# DOING STATISTICS
# FOR BUSINESS *with Excel*

## Data, Inference and Decision Making

**MARILYN K. PELOSI, PH.D.**
*Western New England College*
*Springfield, MA*

**THERESA M. SANDIFER, PH.D.**
*Southern Connecticut State University*
*New Haven, CT*

**JOHN WILEY & SONS, INC.**

**New York • Chichester • Weinheim • Brisbane • Singapore • Toronto**

This book is dedicated to our students, who taught
us how to teach them so that they could learn.

| | |
|---|---|
| ACQUISITIONS EDITOR | Bradford Wiley II/Deborah Berridge |
| ASSISTANT EDITOR | Mary O'Sullivan |
| MARKETING MANAGER | Audra Silverie |
| FULL SERVICES MANAGER | Jeanine Furino |
| COVER DESIGNER | Karin Gerdes Kincheloe |
| ILLUSTRATION EDITOR | Sigmund Malinowski |
| OUTSIDE PRODUCTION SERVICE | Susan L. Reiland |
| TEXT DESIGNER | Nancy Field |
| COVER ILLUSTRATION | Norm Christiansen |

This book was set in New Baskerville by Progressive Information Technologies and
printed and bound by Courier/Westford. The cover was printed by Lehigh Press.

This book is printed on acid-free paper.

*Library of Congress Cataloging in Publication Data:*
Pelosi, Marilyn K.
    Doing statistics for business: data, inferences, and decision
making/Marilyn K. Pelosi, Theresa M. Sandifer.
        p. cm.
    Includes index.
    ISBN 0-471-12208-4 (pbk.: alk. paper)
    1. Commercial statistics.   2. Statistics.   3. Industrial
management—Statistical methods.    I. Sandifer, Theresa M.
II. Title.
HF1017.P367   2000
519.5—dc21
                                                            98-24484
                                                            CIP

Printed in the United States of America

10 9 8 7 6 5 4 3 2 1

# *PREFACE*

## OUR GOALS IN WRITING *DOING STATISTICS*

Our primary goal in writing this text is to teach students how to *do* statistics and how to *use* statistics as a tool for making intelligent, informed decisions. Statistics is the means to an end, not the end itself. We want students to learn how to analyze data and turn it into useful, inference-based information. At the same time, we want students to develop an intuitive understanding of variability, sampling, probability and inference, and models. While it would be nice if students remembered some of the formulas like the standard deviation or the correlation coefficient, it is not essential. It is much more important that students remember *what* these statistics measure and *why* one might want to use them.

Thus, the pedagogical focus of *Doing Statistics* is data driven, emphasizing statistical reasoning, interpretation and decision making. We want to teach the "practice" of statistics; therefore, the emphasis is on comparison and interpretation rather than rote calculation. To support this approach we include discussions of practice as opposed to theory. For example, we discuss when it is OK to be in violation of an assumption and what to watch out for in applying techniques. In order to focus on decision making, the use of the computer is emphasized to keep students from getting bogged down in the details of the calculations.

All the cases found in *Doing Statistics* are based on real business situations and data that are relevant to student life. Most chapters begin with a motivating example based on real-world data. This example is then threaded throughout the chapter and to a lesser extent throughout the book. The data used in many of the examples and exercises are based on our consulting experiences. In order to protect the confidentiality of the companies involved, the data may be rescaled or altered, which does not hinder the usefulness of the data from a pedagogical perspective.

We have adopted a conversational tone so that students may learn the material using vocabulary that is familiar to them. The explanations have an *informal* flavor in order to allow the student to see that the techniques make logical sense. When necessary, we deviate from "pure" mathematics in the interest of understanding, being sure to note when this occurs. The student should see *why* and *how* the technique works. With this kind of understanding, students will know *what* technique to use and *when* to use it.

This book is intended for use in a one or two semester introductory business statistics course. The first 10 chapters cover topics that are usually found in a one-semester course. The remaining chapters are designed to be added or skipped at the instructor's discretion.

## KEY FEATURES OF OUR ACTIVITY-BASED LEARNING APPROACH

In writing this book, we reflected on our many years of teaching experience and examined how people learn new things. The pedagogical features which follow represent the activity-based model we have developed over the years in our own class-rooms. We have found that students . . .

### • Learn by Trying

***Try It Now! Exercises*** are checkpoints embedded in the text, with spaces provided for students to do them, making sure that they understand the basics before moving on.

 **TRY IT NOW!**

**The Glue Company**
*Selecting a Simple Random Sample*

Select a sample of 5 tubes of glue for the glue company. You can assume that each tube of glue has a 5-digit ID number, which the company uses to track its inventory.

Answers are right there (upside down) so that students will have immediate feedback.

### • Learn by Practicing

***Learning It! Exercises*** allow students to practice the basics of computation and include all the techniques covered in the chapter.

#### 2.2.5 Exercises—Learning It!

2.1 The President of the United States wishes to see how popular he/she is after 2 years in office.
(a) What is the population of interest?
(b) Identify which of the reasons for taking a sample (listed on page 21) apply in this case. (There are more than one.)
(c) Identify two variables or characteristics of the members of this population that you may wish to study.

***Thinking About It! Exercises*** emphasize critical thought and ask students to interpret the results of statistical techniques and make decisions based on their analysis.

#### Thinking About It!

1.1 We have said that quality and statistics go hand in hand. What product or service have you used recently that did not have the quality you thought it should have?
(a) Describe the product or service and the area in which it was lacking in quality.
(b) What corrective action would you recommend?

1.2 Describe a situation in which you had a paradigm shift or change.

***Doing It! Exercises*** use large real-world data sets and ask students to perform the statistical analyses and interpret them to solve a real problem. As students progress in their statistical thinking, these exercises will help them to make judgments about "what to do next." The data sets for these exercises are provided on the accompanying disk in Excel v 4.0 worksheet format.

## Doing It!

4.46   The company investigating the golf balls is not satisfied with the limited analysis that it has done. The managers have collected a good deal of data, but they are not sure how to look at them and interpret the output. They decide to hire you to help them understand what the golf balls are doing and how they compare to each other. In addition to measures of the balls' performance, such as the variable *Carry*, the managers know that other factors, both internal (ball-related) and external (environment-related), could affect performance. They tested 36 of each type of ball at three different times using a machine to launch the balls. Data were recorded on 14 different variables. A portion of the data is shown below:

| Ball | Model | S1 | S2 | S3 | Wgt | Dw | Dd | Head | Temp | Carry | Tot Dist | Date | Time |
|------|-------|----|----|----|-----|------|--------|------|------|-------|----------|------|------|
| 1 | M1 | 81 | 81 | 82 | 45.3 | 0.145 | 0.0110 | 686 | 77 | 257 | 270 | 8/20 | 8:15 |
| 2 | M1 | 83 | 83 | 84 | 45.2 | 0.151 | 0.0111 | 688 | 77 | 255 | 267 | 8/20 | 8:15 |
| 3 | M1 | 81 | 82 | 84 | 45.2 | 0.145 | 0.0105 | 687 | 77 | 256 | 267 | 8/20 | 8:15 |
| 4 | M1 | 81 | 81 | 83 | 45.3 | 0.144 | 0.0117 | 688 | 77 | 255 | 271 | 8/20 | 8:15 |
| 5 | M1 | 83 | 81 | 82 | 45.5 | 0.146 | 0.0108 | 687 | 77 | 255 | 268 | 8/20 | 8:15 |

## • Learn by Using the Computer

We have chosen Microsoft Excel because it is the leading analytical software used in the business world. There is a greater chance that students will practice statistical thinking if we show them how to do statistics in a software package that is most likely to be on their desktop at home or at work.

### 4.6.1 Calculating Summary Statistics in Excel

Suppose that we want to calculate a set of summary statistics for the Golf Ball data. Figure 4.9 shows a portion of that data in an Excel worksheet.

The Data Analysis ToolPak has a function that creates a set of summary statistics for a set of data. To access this function, select **Data Analysis** from the **Tools** menu and choose **Descriptive Statistics** from the list of **A**nalysis tools. The dialog box is

|   | A | B | I | J | K | L | M | N |
|---|------|---------|------|------|-------|----------|------|------|
|   | Ball# | Model # | Head | Temp | Carry | Tot Dist | Date | Time |
| 1 | 1 | M1 | 686 | 77 | 257 | 270 | 8/20 | 8:15 |
| 2 | 2 | M1 | 688 | 77 | 255 | 267 | 8/20 | 8:15 |
| 3 | 3 | M1 | 687 | 77 | 256 | 267 | 8/20 | 8:15 |
| 4 | 4 | M1 | 688 | 77 | 255 | 271 | 8/20 | 8:15 |
| 5 | 5 | M1 | 687 | 77 | 255 | 268 | 8/20 | 8:15 |
| 6 | 6 | M1 | 687 | 77 | 256 | 267 | 8/20 | 8:15 |
| 7 | 7 | M1 | 687 | 77 | 255 | 264 | 8/20 | 8:15 |
| 8 | 8 | M1 | 690 | 78 | 258 | 269 | 8/20 | 8:15 |
| 9 | 9 | M1 | 686 | 78 | 252 | 257 | 8/20 | 8:15 |
| 10 | 10 | M1 | 687 | 78 | 256 | 268 | 8/20 | 8:15 |
| 11 | 11 | M1 | 687 | 78 | 253 | 263 | 8/20 | 8:15 |

**FIGURE 4.9**   The golf ball data

The final section of each chapter in the text describes how to implement the techniques taught in the chapter in Excel. In some instances Excel has built-in features to do the necessary statistical calculations. The sections at the end of each chapter walk students through each step in using these features. While Excel does not perform all of the statistical functions found in the text, we have developed some macros to do these calculations. Macros are included on the accompanying disk and the text explains how to use them.

If students are not familiar with using Excel, we have included a short introduction to the basics in Appendix C.

- ## Learn by Discovery

In the process of trying things, we often discover ideas that eventually lead to understanding. Within most of the chapters students will find *Discovery Exercises*. These exercises are written to achieve a high level of student understanding by directing their line of thinking, leading to an "aha" discovery of a key concept. They are designed to help students see that statistical thinking is logical and that the formulas make sense when students discover where they came from and see why they work.

### Discovery Exercise 2.2
### INTRODUCTION TO SAMPLING

Suppose the data shown below represent an entire population. They show the number of people in 50 families living in a small college town in New England. (If you did Discovery Exercise 2.1 then you will recognize this as the same data set.) Now a 2-digit ID number has also been included.

**New England families: large amount of variability**
**Average number of people in 50 families: 4.50**

| | | | | | | | | | |
|---|---|---|---|---|---|---|---|---|---|
| ID: 01 | 1 | ID: 02 | 4 | ID: 03 | 5 | ID: 04 | 7 | ID: 05 | 8 |
| ID: 06 | 3 | ID: 07 | 9 | ID: 08 | 8 | ID: 09 | 8 | ID: 10 | 8 |
| ID: 11 | 4 | ID: 12 | 9 | ID: 13 | 9 | ID: 14 | 1 | ID: 15 | 6 |
| ID: 16 | 4 | ID: 17 | 1 | ID: 18 | 3 | ID: 19 | 9 | ID: 20 | 7 |
| ID: 21 | 8 | ID: 22 | 2 | ID: 23 | 3 | ID: 24 | 1 | ID: 25 | 9 |
| ID: 26 | 1 | ID: 27 | 7 | ID: 28 | 5 | ID: 29 | 1 | ID: 30 | 1 |
| ID: 31 | 1 | ID: 32 | 6 | ID: 33 | 8 | ID: 34 | 2 | ID: 35 | 9 |
| ID: 36 | 4 | ID: 37 | 1 | ID: 38 | 1 | ID: 39 | 1 | ID: 40 | 3 |
| ID: 41 | 4 | ID: 42 | 2 | ID: 43 | 4 | ID: 44 | 9 | ID: 45 | 4 |
| ID: 46 | 1 | ID: 47 | 3 | ID: 48 | 8 | ID: 49 | 1 | ID: 50 | 1 |

**Step 1:** Select a sample of 5 numbers from this population. Use the table of random numbers to do this. Record your sample in the table at the top of page 36.

Several of the *Discovery Exercises* are designed to help students apply the techniques and interpret the results in terms of making a business decision. Most of these exercises require them to *write* a memo or short business report to explain their decisions and recommendations.

- ## Learn by Pulling It All Together

Many times we learn new material by mastering small chunks at a time, but we don't really "get it" until we step back and put all the small pieces together into a complete whole. Thus, even though the statistical techniques are presented one per chapter, they often need to be used together in solving a problem.

*Really Doing It! Exercises* give students some practice at using several tools in an integrated fashion. These exercises are found after Chapters 5, 10, and 17. The data sets for these exercises are provided on the accompanying disk in Excel v 4.0 worksheet format.

## REALLY DOING IT!
### Who Spends Money—What Do They Buy?: Inferential Statistics

A large marketing research firm has an international client that has business interests in many consumer industries such as food, apparel, alcohol, and automobile fuel. As a person with some statistical experience, you have been hired by this marketing research company. Before spending any of the client's money on surveying consumers, the client has asked you to report on general trends in consumer spending. Specifically, you are asked which business areas have experienced increases, and what are the characteristics of people who are spending money in these areas. Your report will guide future survey design and eventually new product development and advertising expenditures.

Using your Internet surfing skills, you find the web site for the U.S. Census Bureau and find the results of the Consumer Expenditure Survey conducted periodically by the Census Bureau. The most recent survey available online is from 1993 and so you decide to use it.

Each chapter ends with summary tables of key terms and formulas. These features are useful when reviewing the material for quizzes and/or exams. Page references are provided so it is easy to return to the more detailed discussion in the chapter.

### Key Terms

| Term | Definition | Page reference |
|---|---|---|
| Biased sample | A **biased sample** is a sample that does not represent the population. | 31 |
| Continuous data | **Continuous data** are data that can take on any one of an infinite number of possible values over an interval on the number line. | 40 |
| Descriptive statistics | **Tools of descriptive statistics** allow you to summarize the data. | 42 |
| Discrete data | **Discrete data** are data that can take on only certain values. These values are often integers or whole numbers. | 40 |
| Inference | An **inference** is a deduction or a conclusion. | 42 |
| Inferential statistics | The **techniques of inferential statistics** allow us to draw inferences or conclusions about the population from the sample. | 43 |

## The Capstone Chapter 17

*Pulling It All Together* allows students to take an overall look at the techniques they have learned throughout the course, and how the tools can be used, in a systematic way, to solve real business problems. This chapter also includes a discussion of ethical issues in data analysis.

## Acknowledgments

No work of this magnitude can happen without a lot of help. We would like to take this opportunity to thank those people who have made the book a reality. First and most importantly, we want to thank Brad Wiley II, our editor, who believed in us, and what we wanted to accomplish. We could not have done this without the help of Mary O'Sullivan, who kept us sane during a very difficult process, and Susan Reiland, who carried us through the production process without losing her sense of humor. There are many other people at Wiley who deserve our thanks, in particular, Debbie Berridge, Jeanine Furino, Martin Batey, and Peter Janzow.

Part of what makes this book unique is the amount of student input in the development process. We thank all of our students who read and used the book in draft form, finding our errors and making suggestions. In particular, we would like to thank Chris Michaelson and Barbara Dedman.

Of course, we could not have done this without the love and support of our families. Thank you, Dick, Ed, Phil, Chris, Erica, and Victoria, for putting up with so much for so long.

# CONTENTS

# Part I
# GETTING STARTED

CHAPTER *1*

# INTRODUCTION: THE ROLE OF STATISTICAL THINKING IN MANAGEMENT

## QUALITY PRODUCTS

If you asked someone to describe the quality of a product made in Japan in the 1950s, what do you think the response would be? Most likely the person would tell you that the Japanese product made in the 1950s was of low quality, not reliable, and probably junk. Now let's roll forward to the 1990s and ask the same question. The person would most likely tell you that the Japanese product is of high quality, reliable, and of great value.

What happened between 1950 and 1990? During this time period Japanese industries started paying increasingly more attention to the quality of their products. At the same time American companies were enjoying the postwar boom and paying little, if any, attention to quality. Interestingly enough, it was an American statistician named W. Edwards Deming who helped the Japanese focus on quality. Deming tried to get American manufacturers to listen to him but they were not interested at the time. Today, American companies, both service industries and manufacturing concerns, are paying close attention to quality in order to compete internationally.

## 1.1    CHAPTER OBJECTIVES

You may be wondering why a story about product quality is used as an introduction to a book on statistics. Students often feel that taking a course in statistics causes unnecessary pain and that they will never again use the statistical techniques they learn. The objective of this book is to refute both of these points: The course need not be painful and you will certainly be asked to collect and analyze data and use the information to make informed business decisions.

The objective of this chapter is to start you down this path. Specifically this chapter covers the following topics:

- Dispelling the Myths About Statistics
- What Managers Should Know About Statistics
- Statistical Thinking—A New Paradigm for Management
- Some Situations That Call for Statistical Thinking
- Key Components of Statistical Thinking
- Organization of This Book

## 1.2    DISPELLING THE MYTHS ABOUT STATISTICS

We should first dispel some of the common myths about statistics. Listed below are three myths we have run across regularly:

**Myth 1:** *"If I had one hour left to live, I would choose to live it in statistics class because it would seem to last forever!"*

A student's lament

**Myth 2:** *"There are three kinds of lies—lies, damned lies, and statistics."*

Benjamin Disraeli

**Myth 3:** *"If it moves, it's biology; if it changes color, it's chemistry; if it breaks, it's physics; if it puts you to sleep, it's statistics."*

Bob Hogg, University of Iowa

If you are like most people you can relate to one of these myths about the "S-word," *statistics*, or *sadistics* as some people refer to it. Statistics is boring and not useful! This book will lead you to another view of the dreaded S-word—one that sees statistics not as a sleeping pill but as a way in which to view all sorts of exciting, amazing, and valuable things.

 *TRY IT NOW!*

### Myths and Fears
*Identifying Some of Your Own*

Be honest. Write down your myths and fears about this subject right now. Get them out in the open so you can deal with them directly and put them behind you. To get you started a common student fear is listed at the top of the next page.

- *I am worried about all of the math in this course.*

---

    You have taken the first step. It is easier to combat the myths about statistics when we acknowledge them. Good work. Next we need to see why it is important to learn about the tools of statistics.

## 1.3  WHAT MANAGERS SHOULD KNOW ABOUT STATISTICS

The use of statistical techniques has long played an important role in quality control and quality improvement in business and industry. Unfortunately, for too long quality has been relegated to the "quality department" and not integrated into the whole organization. The purchasing department would purchase the raw materials needed at the cheapest price without regard to quality and then would throw the material "over the wall" to the engineers. The engineers would make the product, not worrying about the process, since they knew that the quality department would inspect the product before it went out the door. Businesses in the United States have learned that this way of approaching quality does not work and does not allow them to compete internationally. Quality is everyone's job!

    One of the reasons that people prefer to have the quality department handle quality issues is that thinking about quality leads to thinking about data analysis. Data analysis requires the use of statistical techniques, which are often viewed as difficult. But statistical thinking is not difficult to comprehend. Everything we do can be thought of as a series of steps that are connected. Each time we repeat a step it will not be precisely the same. Reducing how much the step changes from time to time is one way to improve things. Suppose your company manufactures copy machines. You would like to know that the raw materials you use to make the copy machine are always the same, and the customer would like to know that the copy machine will perform the same from day to day. Statistical thinking is logical, uses data, recognizes the interdependence of activities, and looks at how things vary.

*Quality and statistics go hand in hand.*

    Four issues must be addressed if statistics is to become an integral part of management:

- Managers must understand *why* they need to possess statistical knowledge.
- Current and future managers (that's you) must *have* this knowledge.

- Steps must be taken to ensure that the knowledge is *used*.
- The *payoff* from the knowledge and its application must be measured.

To achieve this integration of statistics into business and industry we are going to have to change the way we see the world. We need a new pair of glasses.

## 1.4  STATISTICAL THINKING—A NEW PARADIGM FOR MANAGEMENT

Look at Figure 1.1. What do you see? Ask the person next to you what he or she sees.

**FIGURE 1.1**   Two faces or a vase?

Some of you will see two faces and some of you will see a vase. You probably can see both of these things, but which one did you see first? There is no right or wrong answer here. What you see first depends on your viewpoint, your lens, your glasses. This is your **paradigm;** this is how you *see* things. Each of us has a paradigm or view of the world that has been developed based on our individual experiences. The hard part is changing our lens. There will be many times when you must change your paradigm in order to see the same picture in a new way. This book is designed to help you shift your paradigm so that you can see the world through the lens of statistical thinking.

> A *paradigm* is commonly used today to mean a model, theory, perception, assumption, or frame of reference. It was originally a scientific term.

Did you ever take a class on the great artists of the Italian Renaissance? Even if you did not you might have heard of Michelangelo, Leonardo da Vinci, Raphael, and Donatello. If you study Michelangelo, you would probably study his statue of David, and if you study Leonardo you would certainly learn about his painting of the Mona Lisa. Then you would begin to see David and Mona used in all sorts of advertisements and logos as far away from art as software and tomato sauce! Look at the ones shown in Figure 1.2.

Are these ads new? No, you are just more aware of their existence after having studied about them. This is similar to what happened when you looked at the two faces/vase illustration in Figure 1.1. Once you become aware of the second way of looking at the picture you see both things each time you look at the picture. Similarly, once you study Michelangelo and Leonardo your view is different because you are now aware of their works and begin to see them everywhere. You have a new paradigm or view of the world. Your paradigm has *shifted.*

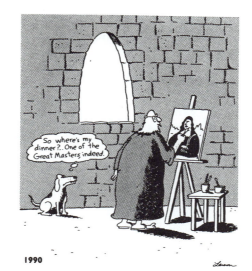

**FIGURE 1.2**   Advertisements using the statue of David and the Mona Lisa

This book is about changing your paradigm to one that allows you to *see* things using statistical thinking. Statistical thinking is public property. Everyone owns it and everyone must use it—not just statisticians. Over the short term, it can improve the quality of decisions; over the long term, it can help turn people into leaders.

You will learn to see data everywhere you turn and to see the information hidden in the data as well. This information will help you make informed investment, marketing, and management decisions and will help you develop quality processes and product designs. How is this possible? By learning and understanding the tools of statistics you will be able to see the world differently. The tools of statistics are actually quite logical and simple to use, yet many people do not use them because of lack of exposure to statistics or dislike of math. Hence, these people often do not make well-informed decisions. You will be different because you will learn the tools of statistics as decision-making tools and not as mathematical manipulation. Think about this book as a management text, not a math text. You will soon see what we mean.

## 1.5   SITUATIONS THAT CALL FOR STATISTICAL THINKING

After finishing this course and graduating from college, you may find yourself facing decisions similar to any one of the following scenarios. In each case you must make decisions based on what you see in the data. A short discussion of possible approaches you might use to investigate each situation is provided. This should help convince you that the tools of statistics make logical sense.

### EXAMPLE 1.1   Tissue Manufacturer

#### Complaints Become Opportunities

Most large companies that manufacture consumer goods receive feedback from their customers. A large manufacturer of paper goods keeps track of consumer complaints for its facial tissue product line. The company receives these complaints through a toll-free number that appears on the product. Next time you pick up a box of tissues see if you can find the toll-free number for making complaints.

The data taken consist of a transcription of the actual complaint and a classification of the complaint into a specific category. The complaint categories are dispensing, foreign material, odor, miscounts, and packaging. There are three subcategories of packaging complaints: damaged, misleading, and defective. The 25 packaging complaints for January 1998 are

| | | | | |
|---|---|---|---|---|
| Damaged | Misleading | Defective | Damaged | Misleading |
| Defective | Defective | Damaged | Damaged | Defective |
| Damaged | Misleading | Misleading | Damaged | Damaged |
| Misleading | Defective | Damaged | Damaged | Damaged |
| Defective | Damaged | Misleading | Defective | Damaged |

You must examine the complaints about packaging and make some recommendations to management. What is your decision?

Possible approaches and things to think about in making a decision:

- Make a table summarizing how many of each type of packaging complaint occurred.
- Make a graph from this table.
- Look to see if there are more complaints of one type than others and if this happens in other months. ■

Let's consider another situation.

## EXAMPLE 1.2   Golf Ball Design

### New Product Design

All companies are currently facing increasing competition at the national and international level. In the face of this competition, American manufacturers are moving to a focus on quality. Customer feedback described in the previous scenario is one aspect of this movement. This change is not simple, but rather encompasses all aspects of the business and all the employees of the company. Managers are learning to listen to the creative ideas of their employees, breaking down the hierarchical management layers, and are monitoring their production processes. In doing so, more and more data are being collected on every aspect of the business. All parts of the organization must work together as a team to achieve Total Quality Management (TQM).

A large manufacturer of golf balls is incorporating some of these concepts. The company is studying two golf balls of equal cost to determine which ball design performs better. Data on how far the ball carried when hit are given (in yards) for 12 balls of each design:

| Design 1 | | | | | | Design 2 | | | | | |
|---|---|---|---|---|---|---|---|---|---|---|---|
| 257 | 259 | 255 | 256 | 260 | 258 | 254 | 252 | 256 | 255 | 255 | 257 |
| 260 | 259 | 259 | 257 | 255 | 260 | 253 | 255 | 254 | 254 | 256 | 255 |

You must make a recommendation on which design to pursue. What is your decision?

Possible approaches and things to think about in making a decision:

- For each design, find the average distance the balls carried.
- Compare these averages.
- Think about whether or not the difference between these averages is large.
- Think about whether the differences that you see in the averages could be caused by something other than the ball design. ■

Here is one more example.

### EXAMPLE 1.3    Dress Down Day

***Impact on Employee Morale***

Many companies have designated one day a week as dress down day. This idea has been implemented even in long-time conservative companies such as IBM. What reason(s) would management have for adopting such a policy? In April 1992, Levi Strauss and Co. conducted the first national survey on business casual dress issues. This phone survey gathered data and opinions from managers from a wide range of industries.

At the company you work for, your boss has conducted a survey of employees. Thirty employees were asked if they strongly agree (SA), agree (A), disagree (D), or strongly disagree (SD) with each of the following two statements:

Statement 1: Casual dress improves morale.

| A | SA | A | A | A | D | SD | SA | A | A | D | D | A | A | SA |
|---|----|---|---|---|---|----|----|---|---|---|---|---|---|----|
| A | A | D | D | A | A | SA | SD | A | D | A | A | D | A | SA |

Statement 2: I do my best work when dressed casually.

| D | SD | A | A | A | A | SA | A | A | A | D | D | A | A | SA |
|---|----|---|---|---|---|----|---|---|---|---|---|---|---|----|
| A | A | A | D | A | D | SD | D | A | D | A | A | D | A | SA |

Your boss would like your recommendation regarding the likely increase in productivity if a dress down day is adopted by your company. What is your decision?

Possible approaches and things to think about in making a decision:

- Summarize the results.
- Examine what percentage of people strongly agreed, agreed, disagreed, or strongly disagreed with the various statements presented to them.
- Consider how productivity was defined in this survey.
- Look to see if the responses are different in different industries. ∎

## *Discovery Exercise 1.1*
## *STARTING TO THINK STATISTICALLY*

This exercise allows you to begin to think statistically. Although you have not yet learned any statistical techniques, you will be surprised to discover that many of the ideas you suggest are the basis for some of the techniques you will learn about later in this book. Do not try to solve the problem, but rather focus on what you might do with the data to analyze the situation and what additional information you would like to know about the data or the situation. Remember that statistical thinking is logical, uses data, recognizes the interdependence of activities, and looks at how things vary.

### Part I.  Reduction in Sick Days Used

It has been a widely held belief that a switch to "participative management" would increase employees' "buy-in" to the company. One of the benefits that should be realized is a reduction in the number of sick days that employees use when they feel that they are an important part of the team. A company that has made the switch in some departments wonders if this has been true. The company decides to sample 25 employees from each of two manufacturing departments. The first department has been using a participative management style for almost two years, and the second is still using a

*(continued)*

traditional management style. The data on the number of sick days used by each employee in the past 12 months are

| Participative | | | | | Traditional | | | | |
|---|---|---|---|---|---|---|---|---|---|
| 1 | 3 | 5 | 5 | 6 | 0 | 5 | 6 | 7 | 9 |
| 1 | 4 | 5 | 6 | 7 | 3 | 5 | 7 | 7 | 9 |
| 2 | 4 | 5 | 6 | 8 | 4 | 6 | 7 | 7 | 10 |
| 2 | 4 | 5 | 6 | 8 | 4 | 6 | 7 | 8 | 11 |
| 3 | 4 | 5 | 6 | 8 | 5 | 6 | 7 | 8 | 11 |

Has the participative management style reduced the number of sick days?

**1.** What else would you like to know about these data sets before comparing them?

**2.** How could you compare the two data sets using graphs?

**3.** What numerical calculations would you perform to compare the two data sets?

## Part II. Are We There Yet?

The amount of time it takes to travel is clearly dependent on how far you travel and the type of vehicle you are driving. The data shown below give information about how long it took to travel the distance (in miles) when the vehicle was either a car or a truck. The travel times (in minutes) are listed as the $Y$ values. The distances traveled are listed as the $X2$ values. The vehicles used are listed as $X1$ values with the following coding scheme: $X1 = 0$ if a truck was driven and $X1 = 1$ if a car was driven.

| Type of vehicle (car = 1, truck = 0) | Distance (miles) | Travel time (min) |
|---|---|---|
| $X1$ | $X2$ | $Y$ |
| 0 | 55 | 35 |
| 0 | 102 | 56 |
| 1 | 33 | 23 |
| 0 | 20 | 12 |
| 1 | 55 | 65 |
| 0 | 48 | 34 |
| 0 | 53 | 23 |
| 0 | 22 | 12 |
| 1 | 45 | 35 |
| 1 | 44 | 46 |
| 1 | 12 | 14 |
| 0 | 45 | 34 |

Look at the data shown and think about how you would analyze the impact that distance and type of vehicle has on travel time.

1. What else would you like to know about the data?

2. How could you display the relationship between the travel time and distance in a graph?

3. How would you predict the time it would take to travel a distance of 90 miles driving a car?

A picture is worth 1000 words. Did you ever hear that saying? Why is this true? Perhaps it is because your eye can process and understand the millions of tiny dots of paint when they are all put together in a meaningful fashion. In this way you see a beautiful picture. Well, that is exactly what the tools of statistics do for you. They put together millions of bits of data in a meaningful fashion. In this way you can make the best decision based on what you see in the data.

## 1.6  KEY COMPONENTS OF STATISTICAL THINKING

It is important to understand the components of this new paradigm that we have called statistical thinking. There are three key components:

- We must use data whenever possible to guide us.
- We must look for connections and relationships.
- We must understand why data values differ from each other.

Now it is time to introduce an official definition of statistics.

*Statistics* is a branch of mathematics dealing with the analysis and interpretation of masses of data.

With statistical thinking as the new paradigm, your way of thinking about numbers has begun to change. You must begin to discipline yourself to think systematically and to collect data systematically. Then you must learn how to see the information contained in the data you have collected. Finally, you must use the information to make informed business decisions. The rest of this book is designed to increase your inventory of specific statistical skills to help you see the information contained in the data and strengthen your decision-making skills. In highly successful organizations, the statistical thinking paradigm will be linked to other important paradigms, such as providing leadership, promoting teamwork, working toward continuous improvement, creating innovative channels of communication, and delighting customers.

## 1.7  ORGANIZATION OF THIS BOOK

Each chapter of this book is designed the same way as this chapter. You can count on seeing the same elements that you have seen in this introductory chapter. Each element of the chapter is designed to help you change your paradigm so that you may see the world through statistical thinking. More specifically each chapter

- Begins with a *business problem* based on an actual situation. The business problem will motivate the material and will be carried throughout the chapter as you learn the techniques needed to solve the problem. Some chapters may have more than one motivating problem.
- Lists the *objectives* of the chapter in the first section.
- Provides you with worked *examples* for each statistical concept, tool, or other important idea. These are always labeled with the word **EXAMPLE** and are visually separated from the text.

- Provides you with *Try It Now!* exercises to give you some practice. These exercises are like checkpoints. Do not skip them. Space is provided for you to work the problems right at that point and the answers are given in the side margin so you may check your work immediately. *Try It Now!* exercises are easily found because the open book graphic shown in the margin is always displayed with them.

- Provides you with one or more *Discovery Exercises*. These exercises are optional and if they are skipped you will not miss any information. They are designed to help you discover some concept or idea. They will help you see that statistical thinking is logical and that the formulas make sense when you discover where they came from and why. They are easily identified by their design. Your instructor will probably use some but not all of these exercises.
- Includes many *Side margin notes,* which contain keywords, hints, and tips. These help you to review the material and locate key formulas, definitions, and concepts.
- Includes *Definitions* in shaded boxes for easy location. Any term that has a definition box will appear in the end-of-chapter summary.
- Identifies *Formulas* with a label in the margin. All such formulas will appear in the end-of-chapter summary.
- Provides you with more practice problems at the end of most sections. (Chapter 1 does not have any of these.) These are like the *Try It Now!* exercises. They are labeled *"Exercises—Learning It"* and by doing them you will master the basics.
- Provides you with detailed instruction on how to use Microsoft Excel to perform most of the statistical analyses described in the chapter. These instructions are found in the last section of each chapter. In many cases, if Excel does not provide a convenient or correct way to do the analysis, we provide an Excel macro, along with directions for using it, that performs the function. These macros are found on the disk with the data files.

- Provides a *Chapter Summary* including tables of key terms and formulas introduced in the chapter.

- Provides you with *End-of-Chapter Exercises.* At the end of most chapters there are three types of exercises: *Learning It!* (similar to the end-of-section exercises), *Thinking About It!,* and *Doing It!* The *Thinking About It!* exercises ask you to go beyond the basics. You are asked to think about the business implications of the computational result and your thinking may be directed beyond the immediate details of the problem. Sometimes a set of data is presented in an exercise in one chapter and then used again in an exercise in a later chapter. In these cases the data are always redisplayed with the problem statement to avoid your flipping back and forth in the text. In addition, if an exercise is built on a previously presented exercise, the exercise number corresponding to the related exercise is noted in the side margin. The *Doing It!* exercises are linked to large data sets that are provided on disk. These data sets are designed to let you use the techniques of the chapter in a more realistic setting, one in which you have lots of data to analyze and not a great deal of structure.

In addition to the exercises found in each chapter, there are three additional groups of exercises, found after Chapters 5, 10, and 17. These exercises are similar to the Doing It! exercises found at the end of each chapter but require you to draw on what you have learned in the previous few chapters. The Exercises after Chapter 5 allow you to use many of the tools of descriptive statistics that you have learned in Chapters 3, 4, and 5 to analyze large data sets and reach some conclusions. The Exercises found after Chapter 10 allow you to use the tools of inferential statistics that you learned in Chapters 8, 9, and 10 to solve a business problem. These Exercises are also preparation for the exercises you will do in the final chapter, entitled "Pulling It All Together."

# CHAPTER 1 SUMMARY

In this chapter you learned how important it is for today's business professional to know how to use the tools of statistics. The business environment is changing rapidly and one of the key issues facing most businesses is the issue of quality. This is true regardless of whether the business is a manufacturer, a restaurant, a hospital, or an entertainment industry. To remain competitive, businesses must pay attention to the quality of the product or service they provide. At the foundation of the quality movement lies the field of statistics.

## Key Terms

| Term | Definition | Page reference |
|------|------------|----------------|
| Paradigm | A **paradigm** is commonly used today to mean a model, theory, perception, assumption, or frame of reference. | 6 |
| Statistics | **Statistics** is a branch of mathematics dealing with the analysis and interpretation of masses of data. | 11 |

# CHAPTER 1 EXERCISES

## Thinking About It!

**1.1**  We have said that quality and statistics go hand in hand. What product or service have you used recently that did not have the quality you thought it should have?

    (a)  Describe the product or service and the area in which it was lacking in quality.

    (b)  What corrective action would you recommend?

**1.2**  Describe a situation in which you had a paradigm shift or change.

**1.3**  There has been an increased use of electronic mail in most organizations. This way of communicating with people changes the dynamics of the exchange of information and how people do business. In fact, you might speculate that e-mail would change the way the organization behaves. It is possible that e-mail is used more by middle management than senior management. It is also possible that e-mail is not used to communicate certain types of decisions. A human resource manager asks 10 senior managers and 10 middle managers how many messages they send each day. The data are

    Senior managers: 10, 15, 2, 21, 14, 35, 19, 12, 18, 19
    Middle managers: 25, 27, 22, 24, 23, 21, 25, 26, 29, 19

    (a)  What might you do with these data to investigate if there is a difference in e-mail usage by level of management?

    (b)  What factors, other than level of management, might influence e-mail usage?

**1.4**  You have just been hired to be the manager of the service department for a car dealership in your town. You begin by examining the service time for all the cars worked on last Friday. The service time is the number of minutes between the time the car arrived at the dealership and the time the service was completed. The data for last Friday are

    15     125     45     65     35     50     20

    (a)  What additional data would you like to examine to get a more complete picture of the current level of service?

    (b)  How would you collect the data identified in part (a)?

# CHAPTER *2*

# *THE LANGUAGE OF STATISTICS*

## THE GLUE COMPANY

All companies are currently facing increasing competition at the national and international level. In the face of this competition, American manufacturers are moving to a focus on quality. Customer feedback is one aspect of this movement. In fact, companies are now building relationships with their vendors to ensure higher quality.

The U.S. automobile industry was hit very hard by international competition. The industry appears to have recovered by paying attention to quality at all levels, right down to the glue that is used in assembling some of the components. One of the large U.S. automobile manufacturers has been building a relationship with the manufacturer of an adhesive product. In doing so, the car company has required a lot of data to support the claims about the performance of the glue. The car company visits the adhesive manufacturer from time to time to further the relationship and discuss product quality problems and solutions.

This type of interaction between the supplier and the customer is much different from the "old days" when the purchasing department bought from the least expensive supplier, which could easily be several different suppliers over a short period of time. There was no long-term commitment on either end of the deal. This short-term mentality led to shortcomings in quality.

## 2.1    CHAPTER OBJECTIVES

In the first chapter you learned that quality and statistics go hand in hand. You also saw that to truly integrate quality into the organization we must view our environment with a statistical thinking paradigm. Recall that a statistical thinking paradigm is a systematic way of thinking about and applying statistical techniques to data and business problems. The first step in changing your view of the world to a statistical thinking paradigm is to develop a language that can be used to describe this new view. This chapter develops the basic language that you need. The following material is covered:

- The Difference Between the Population and a Sample of the Population
- The Difference Between a Parameter and a Statistic
- Factors That Influence Sample Size: Some Sampling and Sample Size Considerations
- Selecting the Sample
- Types of Data
- The Difference Between Descriptive Statistics and Inferential Statistics
- Basic Summation Notation

## 2.2   THE DIFFERENCE BETWEEN THE POPULATION AND A SAMPLE OF THE POPULATION

### 2.2.1 The Population of Interest

In Section 1.5, we saw several situations that called for statistical thinking. If we step back from the details of these situations we see that regardless of the decision that needs to be made, there is always a group of people or things that needs to be studied and understood to make the necessary decision. For the tissue manufacturer it was all the boxes of tissues that the company makes, for the company considering dress down day it was all the employees of the company, and for the golf ball company it was the performance of all the balls made according to each of the two designs. These are examples of what is called the **population** in the statistical thinking paradigm. Thus, *all the boxes of tissues* is the population of interest to the tissue manufacturer, *all the employees* is the population of interest to the company considering dress down day, and *all the golf balls* is the population of interest to the golf ball manufacturer. In each case, the decision maker must learn something about how the population of interest behaves in order to make a recommendation.

The *population* is everything you wish to study.

### EXAMPLE 2.1    The Glue Company

*Identifying the Population*

For the glue company to be a supplier for the car company it must study and understand the performance of all the tubes of glue made for the car company. Call the type of glue used by the car company type C. Thus, the population of interest is all the tubes of type C glue manufactured by the glue company.  ■

## TRY IT NOW!

### The In-line Skate Company
*Identifying the Population of Interest*

There have been a number of failures on the braking device of a new model of roller blades that your company manufactures. What is the population of interest?

ANS. ALL BLADES OF THIS MODEL.

Often, when we study a population, we are really interested in knowing about different characteristics of each member of the population. These characteristics are known as **variables.** For each member of the population we may be interested in knowing about one, two, or even more different variables.

> A *variable* is used to represent a characteristic of each member of the population.

### EXAMPLE 2.2   The Glue Company

*Possible Variables of Interest*

In the case of the glue company, we may wish to know how long the glue will stick but we may also wish to know how the glue smells. In this case we wish to study two variables or characteristics of the members of the population of all the tubes of type C glue manufactured by the glue company. ■

## TRY IT NOW!

### The In-line Skate Company
*Identifying Possible Variables to Study*

There have been a number of failures on the braking device of a new model of roller blades that your company manufactures. Name two variables or characteristics that you might wish to study.

## 2.2.2 The Sample

Now that you understand what a population is, we need to formally define a sample. The statistical definition of the word sample is much like the normal use of the word. If someone tells you to sample the apple pies made by a bakery, then you take a small piece of one and eat it, or you might eat one whole pie out of the many pies made by the bakery. It is the same in statistics. If you take a sample, you take a small piece of the population and look at it or test it. Thus, we have our next definition.

> A *sample* is a piece of the population.

If we think about the population as the big oval (all the pies) shown in Figure 2.1, then a sample from this population might be the small oval shown in the figure. This is clearly not the only sample that could be taken from this population; many different samples may be picked. Some of the samples may overlap and some may be bigger or smaller than the one shown.

**FIGURE 2.1**   The population and a sample

 **TRY IT NOW!**

*Selecting Different Samples*

Draw another oval in Figure 2.1 to represent a different sample that might be selected.

## 2.2.3 Why Pick a Sample at All?

*Reasons why we can't study the entire population*

We have said that the population is everything we wish to study and that the sample is only a piece of the population. A natural question to ask is, Why should we bother to examine a sample when what we really want to know about is the population? Most of the time we cannot study the entire population and must use a sample as a guide.

The main reasons are fairly clear when you think about it for a minute:

- It would take too much time to study the entire population.
- It would take too much money to study the entire population.
- It might not be possible to identify all the members of the population.
- If we test the entire population, we might not have anything left to sell.

Let's consider a couple of populations that we have already talked about and see these reasons in action.

## EXAMPLE 2.3   The Glue Company

***Reason Why the Population Cannot Be Studied***

The glue company needs to study the performance of all tubes of type C glue that it sells to the car company. It would clearly take a great deal of time and money to test each and every tube of type C glue that the company manufactures. In addition, each tube that is tested is opened and then cannot be sold. So, if every tube were tested there would be no glue to sell to the car company.   ■

Here's another example based on one of the situations we looked at in Chapter 1.

## EXAMPLE 2.4   Dress Down Day

***Reason Why the Population Cannot Be Studied***

A company needs to study the productivity of the employees before and after a dress down day is adopted. If there are 5000 employees in this company all doing different jobs, it would be virtually impossible to measure productivity for all of them.   ■

## 2.2.4 The Difference Between the Population and the Sample: An Introduction to Sampling Error

At this point it might look like we should throw up our hands and go home, as we will never have the time, money, or ability to study an entire population. The good news is we don't need to study the entire population. By studying the behavior of a sample we can get a good idea of the behavior of the population. It will not be a perfect picture of the population, but it will be good enough to guide us in our decision making. It is just like eating a piece of the apple pie. You don't know precisely how all the pie tastes but you have a pretty good idea that it is delicious!

*It is not necessary to study the entire population.*

Remember, it is really the population that we wish to understand and study. The sample is a means to this end. By understanding and studying a sample, we gain insight and knowledge about the behavior of the population. The amount of information in the sample is not perfect but adequate.

Imagine that you took a picture of the population but let it develop for only a few seconds. You would be able to get an overall sense of what the population looked like but you would not have all of the details. This is like taking a sample and using the sample to determine how the population behaves. If you let this picture develop a few seconds longer, you would get a little clearer view of the population. This is equivalent to taking a bigger sample. Finally, if you let the picture develop completely, then you would have complete information. This is equivalent to studying the entire population. If you study the entire population, then you have taken what is called a **census.**

A ***census*** is a study of the entire population.

Since the sample is an imperfect snapshot of the population, you know that there will be differences between the sample and the population. This is a bit disconcerting at first. Unless we study the entire population, which we usually cannot do, we will always have incomplete information about the population. That is, unless we study the entire population, we cannot eliminate what is known as **sampling error.**

> ***Sampling error*** is the difference between the behavior of the entire population and a sample of that population.

*The size of the sample influences the size of the sampling error.*

The size of the sampling error is determined by two factors. The first of these factors is the *size of the sample.* Clearly, the larger the sample you take, the more similar the sample will be to the population, thus decreasing the sampling error. The second factor that influences the size of the sampling error is the amount of **variation** that exists in the population. Variation in statistics has the same general meaning that it does in typical language usage.

> The amount of ***variation*** refers to how different the members of the population are from each other with regard to the variable being studied.

*The amount of variability in the population influences the size of the sampling error.*

For example, suppose your population of interest is all students taking classes at your university or college. The variable of interest is the age of these students. If all students were exactly the same age then you would say that there was no variation in the age of the members of the population. In this extreme case you would need to take a sample of only one student to have perfect information about the age of members of the population.

However, you will almost never be studying a characteristic that has no variability. Suppose that the ages of the students have a small amount of variability. Let's say the student ages range from 18 to 22 years old. Clearly, you have to take a sample of more than one student to understand how the ages vary in this population. Now suppose the ages of the students range from 17 to 60 years. You have to take a much larger sample to understand how the ages vary in this population. As the amount of variability in the population increases, the sampling error also increases.

## 2.2.5 Exercises—Learning It!

2.1   The President of the United States wishes to see how popular he/she is after 2 years in office.

   (a)  What is the population of interest?

   (b)  Identify which of the reasons for taking a sample (listed on page 21) apply in this case. (There are more than one.)

   (c)  Identify two variables or characteristics of the members of this population that you may wish to study.

2.2   Tasty Ice Cream Corporation wishes to be sure that all of the half-gallon ice cream cartons do contain one-half gallon of ice cream.

   (a)  What is the population of interest?

   (b)  Identify which of the reasons for taking a sample (listed on page 21) apply in this case. (There are more than one.)

   (c)  Identify two variables or characteristics of the members of this population that you may wish to study.

2.3  A company that manufactures electronic switches wishes to provide the customer with defect-free switches.
  (a) What is the population of interest?
  (b) Identify which of the reasons for taking a sample (listed on page 21) apply in this case. (There are more than one.)
  (c) Identify two variables or characteristics of the members of this population that you may wish to study.

2.4  A university wishes to know how students view the new athletic center on campus.
  (a) What is the population of interest?
  (b) Identify which of the reasons for taking a sample (listed on page 21) apply in this case. (There are more than one.)
  (c) Identify two variables or characteristics of the members of this population that you may wish to study.

2.5  The U.S. government wishes to know how many people are unemployed.
  (a) What is the population of interest?
  (b) Identify which of the reasons for taking a sample (listed on page 21) apply in this case. (There are more than one.)
  (c) Identify two variables or characteristics of the members of this population that you may wish to study.

2.6  The manufacturer of disposable diapers wishes to know how much fluid the diapers can hold before they leak.
  (a) What is the population of interest?
  (b) Identify which of the reasons for taking a sample (listed on page 21) apply in this case. (There are more than one.)
  (c) Identify two variables or characteristics of the members of this population that you may wish to study.

2.7  The Coca-Cola Company wishes to know the proportion of people who prefer Coke over Pepsi.
  (a) What is the population of interest?
  (b) Identify which of the reasons for taking a sample (listed on page 21) apply in this case. (There are more than one.)
  (c) Identify two variables or characteristics of the members of this population that you may wish to study.

2.8  Before selecting a major, a student decides to study the salaries of individuals working as accountants.
  **(a)** What is the population of interest?
  (b) Identify which of the reasons for taking a sample (listed on page 21) apply in this case. (There are more than one.)
  (c) Identify two variables or characteristics of the members of this population that you may wish to study.

## *Discovery Exercise 2.1*

## *INTRODUCTION TO SAMPLING AND VARIABILITY*

Suppose that each set of data in this exercise represents an entire population. Since we don't yet have a way to quantify the amount of variability in a population, the data sets are labeled as having a small amount of variability or a large amount of variability. The first data set shows the number of people in 50 families living in a small college town in New England. The second set of data shows the number of people in 50 families living in a large city in the South.

| New England families: large amount of variability Average number of people in 50 families: 4.50 | | | | | | Southern families: small amount of variability Average number of people in 50 families: 4.36 | | | | |
|---|---|---|---|---|---|---|---|---|---|---|
| 1 | 4 | 5 | 7 | 8 | | 1 | 3 | 4 | 3 | 5 |
| 3 | 9 | 8 | 8 | 8 | | 3 | 4 | 6 | 6 | 3 |
| 4 | 9 | 9 | 1 | 6 | | 6 | 4 | 3 | 4 | 6 |
| 4 | 1 | 3 | 9 | 7 | | 7 | 5 | 4 | 5 | 5 |
| 8 | 2 | 3 | 1 | 9 | | 4 | 4 | 5 | 5 | 6 |
| 1 | 7 | 5 | 1 | 1 | | 3 | 4 | 4 | 7 | 4 |
| 1 | 6 | 8 | 2 | 9 | | 6 | 3 | 5 | 5 | 5 |
| 4 | 1 | 1 | 1 | 3 | | 5 | 4 | 4 | 4 | 5 |
| 4 | 2 | 4 | 9 | 4 | | 4 | 5 | 4 | 3 | 4 |
| 1 | 3 | 8 | 1 | 1 | | 5 | 4 | 2 | 4 | 4 |

**Step 1.**    Select a sample of 5 numbers from each population.

| | Sample from NE families | Sample from Southern families |
|---|---|---|
| Selection 1 | | |
| Selection 2 | | |
| Selection 3 | | |
| Selection 4 | | |
| Selection 5 | | |
| **Average** | | |

**Step 2.**    Calculate the sample average of the 5 numbers by adding them together and dividing by 5.

**Step 3.**    Calculate how far away the sample average is from the true population average by subtracting the sample average from the population average that is provided for you.

**Step 4.**    Which data set had the greater error—the one with a small amount of variability or the one with a large amount of variability?

Repeat steps 1–4 with a sample of 10 numbers.

**Step 1.**   Select a sample of 10 numbers from each population.

|  | Sample from NE families | Sample from Southern families |
|---|---|---|
| Selection 1 |  |  |
| Selection 2 |  |  |
| Selection 3 |  |  |
| Selection 4 |  |  |
| Selection 5 |  |  |
| Selection 6 |  |  |
| Selection 7 |  |  |
| Selection 8 |  |  |
| Selection 9 |  |  |
| Selection 10 |  |  |
| **Average** |  |  |

**Step 2.**   Calculate the sample average of the 10 numbers by adding them together and dividing by 10.

**Step 3.**   Calculate how far away the sample average is from the true population average by subtracting the sample average from the population average that is provided for you.

**Step 4.**   Which data set had the greater error—the one with a small amount of variability or the one with a large amount of variability?

**Step 5.**   Record the errors in the following table:

|  | Sample of size 5 | Sample of size 10 |
|---|---|---|
| NE families (large variability) |  |  |
| Southern families (small variability) |  |  |

What happened to the error when you increased the sample size?

# 2.3   THE DIFFERENCE BETWEEN A PARAMETER AND A STATISTIC

In the last section we saw that by understanding and studying the sample, we can gain insight and knowledge about the behavior of the population. Next we must think about how to describe the behavior of the population to someone else. Suppose you were asked to describe your instructor's behavior in the classroom. You might say that he/she is informative, humorous, organized, and helpful (hopefully)! These are verbal descriptors of your instructor's behavior. If you were asked to provide a visual description of your instructor you might say that she is about 5 feet 2 inches tall, weighs approximately 110 lb, and is probably about 40 years old. These are numerical descriptors of the person. To paint a picture of a population and a sample, we use descriptors that are both verbal and numerical, just like the ones that we used to describe your instructor.

## 2.3.1 Parameters: Numerical Descriptors of the Population

There are many different ways to describe the behavior of a population. One of them is to use numerical values to paint a picture of the population. For instance, if you are told that all the values in the population fall between 0 and 10, you form a mental image of the population that is quite different from the picture that is conjured up if you learn that all the values in the population fall between 0 and 1000. This is just one example of a numerical way to describe the population. Chapter 4 shows you the traditional numerical measures that are used to describe the population. They are all examples of what are known as **parameters.**

A *parameter* is a number that describes a characteristic of the population.

The following example gives three different parameters that might be of interest to the glue company.

### EXAMPLE 2.5   The Glue Company

*Parameters of Interest*

The glue company may wish to know the average number of days that the glue will stick. In this case the parameter of interest is an average value. The company may also wish to know the longest time the glue will stick. In this case the parameter of interest is the maximum stick time. Likewise, the company might wish to know the earliest time that the glue will fail (come unglued). So the shortest stick time is also a parameter of interest. ∎

The next example shows that the parameter of interest might be a percentage.

## EXAMPLE 2.6  Dress Down Day

*Parameter of Interest*

The company thinking of adopting the dress down day would like to know the percentage of employees that had increased productivity as a result of this policy change. The parameter of interest here is a percentage or a proportion. ∎

Remember that generally we do not know much about the population. If we did we would not need to take a sample and would not need the tools of statistics (and we could all go home now!). We are usually trying to discover information about these parameters. In particular, we are often trying to *estimate the value of these parameters*. This is the job of statistics.

## 2.3.2 Statistics: Numerical Descriptors of the Sample

Since we most likely will not know the value of the parameters needed to describe the population, we must resort to using the information contained in the sample. It seems logical that a numerical descriptor for the sample might somehow be used to estimate the corresponding measure for the population. This is the right idea.

So we need numbers to describe the behavior of the sample for two reasons: (1) to paint a picture of the sample and (2) to help us estimate the corresponding population parameter. There is nothing difficult here—in fact, it is quite simple. A statistic is nothing more than a number that describes the behavior of the sample. Let's put that into a definition box.

> A *statistic* is a number that describes the behavior of a sample.

According to this definition we could dream up any formula that we want to to describe the sample and it would count as a statistic. Surprisingly enough, this is exactly right. There are, however, a few measures that are typically used because they convey some fairly standard type of information. These are the subject of Chapter 4.

Let's look at two examples: the glue company and the company considering the dress down day policy. In each case we will see what statistics might be calculated.

## EXAMPLE 2.7   The Glue Company and Dress Down Day

*Statistics That Could Be Calculated*

The glue company might calculate the average stick time from the sample data. This would be one statistic. The company might also record the shortest and the longest stick times in the sample. These are two more statistics.

Let's assume that the company thinking of adopting the dress down day policy has adopted the policy for one department in the company. The company has taken a sample of employees in this department and measured the productivity of each employee in the sample before and after this policy was adopted. Should the policy be adopted companywide? The statistic to calculate is the percentage of employees in the sample who increased their productivity after the dress down policy was adopted.

∎

Notice that the statistics that might be calculated from the sample data are closely related to the parameters of interest. This is often the case. Remember that when you calculate the average stick time for the sample it is exactly that: a description of the sample. Is the average stick time for the population of all tubes of type C glue the same as the average you found in the sample? Probably not. Is it close? That depends on how well your sample reflects the population. This is the subject of the next two sections.

### 2.3.3 Exercises—Learning It!

2.9   The President of the United States wishes to see how popular he/she is after 2 years in office.
(a)  What parameter is of interest?
(b)  What statistic might you calculate from the sample?

2.10  Tasty Ice Cream Corporation wishes to be sure that all of the half-gallon ice cream cartons do contain one-half gallon of ice cream.
(a)  What parameter is of interest?
(b)  What statistic might you calculate from the sample?

2.11  A company that manufactures electronic switches wishes to provide the customer with defect-free switches.
(a)  What parameter is of interest?
(b)  What statistic might you calculate from the sample?

2.12  The university wishes to know how students view the new athletic center on campus.
(a)  What parameter is of interest?
(b)  What statistic might you calculate from the sample?

2.13  The U.S. government wishes to know how many people are unemployed.
(a)  What parameter is of interest?
(b)  What statistic might you calculate from the sample?

2.14  The manufacturer of disposable diapers wishes to know how much fluid the diapers can hold before they leak.
(a)  What parameter is of interest?
(b)  What statistic might you calculate from the sample?

2.15  The Coca-Cola Company wishes to know the proportion of people who prefer Coke over Pepsi.
(a)  What parameter is of interest?
(b)  What statistic might you calculate from the sample?

2.16  Before selecting a major, a student decides to study the salaries of individuals working as accountants.
(a)  What parameter is of interest?
(b)  What statistic might you calculate from the sample?

## 2.4  FACTORS THAT INFLUENCE SAMPLE SIZE: SOME SAMPLING AND SAMPLE SIZE CONSIDERATIONS

The next question is how big does the sample need to be? Do we need to eat half of one pie to know how all the pies in a batch taste? What factors will influence our decision in this matter? We touched on this question when we observed the impact of the sample size and the amount of variability on the size of the sampling error. This section reinforces the connection between the sample size and the two factors we have already studied: amount of variability in the population and the sampling error. In addition, two other factors are introduced.

### 2.4.1 Size of the Population

The first factor is the **size of the population.** How many tubes of glue are manufactured each day? How many employees are there in the company? How many golf balls are manufactured each hour? How many boxes of tissues are produced in a day? These numbers are the size of the population. Intuitively it should seem that it would take a different sample size to learn about a population of 1000 compared to a population of 100,000. The population size is a factor, but it turns out not to be as important as some of the other factors we have identified.

> The *size of the population* is the number of members of the population. It will be referred to as *N*.

### 2.4.2 Extent of Resources Available

The next factor is the amount of time, money and other resources that you have available. If you need a decision in six months rather than next week, the amount of data you could possibly collect and analyze is different. Also, remember that it is expensive to collect and process data, so cost must be a factor. However, we still have not identified the two most important factors.

*The amount of resources available impacts sample size.*

### 2.4.3 Amount of Error That Can Be Tolerated

In Section 2.2.4 we learned that whenever you use a sample to draw conclusions about the population you will have some sampling error. And the bigger the sample we pick, the less error we will have in our conclusions. Returning to the ovals in Figure 2.1, you can see that if you take a bigger sample, then you have captured more of the population. Therefore, the conclusions you draw about the population based on the sample will be more accurate. This is shown by the increasing sample sizes in Figure 2.2. The numbers shown in the large oval are the ages of members of the population being studied. You can see that the smallest oval represents a sample size of 5, since there are 5 members of the population in it. The next larger oval represents a sample of size 10. In addition to the 5 members contained in the small oval, it contains 5 more members of the population.

*Remember! A sampling error does not imply that you did anything wrong. It results simply because you have an incomplete picture of the population.*

If we continue taking increasingly bigger sample sizes, eventually we will end up taking a census, or studying the entire population. This is represented by the largest

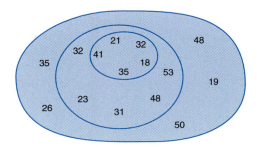

**FIGURE 2.2**    Bigger and bigger samples

oval. When this happens, the sample size is as big as the population size, *N*, and there is no sampling error.

*The amount of error that we can afford is an important factor in determining sample size.*

It seems, then, that if we can live with a little more error, then we can get by with a smaller sample size. The more costly the error, the larger our sample size will need to be. Thus, the *amount of sampling error* (dictated by the cost of the error) we can live with is a factor in determining sample size. We return to this matter in greater detail in Chapters 7–9.

### 2.4.4 Amount of Variation in the Population

*The amount of variation in the population is a key factor in determining sample size.*

Even if we can tolerate only a small amount of error we still may not need a really large sample size. One other factor is very important: the *amount of variability* that exists in the population. Suppose for a moment that I wish to study all the students in your college. I wish to be very accurate in my conclusions but all of the students feel exactly the same about the issue I am studying. How many students do I need to talk to in order to get very precise information about the population? The correct answer is that only 1 student is needed. A sample size of 1 is adequate despite the need to be very accurate and despite the large size of the population. This is because there is no variation in the population; that is, everyone feels the same way about the issue. Granted, this is not likely to happen, but you can see the impact that this factor has on the sample size needed. The more similar the members of the population are to each other, the less variation there is within the population. So the sample size can be smaller. If the population is highly diversified then you will need to talk to more people to get a sense of how the population feels about the issue.

*How will we know the amount of variability in the population?*

A natural question to ask is, How will we know the amount of variability of the population to use in determining the sample size? Considering that we are taking a sample to estimate such parameters as variability in the population, it seems a bit of a problem if we need to know the variability to determine the sample size. We are locked in a loop at this point, a classic "catch 22" situation. We could estimate the amount of variability using information about the amount of variability in a population similar to the one we are studying; we could use previous studies of the same population to give us an idea of the amount of variability; or we could do a pilot study. These are some of the ideas for dealing with this dilemma and they are addressed more completely in Chapter 7.

### 2.4.5 Summary of Factors Influencing Sample Size

We have identified the following factors that are important in determining the size of the sample needed:

- The amount of variation in the population
- The amount of error that can be tolerated
- The amount of resources available for the project
- The size of the population, *N*

You will see that the first two factors show up in the formula for determining the size of the sample. These are developed in Chapter 7. For now it is sufficient for you to have a general understanding of the impact that each of these factors has on the sample size determination.

As a final note in this section, we should agree on a label for the sample size. No matter what statistics book you pick up you will always see the **sample size** referred to as ***n***. Remember that the size of the population is labeled *N*.

> The *size of the sample* will be referred to as *n*.

## 2.5   SELECTING THE SAMPLE

Now that you have a general idea of which factors influence the size of the sample to be selected, we are ready to think about how we could select a sample. In the *Try It Now!* exercise on page 20 you easily drew another oval to represent another sample that could be picked from the population. Let's think for a moment about what qualities we would like our sample to have.

### 2.5.1 Selecting an Unbiased Sample

Ideally, we would like the sample to be a mini version of the population. Remember that we will be using the sample to understand the population. The sample should thus contain all the key features found in the population. It is easiest to understand what this means by looking at some examples of samples that may *not* be a mini version of the population.

### EXAMPLE 2.8   The Glue Company

*Selecting a Sample*

Suppose the glue company decided to use the first 50 tubes of glue made on a Monday morning as the sample.

Why is this not a good idea? It is possible that the manufacturing process varies as the machine runs for a long period of time. It is also possible that different people operate the machine at different times of the day and on different days of the week. And it is possible that a different batch of raw material might lead to differences in the glue performance. If our sample consists of the first 50 tubes made on Monday morning, we may not see the impact of any of these factors on the glue performance. ■

### EXAMPLE 2.9   Dress Down Day

*Selecting a Sample*

Suppose the company considering a dress down day decides to use the recently hired female employees as their sample.

Again the problem is that this sample may not truly represent the population of employees of the company. Why not? First of all, the people in this sample are all women. Maybe women in this company are consistently more or less productive than men. Secondly, they are all new hires and are likely to be putting their best effort into the job. At the same time, they are in training and may make more than the average number of errors. For all of these reasons and probably others this group is not a mini version of the employees of this company. ■

In both of these situations the proposed sample may not let us see all of the variation that exists in the population. Another way to express this is to say that these samples might be **biased.** The word bias in statistics has the same meaning that it has in ordinary usage. It means that the sample is somehow not a fair reflection of the reality we would see in the population. A biased sample would give us an unfair or prejudiced view of the population. This is precisely what we wish to avoid. Let's capture this in a definition.

A *biased sample* is a sample that does not represent the population.

*Try It Now!*

## Stress Relief
### *Identifying Possible Biases in a Sample*

You are studying the methods that students at your school use to relieve stress. You decide to use your statistics class as your sample. Why might this be a biased sample?

## 2.5.2 Selecting a Simple Random Sample

In selecting any sample, we wish to pick an unbiased or fair sample. We also wish to pick a sample that contains as much information as possible. There are many ways to pick a sample but for the purposes of an introductory course we are going to use what is known as a **simple random sample (srs).** This means that each member of the population has an equal chance of being selected as a member of the sample. This is equivalent to placing the names of all the members of the population in a hat and reaching in and picking out members to be in the sample.

> A *simple random sample* is a sample that has been selected in such a way that all members of the population have an equal chance of being picked.

Simple random sampling is the most obvious way to select a sample. In fact, you might wonder how else you could do it. Suppose you wished to learn about prices charged for advertising in newspapers. Your population would be all companies located in the United States that publish newspapers. If you picked a simple random sample of newspaper companies you might end up with all small companies purely by chance. You might not get any of the "biggies" such as the *New York Times,* the *Boston Globe,* the *Chicago Tribune,* or the *Los Angeles Sun* in the sample. Is this bad? Well, that depends on what variable you are studying. We are interested in prices for advertising in these papers and the larger companies might set the pace for the smaller companies with regard to pricing. In this case you might choose to use something other than simple random sampling in order to guarantee that you had some "biggies" in your sample.

As you learn more about statistics you will see that simple random sampling does not always provide you with the most information for your money. All of the concepts that you will learn in this book carry over to the other kinds of selection methods, but the formulas might have to change a bit if you use a procedure other than simple random sampling. In Chapter 14 you will learn how to select a sample by designing an experiment.

## 2.5.3 The Sampling Frame

Now that we have decided that we want to select a simple random sample, we must look at what is needed to pick the sample. Returning to the example of the newspapers from the previous section, we can see that we need a list of all the newspaper publishing companies in the United States. Such a list is known as the **sampling frame.**

> A *sampling frame* is a list of all members of the population.

There are government agencies and private companies, such as Dun & Bradstreet, whose function is to collect data for sampling frames. However, creating a sampling frame may take a fair amount of time, energy, and money. Thus, the list from which you pick your sample may not be accurate or complete. The difficulty in creating these lists is not the focus of this course or this book. However, you should know what a sampling frame is and know enough to watch out for potential problems.

## 2.5.4 Using a Table of Random Numbers to Select the Sample

Once you have the sampling frame and you have determined your sample size (using the formula developed in Chapter 7), you are ready to select your sample. Although it is easy to understand the idea of placing the name of each member of the population in a hat and selecting the sample by picking names from the hat, this is not a practical way to select the sample. To imitate this process you can use a **table of random numbers.**

> A *table of random numbers* is a table that consists of a list of numbers randomly generated and listed in the order in which they were generated.

Such a table is provided in Table 1 in Appendix A. A portion of this table is shown in Figure 2.3.

| | | | | Column | | | |
|---|---|---|---|---|---|---|---|
| **Row** | **1** | **2** | **3** | **4** | **5** | **6** | **7** |
| 1 | 094632795 | 711501513 | 537971597 | 562758635 | 410398128 | 182794408 | 773761503 |
| 2 | 033413186 | 653475420 | 289063704 | 485441982 | 460744361 | 328703833 | 289612212 |
| 3 | 297556368 | 658953044 | 738968017 | 414437050 | 296126017 | 075254187 | 702140315 |
| 4 | 472960570 | 785645638 | 574817322 | 817883255 | 976076280 | 843373358 | 118284363 |
| 5 | 256883707 | 716249997 | 378236162 | 467694224 | 193707682 | 380141891 | 605807481 |
| 6 | 179451522 | 878902420 | 602450872 | 987686989 | 686677180 | 242196303 | 517640224 |
| 7 | 894964682 | 704841116 | 241902107 | 750429362 | 794778197 | 693242123 | 316755091 |
| 8 | 738120861 | 744470405 | 873393138 | 758824215 | 394004646 | 496696605 | 006936567 |
| 9 | 803156944 | 653387115 | 716335974 | 835667154 | 066959782 | 908783760 | 165946696 |
| 10 | 187636922 | 321421098 | 638210137 | 055734541 | 493193305 | 566923120 | 435549770 |

**FIGURE 2.3**   Portion of a table of random numbers

A random number table should contain roughly as many zeros as ones as twos and so forth. There is a way to check to see that such a table does in fact contain random numbers, but for our purposes we will assume that the random number table has been correctly generated.

To use this table to identify which members of our population will be selected to be in the sample, we must first assign each member of the population an identification (ID) number. Suppose that you are studying the students at your college or university. Each student probably has a student ID number. Since many schools use the student's social security number as the ID number, let's suppose that the ID number is a 9-digit number. To use a table of random numbers to select a sample of 30 students from this population, you would first select where in the random number table you should start reading. One way to do this is to close your eyes and point to a spot in the table. Suppose that you did this and you selected row 10, column 2 as your starting point. Then you would read the next nine digits from Figure 2.3 as 321421098. This is the ID number of the first member of your sample. The next 9-digit number from the table is 638210137 and this is the ID number of the second member of your sample. You would continue this process until you have selected 30 ID numbers.

The sample obtained in this manner is a simple random sample. However, you would certainly need to read more than 30 ID numbers from the table to get 30 usable ID numbers. Most of the ID numbers you read from the table of random numbers will not correspond to anyone at your school. This would certainly happen, especially if the school is using the student's social security number as the ID number. Although you will eventually find 30 usable ID numbers, you will waste an incredible amount of time.

This method works a little better when you can assign the ID codes to the members of your population in such a way that they are sequential, so that you will not select any nonusable ID numbers. Let's see how this works in the following example.

### EXAMPLE 2.10   Student Views

#### Selecting a Sample Using a Table of Random Numbers

Each student at your college has a mailbox on campus. The mailboxes are numbered from 0000 to 9000. To select a simple random sample of 10 students we can select 10 mailbox numbers at random using the random number table. Suppose we choose to start at row 7, column 3 of the table shown in Figure 2.3. The first student selected has mailbox number 2419, which is a valid number. Continuing to read off 4-digit numbers from this table, the second number selected has mailbox number 0210. The list of all 10 mailbox numbers selected is:

| 2419 | 0210 | 7750 | 4293 | 6279 | 4778 | 1976 | 2123 | 3167 | 5509 |

Since the numbers are organized in 9-digit blocks and we need only 4-digit mailbox numbers, we just kept reading the numbers sequentially, wrapping down to the next row upon reaching the end of column 7. Notice that the selection after 1976 would have been 9324, which is not a valid mailbox number, so it was simply skipped.

The students who have these mailboxes are the students in the sample.   ■

For the most part, this course and this book assume that the sample has been selected and the data have been collected. Your job is to see the information in the data and make some recommendations and/or decisions based on the data. We do not focus on the actual selection of the sample or the job of collecting the data. This is the subject of another course.

## *TRY IT NOW!*

### The Glue Company
*Selecting a Simple Random Sample*

Select a sample of 5 tubes of glue for the glue company. You can assume that each tube of glue has a 5-digit ID number, which the company uses to track its inventory.

## *Discovery Exercise 2.2*
## *INTRODUCTION TO SAMPLING*

Suppose the data shown below represent an entire population. They show the number of people in 50 families living in a small college town in New England. (If you did Discovery Exercise 2.1 then you will recognize this as the same data set.) Now a 2-digit ID number has also been included.

**New England families: large amount of variability**
**Average number of people in 50 families: 4.50**

| | | | | | | | | | |
|---|---|---|---|---|---|---|---|---|---|
| ID: 01 | 1 | ID: 02 | 4 | ID: 03 | 5 | ID: 04 | 7 | ID: 05 | 8 |
| ID: 06 | 3 | ID: 07 | 9 | ID: 08 | 8 | ID: 09 | 8 | ID: 10 | 8 |
| ID: 11 | 4 | ID: 12 | 9 | ID: 13 | 9 | ID: 14 | 1 | ID: 15 | 6 |
| ID: 16 | 4 | ID: 17 | 1 | ID: 18 | 3 | ID: 19 | 9 | ID: 20 | 7 |
| ID: 21 | 8 | ID: 22 | 2 | ID: 23 | 3 | ID: 24 | 1 | ID: 25 | 9 |
| ID: 26 | 1 | ID: 27 | 7 | ID: 28 | 5 | ID: 29 | 1 | ID: 30 | 1 |
| ID: 31 | 1 | ID: 32 | 6 | ID: 33 | 8 | ID: 34 | 2 | ID: 35 | 9 |
| ID: 36 | 4 | ID: 37 | 1 | ID: 38 | 1 | ID: 39 | 1 | ID: 40 | 3 |
| ID: 41 | 4 | ID: 42 | 2 | ID: 43 | 4 | ID: 44 | 9 | ID: 45 | 4 |
| ID: 46 | 1 | ID: 47 | 3 | ID: 48 | 8 | ID: 49 | 1 | ID: 50 | 1 |

**Step 1:** Select a sample of 5 numbers from this population. Use the table of random numbers to do this. Record your sample in the table at the top of page 36.

| | Sample from NE families |
|---|---|
| Selection 1 | |
| Selection 2 | |
| Selection 3 | |
| Selection 4 | |
| Selection 5 | |
| **Average** | |

**Step 2:**   Calculate the sample average of the 5 numbers by adding them together and dividing by 5.

**Step 3:**   Calculate how far away the sample average is from the true population average by subtracting the sample average from the population average that is provided for you.

**Step 4:**   If you did Discovery Exercise 2.1, compare the sample you selected using the random number table to the sample you selected without the use of the table. Which one had a sample average closer to the population average?

Repeat steps 1–4 with a sample of 10 numbers.

**Step 1:**   Select a sample of 10 numbers from each population. Use the table of random numbers.

| | Sample from NE families | | Sample from NE families |
|---|---|---|---|
| Selection 1 | | Selection 7 | |
| Selection 2 | | Selection 8 | |
| Selection 3 | | Selection 9 | |
| Selection 4 | | Selection 10 | |
| Selection 5 | | **Average** | |
| Selection 6 | | | |

**Step 2:**   Calculate the sample average of the 10 numbers by adding them together and dividing by 10.

**Step 3:** Calculate how far away the sample average is from the true population average by subtracting the sample average from the population average that is provided for you.

**Step 4:** If you did Discovery Exercise 2.1, compare the sample you selected using the random number table to the sample you selected without the use of the table. Which one had a sample average closer to the population average?

## 2.5.5 Exercises—Learning It!

**2.17** The President of the United States wishes to see how popular he/she is after 2 years in office. A sample of 1000 voters is taken from the state of California.
   (a) Why might this be a biased sample?
   (b) How could you get a simple random sample?

**2.18** Tasty Ice Cream Corporation wishes to be sure that all of the half-gallon ice cream cartons do contain one-half gallon of ice cream. A sample of 30 cartons are measured. All of the cartons in the sample were filled on Friday.
   (a) Why might this be a biased sample?
   (b) How could you get a simple random sample?

**2.19** A company that manufactures electronic switches wishes to provide the customer with defect-free switches. A sample of 5 switches is selected every hour throughout the day. Explain why this is most likely an unbiased sample.

**2.20** The university wishes to know how students view the new athletic center on campus. A questionnaire is distributed to people as they enter the building.
   (a) Why might this be a biased sample?
   (b) How could you get a simple random sample?

**2.21** The U.S. government wishes to know how many people are unemployed. A sample of 1000 individuals over age 18 is selected from a national listing. Explain why this is probably an unbiased sample.

**2.22** The manufacturer of disposable diapers wishes to know how much fluid the diapers can hold before they leak. The diapers are put on children who are playing and fluid is injected every 15 minutes until the diaper leaks. Explain why this is probably an unbiased sample.

**2.23** The Coca-Cola Company wishes to know the proportion of people who prefer Coke over Pepsi. A sample of 100 people at a county fair is taken.
   (a) Why might this be a biased sample?
   (b) How could you get a simple random sample?

**2.24** Before selecting a major, a student decides to study the salaries of individuals working as accountants. A sample of accountants is selected from the list of alumni of the school that the student is attending.
   (a) Why might this be a biased sample?
   (b) How could you get a simple random sample?

## 2.6   TYPES OF DATA

Although we will not study the whole area of data collection, we do need to think about the different kinds of data or variables that we might encounter. The kind of statistical analysis that we do depends on the type of data we have. There are 2 major types of variables, **qualitative** and **quantitative.** In this section we concentrate on being able to identify the different types of data. Since different statistical techniques are used with different types of data, it is important to identify the type of data you have before you analyze it so that you don't use the wrong technique.

### 2.6.1 Qualitative Data

**Qualitative data,** also known as **nominal** or **categorical** data, are the simplest form of data. Examples of qualitative data are variables such as gender (male or female) or the expected grade in a course (A, B, C, D, or F). Each item in the sample falls into one of a finite number of possible categories.

> *Qualitative data* describe a particular characteristic of a sample item. They are most often non-numerical in nature.

Suppose you are interested in learning about the length of time that a certain glue, type C, adheres. You collect the data from a sample of glue tubes and after analyzing it you find that there appear to be two groups in the data. Each group has sticking times that cluster around a different value. Why would this occur? When you ask some questions about how the data were collected you learn that some of the tubes came from machine A and some came from machine B. When you return to the sticking times you find that all of the machine A values are clustered at the lower number while those of machine B are clustered at the higher number. The variable *machine* is an example of qualitative data. Knowing which machine produced the tube helps to explain the differences in the glue sticking times that you saw in the data.

 *TRY IT NOW!*

### Dress Down Day
*Possible Qualitative Variables*

The company considering the dress down day tries this policy out with a sample of employees. After the policy has been in effect for some time, the company decides to measure the change in productivity for the employees. It appears that for some employees productivity has increased, while for others it has decreased. What qualitative data might have been collected to help the company understand the differences observed?

The statistical techniques that are used to analyze qualitative data are limited, but are often critical to our understanding of the results of statistical analyses. When you are collecting data it is important to think of qualitative data that may be relevant to the problem to be solved. Qualitative data are easily collected at the time, but almost impossible to reconstruct after the fact. When in doubt, qualitative data should always be collected.

Sometimes numbers are used to classify qualitative data. For example, in surveys that ask for gender we often find that a 0 is used to denote "male" and a 1 is used to denote "female." When this is the case the data are referred to as **nominal** data. The numbers are used simply to represent different categories and have no real meaning as numbers.

> Data that are created by assigning numbers to different categories when the numbers have no real meaning are called *nominal data.*

Nominal data are treated the same way as ordinary qualitative data.

The order in which numbers are assigned to qualitative data may have some meaning. When numbers are used to name ordered categories the data are called **ordinal.** In the case of gender, the numbers 0 and 1 could very easily be reversed and so they have no intrinsic meaning. But suppose you asked a group of people to rank five different versions of a new soft drink. The resulting data might look like this:

| Version | 3 | 2 | 5 | 1 | 4 |
|---|---|---|---|---|---|
| Ranking | 1 | 2 | 3 | 4 | 5 |

In this case the numbers are not entirely meaningless since they indicate a relative position for each version of the product on a scale. However, you cannot tell from this scale whether the person doing the ranking liked version 3 a lot and really hated the other four or whether versions 3 and 2 were similar and much superior to the remaining three versions. The distances between the assigned numbers are not necessarily equal.

Another example of ordinal data would be when a characteristic of a sample item, such as income, is classified as 1 = low, 2 = medium, and 3 = high. Here, the numbers have a relative ordering, but there is no way to compare 1 to 2 to 3 numerically.

> Data that are created by assigning numbers to categories where the order of assignment has meaning are called *ordinal data.*

Most often, ordinal data are analyzed using the same graphical techniques that apply to other qualitative data. There are some limited statistical techniques that can be used to further analyze ordinal data. These techniques are called nonparametric analyses.

## 2.6.2 Quantitative Data

Data that are inherently numerical in form are called **quantitative** data. This type of data falls into two different categories: **discrete** and **continuous.**

Discrete data usually result when the data being collected involve *counting* or *enumerating.* The only possible values are positive integers or whole numbers. Examples of discrete data are the number of prior convictions for a person who has been arrested, or the number of defective items in a sample.

> ***Discrete data*** are data that can take on only certain values. These values are often integers or whole numbers.

Discrete data that result from counting the number of times that something occurs are important. They are the type of data used in analyzing many public opinion polls and other surveys. You will encounter this kind of data many times in the text.

> ***Continuous data*** are data that can take on any one of an infinite number of possible values over an interval on the number line. These values are most often the result of measurement.

Numerical data that are not discrete are called **continuous.** Continuous data occur most often as the result of measuring and are sometimes referred to as *measurement data*. Continuous data usually consist of real numbers, which, as far as we are concerned, will be in decimal form.

### EXAMPLE 2.11   Study Time

***An Example of Continuous Data***

Suppose you are interested in learning about the length of time that a typical student spends studying statistics on any given night. Initially, there are an infinite number of possibilities for values, ranging from 0 minutes up. These data are continuous. Once you decide on a measuring device the number of possibilities is limited, but the limitation is caused by the measuring device and not the variable itself. For example, if you measure the time spent studying to the nearest hour, the possible values are 0, 1, 2, . . . .   If, however, you decide to measure to the nearest one-tenth of an hour (6 minutes), the possible values are 0.0, 0.1. 0.2, . . . , 3.0, 3.1, . . . . . The more precise your measuring device, the more possible values your data can assume.   ■

### 2.6.3 Exercises—Learning It!

**2.25**   The President of the United States wishes to see how popular he/she is after 2 years in office.
  (a)  What is one variable of interest?
  (b)  Is the variable qualitative or quantitative?
  (c)  If it is qualitative, is it nominal or ordinal data? If it is quantitative, is it discrete or continuous?

**2.26**   Tasty Ice Cream Corporation wishes to be sure that all of the half-gallon ice cream cartons do contain one-half gallon of ice cream.
  (a)  What is one variable of interest?
  (b)  Are the data qualitative or quantitative?
  (c)  If the data are qualitative, are they nominal or ordinal? If they are quantitative, are they discrete or continuous?

**2.27**   A company that manufactures electronic switches wishes to provide the customer with defect-free switches.
  (a)  What is one variable of interest?
  (b)  Are the data qualitative or quantitative?
  (c)  If the data are qualitative, are they nominal or ordinal? If they are quantitative, are they discrete or continuous?

**2.28**   The university wishes to know how students view the new athletic center on campus.
  (a)  What is one variable of interest?

(b) Are the data qualitative or quantitative?

(c) If the data are qualitative, are they nominal or ordinal? If they are quantitative, are they discrete or continuous?

**2.29** The U.S. government wishes to know how many people are unemployed.

(a) What is one variable of interest?

(b) Are the data qualitative or quantitative?

(c) If the data are qualitative, are they nominal or ordinal? If they are quantitative, are they discrete or continuous?

**2.30** The manufacturer of disposable diapers wishes to know how much fluid the diapers can hold before they leak.

(a) What is one variable of interest?

(b) Are the data qualitative or quantitative?

(c) If the data are qualitative, are they nominal or ordinal? If they are quantitative, are they discrete or continuous?

**2.31** The Coca-Cola Company wishes to know the proportion of people who prefer Coke over Pepsi.

(a) What is one variable of interest?

(b) Are the data qualitative or quantitative?

(c) If the data are qualitative, are they nominal or ordinal? If they are quantitative, are they discrete or continuous?

**2.32** Before selecting a major, a student decides to study the salaries of individuals working as accountants.

(a) What is one variable of interest?

(b) Are the data qualitative or quantitative?

(c) If the data are qualitative, are they nominal or ordinal? If they are quantitative, are they discrete or continuous?

# 2.7 THE DIFFERENCE BETWEEN DESCRIPTIVE STATISTICS AND INFERENTIAL STATISTICS

So far, we have looked at the concept of population and recognized the usefulness of the sample in gleaning information about the population. We have discussed factors that influence how large the sample needs to be and how we go about picking the sample. And we have looked at the types of data that might be collected from the sample. Now we need an answer to the question, What do I do with the data? The right answer is, It depends. It depends on what type of data you have and on what questions you want answered about the behavior of the population. The rest of this book is devoted to providing tools to answer the question, What do I do with the data? These tools fall into two main categories: the tools of descriptive statistics and the tools of inferential statistics. Although these tools support each other, the jobs they do are quite different.

## 2.7.1 Descriptive Statistics

The tools of descriptive statistics are usually the first ones encountered in any data analysis. They allow you to *describe* the sample. However, that is all they do. This is generally the first step in any data analysis, once the data have been collected, of course. If you were given the results of a survey on customer satisfaction from 100 customers, the first thing you would want to do is to get a handle on the data by summarizing them. This is precisely what the tools of descriptive statistics do for you: summarize the data.

*Tools of descriptive statistics* allow you to summarize the data.

There are *graphical or visual descriptive tools,* which generally include bar charts, pie charts, and histograms. These tools are discussed in Chapter 3. Graphical tools help you to see how the data behave and to summarize the data visually. These tools are used all the time. You frequently will see them in newspapers and in reports. Some examples are shown in Figure 2.4.

There are also *numerical descriptive tools,* which allow you to summarize the data numerically. Typically, a numerical summary would provide you with such *statistics* as the average, the median, the mode, and the largest and smallest data values. We used the word statistics in the last sentence because that is precisely what these values are. Remember that a statistic is simply a number that describes the behavior of the sample. These numbers are covered in more detail in Chapter 4.

Descriptive tools would be adequate if all we wanted to do was describe the sample. But remember that the sample is a means to an end. At the end of the day it is not the sample that we care about but the population. Somehow we must make the leap from the information contained in the sample to the population behavior. To do this we need the tools of inferential statistics.

## 2.7.2 Inferential Statistics

Let's start by defining the word **inference.** If your friend told you not to infer that all business students are brilliant, then your friend is telling you not to reach that *conclusion!*

An *inference* is a deduction or a conclusion.

In our case we wish to draw conclusions about the behavior of the population based on the information in the sample. We have already seen that the sample is only a piece of the population, so we will not have complete information. We have also seen that if the sample is properly selected, then the information contained in the sample will give us reliable information about the population.

How are we going to make this leap from describing the sample data to drawing an inference or conclusion about the behavior of the population? The answer is found in the subject matter of probability. Indeed, without the tools of probability, we could not do anything more than describe the sample data. We could do no inferential statistics.

Think about the population and the sample as the circles shown in Figure 2.5. The larger circle represents the population and the smaller circle represents the sample. This figure is being used to help you see how probability is needed in order to draw inferences about the population from the sample. In reality, the sample circle

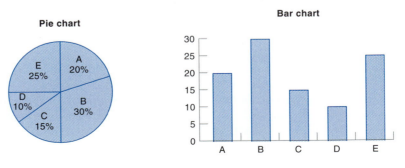

**FIGURE 2.4**   Pie chart and bar chart

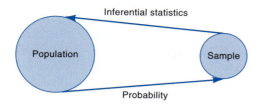

**FIGURE 2.5**   Relationship between probability and inferential statistics

should sit inside the population circle, because it is a piece of it. But for now let us continue with this picture. The line that moves from the sample to the population is labeled "Inferential statistics." The tools of inferential statistics are used to move from sample information to population information. Let's save that thought as a definition.

> The *techniques of inferential statistics* allow us to draw inferences or conclusions about the population from the sample.

To understand the line that moves from the population to the sample, consider the following situation. You and I have decided to play poker and I have dealt us each 5 cards. You see that you have a full house and are wildly excited. However, your excitement is quickly contained as I reveal to you that I have not one, not two, not three, not four, but *five* aces! Immediately you accuse me of cheating (imagine that!). You quickly conclude that I have not used a standard deck.

Let's take a closer look at how you reached that conclusion. First of all, the deck of cards is our population. You had a sample of 5 cards selected from that population and I had a sample of 5 cards from that population. On the basis of the information in my sample you calculated the probability of observing 5 aces if I was using a standard deck and found it to be zero. There is no chance of getting 5 aces if the population has the behavior of a standard deck. So you quickly rejected the notion that I was using a standard deck. This is precisely what we need to do in general. We need to draw conclusions about a population based on the observed sample and the theory of probability. The conclusions will not always be so obvious as this example. So we have a definition for the line labeled "probability," which leads from the population to the sample.

*Reaching a conclusion based on a sample. Did I cheat?*

> We will use *probability* theory to calculate the likelihood of observing or selecting a particular sample from a population.

Although the branch of mathematics known as probability theory is interesting in its own right, we present only those particular parts of probability that will help us in our ultimate goal: to make inferences about the population from the sample.

Let's summarize Figure 2.5. If we make the trip across the bottom line we would be making some assumptions about the population behavior and determining the likelihood of various samples that might come from this population. If we make the trip across the top line we are taking the information in the sample and drawing conclusions about the population from which it came. The tools of probability ultimately allow us to accomplish what we want, which is to do inferential statistics.

### 2.7.3 Upcoming Chapters

Based on our discussions in this section, it is clear that the first thing we should do is learn the tools of descriptive statistics. These are covered in Chapters 3, 4, and 5.

Chapter 6 presents the necessary probability theory that helps us move beyond simply describing the sample data. Chapters 7–16 present various techniques of inferential statistics. The last chapter (Chapter 17) provides a synthesis of the book.

In all of the material that follows this chapter, you will need to be familiar with basic summation notation. If you are familiar with this notation, you may skip the next section. But you may wish to use it as a quick review and refer to it as you proceed through the rest of the book.

# 2.8   BASIC SUMMATION NOTATION

Before we go much further, it is useful to introduce some shorthand notation for writing statistical formulas. This notation is called the **sigma notation.**

> *Sigma notation* is shorthand notation used to write formulas. It is so named because it uses the Greek capital letter sigma, written as $\Sigma$.

## 2.8.1  Summing the Data

*Remember that we agreed to use the letter* n *to represent the size of the sample.*

If we have a sample of size $n$, we can refer to the individual data values as $x_1$, $x_2$, $x_3$, . . ., $x_n$, where $x_1$ represents the first data value, $x_2$ the second data value, and so on. Many of the formulas you encounter will require that you sum the data values. We could write the sum as $x_1 + x_2 + x_3 + \cdots + x_n$ but this will get cumbersome and be awkward to use.

Using the sigma notation, we can write the sum $x_1 + x_2 + x_3 + \cdots + x_n$ as

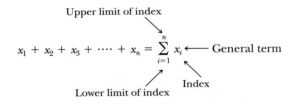

The sigma notation on the right of the equation is read as "the sum of $x_i$ as $i$ goes from 1 to $n$." The letter under the $\Sigma$ is called the index and it really doesn't matter what letter you use since you use the same letter in describing the general term. The value of the index indicated on the bottom of the $\Sigma$ is the starting value for the index and is usually 1, meaning that we are going to start with $x_1$. The value on the top of the $\Sigma$ is the ending value for the index and is usually $n$, the sample size. When the index goes from $i = 1$ to $n$, this indicates that all of the data values in the sample are being used.

Let's try some examples.

## EXAMPLE 2.12   The Glue Company

### *Using the Summation Notation*

The glue company studies the performance (i.e., the stickiness) of 10 tubes of type C glue. The data are recorded in terms of number of days until the glue failed. This is what the company found: 233, 167, 289, 221, 254, 198, 210, 225, 240, 203. Thus, $x_1$ is 233, the first member of the sample. In a similar fashion we can label the rest of the data values: $x_2 = 167$, $x_3 = 289$, $x_4 = 221$, $x_5 = 254$, $x_6 = 198$, $x_7 = 210$, $x_8 = 225$, $x_9 = 240$, and $x_{10} = 203$. The value of $n$ is 10, the size of the sample.

To find $x_1 + x_2 + x_3 + \cdots + x_{10}$ we will first write it as $\sum_{i=1}^{10} x_i$. This means that we start the value of $i$ at 1 and start with $x_1$, which we know is 233. Check to see if the value for $i$ is at the upper limit for $i$, which in this case is 10. Since $i$ is only 1 we continue. Next put a plus sign and change $i$ to be 2 and add in $x_2$, which we know to be 167. So far we have $233 + 167$. Again we check to see if the value for $i$ is bigger than the upper limit of 10. It is not, so we continue by putting a plus sign and changing $i$ to 3 and adding $x_3$, which we know is 289. Now we have $233 + 167 + 289$. This process continues until the value of $i$ is 10 and we have written down $233 + 167 + 289 + 221 + 254 + 198 + 210 + 225 + 240 + 203$ and obtained 2240.

■

## TRY IT NOW!

### The Mail-Order Company

*Using Sigma Notation*

A mail-order company wants some information on the daily demand for a product that has been heavily advertised. The company looks at the orders for an 8-day period and obtains the following data:

| Demand | 31 | 28 | 29 | 32 | 30 | 31 | 29 | 30 |
|--------|----|----|----|----|----|----|----|----|

Use the sigma notation to write down the expression which means to add up all the data values.

## 2.8.2 Summing Differences

When we analyze data we often need to look at how *far away* data values are from some number. Clearly, we could state how far away any one of the data values was from a number by simply subtracting the two numbers. For the glue company data shown in Example 2.12 we could say that $x_1$ is 9 days away from a number, for instance 224. Remember that $x_1$ is 233. In addition to being interested in how far away $x_1$ is from 224, we often have to state how far away the whole data set is from 224. To do this we have to subtract 224 from each value in the data set, each $x_i$, and then add up all these differences.

### EXAMPLE 2.13   The Glue Company

*Use of Sigma Notation to Calculate Differences*

Suppose we wish to quantify how far away the whole data set is from 224 for the glue company. The calculation we want is

$$(233 - 224) + (167 - 224) + (289 - 224) + (221 - 224) + (254 - 224) + (198 - 224)$$
$$+ (210 - 224) + (225 - 224) + (240 - 224) + (203 - 224)$$
$$= (9) + (-57) + (65) + (-3) + (30) + (-26) + (-14) + (1) + (16) + (-21)$$
$$= 0$$

Using the notation $x_1$ to represent the first data value and $x_2$ to represent the second data value, we can write this expression more generally as

$$(x_1 - 224) + (x_2 - 224) + (x_3 - 224) + (x_4 - 224) + (x_5 - 224)$$
$$+ (x_6 - 224) + (x_7 - 224) + (x_8 - 224) + (x_9 - 224) + (x_{10} - 224)$$

Taking this one step further, we can use the sigma notation to make this expression less cumbersome:

$$\sum_{i=1}^{10} (x_i - 224) \qquad\blacksquare$$

This shorthand way of writing the long summation will help us when we get to Chapter 4. Now you try one of these.

## Try It Now!

### The Mail-Order Company
*Using Sigma Notation to Sum Differences*

A mail-order company wants some information on the daily demand for a product that has been heavily advertised. The company looks at the orders for an 8-day period and obtains the following data:

Demand     31     28     29     32     30     31     29     30

Use the sigma notation to write down the expression that means to add up all differences between the data values and the number 30.

## 2.8.3 Exercises—Learning It!

**2.33**  Tasty Ice Cream Corporation wishes to be sure that all of the half-gallon ice cream cartons do contain one-half gallon of ice cream. A sample of 30 cartons are inspected. All of the cartons in the sample were filled on Friday. The data (in gallons) are

0.51   0.50   0.49   0.48   0.51   0.50   0.49   0.50   0.50   0.48   0.53   0.50   0.49   0.49   0.51
0.50   0.50   0.51   0.52   0.50   0.48   0.49   0.50   0.47   0.49   0.51   0.50   0.50   0.50   0.49

(a) Write the sigma notation to represent the sum of the 30 data values.
(b) Tasty would like to know how far away from 0.5 (one-half gallon) the data are on the whole. Write the sigma expression to find the sum of the difference between each data value and 0.5.
(c) Evaluate the expression you constructed in part (b).

**2.34**  The manufacturer of disposable diapers wishes to know how much fluid the diapers can hold before they leak. The diapers are put on 10 children who are playing and fluid is injected every 15 minutes until the diaper leaks. The weight of the diaper (in grams) when it fails is recorded. The data are

503    513    489    499    520    511    525    494    498    501

Write the sigma notation to represent the sum of these data values.

2.35  Before selecting a major, a student decides to study the salaries of individuals working as accountants. A sample of 15 accountants is selected from the list of alumni of the school that the student is attending. The data are

$25,100   $33,000   $27,500   $29,000   $35,000   $26,400   $29,100
$31,050   $32,100   $29,100   $40,000   $21,000   $25,000   $26,500   $33,000

(a) Write the sigma notation to represent the sum of all the salaries in the sample.
(b) Write the sigma notation to represent the sum of the first 5 salaries in the sample.
(c) Write the sigma notation to represent the sum of the last 5 salaries in the sample.

## 2.9  SELECTING A SAMPLE IN EXCEL

There are two ways to use Excel to select a simple random sample from a population. We will look at the method that assumes you have the population values already in a worksheet. Suppose that you have a population with 50 two-digit ID numbers, a portion of which is shown in Figure 2.6.

**FIGURE 2.6**  Population of 50 ID codes

### 2.9.1 Selecting a Random Sample from a Population

You can use the **Data Analysis** tools in Excel to select a random sample. From the **Tools** menu, select **Data Analysis;** the dialog box shown in Figure 2.7 (page 48) opens. These steps will allow you to select a random sample from the population:

1. Scroll down the **Analysis Tools** list and select **Sampling.** The dialog box shown in Figure 2.8 opens. You must tell Excel three things to obtain the sam-

*Note: Make sure the* **Tools** *menu contains the* **Data Analysis** *option. If not, you may have to run the* **MS Office Setup** *program to include the* **Analysis ToolPak** *and then execute the* **Tools**|**Add-Ins** *command.*

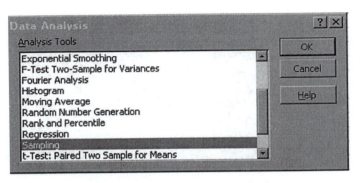

**FIGURE 2.7**   Data Analysis dialog box

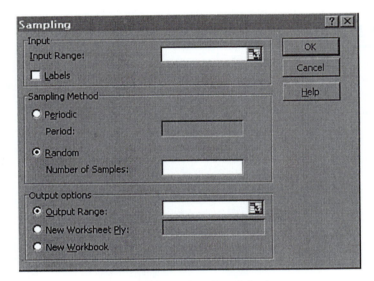

**FIGURE 2.8**   Sampling dialog box

ple: (1) the location of the population, (2) the type of sampling method and the number of samples, and (3) where you want the sample placed.

2. Position the cursor in the box labeled **Input Range** and then highlight the range in the worksheet that contains the data, in this case A1:A51. Since the first row of the population is a title, ID Codes, check the box marked **Labels.**

3. In the section labeled **Sampling Method,** click the radio button for the **Random** sampling method and type "5" in the text box for **Number of Samples:.**

4. In the section labeled Output Options, you can specify that the sample be located in a section of the current worksheet, a new worksheet in the same workbook, or a new workbook. In this example, click the radio button for **Output Range:** and position the cursor in the textbox. Then highlight the cell in the worksheet where you want the output to start. The finished dialog box should look like the one in Figure 2.9.

5. Click **OK;** the random sample of 5 ID codes will appear in the location you specified, in this case cells C1:C5, as shown in Figure 2.10. This tells you to sample the items with ID codes 42, 11, 10, 40, and 46.

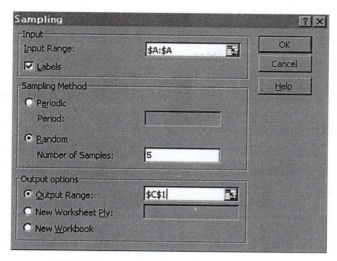

**FIGURE 2.9**   Completed sampling dialog box

|   | A | B | C |
|---|---|---|---|
| 1 | ID Codes | | 42 |
| 2 | 1 | | 11 |
| 3 | 2 | | 10 |
| 4 | 3 | | 40 |
| 5 | 4 | | 46 |
| 6 | 5 | | |
| 7 | 6 | | |
| 8 | 7 | | |

**FIGURE 2.10**   Random sample of 5 ID codes

# CHAPTER 2 SUMMARY

In this chapter you learned the basic language of statistics. You learned that the complete group that you wish to study is called the population. Typically there are several characteristics of each member of the population that you wish to know about. These are called variables. There are two types of variables: quantitative and qualitative. Ideally you would like to study the entire population, or take what is known as a census, but most of the time a census is too expensive and too time-consuming. For these reasons and others you select a sample from a list of the members of the population. In doing so you end up with sampling error, because you are not studying every member of the population. The only way to eliminate sampling error is to study the entire population.

We always want our sample to be a fair representation of the population. This is known as an unbiased sample. There are many ways to select the sample but we will use simple random samples in this book. You learned how to select a simple random sample using a table of random numbers or a software tool such as Excel. You were also introduced to some shorthand notation, known as sigma notation, which is used to write equations throughout the remainder of this book.

## Key Terms

| Term | Definition | Page reference |
|---|---|---|
| **Biased sample** | A **biased sample** is a sample that does not represent the population. | 31 |
| **Continuous data** | **Continuous data** are data that can take on any one of an infinite number of possible values over an interval on the number line. | 40 |
| **Descriptive statistics** | **Tools of descriptive statistics** allow you to summarize the data. | 42 |
| **Discrete data** | **Discrete data** are data that can take on only certain values. These values are often integers or whole numbers. | 40 |
| **Inference** | An **inference** is a deduction or a conclusion. | 42 |
| **Inferential statistics** | The **techniques of inferential statistics** allow us to draw inferences or conclusions about the population from the sample. | 43 |

*(continued)*

| Term | Definition | Page reference |
|------|-----------|----------------|
| Nominal data | Data that are created by assigning numbers to different categories when the numbers have no real meaning are called **nominal data.** | 39 |
| Ordinal data | Data that are created by assigning numbers to categories where the order of assignment has meaning are called **ordinal data.** | 39 |
| Parameter | A **parameter** is a number that describes a characteristic of the population. | 26 |
| Population | The **population** is everything you wish to study. | 18 |
| Population size | The **size of the population** is the number of members of the population. It will be referred to as $N$. | 29 |
| Probability | We use **probability** theory to calculate the likelihood of observing or selecting a particular sample from a population. | 43 |
| Qualitative data | **Qualitative data** describe a particular characteristic of a sample item. They are most often non-numerical in nature. | 38 |
| Quantitative data | Data that are inherently numerical in form are called **quantitative** data. | 39 |
| Sample | A **sample** is a piece of the population. | 20 |
| Sample size | The **size of the sample** will be referred to as $n$. | 30 |
| Sampling error | **Sampling error** is the difference between the behavior of the entire population and a sample of that population. | 22 |
| Sampling frame | A **sampling frame** is a list of all members of the population. | 33 |
| Sigma notation | **Sigma notation** is shorthand notation used to write formulas. It is so named because it uses the Greek capital letter sigma, written as $\Sigma$. | 44 |
| Simple random sample | A **simple random sample** is a sample that has been selected in such a way that all members of the population have an equal chance of being picked. | 32 |
| Statistic | A **statistic** is a number that describes the behavior of a sample. | 27 |
| Table of random numbers | A **table of random numbers** is a table that consists of a list of numbers randomly generated and listed in the order they were generated. | 33 |
| Variable | A **variable** is a characteristic of each member of the population. | 19 |
| Variation | The amount of **variation** refers to how different the members of the population are from each other with regard to the variable being studied. | 22 |

# CHAPTER 2 EXERCISES

## Learning It!

**2.36**  A golf ball manufacturer wishes to know how far its newly designed super fly ball will go.
   (a)  What is the population of interest?
   (b)  Identify which of the reasons for taking a sample (listed on page 21) apply in this case. (There are more than one.)
   (c)  What parameter is of interest?

    (d) What statistic might you calculate from the sample?

    (e) What type of data would you collect?

**2.37** A mall vendor wishes to know how people like his new chocolate chip cookies.

    (a) What is the population of interest?

    (b) Identify which of the reasons for taking a sample (listed on page 21) apply in this case.

    (c) What parameter is of interest?

    (d) What statistic might you calculate from the sample?

    (e) What type of data would you collect?

**2.38** A golf ball manufacturer wishes to know how far its newly designed super fly ball will go. A sample of 50 golf balls is hit and the distance is measured.

    (a) Why might this be a biased sample?

    (b) How could you get a simple random sample?

**2.39** A mall vendor wishes to know how people like his new chocolate chip cookies. Fifty shoppers are stopped and given free samples of the new cookies. Do you think this sample is representative of the population of mall shoppers? Explain why or why not.

**2.40** A golf ball manufacturer wishes to know how far its newly designed super fly ball will go. A sample of 50 golf balls is hit and the distance is measured in yards. The data are

| | | | | |
|---|---|---|---|---|
| 301 | 299 | 307 | 310 | 291 |
| 300 | 297 | 286 | 297 | 302 |
| 296 | 290 | 291 | 306 | 302 |
| 301 | 296 | 287 | 311 | 309 |
| 309 | 290 | 292 | 287 | 276 |
| 299 | 302 | 304 | 300 | 309 |
| 290 | 281 | 308 | 306 | 307 |
| 317 | 277 | 301 | 298 | 285 |
| 289 | 288 | 306 | 311 | 293 |
| 297 | 291 | 285 | 302 | 300 |

    (a) Write the sigma notation to represent the sum of the data values.

    (b) The golf ball manufacturer wishes to know how different the values are from 300. Write the sigma notation to subtract 300 from each data value and sum the differences.

## Thinking About It!

**2.41** Would your information be accurate if you studied every member of the population? Why or why not?

**2.42** In trying to understand the population, what is the only way to eliminate the possibility of reaching the wrong conclusion?

**2.43** Is it likely that the value of the statistic calculated from the sample is exactly the same as the value of the population parameter?

**2.44** Remembering that the percentage of voters who support the President of the United States is an unknown number, what might you suggest using as a best guess for this percentage?

**2.45** There is a classic story about how the wrong inferences can be reached if the sample is biased. Just prior to the presidential election between Truman and Dewey the *Chicago Tribune* selected a national sample of voters from the telephone book. These voters overwhelmingly supported Dewey in the presidential race and so the newspaper confidently printed the headline "Dewey Wins" before the results were finally tallied. As we know, the headline was wrong: Truman became the 33rd President in 1945. What mistake did the *Chicago Tribune* make?

## Doing It!

**2.46** The data in this problem have been collected over the past 11 years of futures trading of Frozen Concentrate Orange Juice (FCOJ). The FCOJ is traded on the New York Cotton Exchange. The unit of trading is 15,000 pounds (one contract = 15,000 lb). The price quoted is in cents per pound. The data show the average price for the month.    *Data file: FCOJ.XXX*

(a) Consider all the months of data to be the population and find the population average.

(b) Select 50 simple random samples of 30 months and find the sample average for each of the samples. Which sample average is closest to the population average? Which sample average is the farthest from the population average?

(c) Select 50 random samples of 40 months and find the sample average for each of the samples. Which sample average is closest to the population average? Which sample average is the farthest from the population average?

(d) Select 50 random samples of 50 months and find the sample average for each of the samples. Which sample average is closest to the population average? Which sample average is the farthest from the population average?

(e) What happens to the accuracy of the sample average as you increased your sample size from 30 to 40 to 50?

# Part II
## EXPLORATORY DATA ANALYSIS

# GRAPHICAL DISPLAYS OF DATA

## ALUACHA BALACLAVA COLLEGE

Aluacha Balaclava College (ABC) is having trouble with faculty compensation. The faculty claim that they have not received pay raises for several years and that new faculty are being hired in at much higher salaries than existing faculty members are getting. The former Provost has just resigned under pressure and the President has hired a new Provost to design a new compensation package that is fair and reasonable. The new Provost knows that, prior to designing anything new, she must first understand the current system and some of the history of the system. She decides that her first step must be to collect some data on faculty salaries.

After some thought, she decides that faculty salaries depend on other factors, so she has the Management Information Systems (MIS) department gather data for each faculty member on Current Rank (Professor, Associate Professor, Assistant Professor, and Instructor), number of years of service to the college, and current salary. The data, which they provide to her in the form of a list, are not very enlightening:

| Rank | Years of service | 1996–97 salary |
|------|------------------|----------------|
| ASST | 22 | 53316 |
| PROF | 11 | 64375 |
| ASSO | 7  | 63501 |
| ASSO | 6  | 59426 |
| ASSO | 20 | 49058 |
| PROF | 4  | 94969 |
| ASST | 21 | 54762 |
| ASSO | 9  | 55516 |

To gain any information from the data she will need to summarize them, probably with some graphical displays.

# 3.1    CHAPTER OBJECTIVES

In Chapter 2 you learned about the different types of data and about how data can and should be collected. You learned that people collect data to provide information about something they are interested in understanding, such as why brakes on rollerblades fail or why customers complain about tissues. Usually there is some problem that needs to be solved, like a new brake design, when this understanding is gained.

In this chapter we look at ways to display different types of data. The chapter covers the following material:

- Graphical Methods for Qualitative Data: frequency tables, bar charts, and pie charts
- Graphical Methods for Quantitative Data: frequency tables and histograms
- Other Graphical Methods: dotplots and Pareto diagrams

# 3.2    ORGANIZING DATA

If one objective of statistics is to obtain information about a set of data, then we need to organize the data in some way. When data are collected the initial result is usually a *list* of the observations for each variable. This is referred to as the *raw data*. Raw data, such as those given to the Provost at the college, provide almost no information. Statistics provides some tools or techniques for turning raw data into information.

## 3.2.1 The Frequency Distribution

One way to organize data is to consolidate them by determining how many times each value in the data occurs and making a table that summarizes this information. This is called a **frequency table** or **frequency distribution.**

> A *frequency table* or *frequency distribution* is a table containing each category, value, or class of values that a variable might have and the number of times that each one occurs in the data. The frequency of the $i$th class is denoted $f_i$.

## *Frequency Tables for Qualitative Data*

Creating a frequency table for qualitative data is not difficult. A basic frequency table has two columns. In the first column of the table each row lists one of the values for the variable of interest. The second column lists the corresponding number of times that the value occurred in the set of data. Later on we will add some additional columns to the table, but right now we will work with only two. Figure 3.1 shows the setup for a typical frequency table:

| Category | Frequency |
|----------|-----------|
| *Category 1* | $f_1$ |
| *Category 2* | $f_2$ |
| **Total** | $n$ |

**FIGURE 3.1**   A frequency table

## EXAMPLE 3.1  ABC Faculty Salaries

*Creating a Frequency Table*

The Provost at ABC knows that salary will differ for each faculty member, depending on several factors. One such factor is the current rank of the faculty member. She decides to look at the faculty by rank to see how the faculty is made up as a whole. She puts the data in the form of a frequency table as shown below:

| Rank | Frequency |
|---|---|
| Professor | 55 |
| Associate Professor | 67 |
| Assistant Professor | 77 |
| Instructor | 8 |
| **Total** | **207** |

From the frequency table she sees that while Assistant Professor has the highest frequency, the three highest ranks are not all that different. There are obviously not many faculty at the rank of Instructor. ∎

The order for the categories in the frequency table is not important. If there is a logical order, or if you know what the classifications will be before you fill out the table, then you might use a particular order. If you are creating the categories as you make the table, you might list them in the order in which they first appear in the raw data.

## EXAMPLE 3.2  Student Distribution

*Creating a Frequency Table*

The faculty in the School of Business at a university are concerned that many students are delaying taking the Introductory Statistics course required by all majors. The course can be taken as early as the freshman year, but traditionally is taken by sophomores. To determine if the concerns are justified, a section of Introductory Statistics is selected at random and each student is classified as a freshman (F), sophomore (S), junior (J) or senior (Sr) according to the number of credits completed to date. The raw data for the 28 students are shown below:

| Sr | Sr | S | S | S | Sr | Sr |
|---|---|---|---|---|---|---|
| S | S | S | S | S | F | F |
| Sr | S | Sr | Sr | Sr | F | F |
| J | S | J | S | S | Sr | Sr |

To use the data to answer their questions, the faculty organize the data in the form of a frequency table:

| Classification | Frequency |
|---|---|
| Freshman | 4 |
| Sophomore | 12 |
| Junior | 2 |
| Senior | 10 |
| **Total** | **28** |

From the table they see that very few students taking the course are freshmen or juniors, and that the rest are about evenly divided between sophomores and seniors. ∎

While the information obtained from the frequency table is certainly better than that from the raw data, it is still not quite as informative as it might be. The values in the frequency table are influenced by the sample size. For large samples the individual frequencies will be much larger numbers than for small samples. This makes comparisons of different samples difficult. To solve this problem we must find a way to express the frequency so that the sample size does not matter. One way to do this is to use the **relative frequency.**

> The *relative frequency* of a classification is the number of times an observation falls into that classification represented as a portion of the total number of observations. It can be expressed as a *fraction, decimal,* or *percentage.*

Thus, to find the relative frequency for the $i$th classification, $rf_i$, we use

$$rf_i = \frac{\text{Frequency of } i\text{th classification}}{\text{Sample size}} = \frac{f_i}{n}$$

### EXAMPLE 3.3   ABC Faculty Salaries

*Finding Relative Frequency*

To get a better idea of how the faculty were distributed among the ranks, the provost decided to calculate the relative frequency for each rank. The calculations and results are shown in the table below:

| Rank | Frequency | Relative frequency | Relative frequency (%) |
|---|---|---|---|
| Professor | 55 | 55/207 | 26.6 |
| Associate Professor | 67 | 67/207 | 32.4 |
| Assistant Professor | 77 | 77/207 | 37.2 |
| Instructor | 8 | 8/207 | 3.9 |
| **Total** | **207** | **207/207** | **100.0** |

*Note: If you add the percentages for relative frequency, you will not get 100% here, because the percentages have been rounded off.*

*Note: Use the actual variable name for the classification to provide the most information.*

It appears from the relative frequencies that the three largest classes are not quite as different as she had thought, but no single class constitutes a majority. ■

It does not matter whether you use fraction, decimal, or percentage to calculate the relative frequency. Percentage is easiest for most people to understand. When using percentage you need to be careful about rounding. When you report relative frequency to the nearest percent or even one decimal place, the values may not sum to exactly 100%. This is not a problem.

### EXAMPLE 3.4   Student Distribution

*Calculating the Relative Frequency*

To get a better picture of the students in the Introductory Statistics course, the person studying the data decides to look at the relative frequency for each classification.

*If these calculations are not readily apparent to you, then you should get out a calculator and practice them until they are!*

**Freshmen:**

$$rf_1 = \tfrac{4}{28} = 0.143 \quad \text{or, as a percentage} = 0.143 \times 100 = 14.3\%$$

**Sophomores:**

$$rf_2 = \tfrac{12}{28} = 0.429 \quad \text{or, as a percentage} = 0.429 \times 100 = 42.9\%$$

**Juniors:**

$$rf_3 = \tfrac{2}{28} = 0.071 \qquad \text{or, as a percentage} = 0.071 \times 100 = 7.1\%$$

**Seniors:**

$$rf_4 = \tfrac{10}{28} = 0.357 \qquad \text{or, as a percentage} = 0.357 \times 100 = 35.7\%$$

Adding a column to the frequency table gives

| Year in school | Frequency | Relative frequency (%) |
|---|---|---|
| Freshman | 4 | 14.3 |
| Sophomore | 12 | 42.9 |
| Junior | 2 | 7.1 |
| Senior | 10 | 35.7 |
| **Total** | **28** | **100.0** |

From this table the School of Business can see that 35.7% of the students in the class are seniors, and 35.7% + 7.1% = 42.8% of the students delayed the course beyond the sophomore year. ■

 **TRY IT NOW!**

## Student Grades

### *Creating a Frequency Table*

A professor in an Introductory Statistics course knows that while students dread taking the course, they also have unusually high expectations for their performance. She surveys (anonymously, of course) her students and asks them what grade they expect to get in the course. The raw data are

| | | | | | | |
|---|---|---|---|---|---|---|
| A | C | B | A | A | B | B |
| B | A | B | A | A | C | B |
| B | A | F | C | B | D | C |
| B | B | D | B | A | B | A |

Make a frequency table for the data. Include both frequencies and relative frequencies.

## *Frequency Tables for Integer Data*

Remember from Chapter 2 that one type of quantitative data you might collect results from *counting* the number of times that something occurs in a set of data or from *ranking* or *rating* objects. This type of data takes on integer values.

Creating a frequency table for this type of data is exactly the same as creating one for qualitative data. Each value of the variable represents one category or classification in the table. The only real difference is that since the numbers for these data have some meaning, the categories must be in numerical order. For qualitative data the order in which you list the categories is not rigid.

### EXAMPLE 3.5    Student Attitudes

***Frequency Table for Integer Data***

The statistics professor who has been trying to learn about her class is also interested in her students' attitudes toward the study of statistics. She is aware that most of the people are enrolled in the course because it is required, and she wonders if this is reflected in their attitude toward the subject. As part of the last anonymous survey on grades, she asks the class to rank "the importance of statistics in my life" on a scale of 0 to 10, where 0 = of no importance whatsoever and 10 = the most important thing I will ever study. The raw data from the study are

| | | | | | | |
|---|---|---|---|---|---|---|
| 0 | 10 | 8 | 0 | 0 | 8 | 6 |
| 5 | 1 | 6 | 1 | 4 | 10 | 8 |
| 7 | 0 | 10 | 9 | 4 | 10 | 10 |
| 7 | 6 | 10 | 9 | 1 | 5 | 1 |

Since this is not very informative she decides to create a frequency table for the data:

| Rating | Frequency | Relative frequency |
|:---:|:---:|:---:|
| 0 | 4 | 4/28 = 14.3% |
| 1 | 4 | 4/28 = 14.3% |
| 2 | 0 | 0/28 = 0.0% |
| 3 | 0 | 0/28 = 0.0% |
| 4 | 2 | 2/28 = 7.1% |
| 5 | 2 | 2/28 = 7.1% |
| 6 | 3 | 3/28 = 10.7% |
| 7 | 2 | 2/28 = 7.1% |
| 8 | 3 | 3/28 = 10.7% |
| 9 | 2 | 2/28 = 7.1% |
| 10 | 6 | 6/28 = 21.4% |
| **Total** | **28** | **28/28 = 100%** |

From the frequency table she sees that most people seemed to have attitudes on either end of the scale, with fewer in the center, neutral, zone.    ■

Most data that are truly integer in nature occur from counting, rankings, or ratings. Sometimes data that are really measurements are recorded as integers. In this case it is important to think about where the data came from and the number of possible values before you pick the classes for your frequency table. Since the data are numbers, you cannot leave out a class (value) if there are no observations in it. If the number of possible values is much larger than 15 or 20 you probably do not want to treat the data as integer data. Instead you should use the methods for continuous data shown later in this section.

## EXAMPLE 3.6   ABC Faculty Salaries

*Looking at Integer Data*

The provost looking at faculty salaries also knows that salary depends on the length of time that the faculty member has been at the College. She decides to summarize these data using a frequency table, and notices that the data are integer data.

She also notices that the shortest length of time involved is 0 years and the longest is 41 years. She realizes that time is a measurement and that it is integer data because it was recorded to the nearest whole year. She sees that treating these data as integer data will not be appropriate.                                                                    ■

Since quantitative data are *ordered* by nature, it is also interesting to ask questions like "What percentage of the class rates statistics on the bottom half of the scale? neutral? the top half?" You can answer the questions by summing the relative frequencies, or by using the **cumulative relative frequency.**

> The *cumulative relative frequency* of a class is the sum of the relative frequencies of all classes at or below that class represented as a portion of the total number of observations. It can be expressed as a fraction, decimal, or percentage.

## EXAMPLE 3.7   Student Attitudes

*Cumulative Frequencies*

For the data from the Student Attitudes example, the professor calculated the cumulative frequency for each class in the table:

| Rating | Frequency | Cumulative relative frequency |
|---|---|---|
| 0 | 4 | 4/28 = 14.3% |
| 1 | 4 | 8/28 = 28.6% |
| 2 | 0 | 8/28 = 28.6% |
| 3 | 0 | 8/28 = 28.6% |
| 4 | 2 | 10/28 = 35.7% |
| 5 | 2 | 12/28 = 42.9% |
| 6 | 3 | 15/28 = 53.6% |
| 7 | 2 | 17/28 = 60.7% |
| 8 | 3 | 20/28 = 71.4% |
| 9 | 2 | 22/28 = 78.6% |
| 10 | 6 | 28/28 = 100.00% |
| **Total** | **28** | |

From the table, the professor can easily see that 35.7% of the students had negative attitudes toward statistics. To find the percentage of students that have neutral attitudes, she must do some arithmetic first. From the table she can see that 42.9% of the students were in the classes from 5 down. Since she knows that 35.7% were in the classes from 4 down, it seems that 42.9% − 35.7% = 7.2% have a neutral attitude toward statistics. This agrees with the relative frequency from Example 3.5.

To find the percentage with positive attitudes, again a little arithmetic is needed. Since she knows that 42.9% are in the classes from 5 down, the remainder must be in the classes from 6 up. She finds that 100% − 42.9% = 57.1% have positive attitudes toward statistics. Teaching this class may be an uphill battle!                      ■

At this point you must be thinking that it would be simpler to use the relative frequencies and just add up the ones you are interested in, like you did for the qualitative data. Sometimes this is true, but when there are a lot of classes the procedure becomes very tedious. There are two other reasons for learning how to deal with cumulative relative frequencies: (1) When you are using statistics or spreadsheet software to analyze data, the cumulative frequencies are part of the output, which makes it a much easier job; and (2) certain statistical tables such as probability tables are cumulative and there is just no other way to use them.

 **TRY IT NOW!**

## New Product Survey
### *Cumulative Relative Frequencies*

A marketing research firm conducted a survey of consumers who invariably use a particular brand of bath soap. The consumers were given a competitor's version of the same product with nonallergenic enhancements and asked whether they would consider buying the new product. Their answers were given on a scale of 1 to 5 where 1 = would not ever buy this product and 5 = will buy this product immediately. The raw data from the survey are given below:

| | | | | | |
|---|---|---|---|---|---|
| 5 | 4 | 1 | 4 | 2 | 2 |
| 3 | 1 | 4 | 4 | 2 | 4 |
| 3 | 1 | 3 | 3 | 5 | 4 |
| 4 | 4 | 2 | 3 | 2 | 5 |
| 4 | 3 | 4 | 4 | 1 | 5 |

Create a frequency table for the data that includes both frequency and cumulative relative frequency.

Which ratings indicate a negative attitude toward the new product? What percentage of the people surveyed had a negative attitude?

Which rating indicates a neutral attitude? What percentage of the people surveyed had a neutral attitude?

Which ratings indicate a positive attitude? What percentage had a positive attitude?

## Frequency Tables for Continuous Data

Another type of quantitative data that is frequently encountered in statistics is *measurement* data. This type of data can assume many different values in a sample. For this reason, we cannot use each value in the sample as a category for the frequency table. It is possible that each value will occur only once in the sample! This would result in entirely too many classes for the frequency table and no real information. Thus, before we can create a frequency table for measurement data we must figure out how to define the categories or classes for the data.

Two questions must be answered to create the frequency table for continuous data:

> *Question 1:* How many classes should there be in the table?
>
> *Question 2:* How large (wide) should each class be?

While there are no absolute rules for creating the frequency classes for continuous data, there are some guidelines that you can apply to answer the questions.

*Question 1:* There are two basic rules for determining the number of classes for the frequency table.

**Rule 1:** The number of classes in the frequency table should be approximately equal to the square root of the sample size, $n$. That is,

$$\text{Number of classes} = \sqrt{n}$$

You can see already that this is just a guideline, since the result of the calculation may not be a whole number. When this happens we usually just take the integer part of the answer. For example, if our sample size were $n = 30$, then $\sqrt{30} = 5.477$ so we would plan on 5 classes.

**Rule 2:** The number of classes should not be less than 5 or more than 20. The reasons for this rule have to do with the kind of information that we are looking for in the frequency table. The reasons for it will be more obvious when we talk about the graphical methods for displaying data.

Once you have decided on the number of classes for the frequency table you can answer the second question.

*Question 2:* To determine how large each class should be we need to think about what the table must accomplish. Once the classes have been determined, each piece of data in the sample must be able to be counted in *one and only one* of the classes. This tells you that

1. The classes must span the entire data set.

2. The classes cannot overlap.

Think about the number of classes as the number of steps you can take to get from the smallest data value (minimum) to the largest data value (maximum). Then the size of each class should be

$$\frac{\text{Maximum} - \text{Minimum}}{\text{Number of classes}}$$

Figure 3.2 (page 66) gives an illustration of the rules.

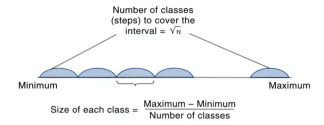

**FIGURE 3.2**   Number of classes and class interval for continuous data

Once the number of classes and size of each class are determined you can define each class precisely. That is, you can specify the smallest and largest data values that will fall in each class. At this point it is probably time for an example.

## EXAMPLE 3.8   ABC Faculty Salaries

*Setting up the Class Intervals for Continuous Data*

The provost of ABC decides to create frequency tables for the years of service data. Although she is going to use a computer package to actually create the table, she knows that some software asks the user for input on the number of classes to use as well as the widths or starting points.

She knows that she has 207 faculty members in the sample, so she calculates $\sqrt{207} = 14.39 \approx 14$ and decides that she will use about 14 classes. Since the highest data point is 41 years and the lowest is 0, she calculates that the width of each class should be

$$\text{Width} = \frac{41 - 0}{14} = 2.93 \text{ years}$$

## EXAMPLE 3.9   On-Time Rates

*Creating Class Intervals for Continuous Data*

A major commuter railroad has been looking at the on-time rates for service to a particular city. The railroad collects data for 30 randomly selected days and finds the percent of trains on time each day:

| | | | | |
|---|---|---|---|---|
| 60.4 | 92.9 | 67.0 | 91.6 | 93.7 |
| 71.9 | 66.4 | 48.0 | 66.6 | 94.6 |
| 83.2 | 68.0 | 74.3 | 78.6 | 40.9 |
| 76.4 | 62.9 | 67.1 | 68.5 | 29.5 |
| 69.2 | 45.8 | 49.1 | 34.7 | 65.1 |
| 72.1 | 40.6 | 35.0 | 60.8 | 54.3 |

To make some sense of the data the railroad decides to create a frequency table. The first task is to decide how many classes the table should have. The railroad first calculates

$$\sqrt{30} = 5.48$$

and so decides to use 5 classes.

Next the railroad must decide how large each class should be. The data show that the maximum value is 94.6% and the minimum value is 29.5%. The total distance that needs to be covered by the classes is

$$94.6 - 29.5 = 65.1$$

and the size of each class must be

$$\frac{65.1}{5} = 13.02$$

∎

To determine the starting and ending points (boundaries) for each class we must remember that the classes cannot overlap and that there cannot be any gaps in the interval. There are several ways to accomplish this, but we will define a class as containing all observations *more than* the lower boundary, *up to and including* the upper boundary:

Lower boundary $< x \le$ Upper boundary

where $x$ represents the value of the variable being studied.

Different texts will use different methods for defining intervals, but they will all yield similar results. The interval definition used here is consistent with the major spreadsheet software such as Excel and Lotus 1-2-3. Most statistical packages, such as Minitab and Data Desk, define the classes to *include* the lower boundary and to go up to but *not include* the upper boundary. As we will see later in the chapter, this causes some differences in the final output, but it is not a problem.

Remember that the intervals have been defined so that they contain the values *above* the lower boundary. To make sure that the classes you define include the minimum, you need to start just *below* the minimum value in the data set and add the class width. The lower boundary for the next class is the upper boundary for the previous class.

### EXAMPLE 3.10  ABC Faculty Salaries

*Creating the Class Intervals*

The provost decides that for the data on the number of years of service she will use a computer software package. She will let the software create the frequency distribution for her and see what it comes up with:

| Class | Frequency |
|---|---|
| $-2.928571 < x \le 0.000000$ | 15 |
| $0.000000 < x \le 2.928571$ | 19 |
| $2.928571 < x \le 5.857143$ | 19 |
| $5.857143 < x \le 8.785714$ | 34 |
| $8.785714 < x \le 11.714286$ | 16 |
| $11.714286 < x \le 14.642857$ | 14 |
| $14.642857 < x \le 17.571429$ | 13 |
| $17.571429 < x \le 20.500000$ | 17 |
| $20.500000 < x \le 23.428571$ | 33 |
| $23.428571 < x \le 26.357143$ | 20 |
| $26.357143 < x \le 29.285714$ | 4 |
| $29.285714 < x \le 32.214286$ | 2 |
| $32.214286 < x \le 35.142857$ | 0 |
| $35.142857 < x \le 38.071429$ | 0 |
| $38.071429 < x \le 41.000000$ | 1 |

The results are not quite what she expected, but she sees that there are 15 classes and that the width of each class is 2.92857 years. ∎

In Example 3.10 we saw that the frequency table generated by a software package used intervals that had six decimal places! While this is not bad, it might be more appealing to use intervals that look more like the original data, with the same precision.

**EXAMPLE 3.11**   **On-Time Rates**

*Creating Class Intervals for Continuous Data*

In the On-Time Rates example we found that the width of each class should be 13.02. The data are measured only to the nearest tenth, so we will use increments of 13.0. Since the smallest value in the data is 29.5, we will start just below that value, at 29.4, and add 13.0. That is, the first class interval contains all values, $x$, of the variable such that

$$29.4 < x \leq 42.4$$

The next class is formed by starting at 42.4 and again adding 13.0:

$$42.4 < x \leq 55.4$$

The remainder of the classes are

$$55.4 < x \leq 68.4$$
$$68.4 < x \leq 81.4$$
$$81.4 < x \leq 94.4$$
$$94.4 < x \leq 107.4$$

*It is okay to wind up with one more class than you planned on. Remember the rules are only guidelines.*

We need this last class for the maximum observation, 94.6%.

To create the frequency table for the data we count the number of data values that fall in each interval and complete the table.

| On-time rates | Frequency | Relative frequency (%) | Cumulative relative frequency (%) |
|---|---|---|---|
| $29.4 < x \leq 42.4$ | 5 | 16.7 | 16.7 |
| $42.4 < x \leq 55.4$ | 4 | 13.3 | 30.0 |
| $55.4 < x \leq 68.4$ | 9 | 30.0 | 60.0 |
| $68.4 < x \leq 81.4$ | 7 | 23.3 | 83.3 |
| $81.4 < x \leq 94.4$ | 4 | 13.3 | 96.6 |
| $94.4 < x \leq 107.4$ | 1 | 3.3 | 100.0 |

From the frequency table the railroad management can obtain information such as, "On 83.3% of the days the on-time rate was no greater than 81.4%" or "The on-time rate was greater than 94.4% on only 3.3% of the days." This allows the railroad to judge its performance and take appropriate action.                      ■

You may be wondering why you need to know how to do this, when you can use a computer to generate the frequency distribution for you. There are two reasons that this knowledge is necessary. First, some software packages require input from the user about how the frequency table is set up. Knowledge of some basic rules will help to answer those questions. The second reason is that even if there is a default setup for the frequency table, the result may not be acceptable. Remember the frequency table for the faculty salaries!

What are we looking for in a frequency table? How can we tell if the one we have created is good? We already know that we need the classes to be the same size and that they must cover all of the data values. What else is important?

Many times you may want the intervals to be defined in nice, round, numbers that are integer data or that end in 0.5 or some other convention. Certainly, it is more appealing to look at classes with numbers like 25, 35, 45, . . . , but software packages follow rules very similar to the ones outlined here and the results are very often not what you want.

Figure 3.3 shows the frequency table produced by Microsoft Excel for the **On-Time Rate** data.

| Bin | Frequency | Cumulative % |
|---|---|---|
| 29.50 | 1 | 3.33 |
| 42.52 | 4 | 16.67 |
| 55.54 | 4 | 30.00 |
| 68.56 | 10 | 63.33 |
| 81.58 | 6 | 83.33 |
| More | 5 | 100.00 |

*In Excel, the number in the Bin column is the highest number that is included in the interval, so the first bin is all numbers less than and including 29.50, and the last bin is all numbers greater than 81.58.*

**FIGURE 3.3**   Microsoft Excel frequency table

It is possible to vary both the starting point of the first class and the interval width, within reason. Remember that the decisions for the number of classes and interval widths are guidelines based on the sample size and the range of the data. As long as your final results do not vary too much from the guidelines you can adjust them.

## EXAMPLE 3.12   ABC Faculty Salaries

### Adjusting Class Boundaries

After looking at the results given by the statistical software, the provost feels that the class intervals do not use numbers with which people will identify. There is no reason to use so many decimal places when the data are measured to the nearest year. Perhaps she should adjust the class intervals a little bit so that they use nice, round numbers. After looking at the data she decides to start just below 0, at $-1$ rather than at the $-3$ suggested by the software. She is less sure about the width of each class. Using a class width of 3 years would not be too different from the 2.93 calculated value, but she feels that a width of 5 years might be more appealing. She decides to try both ways and obtains the following frequency tables.

| Class | Frequency | Class | Frequency |
|---|---|---|---|
| $-1 < x \le 2$ | 34 | $-5 < x \le 0$ | 15 |
| $2 < x \le 5$ | 19 | $0 < x \le 5$ | 38 |
| $5 < x \le 8$ | 34 | $5 < x \le 10$ | 42 |
| $8 < x \le 11$ | 16 | $10 < x \le 15$ | 23 |
| $11 < x \le 14$ | 14 | $15 < x \le 20$ | 29 |
| $14 < x \le 17$ | 13 | $20 < x \le 25$ | 51 |
| $17 < x \le 20$ | 17 | $25 < x \le 30$ | 8 |
| $20 < x \le 23$ | 33 | $30 < x \le 35$ | 0 |
| $23 < x \le 26$ | 20 | $35 < x \le 40$ | 0 |
| $26 < x \le 29$ | 4 | $40 < x \le 45$ | 1 |
| $29 < x \le 32$ | 2 | | |
| $32 < x \le 35$ | 0 | | |
| $35 < x \le 38$ | 0 | | |
| $38 < x \le 41$ | 1 | | |

The first set of changes does not alter the frequency distribution very much. There are 14 classes instead of 15, and the end classes are fairly empty. The second set of changes reduces the number of classes to 10 and there are fewer empty or nearly empty classes. She is not sure if this matters, but she will keep it in mind.   ■

### EXAMPLE 3.13   On-Time Rates

*Adjusting Class Boundaries*

After examining the frequency table for the On-Time Rates problem, the railroad managers decide that they would prefer to see the data classified with rounder, more understandable numbers. They look at the data and decide to start the first interval at just above 25.0% and use a width of 10%. The new frequency table would look like this:

| On-time rates | Frequency | Relative frequency (%) | Cumulative relative frequency (%) |
|---|---|---|---|
| $25.0 < x \leq 35.0$ | 3 | 10.0 | 10.0 |
| $35.0 < x \leq 45.0$ | 2 | 6.7 | 16.7 |
| $45.0 < x \leq 55.0$ | 4 | 13.3 | 30.0 |
| $55.0 < x \leq 65.0$ | 3 | 10.0 | 40.0 |
| $65.0 < x \leq 75.0$ | 11 | 36.7 | 76.7 |
| $75.0 < x \leq 85.0$ | 3 | 10.0 | 86.7 |
| $85.0 < x \leq 95.0$ | 4 | 13.3 | 100.0 |
| **Totals** | **30** | **100.0** | **100.0** |

The changes result in seven classes rather than six because the first interval starts at a lower number and the interval widths are not as large.

Another possibility for class intervals would have been to start at just above 25% and use an interval width of 15%. The frequency table for that arrangement would be

| On-time rates | Frequency | Relative frequency (%) | Cumulative relative frequency (%) |
|---|---|---|---|
| $25.0 < x \leq 40.0$ | 3 | 10.0 | 10.0 |
| $40.0 < x \leq 55.0$ | 6 | 20.0 | 30.0 |
| $55.0 < x \leq 70.0$ | 11 | 36.7 | 66.7 |
| $70.0 < x \leq 85.0$ | 6 | 20.0 | 86.7 |
| $85.0 < x \leq 100.0$ | 4 | 13.3 | 100.0 |
| **Totals** | **30** | **100.0** | **100.0** |

This time the frequency table has only five classes because we used an interval width of more than 13.0%.   ■

  *TRY IT NOW!*

### Assignment Times

*Creating a Frequency Table for Continuous Data*

The instructor for an introductory statistics class wonders about the complaints that she is hearing about the time it takes to complete a computer assignment. The assignments are designed to be done in about 25 minutes. She asks the members of the class to time how long it takes to do the next assignment and to hand the data in with the assignment. The data, in minutes, she obtains are

| | | | | |
|---|---|---|---|---|
| 22.8 | 27.0 | 27.9 | 30.4 | 33.4 |
| 24.8 | 27.2 | 27.9 | 31.1 | 33.9 |
| 24.8 | 27.4 | 28.2 | 31.4 | 35.3 |
| 26.0 | 27.4 | 29.4 | 32.4 | 35.7 |
| 26.0 | 27.4 | 29.6 | 33.1 | 36.3 |
| 26.1 | 27.6 | 29.8 | 33.2 | 40.4 |

Approximately how many classes should the frequency table have?

What should the class width be?

Create a frequency table for the data.

## 3.2.2 Exercises—Learn It!

3.1   The administrators of a local university are trying to determine whether the amount of parking that they have is adequate. They take a random sample of 25 entering freshmen and ask them what mode of transportation they will use to travel to classes. The responses are

| | | | | |
|---|---|---|---|---|
| Car | Car | Car | Car | Car |
| Car | Walk | Car | Car | Car |
| Public | Car | Walk | Car | Car |
| Other | Car | Bicycle | Car | Car |
| Public | Car | Car | Public | Walk |

(a)  Create a frequency table for the responses.

(b)  What mode of transportation was the most popular? the least popular?

(c)  Does any class in the frequency table constitute a majority? If so, which one?

3.2   After the incredible advertising blitz that accompanied the release of Microsoft's Windows 95 operating system, a local computer society wondered what operating platform its members were using. The society surveyed 30 members and obtained the following data:

| | | | | |
|---|---|---|---|---|
| Macintosh | Macintosh | Macintosh | W 3.1* | W 3.1 |
| W 3.1 | W 3.1 | W 3.1 | OS /2 | W 3.1 |
| W 3.1 | W 3.1 | W 3.1 | W 3.1 | W NT |
| W 3.1 | W 3.1 | W 3.1 | W NT | W 3.1 |
| W 3.1 | W 95 | Macintosh | W 95 | W 3.1 |
| OS /2 | DOS | W 3.1 | W 3.1 | W 3.1 |

* The W 3.1 category includes Workgroups for Windows 3.11 users.

(a)  Create a frequency table for the data on operating systems.

(b)  What percentage of the people surveyed used Windows 95?

(c) What percentage of the people surveyed did not use a Microsoft Windows operating system?

(d) Was any operating system used by a majority of those surveyed?

**3.3** A manufacturing company has been receiving complaints from its warehouse operators about the condition of the product cases that are being stored. The complaints indicate that most of the cases are defective in some way and that it is creating logistics problems for the warehouse employees. The managers of the company decide to take a random sample of the cases produced and examine them for various defects. The data they obtain are

| | | | | |
|---|---|---|---|---|
| Dented | No defect | No defect | Dented | No defect |
| No defect | Torn | Crushed | No defect | No defect |
| Unsealed | No defect | No defect | No defect | No defect |
| No defect | No defect | No defect | No defect | No defect |
| No defect | No defect | No defect | Dented | No defect |
| No defect | Unsealed | No defect | No defect | No defect |

(a) Create a frequency table for the defect data.

(b) Based on the data, does the warehouse claim appear justified? Why or why not?

**3.4** A survey in a computer newsletter for employees at a large corporation asked readers how many times per week they logged onto the Internet using any one of the available services at the company. The responses from 50 employees are shown below:

| | | | | |
|---|---|---|---|---|
| 1 | 7 | 3 | 4 | 1 |
| 0 | 7 | 0 | 1 | 8 |
| 3 | 3 | 4 | 2 | 4 |
| 3 | 0 | 3 | 3 | 5 |
| 3 | 3 | 7 | 4 | 1 |
| 7 | 5 | 3 | 3 | 4 |
| 4 | 3 | 3 | 3 | 7 |
| 6 | 8 | 0 | 7 | 2 |
| 7 | 0 | 3 | 9 | 3 |
| 4 | 4 | 4 | 4 | 3 |

(a) Create a frequency table for the data.

(b) What percentage of those surveyed logged onto the Internet exactly 5 times per week? more than 5 times per week?

(c) What percentage of those surveyed logged on 4, 5, or 6 times per week? less than twice per week?

**3.5** As part of a cost/benefit analysis on service contracts for photocopiers, a large company surveyed all departments with copy machines and asked how many times per week their machines needed to be serviced. The data obtained are

| | | | | |
|---|---|---|---|---|
| 3 | 2 | 3 | 0 | 3 |
| 0 | 2 | 2 | 2 | 1 |
| 3 | 2 | 2 | 1 | 0 |
| 2 | 3 | 2 | 1 | 2 |
| 3 | 4 | 2 | 7 | 1 |
| 2 | 2 | 1 | 3 | 1 |
| 0 | 3 | 1 | 1 | 2 |

(a) Create a frequency table for the data. Include relative frequency and cumulative relative frequency.

(b) What percentage of the copiers need repairs more than twice per week?

(c) What percentage need repairs at least four times per week?

(d) Do a majority of the copiers require repairs more than three times per week?

**3.6** To determine whether it was reasonable to request that students use computers to write reports, a school surveyed a typical fifth-grade class and asked the students how many computers they had at home. The responses are

```
1   0   1   1   2
2   2   0   1   0
0   0   3   3   0
0   1   1   2   1
1   2   1   1   0
```

(a) Create a frequency table for the data. Include relative frequency and cumulative relative frequency.

(b) What percentage of the students did not have access to computers at home?

(c) What percentage of the students had easy access (2 or more) to computers at home?

(d) Based on the data collected, do you think that asking students to use computers for reports is reasonable? Why or why not?

**3.7** As part of a study to decide whether to renew the contract of a food service company at a local university, the administration surveyed a group of students who used the service on a regular basis. The students were asked to rate the food quality on a scale of 1 to 5, where 1 was extremely bad and 5 was extremely good. The results are

```
2   4   3   4   1
3   2   3   1   3
1   1   4   1   4
1   3   3   4   1
4   2   3   4   3
3   3   3   1   3
```

(a) Create a frequency table for the data.

(b) What rating had the highest frequency? The lowest?

(c) Based on the data, do you think that the students like the food provided by the company? Why or why not?

**3.8** A local computer group surveyed high school students to find out how many hours per week the students spent logged on to various on-line services. The data are

```
0.2   4.1   11.9   15.5   17.8
0.8   4.2   13.0   15.5   17.9
1.0   6.9   13.4   15.5   18.4
1.4   8.0   13.5   15.5   18.7
2.5   8.2   14.9   16.7   19.5
2.9   8.6   15.1   16.8   19.7
```

Create a frequency table for the amount of time spent on-line by students. Include relative frequency and cumulative relative frequency.

**3.9** A large corporation that relies on a secretarial service group to do much of its work is trying to decide whether it needs more workers in the group. One of the criteria that will be used to make the decision is the turn-around time experienced by users of the service. The corporation surveys a number of employees who use the service on a regular basis and asks them for the turnaround time (in hours) for the last job they submitted. The following data are obtained:

```
14   19   21   25   26   29
15   19   22   25   26   29
16   20   22   25   26   30
16   20   23   25   26   31
18   20   23   25   27   31
18   20   23   25   28   35
18   21   24   26   29   40
```

(a) Create a frequency table for the data. Include relative frequency and cumulative relative frequency.

(b) What percentage of the people who use the service wait more than one day (24 hours) for their jobs to be done?

(c) Modify the class intervals for your frequency table to allow you to easily answer the questions, What percentage of those surveyed wait 20 hours or less? more than 30 hours?

(d) *Approximately* what percentage waited between 20 and 30 hours?

3.10 The Bureau of Weights and Measures conducts random checks of products in various supermarkets. The Bureau decides to check half-gallon containers of a particular brand of orange juice. The contents of each of the containers tested (measured to the nearest 0.1 oz) are

| | | | | |
|---|---|---|---|---|
| 64.8 | 65.2 | 65.6 | 65.7 | 65.9 |
| 64.8 | 65.3 | 65.7 | 65.7 | 66.0 |
| 64.9 | 65.3 | 65.7 | 65.7 | 66.1 |
| 65.1 | 65.4 | 65.7 | 65.8 | 66.1 |
| 65.2 | 65.4 | 65.7 | 65.8 | 66.3 |

(a) Create a frequency table for the data.

(b) If the containers are subject to bursting when filled with more than 66 oz. of fluid, what percentage of the containers are in danger of bursting?

3.11 The local refuse company in charge of recycling in a town is receiving complaints from the people at the end of the pickup route. The complaints indicate that their recycling is not being picked up at the end of the route because the drivers say that the trucks are too full. In an effort to determine how much capacity is really needed for the route, the refuse company randomly samples the newspaper recycling that is put out and measures each pile to the nearest 0.1 inch. The data obtained are

| | | | | |
|---|---|---|---|---|
| 8.3 | 10.2 | 12.2 | 13.1 | 13.9 |
| 8.5 | 10.4 | 12.3 | 13.3 | 14.1 |
| 8.8 | 11.0 | 12.3 | 13.3 | 14.3 |
| 9.1 | 11.0 | 12.3 | 13.4 | 14.3 |
| 9.7 | 11.2 | 12.3 | 13.4 | 14.4 |
| 9.8 | 11.4 | 12.5 | 13.4 | 14.4 |
| 9.9 | 12.0 | 12.7 | 13.6 | 14.4 |
| 9.9 | 12.0 | 12.8 | 13.6 | 14.8 |
| 10.1 | 12.2 | 12.9 | 13.8 | 15.2 |

(a) Create a frequency table for the data. Include relative frequency and cumulative relative frequency.

(b) Modify the frequency table you created in part (a) so that you can easily use it to answer the following questions:

(c) What percentage of the newspaper recycling piles are more than a foot in height?

(d) What percentage of the recycling piles are 10 inches or less?

(e) *Approximately* what percentage of the recycling piles are between 10 and 14 inches?

## 3.3    GRAPHICAL DISPLAYS OF DATA

While the frequency distribution does a good job of organizing and summarizing a set of data, it does not have much of a visual impact. To create a more immediate impression from a data set we need to create a pictorial (graphical) representation of the information in the frequency distribution. Most of the work is already done in creating the frequency table. The displays have the same basic structures with some changes to accommodate the type of data being displayed.

### 3.3.1 Graphical Displays for Qualitative Data

There are two methods that you can use to display qualitative data, a **bar chart** and a pie chart.

A *bar chart* represents the frequency or relative frequency from the table in the form of a rectangle or bar.

## Creating a Bar Chart for Qualitative Data

In a bar chart, one of the axes is used to represent the categories from the frequency table and the other axis is used to represent the frequency or relative frequency for the categories. For qualitative data, assignment of the axes is a matter of preference. We will use the *x* axis for the categories and the *y* axis for the frequencies and relative frequencies to be consistent with the graphs for the other types of data.

**Step 1:** Draw a pair of axes, *x* and *y*.

**Step 2:** At evenly spaced intervals on the *x* axis put tick marks and label them with the categories from the frequency table.

**Step 3:** Scale the *y* axis so that the category with the highest frequency or relative frequency can be graphed. Choose the scale so that you can distinguish different frequencies or relative frequencies from each other.

**Step 4:** At each category on the *x* axis, draw a rectangle (bar) whose height is equal to the frequency or relative frequency for the category. The bases of the rectangles must be the same width and the bars should not touch each other.

**Step 5:** Label the axes and give the graph an appropriate title.

*Steps for creating a bar chart*

Often there is no predetermined order for the categories for qualitative data, but you might want to think about the order in which you place them, since the order may affect the impact of the graph. Some common choices are by descending or ascending frequency or alphabetical order.

## EXAMPLE 3.14   ABC Faculty Salaries

*Creating a Bar Chart*

The Provost at the college decides to create a bar chart for the data she has collected on faculty rank. The frequency table for the data was

| Rank | Frequency | Relative frequency (%) |
|------|-----------|------------------------|
| Professor | 55 | 26.6 |
| Associate Professor | 67 | 32.4 |
| Assistant Professor | 77 | 37.2 |
| Instructor | 8 | 3.9 |
| **Total** | **207** | **100.0** |

She decides to use increasing rank order for the bars on the chart and the resulting bar chart looks like the one below:

Distribution of Faculty by Rank

You may wonder why you would want to know how to create bar charts by hand when it is very likely that you will use a computer to make the charts. Most computer software requires the user to make decisions about how the chart should be set up. Knowing the basics of how the charts are made helps you to make these decisions.

### EXAMPLE 3.15   Student Distribution

***Creating a Bar Chart***

The School of Business would like to create a bar chart of the data that it collected on the year of students who are taking the Introductory Statistics course. The frequency table for the data is

| Year in school | Frequency | Relative frequency (%) |
|---|---|---|
| Freshman | 4 | 14.3 |
| Sophomore | 12 | 42.9 |
| Junior | 2 | 7.1 |
| Senior | 10 | 35.7 |
| **Total** | **28** | **100.0** |

Since in this case there is a natural order for the categories we will use it when labeling the $x$ axis. The $y$ axis needs to accommodate a frequency as high as 12, so the scale for that axis will go from 0 to 12 by twos. The completed bar chart is shown in the figure.

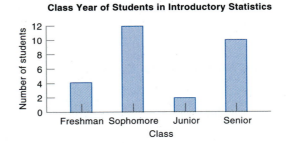

One thing to remember when creating a bar chart is that there are no absolute rules. The idea is to create a chart that conveys information to the viewer. In general, it is desirable to keep the number of categories in a bar chart to no more than ten. When there are too many categories, the axis becomes crowded and difficult to read and the viewer is distracted. To accomplish this, categories with very low frequencies are often consolidated into a single category labeled "Other." Even though the "Other" category might have a higher frequency than some of the categories, it must appear last in the bar chart.

 **TRY IT NOW!**

### Student Grades
***Creating a Bar Chart***

The instructor who surveyed her students about expected grades wants to create a bar chart from the data. The frequency table for the data is

| Grade | Frequency | Relative frequency (%) |
|-------|-----------|------------------------|
| A | 9 | 32.1 |
| B | 12 | 42.9 |
| C | 4 | 14.3 |
| D | 2 | 7.1 |
| F | 1 | 3.6 |
| **Total** | **28** | **100.0** |

Create a bar chart for the data using relative frequency on the *y* axis. Be sure to label the axes and include an appropriate title.

## Pareto Analysis

An important application of bar charts in the area of management is in Pareto analysis. Very often, managers are faced with numerous situations and problems that are not of equal importance. This principle, known as the principle of "the vital few and the trivial many" or the "80–20 rule," was named the Pareto principle in the 1940s by J. M. Juran, one of the major contributors in the field of Total Quality Management (TQM).

The idea behind the principle is that, in most cases, 80% of the phenomena being observed fall into only 20% of the categories. For example, in quality control, most customer complaints can be attributed to only a few of the defect categories; in marketing, a small number of customers account for most of the sales; and in human services and employee relations, only a few of the employees account for most of the personnel problems. If managers can identify the key few categories, they can improve quality, increase sales, and improve employer/employee relations.

A tool that helps accomplish this is known as a **Pareto diagram.**

A *Pareto diagram* is a bar chart in which the categories are plotted in order of decreasing relative frequency. In addition to the bars, the cumulative relative frequency of the categories is plotted on the same graph.

## EXAMPLE 3.16    Tissue Defects

*Pareto Diagram*

A large consumer products company that manufactures facial tissues has been keeping track of customer complaints and classifying them according to the type of complaint. The company has created a frequency table for the complaints for the previous month. The data, in order of decreasing frequency, are

| Complaint category | Frequency | Relative frequency (%) | Cumulative relative frequency (%) |
|---|---|---|---|
| Dispensing | 127 | 58.0 | 58.0 |
| Packaging | 44 | 20.1 | 78.1 |
| Miscounts | 32 | 14.6 | 92.7 |
| Softness | 12 | 5.5 | 98.2 |
| Odor | 3 | 1.4 | 99.6 |
| Pricing | 1 | 0.5 | 100.0 |
| **Total** | **219** | **100.0** | |

To create a Pareto diagram for these data, first construct a bar chart using the relative frequency (see figure).

**Bar Chart – Customer Complaints**

To make the bar chart a Pareto diagram we must add a new, different scale to the right side of the graph. This second y-axis scale must allow us to graph the cumulative relative frequency, so it must go from 0 to 100%. At the location of each bar, a point is plotted that represents the cumulative relative frequency for that bar. At the end, the points are connected to create a line. The Pareto diagram for the tissue complaint data is shown below:

**Pareto Diagram – Customer Complaints**

From the Pareto diagram the company can see that almost 80% (78.1%) of the customer complaints that they received in the last month were about dispensing or packaging. If they concentrate on fixing the defects that generate those complaints they can go a long way toward reducing customer complaints. ■

## *Other Uses for Bar Charts*

Bar charts are not limited to situations where the data are frequencies. Very often bar charts are used to display data for different categories where the data are some kind of quantitative measure for each category. For example, a bar chart could be used to display and compare sales revenues for a sample of different software products. In this case, the *y* axis of the chart would represent the value of the variable being studied.

## EXAMPLE 3.17 Food Sales

### Bar Charts for Nonfrequency Data

A company did a study of sales for a group of foods that are advertised as having "added nutrients." The company looked at supermarket sales for a 52-week period and found the following:

| Product | Manufacturer | Sales ($ million) |
|---|---|---|
| Hawaiian Punch | Procter & Gamble | 125.7 |
| Yoo Hoo Chocolate Flavored Drink | Austin Nichols | 32.9 |
| Life Savers Flavor Pops | Agway | 9.4 |
| Wonder Calcium Enriched White Bread | Ralston-Continental Baking | 1.0 |
| Vicks Vitamin C Lemon Drops | Procter & Gamble | 0.5 |

The company created a bar chart for the data using the product for the *x* axis and the sales (in millions of dollars) for the *y* axis. The chart obtained is shown below:

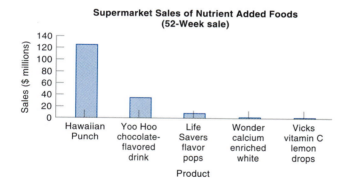

## Creating a Pie Chart for Qualitative Data

Qualitative data can also be displayed using a **pie chart.** This type of chart is often used when the categories of the data represent some part or portion of a whole.

> A *pie chart* represents data in the form of slices or sections of a circle. Each slice represents a category and the size of the slice is proportional to the relative frequency of the category.

Drawing a pie chart by hand requires the use of a protractor to accurately measure the sizes of the slices and is fairly tedious. Since the number of degrees in a circle is 360, you can use the relative frequency to determine how many degrees should be allocated to each slice. Because of this it is usually best to use computer software to create one.

## EXAMPLE 3.18 Class Year

### Creating a Pie Chart

The School of Business wants to use the data on class year for a presentation to the faculty. They think that a pie chart might have the best impact. From the relative frequency table they determine the following:

| Class year | Relative frequency (%) |
|---|---|
| Freshman | 14.3 |
| Sophomore | 42.9 |
| Junior | 7.1 |
| Senior | 35.7 |
| **Totals** | **100.0** |

They use Microsoft Excel to create a pie chart from the relative frequency table:

### 3.3.2 Graphical Displays for Quantitative Data

The tool used to display quantitative data is called a **histogram.** A histogram is very similar to a bar chart, but since numbers are naturally ordered, the $x$ axis of the graph must be scaled to reflect this. There are slight differences for dealing with integer and continuous data.

### Histograms for Integer Data

As in a bar chart, you use a rectangle to represent each possible data value, with the height of the bar corresponding to the frequency or relative frequency for that value. Remember that integers are numbers and they have a definite ordering. The $x$ axis must accommodate all of the possible values, whether or not there were any observations of the value. The rectangles are centered on the data values as in a bar chart, but the bars are contiguous; that is, they touch each other.

**EXAMPLE 3.19**   Student Attitudes

*Creating a Histogram for Integer Data*

The instructor who is interested in student attitudes toward statistics would like to create a graphical display of the data. She feels that this might provide another view of the situation. The frequency table for the data is

| Rating | Frequency | Relative frequency (%) |
|---|---|---|
| 0 | 4 | 14.3 |
| 1 | 4 | 14.3 |
| 2 | 0 | 0.0 |
| 3 | 0 | 0.0 |
| 4 | 2 | 7.1 |
| 5 | 2 | 7.1 |
| 6 | 3 | 10.7 |
| 7 | 2 | 7.1 |
| 8 | 3 | 10.7 |
| 9 | 2 | 7.1 |
| 10 | 6 | 21.4 |
| **Total** | **28** | **100.0** |

To create a relative frequency histogram for the data the *x* axis will be scaled so that the numbers 0 through 10 appear at the center of the bars. The *y* axis will be scaled to accommodate the relative frequencies, the highest of which is 21.4%. The histogram is shown below:

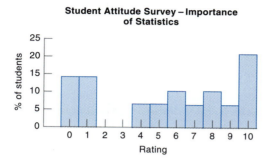

From the histogram the instructor sees that the really negative attitudes are separate from the rest of the group and that the most typical response was a 10. The remainder of the responses were evenly spread out over the 4 through 9 ratings. Although there is no new information in the graph, the visual display does add another dimension to the information. ■

 **TRY IT NOW!**

## New Product Survey

### Cumulative Relative Frequencies

The marketing research firm that is conducting a survey about customer attitudes toward a new brand of soap would like to look at its data graphically. The frequency table for the data is shown below:

| Rating | Frequency | Relative frequency (%) |
|--------|-----------|------------------------|
| 1 | 4 | 13.3 |
| 2 | 5 | 16.7 |
| 3 | 6 | 20.0 |
| 4 | 11 | 36.7 |
| 5 | 4 | 13.3 |
| **Total** | **30** | **100.0** |

Create a relative frequency histogram for the data.

### Histograms for Continuous Data

A histogram for continuous data differs from the one for integer data in that each rectangle represents a class interval, which is a *range of values*. For this reason, the rectangles are not centered on values, but begin and end at each of the class boundaries.

### EXAMPLE 3.20    ABC Faculty Salaries

*Creating a Histogram for Continuous Data*

The provost at ABC wants to make a histogram of the data on number of years of service. She decides to use the frequency distribution that uses intervals of 5 years:

| Class | Frequency |
|-------|-----------|
| $-5 < x \le 0$ | 15 |
| $0 < x \le 5$ | 38 |
| $5 < x \le 10$ | 42 |
| $10 < x \le 15$ | 23 |
| $15 < x \le 20$ | 29 |
| $20 < x \le 25$ | 51 |
| $25 < x \le 30$ | 8 |
| $30 < x \le 35$ | 0 |
| $35 < x \le 40$ | 0 |
| $40 < x \le 45$ | 1 |

The histogram is shown below:

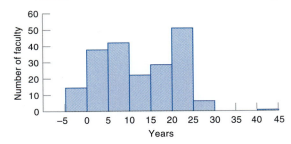

**Faculty Years of Service – Aluacha Balaclava College**

She sees at first glance that the data are certainly heavily concentrated at the bottom half of the class intervals and that the upper half of the intervals are very sparse. ∎

Sometimes you will see histograms that do not use the endpoints of the intervals to label the *x* axis. In this case, the bars of the histogram are centered on the axis tick mark and the *midpoint* of the interval is used instead.

### EXAMPLE 3.21    On-Time Rates

*Creating a Histogram for Continuous Data*

The railroad company interested in on-time rates would like to look at a graphical display of the data that have been collected. The frequency table for the data was

| On-time rates | Frequency | Relative frequency (%) |
|---|---|---|
| $25.0 < x \leq 35.0$ | 3 | 10.0 |
| $35.0 < x \leq 45.0$ | 2 | 6.7 |
| $45.0 < x \leq 55.0$ | 4 | 13.3 |
| $55.0 < x \leq 65.0$ | 3 | 10.0 |
| $65.0 < x \leq 75.0$ | 11 | 36.7 |
| $75.0 < x \leq 85.0$ | 3 | 10.0 |
| $85.0 < x \leq 95.0$ | 4 | 13.3 |
| **Totals** | **30** | **100.0** |

The railroad decides to create a relative frequency histogram. To do this the $y$ axis must be scaled so that it can accommodate percentages from 0 to 36.7%. So the railroad decides to go from 0 to 40% in increments of 5%. The tick marks on the $x$ axis will be the values at the beginning of each of the class intervals. The last tick mark will be at the end of the last interval. The histogram is shown below:

From the histogram the railroad company sees that the most typical on-time rates were in the 65–75% range, and that they were evenly distributed among the other classes. ∎

### 3.3.3 Displaying Small Data Sets

The rules for creating histograms are really not suitable for data sets with less than 25 observations. This is because the number of classes should not be less than 5 and we determine the number of classes by taking the square root of the sample size. Often when we collect data we do not have more than 25 observations. Is there a way to display these types of data sets graphically? The answer is a graphical method called a **dotplot.**

> In a **dotplot,** each observation is plotted as a point on a single, horizontal axis. The axis is scaled so that each of the data points can be located uniquely on the axis. When there is more than one observation with the same value the points are "stacked" on top of each other.

A dotplot can show many of the same features of the data as a histogram.

### EXAMPLE 3.22   Bank Customers

*Creating a Dotplot for a Data Set*

A bank is interested in understanding the number of customers that arrive hourly on Fridays preceding holiday weekends. It takes data for eight hours on each of two different Fridays and obtains the following:

| 10 | 14 | 15 | 14 | 19 | 12 | 11 | 14 |
| 15 | 14 | 20 | 19 | 11 | 12 | 17 | 17 |

After looking at the data the bank decides to use a dotplot to display them because there are not very many observations. An axis is created that is scaled from 10 (the smallest value) to 20 (the largest value) in increments of one, since the data are integer data:

People

The dotplot shows that the number of customers per hour varies widely over the interval, although in 25% (4/16) of the hours there were 14 customers, and in 69% (11/16) there were at least 14 (14 or more) customers. The bank decides to use these data to plan the number of tellers that should be working on these Fridays. ■

### 3.3.4 Using the Computer to Create Graphical Displays

You may be wondering why it is necessary to learn how to create bar charts, pie charts, and histograms by hand, when you will almost certainly be using a computer to create them. The graphs produced by computer software are superior to those produced by hand, but a good deal of critical thought goes into creating a graph that conveys *information* to the viewer. It is not so important that you *create* the charts by hand as it is that you know *how* they are created.

Most computer software packages ask the user for input when creating graphs. Knowing how the charts are created will help you make decisions about how the charts look. In addition, not all software allows the user to make the same decisions. The default graphs produced by computer software offer great starting points for the final graphical display, but they are not usually the end product that you are looking for.

When you create a bar chart by hand you have the freedom to put the bars in any order you want and to combine categories as you wish. Some software will allow you to choose ascending or descending order of frequency for the bars, others will require you to sort the frequency table first, either by frequency or alphabetical order of categories. The same is true about combining categories and using frequency or relative frequency. Some computer software packages such as Excel will not create frequency distributions for categorical data and will make bar and pie charts only from already existing frequency tables. If you did not know what the chart should contain you would not know how to process the raw data that you collect.

Creating histograms differs among software packages. There are differences in the way you go about defining and creating the chart. Almost all software packages will create a histogram using a default set of class intervals, but they differ widely in what they will allow the user to specify. You saw in the previous section that when you are trying to make graphical displays to compare different samples, you need to be able to control the x axis of the graph. Microsoft Excel will allow you to specify the ending values for each of the classes, but when it creates the histogram from the data, it puts the endpoints in the *center* of the bar. Minitab and Data Desk will allow you to specify endpoint values or midpoint values, or the number of classes you want to use, but not both.

There are also differences in the way the final chart looks. Remember that Minitab and Data Desk define their intervals to include the lower value and go up to

**FIGURE 3. 4** Histograms from Minitab and Data Desk for ABC years of service

but not include the upper value. The differences in processing causes the histograms from the different software packages to look a little different. Figure 3.4 shows the histograms produced by each package for the data on faculty years of service for Aluacha Balaclava College.

If you compare these histograms to the one shown on page 82, you see that the major difference is in the first class. In the original histogram, this class contains only those people with 0 years of service $(-5 < x \le 0)$, while in Minitab and Data Desk it contains all people with 0, 1, 2, 3, or 4 years of service $(0 \le x < 5)$. This difference does not really change the general appearance of the histograms. Knowing the basics of creating histograms enables you to make smarter, more informed decisions, no matter what software you might use.

## 3.3.5 Exercises—Learning It!

**3.12** The data that the computer club collected on personal computer operating systems is given again here:

*Requires Exercise 3.2*

| | | | | |
|---|---|---|---|---|
| Macintosh | Macintosh | Macintosh | W 3.1* | W 3.1 |
| W 3.1 | W 3.1 | W 3.1 | OS /2 | W 3.1 |
| W 3.1 | W 3.1 | W 3.1 | W 3.1 | W NT |
| W 3.1 | W 3.1 | W 3.1 | W NT | W 3.1 |
| W 3.1 | W 95 | Macintosh | W 95 | W 3.1 |
| OS /2 | DOS | W 3.1 | W 3.1 | W 3.1 |

* The W 3.1 category includes WFW 3.11 users.

(a) Suppose you were asked to create a bar chart for the data. What order would you pick for the categories? Why?

(b) Make the bar chart for the data.

(c) Does the bar chart indicate that the majority of the people surveyed had switched to Windows 95? Were a majority of those surveyed users of some Windows operating system?

**3.13** An annual survey by J. D. Power and Associates reported in *The Connecticut Post,* 27 May 1994, rated vehicles on the number of problems reported by owners per 100 vehicles. The data for the top 11 cars are shown below:

| Car | Reported Problems |
|---|---|
| Lexus LS 400 | 32 |
| Lexus GS 300 | 48 |
| Lexus SC 300/400 | 52 |
| Geo Prizm | 56 |
| Acura Legend | 57 |
| Infiniti J-30 | 61 |
| Mercury Grand Marquis | 61 |
| Toyota Camry | 63 |

*(continued)*

| Car | Reported Problems |
|---|---|
| Toyota Tercel | 67 |
| Lexus ES 300 | 68 |
| Lincoln Town Car | 68 |

(a) Create a bar chart for the number of reported problems per 100 vehicles.

(b) Comment on any interesting features of the data.

**3.14**  Columbia University conducted a survey of 637 members of the Authors Guild and the Dramatists Guild, all of whom have published at least one book, one play, or three magazine articles. As part of the survey, the members were asked to classify the amount of money that they made from writing. The frequency table below gives the results of the survey:

| Revenue from Writing | Number of Members |
|---|---|
| More than $50,000 | 57 |
| More than $20,000 but less than $50,000 | 127 |
| More than $1000 but less than $20,000 | 192 |
| More than $0 but less than $1000 | 102 |
| No money at all | 159 |
| **Total** | **637** |

(a) Create a relative frequency bar chart for the data.

(b) From the data would you conclude that a majority of those surveyed can or cannot earn a living solely from writing?

(c) Although the data are numerical, why did you have to make a bar chart rather than a histogram?

(d) What problem do you see with the way that the classes were defined?

*Requires Exercise 3.5*   **3.15**  The company that collected data on the number of weekly repairs needed by copy machines decided that it wanted a graphical display of the data to include in a budget report. The data are repeated below:

| | | | | |
|---|---|---|---|---|
| 3 | 2 | 3 | 0 | 3 |
| 0 | 2 | 2 | 2 | 1 |
| 3 | 2 | 2 | 1 | 0 |
| 2 | 3 | 2 | 1 | 2 |
| 3 | 4 | 2 | 7 | 1 |
| 2 | 2 | 1 | 3 | 1 |
| 0 | 3 | 1 | 1 | 2 |

Create a histogram for the data.

*Requires Exercise 3.3*   **3.16**  After collecting the data on packaging defects, the company decided to ask you how it could make a graph of the data to use in a brainstorming session with the employees in the department. The data are repeated here:

| | | | | |
|---|---|---|---|---|
| Dented | No defect | No defect | Dented | No defect |
| No defect | Torn | Crushed | No defect | No defect |
| Unsealed | No defect | No defect | No defect | No defect |
| No defect | No defect | No defect | No defect | No defect |
| No defect | No defect | No defect | Dented | No defect |
| No defect | Unsealed | No defect | No defect | No defect |

(a) What type of graph would be appropriate for these data?

(b) Create a graphical display of the data.

(c) What order did you select for the categories? Why?

**3.17**  A company has recently become concerned about the number of young engineers that seem to be leaving the company and decides to collect some data to try to understand why the engineers are leaving. During the exit interviews with employees the human resources department asks several questions about the reasons the employee has for leaving the company. A table showing the reasons for leaving and the appropriate codes is shown at the top of the next page:

| Dislike engineering | DE |
|---|---|
| Relocation | L |
| Raise in pay | P |
| More responsibilities | R |
| Relations with manager | M |
| Colleagues | C |
| Type of work | B |

The human resources department looks at the results from the exit interview process for 40 engineers that have left the company in the last two years and find the following:

| DE | P | B | R | R | R | R | DE |
|---|---|---|---|---|---|---|---|
| P | L | M | DE | M | M | R | P |
| R | P | L | P | B | P | DE | L |
| R | R | P | P | P | R | C | C |
| M | P | R | P | P | M | P | L |

(a) Create a Pareto diagram for the data.

(b) What category has the highest relative frequency?

(c) Does it appear that the 80–20 rule is true for these data? Why or why not?

**3.18** After a change in the channels provided by the cable television service in a town, subscribers were asked to rate the new service on a scale of 1 = worst service I have ever had to 10 = best service I have ever had. The responses from 40 random subscribers are

| 1 | 4 | 5 | 7 | 8 |
|---|---|---|---|---|
| 2 | 4 | 5 | 7 | 9 |
| 2 | 4 | 5 | 7 | 9 |
| 3 | 4 | 6 | 7 | 9 |
| 3 | 4 | 6 | 7 | 9 |
| 3 | 4 | 6 | 7 | 9 |
| 3 | 4 | 6 | 8 | 9 |
| 3 | 5 | 7 | 8 | 10 |

(a) Create a relative frequency histogram of the ratings.

(b) Are the majority of the ratings favorable, unfavorable, or neither?

**3.19** The company that is worried about losing its engineering staff also looked at the length of time, in months, that the engineers in the sample worked for the company:

| 16.7 | 26.0 | 28.0 | 29.6 | 30.6 | 32.6 | 34.0 | 35.8 |
|---|---|---|---|---|---|---|---|
| 20.9 | 26.6 | 28.1 | 29.7 | 30.8 | 32.6 | 34.6 | 37.4 |
| 25.2 | 26.6 | 28.5 | 30.2 | 31.1 | 33.2 | 34.8 | 37.8 |
| 25.6 | 27.4 | 28.8 | 30.2 | 31.7 | 33.5 | 34.9 | 38.3 |
| 25.8 | 27.7 | 28.8 | 30.3 | 32.1 | 33.6 | 35.2 | 44.6 |

Create a relative frequency histogram for the data.

**3.20** The refuse company decides to use the data it collected in a presentation to the town. *Requires Exercise 3.11* The data are shown again here:

| 8.3 | 10.2 | 12.2 | 13.1 | 13.9 |
|---|---|---|---|---|
| 8.5 | 10.4 | 12.3 | 13.3 | 14.1 |
| 8.8 | 11.0 | 12.3 | 13.3 | 14.3 |
| 9.1 | 11.0 | 12.3 | 13.4 | 14.3 |
| 9.7 | 11.2 | 12.3 | 13.4 | 14.4 |
| 9.8 | 11.4 | 12.5 | 13.4 | 14.4 |
| 9.9 | 12.0 | 12.7 | 13.6 | 14.4 |
| 9.9 | 12.0 | 12.8 | 13.6 | 14.8 |
| 10.1 | 12.2 | 12.9 | 13.8 | 15.2 |

(a) Create a relative frequency histogram for the data.

(b) From the histogram, does it appear that a majority of the newspaper piles put out for recycling exceed 12 inches?

*Requires Exercise 3.10*     **3.21**   The Bureau of Weight and Measures that looked at orange juice containers from super-markets wants to create a graphical display of the data it obtained.

| | | | | |
|---|---|---|---|---|
| 64.8 | 65.2 | 65.6 | 65.7 | 65.9 |
| 64.8 | 65.3 | 65.7 | 65.7 | 66.0 |
| 64.9 | 65.3 | 65.7 | 65.7 | 66.1 |
| 65.1 | 65.4 | 65.7 | 65.8 | 66.1 |
| 65.2 | 65.4 | 65.7 | 65.8 | 66.3 |

Create an appropriate graphical display of the data.

## 3.4   DESCRIBING AND COMPARING DATA

Since the reason for displaying data is to gain understanding, it is important to know what we can learn from the graph of a data set. For quantitative data we usually want to know what a typical observation might be, and how the actual observations differ from the typical values. We would also like to be able to compare different data sets and make some decisions based on the comparisons.

### 3.4.1 Describing Quantitative Data

In statistics, the features of interest for a set of numerical data can be classified as **center, shape,** and **variability.**

> The *center* of a set of data describes where, numerically, the data are centered or concentrated.

> The *shape* of a set of data describes how the data are spread out around the center with respect to the symmetry or skewness of the data.

> The *variability* of a set of data describes how the data are spread out around the center with respect to the smoothness and magnitude of the variation.

Together these three features describe the distribution of the data. To describe the data it is useful to picture the distribution of the data as being represented by a smooth curve that captures the "shape" of the histogram. One way to achieve this curve is to plot a point at the top of each bar of the histogram and then connect the dots as shown in Figure 3.5.

**FIGURE 3.5**   Histogram with line showing the shape of the distribution

**FIGURE 3.6**    Histogram with smoothed curve representing the distribution

If you smooth out the plotted line you get a curve that looks like the one shown in Figure 3.6. From the curve you see that the distribution is centered at the bump or high point of the curve. Since the bump in the distribution is the location of the class with the highest frequency, it locates the most typical data points.

The shape of a distribution concerns how the data are spread out on either side of the center, that is, whether they are **symmetric** or **skewed.**

> When data are evenly spread out on both sides of the center, we describe the distribution of the data as *symmetric.*

A typical symmetric distribution is shown in Figure 3.7.

**FIGURE 3.7**    Typical symmetric distribution

> When the data are not evenly spread out on either side of the center then we refer to the distribution as being *skewed.*

Skewness has a direction associated with it, either left (negative) or right (positive). The direction of the skew describes the side on which the distribution of the data covers a larger distance, that is, the direction in which the distribution "tails off" more slowly. Figure 3.8 shows both right and left skewed distributions.

When data are skewed, either left or right, the tailing off of the data is continuous and gradual as shown in Figure 3.9a. When the tailing off involves a gap in the data—a place where classes in the frequency histogram have no observations—as

**FIGURE 3.8**    Typical skewed distributions

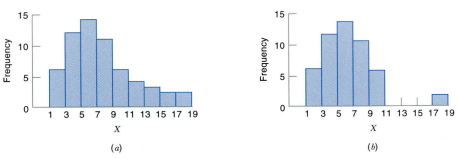

**FIGURE 3.9**    Histograms with (*a*) skewed data and (*b*) extreme values

shown in Figure 3.9*b*, the data are not really skewed. More likely the data you have contain some extreme or unusual observations. We talk about these extreme values more in Chapter 4 when we discuss outliers.

In addition to center and shape we would like to describe how much the data differ from the center or typical values. This is not easy to do without some way to measure the differences. We examine measures of variability in Chapter 4. At this point we can describe the variability of the distribution in two ways: in terms of (1) the "smoothness" of the curve and (2) the total spread of the data.

When data are not very variable, the frequency of observations decreases steadily as you move away from the center. Sometimes when data are highly variable, the distribution will be jagged. That is, the frequency of data values will not decrease steadily as you move away from the center. Another way to describe variability is to describe the distance from one end of the data to the other. Figures 3.10 and 3.11 show histograms with different degrees of variability.

Dotplots can also be used to describe data distributions, although because they represent small data sets it is often more difficult to describe the data.

**FIGURE 3.10**    Histograms with different variability showing spread of data

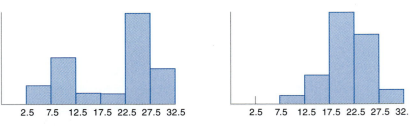

**FIGURE 3.11**    Histograms with different variability showing smoothness

**EXAMPLE 3.23    ABC Faculty Salaries**

*Describing Data Distributions*

The ABC provost looked at the distribution of years of faculty service to see if she could determine how the data are distributed.

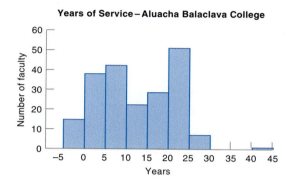

Years of Service – Aluacha Balaclava College

She sees that the data are highly variable and there is no smooth shape to the distribution. In addition, there does not seem to be a unique center or shape. She wonders why this might be. Perhaps another variable is affecting the picture. It might make more sense to look at years of service for each rank separately. ∎

Sometimes when you collect data and make a histogram there is no cohesive picture to be seen. This often happens when there are other, underlying variables that have an effect on the variable you are studying. When this happens it is useful to separate the data into groups on the basis of another variable and look at them again.

**EXAMPLE 3.24    On-Time Rates**

*Describing Data Distributions*

The railroad company that is looking at on-time rates obtained the histogram below:

On-Time Rates

From the histogram we see that the data have a distinct center and that, typically, the on-time rate was between 65% and 75%. The frequency of the data seems to drop off rapidly from the center, and the data are rather uniformly distributed among the other classes. The data seem a little variable. The distribution is skewed to the left, which means that the data are more spread out on the low side (four classes) than on the high side (two classes). ∎

The choice of the number of class intervals for a histogram is very important. If your histogram has too many intervals for the number of data values, then the data might appear to have a lot of variability when, in fact, classes are empty or have few observations because of the lack of data. If your histogram has too few intervals, then the distribution will look like a lump and important features of the data might be hidden. It is often important to look at the data in several different ways to make sure that you are getting a true and consistent picture. We see later on that it is possible to distort graphical displays and bias an analysis by playing with the scale.

 **TRY IT NOW!**

### Assignment Times

*Creating a Frequency Table for Continuous Data*

The instructor for the introductory statistics class wants to see, in graphical form, the data she has collected on the amount of time it took the students to do the assignment. The frequency distribution for the data is

| Time | Frequency | Relative frequency (%) |
|---|---|---|
| $22.70 < x \leq 26.22$ | 6 | 20.0 |
| $26.22 < x \leq 29.74$ | 11 | 36.7 |
| $29.74 < x \leq 33.26$ | 7 | 23.3 |
| $33.26 < x \leq 36.78$ | 5 | 16.7 |
| $36.78 < x \leq 40.30$ | 0 | 0.0 |
| $40.30 < x \leq 43.82$ | 1 | 3.3 |

Create a relative frequency histogram for the data.

Use the histogram to describe the distribution of the times the students took to complete the assignment.

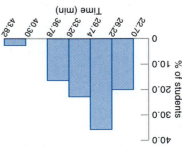

## 3.4.2 Comparing Data Distributions

One of the major reasons for doing statistical analyses of data is to obtain facts for making *informed* decisions. As a result, we often need to make comparisons between or among samples taken from different populations. When comparing different data sets, we make those comparisons based on the qualities of the data you just learned: center, shape, and variability.

To make valid comparisons it is critical that the data be displayed in the same way. Since the graphical displays have a *visual* impact, it is important that the graphs used are comparable in scale so that the viewers are not misled.

To make visual comparisons about center, shape, and variability, the class intervals for the different graphs should be the same. When this is not the case it is difficult if not impossible to compare these qualities in a meaningful way. The following example shows how using different scales can mask the visual impact of the display.

### EXAMPLE 3.25  Loan Application Times

*Comparing Data Distributions*

A large bank with many branch offices has developed two different procedures for filling out mortgage loan applications. Each procedure is being used in a different branch of the bank in comparable locations. After the procedures are used for several months, some data are collected on the amount of time (in minutes) it took for loan candidates to complete the application. The data for each branch are

| Branch 1 | | | | | | Branch 2 | | | | |
|---|---|---|---|---|---|---|---|---|---|---|
| 37.0 | 39.3 | 42.1 | 44.3 | 47.8 | | 40.4 | 42.8 | 43.4 | 44.3 | 46.1 |
| 37.2 | 39.9 | 42.4 | 44.5 | 49.3 | | 40.8 | 42.8 | 43.5 | 44.4 | 46.4 |
| 37.6 | 41.2 | 42.6 | 45.3 | 49.5 | | 41.5 | 42.8 | 43.5 | 44.7 | 46.5 |
| 38.1 | 41.5 | 42.7 | 45.7 | 50.3 | | 41.5 | 43.1 | 43.8 | 44.7 | 47.1 |
| 38.8 | 41.6 | 43.0 | 46.1 | 52.1 | | 41.6 | 43.2 | 44.0 | 45.1 | 47.9 |
| 39.2 | 41.9 | 43.4 | 46.2 | 54.7 | | 42.3 | 43.3 | 44.2 | 45.4 | 48.6 |

To better understand the data and to compare the loan application times for the two branches a computer software package is used to create histograms for each of the data sets. The histograms show that the application times for Branch 1 are typically between 37.5 and 42.5 minutes and that the data are slightly skewed to the right. The loan application times for Branch 2 are typically between 43 and 45 minutes and the data are more symmetric than for Branch 1. On first glance it appears that the first bank is using the better method. Or is it?

The bank decides to re-create the histograms, this time using a common scale for the *x* axis. The results are shown at the top of page 94. From the second set of histograms it appears that the times for Branch 1 are much more variable than those for

Branch 2. It also seems that the typical times for Branch 1 are really not that much different than those for Branch 2. They can see from the second set of graphs that the method used for Branch 2 is more consistent than that for Branch 1 and it has comparable application times. ∎

In the previous example it does not really matter that the *y* axis of the histograms is frequency rather than relative frequency, because the number of observations is the same for each sample. If this is not the case it is necessary to use relative frequency in order to make valid and meaningful comparisons.

## *Discovery Exercise 3.1*
## *THINKING ABOUT VARIABILITY*

A manufacturer of compact discs uses two different suppliers for the jewel boxes used to hold the discs. There have been problems with these boxes in the past. The inside width of the jewel box has critical specifications of $119.0 \pm 0.2$ mm. If the case is too narrow the disc will not fit in it and if it is too wide, the front label insert slips around. Because it is time to renew the purchasing contracts for the jewel boxes, the CD manufacturer decides to take a look at a sample of the boxes from each supplier. The data (in mm) for each source are

| Supplier A | | | | | Supplier B | | | | |
|---|---|---|---|---|---|---|---|---|---|
| 118.7 | 118.9 | 119.0 | 119.1 | 119.2 | 118.8 | 118.8 | 118.9 | 118.9 | 118.9 |
| 118.8 | 119.0 | 119.0 | 119.1 | 119.2 | 118.8 | 118.8 | 118.9 | 118.9 | 119.0 |
| 118.8 | 119.0 | 119.0 | 119.1 | 119.2 | 118.8 | 118.8 | 118.9 | 118.9 | 119.0 |
| 118.9 | 119.0 | 119.1 | 119.2 | 119.2 | 118.8 | 118.8 | 118.9 | 118.9 | 119.0 |
| 118.9 | 119.0 | 119.1 | 119.2 | 119.3 | 118.8 | 118.9 | 118.9 | 118.9 | 119.1 |

Make a relative frequency histogram of the data for each supplier.

Describe the distribution of jewel box widths for each supplier and compare them.

Your company has decided to single source its supply of jewel boxes. The purchasing agent in charge of the accounts argues that supplier B should not get a renewed contract since he observes that the jewel boxes from that source are not centered at the target specification of 119.0 mm while the jewel boxes from supplier A are right on target. Can you explain to him why, although his observation is true, his decision to use supplier A is not necessarily correct? What factor has he failed to consider?

Which supplier would you recommend that your company use? Write a short memo to the manager with your recommendation and your supporting reasons.

## 3.4.3 Exercises—Learning It!

**3.22** A university collected some data on the amount of money ($) that students spend on textbooks in a typical semester:

| | | | | |
|---|---|---|---|---|
| 214 | 241 | 248 | 258 | 269 |
| 233 | 244 | 249 | 260 | 274 |
| 234 | 245 | 250 | 262 | 276 |
| 236 | 247 | 253 | 262 | 277 |
| 239 | 248 | 254 | 263 | 277 |
| 241 | 248 | 254 | 265 | 281 |

(a) Create a histogram of the data.

(b) Where is the center of the distribution located?

(c) Are the data symmetric or skewed? If they are skewed, are they left or right skewed?

(d) Describe the variability of the data.

**3.23** The company that is looking at the length of time that engineers stayed at the company wants to summarize its data, shown below:

*Requires Exercise 3.19*

| | | | | | | | |
|---|---|---|---|---|---|---|---|
| 16.7 | 26.0 | 28.0 | 29.6 | 30.6 | 32.6 | 34.0 | 35.8 |
| 20.9 | 26.6 | 28.1 | 29.7 | 30.8 | 32.6 | 34.6 | 37.4 |
| 25.2 | 26.6 | 28.5 | 30.2 | 31.1 | 33.2 | 34.8 | 37.8 |
| 25.6 | 27.4 | 28.8 | 30.2 | 31.7 | 33.5 | 34.9 | 38.3 |
| 25.8 | 27.7 | 28.8 | 30.3 | 32.1 | 33.6 | 35.2 | 44.6 |

(a) Describe the center of the data.

(b) Describe the shape of the data.

(c) Describe the variability of the data.

**3.24** The refuse company decides to use the data it collected in a presentation to the town. The data are shown again:

*Requires Exercises 3.11, 3.20*

| | | | | |
|---|---|---|---|---|
| 8.3 | 10.2 | 12.2 | 13.1 | 13.9 |
| 8.5 | 10.4 | 12.3 | 13.3 | 14.1 |
| 8.8 | 11.0 | 12.3 | 13.3 | 14.3 |
| 9.1 | 11.0 | 12.3 | 13.4 | 14.3 |
| 9.7 | 11.2 | 12.3 | 13.4 | 14.4 |
| 9.8 | 11.4 | 12.5 | 13.4 | 14.4 |
| 9.9 | 12.0 | 12.7 | 13.6 | 14.4 |
| 9.9 | 12.0 | 12.8 | 13.6 | 14.8 |
| 10.1 | 12.2 | 12.9 | 13.8 | 15.2 |

Describe the distribution of the data. Remember to include center, shape and variability.

**3.25** The company that is looking at turnaround time from the Office Services department wants to summarize its data:

*Requires Exercise 3.9*

| | | | | | |
|---|---|---|---|---|---|
| 14 | 19 | 21 | 25 | 26 | 29 |
| 15 | 19 | 22 | 25 | 26 | 29 |
| 16 | 20 | 22 | 25 | 26 | 30 |
| 16 | 20 | 23 | 25 | 26 | 31 |
| 18 | 20 | 23 | 25 | 27 | 31 |
| 18 | 20 | 23 | 25 | 28 | 35 |
| 18 | 21 | 24 | 26 | 29 | 40 |

(a) How long is a typical turnaround time for a job?

(b) What shape is the distribution of turnaround times?

(c) How would you describe the variability in turnaround times?

# 3.5 CREATING GRAPHICAL DISPLAYS USING EXCEL 97

The graphical displays that you learned about in this chapter can be created easily using Microsoft Excel 97. The power of using a computer package such as Excel to create the graphs is that you can modify them easily. This leads to better graphical displays because it eliminates much of the drudgery. This section will cover the basics of creating a chart using Microsoft Excel, as well as the specific methods for creating bar charts, pie charts, and histograms.

## 3.5.1 The Basics of Creating a Chart in Excel

In most cases, graphs and charts in Excel are easily created by using a special assistant called the *Chart Wizard*. The steps for creating a chart or graph using the Chart Wizard are:

*Chart Wizard belongs to the Standard Toolbar.*

1. Highlight the frequency table that you want to graph.
2. Invoke the Chart Wizard by clicking on the icon on the toolbar.
3. Follow the directions and hints from the Chart Wizard.
4. Edit the graph to include any other features or changes that you want.

## 3.5.2 Creating a Bar Chart in Excel

Excel differentiates between *bar* and *column* charts. The only real difference is the direction (horizontal vs. vertical) of the bars. What we call a bar chart in statistics is called a column chart in Excel. Using the *Chart Wizard* to create a bar or column chart in Excel assumes that the data have already been tabulated in the form of a frequency table. Once this is done, it is relatively easy to create the chart.

A Bar Chart

Suppose we want to create a bar chart for the faculty rank data for Aluacha Balaclava College. Figure 3.12 shows the frequency table as part of an Excel worksheet.

To use the *Chart Wizard* to create a bar or column chart, select the range that contains the frequency table, in this case **A2:B6,** and click the *Chart Wizard* icon on the Standard Toolbar.

A Column Chart

There are four steps to creating the chart:

1. In the first step you select the *Chart Type*. Select **Column Chart** as the *type* and **Clustered Chart** as the *sub-type,* and then click the **Next>** button.
2. The second step of the process tells Excel where the frequency table is located and how the data are arranged, in columns or in rows. The default setting in this box is usually correct. You can see what your chart will look like. If it does

|   | A | B | C |
|---|---|---|---|
| 1 |  |  |  |
| 2 | Rank | Total |  |
| 3 | Associate | 67 |  |
| 4 | Assistant | 77 |  |
| 5 | Instructor | 8 |  |
| 6 | Professor | 55 |  |
| 7 | Grand Total | 207 |  |
| 8 |  |  |  |

**FIGURE 3.12**  Excel worksheet with frequency table

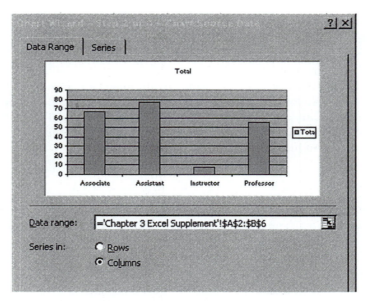

**FIGURE 3.13**    The Chart Source Data dialog box

not appear the way you expect, experiment with the settings. Figure 3.13 shows what you should see at this point in the process.

3. Once you have indicated where the data are located, you can click on the Next button and go to the third step of the process. In this step you will be making decisions about the format of the chart. The dialog box for this step has six tabs—Titles, Axes, Gridlines, Legend, Data Labels, and Data Table. In the **Titles** section, click the **Chart title** box and type **"Aluacha Balaclava College—Faculty Rank"**. Next, press TAB and type **"Rank"**. Finally, press TAB and type **Number of Faculty"**. Figure 3.14 shows a portion of the dialog box at this point. Click the **Gridlines** tab and clear the **Major gridlines** check box. Then Click the **Legend** tab and clear the **Show legend** check box. Finally, click the Data Labels tab and click the **Show value** check box. Notice that while you are making these choices, you can see what your chart will look like. When you are finished with this step, click the Next button and begin the fourth and last step.

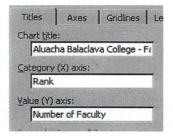

**FIGURE 3.14**    Adding titles to the chart

4. The fourth step of the process lets you select where you want the chart to appear. You can either put the chart on a new worksheet or locate the chart on an existing sheet. To have the chart located on the current sheet, check the **As object in** radio button and click **Finish.** Excel will place the chart on the current worksheet. Figure 3.15 shows the chart.

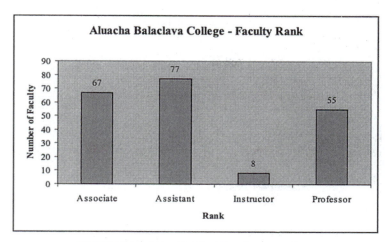

**FIGURE 3.15**     Finished Excel column chart

### 3.5.3 Creating a Pie Chart in Excel

Creating a pie chart in Excel uses the same procedure as that for creating a bar chart. The major difference occurs in the first dialog box of the Chart Wizard, where you select a different type of chart: In the first dialog box of the Chart Wizard, select Pie as the type of chart you want. The second step defines the location of the data, just as it did for the bar chart. The real differences are in step 3, where you determine the format of the chart. The dialog box for a pie chart has three tabs—Title, Legend, and Data Labels. The Title and Legend boxes are the same as those for the bar chart. The Data Labels tab allows you to select how the pie slices are labeled. Figure 3.16 shows the dialog box with the selections to label each slice with the category and the percent. The chart to the right updates to show how your selections will look. Note that if you are going to label the slices of the pie, it is not necessary to have a legend show as well.

**FIGURE 3.16**     Formatting the pie chart

### 3.5.4 Creating a Pareto Diagram in Excel

As you learned in this chapter, a Pareto diagram is a special kind of bar chart used to identify trouble spots in a process. It is really a bar chart with the bars sorted by decreasing frequency. It also has a line that plots the cumulative frequency for the categories.

Sort Descending

To create a Pareto diagram in Excel, you must first sort your frequency table by using the Sort command. You can do this in two ways. The quick way is to highlight the range that you want to sort—in this case, it is the data in the frequency table for Total—and click on the *Sort Descending* button on the Standard Toolbar. This will sort the range you have selected and will also sort the adjacent data accordingly. You can also highlight the entire table you want to sort and select **Sort** from the **Data** menu. In the dialog box, select the field you want to sort by and indicate whether you want to sort in ascending or descending order.

To create a Pareto diagram, you will also need a column for cumulative relative frequency. You can do this using the SUM function to keep a running total similar to the way you would update a checkbook register. The steps are:

1. Label the cell adjacent to Total in the frequency table Cumulative and copy the frequency of the first category to the adjacent cell.

2. In the cell directly below this, you will enter a formula using the SUM function. You tell Excel that you are entering a formula or function by typing an equal sign "=" and entering the formula or function name. In this case, the function you want to use is SUM, so you type this directly after the equal sign. The SUM function has two parts here, the location of the current subtotal and the location of the data you want to add to that subtotal. For example, looking at the data table in Figure 3.17, you see that the current subtotal is in cell C3 and you want to add the data in cell B4. You can see the formula in Cell C4. When you hit **Enter**, the formula will calculate the value you have indicated.

|   | A | B | C |
|---|---|---|---|
| 1 |   |   |   |
| 2 | Rank | Total | Cumulative |
| 3 | Associate | 67 | 67 |
| 4 | Assistant | 77 | =SUM(C3,B4) |
| 5 | Instructor | 8 |   |
| 6 | Professor | 55 |   |

**FIGURE 3.17**   Formula for Cumulative Frequency

3. Copy this formula to the rest of the cells in the column labeled Cumulative and you will see the cumulative frequencies calculated.

4. Now you need to calculate the cumulative relative frequencies. Using the next column (you can label it Cumulative Relative Frequency) in the worksheet, you will enter the formula—that is, the cumulative frequency divided by the total. In cell D4 type "=C4/207" and press enter. The cumulative relative frequency for the first category is calculated. Copy the formula to the remainder of the cells in this column.

Creating the Pareto diagram is easy using one of the custom charts in the Chart Wizard. After starting the Chart Wizard, the steps are as follows:

*Note: If the ranges are not adjacent, hold down the CTRL button while you highlight the ranges.*

1. In the first dialog box for chart type, click on the tab labeled Custom Types and scroll down until you see a chart called Line–Column on 2 Axes as shown in Figure 3.18. This will plot the first data series in your table as columns and the second as a line. The left axis corresponds to the bars and the right to the line. Click **Next>** to proceed.

2. The second step lets you specify the location of the data and see that your chart is set up correctly. When this is done, click **Next>** to continue.

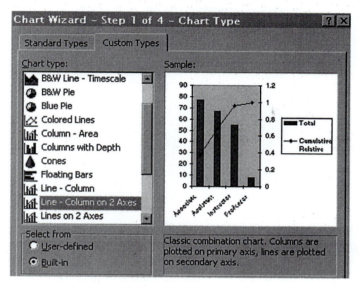

**FIGURE 3.18**   Custom Chart Type menu

3. The third step allows you to format the chart by adding labels. You might want to use the Legend tab to relocate the legend to the bottom of the chart so that it does not interfere with the right axis. When the graph is the way you want it, click **Next>** to continue.

4. In the fourth step you can specify the location of the chart and click **Finish** to see your result. The finished chart is shown in Figure 3.19.

**FIGURE 3.19**   Pareto diagram

## 3.5.5 Creating a Frequency Table in Excel

In the previous section, you learned how to use the Chart Wizard to create a bar chart in Excel when you have the frequency table available. If you have the raw data instead of the frequency table, then you need to use **Pivot Table** to create it.

A **Pivot Table** in Excel is a tool that you will use often to manipulate raw data that are qualitative. You can create a pivot table by using the Pivot Table Wizard in Excel. The steps for creating the frequency table are:

1. Highlight the data for which you want to create the frequency table.

2. Invoke the Pivot Table Wizard in Excel by using the **Data > Pivot Table Report** menu.

|    | A | B | C |
|----|---|---|---|
| 1  | Rank | Service (Yr) | Salary |
| 2  | ASST | 22 | 53316 |
| 3  | PROF | 11 | 64375 |
| 4  | ASSO | 7 | 63501 |
| 5  | ASSO | 6 | 59426 |
| 6  | ASSO | 20 | 49058 |
| 7  | PROF | 4 | 94969 |
| 8  | ASST | 21 | 54762 |
| 9  | ASSO | 9 | 55516 |
| 10 | ASSO | 18 | 45932 |

**FIGURE 3.20**   Raw data for faculty rank

3. Follow the directions that the Pivot Table Wizard gives you.

4. Copy or edit the pivot table to include any changes you want.

Remember that the Provost at Aluacha Balaclava College received the data in raw form; that is, the data had not been summarized. A portion of the data for Faculty Rank is shown in Figure 3.20.

The frequency table for these data will have a column for each rank that appears and a column for the frequency, or number of times that it appears. You can also add columns for the relative frequency and cumulative relative frequency. Highlight the range that contains the data you want to summarize—in this case, the data are in the range **A1:A208**—and follow these steps:

1. From the **Data** menu select **Pivot Table Report.** The Pivot Table Wizard will start by asking what type of data you want to work with. Since you are using data that are already in an Excel worksheet, simply leave the default radio button checked and click on **Next>** to continue.

2. Since you have already highlighted the range that contains your data, you can proceed by clicking **Next>.** If you have not indicated where your data are located, you can do that now by highlighting the range in the worksheet. The next step of the process will define the table you want to create. The dialog box is shown in Figure 3.21.

3. At the right of the dialog box, there are small rectangles called field buttons, which represent all of the variables you have highlighted. In this case, Rank is

**FIGURE 3.21**   Creating the table

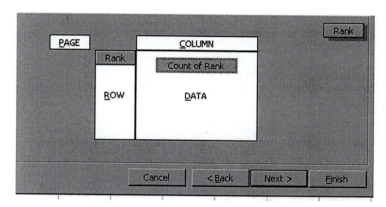

**FIGURE 3.22**    Defining the table

the only one. You create the table by dragging the variable names to the places you want them to appear. Since we want the ranks to appear as rows in the table, click on the Rank box and drag it to the area marked **ROW,** as indicated at the top of the dialog box. To get the frequencies, you must tell the Pivot Wizard how to process the data. To do this, drag the field button for Rank into the area of the dialog box marked **DATA.** The dialog box should now look like the one in Figure 3.22.

The default operation in a Pivot Table is to provide a count (frequency) of each value. If you want to change this, or if you want to provide additional information in the table, you can double click on the field in the DATA area and select a different data operation such as Sum or Average. When you have finished this step, click **Next>** to continue.

4. The last step of the process is to tell Excel where you want the Pivot Table to appear. You can have it appear in the current worksheet or on a new worksheet. To have it appear in the current sheet, select the **Existing worksheet** radio button and then click on the cell in the worksheet where you want the left corner of the table to appear. Click **Next>** and the pivot table shown in Figure 3.23 will appear in the worksheet location you indicated.

If you want to change the pivot table so that it looks better, you can click on the cell and type in exactly what you want to appear. For example, if you want the ranks to appear as Associate Professor, Assistant Professor, etc., simply edit those cells in the table.

| Count of Rank | |
|---|---|
| Rank | Total |
| ASSO | 67 |
| ASST | 77 |
| INST | 8 |
| PROF | 55 |
| Grand Total | 207 |

**FIGURE 3.23**    Finished pivot table

It is possible to change other features of the table by double clicking on the cell in the table with the field label and clicking on the Advanced button. This will allow you to change the order of the categories and several other options.

Once you have the frequency table, you can create a bar chart, pie chart, or Pareto diagram as explained previously.

*Note: Make sure the* **Tools** *menu contains the* **Data Analysis** *option. If not, you may have to run the* **MS Office Setup** *program to include the* **Analysis ToolPak** *and then execute the* **Tools\Add-Ins** *command.*

### 3.5.6 Creating Histograms in Excel

Quantitative data are easier to process using Excel because there are built-in functions for many of the statistical tools. Selecting **Data Analysis** from the **Tools** menu accesses these functions, as shown in Figure 3.24. Macros for other types of analyses are available on the disk that comes with this book.

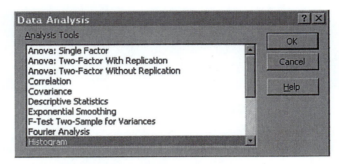

**FIGURE 3.24**   Data Analysis dialog box

The basics of creating a histogram in Excel are quite simple. Suppose you want to create a histogram for the data on Years of Service for Aluacha Balaclava College. Figure 3.20 shows the data as part of an Excel file.

To produce a basic histogram, simply follow these steps:

1. From the Data Analysis dialog box, select Histogram and click **OK.** The **Histogram** dialog box will open.
2. Position the cursor in the **Input Range** text box and highlight the range that contains the data, **B1:B208.**
3. Click the checkbox for **Labels** since the range contains the variable name.
4. Excel calls the class intervals for continuous data Bins. Leave the Bin text box empty this time and let Excel pick the class intervals.
5. Click the radio button for **Output Range,** position the cursor in the text box, and click on the cell where you want the top left corner of the output to appear.
6. Finally, click the checkbox for **Chart Output.** The completed dialog box appears in Figure 3.25.
7. Click **OK** and the output will appear as shown in Figure 3.26.

**FIGURE 3.25**   Completed histogram dialog box

| Bin | Frequency |
|---|---|
| 0 | 15 |
| 2.928571 | 19 |
| 5.857143 | 19 |
| 8.785714 | 34 |
| 11.71429 | 16 |
| 14.64286 | 14 |
| 17.57143 | 13 |
| 20.5 | 17 |
| 23.42857 | 33 |
| 26.35714 | 20 |
| 29.28571 | 4 |
| 32.21429 | 2 |
| 35.14286 | 0 |
| 38.07143 | 0 |
| More | 1 |

**FIGURE 3.26**    Output from histogram tool

Clearly this is not the output you hoped for, but it is a starting point. There are two basic problems with this output. First, Excel used rules very similar to the ones you learned in this chapter for creating the class intervals or Bins. As a result, the numbers are not particularly friendly. The number that shows in each bin is the *upper* boundary for that class interval. A bin represents all of the data values that are greater than the previous bin number, up to and including the current bin number. That is, the number 2.928571 represents all of the data greater than 0 and less than or equal to 2.928571 ($0 < x \le 2.928571$). The first class is all data values that are less than or equal to zero, and the last is all data values greater than 38.07143.

The second problem is with the format of the graph. No labels can be seen on the axis, the graph is squashed, and the titles and axis labels are not very useful. This can be fixed by clicking on the parts of the chart that you do not like and editing them. A set of steps for editing the chart in Figure 3.26 is given below:

- Click on the chart to select it. Handles appear on the corners and sides.
- Click on the bottom handle and drag it down to increase the height of the chart, as shown in Figure 3.27. You see that the axis labels become visible.

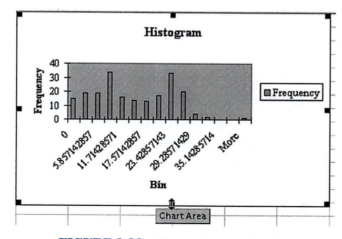

**FIGURE 3.27**    Changing the size of the chart

To change any portion of the chart, you simply double click on it and the appropriate dialog box opens. There are many changes that can be made to the chart. We will focus on a few that will produce an acceptable finished product.

The major problems with the chart are the way the *x* axis is labeled, the titles, and the gaps between the bars. We will fix some of these easily by making some changes to the bins in the frequency table. The frequency table and the chart are *linked* so that making changes to the table will update the chart.

- To get a little more room for the chart, click on the legend box and press the Delete key to remove it.
- Since the original data are integers, it really does not make sense to have the classes use six decimal places. Highlight the range in the worksheet that contains the frequency table and select **Cells** from the **Format** menu. The Format Cells dialog box opens. The box has six tabs: Number, Alignment, Font, Border, Patterns, and Protection. Click on the tab for Number and select Number from the list under **Category.** In the textbox for **Decimal places:** type 0, because you want integers. The dialog box is shown in Figure 3.28. Click **OK** to continue. The bins will change to integer values and you will see the changes on the chart.

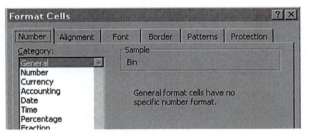

**FIGURE 3.28**   Format Cells dialog box

*When you have an extreme outlier in the data, Excel will include it in that last interval regardless of the width of the other intervals.*

- Before we proceed, we really need to change the last bin label. Excel automatically labels the bin More, but we know it should be a number. From the updated frequency table, you can see that the class width is 3, so the last number in the bin column should be $38 + 3 = 41$. Click on the cell that contains the word "More" and change it to 41. The worksheet should look like the one in Figure 3.29.

Notice that this changed the orientation of the text on the axis. That happened because the word "More" was too long to fit horizontally. If this change

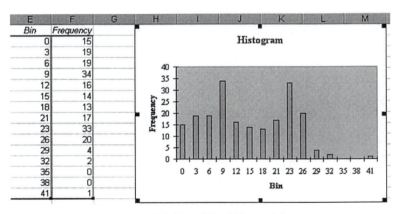

**FIGURE 3.29**   Edited bins and chart

**FIGURE 3.30**    Changing the gap width

had not occurred, you could have changed the orientation of the text as follows:

- Move the cursor to the chart and click on the *x* axis labels. The ToolTips box will say "Category Axis" when your cursor is in the right location. The Format Axis dialog box will open. The box has five tabs: Patterns, Scale, Font, Number, and Alignment. To change the orientation of the text on the axis, click on the Alignment tab. Click on the red diamond to orient the text horizontally.

*Note: Clicking and dragging the red diamond will change the orientation of the text to any angle you want.*

We still need to change the title and axis labels and to remove the gaps between the bars.

- To change any title or label, simply click on the title and type the new text.
- To remove the gaps, double click on any of the histogram bars and the **Format Data Series** dialog box will open. This box allows you to change the way the bars in the histogram look. For example, from the Patterns box you can change the color of the bars or from the Data Labels box you can have the frequency for each bar print above the box. To change the spacing between the bars, click on the **Option** tab, and in the text box for **Gap width,** type 0 as shown in Figure 3.30.

Click **OK,** and the chart should look like the one in Figure 3.31.

Even though Excel defines the bin numbers as the endpoints of the class intervals, it prints them in the center of the histogram bars. There is no way to change this. It is one of the problems that arise because Excel is not a statistical package.

There are many other changes that you can make to a chart in Excel. Experiment with them and see what improvements you can make to the histogram.

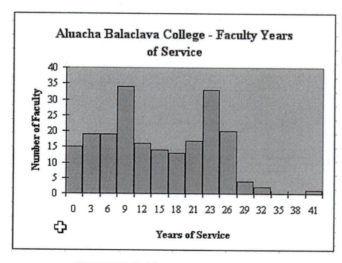

**FIGURE 3.31**    Completed histogram

# *CHAPTER 3 SUMMARY*

Raw data are nothing more than a list of words or numbers. The purpose of descriptive statistics is to turn *data* into *information.* One way to do this is by summarizing the data with a frequency distribution and then using a graphical display that is appropriate for the type of data. Qualitative data can be displayed using *bar charts* or *pie charts,* while quantitative data are usually displayed using a *histogram* or *dotplot.*

Graphical displays of data can be used to describe data in terms of *center, shape,* and *variability.* They can also be used to compare different data sets.

Although it is possible to create graphs by hand, using a computer software package such as Minitab or Excel greatly enhances the quality of the finished product. It is important, however, to know how each graphical technique works so that you can use the software intelligently and make useful (and not just flashy) graphs.

## Key Terms

| Term | Definition | Page reference |
|---|---|---|
| Bar chart | A **bar chart** is a graph that represents the frequency or relative frequency from a frequency table in the form of a rectangle or bar. | 75 |
| Center | The **center** of a distribution tells where on the number line the data are centered or concentrated. | 88 |
| Cumulative relative frequency | The **cumulative relative frequency** of a class is the sum of the relative frequencies of all classes at or below that class represented as a portion of the total number of observations. It can be expressed as a fraction, decimal, or percentage. | 63 |
| Dotplot | A **dotplot** is a graph used for small data sets, in which each observation is plotted as a point on a single, horizontal axis. | 83 |
| Frequency table | A **frequency table** or frequency distribution is a table containing each category, value, or class of values that a variable might have and the number of times that each one occurs in the data. | 58 |
| Histogram | A **histogram** is very similar to a bar chart, but since numbers are naturally ordered, the $x$ axis of the graph must be scaled to reflect this. | 80 |

*(continued)*

| Term | Definition | Page reference |
|------|-----------|----------------|
| **Pareto diagram** | A **Pareto diagram** is a bar chart in which the categories are plotted in order of decreasing relative frequency. The cumulative relative frequency of the categories is plotted on the same graph. | 77 |
| **Pie chart** | A **pie chart** is a graph in which a circle is used to represent the whole, and each "slice" is used to represent one of the categories. The size of the slice is proportional to the relative frequency of the category. | 79 |
| **Relative frequency** | The **relative frequency** of a classification is the number of times an observation falls into that classification represented as a portion of the total number of observations. It can be expressed as a *fraction, decimal,* or *percentage.* | 60 |
| **Shape** | The **shape** of a distribution describes how the data are spread out around the center with respect to symmetry or skewness. | 88 |
| **Skewed** | When the data are not evenly spread out on either side of the center then we refer to the distribution as being **skewed.** | 89 |
| **Symmetric** | When data are evenly spread out on both sides of the center, we describe the distribution of the data as **symmetric.** | 89 |
| **Variability** | The **variability** of a distribution describes how the data are spread out around the center with respect to the magnitude of the variation. | 88 |

# CHAPTER 3 EXERCISES

## Learning It!

**3.26**   In a study on housing costs in major international cities, data were collected on the construction cost ($ per square meter) for housing in 23 capital cities. Here are the data:

| | | | | | |
|---|---|---|---|---|---|
| Jakarta | 65 | Kingston | 157 | London | 560 |
| Dar Es Salaam | 67 | Bogota | 171 | Toronto | 608 |
| Karachi | 87 | Johannesburg | 192 | Seoul | 617 |
| Beijing | 90 | Rio de Janeiro | 214 | Hong Kong | 641 |
| New Delhi | 94 | Melbourne | 383 | Singapore | 749 |
| Istanbul | 110 | Algiers | 500 | Paris | 990 |
| Manila | 148 | Washington, D.C. | 500 | Tokyo | 2,604 |
| Bangkok | 156 | Madrid | 510 | | |

(a)  What kind of graphical display is appropriate for these data? Why?

(b)  Create a dotplot of the data.

(c)  Use the data to describe housing construction costs for these cities.

(d)  Do any of the data values appear to be unusual? If so, which one(s)?

**3.27**   An article in *Dartmouth Life* (15 September 1994) reported on the number of students from different classes who enrolled in various majors as freshmen. The data for the class of 1986 are:

| Major | 1986 |
|-------|------|
| Anthropology | 10 |
| Art History | 28 |
| Asia | 11 |
| Biology | 31 |
| Chemistry | 28 |

*(continued)*

| Major | 1986 |
|---|---|
| Classics | 2 |
| Comparative Literature | 5 |
| Drama | 13 |
| Earth Science | 8 |
| Economics | 90 |
| English | 156 |
| French | 23 |
| Geography | 23 |
| Government | 147 |
| History | 102 |
| Mathematics | 35 |
| Music | 10 |
| Philosophy | 21 |
| Physics | 13 |
| Psychology | 70 |
| Religion | 22 |
| Sociology | 17 |

(a) Create a bar chart for these data.

(b) What percentage of the students majored in the sciences?

(c) What order did you pick for the categories on the $x$ axis? Why did you choose this order? Can you think of any other orders that might also be appropriate? If so, what are they?

**3.28** The *Chronicle of Higher Education* (25 May 1994) reported on the distribution of college and university faculty by rank. The data are summarized below:

| Academic rank | Relative frequency (%) |
|---|---|
| Professor | 36.3 |
| Associate Professor | 27.8 |
| Assistant Professor | 26.5 |
| Instructor | 6.1 |
| Lecturer | 2.3 |
| No Rank | 1.0 |

(a) Create a bar chart for these data.

(b) What do the data tell you about the distribution of faculty over the various ranks?

**3.29** A report by the Bureau of Labor Statistics says that in 1992, the last year for which such data are available, 1004 homicides occurred while people were on the job. These homicides were classified by the type of job that the victim had:

| Job | People killed on job |
|---|---|
| Managerial and professional | 177 |
| Sales | 335 |
| Service industry | 225 |
| Drivers/factory workers | 202 |
| Other | 65 |

SOURCE: *Connecticut Post,* 11 June 1994.

(a) Create a bar chart for these data.

(b) Describe any interesting features of the data.

**3.30** The publishing industry has long felt that the field of food magazines is in a recession. They looked at paid circulations for all of the magazines in the field and obtained the following data:

| Magazine | Paid circulations |
|---|---|
| *Bon Appetit* | 1,294,945 |
| *Cooking Light* | 1,119,811 |
| *Gourmet* | 906,299 |
| *Food & Wine* | 734,831 |

(*continued*)

| | |
|---|---|
| *Eating Well* | 648,697 |
| *Cook's Illustrated** | 160,000 |
| *Fine Cooking** | 100,000 |
| *Saveur** | 100,000 |

* Publishers' current estimates.

(a) Create a bar chart for the data.

(b) Does one of the magazines account for a majority of the paid circulations in food magazines? If so, which one?

(c) Do you think that *Bon Appetit, Food & Wine,* and *Gourmet* deserve their reputation as the Big Three? Why or why not?

**3.31** As part of a survey conducted by the Bureau of Transportation in 1992, drivers were asked questions about the age of the vehicle that they normally drive. The survey recorded the year that the vehicle was manufactured. Data for 50 drivers surveyed are:

| | | | | |
|---|---|---|---|---|
| 1955 | 1978 | 1983 | 1986 | 1988 |
| 1966 | 1979 | 1983 | 1986 | 1988 |
| 1970 | 1979 | 1984 | 1986 | 1988 |
| 1973 | 1980 | 1984 | 1987 | 1988 |
| 1974 | 1980 | 1984 | 1987 | 1988 |
| 1975 | 1981 | 1984 | 1987 | 1988 |
| 1976 | 1982 | 1984 | 1987 | 1988 |
| 1977 | 1982 | 1985 | 1987 | 1989 |
| 1977 | 1982 | 1985 | 1987 | 1989 |
| 1978 | 1982 | 1986 | 1987 | 1990 |

(a) Create a histogram for these data.

(b) Use the histogram to describe the distribution of the age of vehicles driven by the people in the sample.

(c) These data are really integer data. Why would a histogram using each year as a separate class not work for these data?

**3.32** In addition to information about vehicles, the Personal Transportation Survey conducted by the Bureau of Transportation also asked questions about households, in particular, the number of people who lived in the household. The data for 45 random respondents are:

| | | | | |
|---|---|---|---|---|
| 1 | 2 | 2 | 3 | 4 |
| 1 | 2 | 2 | 3 | 4 |
| 1 | 2 | 2 | 3 | 4 |
| 1 | 2 | 2 | 3 | 4 |
| 1 | 2 | 2 | 3 | 4 |
| 1 | 2 | 3 | 4 | 4 |
| 2 | 2 | 3 | 4 | 4 |
| 2 | 2 | 3 | 4 | 5 |
| 2 | 2 | 3 | 4 | 5 |

(a) Create a relative frequency histogram for the data.

(b) What percentage of the households surveyed had 3 or more people in them?

(c) What percentage of the households had fewer than 4 people?

**3.33** A manufacturer of breakfast cereals packs the cereal in 15-ounce boxes. Periodic quality checks must be made to ensure that the boxes are being filled adequately. A random sample of 35 boxes of the cereal are selected and weighed. The data are

| | | | | | | |
|---|---|---|---|---|---|---|
| 14.91 | 15.23 | 15.34 | 15.40 | 15.48 | 15.59 | 15.62 |
| 14.96 | 15.24 | 15.34 | 15.42 | 15.49 | 15.60 | 15.62 |
| 15.11 | 15.28 | 15.36 | 15.45 | 15.50 | 15.60 | 15.63 |
| 15.21 | 15.30 | 15.38 | 15.46 | 15.52 | 15.61 | 15.64 |
| 15.22 | 15.30 | 15.39 | 15.48 | 15.58 | 15.61 | 15.67 |

(a) Create a relative frequency histogram for the cereal box weights.

(b) Use the histogram to describe the distribution of cereal weights.

**3.34** A large company buys computer storage media (diskettes) from one supplier. The company wants to know something about how long the diskettes last before they show signs of wear, so it collects data (in hours) for 20 diskettes from different boxes:

| | | | |
|---|---|---|---|
| 486 | 494 | 502 | 508 |
| 490 | 496 | 504 | 510 |
| 491 | 498 | 505 | 514 |
| 491 | 498 | 506 | 515 |
| 494 | 498 | 507 | 517 |

(a) Create a dotplot for the data.

(b) Use the dotplot to describe the distribution of lifetimes for the 20 diskettes.

**3.35** A survey of 40 households asked the head of the household how much was spent on clothing ($) during the previous three months. The data are

| | | | | |
|---|---|---|---|---|
| 0.00 | 0.00 | 0.00 | 35.00 | 210.00 |
| 0.00 | 0.00 | 0.00 | 36.75 | 221.00 |
| 0.00 | 0.00 | 0.00 | 60.89 | 224.00 |
| 0.00 | 0.00 | 0.00 | 64.66 | 303.33 |
| 0.00 | 0.00 | 10.62 | 96.00 | 310.55 |
| 0.00 | 0.00 | 15.00 | 114.61 | 365.95 |
| 0.00 | 0.00 | 17.21 | 149.27 | 365.95 |
| 0.00 | 0.00 | 21.33 | 165.37 | 702.60 |

(a) Create a frequency table for the data.

(b) Use the frequency table to make a histogram of the data.

(c) Describe the distribution of the amount of money that a household spends on clothing in a three-month period.

## Thinking About It!

**3.36** The problem of piracy has long been a concern for the computer software industry. Each year, the amount of revenue lost from piracy is staggering. Since the Internet has become more developed, the problems have become even more severe. In 1994 piracy losses were $15.2 billion. A study showed that the losses were distributed over the world as follows:

| Region | % of total |
|---|---|
| Europe | 39 |
| Asia | 29 |
| Africa/Middle East | 2 |
| Latin America | 9 |
| North America | 21 |

Source: *The Economist,* 27 July 1996.

(a) Create a bar chart for these data.

(b) Create a pie chart for these data.

(c) Which chart do you think does a better job of displaying the data and what they mean? Why?

**3.37** Consumers are increasingly complaining that it is not worth the effort to clip coupons any more. One of the reasons they cite for this is that the life of the coupons is decreasing and it takes more time to sort and discard expired coupons. In an effort to substantiate these claims a consumer group looked at samples of 500 coupons from both 1995 and 1996. The data are summarized as follows:

| Coupon life | Number of Coupons 1995 | 1996 |
|---|---|---|
| Less than 1 month | 28 | 123 |
| 1–4 months | 88 | 140 |
| 5–8 months | 204 | 122 |
| 9–12 months | 141 | 88 |
| More than 1 year | 39 | 27 |

(a) Create bar charts for 1995 and 1996.

(b) Compare and describe the data for the two years. Do you think that there has been a change in coupon life? Why or why not?

**3.38** The problem of workplace violence is growing. A survey of 600 full-time American workers who were the victims of violence while working reports the following:

*Workplace violence survey*

| Type of Violence | % Who Reported |
|---|---|
| Harassment | 19 |
| Threat of physical harm | 7 |
| Physical attack | 3 |

| Major effect on worker | Type of violence (%) | | |
|---|---|---|---|
| | Harassment | Threat | Physical |
| Psychologically | 49 | 53 | 49 |
| Disrupted work life | 34 | 25 | 25 |
| Physically injured or sick | 13 | 9 | 17 |
| No negative effect | 4 | 13 | 9 |

Create graphical displays of these data. Use any techniques you feel are appropriate to convey the information in the data.

**3.39** The article that looked at majors of students at Dartmouth College also reported data for the classes of 1991 and 1996: *Requires Exercise 3.27*

| Major | 1991 | 1996 |
|---|---|---|
| Anthropology | 19 | 17 |
| Art History | 20 | 15 |
| Asia | 15 | 22 |
| Biology | 44 | 108 |
| Chemistry | 15 | 55 |
| Classics | 1 | 8 |
| Comparative Literature | 6 | 6 |
| Drama | 8 | 5 |
| Earth Science | 15 | 7 |
| Economics | 66 | 93 |
| English | 143 | 92 |
| French | 15 | 19 |
| Geography | 28 | 28 |
| Government | 172 | 142 |
| History | 125 | 115 |
| Mathematics | 23 | 14 |
| Music | 3 | 11 |
| Philosophy | 28 | 25 |
| Physics | 10 | 16 |
| Psychology | 74 | 87 |
| Religion | 31 | 8 |
| Sociology | 26 | 15 |

(a) Create bar charts for the 1991 and 1996 data. Design the charts so that you will be able to compare the two classes and also compare them to the data for the class of 1986.

(b) Did the fact that you needed to make comparisons change the way you might have arranged the categories? Why or why not?

(c) Did the fact that you needed to make comparisons affect the way you scaled the *y* axis?

(d) Do you think that there is any basis for the claim that there has been an increase in the number of students who are interested in the sciences? Why or why not?

(e) Make comparative bar charts for only those majors that are sciences. Do you think that these charts make trends in the sciences easier to see? Why or why not?

(f) Compare the distribution of majors in general for each of the classes. Do you see any other changes or trends?

**3.40**  A study in 1992 by the National Center for Health Statistics looked at the death rates for a group of diseases. The data were grouped by sex and race:

| Disease | Deaths (per 100,000 people) | | | |
|---|---|---|---|---|
|  | **White Males** | **Black Males** | **White Females** | **Black Females** |
| Heart disease | 190.3 | 264.1 | 98.1 | 162.4 |
| Prostate cancer | 15.7 | 35.8 | NA | NA |
| Breast cancer | NA | NA | 21.7 | 27.0 |
| Diabetes | 11.6 | 24.2 | 9.6 | 25.8 |
| HIV infection | 18.1 | 61.8 | 1.6 | 14.3 |
| Chronic liver disease | 11.1 | 17.2 | 4.6 | 6.9 |

(a) Create a bar chart for the total deaths for each disease. What information do you get from this chart?

(b) Create separate bar charts for each of the four groups.

(c) Compare the deaths for the groups to each other and to the chart for the total deaths. What interesting features do you see in the data?

**3.41**  The company looking at the life of its diskettes, decides to buy some diskettes from a different vendor and do a comparative test. The company selects 45 diskettes from each supplier and distributes them to a random sample of employees. The employees are asked to use the diskettes as they normally would, but to check them regularly for signs of wear using a disk-checking utility. They are to record how long (in hours) their disk lasts before it exhibits signs of wear. The data, for each supplier, are:

| Supplier A | | | | | | Supplier B | | | | |
|---|---|---|---|---|---|---|---|---|---|---|
| 474 | 492 | 498 | 504 | 511 | | 487 | 492 | 495 | 497 | 499 |
| 486 | 492 | 500 | 505 | 511 | | 488 | 492 | 495 | 497 | 499 |
| 489 | 494 | 501 | 506 | 512 | | 489 | 492 | 495 | 497 | 499 |
| 490 | 494 | 501 | 507 | 513 | | 489 | 492 | 495 | 497 | 501 |
| 490 | 494 | 501 | 508 | 514 | | 489 | 493 | 495 | 498 | 502 |
| 490 | 495 | 502 | 508 | 515 | | 491 | 493 | 496 | 498 | 503 |
| 491 | 496 | 502 | 509 | 517 | | 491 | 494 | 496 | 498 | 503 |
| 491 | 498 | 504 | 509 | 519 | | 491 | 494 | 496 | 498 | 505 |
| 491 | 498 | 504 | 510 | 528 | | 492 | 494 | 496 | 499 | 506 |

(a) Create histograms for the lifetimes of diskettes from each supplier. Be sure to choose your scales so that you can make meaningful comparisons.

(b) Describe the distribution of diskette lifetimes for each supplier.

(c) Compare the lifetimes for the diskettes from supplier B, to those from the company's current supplier, supplier A. Which diskettes would you recommend the company use? Why?

**3.42**  *The New York Times* (28 October 1994) published a state by state report on the number of people incarcerated per 100,000 people in the population. The data are shown in the table below:

| | | | | | | | |
|---|---|---|---|---|---|---|---|
| District of Columbia | 1578 | Mississippi | 385 | Tennessee | 278 | Wisconsin | 172 |
| | | California | 382 | Colorado | 272 | Hawaii | 170 |
| Texas | 545 | Virginia | 374 | Alaska | 256 | Oregon | 169 |
| Louisiana | 514 | United States (Federal) | 373 | Indiana | 256 | New Hampshire | 167 |
| South Carolina | 504 | | | Idaho | 253 | Massachusetts | 165 |
| Oklahoma | 501 | Ohio | 369 | Wyoming | 247 | Utah | 154 |
| Nevada | 456 | New York | 361 | Kansas | 239 | Nebraska | 148 |
| Arizona | 448 | Arkansas | 355 | South Dakota | 227 | Vermont | 138 |
| Alabama | 439 | Connecticut | 331 | Pennsylvania | 224 | Maine | 113 |
| Michigan | 423 | Missouri | 321 | New Mexico | 216 | West Virginia | 106 |
| Georgia | 417 | North Carolina | 314 | Washington | 198 | Minnesota | 100 |
| Florida | 404 | New Jersey | 307 | Montana | 192 | North Dakota | 75 |
| Maryland | 392 | Illinois | 302 | Rhode Island | 185 | | |
| Delaware | 391 | Kentucky | 281 | Iowa | 180 | | |

(a) Create a histogram for the data.

(b) Describe the number of people incarcerated (per 100,000) in the United States.

(c) Do you think that the observation from the District of Columbia has an effect on the way your histogram turned out? If so, what is it?

(d) Recreate the histogram without the data from the District of Columbia.

(e) Use the new histogram to describe the number of people incarcerated (per 100,000) in the United States.

(f) Has your description changed from that of part (b). If so, what is different?

**3.43** The cable company that is looking at its ratings after changing channel selection wonders if there has been a change in customer feelings about the service. The company looks at the responses from a survey done six months prior to the change and finds the following data:

*Requires Exercise 3.18*

| 2 | 3 | 3 | 4 | 5 |
|---|---|---|---|---|
| 2 | 3 | 3 | 4 | 6 |
| 2 | 3 | 3 | 4 | 6 |
| 2 | 3 | 3 | 4 | 6 |
| 2 | 3 | 4 | 5 | 6 |
| 2 | 3 | 4 | 5 | 7 |
| 3 | 3 | 4 | 5 | 8 |

(a) Create a graphical display for the ratings data before the service changes were made.

(b) Do you think that the company can conclude that the ratings have improved? Why or why not?

**3.44** It has been a widely held belief that the switch to participative management would increase employee "buy-in" to the company. One of the benefits that should be realized is a reduction in the number of sick days that employees use. A company that has made the switch in some departments wonders if this has been true. The company decides to sample 25 employees from each of two manufacturing departments. The first has been using participative management for almost two years and the second is still using a traditional management style. The data on the number of sick days used by each employee in the past 12 months are

| Participative Management | | | | | Traditional | | | | |
|---|---|---|---|---|---|---|---|---|---|
| 1 | 3 | 5 | 5 | 6 | 0 | 5 | 6 | 7 | 9 |
| 1 | 4 | 5 | 6 | 7 | 3 | 5 | 7 | 7 | 9 |
| 2 | 4 | 5 | 6 | 8 | 4 | 6 | 7 | 7 | 10 |
| 2 | 4 | 5 | 6 | 8 | 4 | 6 | 7 | 8 | 11 |
| 3 | 4 | 5 | 6 | 8 | 5 | 6 | 7 | 8 | 11 |

(a) Create a graphical display for each of the samples.

(b) Use the graphical display to describe the distribution of sick days for each department.

(c) Based on the data, do you think there is a difference in the number of sick days taken by each group? Why or why not?

(d) Do you think, on the basis of these data, that the company can conclude that participative management reduces the number of sick days taken by employees? Why or why not?

## Doing It!

**3.45** A large manufacturer of paper goods keeps track of consumer complaints for its facial tissue product line. The company receives these complaints through a toll-free number that appears on the product. The data taken consist of a transcription of the actual complaint language and a classification of the complaint into a specific category. When a complaint is received, the company generally asks for additional information about the specific package. The information is then used to relate the complaints to manufactur-

*Datafile:*
*COMPLAIN.XXX*

ing data so that in the future similar problems can be prevented. The categories and subcategories used for classifying complaints are given below:

| Category | Subcategory |
|---|---|
| Dispensing | Sheets tear on removal |
| | Reach in/fallback |
| Foreign material | Lint/dust |
| | Other |
| Odor | |
| Miscounts | |
| Packaging | Defective |
| | Misleading |
| | Damaged |
| | Other |

For example, if a consumer calls the company and says, "The box of tissues I bought was not full," the customer service representative classifies this as a miscount. Sometimes complaints can be tricky to classify. If a customer calls, for example, after a product change and says, "I eat tissues and I prefer the taste of the old product," into what category should this complaint be placed?

The company is looking at data for 3 different years. It compiles data for five different variables:

| Variable name | Description |
|---|---|
| *Month* | Month in which complaint occurred |
| *Year* | Year of the complaint |
| *Category* | Category of complaint |
| *Subcategory* | Subcategory of complaint |
| *Number* | Number of complaints of the particular category and subcategory in the time period |

A portion of the data file is shown below:

| Month | Year | Category | Subcategory | Number |
|---|---|---|---|---|
| JANUARY | 1989 | DISPENSING | FALLBACK | 80 |
| JANUARY | 1989 | DISPENSING | SHEETS_TEAR | 118 |
| JANUARY | 1989 | FOREIGN MATL. | LINT/DUST | 53 |
| JANUARY | 1989 | FOREIGN MATL. | OTHER | 1 |
| JANUARY | 1989 | MISCOUNTS | MISCOUNTS | 37 |
| JANUARY | 1989 | ODOR | ODOR | 5 |
| JANUARY | 1989 | PACKAGING | ADVERTISING | 1 |

(a) Create a bar chart for customer complaints by subcategory for January 1989. What subcategory has the largest number of complaints? Does any subcategory constitute a majority?

(b) Create similar charts for January 1990 and January 1991. Compare the three months. Do you notice any similarities or differences?

(c) Create bar charts for the complaint categories for the same three months. Describe and compare the bar charts.

(d) Look at the data for all of 1989. Create a bar chart that displays the total number of complaints by month. Do you notice any trends or interesting features? Does any month or season have considerably more or less complaints than others? Can you offer any explanation for this?

(e) Create similar bar charts of total complaints for 1990 and 1991. Compare these years to 1989 and to each other. Are they similar? Would you expect them to be? Why or why not?

(f) Take the bar chart that you made for January 1989 and modify it so that it is a Pareto chart. Do the same thing for January 1990 and January 1991. As manager of the company, where would you concentrate your resources to eliminate complaints? Why?

(g) Look at the data for July 1989 and make bar charts for both complaint category and complaint subcategory. Compare July 1989 to January 1989. How are they similar? How are they different?

(h) Look at the same data as in part (h), but this time create pie charts. Which chart does a better job at conveying information about customer complaints? Why?

(i) Look at the data by category for each year and make pie charts comparable to the bar charts from parts (d) and (e). Have there been any changes over the three-year period in the type and number of complaints? If so, what are they?

(j) The management of this company asks for your assessment of the complaint data. In particular, the managers would like to know where their biggest problems are and what the trends seem to be over the past three years. Create any additional graphs that will help you answer their questions and prepare a report for them offering your analysis and your suggestions.

**3.46** The provost at Aluacha Balaclava College would like to look at the data she has collected in a different way. She has collected data on salary, rank, school, gender, and tenure. A sample of the data is shown below:

*Datafile: FACULTY.XXX*

| Salary | Years of Service | Rank | School | Gender M/F | Tenure Y/N |
|--------|------------------|------|--------|------------|------------|
| 48,000 | 22 | ASST | BUSINESS | F | Y |
| 57,956 | 16 | PROF | BUSINESS | M | Y |
| 57,170 | 7 | ASSO | BUSINESS | M | Y |
| 53,500 | 11 | ASSO | BUSINESS | M | N |
| 44,166 | 20 | ASSO | BUSINESS | M | Y |
| 85,500 | 23 | PROF | BUSINESS | M | N |
| 49,302 | 21 | ASST | BUSINESS | M | Y |
| 49,981 | 10 | ASSO | BUSINESS | M | Y |

(a) Create histograms of salary for each rank separately.

(b) Compare the distribution of salaries for the four ranks.

(c) Create histograms of salary for each school separately.

(d) Compare the distributions of salary for the different schools.

(e) Create separate histograms of salary by gender and for tenured/untenured faculty and compare them.

(f) Write a report to the provost that summarizes your observations.

# NUMERICAL DESCRIPTORS OF DATA

## THE GOLF BALL COMPANY

A company that manufactures golf balls is preparing for an advertising campaign. The company wants to compare the ball that it manufactures to a competitor's product to see how they differ. In preparation for the study, sample data are collected on a number of different variables for the two types of balls and you are asked to summarize the data and report on what you find. A portion of the data is shown below:

| Ball number | Model number | S1 | S2 | S3 | Weight (g) | Dimple width (mm) | Dimple depth (mm) | Head | Temper- ature (°F) | Carry (yd) | Total distance (yd) | Date | Time |
|---|---|---|---|---|---|---|---|---|---|---|---|---|---|
| 1 | M1 | 81 | 81 | 82 | 45.3 | 0.1450 | 0.0110 | 686 | 77 | 257 | 270 | 8/20 | 8:15 |
| 2 | M1 | 83 | 83 | 84 | 45.2 | 0.1510 | 0.0111 | 688 | 77 | 255 | 267 | 8/20 | 8:15 |
| 3 | M1 | 81 | 82 | 84 | 45.2 | 0.1450 | 0.0105 | 687 | 77 | 256 | 267 | 8/20 | 8:15 |
| 4 | M1 | 81 | 81 | 83 | 45.3 | 0.1440 | 0.0117 | 688 | 77 | 255 | 271 | 8/20 | 8:15 |
| 5 | M1 | 83 | 81 | 82 | 45.5 | 0.1460 | 0.0108 | 687 | 77 | 255 | 268 | 8/20 | 8:15 |
| 6 | M1 | 83 | 83 | 82 | 45.3 | 0.1560 | 0.0111 | 687 | 77 | 256 | 267 | 8/20 | 8:15 |
| 7 | M1 | 81 | 81 | 82 | 45.2 | 0.1495 | 0.0111 | 687 | 77 | 255 | 264 | 8/20 | 8:15 |
| 8 | M1 | 83 | 81 | 82 | 45.1 | 0.1505 | 0.0110 | 690 | 78 | 258 | 269 | 8/20 | 8:15 |

The company wants to know things like how far most of the balls go when they are hit, how much variation there is in the distance, and what percentage of the balls go beyond a certain distance. You need to figure out which numerical measures will provide the company with the most information.

## 4.1   CHAPTER OBJECTIVES

Remember that when you looked at different graphical techniques for displaying numerical data you were interested in three characteristics: center, spread or dispersion, and shape. In this chapter we look at different numerical measures that can be used to describe the same features of the data. The chapter covers the following material:

- Numerical measures of center: the mean, the median, and the mode
- Numerical measures of variability: the range and the standard deviation
- Describing a set of data: the empirical rule and boxplots
- Measures of relative standing: percentiles and percentile rank
- Identifying outliers: z-scores and boxplots

## 4.2   DESCRIBING DATA NUMERICALLY

Although we may say that a picture is worth a thousand words, it is useful to find numerical quantities to describe the data as well. You may wonder why you need to do this, when the graphs let you *see* the data. There are two reasons. First, while you can certainly see the data using histograms and bar charts, it is difficult to *talk* or *write* about pictures. We often need other references to describe the data. A second reason for using numerical descriptors is that we may want to make inferences based on the sample data. To make statistical inferences you need to use *numerical measures*.

When you collect data you may have either a *population* or a *sample* from the population. Numerical measures calculated from the data are known as either **statistics** or **parameters.**

> A *statistic* is a numerical descriptor that is calculated from sample data and is used to describe the sample. Statistics are usually represented by Roman letters.

> A *parameter* is a numerical descriptor that is used to describe a population. Parameters are usually represented by Greek letters.

Most of the time in statistics you will be working with sample data, but you may sometimes have the entire population available for study. We usually use Greek letters to denote parameters and Roman letters to describe statistics.

When you look at numerical descriptors for a set of data, you want to describe the same properties of the data that you described from the graphical displays. You will find, however, that there are several different statistics that you can use to describe each property and that the choice of the statistic is dependent on the problem you are trying to solve.

## 4.3   MEASURES OF CENTRAL TENDENCY

The golf ball company would like to know about values that represent a "typical" golf ball. The company would like to measure the *center* of its data. We will look at three different statistics that measure central tendency: the *sample mean,* the *median,* and the *mode.*

## 4.3.1 The Arithmetic Mean

You are probably already familiar with the most common measure of center, the **sample mean.** The mean, or average, as it is commonly known, is calculated by adding all of the data values in the sample and then dividing by the number of values. The symbol for the sample mean is $\overline{X}$ (this is read as "X bar").

$$\overline{X} = \frac{\text{Sum of all the values in the sample}}{\text{Total number of observations}}$$

> The *sample mean* is the center of balance of a set of data, and is found by adding up all of the data values and dividing by the number of observations.

The population parameter that corresponds to the sample mean is the **population mean,** $\mu$ (mu).

> The *population mean* is represented by the Greek letter $\mu$ (mu).

Using the $\Sigma$ notation that you saw in Chapter 2, we can write the formula for the sample mean as:

$$\overline{X} = \frac{\sum_{i=1}^{n} x_i}{n} \quad \text{or} \quad \frac{\Sigma x}{n}$$

*SAMPLE MEAN*

When we talk about a variable or a statistic in general we use *capital* letters, such as $X$ or $\overline{X}$. When we talk about a *specific value* of a variable we use *lowercase* letters such as $x$. For example, we write, "The sample mean, $\overline{X}$, is calculated by . . . ," or "The third value in the sample is $x = 27.2$."

*Most of the time in statistics, it is understood that the sum is over the entire sample and so we can leave out the index on the summation sign.*

### EXAMPLE 4.1    The Mail-Order Company

*Calculating the Sample Mean*

A mail-order company wants some information about the daily demand for a product that has been heavily advertised. The company wants a measure of what it might typically expect the demand to be. The company looks at the number of orders for a 10-day period and obtains the following data:

| Demand | 29 | 28 | 29 | 31 | 30 | 31 | 27 | 29 | 30 | 32 |
|---|---|---|---|---|---|---|---|---|---|---|

Since it is interested in a *typical* value for the demand, the company decides to calculate the sample mean, $\overline{X}$:

$$\overline{X} = \frac{29 + 28 + 29 + 31 + 30 + 31 + 27 + 29 + 30 + 32}{10} = \frac{296}{10} = 29.6 \text{ orders} \quad \blacksquare$$

### What Does the Sample Mean Really Measure?

You can think of the sample mean as the *balance point* of the data. The value of $\overline{X}$ balances the higher values against the lower ones. It is easiest to see this when you look at the mean on a dotplot of the data. In Figure 4.1 (page 122) you can see that the mean, 29.6, is in the center of the data and that the data are fairly evenly spread out on both sides of the mean.

Suppose that the data on the right (high) side are more spread out than those on the left. What will happen to the value of the sample mean? Remember, we said

*You want to use a dotplot for these data because the sample size is so small.*

**FIGURE 4.1**    Dotplot of the data for the mail order company

that the sample mean is the balance point of the data values. When there are a few data points on one side that are far from the bulk of the data (the bump), the sample mean moves toward them in order to maintain balance with the data on the other side. The next example shows what happens.

**EXAMPLE 4.2   The Mail-Order Company**

*The Sample Mean as a Balance Point*

Suppose that when the mail-order company looks at its data it finds the following:

| Demand | 40 | 28 | 29 | 31 | 30 | 31 | 27 | 29 | 39 | 36 |
|---|---|---|---|---|---|---|---|---|---|---|

Most of the data values are still around 30, but three of them are quite a bit higher. If you calculate the sample mean from the data,

$$\overline{X} = \frac{40 + 28 + 29 + 31 + 30 + 31 + 27 + 29 + 39 + 36}{10} = \frac{320}{10} = 32.0 \text{ items}$$

you see that the sample mean has changed from 29.6 to 32.0. This is not really where the bulk of the data are located. You can see that although the bulk of the data are still located around 30, the value of the sample mean has changed to 32.0 to balance the three data values at the high end.

# TRY IT NOW!

## Restaurant Table Times
### *Calculating the Sample Mean*

A restaurant is trying to decide whether it has an adequate number of tables available. The restaurant owner would like some information on the amount of time a table is occupied by a customer. She collects data on the length of time a customer occupies a table for a random sample of 10 customers:

| Customer | 1 | 2 | 3 | 4 | 5 | 6 | 7 | 8 | 9 | 10 |
|---|---|---|---|---|---|---|---|---|---|---|
| Time (min) | 59.3 | 58.6 | 62.7 | 65.4 | 59.0 | 67.3 | 62.8 | 68.1 | 59.4 | 63.7 |

Calculate the sample mean for the length of time a table is occupied.

### EXAMPLE 4.3    The Golf Ball Problem

*The Sample Mean*

The company that manufactures golf balls is interested in describing the way the two different golf balls behave so that an advertising campaign can be planned. One way to describe the distance that the balls travel is to use the variable *Carry,* which measures the distance (yd) from the point where the ball was hit to the point where it hit the ground. Since the company is interested in *comparing* the two different designs, it will want to look at the designs separately. Using a computer package, you can calculate a set of descriptive statistics for the two different ball designs. From the output you find the following information:

|  | Type M1 | Type M2 |
|---|---|---|
| **Sample size** | 36 | 36 |
| **Sum of data** | 9267 | 9244 |
| **Sample mean** | 257.4 | 256.8 |

It would appear from these values that there is not much difference in the way a "typical" ball behaves, but at this point that is just conjecture.

To better understand what the numbers really mean you can locate the values obtained for the mean on a histogram of the data. Perhaps the numbers and graphs together will provide more information.

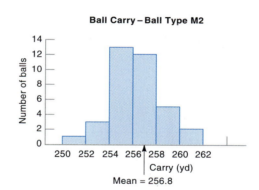

When you locate the mean *Carry* for each ball type on the appropriate histogram you see that the mean appears to be a good measure of the center of the data and that it does not appear to be influenced by extreme values. However, you notice that the sample mean does not provide any information about the number of golf balls that went more than that number of yards or less than that number of yards. This could be useful information, and so it seems that you need another measure of center. ■

## 4.3.2 The Sample Median

While the sample mean does measure the center of the data, its value might be influenced by unusually high or low values in the sample and might not present a true picture of the sample data. For this reason we often look at other measures of center in addition to the sample mean so that we will be able to see a better picture of the data.

Another measure of central tendency that is often used is the **sample median.** While the sample mean is a measure of the center of balance of the data and is sensitive to the actual values, the sample median is a measure of the middle of the data after it is sorted from lowest to highest.

> The *sample median* is the value of the middle observation in an ordered set of data.

Finding the sample median requires sorting the data set first. Once this is done, the *sample median* is the value of the observation that is in the middle of the data. The exact location of the middle will depend on whether the number of observations in the sample is even or odd.

*STEPS FOR LOCAT-*
*ING THE MEDIAN*

**Step 1:**  If the number of observations in the sample, $n$, is odd, then the median is the value of the observation in the $(n + 1)/2$ position.

**Step 2:**  If $n$ is even, then the median is the average of the values in the $n/2$ and $n/2 + 1$ positions.

## EXAMPLE 4.4   Mortgage Waiting Times

*Calculating the Sample Median*

A bank has been receiving complaints from real estate agents that their customers have been waiting too long for mortgage confirmations. The bank prides itself on its mortgage application processing and decides to investigate the claims. The bank takes a random sample of 15 customers whose mortgage applications have been processed in the last six months and finds the following data:

**Wait (days)**    8    10    12    12    6    10    6    15    8    7    13    9    6    12    14

What is the median number of days required for mortgage confirmation?

**Step 1.**  Put the data in numerical order from lowest to highest.

| Position | 1 | 2 | 3 | 4 | 5 | 6 | 7 | 8 | 9 | 10 | 11 | 12 | 13 | 14 | 15 |
|---|---|---|---|---|---|---|---|---|---|---|---|---|---|---|---|
| Wait (days) | 6 | 6 | 6 | 7 | 8 | 8 | 9 | 10 | 10 | 12 | 12 | 12 | 13 | 14 | 15 |

**Step 2.**  Locate the middle observation.

Since $n = 15$ is odd, the middle position is the $(15 + 1)/2 = 16/2 = $ 8th observation. The median of the sample is the *value* of the eighth observation, 10 days.

Since the median represents the middle of the data set, it tells the company that about half of the customers waited less than 10 days for a mortgage confirmation and about half of the customers in the sample waited more than 10 days.   ■

## EXAMPLE 4.5   The Mail-Order Company

*Calculating the Sample Median*

Look at the data from the mail-order company in Example 4.1. If we sort the data in numerical order from highest to lowest we obtain:

| Position | 1 | 2 | 3 | 4 | 5 | 6 | 7 | 8 | 9 | 10 |
|---|---|---|---|---|---|---|---|---|---|---|
| Demand | 27 | 28 | 29 | 29 | 29 | 30 | 30 | 31 | 31 | 32 |

Since $n = 10$ is even, the sample median is the average of the observations in positions $10/2 = 5$ and $10/2 + 1 = 6$. So we find that the median $= (29 + 30)/2 = 29.5$ orders.   ■

## *TRY IT NOW!*

### Town Hall Traffic
#### *Calculating the Sample Median*

In the past few years the town council of a small town has received complaints that it has become increasingly difficult to cross the main street in town near the library. The council decides to look at traffic flow on the street. It selects a site directly in front of the library where most people try to cross the road and records the number of cars that pass the point in a two-minute period. This is done for 10 two-minute periods at 3:00 P.M. over several weeks and the following data are obtained.

| Number of cars | 20 | 27 | 29 | 28 | 37 | 23 | 21 | 28 | 29 | 28 |
|---|---|---|---|---|---|---|---|---|---|---|

Find the median number of cars that pass the site in two minutes.

*Remember to SORT the data before you locate the median!*

### *Why Use Two Different Measures?*

At this point, you still may be wondering why we need to have two different measures of center. For most sets of sample data, the mean and the median will be very close to each other in value. If you need to decide on a single measure, then the choice will depend on the problem you are trying to solve and what information you need. The median tells you that half of the observations in the sample are above that value and half of the observations are below it. Because it is a measure of *location* it ignores the actual values of the observations and may not fully reflect the sample data. The mean uses all of the data values in its calculation and measures the center of balance of the data. While it can be shifted by extreme values, it does reflect all of the data values equally.

### 4.3.3 Comparing the Mean and the Median

We know that the mean and the median provide different information about the center of a set of sample data, but can anything additional be gained from knowing both of them?

   If we start out with a symmetric, mound-shaped distribution, then the mean and the median are both located at the center of the distribution, at the bump. This is illustrated in Figure 4.2. We know that when the data are more spread out in one di-

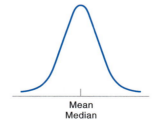

Mean
Median

**FIGURE 4.2**   Mean and median for a symmetric distribution

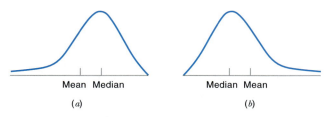

Mean  Median                          Median  Mean

(*a*)                                        (*b*)

**FIGURE 4.3**   Mean and median for skewed distributions

rection (that is, when the data are skewed), then the mean is pulled toward these values, in the direction of the skew. This is illustrated in Figure 4.3. If we compare the mean and the median then we can learn about the shape of the distribution. In particular,

| If | Distribution is | Illustration |
|---|---|---|
| Mean = median | Symmetric | Figure 4.2 |
| Mean < median | Skewed left | Figure 4.3a |
| Mean > median | Skewed right | Figure 4.3b |

At this point you might be saying, "Hey, hold on here. The mean and the median will almost never be exactly equal. How different do they have to be to say that the data are skewed?" This is a really good question and it is one that is often ignored. There is no exact answer, but there are several rules of thumb that can be used. One rule is that the mean and the median should differ by at least the width of a class in a histogram of the data. For small data sets or data values that do not have large magnitudes another rule is that they should differ by at least 10% of either measure. Because these are rules of thumb they will not work in all situations. Knowledge of the data, intuition, and experience are always important in interpreting statistics.

**EXAMPLE 4.6**   **College Graduate Salaries**

*Comparing the Mean and the Median*

A college wonders about the numbers reported by its placement office on graduates' salaries. The college thinks that the mean seems inflated and wonders whether it gives a good picture of what is really happening. It decides to collect some data on the salaries earned by students graduating from the School of Business and to calculate both the mean and the median. The college takes a random sample of 100 graduates from the past year and obtains the following data:

| | | | | | | | | | |
|---|---|---|---|---|---|---|---|---|---|
| $25,000 | $25,400 | $25,600 | $25,800 | $25,900 | $26,200 | $26,400 | $26,600 | $27,000 | $28,000 |
| 25,100 | 25,400 | 25,600 | 25,800 | 25,900 | 26,200 | 26,400 | 26,600 | 27,100 | 28,200 |
| 25,100 | 25,400 | 25,600 | 25,800 | 26,000 | 26,200 | 26,400 | 26,600 | 27,100 | 28,300 |
| 25,200 | 25,500 | 25,700 | 25,800 | 26,000 | 26,200 | 26,400 | 26,600 | 27,100 | 28,300 |
| 25,200 | 25,500 | 25,700 | 25,800 | 26,000 | 26,200 | 26,400 | 26,600 | 27,200 | 28,400 |
| 25,200 | 25,500 | 25,700 | 25,800 | 26,100 | 26,200 | 26,400 | 26,700 | 27,300 | 28,500 |
| 25,200 | 25,500 | 25,700 | 25,900 | 26,100 | 26,200 | 26,500 | 26,700 | 27,400 | 28,600 |
| 25,300 | 25,600 | 25,700 | 25,900 | 26,100 | 26,300 | 26,500 | 26,800 | 27,600 | 29,400 |
| 25,300 | 25,600 | 25,700 | 25,900 | 26,100 | 26,300 | 26,500 | 26,900 | 27,700 | 30,700 |
| 25,300 | 25,600 | 25,800 | 25,900 | 26,100 | 26,300 | 26,500 | 26,900 | 27,700 | 30,800 |

The mean of the salaries is

$$\overline{X} = \frac{2,638,500}{100} = \$26,385$$

The median is the average of the 50th and 51st observations in the data set so

$$\text{Median} = \frac{26{,}100 + 26{,}200}{2} = \$26{,}150$$

In this case, the actual difference between the mean and the median is

$$\$26{,}385 - \$26{,}150 = \$235$$

Clearly the mean is larger than the median, but are they different enough to indicate that the data are skewed? If you apply the 10% rule, you find

$$10\% \text{ of } \$26{,}385 = \$2638.50$$

The actual difference of $235 is much less than $2638.50, and, so, using the 10% rule, the salaries would not appear to be skewed.

*Remember! No **one** statistical tool will provide all of the information in a sample.*

The administrators decide to use a statistical software package to make a histogram to see if this conclusion is consistent with what they see. From the histogram we see that, in fact, the data are skewed right. This is not what the administrators expected.

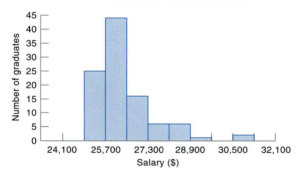

**Salaries of Business School Graduates**

They see that the width of the classes in the histogram is $800 and that the difference between the mean and median, $235, is not more than this either. They understand that this is why it is important to look at data in more than one way.

Knowing that the data are skewed, the college officials decide that they should look at other variables to see why this might be true. This way the placement office will be able to prepare a report that better represents the true situation. ∎

 ## *Try It Now!*

### Airline Cancellations
#### *Comparing the Mean and the Median*

An airline company is wondering about the number of cancellations that it receives for a particular business commuter flight. The airline takes a random sample of 15 days from the first quarter of the year and obtains the following data:

| Number of cancellations | 4 | 9 | 9 | 12 | 12 | 13 | 14 | 14 | 15 | 15 | 16 | 16 | 17 | 17 | 24 |

Find the mean and median for the number of cancellations for the commuter flight.

*Note: The data have been sorted for you.*

*(continued)*

When compared, do the data appear symmetric or skewed?

Make a dotplot of the data.

From the dotplot, do the data appear symmetric or skewed?

---

### EXAMPLE 4.7   The Golf Ball Problem

***Comparing the Mean and the Median***

The golf ball company needs another number to describe a "typical" golf ball. Although the mean is a good statistic, you know you can give the company more information if you look at the median, too. You know that the value of the sample median will represent the distance that had half of the golf balls above it and half below it. This will tell the company a little more about how the golf balls behaved.

From the same computer output that gave us the mean, we learn that the median for the M1 balls is 257.5 yd and the median for the M2 balls is 257.0 yd. This information tells the company that 50% of the M1 balls in the sample went farther than 257.5 yards and that 50% of the M2 balls in the sample went farther than 257 yards.

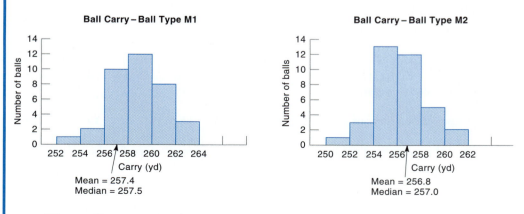

The medians are very close to the sample means. ■

## Discovery Exercise 4.1
## THE TRIMMED MEAN

### Part I. Investigating the Data

In a report to the administration of a large university, the Psychology Department states that the average class size is greater than the 35 students per class allowed by the university charter. The report indicates that the mean class size is 39.4.

No data are appended to the report, but you can obtain the current enrollments easily. The data you find are

| | | | | |
|---|---|---|---|---|
| 3 | 14 | 22 | 26 | 42 |
| 3 | 15 | 23 | 27 | 45 |
| 5 | 15 | 24 | 28 | 45 |
| 9 | 17 | 24 | 28 | 190 |
| 11 | 21 | 25 | 36 | 193 |
| 13 | 22 | 26 | 38 | 193 |

(a) Do you think that the mean is a good measure of center for these data? Why or why not?

(b) By simply studying the data, what do you think a typical class size for the Psychology Department is?

(c) What is the median of the data? Is this close to what you thought?

(continued)

(d) Compare the mean and the median. What does the comparison lead you to believe about the data?

(e) Display the data graphically. Do you still think the same thing?

## Part II.  Solving the Problem

In the first part of this exercise you saw that neither the mean nor the median give a very good measure of a typical class size. In addition, a comparison of the mean and the median leads you to believe that the data are skewed right! From the histogram you can see that this is not the case.

## How can we measure the center of the data when we have extreme values that influence the statistics we usually rely on?

You really can't just discard the extremes without careful investigation of the causes. One way to do this is to use a measure other than the median that removes the effect of the extremes. The ***trimmed mean*** is a statistic that does this. The trimmed mean allows you to drop a specified percentage of the observations in the data set from ***each end.*** By doing this you are not simply dropping outliers, but are looking at the center of the data, which may prove to be more reliable. Typically the trimmed mean is used with a percentage of 10%, but the percentage can be varied to suit the specific circumstances.

## Computing the Trimmed Mean

To compute a 10% trimmed mean you must first determine how many values will be dropped from each end of the data set.

(a) Find 10% of the sample size.

(b)  Do you think that dropping this many values from each end will be effective? Why or why not? If not, how many values do you think would be effective?

(c)  Drop the top and bottom 3 observations from the data and recalculate the mean.

(d)  Compare the trimmed mean and the median. What do you think about the data? Do you think this is a more accurate representation?

By using the trimmed mean together with the mean and the median you find that the mean was not influenced by the *skewness* of the data, but by extreme values on the high end.

(e)  As an administrator at this university, do you think that the Psychology Department can claim that it exceeds the class size specification of 35 students per section? Use all of the information you have gathered to write a memo explaining your decision.

## 4.3.4 The Sample Mode

There is another measure of center that is used in statistics to measure the center of the data. This measure is the **mode.** In a bar chart the mode is analogous to the bar with the highest frequency.

> The sample *mode* is the data value that has the highest frequency of occurrence in the sample.

It would appear that the mode would be a very good measure of a typical value, but there are some obvious reasons why it will not always provide useful information. One is that, depending on the size of the sample and the number of possible data values, there may not be any repeated values in the sample. *That is, for some samples, the mode may not exist.* For continuous data, where there are many different possible values, we do not usually talk about the mode because of the problems just mentioned. In these cases, we often refer to the **modal class** in a frequency distribution or histogram.

> The *modal class* is the class interval in a frequency distribution or histogram that has the highest frequency.

Another problem with the mode is that there may appear to be more than one mode for a sample. This frequently happens with small samples. When this occurs it is not necessarily a case of two or three values that may occur much more frequently than any others. Rather, it is that most of the values happen to occur more than once.

For these reasons the mode is considered by many people to be an unreliable measure of central tendency. However, sometimes a sample does have more than one distinct mode. This indicates a *bimodal* (two modes) or *multimodal* (many modes) sample and should raise a number of questions in your mind.

1. Is it likely that these data could have two or more distinct centers? What would cause such a phenomenon?
2. Is it possible that the sample represents two or more different populations that were not understood when the data were taken?

## EXAMPLE 4.8   The Clothing Store

*Calculating the Sample Mode*

A large retailer of women's clothing is trying to obtain some information that will help it formulate an ordering policy for clothing sizes. The retailer decides to look at a single line of apparel and collect data on the sizes of the items sold in a two-week period. The data after sorting in size order are

| | | | | | |
|---|---|---|---|---|---|
| 6 | 10 | 10 | 12 | 12 | 14 |
| 8 | 10 | 10 | 12 | 12 | 14 |
| 8 | 10 | 10 | 12 | 12 | 14 |
| 10 | 10 | 10 | 12 | 12 | 14 |
| 10 | 10 | 10 | 12 | 12 | 16 |

A frequency table of the data is shown below:

| Size | Frequency |
|------|-----------|
| 6 | 1 |
| 8 | 2 |
| 10 | 12 |
| 12 | 10 |
| 14 | 4 |
| 16 | 1 |

From the frequency table you can see that the mode of the sample is 10. That is, size 10 was sold the most often during the period of study.

What would the sample mean and median be for these data?

Sample mean: $\overline{X} = \dfrac{334}{30} = 11.1$

Sample median: Average of the $\dfrac{30}{2} = 15$th and $\dfrac{30}{2} + 1 = 16$th observations

so

$$\text{Median} = \dfrac{10 + 12}{2} = 11$$

You can see that neither the mean nor the median is a real size. The mean measures the balance point of the sizes and the median simply tells you the size that had half the sales below it and half above. Neither supplies information that helps to develop an ordering policy. ∎

You might wonder why, in the previous example, we did not say that the data are bimodal, since size 12 has a frequency that is very close to the frequency for size 10. Usually when we refer to bimodal or multimodal data we are talking about data with very distinct, different centers. In a histogram this would mean that the two modal classes are not adjacent classes. Figure 4.4 shows a histogram that is bimodal.

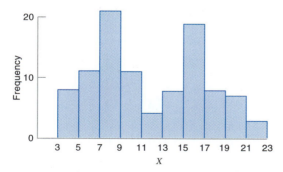

**FIGURE 4.4**   Histogram of bimodal data

## 4.3.5 Exercises—Learn It!

4.1    The management of Disney World would like to know something about the amount of time that visitors spend waiting for the monorail at one of the hotels at the resort. They take some sample data of times (minutes) and find the following:

    5.5    9.6    5.1    13.6    6.5    8.6    9.3    9.1    9.5    15.0    9.7    14.1

(a) What is the mean time spent waiting for a monorail?
(b) What is the median time spent waiting for a monorail?

*In many exercises, the data are already sorted for you.*

**4.2**  A company that sells mail-order computer systems needs to know something about typical weekly sales in order to plan inventory and staffing. The company selects 15 weeks at random from the past year and obtains the data (thousands of dollars) shown below:

Weekly sales  191  222  222  223  223  225  227  228  229  232  234  234  236  244  253

(a)  What are the mean and median weekly sales for the company?

(b)  What is the mode for the data? Do you think the mode is a good measure of typical weekly sales? Why or why not?

**4.3**  In manufacturing, knowing the amount of material that is wasted or lost during a process is very important. A company looks at the percent material lost per day for a new manufacturing process and finds the following data:

Daily loss   10    12    12    13    14    14    18    19    19    20

(a)  What is the mean percent material loss for the process?

(b)  What is the median percent material loss for the process?

(c)  Do you think that either one of these two statistics is preferable for this data set? Why or why not?

**4.4**  To revise an existing inventory system, a company needs to know something about the lead time for orders it places for a critical part. The company looks at the last 20 orders placed and finds that the lead times (days) are:

| | | | |
|---|---|---|---|
| 1 | 4 | 6 | 7 |
| 4 | 5 | 6 | 8 |
| 4 | 5 | 7 | 8 |
| 4 | 6 | 7 | 10 |
| 4 | 6 | 7 | 11 |

(a)  What is the mean lead time for an order?

(b)  What is the median lead time for an order?

(c)  What is the mode lead time for an order?

(d)  Which statistic do you think best represents a typical lead time? Why?

**4.5**  In an effort to understand the cost overruns (the amount by which actual spending exceeds the estimated or budgeted cost) by a particular department in a company, data were collected on the amount of the overrun ($ thousand) for 24 different days:

| | | | | | |
|---|---|---|---|---|---|
| 87.3 | 93.7 | 96.8 | 98.4 | 100.9 | 107.8 |
| 89.9 | 94.9 | 97.0 | 99.6 | 101.3 | 109.7 |
| 91.5 | 96.5 | 97.1 | 100.3 | 105.7 | 111.5 |
| 93.6 | 96.7 | 97.3 | 100.4 | 107.7 | 114.2 |

(a)  Find the mean and median daily overrun by the department.

(b)  Compare the two statistics. Do you think the data are skewed or symmetric? Why?

**4.6**  To assess the effectiveness of the Total Quality Management (TQM) a company that manufactures paper products looks at daily production. Data on the number of items per minute that are produced are taken hourly. The sorted data are:

| | | |
|---|---|---|
| 312 | 323 | 327 |
| 313 | 325 | 331 |
| 315 | 325 | 332 |
| 317 | 326 | 332 |
| 321 | 326 | 335 |

(a)  What is the mean number of items produced during the 15-minute period?

(b)  What is the median number of items produced during the 15-minute period?

(c)  Compare the mean and the median and decide if the data are skewed or symmetric.

**4.7**    The transportation department of a small city has received numerous complaints about the length of the light cycle at a busy intersection. Drivers complain that there are so many cars backed up at the light that they cannot all get through when it turns green. In an effort to determine the validity of the complaints the transportation engineer decides to collect some data on the number of cars that are waiting at the light when it turns green. The data for 10 randomly selected light cycles are

4    4    5    6    7    8    9    10    11    12

(a)  Find the mean and median of the number of cars waiting at the light.

(b)  Are the data skewed or symmetric?

**4.8**    A study on the effects of television on behavior in adolescents, uses, as part of the data, the number of hours per day that the television set is turned on in a household. Twenty-six households are randomly selected and the data are

| 3.6 | 3.7 | 3.7 | 3.8 | 3.9 | 3.9 | 3.9 | 3.9 | 3.9 | 3.9 | 4.2 | 4.3 | 4.6 |
|-----|-----|-----|-----|-----|-----|-----|-----|-----|-----|-----|-----|-----|
| 5.0 | 5.3 | 5.6 | 5.7 | 5.8 | 6.0 | 6.0 | 6.0 | 6.0 | 6.0 | 6.0 | 6.3 | 6.9 |

(a)  Find the mean and median for the number of hours that the television is turned on in a household.

(b)  Do you think that these statistics are good measures for a typical household?

(c)  What is the mode or modal class for the data? Do you think that this statistic provides any information that is not provided by the mean or median? Why or why not?

## *Discovery Exercise 4.2*
## *INVESTIGATING VARIABILITY*

The table contains air-quality data collected by the Environmental Protection Agency. The data show the number of unhealthy days for 14 major U.S. cities in 1989.

| City | Number of unhealthy days |
|------|--------------------------|
| Atlanta | 3 |
| Boston | 1 |
| Chicago | 2 |
| Dallas | 3 |
| Denver | 10 |
| Houston | 12 |
| Kansas City | 2 |
| Los Angeles | 206 |
| New York | 9 |
| Philadelphia | 18 |
| Pittsburgh | 11 |
| San Francisco | 0 |
| Seattle | 4 |
| Washington, DC | 7 |

### Part I

(a)  Display these data using a dotplot.

*(continued)*

**(b)** Find the typical number of unhealthy days by calculating the average value.

**(c)** Can you expect every observation to be typical? Why not?

## Part II

**(a)** What you noticed in Part I is that although we know how to measure a typical value we don't have any way to describe the differences from this typical value that we see in the data. This is called *variability*. Using the dotplot, decide if you think this data set has a lot of variability, a little variability, or something in between a lot and a little.

**(b)** How can we measure this variability?

**(c)** You might have tried subtracting the smallest value from the largest value in an attempt to measure this variability. If you did so, you calculated what is known as the *range*. If you did not do this already, calculate the range.

**(d)** In this case the range is a large number, which presumably tells you that there is a large amount of variability in the data. Do you agree? Based on this, do you think that the range is a good measure of variability?

## Part III

**(a)** Hopefully you concluded that the range can, in fact, give a very misleading picture of the amount of variability in the data. The range is quite large for this data set. Why?

**(b)** What about if we tried measuring how far away each data point is from the middle (i.e., the average) of the data? Let's do this by filling in the table below. Remember that the average is 20.57. The first measurement has been done for you.

| City | Number of unhealthy days | Distance from middle |
|---|---|---|
| Atlanta | 3 | $3 - 20.57 = -17.57$ |
| Boston | 1 | |
| Chicago | 2 | |
| Dallas | 3 | |
| Denver | 10 | |
| Houston | 12 | |
| Kansas City | 2 | |
| Los Angeles | 206 | |
| New York | 9 | |

*(continued)*

| City | Number of unhealthy days | Distance from middle |
|------|--------------------------|----------------------|
| Philadelphia | 18 | |
| Pittsburgh | 11 | |
| San Francisco | 0 | |
| Seattle | 4 | |
| Washington, DC | 7 | |
| **Average or typical** | **20.57** | |

**(c)** This is still not informative because we want a single number that will tell us what the typical deviation from the middle is. What is one way to measure "typical"?

Now, calculate the typical deviation.

**(d)** Does this value give us a good idea of how much variation there is in the data? Why not?

**(e)** What caused the typical variation just calculated to be such a small number? How can we fix this?

## Part IV

**(a)** There are two ways to handle this. One way is to convert all the numbers to positive values by squaring them. The other way is to take the absolute value of the numbers. Fill in the table below. Then calculate the typical or average of the values in column 4 and the average of the values in column 5.

| City | Number of unhealthy days | Distance from middle | Absolute value of distance from the middle | Distance from the middle squared |
|---|---|---|---|---|
| Atlanta | 3 | $(3 - 20.57) = -17.57$ | 17.57 | $(-17.57)^2 = 308.7$ |
| Boston | 1 | | | |
| Chicago | 2 | | | |
| Dallas | 3 | | | |
| Denver | 10 | | | |
| Houston | 12 | | | |
| Kansas City | 2 | | | |
| Los Angeles | 206 | | | |
| New York | 9 | | | |
| Philadelphia | 18 | | | |
| Pittsburgh | 11 | | | |
| San Francisco | 0 | | | |
| Seattle | 4 | | | |
| Washington, DC | 7 | | | |
| **Average** | 20.57 | | | |

**(b)** Why are these two averages not close in magnitude?

**(c)** The average of the absolute distances (column 4) is called the *mean absolute deviation* (MAD) and the average of the squared distances (column 5) is called the *variance*. What are the units for the MAD and what are the units for the variance?

*(continued)*

**(d)** The MAD is actually easier to interpret but is not often used because absolute values do not "behave well." What can we do to the variance to get the magnitude to be about the same as the MAD? Do it.

What you have calculated is called the standard deviation.

## 4.4   MEASURES OF DISPERSION OR SPREAD

In the previous chapter you saw that simply describing the center of the data or a typical data value does not provide complete information about the data set. In addition to knowing what a typical value for the sample is, it is important to know how diverse the values in the sample can be. That is, we need to know how *spread out* or *dispersed* the data values are relative to the typical values. Understanding the *variation* in a set of data is of critical importance in statistics. When people use statistics to make decisions, it is important to understand not only a typical outcome, but all possibilities as well. We will look at two different measures of dispersion, the *sample range* and the *sample standard deviation*.

### 4.4.1 The Sample Range

The simplest measure of dispersion, the **sample range,** involves looking at the two extreme values in the sample: the highest (maximum) and the lowest (minimum) values.

> The *sample range, R,* is the difference between the maximum and minimum observations in the sample.

The sample range is very easy to calculate and understand. It gives information about the distance from one end of an ordered data set to the other. If the sample data are symmetric, then it also gives information about the spread of the data relative to the measures of central tendency.

## EXAMPLE 4.9   The Mail-Order Company

### Calculating the Sample Range

Look at the data on demand from the mail-order company. In Example 4.1 we had the sample data:

| Position | 1 | 2 | 3 | 4 | 5 | 6 | 7 | 8 | 9 | 10 |
|---|---|---|---|---|---|---|---|---|---|---|
| Demand | 27 | 28 | 29 | 29 | 29 | 30 | 30 | 31 | 31 | 32 |

The sample range for this data is

$$R = 32 - 27 = 5$$

This tells the company that the demand for the product has a range or spread of 5 units around its center. In this case the range is a reliable measure of how spread out the demands are.  ■

Information on the range *along with a measure of central tendency* gives you a mental image of the data. If the data are symmetric, then the company would expect that a typical demand for the product is 30 units, and that the demand is evenly spread out on either side of the center. The actual demand might be as low as 27.1 $(29.6 - 2.5)$ or as high as 32.1 $(29.6 + 2.5)$. In Example 4.9 these bounds agree very well with the actual data.

It would seem that the range is a good statistic because it gives a clear picture of the spread and is easy to calculate and understand. The next example illustrates why this is not always the case.

## EXAMPLE 4.10   Hold Times

### Understanding the Sample Range

A company is wondering whether complaints about the amount of time customers spend on hold for technical service are justified or whether the complaints are a result of the "squeaky wheel" phenomenon. The company takes a sample of 15 customer calls to the technical service phone line and records the amount of time each customer spends on hold.

| Customer | 1 | 2 | 3 | 4 | 5 | 6 | 7 | 8 | 9 | 10 | 11 | 12 | 13 | 14 | 15 |
|---|---|---|---|---|---|---|---|---|---|---|---|---|---|---|---|
| Wait (min) | 5.6 | 10.2 | 6.6 | 6.9 | 9.4 | 6.7 | 0.6 | 9.2 | 7.6 | 10.7 | 9.6 | 2.9 | 6.0 | 8.6 | 4.6 |

To find the range, find the minimum and maximum observations and subtract them:

$$R = 10.7 - 0.6 = 10.1 \text{ min}$$

What does the range tell the company?

At first glance, the company would think that the length of time that a customer spends on hold has a spread of 10.1 minutes. If the company considers that the largest value in the sample is 10.7 minutes, it could conclude that the data are quite variable. Or are they?

A dotplot of the data shows that the value of 0.6 minute is really quite a distance from the next nearest value, while there are quite a few observations in the

Waiting time on hold (min)

9- to 10-minute range. The image that the company receives from the data is actually quite distorted. If it calculates the average hold time

$$\overline{X} = \frac{105.2}{15} = 7.0 \text{ min}$$

then the company would think that a typical caller is on hold for 7.0 minutes and that the time a customer spends on hold might be as low as 1.95 (7.0 − 5.05) minutes or as high as 12.05 (7.0 + 5.05) minutes. This is certainly not an accurate picture of the data since the mean is biased downward by the few extremely low values and the range was biased upward by the same values! ■

## TRY IT NOW!

### Restaurant Table Time
*Calculating the Sample Range*

The restaurant looking at the turnaround time for its tables, wonders how variable the occupation time for a table really is. The data the restaurant had collected are

| Time (min) | 59.3 | 58.6 | 62.7 | 65.4 | 59.0 | 67.3 | 62.8 | 68.1 | 59.4 | 63.7 |
|---|---|---|---|---|---|---|---|---|---|---|

What is the range of turnaround times?

Previously (page 122) you calculated the mean turnaround time to be 62.6 minutes. Using this information and the value for the range, what would the restaurant expect as its lowest turnaround time? its highest turnaround time?

The range is heavily influenced by unusual or extreme values in the sample. We saw that the mean is also influenced by extremes, but the amount of bias introduced is much less, since when you calculate the sample mean you use all of the values in the sample. When you calculate the range you use only the extreme values, so when these are unusual they have a large impact on the statistic. In fact, the sample range is one of the only statistics that gets more *unreliable* as the sample size gets larger. As a rule, when the sample size is more than 25, the sample range should not be used as a measure of variability.

ANS. RANGE = 9.5 MIN; LOWEST = 57.9 MIN; HIGHEST = 67.4 MIN

**EXAMPLE 4.11    The Golf Ball Company**

*Looking at Variation*

The company investigating the behavior of the two golf ball designs will need to know about the variation in the characteristics as well as the measure of typical distance. Unless the distance that its golf balls carry is pretty consistent, the company will have customer complaints about any claims it makes in advertising. From the histogram, you saw that the data did not appear to have any unusual values and so you decide to use the sample range as a measure of variation. You look at the computer printout and find the following information:

|  | **Ball type** | |
|---|---|---|
|  | **M1** | **M2** |
| Minimum | 252 | 251 |
| Maximum | 262 | 262 |
| Range | 10 | 11 |
| Average | 257.4 | 256.8 |

From the data it appears that the range is about the same for each of the ball types.

Since we also know that the data are symmetric, by using the sample ranges and sample means together, we can provide the company with a better picture of how the golf balls behave. The M1 balls would appear to carry between a low of 252.4 yards and a high of 262.4 yards. The M2 balls have a low carry of 251.3 yards and a high of 262.3 yards. The fact that these numbers agree with the observed minimums and maximums for the data support the idea that the range is a good measure for this data.

Still, the range does not answer all of the questions. Although you know that an M2 golf ball *could* travel between 251.3 and 262.3 yards, you do not have a measure of how the actual values are spread out within that range. What you would like to know, really, is what the typical variation in travel is.  ■

## 4.4.2 The Sample Standard Deviation

We have seen that the sample range proves unreliable in the presence of extreme data values. As the size of the sample increases, the chance that the sample will contain an extreme value increases. So for large sample sizes another measure of dispersion or spread must often be used.

If you think about what we are really trying to describe when we measure variability, it seems logical to try to define a measure of how far away from the center of the data a value might be. In fact, what we would probably like to know is, on the average, or typically, how far the values *vary* (differ) from the center. This measure is the **sample standard deviation.** The standard deviation is most often defined relative to another measure of dispersion called the **sample variance.** In practice, the measure that is used is the standard deviation because its units and order of magnitude are the same as those of the actual data.

> The *sample variance, $s^2$,* is the average of the squared deviations of the data values from the sample mean.

> The *sample standard deviation, s,* is the positive square root of the sample variance.

By their definition alone the sample variance and standard deviation seem overly complicated, especially when compared to the sample range. To calculate the sample standard deviation, you calculate the sample variance first, using the formula

SAMPLE VARIANCE
*Note: The sum for the
variance still includes
all n sample values even
though we divide by
n − 1.*

SAMPLE STANDARD
DEVIATION

$$s^2 = \frac{\sum_{i=1}^{n}(x_i - \bar{x})^2}{n - 1}$$

To obtain the sample standard deviation, $s$, you take the positive square root of the sample variance to obtain

$$s = \sqrt{s^2}$$

> The population variance and standard deviation are represented by the Greek letter $\sigma$ (sigma), where $\sigma^2$ is the *population variance* and $\sigma$ is the *population standard deviation.*

### EXAMPLE 4.12   Hold Times

*Calculating the Sample Standard Deviation Using the Definition*

The company looking at customer hold times knows that a measure of a typical hold time will not provide enough information about what its customers might encounter. The managers of the company need to know how much the time of a call might vary from the center. They decide to calculate the sample standard deviation for their data on customer hold times:

| Customer | 1 | 2 | 3 | 4 | 5 | 6 | 7 | 8 | 9 | 10 | 11 | 12 | 13 | 14 | 15 |
|---|---|---|---|---|---|---|---|---|---|---|---|---|---|---|---|
| Wait (min) | 5.6 | 10.2 | 6.6 | 6.9 | 9.4 | 6.7 | 0.6 | 9.2 | 7.6 | 10.7 | 9.6 | 2.9 | 6.0 | 8.6 | 4.6 |

To calculate the sample standard deviation the managers first need to calculate the sample mean, since the standard deviation measures the average of the squared distances from the sample mean. For this set of data we found previously that the mean hold time is 7.0 min.

The next part of the calculation for the sample variance finds the distance of each data value, $X$, from the sample mean, $\bar{X}$, squares each of them, and then adds them. The table below gives the details of each part of the calculation. Each column of the table below represents one part of the calculation. The last row of the table is the sum of all of the previous rows.

| Customer | Wait | $(X - \bar{X})$ | $(X - \bar{X})^2$ |
|---|---|---|---|
| 1 | 5.6 | − 1.4 | 1.96 |
| 2 | 10.2 | 3.2 | 10.24 |
| 3 | 6.6 | − 0.4 | 0.16 |
| 4 | 6.9 | − 0.1 | 0.01 |
| 5 | 9.4 | 2.4 | 5.76 |
| 6 | 6.7 | − 0.3 | 0.09 |
| 7 | 0.6 | − 6.4 | 40.96 |
| 8 | 9.2 | 2.2 | 4.84 |
| 9 | 7.6 | 0.6 | 0.36 |
| 10 | 10.7 | 3.7 | 13.69 |
| 11 | 9.6 | 2.6 | 6.76 |
| 12 | 2.9 | − 4.1 | 16.81 |
| 13 | 6.0 | − 1.0 | 1.00 |
| 14 | 8.6 | 1.6 | 2.56 |
| 15 | 4.6 | − 2.4 | 5.76 |
| **Sum** | **105.2** | **0.2** | **110.96** |

The last part of the calculation divides the sum of the squared deviations by $n - 1$, in this case, 14. So we have the sample variance

$$s^2 = \frac{110.96}{14} = 7.93 \text{ min}^2$$

and the sample standard deviation

$$s = \sqrt{7.93} = 2.82 \text{ min}$$

This tells the company that while an average call lasts 7.0 minutes, the actual call times will vary from that value. Typically, the variations or differences from the average will be about 2.82 minutes.    ■

By now you must be convinced that there has to be a really good reason to use the standard deviation instead of the range! In fact, almost nobody actually *calculates* the sample variance and standard deviation using the definition. There is a shortcut formula that reduces the number of calculations you have to perform considerably:

$$s^2 = \frac{n \sum x^2 - (\sum x)^2}{n(n - 1)}$$

*SHORTCUT FORMULA FOR SAMPLE VARIANCE*

To calculate the sample variance and sample standard deviation this way you need to sum all of the data values and also to square each value and sum the squares. The next example recalculates both the sample variance and standard deviation from the previous example using the new, shorter method.

## EXAMPLE 4.13    Hold Times

### Calculating s Using the Shortcut Formula

Using the same data, you would need to sum both the values and the values squared. The table below gives the details of the calculations:

| Customer | Wait (X) | $X^2$ |
|----------|----------|-------|
| 1 | 5.6 | 31.36 |
| 2 | 10.2 | 104.04 |
| 3 | 6.6 | 43.56 |
| 4 | 6.9 | 47.61 |
| 5 | 9.4 | 88.36 |
| 6 | 6.7 | 44.89 |
| 7 | 0.6 | 0.36 |
| 8 | 9.2 | 84.64 |
| 9 | 7.6 | 57.76 |
| 10 | 10.7 | 114.49 |
| 11 | 9.6 | 92.16 |
| 12 | 2.9 | 8.41 |
| 13 | 6.0 | 36.00 |
| 14 | 8.6 | 73.96 |
| 15 | 4.6 | 21.16 |
| **Sum** | **105.2** | **848.76** |

Substituting the values in the last row of the table into the formula you get

$$s^2 = \frac{(15)(848.76) - (105.2)^2}{(15)(14)} = \frac{1664.36}{210} = 7.93 \text{ min}^2$$

and

$$s = \sqrt{7.93} = 2.82 \text{ min}$$

which is the same answer we got using the original formula.    ■

*CAUTION! Calculators and spreadsheets can calculate more than one type of standard deviation. Be sure that you are finding the **sample** standard deviation.*

It is probably hard to convince yourself that this is actually any better than the first set of calculations. Actually it is, but unless you do it several times it is hard to see. In fact, there is really no reason to calculate the sample variance and sample standard deviation by hand more than once or twice in your lifetime. Many calculators available today include statistical calculations like the sample mean and sample standard deviation, and these functions are also built into every spreadsheet package. What is important is understanding what the standard deviation measures and how it can be used to interpret sample data.

 **TRY IT NOW!**

### Town Hall Traffic Flow
*Calculating the Sample Variance and Standard Deviation*

The town council looking at the traffic flow problem has seen reports that use the standard deviation, and wants to use it to describe the variability of traffic flow. The data are

| **Number of Cars** | 20 | 27 | 29 | 28 | 37 | 23 | 21 | 28 | 29 | 28 |
|---|---|---|---|---|---|---|---|---|---|---|

What is the sample standard deviation of the traffic flow?

*Use whatever method you feel most comfortable with. If you have a statistical calculator learn how to use it NOW!*

### 4.4.3 Interpreting the Standard Deviation— The Empirical Rule

Admittedly, the standard deviation is not as intuitive or appealing as the sample range. From the sample range you get an immediate (although sometimes false) picture of how far the data spread out around the center. The sample standard deviation does not give you the same intuitive response.

One way to understand what information the standard deviation gives is to use the **empirical rule.**

The *empirical rule* says that for a mound-shaped, symmetric distribution

- about 68% of all observations are within one standard deviation of the mean
- about 95% of all observations are within two standard deviations of the mean
- almost all (more than 99%) of the observations are within three standard deviations of the mean.

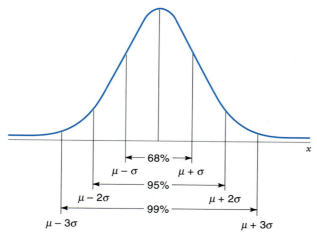

**FIGURE 4.5** The empirical rule

The empirical rule (Figure 4.5) is defined for large data sets and distributions that are symmetric and mound-shaped, often called bell-shaped or normal curves. For distributions that are only slightly skewed the empirical rule is surprisingly accurate as well. It provides sets of bounds for data values from a given population.

In general, if we are talking about sample data we will not know the population mean and standard deviation, so it is necessary to substitute $\overline{X}$ and $s$ for $\mu$ and $\sigma$. To really understand the empirical rule an example is necessary.

## EXAMPLE 4.14   Hold Times

### *The Empirical Rule*

Suppose that the company looking at hold times had collected a total of 50 observations on customer hold times. The sorted data are shown below:

| | | | | | | | | | |
|---|---|---|---|---|---|---|---|---|---|
| 0.6 | 4.6 | 5.6 | 6.3 | 6.8 | 7.5 | 7.8 | 8.3 | 8.9 | 9.6 |
| 2.9 | 4.7 | 6.0 | 6.3 | 6.9 | 7.5 | 7.9 | 8.4 | 9.2 | 10.1 |
| 3.4 | 5.2 | 6.0 | 6.6 | 6.9 | 7.6 | 8.0 | 8.4 | 9.2 | 10.2 |
| 3.8 | 5.5 | 6.1 | 6.6 | 7.0 | 7.6 | 8.1 | 8.6 | 9.4 | 10.7 |
| 4.5 | 5.5 | 6.1 | 6.7 | 7.2 | 7.8 | 8.2 | 8.6 | 9.4 | 11.1 |

*For a data set this large you should use your calculator or a computer software package to find $\overline{X}$ and $s$.*

To investigate the empirical rule we need to find the sample mean and sample standard deviation of the data. For this data set $\bar{x} = 7.12$ minutes and $s = 2.08$ minutes. We should also verify that the empirical rule applies—that is, that the data we have follow a normal curve. To do this we can make a histogram of the data:

The histogram shows that the data are reasonably symmetric and bell-shaped, so we can use the empirical rule:

| | |
|---|---|
| $\bar{x} + s = 7.12 + 2.08 = 9.20$ min | *About 68% of the data values should be* |
| $\bar{x} - s = 7.12 - 2.08 = 5.04$ min | *between 5.04 and 9.20 min.* |
| $\bar{x} + 2s = 7.12 + 2(2.08) = 11.28$ min | *About 95% of the data values should be* |
| $\bar{x} - 2s = 7.12 - 2(2.08) = 2.96$ min | *between 2.96 and 11.28 min.* |
| $\bar{x} + 3s = 7.12 + 3(2.08) = 13.36$ min | *Almost all of the data (more than 99%) should* |
| $\bar{x} - 3s = 7.12 - 3(2.08) = 0.88$ min | *be between 0.88 and 13.36 min.* |

To see how well the data we collected agrees with this rule we can calculate the actual percentage of data values that fall within each interval. The table below contains the number of data values and the percentage that fall within each of the three intervals.

| Interval | Number of data values | Percentage of data values | Empirical rule (%) |
|---|---|---|---|
| 5.04 to 9.20 | 36 | 72 | 68 |
| 2.96 to 11.28 | 48 | 96 | 95 |
| 0.88 to 13.36 | 49 | 98 | 99 |

Looking at the table we see that the actual percentages are slightly different from what the empirical rule predicted. Remember that variation in data is to be expected. In statistics it is important to understand when the variation is within the limits of what is expected or typical. When the variation from what is expected is too large, something may be happening to cause it.

In this case actual percentages are fairly close to those predicted by the empirical rule. The largest difference occurs in the first interval. If the percentages are very different from what is expected then the cause is that the data are probably from a distribution that is not symmetric. The more skewed a data set is, the more it will deviate from the empirical rule in the first two intervals. ■

 **TRY IT NOW!**

**Loan Processing**

*The Empirical Rule*

Errors in filling out loan applications can lead to delays in having the loans approved. Bank employees must contact the applicants to correct the errors. This sometimes requires multiple contacts. To understand the extent to which the errors affect the application process a bank collected data on the number of follow-up contacts required before a loan could be processed. The bank looked at 25 different applications and found

| | | | | |
|---|---|---|---|---|
| 0 | 1 | 2 | 3 | 4 |
| 0 | 2 | 2 | 4 | 4 |
| 1 | 2 | 3 | 4 | 5 |
| 1 | 2 | 3 | 4 | 5 |
| 1 | 2 | 3 | 4 | 7 |

Make a dotplot of the data.

From the dotplot, do you think that the assumption that the data have a symmetric, bell-shaped distribution is a reasonable one?

Find the mean and standard deviation of the data.

According to the empirical rule, between what two values should 68% of the observations fall?

Between what two values should 95% of the observations fall?

Between what two values should more than 99% of the observations fall?

### 4.4.4 z-Scores

In the previous section we saw that the standard deviation can be used to measure how likely it is for a data value to occur. Since, for a symmetric, bell-shaped distribution, 68% of the data values fall within one standard deviation of the mean, it is reasonable to assume that data values in that interval will make up about 68% of the sample values. Certainly, most (more than 99%) of the data values in the sample should be within three standard deviations of the mean! We could use this rule as a measure of how "usual" that data value is. This measure is called the **z-score.**

> A **z-score** measures the number of standard deviations that a data value is from the mean.

To calculate the z-score of a data value we first find the *distance* that the data value is from the mean and then divide by the standard deviation:

$$z = \frac{\text{Distance between the data value and the mean}}{\text{Standard deviation}} = \frac{X - \mu}{\sigma}$$

*Remember! When we substitute $\overline{X}$ and s for $\mu$ and $\sigma$ we are relying on large sample sizes to ensure that they are good estimates.*

As in the empirical rule, for sample data we substitute $\overline{X}$ and $s$ for $\mu$ and $\sigma$. A positive z-score indicates that the data value is *above* the mean, while a negative z-score indicates that the data value is *below* the mean.

If you think about the empirical rule together with the z-score you can begin to make some inferences on how data values compare to what is expected from a random sample. Table 4.1 gives you an idea of how this can be used.

#### TABLE 4.1   Using the Empirical Rule and z-Scores Together

| If the z-score is . . . | The empirical rule says it will occur . . . | You can conclude that . . . |
| --- | --- | --- |
| Less than −2 or more than 2 | About 5% of the time | It is unusual and possibly an outlier |
| Less than −3 or more than 3 | Less than 1% of the time | It is very unusual and probably an outlier |

*Just because a data value is identified as an outlier does not mean you can discard it from the sample!*

It is possible to use the z-score of a data value to identify *outliers*. The problem with doing this is that everyone's definition of unusual is not the same and so inconsistencies in data analysis can result. z-scores are discussed much more extensively in Chapters 6 and 8.

### EXAMPLE 4.15   The Golf Ball Company

*z-scores*

From the histograms, the company investigating the two designs of golf balls did not think that any of the data are potential outliers. Still, it is better to be sure, so the company asks you to calculate z-scores. Rather than calculate the z-score for every data value you know, it will be easier to just look at the maximum and minimum values first. If neither of these have unusual z-scores, then you know that none of the other values will either. From the tables of summary statistics you find the information you need and compute the z-scores shown in the table. The z-scores are indeed in the 2−3 range, but they do not indicate that the extreme sample values are outliers.

| Ball type | Carry (yd) | | Carry (yd) | | z-score |
| --- | --- | --- | --- | --- | --- |
| M1 | Max | 262 | $\overline{x} =$ | 257.4 | 1.92 |
| | Min | 252 | $s =$ | 2.4 | −2.25 |
| M2 | Max | 262 | $\overline{x} =$ | 256.8 | 2.26 |
| | Min | 251 | $s =$ | 2.3 | −2.52 |

## *TRY IT NOW!*

### Town Hall Traffic
*Calculating z-Scores*

The town that was looking at traffic flow in front of the town hall wonders if the observation of 37 cars is unusual. Although the town officials know that their sample size of 10 cars is not large enough to ensure accuracy, they want to use z-scores to look at the data:

| **Number of cars** | 20 | 27 | 29 | 28 | 37 | 23 | 21 | 28 | 29 | 28 |
|---|---|---|---|---|---|---|---|---|---|---|

What is the z-score for the observation of 37 cars?

Comparing the z-score to the empirical rule, do you think that the value is unusual?

## 4.4.5 Exercises—Learn It!

**4.9**  In addition to knowing the typical time visitors to Disney World wait for a monorail, management wants to know about the variation in waiting times. The data for the waiting times (min) are    *Requires Exercise 4.1*

  5.5  9.6  5.1  13.6  6.5  8.6  9.3  9.1  9.5  15.0  9.7  14.1

   (a) What is the range of the waiting times?
   (b) What are the variance and the standard deviation of the waiting times?

**4.10**  To understand its customer base, a mail-order book club needs to know something about the frequency with which customers make subsequent purchases. They collect data on the number of purchases that customers make during their initial three-year membership. From a sample of 15 randomly selected customers they obtain the following data:

  12  12  14  15  14  11  13  10  14  14  13  14  14  11  15

   (a) Find the range of the number of purchases in a three-year period.
   (b) Find the variance and the standard deviation of the number of purchases.
   (c) Do you think that the range or the standard deviation is a better measure of variation for these data? Why?

**4.11** A professor in an introductory statistics course is interested in the number of hours that students spend doing homework during the week. A random selection of 12 students yields the following:

4.1   2.8   6.1   4.9   4.2   5.5   3.2   5.9   2.7   5.4   6.9   3.7

(a) Find the range and the standard deviation of the number of hours spent doing homework.

(b) According to the empirical rule, between what two amounts of time will 68% of the students spend on homework? 95%? more than 99%?

(c) How do the actual data compare to the predictions from the empirical rule?

**4.12** A manager at XYZ Corporation thinks that the number of travel miles claimed in expense reports of its sales personnel has increased in recent months. To substantiate his idea he collects historical data from expense reports filed in the previous year:

2171   1709   2062   2075   1758   1733   1716   1963   1655   1558   1617   1908

(a) Find the range of the number of miles claimed.

(b) Find the standard deviation of the number of miles claimed.

(c) What is the z-score of the smallest data value? Do you think this value is unusual? Why or why not?

(d) What is the z-score of the largest data value? Do you think this value is unusual? Why or why not?

*Requires Exercise 4.2*   **4.13** The company that sells mail-order computer systems wants to further analyze its typical weekly sales. The data (in thousands of dollars) for the 15 weeks are

Weekly sales  191  222  222  223  223  225  227  228  229  232  234  234  236  244  253

(a) What are the range and standard deviation of the weekly sales?

(b) According to the empirical rule, between what two values will 68% of the weekly sales fall? 95%? more than 99%?

(c) Do you think that the empirical rule is appropriate for these data? Why or why not?

(d) What is the z-score for the week that had $228,000 in sales? What does this tell you about that observation?

**4.14** To make a decision about replacing the cars in its current fleet, a company looks at the amount of money spent on repairs in the past 12 months, for a random sample of ten cars:

472   472   603   459   538   601   449   588   539   521

(a) Find the range, variance, and standard deviation of the amount spent on repairs.

(b) Use the empirical rule to describe the distribution of the amount spent on repairs.

(c) Do you think that the use of the empirical rule was appropriate here? Why or why not?

(d) Find the z-score for each data value. Are any of the values unusual? If so, why?

*Requires Exercise 4.3*   **4.15** Look at the data on percent material lost per day in manufacturing:

Daily loss   10   12   12   13   14   14   18   19   19   20

(a) What is the range of percent material loss for the process?

(b) What is the standard deviation of the percent material loss for the process?

(c) Find the z-score for each data value. Do you think that any of the values are unusual? Why or why not?

**4.16** A company that buys blank VHS tapes for video recording is concerned about the actual amount of time that the tapes are able to record. The tapes are rated at 120 minutes, but the company knows that there is variation in the actual recording time. The company collects data on 25 randomly selected tapes and finds the actual recording times (min) listed below:

| | | | | |
|---|---|---|---|---|
| 116 | 118 | 119 | 119 | 120 |
| 117 | 118 | 119 | 119 | 121 |
| 117 | 118 | 119 | 120 | 121 |
| 117 | 119 | 119 | 120 | 121 |
| 117 | 119 | 119 | 120 | 121 |

(a) What is the range of actual recording times for the tapes? What are the variance and standard deviation of the recording times of the tapes?

(b) Make a dotplot or a histogram of the data.

(c) Use the empirical rule to find the intervals $\overline{X} \pm 1s$, $\overline{X} \pm 2s$, $\overline{X} \pm 3s$ and mark them on the graphical display.

(d) Find the actual number of data points that fall in each interval. Does this agree with the predictions of the empirical rule? Why or why not?

**4.17** A large company that sells software has had complaints from customers lately that the disks provided by the company fail to work properly after extended use. The company decides to investigate the complaint and looks at the disks from its current supplier. The disks are rated with a life of 500 hours of use before failure. After testing 20 disks until they fail, the company obtains the following data on disk life (hours):

| | | | |
|---|---|---|---|
| 486 | 494 | 502 | 508 |
| 490 | 496 | 504 | 510 |
| 491 | 498 | 505 | 514 |
| 491 | 498 | 506 | 515 |
| 494 | 498 | 507 | 527 |

(a) Find the range and standard deviation of the time to failure of the disks.

(b) Make a dotplot of the data. Do any of the data values appear unusual?

(c) Do you think that the range or the standard deviation is more reliable for this set of data? Why?

(d) Find the z-score of the two largest and the two smallest disk lives. What does this tell you about the observations?

**4.18** The company that is looking at cost overruns by a particular department also wants to look at how the overruns vary:    *Requires Exercise 4.5*

| | | | | | |
|---|---|---|---|---|---|
| 87.3 | 93.7 | 96.8 | 98.4 | 100.9 | 107.8 |
| 89.9 | 94.9 | 97.0 | 99.6 | 101.3 | 109.7 |
| 91.5 | 96.5 | 97.1 | 100.3 | 105.7 | 111.5 |
| 93.6 | 96.7 | 97.3 | 100.4 | 107.7 | 114.2 |

(a) Find the range and standard deviation of the daily overrun by the department.

(b) Use the empirical rule to give the company a picture of how the overruns vary.

(c) Find the z-scores of the maximum and minimum overruns. Is either of these two values unusual? Why or why not?

# 4.5 MEASURES OF RELATIVE STANDING

Measures of center and dispersion are certainly important, but they are not the only numerical measures that can be used to obtain information about a set of data. Other measures, called measures of relative standing, or *order statistics,* give information about the position of an observation in the sample. We looked at one measure of relative standing, the median, when we looked at measures of center. Now we look at some additional measures.

## 4.5.1 Percentiles

It is useful in some real situations to know what data value in a sample has a certain percentage of the sample above or below it. This measure is known as the **percentile** of the data.

> The *p*th **percentile** of a data set is the value that has *p*% of the data at or below it.

*The methods used for calculating percentiles vary slightly among software packages, and so the values that you get might differ.*

Two questions can be asked involving percentiles: What value has $p\%$ of the data at or below it? and What is the percentile rank of a particular data value? The first question involves finding either a particular percentile or set of percentiles, such as the *deciles* (10%, 20%, . . . , 90%). It can be tedious to do by hand, but is easily done using most statistical software packages. It is of particular use when the observations in the data set are to be compared to each other or some other norm.

## EXAMPLE 4.16   The Golf Ball Company

### *Finding Percentiles*

The statistics you have provided so far tell the management of the golf ball company a lot about how its golf balls behave when they are hit, but these statistics will not be useful for making statements like "Ninety Percent of Brand X balls go XXX yards when they are hit!" No number that the company has looked at so far seems to provide that information, although the median comes close. The company wonders if there is a more general statistical measure similar to the median.

*Remember! The median gives the data value that has 50% of the values above or below it.*

A statistician is consulted who says that the measures the company wants are called percentiles. These percentiles are easily obtained from a statistical software package:

| Data vector: GOLFDATA.Carry M1 | | | Data vector: GOLFDATA.Carry M2 | | |
|:---:|:---:|:---:|:---:|:---:|:---:|
| **Percentages** | | **Percentiles** | **Percentages** | | **Percentiles** |
| 10 | = | 255 | 10 | = | 254 |
| 20 | = | 255 | 20 | = | 255 |
| 30 | = | 256 | 30 | = | 255 |
| 40 | = | 257 | 40 | = | 256 |
| 50 | = | 257.5 | 50 | = | 257 |
| 60 | = | 258 | 60 | = | 257 |
| 70 | = | 259 | 70 | = | 258 |
| 80 | = | 260 | 80 | = | 258 |
| 90 | = | 260 | 90 | = | 260 |

From this information the golf ball company sees, for example, that 20% of the M1 balls went further than 260 yards, compared to only 10% of the M2 balls. This is really the first difference that the company has seen in the way the two balls behaved, but it might just be part of the natural variation.   ∎

The second question involves finding the **percentile rank** of a particular value in a data set.

> The *percentile rank* of a value is the percentage of the data in the sample that are at or below the value of interest.

This measure allows you to determine the *relative standing* of an observation in a set of data. To find the percentile rank of an observation, the data must be put in numerical order. The percentile rank, $P$, is then found by

*PERCENTILE RANK*

$$P = \frac{b + \frac{1}{2}e}{n}$$

where

$b$ = the number of data values *below* the value of interest

$e$ = the number of data values *equal* to the value of interest

$n$ = the sample size

**EXAMPLE 4.17**   **Starting Salaries**

*Finding the Percentile Rank*

Suppose that in the example where we looked at starting salaries, a person in the group who had a starting salary of $26,200 wanted to know how she ranked relative to her peers. Looking at the data you see that

$$b = 50 \quad \text{(there are 50 salaries below \$26,200)}$$

$$e = 7 \quad \text{(there are 7 salaries equal to \$26,200)}$$

$$n = 100$$

so

$$P = \frac{50 + \frac{1}{2}(7)}{100} = 0.535 = 53.5\%$$

This tells her that 53.5% of the starting salaries were at or below hers. ■

 **TRY IT NOW!**

### Aptitude Test Scores
*Calculating the Percentile Rank*

A group of employees at a manufacturing facility take a test to determine their aptitude for training. The tests are scored on a 400-point scale and are shown here in increasing order:

| 185 | 227 | 241 | 257 | 281 | 299 | 314 | 329 |
| 195 | 228 | 243 | 261 | 283 | 304 | 318 | 333 |
| 196 | 234 | 248 | 269 | 283 | 307 | 319 | 335 |
| 199 | 238 | 250 | 271 | 291 | 309 | 322 | 349 |
| 223 | 241 | 253 | 272 | 297 | 310 | 328 | 353 |

One of the employees who scored 283 wants to know how he stands relative to the other employees who took the exam. What is the percentile rank for the employee's score?

What is the percentile rank of the employee that scored 319?

---

### 4.5.2 Quartiles

Although calculating percentiles in general can be tedious, there are certain percentiles that are used frequently. These percentiles are the 25th percentile and the 75th percentile, also known as the **first and third quartiles.**

> The *first quartile, $Q_1$*, is the value in the sample that has 25% of the data at or below it.

> The *third quartile, $Q_3$*, is the value in the sample that has 75% of the data at or below it.

You may be wondering what happened to the second quartile, but we have already seen it! If you think about the definition of the quartile, it is not hard to see that the second quartile, $Q_2$, must be the median.

Just like finding percentiles, finding the quartiles for a set of data can be tedious, depending on how you choose to do it. Actually, several methods can be used to find the quartiles of a set of data. As with the percentiles, different software packages use different methods and so it is not unusual to get two slightly different answers when using two different packages on the same data set. The method that we use estimates the quartiles and is not really exact. It gives values that are quite close to the more exact methods and it is much simpler.

If you think about the quartiles you see that they divide the data set in fourths. We know that the median divides the data set in half, and that if you take half of one-half you get a quarter. So, it would appear that if we find the median of each of the two halves of the data we will have our quartiles! It may sound complicated, but it really isn't.

Since percentiles and quartiles are order statistics, finding them requires that the data set be sorted from lowest to highest.

*STEPS FOR FINDING THE QUARTILES*

**Step 1:** Put the data set in order and find the median of the data.

**Step 2:** Take the lower half of the data (all of the values that are below the median) and find the median of the lower half of the data. This value will be the first quartile, $Q_1$.

**Step 3:** Take the upper half of the data (all of the values that are above the median) and find the median of the upper half of the data. This value will be the third quartile, $Q_3$.

When the median is actually one of the data values (when $n$ is odd) do not include the median in either half of the data.

**EXAMPLE 4.18**   **Starting Salaries**

*Finding the Quartiles*

The university that was looking at starting salaries of their graduates wants to know among the graduates sampled, what salary has 25% of the students earning less than that salary and what salary has 25% of the students earning more than that salary. In other words, the university wants to know the quartiles of the data. To make this easier the data set is repeated here:

| $25,000 | $25,400 | $25,600 | $25,800 | $25,900 | $26,200 | $26,400 | $26,600 | $27,000 | $28,000 |
|---------|---------|---------|---------|---------|---------|---------|---------|---------|---------|
| 25,100  | 25,400  | 25,600  | 25,800  | 25,900  | 26,200  | 26,400  | 26,600  | 27,100  | 28,200  |
| 25,100  | 25,400  | 25,600  | 25,800  | 26,000  | 26,200  | 26,400  | 26,600  | 27,100  | 28,300  |
| 25,200  | 25,500  | 25,700  | 25,800  | 26,000  | 26,200  | 26,400  | 26,600  | 27,100  | 28,300  |
| 25,200  | 25,500  | 25,700  | 25,800  | 26,000  | 26,200  | 26,400  | 26,600  | 27,200  | 28,400  |
| 25,200  | 25,500  | 25,700  | 25,800  | 26,100  | 26,200  | 26,400  | 26,700  | 27,300  | 28,500  |
| 25,200  | 25,500  | 25,700  | 25,900  | 26,100  | 26,200  | 26,500  | 26,700  | 27,400  | 28,600  |
| 25,300  | 25,600  | 25,700  | 25,900  | 26,100  | 26,300  | 26,500  | 26,800  | 27,600  | 29,400  |
| 25,300  | 25,600  | 25,700  | 25,900  | 26,100  | 26,300  | 26,500  | 26,900  | 27,700  | 30,700  |
| 25,300  | 25,600  | 25,800  | 25,900  | 26,100  | 26,300  | 26,500  | 26,900  | 27,700  | 30,800  |

1. Since there are 100 observations in the sample, the median must be the average of the 50th and 51st data values:

$$Q_2 = \frac{26,100 + 26,200}{2} = \$26,150$$

*Remember! Since 100 is even you average the two middle values to find the median.*

   The lower half of the data set, everything *below* the median, consists of the first 50 values, and the upper half of the data, everything above the median, consists of the second 50 values.

2. To find the first quartile, find the median of 50 values, which will be the average of the 25th and 26th observations:

$$Q_1 = \frac{25,700 + 25,700}{2} = \$25,700$$

3. To find the median of the 51st through 100th values you will need to average the 75th and 76th observations:

$$Q_3 = \frac{26,600 + 26,700}{2} = \$26,650$$

So, the university finds that 25% of the graduates earn less than $25,700 and 25% of the graduates earn more than $26,650. ■

 **TRY IT NOW!**

**Training Aptitude**

*Finding the Quartiles*

The company looking at training aptitude wants to give employees who scored in the top 25% on the test the opportunity to attend a seminar on training. The test scores are

| 185 | 227 | 241 | 257 | 281 | 299 | 314 | 329 |
|-----|-----|-----|-----|-----|-----|-----|-----|
| 195 | 228 | 243 | 261 | 283 | 304 | 318 | 333 |
| 196 | 234 | 248 | 269 | 283 | 307 | 319 | 335 |
| 199 | 238 | 250 | 271 | 291 | 309 | 322 | 349 |
| 223 | 241 | 253 | 272 | 297 | 310 | 328 | 353 |

*(continued)*

In the sample, what is the cutoff score for those people who will be able to attend the seminar?

*Hint: The value that defines the top 25% is the same as the value that defines the bottom 75%.*

Suppose that the company decides that the employees who scored in the bottom 25% need some additional classes on team building. What is the cutoff score for those employees who need the classes on team building?

### 4.5.3 Displaying the Data Using Boxplots

In Chapter 3 we looked at several different methods for displaying quantitative data. These methods allowed us to look at the data and describe central tendency, shape, variability, and outliers. Both methods, histograms and dotplots, provide displays of the actual data in the data set.

There is another method for displaying a set of data which uses not the individual data values, but rather a set of summary statistics taken from the data. The plot is called a **boxplot** or a **box and whisker diagram.**

> A *boxplot* or *box and whisker diagram* is a graphical display that uses summary statistics to display the distribution of a set of data.

A boxplot summarizes a sample using the quartiles and the median. You know that the median is a measure of center, but how can these statistics be used to show shape and variability? How can they identify outliers?

If you look at the first and third quartiles of a sample, $Q_1$ and $Q_3$, you see that 50% of the data in the sample fall between these two values. The distance between these two values is called the **interquartile range (IQR).**

> The *interquartile range* (**IQR**) is the difference between the third and first quartiles, $Q_3 - Q_1$.

We found that one way of describing the variability of a set of data is to use the empirical rule. Remember, the empirical rule says that approximately 68% of the data in the sample should be within one standard deviation of the mean (in the interval $\mu - 1\sigma, \mu + 1\sigma$). The interval from $Q_1$ to $Q_3$ is similar to the first interval of the empirical rule.

The main part of the boxplot consists of a rectangle (box) constructed from the median and the quartiles, as shown in Figure 4.6. This provides a partial picture of

*Note: In drawing the box the height of the vertical lines is arbitrary. The scale on the x axis should span the values of the data similar to a histogram or dotplot.*

**FIGURE 4.6**   Box portion of boxplot

the data set. To complete the description with the empirical rule we used two additional intervals, $\mu \pm 2\sigma$ and $\mu \pm 3\sigma$. The boxplot uses multiples of the IQR instead of the standard deviation. The second interval in a boxplot is described by the **inner fences.** This is similar to the $2\sigma$ interval of the empirical rule. The last interval is described by the **outer fences.**

*Note: The parts of a boxplot are defined from the quartiles, not the median.*

> The *inner fences* of a boxplot are located at $Q_1 - 1.5(\text{IQR})$ *and* $Q_3 + 1.5(\text{IQR}).$

> The *outer fences* of a boxplot are located at $Q_1 - 3(\text{IQR})$ *and* $Q_3 + 3(\text{IQR}).$

The inner and outer fences are not drawn on the boxplot but are used to define the "whisker" portion of the plot. The whiskers are constructed by drawing horizontal lines from the quartiles to the smallest (and largest) observations in the sample that are *within* the inner fences. This is illustrated in Figure 4.7.

**FIGURE 4.7**   Boxplot with whiskers

It is easy to see that the median gives us a picture of the center of the data, but you may be wondering how the boxplot shows the shape of the distribution. You know that when data are symmetric the mean and the median are both located at the bump in the distribution and that the distribution tails off at the same rate in both directions. In a boxplot, if the data are symmetric, then you would expect that the median is located halfway between the two quartiles and that the whiskers are the same length.

If data are skewed the median stays located near the bump in the data and one side of the distribution tails at a slower rate than the other. In a boxplot, the median will be closer to one of the quartiles than the other and/or the whisker on the tail side of the distribution will be longer than the other. Figure 4.8 shows boxplots for data that are skewed.

While all this may seem quite complicated, it really is not. It is actually *easier* to make a boxplot than a histogram because the rules are standard and there is no judgment involved in setting it up. Making a boxplot can be separated into two parts, the calculations and the construction.

**FIGURE 4.8**   Boxplots for skewed data

*CREATING A BOXPLOT* **Calculations**

**Step 1:** Find the median and the first and third quartiles for the data.

**Step 2:** Calculate the interquartile range (IQR) by finding $Q_3 - Q_1$.

**Step 3:** Find the values for locating the inner and outer fences:

| | |
|---|---|
| Lower inner fence (LIF): | $Q_1 - 1.5(\text{IQR})$ |
| Upper inner fence (UIF): | $Q_3 + 1.5(\text{IQR})$ |
| Lower outer fence (LOF): | $Q_1 - 3(\text{IQR})$ |
| Upper outer fence (UOF): | $Q_3 + 3(\text{IQR})$ |

**Construction**

**Step 1:** Draw the $x$ axis and select a scale that will allow you to locate the largest and smallest values in the sample.

**Step 2:** Draw three vertical lines, all of the same height, one at $Q_1$, one at $Q_3$, and one at the median.

**Step 3:** Construct a box using the quartiles as the vertical edges and enclosing the median.

**Step 4:** Locate the inner fences on the axis. Draw a horizontal line from $Q_1$ to the smallest value in the sample that is within (larger than) the lower inner fence. Draw a similar horizontal line from $Q_3$ to the largest sample value that is within (less than) the upper inner fence.

It is probably time for an example!

### EXAMPLE 4.19  Starting Salaries

*Constructing a Boxplot*

Consider the data about starting salaries. We have already summarized the data by constructing a histogram and by looking at sample statistics. What can we learn from a boxplot?

The first step in the calculation portion is done, since in Example 4.18 we found the median and the quartiles for the data to be

| | |
|---|---|
| Median: | $26,150 |
| $Q_1$: | $25,700 |
| $Q_3$: | $26,650 |

The interquartile range is

$$Q_3 - Q_1 = \$26{,}650 - \$25{,}700 = \$950$$

For the inner fences we find that

$$\text{Lower inner fence} = 25{,}700 - 1.5(950) = \$24{,}275$$

$$\text{Upper inner fence} = 26{,}650 + 1.5(950) = \$28{,}075$$

and for the outer fences we find that

$$\text{Lower outer fence} = 25{,}700 - 3(950) = \$22{,}850$$

$$\text{Upper outer fence} = 26{,}650 + 3(950) = \$29{,}500$$

We now have all of the information needed to draw the boxplot.

Select the scale for the $x$ axis so that it will accommodate the smallest and largest values in the sample, in this case $25,000 and $30,800. We will let the axis go from $24,000 to $31,000 in $1000 increments.

Locating the box portion of the plot is not difficult. To determine how far to extend the whiskers, compare the minimum and maximum values in the data set to the lower and upper inner fences:

| | |
|---|---|
| Minimum: $25,000 | Lower inner fence: $24,275 |
| Maximum: $30,800 | Upper inner fence: $28,075 |

You see that the lower whisker can extend all the way to the minimum data value, but the upper whisker cannot go to the maximum because it is beyond the inner fence. Looking back at the data you see that the largest data value that is less than $28,075 is $28,000. The boxplot will look like the one shown below:

Starting salary ($)

From the boxplot it would appear that the data are only slightly skewed to the right, since the median is located only a little closer to the lower quartile and the right whisker is only slightly longer than the left. ■

## 4.5.4 Using a Boxplot to Identify Outliers

At this point you may have some questions such as Why did we find the outer fences if we never use them? and What happens to the values from the sample that are beyond the inner fences? Both of these questions are quite reasonable and we are ready to answer them.

When we started talking about boxplots we said that they could be used to identify outliers. The intervals defined by the fences are similar to the $2\sigma$ and $3\sigma$ intervals from the empirical rule. Sample data that fall between the inner and outer fences are called *possible outliers,* while data values that fall beyond the outer fences are called *probable outliers.* If you are having trouble figuring out the difference between *probable* and *possible,* think about the difference in your reaction when your instructor tells you, "It is *possible* that you will pass this course" vs. "It is *probable* that you will pass this course."

The observations that fall between the inner and outer fences, the possible outliers, correspond to those that fall between two and three standard deviations from the mean using the empirical rule. Remember that less than 5% of the data should fall in this region, so data values that fall there are fairly unlikely events.

The observations that fall beyond the outer fences, the probable outliers, correspond to those that are more than three standard deviations from the mean in the empirical rule. Less than 1% of the data should fall in this region, so data values that fall there are very unlikely events.

The possible and probable outliers are plotted individually on the boxplot using special symbols, such as an open dot for possible outliers and a closed dot for probable outliers.

## EXAMPLE 4.20    Starting Salaries

*Locating Outliers on a Boxplot*

Let's finish up the boxplot for the starting salaries data by identifying the possible and probable outliers. To do this we need to compare the data values that were not included in the whiskers to the inner and outer fences. You remember that nine observations were beyond the inner fence on the upper side of the boxplot. Comparing these values to the values of the inner and outer fences, we find

|  |  |
|---|---|
| Upper inner fence: | $28,075 |
| Possible outliers: | $28,200; 28,300; 28,300; 28,400; 28,500; 28,600; 29,400 |
| Upper outer fence: | $29,500 |
| Probable outliers: | $30,700; 30,800 |

The completed boxplot is shown below:

When the outliers are added to the boxplot the extreme skewness shown in the histogram is evident. Further, now we know that it is not just a case of skewed data but that in fact at least two of the data points are very different from the rest of the data. ■

Now that we know that some of the data might be outliers, you might wonder what we will do about it. You remember that when we sample from populations we want our sample to represent the population of interest. An outlier is a data value that has a large variation from the center. The first thing to do when we have identified outliers is to search for the *cause* of the variation. If there is some identifiable, assignable cause that makes the data value *not* representative of the data, then it is reasonable to drop the observation from the sample and redo the analysis. You cannot drop the observation without cause, but you can note in your report or analysis that an observation might have influenced the results.

Why do we need *another* method for displaying sample data and for identifying outliers? One important reason for using the boxplot is that it relies on statistics that are *invariant* (do not change) to the outliers themselves. When we use *z*-scores to identify outliers we use the mean and the standard deviation, both of which are sensitive to extreme values. If we have cause to actually delete the extreme observations the mean and standard deviation will change and *so will the definition of an outlier!* In a way, using *z*-scores to identify outliers is like trying to hit a moving target. The median and quartiles, on the other hand, remain relatively stable. Removing outliers does not really affect the locations of the inner and outer fences and so the definition of an outlier stays the same.

Another reason for using boxplots to display data is that they really do give an excellent picture of the distribution and they make it easy to compare samples from the same or different populations. Multiple boxplots lend themselves to being put on the same axis and make comparisons easier than multiple histograms, each of which requires a separate graph.

### EXAMPLE 4.21  The Golf Ball Company

*Comparing Data Using Boxplots*

As the statistical consultant, you decide to give the management of the golf ball company a final picture by looking at boxplots for each sample. A statistical software package is used to create the plots and the results are shown below:

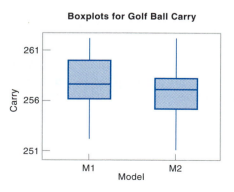

**Boxplots for Golf Ball Carry**

This is the first time that the two ball types have actually been compared side by side and the boxplots provide new information. From the plots it appears that while the middle 50% of the M1 balls carry between 256 and 259 yards, the middle 50% of the M2 balls only carry between 255 and 258 yards. In addition, based on the size of the plot, it appears that the M2 balls are slightly more variable, but, again, this may be due entirely to the samples selected. ■

## TRY IT NOW!

### Training Aptitude

*Finding the Quartiles*

The company that administered the training aptitude test to its employees would like a better picture of how the employees performed on the test. The data are

| | | | | | | | |
|---|---|---|---|---|---|---|---|
| 185 | 227 | 241 | 257 | 281 | 299 | 314 | 329 |
| 195 | 228 | 243 | 261 | 283 | 304 | 318 | 333 |
| 196 | 234 | 248 | 269 | 283 | 307 | 319 | 335 |
| 199 | 238 | 250 | 271 | 291 | 309 | 322 | 349 |
| 223 | 241 | 253 | 272 | 297 | 310 | 328 | 353 |

Previously (pages 157–158), you found the first and third quartiles of the data set. Use these values to complete the calculations needed for a boxplot.

Draw a complete boxplot of the data.

Were there any outliers? If so, which data values were they?

## 4.5.5 Exercises—Learn It!

**4.19**  The management of Disney World thinks that between 5 and 10 minutes is about the right amount of time for someone to wait for a monorail. Looking at the sample data, we find

5.1    5.5    6.5    8.6    9.1    9.3    9.5    9.6    9.7    13.6    14.1    15.0

(a)  What is the percentile rank of the person who waited 9.7 minutes?
(b)  What is the percentile rank of the person who waited 5.1 minutes?
(c)  Approximately what percentage of the people in the sample waited between 5 and 10 minutes?

*Requires Exercise 4.7*     **4.20**  The transportation department of the small city wants to know what percentage of time there were 10 cars backed up at the light. Use the data below:

4    4    5    6    7    8    9    10    11    12

(a)  What is the percentile rank of the observation of 10 cars waiting at the light?
(b)  What is the percentile rank of the observation of 7 cars waiting at the light?
(c)  What information does this give the transportation department?

*Requires Exercise 4.10*     **4.21**  The mail-order book club wants to know about the percentage of its customers that make subsequent purchases. Use the sample of 15 randomly selected customers given below:

10    11    11    12    12    13    13    14    14    14    14    14    14    15    15

(a)  What is the percentile rank of a person who makes 13 subsequent purchases?
(b)  What are the first and third quartiles of the data?
(c)  What information do the quartiles give the book club?

*Requires Exercise 4.12*     **4.22**  The manager looking at the travel expenses of sales representatives would like to look at the data in terms of percentiles:

1558    1617    1655    1709    1716    1733    1758    1908    1963    2062    2075    2171

(a)  Find the first and third quartiles of the data and explain what they mean.
(b)  What is the interquartile range? How would you explain what that tells the manager?
(c)  Make a boxplot of the data.
(d)  Use the boxplot to describe the distribution of travel miles claimed.
(e)  Are any of the observations unusual? Why or why not?

*Requires Exercise 4.2*     **4.23**  The company that sells mail-order computer systems wants to look at typical weekly sales in terms of quartiles. The data (in thousands of dollars) for the 15 weeks are given below:

191    222    222    223    223    225    227    228    229    232    234    234    236    244    253

(a)  Find the first and third quartiles of the data and explain what they mean.
(b)  What is the interquartile range? How would you explain what it means?
(c)  Make a boxplot of the data.
(d)  Use the boxplot to describe the distribution of weekly sales.
(e)  Are any of the observations unusual? Why or why not?

**4.24** *The New York Times* (28 October 1994) published a state-by-state report on the number of people incarcerated per 100,000 people in the population. The data are shown in the table below:

| | | | | | | | |
|---|---|---|---|---|---|---|---|
| District of Columbia | 1578 | Mississippi | 385 | Tennessee | 278 | Hawaii | 170 |
| Texas | 545 | California | 382 | Colorado | 272 | Oregon | 169 |
| Louisiana | 514 | Virginia | 374 | Alaska | 256 | New Hampshire | 167 |
| South Carolina | 504 | United States (Federal) | 373 | Indiana | 256 | Massachusetts | 165 |
| Oklahoma | 501 | Ohio | 369 | Idaho | 253 | Utah | 154 |
| Nevada | 456 | New York | 361 | Wyoming | 247 | Nebraska | 148 |
| Arizona | 448 | Arkansas | 355 | Kansas | 239 | Vermont | 138 |
| Alabama | 439 | Connecticut | 331 | South Dakota | 227 | Maine | 113 |
| Michigan | 423 | Missouri | 321 | Pennsylvania | 224 | West Virginia | 106 |
| Georgia | 417 | North Carolina | 314 | New Mexico | 216 | Minnesota | 100 |
| Florida | 404 | New Jersey | 307 | Washington | 198 | North Dakota | 75 |
| Maryland | 392 | Illinois | 302 | Montana | 192 | | |
| Delaware | 391 | Kentucky | 281 | Rhode Island | 185 | | |
| | | | | Iowa | 180 | | |
| | | | | Wisconsin | 172 | | |

(a) Make a boxplot of the data.

(b) Use the boxplot to describe the number of people incarcerated in the United States.

(c) Are any of the data values unusual? Why or why not?

## 4.6  NUMERICAL DESCRIPTORS IN EXCEL

This section covers the basics of using Excel to calculate summary statistics and to create boxplots.

### 4.6.1 Calculating Summary Statistics in Excel

Suppose that we want to calculate a set of summary statistics for the Golf Ball data. Figure 4.9 shows a portion of that data in an Excel worksheet.

The Data Analysis ToolPak has a function that creates a set of summary statistics for a set of data. To access this function, select **Data Analysis** from the **Tools** menu and choose **Descriptive Statistics** from the list of **A**nalysis tools. The dialog box is

| | A | B | I | J | K | L | M | N |
|---|---|---|---|---|---|---|---|---|
| 1 | Ball# | Model # | Head | Temp | Carry | Tot Dist | Date | Time |
| 2 | 1 | M1 | 686 | 77 | 257 | 270 | 8/20 | 8:15 |
| 3 | 2 | M1 | 688 | 77 | 255 | 267 | 8/20 | 8:15 |
| 4 | 3 | M1 | 687 | 77 | 256 | 267 | 8/20 | 8:15 |
| 5 | 4 | M1 | 688 | 77 | 255 | 271 | 8/20 | 8:15 |
| 6 | 5 | M1 | 687 | 77 | 255 | 268 | 8/20 | 8:15 |
| 7 | 6 | M1 | 687 | 77 | 256 | 267 | 8/20 | 8:15 |
| 8 | 7 | M1 | 687 | 77 | 255 | 264 | 8/20 | 8:15 |
| 9 | 8 | M1 | 690 | 78 | 258 | 269 | 8/20 | 8:15 |
| 10 | 9 | M1 | 686 | 78 | 252 | 257 | 8/20 | 8:15 |
| 11 | 10 | M1 | 687 | 78 | 256 | 268 | 8/20 | 8:15 |
| 12 | 11 | M1 | 687 | 78 | 253 | 263 | 8/20 | 8:15 |

**FIGURE 4.9**  The golf ball data

**FIGURE 4.10**   The Descriptive Statistics dialog box

shown in Figure 4.10. You will see that it is similar to the one you used to create a histogram.

The steps for creating the summary statistics are:

1. Position the cursor in the textbox labeled **Input Range** and highlight the range of data for which you want to calculate summary statistics. If your first row contains the variable name, be sure to check the box labeled **Labels in First Row.**

2. You must specify a location for the output, either a section of the current worksheet, or a new worksheet or workbook. Click on the radio button for your choice. If you select **Output Range,** you must specify a location on the worksheet. Position the cursor in the textbox for Output Range and click on the cell where you want the upper left corner of the results to appear. If you want to put the results in a new worksheet, you have the option of giving the sheet a name in the textbox or just letting Excel create a new, numbered sheet.

3. The last thing you must do is specify what kind of descriptive statistics you want. Click on the box labeled **Summary statistics** and finally click on **OK.** The output will appear in the location you specified and should look like Figure 4.11. The output includes most of the summary statistics that you learned in this chapter, as well as some that you have not yet encountered. The output does not include the quartiles. When a statistic (such as the mode) cannot be computed, the output will read N/A (Not Available).

| | AE | AF |
|---|---|---|
| | *Carry* | |
| Mean | | 257.0972 |
| Standard Error | | 0.274902 |
| Median | | 257 |
| Mode | | 258 |
| Standard Deviation | | 2.332621 |
| Sample Variance | | 5.441119 |
| Kurtosis | | -0.18199 |
| Skewness | | -0.04536 |
| Range | | 11 |
| Minimum | | 251 |
| Maximum | | 262 |
| Sum | | 18511 |
| Count | | 72 |

**FIGURE 4.11** Output from **Tools> Data Analysis> Descriptive Statistics**

### 4.6.2 Making a Boxplot in Excel

Excel does not include boxplots as part of the graphs it can create. However, it is possible to create them if you know how Excel creates graphs. We have provided a macro for creating boxplots on the data disk that comes with this book.

To access the macros, you must open the file MacDoIt.xls either from your hard drive or from the floppy disk. This makes the macros available to you during your Excel session. Once you have the file open, you can use the macros in any workbook or worksheet. To use the boxplot macro, follow these instructions:

*Note: The MacDoIt workbook must be open to make the macros available.*

1. From the **Tools** menu select **Macro>Macros,** as shown in Figure 4.12. The macro dialog box shown in Figure 4.13 opens and you are presented with a list of macros for different statistical tools.

**FIGURE 4.12** Menu selection to run a macro

**FIGURE 4.13** The Macro dialog box

**2.** From the list of available macros, highlight **MacDoIt.xls!BoxPlot** and click on **<u>R</u>un.** A small dialog box will open, prompting you to enter the location of your data. You can either enter the name of a range (if you have done this) or just use the cursor to highlight the data range in the worksheet. Click **OK** and the boxplot will be created in a new worksheet. In addition to the box-plot, the macro prints the Min and Max values of the data, and $Q_1$, $Q_3$, and the median.

The boxplot and summary statistics for the distance that the golf ball traveled when hit (Carry) for the M1 ball is shown in Figure 4.14.

| | A | B | C | D | E | F | G | H |
|---|---|---|---|---|---|---|---|---|
| 1 | Min | Q1 | Q2 | Q3 | Max | | | |
| 2 | 252 | 256 | 257.5 | 259.25 | 262 | | | |

Box Plot For 'Golf Ball Data'!$Q$1:$Q$37

**FIGURE 4.14**   Finished boxplot for golf ball data

# CHAPTER 4 SUMMARY

There are many ways to describe a set of data using sample statistics. No single number will do the job, nor is there any standard way to proceed. The measures that you choose must reflect the characteristics of the data itself. Most of the time the best descriptions come from the use of multiple measures and the conclusions that can be reached by comparing them.

Rather than summarize a sample with a list of numbers it is often useful to create images of the data using combinations of different statistics. An example of this is the *empirical rule*, which gives a picture of the distribution of the data. Another example of using summary statistics together to get a picture of the distribution is the *boxplot*.

Data analysis is not a static tool. You will need to look at a set of data in every way possible to obtain all of the information that it contains. Sometimes different methods will all lead you to the same conclusions and sometimes one method will yield an insight that is hidden in every other method.

## Key Terms

| Term | Definition | Page reference |
|---|---|---|
| **Boxplot** | A graphical tool that summarizes a sample using the quartiles and the median. | 158 |
| **Empirical rule** | For a symmetric distribution<br>• about 68% of all observations are within one standard deviation of the mean.<br>• about 95% of all observations are within two standard deviations of the mean.<br>• almost all (more than 99%) of the observations are within three standard deviations of the mean. | 146 |
| **First quartile, $Q_1$** | The value in the sample that has 25% of the data at or below it. | 156 |

*(continued)*

| Term | Definition | Page reference |
|---|---|---|
| **Interquartile range, IQR** | The difference between the third and first quartiles, $Q_3 - Q_1$. | 158 |
| **Median** | The value of the middle observation in an ordered set of data. | 124 |
| **Modal class** | The class interval in a frequency distribution or histogram that has the highest frequency. | 132 |
| **Mode** | The data value that has the highest frequency of occurrence in the sample. | 132 |
| **Parameter** | A numerical descriptor that is used to describe a population. | 120 |
| **pth percentile** | The value in the data that has $p\%$ of the data at or below it. | 153 |
| **Population mean** | $\mu$ (mu). | 121 |
| **Possible outlier** | Observations that fall between the inner and outer fences. | 161 |
| **Probable outlier** | Observations that fall beyond the outer fences. | 161 |
| **Sample mean** | The center of balance of a set of data, and is found by adding up all of the data values and dividing by the number of observations. | 121 |
| **Sample range** | The difference between the maximum and minimum observations in the sample. | 140 |
| **Sample standard deviation, s** | The positive square root of the sample variance. | 143 |
| **Sample variance, $s^2$** | The average of the squared deviations of the data values from the sample mean. | 143 |
| **Statistic** | A numerical descriptor that is calculated from sample data and is used to describe the sample. | 120 |
| **Third quartile, $Q_3$** | The value in the sample that has 75% of the data at or below it. | 156 |
| **z-score** | A **z-score** measures the number of standard deviations that a data value is from the mean. | 150 |

## Key Formulas

| Term | Formula | Page reference |
|---|---|---|
| **Inner fences** | $Q_1 - 1.5(\text{IQR})$ <br> $Q_3 + 1.5(\text{IQR})$ | 159 |
| **IQR** | $Q_3 - Q_1$ | 158 |
| **Outer fences** | $Q_1 - 3(\text{IQR})$ <br> $Q_3 + 3(\text{IQR})$ | 159 |
| **Percentile rank** | $P = \dfrac{b + \frac{1}{2}e}{n}$ | 154 |
| **Sample mean, $\overline{X}$** | $\overline{X} = \dfrac{\sum_{i=1}^{n} x_i}{n}$   or   $\dfrac{\sum x}{n}$ | 121 |
| **Sample range, $R$** | $\text{Max} - \text{Min}$ | 140 |
| **Sample standard deviation, s** | $s = \sqrt{s^2}$ | 144 |
| **Sample variance, $s^2$** | $s^2 = \dfrac{\sum_{i=1}^{n}(x_i - \overline{x})^2}{n-1}$ | 144 |
| **z-score** | $\dfrac{X - \mu}{\sigma}$ | 150 |

# CHAPTER 4 EXERCISES

## Learning It!

**4.25**  A manufacturer of pain relievers is interested in studying the amount of time it takes a person to be relieved of headache pain after taking the medication. The manufacturer selects a random sample of 12 people and conducts a study. The data (minutes) are

13.0   12.9   13.2   12.7   13.1   13.0   13.1   13.0   12.6   13.1   13.0   13.1

(a) Find the mean, median, and mode of the relief time.
(b) Find the range and standard deviation of the relief time.
(c) What is the percentile rank of the person who took 12.9 minutes to be relieved of their headache pain?
(d) Find the quartiles for the data set and interpret them.

**4.26**  A group of elderly people in a town filed a grievance with the town council, saying that the length of the walk signals in the town was inadequate for many elderly people to cross safely. In an attempt to investigate the claim, the town collected data on street crossing times for 12 elderly persons. The times, to the nearest tenth of a second, are

21.4   15.1   13.6   16.0   15.0   19.1   21.0   14.2   15.6   20.1   21.1   22.2

(a) Find the mean, median, and mode for the crossing times.
(b) Find the range and standard deviation of the crossing times.
(c) Compare the mean and the median. Do you think the data are skewed?
(d) What is the percentile rank of the person who took 20.1 seconds to cross the street?
(e) Find the quartiles and interpret them.
(f) Make a boxplot of the data.
(g) Describe the distribution of the crossing times.

**4.27**  A manufacturer of compact discs decides to take a random sample of the discs and measure the diameter. The diameter of the discs is a critical measurement, since if the discs are too large they will not fit in the players and if they are too small then they will not be read properly. The sample of 10 discs yields the following data (in.):

4.74   4.72   4.76   4.72   4.73   4.72   4.76   4.74   4.75   4.75

(a) Find the mean diameter of the compact discs.
(b) Find the median diameter of the compact discs.
(c) Compare the median and the mean. Do you think the distribution of the diameters is symmetric? Why or why not?
(d) Would it make sense to use the mode as a measure of center for this sample? Why or why not?
(e) Find the range of the diameters of the compact discs.
(f) Find the standard deviation of the diameters of the compact discs.

**4.28**  The vice president of marketing at a large corporation is wondering how many of the company's employees arrive at work before he does. He decides to collect some data by counting the number of cars that are in the parking lot when he arrives at 6:30 A.M.

23   24   24   25   25   25   25   26   26   37

(a) Find the sample mean of the data.
(b) Find the median of the data.
(c) Find the range of the data.
(d) Find the standard deviation of the data.

**4.29**  A company that sells personal computers via mail order would like to know something about the amount of time that a customer spends on the phone with the Technical Support department during a call. The company collects information on the length of a call in minutes for 20 different customers:

| | | | | |
|---|---|---|---|---|
| 15 | 23 | 34 | 38 | 41 |
| 16 | 25 | 37 | 40 | 43 |
| 19 | 25 | 38 | 40 | 43 |
| 20 | 28 | 38 | 41 | 44 |

(a) What is the average length of a call to Technical Support? What is the median length of a call?

(b) What is the range of call length? What is the standard deviation of the call length?

(c) Compare the mean and the median. Do you think the distribution of call times is symmetric or skewed? If it is skewed, which way is it skewed?

(d) Use one of the graphical methods you learned in Chapter 3 to display the data. Does the display agree with your answer to part (c)? If not, why not?

**4.30**  A manufacturer of breakfast cereals packs the cereal in 15-ounce boxes. Periodic quality checks are made to ensure that the boxes are being filled adequately. A random sample of 35 boxes of the cereal are selected and weighed. The data are

| | | | | | | |
|---|---|---|---|---|---|---|
| 14.91 | 15.23 | 15.34 | 15.40 | 15.48 | 15.59 | 15.62 |
| 14.96 | 15.24 | 15.34 | 15.42 | 15.49 | 15.60 | 15.62 |
| 15.11 | 15.28 | 15.36 | 15.45 | 15.50 | 15.60 | 15.63 |
| 15.21 | 15.30 | 15.38 | 15.46 | 15.52 | 15.61 | 15.64 |
| 15.22 | 15.30 | 15.39 | 15.48 | 15.58 | 15.61 | 15.67 |

(a) Find the mean, median, and mode for the weight of the cereal boxes.

(b) Find the range and the standard deviation of the cereal box weights.

(c) Describe the distribution of the data using the information from parts (a) and (b).

(d) Find the quartiles of the data and interpret them.

(e) Make a boxplot of the data.

**4.31**  A university collected some data on the amount of money ($) that students spend on textbooks in a typical semester:

| | | | | |
|---|---|---|---|---|
| 214 | 241 | 248 | 258 | 269 |
| 233 | 244 | 249 | 260 | 274 |
| 234 | 245 | 250 | 262 | 276 |
| 236 | 247 | 253 | 262 | 277 |
| 239 | 248 | 254 | 263 | 277 |
| 241 | 248 | 254 | 265 | 281 |

(a) Find the mean and the standard deviation of the amount spent on textbooks.

(b) Find the intervals $\bar{X} \pm 1s$, $\bar{X} \pm 2s$, and $\bar{X} \pm 3s$.

(c) Make a histogram of the data and mark the intervals on the histogram.

(d) Calculate the actual percentage of the data that fall within each interval.

(e) Compare the actual percentages with those predicted by the empirical rule.

**4.32**  To begin a study comparing the monthly salaries of the auditors for a particular company to accountants' salaries in the industry in general, a random sample of twenty auditors was taken and their monthly salaries recorded. The data are

| | | | |
|---|---|---|---|
| 2832 | 3050 | 3231 | 3286 |
| 2843 | 3087 | 3233 | 3302 |
| 2978 | 3096 | 3237 | 3308 |
| 3016 | 3193 | 3239 | 3396 |
| 3032 | 3223 | 3252 | 3429 |

(a) Find the mean and the standard deviation of the monthly salaries.

(b) Find the intervals $\bar{X} \pm 1s$, $\bar{X} \pm 2s$, and $\bar{X} \pm 3s$.

(c) Calculate the actual percentage of the data that fall within each interval.

(d) Compare the actual percentages with those predicted by the empirical rule.

(e) Create a boxplot of the monthly salaries.

(f) Use the boxplot to describe the monthly salaries of auditors in the company.

**4.33** It has been a widely held belief that the switch to participative management would increase employees' "buy-in" to the company. One of the benefits that should be realized is a reduction in the number of sick days that employees use. The company decides to sample 25 employees from each of two manufacturing departments. The first department has been participative for almost two years and the second is still using a traditional management style. The data on the number of sick days used by each employee in the past 12 months are

| Participative Management | | | | | | Traditional | | | | |
|---|---|---|---|---|---|---|---|---|---|---|
| 1 | 3 | 5 | 5 | 6 | | 0 | 5 | 6 | 7 | 9 |
| 1 | 4 | 5 | 6 | 7 | | 3 | 5 | 7 | 7 | 9 |
| 2 | 4 | 5 | 6 | 8 | | 4 | 6 | 7 | 7 | 10 |
| 2 | 4 | 5 | 6 | 8 | | 4 | 6 | 7 | 8 | 11 |
| 3 | 4 | 5 | 6 | 8 | | 5 | 6 | 7 | 8 | 11 |

(a) Find the mean, median, and mode for the number of sick days used for each group of employees.

(b) Find the range and standard deviation for the two sets of data.

(c) Make a boxplot for each set of data.

## Thinking About It!

**4.34** Consider the data about the number of cars that are in the parking lot when the vice president arrives:

*Requires Exercise 4.28*

$$23 \quad 24 \quad 24 \quad 25 \quad 25 \quad 25 \quad 25 \quad 26 \quad 26 \quad 37$$

(a) If the vice president wanted an estimate of the typical number of employees who arrive at work before he does, would the mean or the median be a better statistic to use? Why?

(b) Is the standard deviation or the range more reliable for these data? Why?

(c) Find the z-score of the data value of 37 cars. Does it confirm your suspicion that the data value is unusual?

**4.35** Grading homework is a real problem. It takes an enormous amount of time and many students do not do a very good job or copy answers from other students or the back of the book. A teacher of Elementary Statistics decided to conduct a study to determine what effect grading homework had on her students' exam scores. She taught three sections of Elementary Statistics and randomly assigned each class one of three conditions: (1) no homework given, (2) homework given but not collected, and (3) homework given, collected, and graded. After the first exam, she collected the data (exam scores) and made histograms of the data and calculated some numerical measures. The histograms and summary data for each group are:

(a) Describe the distribution of the data for each of the groups and use the information to compare the three data sets and come to any conclusions you can about the effects of homework on exam grades.

(b) Do you think that this was a valid way for the instructor to examine the effects of homework on exam scores? Why or why not? What other factors might need to be considered? Based on the data, if you were the instructor, what would you do about assigning homework?

*Requires Exercise 4.29*    **4.36**  The mail-order computer firm that collected data on the length of a call to Technical Support is unsure about which sample statistics should be used to summarize the data and what they really mean. The firm calls on you to be its expert. What can you tell the firm about the length of phone calls to its Technical Support lines?

**4.37**  A company has been recording data on the number of defective items found in the daily production lots after 100% inspection. The data recorded for the last five days had an average of 20 defectives. One of the people on the inspection team thought this seemed strange and realized that on one of the days the number of defectives was really 10 and had been miscopied. When the correct value was written down the average number of defectives changed from 20 to 15. What value had been written down instead of the 10?

*Requires Exercises 4.28 and 4.34*    **4.38**  Create a boxplot for the data on the number of cars in the parking lot when the vice president arrives:

23    24    24    25    25    25    25    26    26    37

(a) What does the boxplot indicate about outliers?

(b) Does this agree with the information obtained from the z-score? Why or why not?

(c) Which do you think is a more reliable indicator in this case?

*Requires Exercise 4.17*    **4.39**  The software company that is looking at the time to failure of the diskettes it uses decides to look at an alternative supplier of the product. The data for its current supplier and for the new supplier are

| Current Supplier | | | | Alternative Supplier | | | |
|---|---|---|---|---|---|---|---|
| 486 | 494 | 502 | 508 | 489 | 492 | 495 | 498 |
| 490 | 496 | 504 | 510 | 489 | 492 | 496 | 499 |
| 491 | 498 | 505 | 514 | 491 | 493 | 497 | 502 |
| 491 | 498 | 506 | 515 | 492 | 493 | 497 | 503 |
| 494 | 498 | 507 | 527 | 492 | 494 | 497 | 505 |

(a) Find the mean, median, range, and standard deviation for each supplier.

(b) Make a boxplot of the data for each supplier.

(c) Based on these statistics, describe the diskettes for each supplier and make a recommendation to the company on which supplier to choose. Be sure to support your recommendation with references to the data.

*Requires Exercise 4.31*    **4.40**  The university that collected the data on the amount that students spend on textbooks wants to include the information in their catalog on typical semester expenses. The current catalog estimates that students spend $225 on books in a semester.

(a) Do you think the catalog should be changed?

(b) If you think the catalog should be changed, what amount should be used? Justify your choice. If you think the catalog does not need to be changed, explain why.

(c) Suppose that changing catalog copy is costly considering the number of catalogs that are currently on hand. Does this change your answer to part (b)? If so, why? If not, why not?

*Requires Exercise 4.32*    **4.41**  The company that wants to compare its auditors' salaries to the industry in general finds that the average monthly salary of accountants in the industry is $3400 with a standard deviation of $50. The company asks you to interpret the data and to show how its auditors' salaries compare to those of the industry in general.

**4.42**  As part of an annual program to calibrate their quality inspectors, each inspector in a company is asked to do a 100% inspection of a lot of 500 items. The data on the number of defectives found by each of the 35 inspectors are

| 8 | 10 | 13 | 14 | 15 | 16 | 17 |
|---|----|----|----|----|----|----|
| 9 | 11 | 13 | 15 | 15 | 16 | 18 |
| 9 | 12 | 13 | 15 | 15 | 17 | 18 |
| 9 | 12 | 14 | 15 | 15 | 17 | 19 |
| 10 | 12 | 14 | 15 | 15 | 17 | 20 |

(a) If you did not know the actual number of defectives in the lot, would you use the sample mean, median, or mode as your estimate of the number of defectives in the lot? Justify your decision.

(b) Suppose that you were the training coordinator for the quality inspectors. How would you interpret the data? What would you do as a result?

**4.43** The company that is looking at employee sick days asks you to examine the summary statistics and compare the number of sick days taken by the two groups.

*Requires Exercise 4.33*

(a) Based on the data, what would you tell the company?

(b) Are the data sufficient for you to conclude that participative management reduces the number of sick days taken by employees?

(c) If your answer to part (b) is yes, why can you reach this conclusion? If your answer to part (b) is no, why do you think this conclusion is not valid?

(d) As a consultant, would you like the company to collect any additional information for you? If so, what information might be useful?

**4.44** A study was recently completed by an insurance company concerning a particular surgical procedure. The study looked at the hospital records of 40 patients at two hospitals and compared the length of patient stay. The data were analyzed using Minitab. Output showing the descriptive statistics for the data and boxplots of the two samples is shown below:

Descriptive Statistics Output for Patient Stay (Minitab)

**Descriptive Statistics**

| Variable | N | Mean | Median | Tr Mean | StDev | SE Mean |
|----------|----|-------|--------|---------|-------|---------|
| Hospital 1 | 40 | 7.725 | 7.500 | 7.667 | 2.562 | 0.405 |
| Hospital 2 | 40 | 10.350 | 10.000 | 10.222 | 3.340 | 0.528 |

| Variable | Min | Max | Q1 | Q3 |
|----------|-------|--------|-------|--------|
| Hospital 1 | 2.000 | 14.000 | 6.000 | 9.000 |
| Hospital 2 | 5.000 | 18.000 | 8.000 | 13.500 |

Boxplots for Patient Stay Data (Minitab)

(a) Using the computer output, describe each of the data sets. Be sure to include information about central tendency, variability, shape, and outliers.

(b) Compare the length of patient stay for the two hospitals.

(c) Proponents of hospital 1 say that this hospital is better than hospital 2 at the surgical procedure because the patients from hospital 1 recover faster than the patients from hospital 2. Can you conclude, on the basis of these data, that hospital 1 is better at the surgical procedure than hospital 2?

(d) If you think that the conclusion described in part (c) is reasonable, justify it. If you think this conclusion is not possible, what can you conclude? What additional information would you, as the insurance company president, find useful?

*Requires Exercise 4.26*    **4.45**   The town that is studying crossing times consults you about what the data mean. Currently, the walk signals are set for 15 seconds. Write a report telling the town council what you have found.

## Doing It!

*Datafile:*
*GOLFBALL.XXX*

**4.46**   The company investigating the golf balls is not satisfied with the limited analysis that it has done. The managers have collected a good deal of data, but they are not sure how to look at them and interpret the output. They decide to hire you to help them understand what the golf balls are doing and how they compare to each other. In addition to measures of the balls' performance, such as the variable *Carry,* the managers know that other factors, both internal (ball-related) and external (environment-related), could affect performance. They tested 36 of each type of ball at three different times using a machine to launch the balls. Data were recorded on 14 different variables. A portion of the data is shown below:

| Ball | Model | S1 | S2 | S3 | Wgt | Dw | Dd | Head | Temp | Carry | Tot Dist | Date | Time |
|------|-------|----|----|----|-----|-----|------|------|------|-------|----------|------|------|
| 1 | M1 | 81 | 81 | 82 | 45.3 | 0.145 | 0.0110 | 686 | 77 | 257 | 270 | 8/20 | 8:15 |
| 2 | M1 | 83 | 83 | 84 | 45.2 | 0.151 | 0.0111 | 688 | 77 | 255 | 267 | 8/20 | 8:15 |
| 3 | M1 | 81 | 82 | 84 | 45.2 | 0.145 | 0.0105 | 687 | 77 | 256 | 267 | 8/20 | 8:15 |
| 4 | M1 | 81 | 81 | 83 | 45.3 | 0.144 | 0.0117 | 688 | 77 | 255 | 271 | 8/20 | 8:15 |
| 5 | M1 | 83 | 81 | 82 | 45.5 | 0.146 | 0.0108 | 687 | 77 | 255 | 268 | 8/20 | 8:15 |

| Variable | Description |
|----------|-------------|
| Ball | Keeps track of the observation number and goes from 1 to 72 |
| Model | Identifies ball design, M1 and M2 |
| S1, S2, S3 | Measurement of circumference at three different points |
| Wgt | Weight of the ball |
| Dw, Dd | Dimple width and depth |
| Head | Head speed of the ball when it is hit |
| Temp | Environmental temperature in degrees Fahrenheit |
| Carry | Distance from the point the ball was hit to the point where it first hit the ground, measured in yards |
| TotDist | Total distance the ball travels from the point where it was hit to its final position |
| Date | Date of the testing |
| Time | Time of day that the observation took place. There were 3 different time periods. |

(a) For each model number, calculate the mean, median, mode, range, and standard deviation of the variable *TotDist.*

(b) Describe the distribution of *TotDist* for each ball. How does *TotDist* compare with *Carry?* Which measure do you think is better from a statistical point of view?

(c) What can you tell the company about the effects of temperature during the trials?

(d) What can you say about the variable *Head* over the entire trial? Is it consistent for both ball models? Do you think that head speed needs to be considered as a variable in this trial or is it controlled well enough?

(e) Would you expect there to be much difference in performance from one time period to the next? Examine the effects of *Time* for each model separately and describe what you see. Make sure you look at both *Carry* and *TotDist.*

(f) Look at the variables *S1, S2,* and *S3.* The closer these measures are to being equal, the closer to spherical the ball is. Do you think that the balls are spherical? Why or why not?

(g) Look at *Dw* and *Dd.* How do these compare for each ball type?

(h) Think about the problem and analyze the data in any other way that you think might be interesting or important. Prepare a report for the company describing the behavior of each ball type and comparing the two balls. Include any graphs that you think might be useful in illustrating your findings.

**4.47** Remember the complaint data from the tissue manufacturer that we looked at in the last chapter? Recall that one of the largest categories of customer complaints involved sheets tearing on removal. One of the product variables that can cause tissues to tear is tensile strength. The manufacturer has decided to look at the manufacturing process and to investigate the tensile strength of the tissues. Two hundred samples are taken from tissues produced on a single-tissue machine. The samples are taken over three different days. A portion of the data file is shown below:

*Datafile: TISSUES.XXX*

| Day | MDStrength | CDStrength |
|-----|------------|------------|
| 1 | 1006 | 422 |
| 1 | 994 | 440 |
| 1 | 1032 | 423 |
| 1 | 875 | 435 |
| 1 | 1043 | 445 |

*Day* keeps track of the day on which the sample was taken and goes from 1 to 3. *MDStrength* measures the machine-directional strength and is measured in lb/ream. *CDStrength* measures the cross-directional strength and is measured in lb/ream.

(a) Calculate a set of summary statistics for *MDStrength* for the entire three-day period.

(b) Use the summary statistics to describe the *MDStrength* of the tissues. Be sure to comment on central tendency, variability, shape, and outliers.

(c) Display the data for *MDStrength* for the three-day period graphically. How does the graph agree with or disagree with your answer to part (b)?

(d) Calculate a set of summary statistics for each day separately. How does the distribution of *MDStrength* for each of the three days compare to the overall distribution?

(e) Suppose that the *MDStrength* for the product is supposed to have a mean of 1050 and a standard deviation of 50. Do you think that the manufacturing product is meeting these specifications?

(f) Repeat parts (a)–(e) for the variable *CDStrength.* The specifications for *CDStrength* are a mean of 450 and a standard deviation of 25. In addition to a target mean and standard deviation, each variable has defect limits. For example, *MDStrength* should be between 850 and 1075. If the strength is too low the tissue snaps and causes the machine to shut down. If it is too high the tissues are stiff. The *CDStrength* should be between 390 and 480. If the strength is too low then the tissues will tear when you pull on them to remove them from the box. If it is too high, they will be stiff.

(g) Prepare a report for the management of the tissue company telling them how the actual process compares to the specifications. Include any appropriate graphs or tables of data with the report.

# CHAPTER *5*

# *ANALYZING BIVARIATE DATA*

## ALUACHA BALACLAVA COLLEGE

The Provost at Aluacha Balaclava College decides that there is more to the faculty salary picture than can be seen in the set of data she has currently collected. She decides to look at some additional variables that may help her to understand the problem a little more. She wonders if the department or school in which the faculty member is employed has an effect on salary, so she decides to look at the data for both variables.

In addition, she decides to collect data on gender and whether or not the faculty member has tenure at the university. Perhaps these additional variables will give her some insight into how the faculty is paid. A portion of the data is shown below:

| Rank | Years of service | Salary | Gender | Tenure? | Schl |
|------|------------------|--------|--------|---------|------|
| ASST | 22 | 53316 | F | Y | BUSINESS |
| PROF | 11 | 64375 | M | Y | BUSINESS |
| ASSO | 7 | 63501 | M | Y | BUSINESS |
| ASSO | 6 | 59426 | M | N | BUSINESS |
| ASSO | 20 | 49058 | M | Y | BUSINESS |
| PROF | 4 | 94969 | M | N | BUSINESS |
| ASST | 21 | 54762 | M | Y | BUSINESS |

## 5.1   CHAPTER OBJECTIVES

In Chapters 3 and 4 you learned how to summarize data using both graphical and numerical measures. You learned how to use these techniques to obtain *information* about the data and to use that information to help in making decisions. In this chapter we look at some additional methods for summarizing data and also at how to display and analyze *bivariate* data. The chapter covers the following material:

- Bivariate Qualitative Data: contingency tables, clustered and stacked bar charts
- Identifying Quantitative Relationships: scatter plots and the least-squares line

## 5.2   QUALITATIVE BIVARIATE DATA

In the previous chapters of this book we have looked at ways to organize, summarize, and describe data that represent a *single* variable or characteristic of the population under study. Most often when data are collected, multiple variables are involved. The reason for doing this is that very often phenomena that occur in one variable are connected to or can be explained by the values of other, related variables. In this section we discuss methods for summarizing and displaying data for two variables, known as *bivariate data.*

### 5.2.1 Summarizing Two Qualitative Variables

In Chapter 3 you learned that the first step in organizing and summarizing data for a single variable is to create a frequency table. The frequency table contains the number of occurrences of each value of the variable expressed as a count, fraction, decimal, or percentage. When data are collected on two *related* variables, they are organized by using a **cross-classification table** or **contingency table.**

> A *contingency table* is a table whose rows represent the possible values of one variable and whose columns represent the possible values for a second variable. The entries in the table are the number of times that each *pair* of values occurs.

Creating a contingency table for a set of bivariate data is similar to creating a frequency table for a single variable. The basic setup of a contingency table is shown in Table 5.1. Each category for one of the variables is represented by a column, while

**TABLE 5.1   Layout of Contingency Table**

|  |  | Variable 2 | | | | |
|---|---|---|---|---|---|---|
|  |  | Category 1 | Category 2 | . . . | Category $n$ | Totals |
| Variable 1 | Category 1 | $f_{11}$ | $f_{12}$ | . . . | $f_{1n}$ |  |
|  | Category 2 | $f_{21}$ | $f_{22}$ | . . . | $f_{2n}$ |  |
|  | . | . | . | . | . |  |
|  | . | . | . | . | . |  |
|  | . | . | . | . | . |  |
|  | Category $m$ | $f_{m1}$ | $f_{m2}$ | . . . | $f_{mn}$ |  |
|  | Total |  |  |  |  |  |

each category for the other variable is represented by a row. The numbers in the table, the $f_{ij}$'s, represent the number of observations in the data set that were in category $i$ of variable 1 *and* category $j$ of variable 2.

By using a contingency table, we can look at the number, proportion, or percentage of the data that fall into each *pair* of values for variable 1 and variable 2. Adding a second variable to an analysis gives you a different, and sometimes valuable, perspective on the data you have collected.

## EXAMPLE 5.1   ABC Faculty Salary

*Creating a Contingency Table*

Since she is still convinced that rank should be a factor in determining salary, the provost decides to look at the school in which the faculty member is employed in addition to rank. The college has four schools: Business, Liberal Arts, Professional Studies, and Science. The ranks of the faculty are Professor, Associate Professor, Assistant Professor, and Instructor. She creates a contingency table for the data and finds the following:

| Rank | Business | Liberal arts | Professional studies | Science | Total |
|------|----------|--------------|----------------------|---------|-------|
| Professor | 6 | 27 | 7 | 15 | 55 |
| Associate Professor | 11 | 30 | 9 | 17 | 67 |
| Assistant Professor | 10 | 20 | 15 | 32 | 77 |
| Instructor | 0 | 3 | 3 | 2 | 8 |
| Total | 27 | 80 | 34 | 66 | 207 |

Making comparisons with the actual frequencies is difficult, so she decides to use the relative frequencies for each cell. The relative frequency table is shown below:

| Rank | Business (%) | Liberal arts (%) | Professional studies (%) | Science (%) | Totals (%) |
|------|--------------|------------------|--------------------------|-------------|------------|
| Professor | 3 | 13 | 3 | 7 | 27 |
| Associate Professor | 5 | 14 | 4 | 8 | 32 |
| Assistant Professor | 5 | 10 | 7 | 15 | 37 |
| Instructor | 0 | 1 | 1 | 1 | 4 |
| Totals | 13 | 39 | 16 | 32 | 100 |

The provost sees that the distribution of faculty over the schools is not as uniform as the distribution over ranks. The Schools of Liberal Arts and Sciences make up 71% of the faculty, while Business and Professional Studies make up only 29%. She wonders if this has any impact on the distribution of faculty salaries. ■

## EXAMPLE 5.2   Student Distribution

*Creating a Contingency Table*

When the faculty of the School of Business looked at the data on students taking the Introductory Statistics course, they wondered if they could identify a reason why so many seniors were taking the course. They went back to the class and asked again for the information on year, but also asked a few more questions to help them classify the student. One of the questions they asked was whether or not the student was a transfer student. They guessed that students who transferred into the college might not have had the course previously. The data that they collected looked like this:

| | | | | | | |
|---|---|---|---|---|---|---|
| Fr | Transfer | So | Nontransfer | So | Nontransfer | Sr | Transfer |
| Fr | Nontransfer | So | Nontransfer | So | Transfer | Sr | Nontransfer |
| Fr | Nontransfer | So | Nontransfer | Jr | Nontransfer | Sr | Transfer |
| Fr | Nontransfer | So | Nontransfer | Jr | Nontransfer | Sr | Transfer |
| So | Nontransfer | So | Transfer | Sr | Transfer | Sr | Transfer |
| So | Nontransfer | So | Nontransfer | Sr | Transfer | Sr | Transfer |
| So | Nontransfer | So | Nontransfer | Sr | Transfer | Sr | Transfer |

To make sense of the data they decided to create a contingency table using the students' year as the row variable and the transfer status as the column variable. The table is shown below:

| | Status | | |
|---|---|---|---|
| **Year** | **Nontransfer** | **Transfer** | **Total** |
| Fr | 3 | 1 | 4 |
| So | 10 | 2 | 12 |
| Jr | 2 | 0 | 2 |
| Sr | 1 | 9 | 10 |
| Total | 16 | 12 | 28 |

From the table they saw that nine of the seniors who were taking the course were indeed transfer students. To make the table more general, they decided to create the same table expressing each frequency as a percentage of the total:

| | Status | | |
|---|---|---|---|
| **Year** | **Nontransfer** | **Transfer** | **Total (%)** |
| Fr | 11 | 4 | 14 |
| So | 36 | 7 | 43 |
| Jr | 7 | 0 | 7 |
| Sr | 4 | 32 | 36 |
| Total | 57 | 43 | 100 |

From the second table they learned that, in fact, 43% of the students in the class were sophomores, which is supposed to be the norm. They also learned that senior transfer students constituted 32% of the total students enrolled in the course. Since the table also indicated that 43% of the students in the course were transfer students, they decided that establishing a task force to study transfer students was probably a good idea. ■

Creating contingency tables from large data sets can be tedious and time-consuming. Often, when data are collected, many more than two variables are considered and the person doing the analysis wants to look at different pairs of variables. Almost any software package that can be used for statistical analyses can create contingency tables easily. It is the *information* obtained from the tables that is important.

 **TRY IT NOW!**

## Quality Problems
*Creating a Contingency Table*

A company that manufactures cardboard boxes is trying to understand some of its quality problems. The company has analyzed some data and determined that the major defects are printing, color, and skewness (how square the box is). Further attempts to pinpoint the problems have resulted in many opinions and finger pointing about responsibility. The company decides to collect some additional data on defect type and the shift during which production occurred. The data (sorted by defect type) are as follows:

| | | | | | |
|---|---|---|---|---|---|
| Color | 1 | Color | 2 | Printing | 3 |
| Color | 2 | Color | 1 | Printing | 1 |
| Color | 1 | Color | 1 | Printing | 2 |
| Color | 3 | Color | 3 | Printing | 1 |
| Color | 3 | Color | 1 | Printing | 2 |
| Color | 2 | Printing | 3 | Printing | 1 |
| Color | 1 | Printing | 2 | Printing | 2 |
| Color | 1 | Printing | 1 | Printing | 1 |
| Color | 1 | Printing | 2 | Skewness | 2 |
| Color | 2 | Printing | 1 | Skewness | 2 |

Create a relative frequency contingency table for the data.

What percentage of the defects were color?

Does there appear to be any credence to the claim that the majority of the defects occur on the third shift?

ANS.

| Defect | Shift (%) 1 | 2 | 3 | Total % |
|---|---|---|---|---|
| Color | 27 | 13 | 10 | 50 |
| Printing | 20 | 17 | 7 | 43 |
| Skewness | 0 | 7 | 0 | 7 |
| Total | 47 | 37 | 17 | 100 |

## 5.2.2 Displaying Contingency Tables

While there is certainly a great deal of information to be obtained from a contingency table, we learned in Chapter 3 that a visual representation of the data can strengthen our understanding of what is happening. The question then is how can we best represent bivariate data.

If we select one of the variables it would be possible to create bar or pie charts for each of the possible categories of that variable. For the student distribution data in Example 5.2, the corresponding bar charts are shown in Figure 5.1. This accomplishes what we wanted, but it will certainly be cumbersome if the number of possible categories increases! Also, as you learned in Chapter 3, when you are making comparisons on separate charts it is important that the scales of the charts be comparable.

**FIGURE 5.1**   Two bar charts representing a contingency table

## Clustered and Stacked Bar Charts

There is a way to display contingency table data on the same chart, by using what are called *clustered* or *stacked* bar charts. Both of these charts use a different type of bar (color, shading) to represent one of the values of a selected qualitative variable.

## Clustered Bar Charts

**Clustered bar charts** can be used in place of multiple bar charts when you want to compare the data for the values of the selected qualitative variables. An example of a clustered bar chart is shown in Figure 5.2.

> In a *clustered bar chart,* the bars for one variable are grouped according to the values of the other qualitative variable.

**FIGURE 5.2**   Example of a clustered bar chart

You can see that a clustered bar chart really *combines* the results of several bar charts into a single chart. A legend is included to identify each of the categories.

### EXAMPLE 5.3 ABC Faculty Salaries

*Creating a Clustered Bar Chart*

The provost is interested in seeing how the faculty of the different schools are distributed over the ranks. She decides to create a clustered bar chart using rank as the category on the *x* axis and a different bar for each school:

Looking at the bar chart, the provost sees that the faculty in the School of Liberal Arts tends toward the higher ranks, while the School of Science has a higher percentage of the faculty in the lower ranks. The School of Business appears to be similar to Liberal Arts, while Professional Studies is similar to Science. ∎

### EXAMPLE 5.4 Student Distribution

*Creating a Clustered Bar Chart*

In the Student Distribution example the faculty wanted to compare transfer students to nontransfer students. They decided that to get a good picture of the data they would use a clustered bar chart to display the data they collected. Since the variable that they wanted to compare was Student Status, they made a bar for each of the two student types for each of the four years. There are really no new calculations to do. The chart that resulted is shown below:

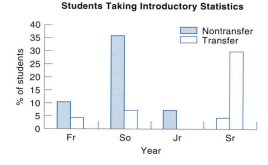

From the chart the faculty could see that transfer students take the statistics course primarily as seniors, while nontransfer students take the course primarily as sophomores. ∎

## TRY IT NOW!

## Quality Problems

### Creating a Clustered Bar Chart

The management of the company that manufactures cardboard boxes decide to have a meeting with the crews about the quality problems. The managers decide that a graphical display of the data is the best way to convey the information they have obtained and that they want to display the crew data for each of the quality defects.

| | | Shift (%) | | |
|---|---|---|---|---|
| **Defect** | **1** | **2** | **3** | **Total (%)** |
| Color | 27 | 13 | 10 | 50 |
| Printing | 20 | 17 | 7 | 43 |
| Skewness | 0 | 7 | 0 | 7 |
| Total | 47 | 37 | 17 | 100 |

Create a clustered bar chart that displays the percent defects for each shift using the quality problems as the categories for the x axis.

Which shift contributes the largest percentage of color defects? the least percentage?

What do you notice about skewness defects?

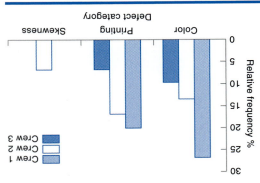

ANS. CREW 1, CREW 3, MADE ONLY BY CREW 2

## Stacked Bar Charts

In Chapter 3 you learned that when you want to look at different categories as part of a whole, a pie chart can be more informative than a bar chart. For bivariate data, instead of using multiple pie charts you can use a **stacked bar chart.**

> In a *stacked bar chart* the data for the selected variable are represented as a percentage of the total for each category of the second variable. Each value of the selected variable is represented in a different way and the bars are "stacked" to total 100%.

An example of a stacked bar chart is shown in Figure 5.3.

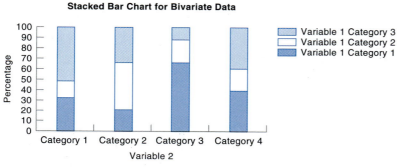

**FIGURE 5.3**   Stacked bar chart

## EXAMPLE 5.5   ABC Faculty Salaries

*Creating a Stacked Bar Chart*

After looking at the clustered bar charts, the provost is interested in how the different schools are made up by rank of faculty. She decides she needs a stacked bar chart to look at this, and so she must modify the original contingency table shown below:

| Rank | Business | Liberal arts | Professional studies | Science | Total |
|------|----------|--------------|----------------------|---------|-------|
| Professor | 6 | 27 | 7 | 15 | 55 |
| Associate Professor | 11 | 30 | 9 | 17 | 67 |
| Assistant Professor | 10 | 20 | 15 | 32 | 77 |
| Instructor | 0 | 3 | 3 | 2 | 8 |
| Total | 27 | 80 | 34 | 66 | 207 |

Since what she wants to know is, for example, what percentage of the School of Business is Professors, she needs to calculate each frequency as a percentage of the column total. This calculation is found to be

$$6/27 = 22.2\% \text{ or approximately } 22\%$$

The next calculation is for the rank of Professor in the School of Liberal Arts. This is found by $27/80 = 33.8\%$ or approximately 34%. The completed table is shown below:

| Rank | Business (%) | Liberal arts (%) | Professional studies (%) | Science (%) | Total (%) |
|---|---|---|---|---|---|
| Professor | 22 | 34 | 21 | 23 | 27 |
| Associate Professor | 41 | 38 | 26 | 26 | 32 |
| Assistant Professor | 37 | 25 | 44 | 48 | 37 |
| Instructor | 0 | 4 | 9 | 3 | 4 |
| Total | 100 | 100 | 100 | 100 | 100 |

The stacked bar chart for this table is shown below:

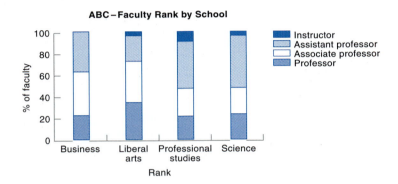

She sees from the graph that the schools of Science and Professional Studies have a higher proportion of Assistant Professors than the other schools and that Liberal Arts has the highest percentage of Professors. ■

### EXAMPLE 5.6   Student Distribution

#### Creating a Stacked Bar Chart

The faculty of the School of Business decide to look at the data one more time. They want to see if they can determine the proportion of transfer students enrolled in the course that are Freshmen, Sophomores, Juniors, and Seniors. The first task is to figure out the percentages for each column of the contingency table. Since they are trying to express the year as a percentage of the type of student rather than as a percentage of the total, they use the column totals instead of the grand total to calculate them. The original contingency table is shown below:

| Year | Status Nontransfer | Transfer | Total |
|---|---|---|---|
| Fr | 3 | 1 | 4 |
| So | 10 | 2 | 12 |
| Jr | 2 | 0 | 2 |
| Sr | 1 | 9 | 10 |
| Total | 16 | 12 | 28 |

The number of nontransfer students who are freshmen was 3. To express this as a percentage of the nontransfer students calculate $3/16 = 18.75\%$, or approximately 19%.

In the same way, to find the percentage of transfer students that are freshmen calculate $1/12 = 8.33\%$ or approximately 8%. The completed table is shown at the top of the next page.

|  | Status (%) | |
|---|---|---|
| **Year** | **Nontransfer** | **Transfer** |
| Fr | 19 | 8 |
| So | 63 | 17 |
| Jr | 13 | 0 |
| Sr | 6 | 75 |
| Total | 100 | 100 |

The stacked bar chart for the data is

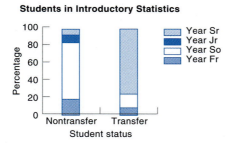

This chart shows that the majority of nontransfer students enrolled in the course are freshmen and sophomores, which is appropriate. The majority of the transfer students are seniors and there are no juniors. This information reinforces the earlier idea that better advising of transfer students might alleviate the problem. ■

 **TRY IT NOW!**

### Quality Problems
*Creating a Stacked Bar Chart*

The company that manufactures cardboard boxes decides to create a graphical display of the data to show to the employees. What the company wants to do is to display type of defect as a percentage of total defects for each shift. It is hoped that these data will help each shift concentrate on its own priorities. The contingency table for the defect data is shown below:

|  | Shift | | | |
|---|---|---|---|---|
| **Defect** | **1** | **2** | **3** | **Total** |
| Color | 8 | 4 | 3 | 15 |
| Printing | 6 | 5 | 2 | 13 |
| Skewness | 0 | 2 | 0 | 2 |
| Total | 14 | 11 | 5 | 30 |

Modify the contingency table to display defect types as a percentage of total defects for each shift.

(*continued*)

Create a stacked bar chart for the data.

Which defect type should each crew concentrate on? Why?

## 5.2.3 Exercises—Learn It!

**5.1** Tracking public opinion about the president of the United States is a favorite pastime of the news media. In August 1995, the *Wall Street Journal* reported on public attitude toward then President Bill Clinton. They reported the following data:

| Attitude | Time Period (%) | | |
|---|---|---|---|
| | **Aug 95** | **Jan 95** | **Oct 94** |
| Very positive | 15 | 15 | 16 |
| Somewhat positive | 34 | 29 | 26 |
| Neutral | 17 | 18 | 16 |
| Somewhat negative | 17 | 20 | 16 |
| Very negative | 17 | 17 | 25 |

(a) Create a clustered bar chart for the data using the variable *Attitude* as the categories for the $x$ axis.

(b) Redo the chart using *Time Period* as the *x*-axis category.

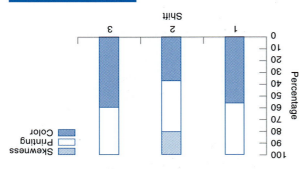

**Defect Data for Cardboard Box Company**

| Defect | Shift (%) | | | Total (%) |
|---|---|---|---|---|
| | **1** | **2** | **3** | |
| Color | 57 | 36 | 60 | 50 |
| Printing | 43 | 45 | 40 | 43 |
| Skewness | 0 | 18 | 0 | 7 |
| Total | 100 | 100 | 100 | 100 |

ANS. CREW 1 COLOR; CREW 2 PRINTING; CREW 3 COLOR

(c) Which chart do you think does a better job of giving information about how attitudes have changed over time? Why?

(d) Using the data, describe the public's attitude toward President Clinton over the time period involved.

**5.2** The following data were published on deaths (per 100,000 people) for various diseases by race and gender:

| Disease | White males | Black males | White females | Black females |
|---|---|---|---|---|
| Heart disease | 190.3 | 264.1 | 98.1 | 162.4 |
| Prostate cancer | 15.7 | 35.8 | NA | NA |
| Breast cancer | NA | NA | 21.7 | 27.0 |
| Diabetes | 11.6 | 24.2 | 9.6 | 25.8 |
| HIV infection | 18.1 | 61.8 | 1.6 | 14.3 |
| Chronic liver disease | 11.1 | 17.2 | 4.6 | 6.9 |

*Note.* NA, data not available

(a) Display the data on a clustered bar chart using disease as the $x$-axis category.

(b) Comment on any interesting features in the data.

(c) Do you think that the data would be better displayed with race/gender as the $x$-axis categories? Why or why not?

(d) Why would a stacked bar chart not be appropriate for these data?

**5.3** In recent years many prescription drugs have shifted to over-the-counter status as a means of increasing profit margins for pharmaceutical companies. A study of data reveals the following:

| | Sales ($ million) | | |
|---|---|---|---|
| Drug | Before switch* | 1993 | 1994 (projected) |
| Actifed | 20 | 41 | 40 |
| Benadryl | 17 | 130 | 155 |
| Imodium A-D | 20 | 82 | 84 |
| Monistat 7 | 80 | 130 | 105 |

\* Last 12 months preceding the switch from prescription to OTC

(a) Create a clustered bar chart using the drugs as the $x$-axis categories.

(b) Comment on the idea that switching from prescription to nonprescription status greatly increased sales.

(c) Why would a chart with the time periods on the $x$ axis not be as informative?

(d) Why is making comparisons among the drugs not appropriate here?

**5.4** The relationship between cities and the economic health of countries is of interest to city planners. The following data show where cities get revenues for eight different developing countries:

| | Source of funding (%) | | |
|---|---|---|---|
| Country | Central transfers | Local taxes | Local fees |
| Indonesia | 84 | 7 | 9 |
| Brazil | 65 | 24 | 11 |
| Mexico | 63 | 12 | 25 |
| Turkey | 63 | 10 | 27 |
| Tunisia | 56 | 34 | 10 |
| Colombia | 42 | 46 | 12 |
| India | 27 | 61 | 12 |
| Kenya | 4 | 38 | 58 |

(a) Use a stacked bar chart to display the data. The $x$-axis categories should be the countries.

(b) What interesting features do you see in the data?

## 5.3   QUANTITATIVE BIVARIATE DATA

In the previous section you learned ways to investigate the relationships among *qualitative* variables. It is also interesting to understand how *quantitative* variables are related.

When we say that two variables are *related* we mean that they vary in some systematic way. For example, as one variable increases, the other might also increase. This would be true for variables such as height and weight, and it might or might not be true for variables such as the amount of time you study and the grade you earn on an exam. Another type of relationship occurs when as one variable increases the other decreases. This might apply to the relationship between the price of a product and sales of that same product.

In statistics, it is often important to investigate the relationships between variables in order to learn how to predict the value of one variable based on the value of the other, related variable. This is called *regression analysis* and is covered in Chapter 11. But, just as in all statistical studies, before any analytical tool is used it is important to look at the data first to get an understanding of what is happening. Viewing the data with different graphical tools helps a statistician to decide what statistical techniques are appropriate. The most important tool for exploring relationships between quantitative variables is a graphical tool called a **scatter plot.**

### 5.3.1 Scatter Plots

Each observation of bivariate quantitative data can be thought of as a data pair of the form $(x, y)$, where $x$ represents the values of the first variable and $y$ represents the value of the second variable. The $x$ variable is often called the **independent** variable, and the $y$ variable is often called the **dependent** variable. The type of graph used to display this kind of data is called a **scatter plot.**

> In a *scatter plot* an axis is used to represent each of the variables and the data are plotted as points on the graph. Typically, the *independent variable* is plotted on the $x$ axis and the *dependent variable* is plotted on the $y$ axis.

An example of a scatter plot is shown in Figure 5.4. The plot gives information about whether or not a relationship exists between two variables and what the form of that relationship might be. From the plot it appears that as the value of $X$ increases, so does the value of $Y$. Of course, what we know about variation in data tells us that the relationship will not be perfect. If there is a strong enough relationship between the two variables, then it should be apparent even in the presence of normal variation.

Creating a scatter plot is not difficult. Once you have selected which variable is the *independent* variable and which is the *dependent* variable, all that needs to be done

**FIGURE 5.4**   Scatter plot

is to draw a set of axes and scale them so that you can plot all of the data. You might have learned in a mathematics class at some time that when you create a graph on a pair of *x-y* axes the scales should both start at 0. While this is often desirable in mathematics, it may not be suitable for a scatter plot in statistics. If your data are very large, then starting both axes at zero might squash the graph into the top portion and actually hide the relationship. For this reason, we usually pick the starting points on the axes to be some reasonable number that is smaller than the smallest *x* and *y* values. When the axes are drawn, the data pairs are plotted as points on the graph.

### EXAMPLE 5.7    ABC Faculty Salaries

#### Looking for Relationships

The provost who is looking at faculty salaries is still convinced that there should be a relationship between the number of years of service to the university and faculty salary. She feels that it is time to look at these two variables together. Since both variables are quantitative, she knows that the first step in understanding the relationship will be to make a scatter plot of salary versus the number of years of service. Since she thinks that salary depends on years of service, she picks salary for the *y* (dependent) variable and years of service for the *x* (independent) variable. The plot she obtains is shown below:

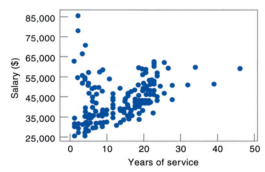

She looks at the plot and knows that she is basically correct—that salary and years of service are related and that the relationship is, for the most part, linear. She sees, however, that many data points do not quite fit the picture. She guesses that some of the other, qualitative variables that she has been looking at might be of interest.    ■

### EXAMPLE 5.8    Sales and Advertising

#### Creating a Scatter Plot

A pharmaceutical company is interested in determining whether there is a relationship between the amount of money spent in advertising and factory sales for a group of over-the-counter products. It seems reasonable to think that spending more on advertising will increase sales, but the company is not sure. The managers collect some data and find the following:

| Advertising (millions of dollars) | 143.8 | 91.7 | 43.8 | 26.7 |
| Factory sales (millions of dollars) | 855 | 360 | 170 | 130 |

They decide to create a scatter plot to look at the data. Since they are expecting the amount of sales to depend on the amount spent on advertising, they pick sales to be the dependent (*y*) variable and advertising expenditures to be the independent (*x*) variable. After looking at the data they choose to scale the *x* axis from $0 to $150 million and the y axis from $100 to $900 million. They then plot each of the four data points on the graph.

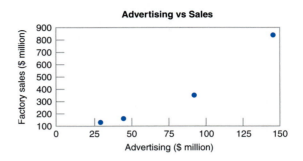

From the graph they see that their idea was correct, and that as advertising expenditures increase so do sales. ■

# TRY IT NOW!

## Airline Fares
### Creating a Scatter Plot

A travel agency was interested in knowing how airline fares were related to the length of the flight in miles. The agency hypothesized that the longer the flight, the more the airfare. The following data were collected:

| Miles | 2375 | 1400 | 1250 | 2325 | 985 | 2025 |
|-------|------|------|------|------|-----|------|
| Airfare ($) | 430 | 272 | 252 | 422 | 207 | 373 |

Which variable is the dependent variable and which is the independent variable?

Make a scatter plot of the data.

(*continued*)

From your plot, do you think that the travel agency's hypothesis was correct? Why or why not?

## 5.3.2 Types of Relationships

Once we determine that two variables are related, we often want to find an equation describing the relationship. This equation shows how the variables are related and can then be used to predict values of the dependent variable for different values of the independent variable. The purpose of a scatter plot is to give a visual clue about whether a relationship exists between two quantitative variables. The scatter plot will also give information about the type of relationship, which can be used to determine the correct method for finding the equation that relates the variables.

When two variables are related, the scatter plot forms some type of pattern that characterizes the form of the relationship. In Figure 5.5 you see some of the different types of relationships that can exist between variables.

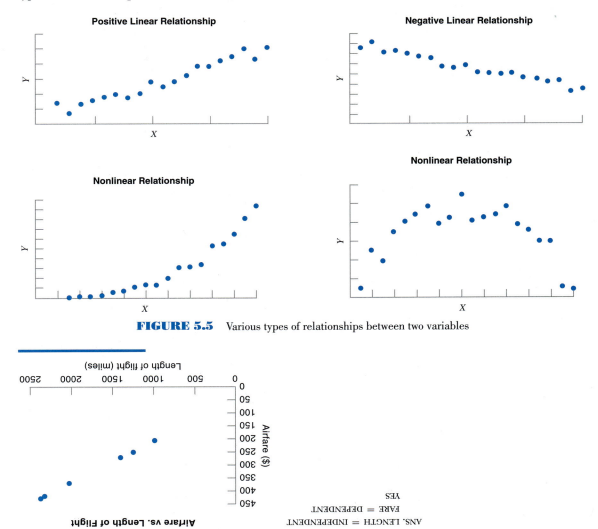

**FIGURE 5.5** Various types of relationships between two variables

When two variables are not related, the scatter plot forms no pattern, as is seen in Figure 5.6.

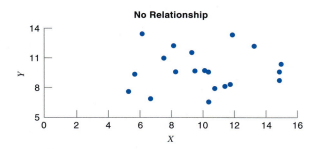

**FIGURE 5.6**   Scatter plot indicating no relationship between two variables

**EXAMPLE 5.9**   **Training and Performance**

*Finding the Least-Squares Line*

A company is interested in determining whether there is a relationship between the number of days that it spends training employees at a particular job and the employees' performance as measured on a standardized test. The managers collect the following data:

| Training days | Score |
| --- | --- |
| 1.0 | 41 |
| 1.5 | 60 |
| 2.0 | 72 |
| 2.5 | 91 |
| 3.0 | 99 |

The first thing they do is display the data on a scatter plot to determine whether the two variables are related. Since they think that the score depends on the number of training days, they plot the score on the $y$ axis and the number of training days on the $x$ axis.

The plot confirms their hypothesis and indicates that a linear relationship is appropriate.

**TRY IT NOW!**

### Starting Salaries and Math Courses

*Finding the Least-Squares Line*

The Career Planning office of a large university is interested in knowing whether there is a relationship between the starting salary of graduates and the number of mathematics courses the graduates had taken as students. The Career Planning office goes through the records for the last year and finds the following data.

| Number of math courses | Starting salary ($) |
|:---:|:---:|
| 1 | 26,284 |
| 1 | 25,470 |
| 2 | 26,777 |
| 3 | 27,269 |
| 4 | 28,553 |
| 6 | 30,054 |

Which variable is the independent variable? the dependent variable?

Use the grid below to create a scatter plot of the data.

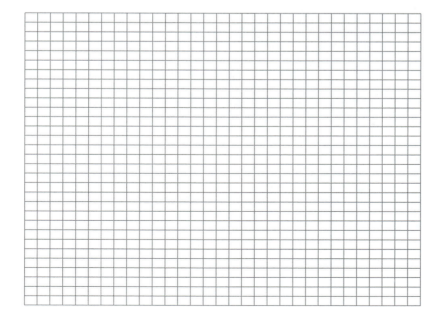

(*continued*)

Do you think that there is a linear relationship between the number of math courses taken and the starting salary? If so, describe the relationship.

 ## *Discovery Exercise 5.1*
## *DISCOVERING RELATIONSHIPS*

### Part I

1. The data below represent data taken on the number of weeks that a student was enrolled in a speed reading program and the speed gain in words per minute that the student has experienced:

| Weeks | 2 | 3 | 4 | 6 | 8 |
|---|---|---|---|---|---|
| Speed gain | 49 | 86 | 109 | 164 | 193 |

2. Plot the data on the graph paper below. Be sure to label your axes and to construct the graph so that it uses most of the paper.

ANS. DEPENDENT—SALARY; INDEPENDENT—NUMBER OF MATH COURSES; YES, AS NUMBER OF MATH COURSES INCREASES, SO DOES SALARY.

3. Draw a straight line through the points that you think best represent the relationship between $x$ and $y$. What criteria did you use for drawing the line you selected?

**Part II**

Use your *line and the graph* to answer the following questions. Do NOT use the data given:

1. What is the speed gain for a person who has spent 4 weeks in the program as predicted by the line?

2. If you were to select a person at random who had spent 3 weeks in the program, what would your prediction of that person's speed gain be?

3. By how much would you predict speed gain will change for a 1-week increase in time in the program?

4. Draw the line whose equation is $y = 10.1 + 23.9x$ on the graph.

(*continued*)

5. Answer the prediction questions you previously answered using the new, "best" line.

6. Agree that my line is better than yours (if it is different).

7. What do you think would happen if you used the line to predict the speed gain of a person who had been in the program for 1 week? 0 weeks? Is this reasonable?

8. What problems can you see with what we have done?

### 5.3.3 Least-Squares Line

If the relationship between the variables looks linear (like a straight line), then a curve-fitting technique called least squares can be used to find the equation of the line that best fits the data. In Chapters 11 and 12 you will use least squares in more depth in a statistical technique called regression analysis. At this point we use least squares to find the equation relating $X$ and $Y$.

In drawing a line to fit a set of points, we would like to satisfy some criteria. First, we would like the line that is drawn to be unique. It would certainly not be very useful if everyone who saw the same set of data used a technique that produced a *different* line! Second, we would like the line to be as close as possible to all of the points. In Figure 5.7 you see the data points and a line drawn to represent the relationship between the variables. The distances from each point to the line are called the *deviations* or *errors*. We will use **least squares** to find the equation of the line that best represents the relationship between the two variables.

> The *least squares* technique finds the equation of the line that minimizes the sum of the squared errors between the actual data points and the line.

The reasons for minimizing the sum of the squared deviations rather than just the deviations or the absolute deviations are mathematical and are discussed in more detail in Chapter 11.

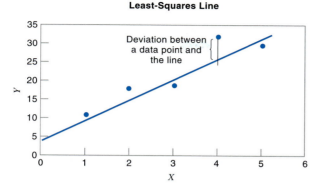

**FIGURE 5.7**   Deviations between the data points and the line

The *least-squares line* is defined by

$$\hat{y} = a + bx$$

The value $\hat{y}$ ($y$-hat) is the **predicted value** of $y$ for a selected value of $x$. We are also interested in the **deviation** of the data from the least-squares line.

The distance between the predicted value of $Y$, $\hat{y}$, and the actual value of $Y$, $y$, is called the *deviation* or *error*. The *deviation, e,* is equal to $\hat{y} - y$.

## Calculating the Least-Squares Line

The actual method of least squares finds the *slope* and the *y intercept* of the line $\hat{y} = a + bx$. The least squares estimates for the slope, $b$, and the $y$ intercept, $a$, are given by

$$b = \frac{n\sum XY - \sum X \sum Y}{n\sum X^2 - (\sum X)^2} \quad \text{and} \quad a = \frac{\sum Y}{n} - b\frac{\sum X}{n}$$

*Slope and Intercept of Least-Squares Line*

While these equations might seem too complicated to be true, if you look at them carefully you will see that really only five quantities have to be calculated. Four of the quantities are sums involving the data values $x$ and $y$, and the fifth is the number of observations in the sample. As an example, look at the quantity $\sum XY$. This quantity is the sum of the products of the $x$ and $y$ values for each observation. The easiest way to calculate all of the quantities involved is to make a table with a column for each sum needed. The table will look like the one in Table 5.2.

## TABLE 5.2   Table for Calculating the Least-Squares Line

| Observation number | X | Y | XY | X² |
|---|---|---|---|---|
| 1 | $x_1$ | $y_1$ | $x_1 * y_1$ | $x_1 * x_1$ |
| 2 | $x_2$ | $y_2$ | $x_2 * y_2$ | $x_2 * x_2$ |
| ⋮ | ⋮ | ⋮ | ⋮ | ⋮ |
| $n$ | $x_n$ | $y_n$ | $x_n * y_n$ | $x_n * x_n$ |
| **Totals** | $\sum X$ | $\sum Y$ | $\sum XY$ | $\sum X^2$ |

In the table each $x_i$ represents the value of $X$ for the $i$th data point, and each $y_i$ represents the $Y$ value of the $i$th data point. The first column keeps track of the observation number, the next two columns are the values of $X$ and $Y$ for the observation, the next column is the product of the $X$ and $Y$ value for each observation, and the last column is the square of the $X$ value of the observation.

### EXAMPLE 5.10   Training and Performance

*Finding the Least-Squares Line*

The company looking at the relationship between training and performance decides to use the least squares method to find the equation of the line that describes the relationship. The first thing the managers of the company do is to create the table shown below:

*When you are substituting the value for b into the equation for a, do NOT round it. Round all of the values AFTER the calculations are done.*

| Observation number | Training days ($X$) | Score ($Y$) | $XY$ | $X^2$ |
|---|---|---|---|---|
| 1 | 1.0 | 41 | 41.0 | 1.00 |
| 2 | 1.5 | 60 | 90.0 | 2.25 |
| 3 | 2.0 | 72 | 144.0 | 4.00 |
| 4 | 2.5 | 91 | 227.5 | 6.25 |
| 5 | 3.0 | 99 | 297.0 | 9.00 |
| Totals | 10.0 | 363 | 799.5 | 22.50 |

Substituting the values in the last row of the table into the formulas, they obtain

$$b = \frac{(5)(799.5) - (10)(363)}{(5)(22.5) - (10^2)} = \frac{367.5}{12.5} = 29.4$$

$$a = \frac{363}{5} - (29.4)\frac{10}{5} = 13.8$$

Using these values, they find the least squares line to be

$$\hat{y} = 13.8 + 29.4x$$

The equation tells them that the base score on the test is 13.8 points and that every day of training increases the score by 29.4 points.   ■

**TRY IT NOW!**

## Starting Salaries and Math Courses

*Finding the Least-Squares Line*

The Career Planning office at the university wants to use least squares to find the equation that relates the number of math courses taken and starting salary. The data are

*Many calculators that do statistical calculations also find the least-squares line. If you have such a calculator it is probably worthwhile to learn how to do this.*

| Number of math courses | Starting salary ($) |
|---|---|
| 1 | 26,284 |
| 1 | 25,470 |
| 2 | 26,777 |
| 3 | 27,269 |
| 4 | 28,553 |
| 6 | 30,054 |

(a) Find the equation for the least-squares line for the data. (*Hint:* Create a table like the one in Example 5.10 or use a computer software package.)

(b) Explain what the least-squares line tells the Career Planning Office about math courses and starting salary.

When you obtain the least-squares line it is often useful to plot the line on the same plot as the original data. This allows you to see how well the line fits the data values. Since two points determine a straight line, the easiest way to plot the line is to substitute two different values for $X$ into the least-squares equation and solve for $\hat{y}$. Once you have two pairs of points you can plot them and draw the line.

### EXAMPLE 5.11   Training and Performance

***Plotting the Least-Squares Line***

The company that is looking at performance scores and training days decides to plot the data and the line it had found on the same graph to see how well the line represents the data. To do this the company chooses two different values for $X$, 1 and 3 training days, and uses the equation to find the predicted scores. The following equation is obtained:

$$\hat{y} = 13.8 + 29.4(1) = 43.2 \quad \text{and} \quad \hat{y} = 13.8 + 29.4(3) = 102$$

Thus, the least-squares line predicts that a person with 1 day of training would score 43.2 points and a person with 3 days of training would score 102 points. The company plots these points on the graph with the original data and draws the line. The resulting plot looks like

**Least-squares line for Training vs Performance**

$Y = 13.8 + 29.4X$

The plot shows that the line fits the data very well.  ■

## Starting Salaries and Math Courses

*Finding the Least-Squares Line*

Once the least-squares equation is obtained, the Career Planning Office wants to see whether the equation does a good job of predicting starting salary for a given number of math courses.

Plot the data and the least-squares line on the same graph.

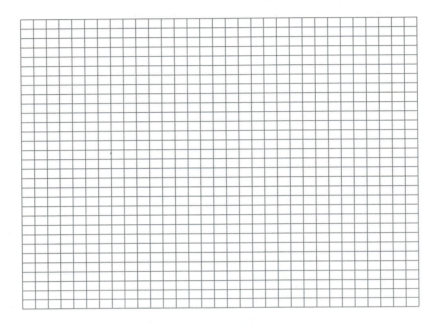

## *Interpolation and Extrapolation*

After finding the least-squares line, it might be reasonable to wonder when the equation can be used. When are the predictions useful? Are there any times when the predictions are not useful?

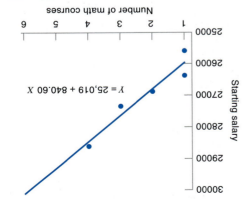

ANS.  $y = 25{,}019 + 840.60\,X$

Least-squares Line for Math Courses vs Salary

In the example on training days and performance score, the company used the equation of the least-squares line to predict scores for a person who was trained for 1 day and a person who was trained for 3 days. Notice that both of the values that they chose for $X$ were *within* the range of $X$ from the data they collected. This is called **interpolation.** There is another kind of prediction called **extrapolation.**

> Predicting values for $Y$ using values of $X$ that are outside the data range is called *extrapolation.* Finding values for $Y$ using $X$ values that are within the data range is called *interpolation.*

Interpolations are valid and they will produce good results if the line is a good fit. Predicting values for $Y$ using values of $X$ that are *outside* the data range, extrapolation, is usually quite dangerous. In reality, you have absolutely no knowledge of the relationship outside the range of the data and you have no business trying to predict it. It might be the case that the true relationship changes completely outside that range—perhaps as the number of days of training goes beyond a certain point, performance scores *decrease!* If you want to find a line to make predictions for certain values of the independent variable, it is necessary to collect data in that range.

 *TRY IT NOW!*

### Starting Salaries and Math Courses
*Finding the Least-Squares Line*

Use the equation you just found to predict the starting salary of a person who takes 2 mathematics courses. Do the same thing for the starting salary of a person who takes 5 mathematics courses.

Do you think the predictions from the least-squares line are useful? Why or why not?

Is what you just did interpolation or extrapolation?

(*continued*)

Now use the equation of the least-squares line to predict the starting salary of a person who takes no math courses. Do the same thing for a person who takes 10 math courses.

Do you think these predictions are valid? Why or why not?

## 5.3.4 Exercises

**5.5**  In a study about postage tariffs and revenue, The European Economic Community wanted to look at the relationship between the number of postal employees in a country and the amount of domestic mail that was processed. They collected data for six different countries and found:

| Country | Number of staff | Domestic traffic (billions of pieces) |
|---------|-----------------|----------------------------------------|
| Germany | 342,413 | 18.32 |
| France | 289,156 | 23.87 |
| Italy | 221,534 | 6.62 |
| Britain | 189,000 | 16.75 |
| Spain | 65,355 | 4.06 |
| Sweden | 52,251 | 4.21 |

Source: *The Economist,* 15 June 1996

(a) Create a scatter plot of the data and determine whether a linear relationship is appropriate.

(b) Find the equation of the least-squares line for the data.

(c) The Netherlands employs 53,560 postal workers. Use the least-squares equation to predict the domestic mail traffic for The Netherlands.

**5.6**  In trying to look at the effects of shopping center expansion, the Commerce Department decided to look at the relationship between the number of shopping centers and the retail sales for different states in the same region. The Department collected the data for the North Central states in the United States and found the following:

| State | Number of shopping centers | Retail sales ($ billion) |
|-------|----------------------------|--------------------------|
| Ohio | 1559 | 34.8 |
| Indiana | 848 | 18.0 |
| Illinois | 1961 | 34.1 |

(*continued*)

| Michigan | 967 | 21.0 |
| Wisconsin | 588 | 12.0 |
| Minnesota | 436 | 11.5 |
| Iowa | 277 | 6.3 |
| Missouri | 834 | 18.9 |
| North Dakota | 84 | 1.8 |
| South Dakota | 51 | 1.1 |
| Nebraska | 245 | 4.8 |
| Kansas | 458 | 9.7 |

Source: *Statistical Abstract of the United States 1995*

(a) Create a scatter plot of the data.

(b) Find the equation of the least-squares line relating retail sales and number of shopping centers.

(c) Plot the least-squares line and the data on the same plot. Do you think the line fits the data well? Why or why not?

(d) Use the least-squares equation to predict retail sales for each state.

5.7 As part of an international study on energy consumption, data were collected on the number of cars in a country and the total travel in kilometers. The data for 12 of the countries are shown below:

| Country | Total cars (million cars) | Travel (billion km) |
|---|---|---|
| United States | 142,352 | 3140 |
| Finland | 1,823 | 35 |
| Denmark | 1,664 | 31 |
| Britain | 21,316 | 353 |
| Australia | 8,534 | 138 |
| Sweden | 3,321 | 53 |
| Netherlands | 5,527 | 84 |
| France | 23,268 | 348 |
| Norway | 1,592 | 24 |
| Italy | 26,117 | 368 |
| Germany | 43,752 | 609 |
| Japan | 40,246 | 439 |

Source: *The Economist*, 22 June 1996

(a) Create a scatter plot of the data. Do you think that there is a linear relationship between number of kilometers traveled and the number of cars?

(b) Find the least-squares line for the data. Interpret the value of the slope.

(c) Does the intercept make sense for these data? Why or why not?

(d) Plot the least-squares line on the same plot with the data. Does the line make you feel confident about predicting travel as a function of the number of cars?

(e) Use the least-squares equation to predict the number of kilometers traveled for Sweden and Japan. How well do the predictions agree with the original data?

5.8 In a study that was looking at why retail stores fail, data were collected on the type of retailer, the rental charges ($ per sq ft), and sales ($ per sq ft) for super regional shopping malls in the United States in 1993. The data are shown in the table below:

| Type of retailer | Rental charges ($/sq ft) | Sales ($/sq ft) |
|---|---|---|
| Department stores | 2.0 | 131 |
| Clothing and accessories | 19.0 | 237 |
| Gift/specialty | 22.5 | 250 |
| Shoes | 22.5 | 259 |
| Food service | 32.0 | 342 |
| Jewelry | 40.0 | 555 |

Source: *The Economist*, 1 March 1997

(a) Create a scatter plot for the data. Do you think that the relationship between rental charges and sales is linear?

(b) Find the equation of the least-squares line for the data and interpret the value of the slope and the intercept.

(c) Does the intercept make sense for these data? Why or why not?

(d) Plot the least-squares line on the same plot with the data. Do you think the line does a good job of predicting sales as a function of rental charges? Why or why not?

## 5.4 BIVARIATE DATA IN EXCEL

This section covers the basics of displaying and analyzing bivariate data in Excel. Many of the steps involved are similar to those discussed in previous chapters.

### 5.4.1 Creating Contingency Tables in Excel

You have already learned the basics of creating tables in Excel, using the **Pivot Table** command. We will use this same tool to create contingency tables for bivariate data. Figure 5.8 shows a portion of the Aluacha Balaclava College data in an Excel worksheet. Suppose that we want to create a contingency table for the variables Rank and School. Highlight the range that contains the data you want to analyze. The steps for accomplishing this are:

*Note: It is not necessary for the two columns you are interested in to be next to each other. You can highlight more than just the columns you want to analyze.*

1. From the **Data** menu, select **Pivot Table Report.** The Pivot Table Wizard will start by asking what type of data you want to work with. Since you are using data that are already in an Excel worksheet, simply leave the default radio button checked and click on **Next>** to continue.

2. Since you have already highlighted the range that contains your data, you can just proceed by clicking **Next>.** If you have not indicated where your data are located, you can do that now by highlighting the range in the worksheet.

The next step of the process will define the table you want to create. The dialog box will show the list of variables (columns) that you highlighted. In this case, we will use

| | A | B | C | D | E | F | G |
|---|---|---|---|---|---|---|---|
| 1 | Rank | Years at ABC | Salary 96-97 | Gender | Tenure? | School | Rank |
| 2 | ASST | 22 | 53316 | F | Y | BUSI | ASST |
| 3 | PROF | 11 | 64375 | M | Y | BUSI | PROF |
| 4 | ASSO | 7 | 63501 | M | Y | BUSI | ASSO |
| 5 | ASSO | 6 | 59426 | M | N | BUSI | ASSO |
| 6 | ASSO | 20 | 49058 | M | Y | BUSI | ASSO |
| 7 | PROF | 4 | 94969 | M | N | BUSI | PROF |
| 8 | ASST | 21 | 54762 | M | Y | BUSI | ASST |
| 9 | ASSO | 9 | 55516 | M | Y | BUSI | ASSO |
| 10 | ASSO | 18 | 45932 | M | Y | BUSI | ASSO |
| 11 | ASSO | 11 | 56871 | M | Y | BUSI | ASSO |

**FIGURE 5.8** ABC Excel Worksheet

the variables Rank and School to create the table, with Rank on the rows and School in the columns.

3. Drag the field box for Rank to the area marked **Row,** and the field box for School to the area marked **Column.** Now drag the field box for *either* Rank or School into the area marked **Data.** The dialog box should look like the one in Figure 5.9.

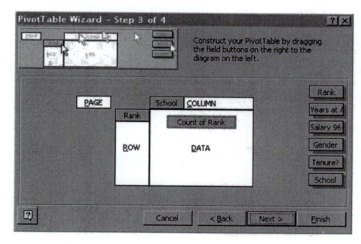

**FIGURE 5.9** Creating the contingency table

4. Click on **Next** and tell Excel where you want the table to appear, either in the current worksheet or in a new worksheet.
5. Finally, click on **Finish;** the table shown in Figure 5.10 should appear.

| Count of Rank | School | | | | |
|---|---|---|---|---|---|
| Rank | BUSI | LIBS | PROS | SCI | Grand Total |
| ASSO | 11 | 30 | 9 | 17 | 67 |
| ASST | 10 | 20 | 15 | 32 | 77 |
| INST | | 3 | 3 | 2 | 8 |
| PROF | 6 | 27 | 7 | 15 | 55 |
| Grand Total | 27 | 80 | 34 | 66 | 207 |

**FIGURE 5.10** Contingency table of Rank and School

You might want to use relative frequencies instead of frequencies for the contingency table, particularly if you are going to use the table to create clustered or stacked bar charts. You can change the way the data are displayed while you are creating the contingency table.

1. After you drag the field to the **Data** section of the table, double click on it and the **Field** dialog box opens. Click on **Options>>** and the box expands, as shown in Figure 5.11.

**FIGURE 5.11**   Changing the data display

2. Click on the drop down menu in the **Show data as** box, and select "% of total," "% of row," or "% of column." Your choice will, of course, be determined by what you want to do with the table later. Figure 5.12 shows the contingency table with the frequencies displayed as a percent of the total number of observations.

| Count of Rank | School | | | | |
|---|---|---|---|---|---|
| Rank | BUSI | LIBS | PROS | SCI | Grand Total |
| ASSO | 5.31% | 14.49% | 4.35% | 8.21% | 32.37% |
| ASST | 4.83% | 9.66% | 7.25% | 15.46% | 37.20% |
| INST | 0.00% | 1.45% | 1.45% | 0.97% | 3.86% |
| PROF | 2.90% | 13.04% | 3.38% | 7.25% | 26.57% |
| Grand Total | 13.04% | 38.65% | 16.43% | 31.88% | 100.00% |

**FIGURE 5.12**   Contingency table with relative frequencies

## 5.4.2 Creating Clustered Bar Charts in Excel

Suppose you want to create a clustered bar chart for the faculty data, using Rank as the axis variable, with a bar for each school. You can do this using either the frequency or relative frequency contingency table we just made.

The steps for creating a clustered bar chart are very similar to those for a simple bar chart. They are:

*Note: If the data in the copy do not appear as percents, use **Format cells** command to change the format to percents.*

1. To highlight just a portion of the pivot table, you will need to copy it to another location in the worksheet using the **Paste Special** command. To do this, click

anywhere on the table and select **Copy** from the **Edit** menu. Then, click on the location for the copy and select **Paste Special.** In the **Paste** section, click on the radio button for **Values** and in the Operation section, select **None,** as shown in Figure 5.13. Click **OK** and a copy of the table will appear in the location you specified.

**FIGURE 5.13**   Paste Special dialog box

2. Highlight the copy of the table, being careful *not* to highlight the Grand Total row or column. Then start the Chart Wizard and select Column as the type of chart and the clustered chart as the sub-type.

3. Click **Next** twice to get to the Options step; add titles and make any desired formatting choices.

4. Click **Next** to indicate where you want the chart to be located and then click **Finish.** A chart similar to the one shown in Figure 5.14 should appear.

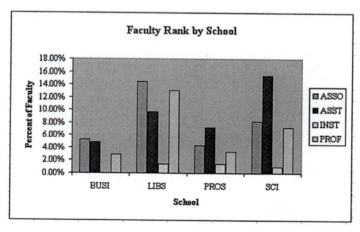

**FIGURE 5.14**   Clustered bar chart

### 5.4.3 Creating Stacked Bar Charts in Excel

To create a stacked bar chart in Excel, you will need to display the data in the contingency table differently. Suppose that you want to look at the different ranks as a percentage of the total for each school. You will need to recreate the contingency table for the data, displaying the data as a percent of the column (since School is the column variable). The new contingency table should look like the one in Figure 5.15.

| Count of Rank | School | | | | |
|---|---|---|---|---|---|
| Rank | BUSI | LIBS | PROS | SCI | Grand Total |
| ASSO | 40.74% | 37.50% | 26.47% | 25.76% | 32.37% |
| ASST | 37.04% | 25.00% | 44.12% | 48.48% | 37.20% |
| INST | 0.00% | 3.75% | 8.82% | 3.03% | 3.86% |
| PROF | 22.22% | 33.75% | 20.59% | 22.73% | 26.57% |
| Grand Total | 100.00% | 100.00% | 100.00% | 100.00% | 100.00% |

**FIGURE 5.15**   Contingency table for stacked bar chart

The steps for creating a stacked bar chart are very similar to those for creating a clustered bar chart. You must make a copy of the table using the **Paste Special** command before you can highlight the data range.

1. Highlight the rows of the pivot table, being careful not to include the Grand Total row and column and start the Chart Wizard.
2. Select Column as the chart type and 100% Stacked Column as the sub-type, as shown in Figure 5.16.

**FIGURE 5.16**   Selecting a stacked bar chart

3. Click **Next** twice to get to the Options step; put in titles and any other formatting changes that you want.
4. Click **Next,** enter the location for the chart, and click **Finish** to display the chart. It should look similar to the one in Figure 5.17.

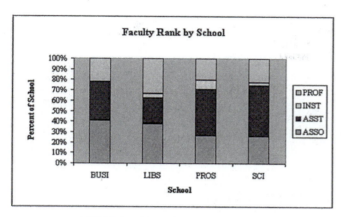

**FIGURE 5.17**   Stacked bar chart

### 5.4.4 Analyzing Quantitative Bivariate Data in Excel

When you learned how to analyze quantitative bivariate data, you learned about scatter plots and the least-squares line. Excel allows you to create a scatter plot and plot the least-squares line on the plot. Suppose that we want to plot the data for the number of training days and score on the aptitude test, and find the least-squares line. The data are shown in Figure 5.18.

When creating a scatter plot in Excel, it is important that the independent variable be in the first column, and the dependent variable in the second column. The steps for plotting the data and finding the least-squares line are as follows:

1. Highlight the range of the data and start the Chart Wizard. Select XY (Scatter) as the chart type and Scatter (no lines) for the subtype, as shown in Figure 5.19.

2. Select **Next>** twice to get to the Chart Options dialog box; put in titles and make any other desired formatting choices.

|   | A | B |
|---|---|---|
| 1 | Training | Score |
| 2 | 1.0 | 41 |
| 3 | 1.5 | 60 |
| 4 | 2.0 | 72 |
| 5 | 2.5 | 91 |
| 6 | 3.0 | 99 |

**FIGURE 5.18**   Training Data        **FIGURE 5.19**   Selecting the scatter plot option

3. Select **Next>** to tell Excel where to locate the chart and **Finish** to display the chart. The chart should look like the one in Figure 5.20.

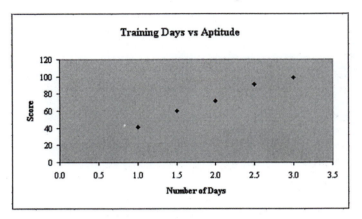

**FIGURE 5.20**    Finished scatter plot

Once you have the scatter plot, you can add the least-squares line as follows:

*Make sure you single click here; double clicking will bring up a different dialog box.*

1. Click on any one of the points in the scatter plot to highlight them all.
2. From the **Chart** menu, select **Add Trendline,** as shown in Figure 5.21. The Add Trendline dialog box will open.

**FIGURE 5.21**    Chart menu

3. In the Trend/Regression section, highlight **Linear,** as shown in Figure 5.22.

**FIGURE 5.22**    Selecting the type of trendline

4. Click on the **Options** tab and click the box labeled **Display equation on chart,** as shown in Figure 5.23. You can also choose to label the line with a title other than the default.

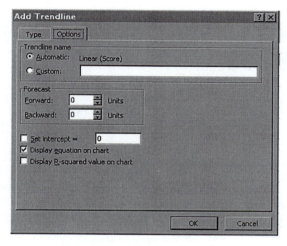

**FIGURE 5.23** Trendline options

5. Click **OK;** the least-squares line and its equation will appear on the chart, as shown in Figure 5.24.

**FIGURE 5.24** Scatter plot with least-squares line

Notice that Excel displays the equation with the slope term first and the intercept term last.

# CHAPTER 5 SUMMARY

In this chapter you learned some methods for looking at how two variables are related. You learned that very often, we are interested not in a single variable, but rather in how variables are *related*. There are several different methods for looking at these relationships. There are graphical methods, such as *stacked bar charts, clustered bar charts,* and *scatter plots,* which allow you to see the nature of the relationships. There are also quantitative methods, such as *least-squares analysis,* which allow you to describe the relationship numerically.

By now you should be more convinced than ever that there is no standard way to analyze a set of data. Rather, various statistical tools are used together to produce the most complete *picture* of the data and the information they contain.

## Key Terms

| Term | Definition | Page reference |
| --- | --- | --- |
| $\hat{y}$ | $\hat{y}$ (**y-hat**) is the predicted value of $y$ for a selected value of $x$. | 201 |
| **Clustered bar chart** | A **clustered bar chart** is used in place of multiple bar charts to make comparisons for one of the qualitative variables. The bars are grouped according to the values of the other variable. | 184 |
| **Contingency table** | A **contingency table** is a table whose rows represent the possible values of one variable and whose columns represent the possible values for a second variable. The entries in the table are the number of times that each pair of values occurs. | 180 |
| **Deviation (e)** | The difference between the data point and the value predicted by the least squares equation is called the **deviation, e.** | 201 |
| **Extrapolation** | Predicting values for $Y$ using values of $X$ that are outside the data range is called **extrapolation.** | 205 |
| **Interpolation** | Finding values for $Y$ using $X$ values that are within the data range is called **interpolation.** | 205 |
| **Least squares** | **Least squares** finds the equation of the line that minimizes the sum of the squared errors between the actual data points and the line. | 200 |

| Term | Definition | Page reference |
|------|-----------|----------------|
| Scatter plot | A **scatter plot** is a graph used to represent bivariate quantitative data. Each of the variables and the data are plotted as points on the graph. | 192 |
| Stacked bar chart | In a **stacked bar chart** the data for the selected variable are represented as a percentage of the total for each category of the second variable. Each value of the selected variable is represented in a different way and the bars are "stacked" to total 100%. | 187 |

## Key Formulas

| Term | Formula | Page reference |
|------|---------|----------------|
| Least-squares equation | $\hat{y} = a + bx$ | 201 |
| $\hat{y}$ (predicted value) | $\hat{y}$ | 201 |
| $e$ (deviation) | $\hat{y} - y$ | 201 |
| $a$ (intercept estimate) | $a = \dfrac{\Sigma Y}{n} - b\dfrac{\Sigma X}{n}$ | 201 |
| $b$ (slope estimate) | $b = \dfrac{n\,\Sigma XY - \Sigma X\,\Sigma Y}{n\,\Sigma X^2 - (\Sigma X)^2}$ | 201 |

# CHAPTER 5 EXERCISES

## Learning It!

**5.9**  Consider the data on the number of students choosing different majors at Dartmouth College that you looked at in Chapter 3.

| Major | 1986 | 1991 | 1996 |
|-------|------|------|------|
| Anthropology | 10 | 19 | 17 |
| Art History | 28 | 20 | 15 |
| Asia | 11 | 15 | 22 |
| Biology | 31 | 44 | 108 |
| Chemistry | 28 | 15 | 55 |
| Classics | 2 | 1 | 8 |
| Comparative Literature | 5 | 6 | 6 |
| Drama | 13 | 8 | 5 |
| Earth Science | 8 | 15 | 7 |
| Economics | 90 | 66 | 93 |
| English | 156 | 143 | 92 |
| French | 23 | 15 | 19 |
| Geography | 23 | 28 | 28 |
| Government | 147 | 172 | 142 |
| History | 102 | 125 | 115 |
| Mathematics | 35 | 23 | 14 |
| Music | 10 | 3 | 11 |
| Philosophy | 21 | 28 | 25 |
| Physics | 13 | 10 | 16 |
| Psychology | 70 | 74 | 87 |
| Religion | 22 | 31 | 8 |
| Sociology | 17 | 26 | 15 |

(a)  What type of bivariate chart would be appropriate for displaying these data?

(b)  What variable would you choose for the *x*-axis categories?

(c)  Create a clustered bar chart for the data.

**5.10**  In an effort to learn whether customers were correct about coupon life being shorter, the following data were collected:

| | Number of coupons | |
|---|---|---|
| **Coupon life** | **1995** | **1996** |
| Less than 1 month | 28 | 123 |
| 1–4 months | 88 | 140 |
| 5–8 months | 204 | 122 |
| 9–12 months | 141 | 88 |
| More than 1 year | 39 | 27 |

(a) Use a clustered bar chart to display the data.

(b) Is the perception that coupon life is getting shorter justified? Why or why not?

**5.11**  Look again at the data on the percentage of people who reported workplace violence that were presented in Chapter 3.

| Workplace violence survey | |
|---|---|
| **Type of violence** | **% who reported** |
| Harassment | 19 |
| Threat of physical harm | 7 |
| Physical attack | 3 |

| | Type of violence (%) | | |
|---|---|---|---|
| **Major effect on worker** | **Harassment** | **Threat** | **Physical** |
| Psychologically | 49 | 53 | 49 |
| Disrupted work life | 34 | 25 | 25 |
| Physically injured or sick | 13 | 9 | 17 |
| No negative effect | 4 | 13 | 9 |

Display these data on a single graph.

**5.12**  Large companies are always looking for expanding markets. Media companies in particular are looking at foreign markets for expansion. To obtain some information about expansion possibilities in Central Europe, a major communications company collected data on revenues from television advertising and the percentage of households with cable or satellite television for seven countries.

| Country | % of households | TV advertising revenues |
|---|---|---|
| Hungary | 55 | 170 |
| Slovakia | 47 | 28 |
| Poland | 33 | 369 |
| Czech Republic | 28 | 105 |
| Romania | 27 | 40 |
| Russia | 15 | 250 |
| Bulgaria | 9 | 23 |

Source: The Economist, 6 July 1996

(a) Create a scatter plot of the data. Does it appear that advertising revenues are related to the percentage of households that have cable or satellite TV?

(b) Find the equation of the least-squares line for the data.

(c) Plot the least-squares line on the same plot as the data. Do you think that the line does a good job of predicting TV advertising revenues? Why or why not?

(d) In the Ukraine, 8% of the households have cable or satellite television. Use the line to predict TV advertising revenues for the Ukraine.

**5.13**  For many countries, tourism contributes a large part of revenues. In trying to predict tourism revenues, one of the independent variables that is considered important is the number of foreign visitors to the country. Data for six different countries were collected:

| Country | Number of visitors (million) | Tourism receipts ($ billion) |
|---------|------------------------------|------------------------------|
| France | 60 | 27.3 |
| Spain | 48 | 25.1 |
| US | 45 | 58.4 |
| Italy | 30 | 27.1 |
| Britain | 23 | 17.5 |
| Germany | 15 | 11.9 |

*Source: The Economist,* July 1996

(a) Create a scatter plot of the data and find the equation of the least-squares line.

(b) Use the least-squares line to predict tourism revenues for Italy and the United States.

(c) For which country does the least-squares line do the best job of predicting tourism receipts?

**5.14** As part of the anti-crime bill passed in 1995, the U.S. government granted money to different cities to hire new beat-patrol officers. Nine cities in New Jersey were given grants and hired officers. In trying to assess the program, government analysts wanted to look at the relationship between the number of officers hired and the amount of the grant. The data are

| Grant ($) | Number of officers |
|-----------|--------------------|
| 150,000 | 2 |
| 375,000 | 5 |
| 471,125 | 6 |
| 70,967 | 1 |
| 450,000 | 6 |
| 525,000 | 7 |
| 375,370 | 7 |
| 750,000 | 10 |
| 1,000,000 | 12 |

*Source:* Justice Department

(a) Create a scatter plot of the data.

(b) Find the equation of the least-squares line.

(c) Plot the least-squares line on the same plot as the data.

(d) How well do you think the line fits the data?

(e) Use the equation to predict the number of officers hired for each city.

(f) Compare the values from the least-squares line to the actual data.

**5.15** As communications and the media change and become more important in the economy, the radio industry has become an area of concern. One of the major communications companies wanted to look at the relationship between the number of radio stations a company owned and the revenues generated by radio. The company collected the following data:

| Company | Number of stations | Revenues ($ billion) |
|---------|--------------------|----------------------|
| Westinghouse/Infinity | 83 | 1.05 |
| Jacor | 57 | 0.31 |
| Clear Channel | 104 | 0.31 |
| Evergreen | 35 | 0.30 |
| Disney/ABC | 21 | 0.29 |
| American Radio Systems | 63 | 0.23 |
| SFX | 67 | 0.22 |
| Chancellor | 41 | 0.21 |
| Cox | 38 | 0.21 |
| Bonneville | 20 | 0.12 |

*Source: The Economist,* 29 June 1996; *Duncan's American Radio*

(a) Create a scatter plot of the data. Do you think that there is a relationship between the number of stations that a company owns and the revenues generated by radio?

(b) Find the least-squares line for the data.

## Thinking About It!

**5.16**  The summer of 1995 is remembered as one of the hottest summers on record. The *New York Times* reported on the amount of electricity used and the temperature for 10 different record breaking days:

| Date | Temperature (°F) | Electrical usage (MW) |
|---|---|---|
| 2 Aug 95 | 95 | 10,805 |
| 23 Jul 91 | 99 | 10,752 |
| 8 Jul 93 | 100 | 10,667 |
| 27 Jul 93 | 93 | 10,654 |
| 27 Jul 95 | 90 | 10,567 |
| 20 Jun 95 | 95 | 10,551 |
| 9 Jul 95 | 101 | 10,398 |
| 26 Jul 95 | 90 | 10,391 |
| 8 Jul 94 | 90 | 10,368 |
| 19 Jul 91 | 96 | 10,349 |

(a) Do you think it is likely that the amount of electrical usage in New York City depends on the temperature?

(b) Create a scatter plot showing the relationship between electrical usage and temperature.

(c) Do you think the data indicate a relationship between the two variables? Why or why not?

(d) Can you think of any other variables or factors that might be hiding a relationship? If so, what might that variable or factor be?

(e) Eliminate the days from years other than 1995 and replot the data. Now what do you see?

(f) Do any of the data points appear to be unusual? Can you offer any explanation for this?

(g) 9 July 1995 was a Sunday (which means that businesses in New York City were closed). Eliminate this point and replot the data. Now what do you see?

(h) Find the least-squares line for the data from the last plot.

**5.17**  The pharmaceutical company that is looking at the relationship between advertising expenditures and sales of over the counter drugs collects some additional data:

| Drug | Advertising ($ millions) | Factory sales ($ millions) |
|---|---|---|
| Tylenol | 143.8 | 855 |
| Advil | 91.7 | 360 |
| Vicks | 26.6 | 350 |
| One Touch | 2 | 220 |
| Robitussin | 37.7 | 205 |
| Bayer Aspirin | 43.8 | 170 |
| Alka-Seltzer | 52.2 | 160 |
| Centrum | 16.5 | 150 |
| Mylanta | 32.8 | 135 |
| Tums | 27.6 | 135 |
| Excedrin | 26.7 | 130 |
| Benadryl | 30.9 | 130 |
| Halls | 17.4 | 130 |
| Metamucil | 12.1 | 125 |
| Sudafed | 28.6 | 115 |

(a) Make a scatter plot of the data.

(b) What explanation can you offer for the fact that there does not appear to be a relationship between advertising expenditures and sales? What additional information do you think would help put the data in perspective?

(c) Separate the drugs listed into four categories: pain relievers, digestive, cold/allergy, and other.

(d) Create four different scatter plots. Does this change anything? If so, how?

(e) For each of the categories that show a relationship between sales and advertising, find the equation of the least-squares line.

(f) Do the equations have any similarities? If so, what are they?

**5.18** Consider the data on revenue and number of radio stations owned by various companies.    *Requires Exercise 5.15*

(a) Use the least-squares line to predict the radio revenues for Chancellor and Westinghouse/Infinity. Which prediction is better?

(b) Leave out the data point for Westinghouse/Infinity and recalculate the least-squares line. Do you think that this line will do a better job of predicting radio revenues? Why or why not?

**5.19** The Commerce Department also has data available on number of shopping centers and retail sales for the South Central States. The data are given below:    *Requires Exercise 5.6*

| State | Number of shopping centers | Retail sales ($ billions) |
|---|---|---|
| Kentucky | 593 | 11.7 |
| Tennessee | 1,137 | 19.1 |
| Alabama | 601 | 13.2 |
| Mississippi | 418 | 7.2 |
| Arkansas | 339 | 6.4 |
| Louisiana | 676 | 15.6 |
| Oklahoma | 556 | 11.4 |
| Texas | 2,824 | 72.7 |

*Source: Statistical Abstract of the United States 1995*

(a) Create a scatter plot of the data. Do you think that the variables are linearly related?

(b) Find the equation of the least-squares line for the data.

(c) How does the equation for the South Central States compare to the one you found for the North Central States?

(d) Would you have expected the least-squares lines for the South Central and North Central States to be exactly the same? Why or why not?

(e) What similarities would you expect them to have? What differences?

**5.20** Combine the data on retail sales and number of shopping centers for the North Central and South Central states. Data for the North Central states are repeated here for convenience.    *Requires Exercises 5.6, 5.19*

| State | Number of shopping centers | Retail sales ($ billions) | State | Number of shopping centers | Retail sales ($ billions) |
|---|---|---|---|---|---|
| Ohio | 1,559 | 34.8 | Iowa | 277 | 6.3 |
| Indiana | 848 | 18.0 | Missouri | 834 | 18.9 |
| Illinois | 1,961 | 34.1 | North Dakota | 84 | 1.8 |
| Michigan | 967 | 21.0 | South Dakota | 51 | 1.1 |
| Wisconsin | 588 | 12.0 | Nebraska | 245 | 4.8 |
| Minnesota | 436 | 11.5 | Kansas | 458 | 9.7 |

(a) Create a scatter plot of the data and find the equation of the least-squares line.

(b) Plot the least-squares line on the scatter plot.

(c) Your boss says that the model for the combined data will be better than the model for the two areas separately because there are more data for the combined equation. Write a memo to your boss explaining your results and which method you think is better. Be sure to back up your opinions with the analyses.

## Doing It!

*Datafile:*
*FACULTY.XXX*

**5.21**   The provost at Aluacha Balaclava College wants to look at the faculty salary data in some other ways. She has collected data on salary, years of service, rank, school, gender, and tenure. A sample of the data is shown below:

| Salary ($) | Years of service | Rank | Schools | Gender (M/F) | Tenure (Y/N) |
|---|---|---|---|---|---|
| 53,316 | 22 | ASST | BUSINESS | F | Y |
| 64,375 | 11 | PROF | BUSINESS | M | Y |
| 63,501 | 7 | ASSO | BUSINESS | M | Y |
| 59,426 | 6 | ASSO | BUSINESS | M | N |
| 49,058 | 20 | ASSO | BUSINESS | M | Y |
| 94,969 | 4 | PROF | BUSINESS | M | N |
| 54,762 | 21 | ASST | BUSINESS | M | Y |
| 55,516 | 9 | ASSO | BUSINESS | M | Y |

(a) Create graphical displays that analyze the relationship between gender and rank and gender and school. Is there anything that the provost might find interesting? Why or why not?

(b) Create graphical displays that look at tenure by school and by rank. What can you say about these data?

(c) Look at salary vs. years of service for each rank separately. Do you think that the relationship between the two variables is still valid? Is it easier to see the relationship when the data are separated like this? Why or why not?

(d) Find least-squares equations that will enable the provost to predict salary based on years of service for each rank. Do you think the equations do a good job of predicting salary? Why or why not?

(e) Are each of the equations you found in the previous part the same? Would you have expected them to be? How do they compare?

(f) Redo the analysis of salary and years of service considering some of the other qualitative variables. Do any of them make the least-squares equations better?

(g) Which qualitative variable, if any, would you suggest that the provost at ABC use to separate the data? Why?

**5.22**   The following data were collected from 10 big fund managers about the ways in which their holdings were distributed:

**Bond Holdings by Currency (%)**

| Currency | Neutral weights | A | B | C | D | E | F | G | H | I | J |
|---|---|---|---|---|---|---|---|---|---|---|---|
| Dollar | 33 | 31 | 54 | 45 | 38 | 32 | 32 | 46 | 27 | 42 | 31 |
| Yen | 19 | 19 | 6 | 15 | 15 | 15 | 18 | 0 | 12 | 11 | 18 |
| Sterling | 5 | 6 | 2 | 6 | 4 | 3 | 4 | 7 | 5 | 6 | 7 |
| DM | 10 | 8 | 12 | 18 | 15 | 13 | 10 | 18 | 32 | 14 | 16 |
| FFr | 7 | 5 | 12 | 8 | 2 | 10 | 8 | 5 | 5 | 4 | 8 |
| Others | 26 | 31 | 14 | 8 | 26 | 27 | 28 | 24 | 19 | 23 | 20 |
| Sum | 100 | 100 | 100 | 100 | 100 | 100 | 100 | 100 | 100 | 100 | 100 |

**Holdings by Instrument (%)**

| Instrument | A | B | C | D | E | F | G | H | I | J |
|---|---|---|---|---|---|---|---|---|---|---|
| Equities | 40 | 78 | 70 | 65 | 50 | 50 | 50 | 36 | 60 | 31 |
| Bonds | 50 | 14 | 20 | 30 | 42 | 45 | 45 | 54 | 40 | 48 |
| Cash | 10 | 8 | 10 | 5 | 8 | 5 | 5 | 10 | 0 | 21 |
| | 100 | 100 | 100 | 100 | 100 | 100 | 100 | 100 | 100 | 100 |

**Equity Holdings by Area (%)**

| Area | Neutral weights | A | B | C | D | E | F | G | H | I | J |
|---|---|---|---|---|---|---|---|---|---|---|---|
| Americas | | | | | | | | | | | |
| United States | 42 | 39 | 46 | 35 | 24 | 26 | 41 | 28 | 28 | 34 | 29 |
| Others | 2 | 4 | 4 | 1 | 5 | 2 | 3 | 2 | 0 | 2 | 2 |
| Europe | | | | | | | | | | | |
| Britain | 9 | 8 | 4 | 7 | 7 | 6 | 6 | 7 | 12 | 6 | 8 |
| Germany | 4 | 4 | 2 | 8 | 7 | 8 | 4 | 9 | 6 | 8 | 4 |
| France | 4 | 3 | 2 | 3 | 4 | 10 | 4 | 7 | 14 | 1 | 6 |
| Others | 11 | 9 | 8 | 1 | 15 | 17 | 8 | 7 | 13 | 15 | 13 |
| Asia | | | | | | | | | | | |
| Japan | 22 | 21 | 23 | 35 | 30 | 25 | 24 | 30 | 17 | 25 | 30 |
| Others | 6 | 12 | 11 | 10 | 8 | 6 | 10 | 10 | 10 | 9 | 8 |
| Sum | 100 | 100 | 100 | 100 | 100 | 100 | 100 | 100 | 100 | 100 | 100 |

*Key:*
FUND

| | | | |
|---|---|---|---|
| A | Merrill Lynch | F | Robeco Group Asset Management |
| B | Lehman Brothers | G | Bank Julius Baer (Zurich) |
| C | Nikko Securities | H | UBS International Investment |
| D | Daiwa Europe | I | Commerz International Capital Management |
| E | Credit Agricole | J | Credit Suisse Asset Management |

**(a)** Create an appropriate graphical display to show how holdings are distributed among the three instruments (equities, bonds, cash) for each of the funds.

**(b)** Create a graphical display of equity holdings both for continent and countries. Compare these to each other and to the neutral weights.

**(c)** Create a graphical summary for bond holdings by currency for each of the funds. Compare them to each other and to the neutral weights.

**(d)** Create any other graphs and displays that you think help to summarize the data. Prepare a report on distribution of holdings for the 10 funds surveyed.

# REALLY DOING IT
## Advising the President: Exploratory Data Analysis

You have taken a job at the White House on the staff of the President's Council of Economic Advisors. As the new kid on the block, you have been given the assignment of providing the President with a view of the country. The President wants you to capture the country with statistics, so that he understands how the different demographic and economic factors vary around the country.

You have data available to you from the U.S. Census Bureau by state and region of the country. The Census Bureau tracks 63 different variables covering nine main categories: population, health, education, income, crime, employment, banking, transportation, and housing. A very small portion of the data is shown at the top of the next page.

| Region division, and state | Resident population\1 | | | Resident population\1 | | |
|---|---|---|---|---|---|---|
| | Total,\2 1994 (1,000) | Percent increase,\3 1990–94 (%) | 65 years old and over, 1994 (%) | Householder 65 years old and over, 1994\3 (%) | In metro areas, 1992\4 (%) | Net domestic immigration, 1990–94\1 |
| United States | 260,341 | 4.7 | 12.7 | 21.8 | 79.7 | 0 |
| Northeast | 51,396 | 1.2 | 14.1 | 23.7 | 89.4 | (1,439,371) |
| New England | 13,270 | 0.5 | 13.9 | 22.9 | 84.1 | (364,734) |
| Maine | 1,240 | 1.0 | 13.9 | 22.8 | 35.7 | (12,109) |
| New Hampshire | 1,137 | 2.5 | 11.9 | 19.5 | 59.4 | (9,912) |
| Vermont | 580 | 3.1 | 12.1 | 20.1 | 27.0 | 2,287 |
| Massachusetts | 6,041 | 0.4 | 14.1 | 23.3 | 96.2 | (183,995) |
| Rhode Island | 997 | − 0.7 | 15.6 | 25.6 | 93.6 | (35,219) |
| Connecticut | 3,275 | − 0.4 | 14.2 | 23.1 | 95.7 | (125,786) |

The data are found in a file named CENSUS.XXX. Looking at the data, you notice that they cover the time period from 1990 to 1994. You wonder about the data in parentheses and remember from your accounting course that this means negative numbers. You also notice that there are summary lines for geographic regions and realize that you had better remove them before you do any type of analysis by state.

Completely overwhelmed, you decide to approach your boss and ask for some direction. She tells you that the President is looking for summaries and relationships among the variables and regions of the country and that reminds you of the popularity of Ross Perot's charts and graphs. In addition, she tells you that the President wants to be prepared for attacks from Capitol Hill on health care and education, and that you should start your analysis there.

1. Prepare a set of graphical displays and statistics that summarize the availability of health care in the country. Do any of the states or regions exhibit unusual behavior? If so, which ones are they? Is any state or region consistently low or high on all aspects of health care?

2. Of particular interest is the cost of health care. Does there appear to be any relationship between the cost of hospital care and the other variables? If so, how are they related?

Now that you have gotten the idea, you understand what you need to do for each of the rest of the categories in order to have your report on the President's desk by the end of the week. Oh, by the way, the President is wondering where is the best place in the country to live?

## REALLY DOING IT
## What's on the Road? Who's Driving It?: Exploratory Data Analysis

Before you can make decisions about the future, you need to understand the current situation. The transportation industry controls the infrastructure of the country, and decisions it makes impact everyone.

One method that the industry uses to understand current trends is the Nationwide Personal Transportation Survey (NPTS). This survey is done approximately every seven years under the sponsorship of the U.S. Department of Transportation.

The NPTS compiles national data on the nature and characteristics of personal travel. NPTS data may be used to describe current travel patterns, and given projections of demographic change, can provide a valuable tool to forecast future travel demand. One aspect of transportation planning is knowing what is currently on the road. The NPTS data include motor vehicle information such as year, make, model, and other vehicle-related information.

You are asked to describe the current vehicle population of the United States. To assist in this, you have been given a portion of the NPTS data from the 1990 survey, in which data were collected on 26,172 households using computer-assisted telephone interviewing (CATI). Specifically, you have been given data from all survey interviews conducted in April 1990. A portion of the data is shown below:

| CMSA | HHFAMINC | HHLOC | HHMSA | HHSIZE | HH_RACE | HOUSEID | LIF_CYC | MAKECODE |
|------|----------|-------|-------|--------|---------|---------|---------|----------|
| 5602 | 05 | 1 | 5640 | 01 | 01 | 00038 | 01 | 014 |
| 5602 | 04 | 1 | 5640 | 06 | 01 | 00039 | 08 | 038 |
| 5602 | 04 | 1 | 5640 | 06 | 01 | 00039 | 08 | 022 |
| 5602 | 04 | 1 | 5640 | 06 | 01 | 00039 | 08 | 021 |
| 5602 | 98 | 1 | 5640 | 05 | 01 | 00041 | 06 | 021 |
| 5602 | 98 | 1 | 5640 | 05 | 01 | 00041 | 06 | 021 |
| 5602 | 98 | 1 | 5640 | 05 | 01 | 00041 | 06 | 035 |
| 5602 | 14 | 2 | 5190 | 02 | 01 | 00045 | 10 | 009 |
| 5602 | 14 | 2 | 5190 | 02 | 01 | 00045 | 10 | 006 |
| 5602 | 99 | 2 | 5190 | 03 | 01 | 00047 | 06 | 019 |

The data are found in a data file called CARS.XXX. As is the case with much publicly available data, the variable names are cryptic and the qualitative variables are coded. Clearly you will need to know what each variable and code mean. This information is detailed in Appendix C at the end of the book.

Of particular interest to automobile manufacturers is information about the age and type of vehicles on the road.

1. Use graphical and numerical summary tools to describe the distribution of vehicles by make and year.
2. Are the distributions affected by various demographics? Who drives foreign cars? Who drives old cars?
3. Is the distance driven annually impacted by where people live, gender, age, or type of vehicle?

Using these questions as a starting point, prepare a report that completely describes the vehicle population of the United States. Be sure to include any limitations of the data.

# Part III
# THE KEYS THAT UNLOCK THE DOOR TO INFERENTIAL STATISTICS

CHAPTER **6**

# *PROBABILITY*

## CREDIT PROBLEMS

One of the problems facing a small business is that of borrowing money. As a result of problems with loan defaults, banks and other lending institutions have imposed more stringent conditions for obtaining loans. As a result, many small businesses find themselves forced to close because they cannot meet the stiffer requirements.

The problems of small businesses affect the cities and towns in which the businesses are located. As a result, the towns would like to know what is going on and how they can help. A medium-sized New England city has been trying to understand the problems facing the small businesses in their town by administering a questionnaire every six months. The questionnaire collects information on the size of the business in terms of annual sales and employees, the type of business, and whether the business is experiencing recession-related problems. A sample of the data is shown below:

| Number | Size | Employee | Nature | Problem | Understand | Concerned | Call | Loan | Collateral | Access |
|--------|------|----------|--------|---------|------------|-----------|------|------|------------|--------|
| 1 | 2 | 2 | 1 | 1 | 2 | 1 | 2 | 2 | 0 | 2 |
| 2 | 1 | 2 | 3 | 1 | 2 | 2 | 0 | 2 | 2 | 1 |
| 3 | 4 | 3 | 1 | 2 | 1 | 2 | 2 | 2 | 2 | 0 |
| 4 | 1 | 1 | 1 | 2 | 1 | 2 | 2 | 2 | 2 | 2 |
| 5 | 1 | 2 | 1 | 2 | 0 | 0 | 0 | 0 | 0 | 0 |
| 6 | 3 | 1 | 5 | 2 | 0 | 2 | 2 | 2 | 2 | 1 |

The town has employed an analyst to help them interpret the results of the study and to determine how the small businesses in the city are being affected by credit problems.

## 6.1    CHAPTER OBJECTIVES

In Chapters 3 through 5 you learned different ways to summarize and describe data that were collected. The data were usually random samples from different populations, but the techniques used really described only the sample data.

In this chapter we look at how *probability theory* can be used to measure and predict what is *likely* to happen when data are collected from different populations. This chapter covers the following topics:

- Basic Probability Rules
- Random Variables and Probability Distributions
- The Binomial Probability Distribution
- The Normal Probability Distribution

## 6.2    BASIC RULES OF PROBABILITY

Probability is a concept that you are most likely familiar with, although perhaps not in a formal sense. Whenever you talk about whether some event is likely to occur, such as whether it will rain tomorrow, you are using the concepts of **probability.**

> *Probability* is a measure of how likely it is that something will occur.

To talk about probability using the language of probability it is necessary to define some terms. In probability and statistics we often speak of **experiments.** These are not experiments in the laboratory sense, but rather are the actions we perform to collect data. As an experiment we might count the number of students who miss statistics class each day or record the color of the car parked next to us in the parking lot.

> An *experiment* is any action whose outcomes are recordable data.

When we perform an experiment, or collect data, we must think about what outcomes might occur so that we know what form our data will take. This is called the **sample space** of the experiment. You remember from the previous chapters that data can be qualitative or quantitative and that we use different techniques when we analyze them.

> The *sample space, S,* is the set of all possible outcomes of an experiment.

*The brackets { } are used to indicate a sample space.*

As a very simple experiment, consider rolling a single, six-sided die and recording the number of spots on the top facing side. You know that there are six possible outcomes of the experiment, so the sample space is

or    S = {1,2,3,4,5,6}

Another example is the sample space for the experiment of tossing a coin. In this case the sample space is

$$S = \left\{ \underset{\text{}}{} \right\}$$

or   S = {H, T}, where H represents a head and T represents a tail

Depending on the experiment and the type of data collected, a sample space might have a *finite* number of elements such as the ones we just described. The word *finite* means that you can count the number of elements. This will most often happen when the experiment results in qualitative data, or in quantitative data that are integer.

When the data from the experiment are quantitative and continuous, then the sample space contains an *infinite* number of possible values and must be described mathematically or in words. For example, suppose that as our experiment we decide to take college students at random and measure their heights. This is an example of *continuous* qualitative data. The number of possible values that we get will be *bounded* (there are probably not any college students shorter than one foot or taller than eight feet) but *infinite* (since height is a measurement and there are an infinite number of measurements between 1 foot and 8 feet). To describe this sample space we might write

S = {all numbers greater than or equal to 1 and less than or equal to 8}

or using mathematical notation

$$S = \{x: 1 \le x \le 8\}$$

Figuring out the sample space for an experiment is important in data collection, too. It helps the experimenter think about all of the possible results they might obtain and to plan for them.

## EXAMPLE 6.1   Flipping Two Coins

### *Writing Out a Sample Space*

An experiment consists of flipping two standard coins and writing down what is on the side facing up for each of the coins. If we let H represent a head and T represent a tail, then the sample space for this experiment is

S = {HH, HT, TH, TT}

If you are thinking that HT and TH are the same outcome, then try to think of tossing the coins one after another or think of the problem as if the coins are different, such as a penny and a quarter. In writing out sample spaces it is best not to combine outcomes, but to write each one separately. ∎

*HT and TH are different even if the coins are both quarters, but this is another way to try and think about it.*

 **TRY IT NOW!**

### The Spinner Problem
#### *Writing out the Sample Space*

An experiment consists of spinning the two different spinners pictured below:

*(continued)*

Write down the sample space for this experiment.

You may be wondering what rolling dice and spinning spinners has to do with collecting data and you are right to wonder. In fact, they have little to do with the types of problems you will encounter in statistics, but they do form an easy basis for discussing and understanding the general rules of probability.

## 6.2.1 Probability of an Event

Several of the sample spaces that you looked at have something in common that make calculating probabilities easy. This feature is that each of the outcomes in the sample space is equally likely to occur on any given trial of the experiment. For example, when you roll a single die, each of the six possible outcomes in **S** is equally likely to occur. The same is true about flipping coins, both one coin and two coins.

Often we are interested in knowing how likely is that that a certain outcome or outcomes of the experiment will occur. We call these outcomes the **events** of interest.

> An *event, A,* is an outcome or a set of outcomes that are of interest to the experimenter.

The likelihood that the event A will occur is called the **probability of A** and is written P(A).

> The *probability of an event A,* P(A), is a measure of the likelihood that an event A will occur.

When each of the outcomes in a sample space *is equally likely*, then the probability of A can be calculated using the following formula:

*Formula for P(A)*

$$P(A) = \frac{\text{Number of ways that A can occur}}{\text{Total number of possible outcomes}}$$

or

$$P(A) = \frac{n_A}{N}$$

*Although you have been taught to reduce fractions and use decimals, when we are calculating probabilities here we will not do this so that you can see the actual number of outcomes of interest.*

where $n_A$ is the number of outcomes that correspond to the event, A, and $N$ is the total number of outcomes in the sample space, S.

If you look at the simple experiment of flipping a coin described earlier, you can see that if S = {H, T} and we look at the event A = coin turns up heads, then P(A) = $\frac{1}{2}$, since there is only one way that the event A can occur and there are two possible outcomes of the experiment. This should not be surprising, since it is one of the ways you know probability intuitively.

Similarly if we look at the experiment of rolling a die where S = {1, 2, 3, 4, 5, 6} and we let A be the event that the number that comes up is even, then

$$P(A) = \tfrac{3}{6} = \tfrac{1}{2}$$

since there are 3 even numbers possible out of the 6 possible outcomes of the experiment. Again, this is not surprising since it is somewhat intuitive.

There are some facts about probabilities that must be true:

1. $0 \leq P(A) \leq 1$. This says that the probability of an event must be a number between 0 and 1 inclusive. Since you know that probabilities are formed by taking the ratio of the number of ways that A can happen to the total number of outcomes, the numerator is a subset of (smaller than or equal to) the denominator.

2. $P(S) = 1$. This says that the sum of the probabilities for the entire sample space must be equal to 1, or that essentially, when you perform an experiment, something must happen!

3. If an event A MUST happen, then $P(A) = 1$, and if the event cannot happen, then $P(A) = 0$.

*These facts may seem trivial, but try to keep them in mind when you calculate probabilities. If you get an answer that is not a number between 0 and 1 then it **cannot be correct!***

## EXAMPLE 6.2    Flipping Two Coins

### Finding Probabilities

Joe and Tom decide to flip coins to determine who will pay for dinner. They agree that if the coins match Joe will pay and if they do not match Tom will pay. What is the probability that Tom pays for dinner?

Remember from the previous example that the sample space for this experiment is given by S = {HH, HT, TH, TT}. If we let A be the event that Tom pays, then we can write

A = the two coins do not match

Looking at S, we see that the outcomes in the sample space are equally likely to occur and that two outcomes correspond to the event A: HT and TH. Thus, the probability that Tom pays for dinner is

*It was important to list HT and TH as different outcomes so that the outcomes in S are all equally likely.*

$$P(A) = \tfrac{2}{4}$$

Using the same approach, if we let B be the event that Joe pays for dinner, or

B = the two coins match

then there are two outcomes in S that correspond to B, HH and TT, so

$$P(B) = \tfrac{2}{4}$$

From this we see that the arrangement is fair to both people.    ■

Finding probabilities when the events are equally likely is not hard if you can write down the sample space, S, without much difficulty. These types of problems are knows as "classical" probability problems. Although the situations where the ideas of classical probability apply are limited, they do make it easy to illustrate the rules of probability.

*TRY IT NOW!*

### The Spinner Problem
*Classical Definition of Probability*

In the previous exercise you found the sample space for the spinner example to be

$$S = \{1A, 1B, 1C, 2A, 2B, 2C, 3A, 3B, 3C\}$$

Let A be the event that the first spinner lands on an odd number. Find P(A).

Let B be the event that the second spinner is a vowel. Find P(B).

---

*The complement of an event A is often referred to as the event, not A.*

When the number of possible outcomes of an experiment is large, calculating probabilities can be tedious. At times it is easier to solve the opposite or **complement** of the problem you are interested in! You may wonder how this can be possible. It is possible because it makes use of one of the three rules of probabilities, the one that says that the probabilities must sum to one.

> The *complement* of an event A, denoted **A′**, is the set of all outcomes in the sample space, S, that do not correspond to the event A.

Since the event A is a set of outcomes in S and the complement of the event A′ is the set of all outcomes in S that do NOT correspond to A, we can see that

*Formula for the probability of the complement of A*

$$P(A) + P(A') = 1$$

and that we can find P(A) by calculating P(A′) and subtracting it from 1. That is,

$$P(A) = 1 - P(A')$$

For the experiment of rolling a single die, the sample space was

$$S = \{1, 2, 3, 4, 5, 6\}$$

If we define the event A to be that the number that comes up is greater than 2 we can calculate $P(A) = \frac{4}{6}$ directly, since there are 4 outcomes that are larger than 2 out of the 6 possible outcomes.

---

ANS. $P(A) = \frac{6}{9}$, $P(B) = \frac{6}{9}$

We notice that the complement of A, $A'$, is defined as all of the numbers that are NOT larger than 2. Since two outcomes correspond to $A'$ we know that $P(A') = \frac{2}{6}$ and we can find $P(A) = 1 - P(A') = 1 - \frac{2}{6} = \frac{4}{6}$.

Although this seems to complicate something that was not very complicated, there are times when using the complement of an event to find a probability is much easier. Usually this is when the number of possible outcomes that correspond to the event of interest is a large part of the sample space, S.

Despite what it may look like, the idea of equally likely outcomes is not limited to dice, coins, and spinners! In fact, the ideas of classical probability apply to random sampling as well. Whenever an experiment consists of selecting an item at random from a group of items, the sample space will consist of all of the possible items. If the sample is truly random, then each item in the group is just as likely to be selected as any other item.

## EXAMPLE 6.3   Credit Problems

### Probabilities and Random Sampling

The Chamber of Commerce in the city looking at credit problems of small businesses administered the questionnaire to 166 random businesses out of 1536 existing small businesses. The Chamber of Commerce was worried that if the surveys were administered by mail the sample might be biased toward companies that were having problems. These companies might be more likely to respond to the survey because it would give them an opportunity to report their problems to a group who might be able to help them. To ensure that the sample was truly representative, the surveys were administered by phone. One of the questions asked was about the size of the company in annual sales. The questionnaire allowed the following responses:

| | |
|---|---|
| 1 | Under $1 million |
| 2 | $1–5 million |
| 3 | $6–10 million |
| 4 | $11–20 million |
| 5 | Over $20 million |

If the business did not respond to the question then the response was coded as a zero (0). The results of the questionnaire are

| Size | Number of responses |
|---|:---:|
| No response | 1 |
| Under $1 million | 60 |
| $1–5 million | 21 |
| $6–10 million | 42 |
| $11–20 million | 21 |
| Over $20 million | 21 |
| Total | 166 |

The Chamber of Commerce decided to select a business at random and interview it along with the bank with which it does business. What is the probability that the company selected will have over $20 million in sales?

This is an experiment in which the sample space is the set of 166 companies that respond. That is, S = {C1, C2, C3, . . . , C166} and each element of S is equally likely to be chosen. Since we are interested in the event that the company has over $20 million in sales (call that event A) we see that 21 elements (companies) in the sample space correspond to A out the total of 166, so

$$P(A) = \frac{21}{166}$$

■

**EXAMPLE 6.4**    **Product Preference**

*Random Sampling*

A market research firm has conducted a product preference survey at a large manufacturing company. A group of 250 production workers were asked to use three different pairs of safety glasses and to select the one that they preferred. The results of the survey showed that 120 preferred product A, 85 preferred product B, and 45 preferred product C. The marketing research firm decides to select a worker at random and interview him more thoroughly about his choice. What is the probability that the worker interviewed prefers product B?

If you think about the problem you will see that the experiment consists of selecting a worker at random. Thus, the sample space of the experiment will consist of the 250 workers,

$$S = \{W1, W2, W3, \ldots, W248, W249, W250\}$$

Let B be the event that the worker chosen preferred product B. Then $P(B) = \frac{85}{250}$, since 85 outcomes (workers) in S corresponded to the event B, and there were 250 possible outcomes. ∎

## 6.2.2 Combinations of Events—OR and AND

Calculating the probability of a single event is not difficult. Much of the time, however, we are interested in looking at more than one event. For example, if an experiment consisted of selecting employees at random from a company, we might be interested in the event A = the employee is an hourly worker, or the event B = the employee participates in the company's stock purchase program. From what we have learned, if we had the data on the number of employees in the company and the number that fell into each category, we could calculate P(A) and P(B) without much trouble. But suppose that the company is interested in how these two events behave *together*? We must look at the events **A OR B** and **A AND B.**

> The event **A OR B** describes the event when either A happens or B happens or they both happen.

> The event **A AND B** is the event that A and B both occur.

Let's look at the problem in terms of one of the smaller problems we studied earlier. When you roll a single die, the sample space is S = {1, 2, 3, 4, 5, 6}. If you let the event A = the number that comes up is even, and the event B = the number that comes up is a 3, then you can easily see that $P(A) = \frac{3}{6}$ and $P(B) = \frac{1}{6}$. What if you were interested in the event that the number that comes up is even OR a 3? The same rules that we used to calculate simple probabilities will apply here.

If we look at the sample space S we can count how many of the possible outcomes correspond to the event A OR B. We see that 2, 4, and 6 correspond to A and that 3 corresponds to B. Thus, using the rules for probability when events are equally likely, we find that $P(A \text{ OR } B) = \frac{4}{6}$, since 4 outcomes out of the 6 possible outcomes correspond to what we are interested in.

You may have noticed two things about this problem. First, the answer, $\frac{4}{6}$, is simply the sum of the two individual probabilities, $\frac{3}{6}$ and $\frac{1}{6}$. This is not a coincidence. Second, the events A and B have no outcomes in common. In probability we refer to these kinds of events as **mutually exclusive.**

Two events, A and B, are said to be *mutually exclusive* if they have no outcomes in common.

When two events are mutually exclusive, then the probability that A occurs or B occurs, **P(A OR B),** is the sum of the individual probabilities. This is known as the **simple addition rule** and is found by

$$P(A \text{ OR } B) = P(A) + P(B)$$

*Formula for simple addition rule: P(A OR B)*

The simple addition rule easily extends to any number of mutually exclusive events. For example, if A, B C, and D are four mutually exclusive events, then

$$P(A \text{ OR } B \text{ OR } C \text{ OR } D) = P(A) + P(B) + P(C) + P(D)$$

## EXAMPLE 6.5    Flipping Two Coins

*Finding* **P(A OR B)**

Remember that Joe and Tom were matching coin flips to see who would pay for dinner. The sample space for the experiment was S = {HH, HT, TH, TT}.

If A is the event that the two coins both come up tails, and B is the event that the two coins do not match, find P(A OR B).

You can see that one outcome, TT, corresponds to the event A and that two outcomes, HT and TH, correspond to the event B, so that

$$P(A \text{ OR } B) = \tfrac{3}{4}$$

You can also see that A and B have no outcomes in common. That is, A and B are mutually exclusive, so that you should be able to find P(A OR B) by adding P(A) and P(B). Since $P(A) = \tfrac{1}{4}$ and $P(B) = \tfrac{2}{4}$ (these should be easy for you by now),

$$P(A \text{ OR } B) = \tfrac{1}{4} + \tfrac{2}{4} = \tfrac{3}{4} \qquad \blacksquare$$

In probability, as in any kind of mathematics, we must be very precise in our use of words. The word OR in probability is *inclusive.* That means when we talk about the event A OR B occurring we mean that A occurs or B occurs or BOTH occur. This was not a problem when we were talking about rolling a die, since there is no way that a number can be both even and a 3!

## EXAMPLE 6.6    Credit Problems

*Simple Addition Rule*

In the example with the Chamber of Commerce and the small businesses, the Chamber of Commerce wants to know the probability that the company selected to be interviewed in depth is one of the largest (sales over $20 million) or one of the smallest (sales under $1 million).

In the last example we defined A to be the event that the company has sales of over $20 million and we found that P(A) was $\tfrac{21}{166}$. We can define event B to be the event that the company chosen has sales under $1 million. What we are trying to determine here is P(A OR B).

It is easy to see that the events A and B are mutually exclusive, that is, that they cannot have any company in common. This is the same as saying that there is no company that has sales under $1 million and over $20 million. We can use the simple addition rule here since the sample space is much too large to work with comfortably:

$$P(A \text{ OR } B) = P(A) + P(B)$$

We need to find P(B) first. The results of the survey showed that 60 companies had sales under \$1 million, so $P(B) = \frac{60}{166}$. Thus,

$$P(A \text{ OR } B) = \tfrac{21}{166} + \tfrac{60}{166} = \tfrac{81}{166} \qquad ■$$

### EXAMPLE 6.7   Product Preference

***Finding the Probability of* A OR B**

In the product preference example, a random sample of 250 production workers were asked to use three different pairs of safety glasses and to select the one that they preferred. The results of the survey found that 120 preferred product A, 85 preferred product B, and 45 preferred product C.

    We defined an experiment where a worker was selected at random for further interviewing. We let the event B = the worker preferred product B and we found that $P(B) = \frac{85}{250}$. Let the event C = the worker preferred product C.

    Find the probability that the worker selected for further interviewing preferred product B or C.

    Since the sample space for this experiment does not help us, we must use the simple addition rule. First we need to find P(C). Since 45 outcomes (workers) in the sample space preferred product C, we know that

$$P(C) = \tfrac{45}{250}$$

To find P(B OR C) we use the simple addition rule and find that

$$P(B \text{ OR } C) = \tfrac{85}{250} + \tfrac{45}{250} = \tfrac{130}{250} \qquad ■$$

    Up to this point we have been expressing probabilities as fractions without reducing them. This is only to illustrate what numbers we use to calculate the probabilities. In fact, probabilities are often expressed as decimals or percentages. In the previous example you could also say that the probability that the worker selected preferred product B or C is 0.52 or 52%. Looking at the probability as a percentage is really quite illuminating in this case. We see that while product A had the largest number of people preferring it, a majority of the workers prefer something else.

## TRY IT NOW!

### The Spinner Problem
***Calculating the Probability of* A OR B**

The sample space for the experiment of spinning the two spinners is

$$S = \{1A, 1B, 1C, 2A, 2B, 2C, 3A, 3B, 3C\}$$

Let A be the event that the first spinner comes up a 1 and let B be the event that it comes up a 3.

    Find the probability that A OR B occurs using the sample space.

Now find the same probability using the simple addition rule.

Why are the two answers the same?

---

The simple addition rule for probability that you just learned applies only when the events of interest have nothing in common, that is, they are mutually exclusive. Sometimes that will always be the case, as in the Product Preference example. It is not physically possible for a single worker to prefer more than one of the products, just as it is not possible for a single roll of a die to result in a 2 and a 3 at the same time. What happens when this is *not* the case, when it is possible that both of the events, A AND B, can occur?

Look at the example of rolling a single die. The sample space for the experiment is

$$S = \{1, 2, 3, 4, 5, 6\}$$

When we let A be the event that the number that came up was even and B be the event that the number that came up was a 3, these events had nothing in common. It is impossible for them to both occur on a *single trial* of the experiment. The law of probability worked just fine and $P(A \text{ OR } B) = \frac{3}{6} + \frac{1}{6} = \frac{4}{6}$.

Suppose that we define a new event, C, where

C = the number that comes up is exactly divisible by 3

What is the probability that the number that comes up is even or exactly divisible by 3? Remember that our definition of the word OR is that one or the other or *both* of the events can occur. This means that we are looking at any outcomes in S that are even, or exactly divisible by 3, or both even AND exactly divisible by 3. Looking at the sample space, you can see that there are four outcomes that fit this description, 2, 3, 4, and 6. Thus, using the formula for probability, $P(A \text{ OR } C) = \frac{4}{6}$. What happens if we try to use the simple addition rule?

*This is the same as asking for the probability that A OR C will occur.*

From the sample space you can see that $P(A) = \frac{3}{6}$ (there are three outcomes that are even) and $P(C) = \frac{2}{6}$ (there are two outcomes that are exactly divisible by 3). If we add these two probabilities we get: $\frac{3}{6} + \frac{2}{6} = \frac{5}{6}$. Wait! This is *not* the answer we got using the definition. What went wrong?

What went wrong is what usually goes wrong when mathematics leads to an incorrect answer—we violated the rules, in this case, the rules for using the simple addition rule. The simple addition rule is valid only if the events of interest are *mutually exclusive*. In this case, events A and C are *not* mutually exclusive; they have an outcome in common, the outcome of a 6.

---

When we calculated P(A), the outcome that the die turned up a 6 was included in that probability. Then, when we calculated P(C), the outcome of a 6 was included again. When we added the two probabilities together, the outcome got included twice, once for each event. Thus, the answer we obtained was too large by $\frac{1}{6}$, which is the probability that the die turns up a 6.

How can we adjust the simple addition rule of probability to work in situations when the events are NOT mutually exclusive? You can see from the example that the problem occurs when an outcome is included in *both* events, that is, in A AND C. The answer obtained by adding the individual probabilities was too large because the probability that both A AND C occur, P(A AND C), is included in both individual probabilities. If we consider this we can come up with a **general addition rule** for probability that considers the probability of A OR B when the events are not mutually exclusive:

*Formula for general addition rule:* P(A OR B)

$$P(A\ OR\ B) = P(A) + P(B) - P(A\ AND\ B)$$

## EXAMPLE 6.8   Flipping Two Coins

### *Calculating* P(A OR B)

In the example where Joe and Tom are matching coins, the sample space is

$$S = \{HH, HT, TH, TT\}$$

We let A be the event that both coins come up tails and we know that $P(A) = \frac{1}{4}$.

Suppose we define the event B = the coins match. What is the probability that the coins match or they both come up tails?

From the sample space we can see that two outcomes (HH and TT) correspond to the event A OR B. Thus, $P(A\ OR\ B) = \frac{2}{4}$. To use the addition rule we need to look at events A and B and decide if they have any elements in common. We see that they do, since the outcome TT satisfies both the definition of A and the definition of B. We will have to use the general rule and remember to subtract P(A AND B). Thus, we find

$$P(A\ OR\ B) = P(A) + P(B) - P(A\ AND\ B)$$

or

$$P(A\ OR\ B) = \frac{1}{4} + \frac{2}{4} - \frac{1}{4} = \frac{2}{4}$$

which we know is the correct answer.                                    ■

In Chapter 5 you learned about collecting data that involve two qualitative variables. You learned how to organize that data into a contingency table and to use the data to describe the sample. Data of this type are also important in probability problems.

## EXAMPLE 6.9   Credit Problems

### *The General Addition Rule*

The size of a company cannot be determined by sales alone. It also depends on the price of the products to a large extent, so the Chamber of Commerce also asked a question about the number of employees that the company employed. The following table is a contingency table based on the questions about sales and number of employees:

| Employees | No response | Under $1 m | $1–5 m | $6–10 m | $11–20 m | Over 20 m | Grand Total |
|-----------|-------------|------------|--------|---------|----------|-----------|-------------|
| 0–5 | 0 | 44 | 7 | 7 | 0 | 0 | 58 |
| 6–10 | 0 | 12 | 9 | 3 | 3 | 0 | 27 |
| 11–50 | 1 | 4 | 4 | 14 | 3 | 3 | 29 |
| 51–150 | 0 | 0 | 1 | 8 | 5 | 7 | 21 |
| 151–250 | 0 | 0 | 0 | 6 | 5 | 6 | 17 |
| Over 250 | 0 | 0 | 0 | 4 | 5 | 5 | 14 |
| Grand total | 1 | 60 | 21 | 42 | 21 | 21 | 166 |

Suppose that the Chamber of Commerce is interested in knowing the probability that the company chosen for further interviewing had sales over $20 million (A) or more than 250 employees (C). How can we find this probability, P(A OR C)?

If we look at the two events we see that $P(A) = \frac{21}{166}$ and $P(C) = \frac{14}{166}$. Now, because we are going to *add* the probabilities, we have to determine whether the two events have any outcomes in common. We need to know whether there are any businesses that have both sales over $20 million and more than 250 employees. Looking at the table, we see that 5 businesses are included in both events. We have to use the general addition rule:

$$P(A \text{ OR } C) = P(A) + P(C) - P(A \text{ AND } C)$$

or

$$P(A \text{ OR } C) = \tfrac{21}{166} + \tfrac{14}{166} - \tfrac{5}{166} = \tfrac{30}{166} \qquad \blacksquare$$

When you are looking for the probability of the event A OR B in a contingency table problem, it is not difficult to recognize when you need to use the general addition rule or when the simple addition rule will work. You can think about the outcomes that the events might have in common as the *intersection* of the two simple events in the table. If one event is represented by a row and the other by a column, then there will be a place where they intersect or overlap and you will need to use the general addition rule. If both of the simple events A and B are rows or both columns, then they cannot intersect, and the simple addition rule applies.

### EXAMPLE 6.10   Student Status

*Using the General Addition Rule*

Remember the School of Business that is concerned about the number of upper level students who are enrolled in the Introductory Statistics course? The school collected data on 28 students and asked the students their year and whether they had transferred to the University. The data are shown in the contingency table below:

| Year | Status Nontransfer | Transfer | Total |
|------|--------------------|----------|-------|
| Fr | 3 | 1 | 4 |
| So | 10 | 2 | 12 |
| Jr | 2 | 0 | 2 |
| Sr | 1 | 9 | 10 |
| Total | 16 | 12 | 28 |

Suppose that the committee decides to select a student at random and look at that student's records more closely. The sample space of the experiment is the twenty-eight students. Since every student is just as likely to be selected, the outcomes in S are equally likely. What is the probability that the student whose records are examined is a sophomore or a transfer student?

*In general, it is helpful to use letters to represent events that are coded to the actual data rather than just A or B.*

Let S represent the event that the student selected is a sophomore and T be the event that the student selected is a transfer student. We are looking for P(S OR T).

To start, we will find P(S) and P(T). From the table we can find that $P(S) = \frac{12}{28}$ and $P(T) = \frac{12}{28}$.

Next, we need to think about whether these events have any outcomes in common. That is, are there any students who are both sophomores AND transfer students? Again, looking at the table, we see that there are 2 students who are classified as sophomores and transfer students. These two students are included in both P(S) and P(T), so we will have to use the general addition rule. Thus,

$$P(S \text{ OR } T) = \tfrac{12}{28} + \tfrac{12}{28} - \tfrac{2}{28} = \tfrac{22}{28}$$

If we look at the table and use the cells that correspond to S OR T instead of the totals we get

$$P(S \text{ OR } T) = \frac{10 + 2 + 1 + 0 + 9}{28} = \frac{22}{28}$$

which is the same answer.   ■

 ## TRY IT NOW!

### Quality Problems
#### *Using the General Addition Rule*

The company that manufactures cardboard boxes collected data on the defect type and production shift. The data are summarized in the contingency table below:

|         |    | Shift |   |       |
|---------|----|----|----|-------|
| Defect  | 1  | 2  | 3  | Total |
| Color   | 8  | 4  | 3  | 15    |
| Printing| 6  | 5  | 2  | 13    |
| Skewness| 0  | 2  | 0  | 2     |
| Total   | 14 | 11 | 5  | 30    |

If a box has more than one defect, then it is classified by the more serious of the defects only.

Suppose that a box from the sample is selected at random and examined more closely. What is the probability that the box has a color defect?

What is the probability that the box was produced during the second shift?

Is it possible for the selected box to have a color defect and to have been produced on the second shift? If so, what is the probability?

What is the probability that the selected box will have a color defect or will have been produced on the second shift?

## 6.2.3 Probabilities as Relative Frequencies

At this point you may wonder why we need to study probability at all. It is a reasonable question if you think of probability only in terms of coins, dice, and spinners, or as random samples taken from known populations. These contexts are good for explaining how probabilities are calculated. The situations serve as models for examining probabilities for other situations.

In the previous chapters of this book you looked at methods for describing sample data taken from some population. We talked about the fact that the descriptions are exact for the sample, but are only estimates when applied to the whole population. How good those estimates are depends on many factors, such as the size of the sample and how well the sample represents the population of interest.

The same is true of probability models. When you collect sample data and calculate relative frequencies you can find, exactly, the probability that an item taken from the sample will have some characteristic(s) of interest. If the data are a good representation of the population, then we can also use the relative frequencies as estimates of the true probabilities for the population. Probabilities calculated in this way are often called **empirical** probabilities. The same rules of probability that you have already learned apply to these problems as well.

> An *empirical probability* is one that is calculated from sample data and is an estimate for the true probability.

## EXAMPLE 6.11   Restaurant Survey

### Probabilities as Relative Frequencies

A local marketing firm took a random sample of people in a large city to learn what kind of restaurants they think the city needs. The firm asked the customers to choose the type from the following list:

- Fast food
- Family restaurant
- Adult economical
- Adult moderate
- Adult upscale

ANS. $P(C) = \frac{15}{30}$, $P(2) = \frac{11}{30}$, YES, $\frac{4}{30}$, $\frac{22}{30}$.

As a result of their research they obtained the following data:

| Type of restaurant | Number of people | Relative frequency (%) |
|---|---|---|
| Fast food | 127 | 20 |
| Family restaurant | 234 | 36 |
| Adult economical | 158 | 24 |
| Adult moderate | 72 | 11 |
| Adult upscale | 56 | 9 |
| Total | 647 | 100.00 |

If we assume that the sample was a good representation of the adult residents of the city, what is the probability that a person selected at random in the city would favor some type of adult restaurant?

This question is really asking for the probability that a person would favor either an adult economical (AE), an adult moderate (AM), or an adult upscale (AU) restaurant. Since the events are mutually exclusive, we can use the simple addition rule to find the answer. What has changed is that our sample space is no longer the 647 original respondents to the survey. Instead, the sample space of the experiment becomes the possible responses that a person can give, or

$$S = \{\text{fast food, family, adult economical, adult moderate, adult upscale}\}$$

and the outcomes are no longer equally likely. We can, however, use the relative frequencies as estimates of the probabilities for each outcome:

$$P(\text{AE OR AM OR AU}) = 24\% + 11\% + 9\% = 44\% \quad\blacksquare$$

## 6.2.4 Exercises—Learning It!

**6.1**  In 1997, the competition for local phone service increased significantly as a result of changes in federal law. A survey of over 10,000 people in a large metropolitan area asked whether they rated their current local phone service provider as Excellent, Very Good–Good, Satisfactory, or Poor–Very Poor. The results were tabulated and are shown in the table below:

| Rating | Excellent | Very Good–Good | Satisfactory | Poor–Very Poor |
|---|---|---|---|---|
| % Responding | 12 | 21 | 25 | 42 |

(a)  What is the probability that a person selected at random from the area will rate his current phone service as Excellent or Very Good–Good?

(b)  What is the probability that a person selected at random from the area will not rate his service as Poor–Very Poor?

**6.2**  A company was interested in looking at the way in which employees used the 2 floating holidays that were part of their benefits packages. They surveyed 300 employees and asked them what type of job they had and how they used the holidays. The answers were tabulated into a contingency table:

| Type of Job | How Days Were Used | | |
|---|---|---|---|
| | Took Actual Holiday | Added to Vacation | Took Random Days |
| Professional | 5 | 17 | 51 |
| Clerical | 13 | 46 | 32 |
| Hourly | 53 | 78 | 5 |

(a)  What is the probability that an employee surveyed took the floating holidays on the actual holiday?

(b)  What is the probability that an employee surveyed was professional?

(c) What is the probability that an employee surveyed was clerical and added the floating holidays to his vacation?

(d) What is the probability that an employee surveyed took the floating holidays as random days off or was an hourly worker?

**6.3** A computer magazine surveyed its readers to determine how likely it was that people who planned to purchase new computers in the near future would buy a portable/notebook or desktop model. The results are tabulated in the table below:

| | *When Purchase Will Be Made* | | |
|---|---|---|---|
| **Type of Computer** | **0–3 Months** | **3–6 Months** | **6–12 Months** |
| Notebook/Portable | 34 | 156 | 258 |
| Desktop | 56 | 346 | 128 |

(a) What is the probability that a person surveyed planned to buy a desktop model in the next 3–6 months?

(b) What is the probability that a person surveyed planned to buy a computer in the next 0–3 months or that the person was planning to buy a desktop model?

(c) What is the probability that a person was planning to buy a notebook computer and that the person planned to make the purchase in the next 0–3 or 3–6 months?

(d) What is the probability that a person did not plan to make a purchase in the next 6 months?

**6.4** A marketing firm in the Northeast was looking at the type of medication that people with allergies took during the autumn allergy season. In particular, the firm wanted to know whether the person took medication daily and whether that medication was prescribed by a physician or purchased over the counter. The results of the survey are in the table:

| | *Type of Medication* | |
|---|---|---|
| **Frequency of Taking Medicine** | **Prescription** | **Over the Counter** |
| Daily | 86 | 43 |
| As needed (sporadic) | 23 | 156 |

(a) What is the probability that an allergy sufferer took prescription medication daily?

(b) What is the probability that an allergy sufferer took over-the-counter medication or that he took the medication daily?

(c) What is the probability that an allergy sufferer took medication daily and that he used over-the-counter medication?

**6.5** A report by the Department of Justice on rape victims reports on interviews with 3721 victims. The attacks were classified by age of the victim and the relationship of the victim to the rapist. The results of the study are given below:

| | *Relationship of Rapist* | | |
|---|---|---|---|
| **Age of Victim** | **Family** | **Acquaintance or Friend** | **Stranger** |
| Under 12 | 153 | 167 | 13 |
| 12 to 17 | 230 | 746 | 172 |
| Over 17 | 269 | 1232 | 739 |

(a) What is the probability that a victim was under 12 years of age?

(b) What is the probability that a victim was between 12 and 17 and that the rapist was a member of the family?

(c) What is the probability that a victim was under 12 or that the rapist was an acquaintance or friend?

(d) What is the probability that the victim was not under 12 years of age?

(e) What is the probability that the rapist was not a family member, acquaintance, or friend?

## 6.3 RANDOM VARIABLES

When you learned the different techniques for summarizing and analyzing data, you learned that while qualitative data are important for understanding and interpreting the information obtained from a set of data, there is not much you can do with the data itself. Most statistical analyses use numerical or quantitative data as their basis. In much the same way, we also prefer to discuss probabilities for experiments whose outcomes are numerical.

You know from algebra that a variable is a quantity whose value can change or vary. In algebra the exact value that the variable takes on depends on the equation that includes it. We have also referred to the different kinds of data that can be collected from a population as variables. The exact value that a statistical variable will take on depends on the laws of chance or probability. When the variables in question are quantitative, they are known as **random variables.**

> A *random variable, X,* is a quantitative variable whose value varies according to the rules of probability.

When the outcome of an experiment is a random variable, then the elements of the sample space are all of the possible values that the variable can have. Random variables, like data, can be discrete (integers) or continuous (real numbers). In our initial discussion of random variables we will consider only the discrete case.

In the experiment of rolling a single die and recording the number of spots that are on the top face the sample space for the experiment can be written as

$$S = \{1, 2, 3, 4, 5, 6\}$$

The outcomes are numerical and the exact value that will turn up varies. We know from the rules of probability that each of the values is equally likely, or has an equal probability of happening. Thus, $X =$ the number of spots on the top face of a die is a random variable.

### EXAMPLE 6.12   Tossing Two Coins

**Defining a Random Variable**

In the previous section we looked at an experiment in which two coins were tossed. The sample space of that experiment was

$$S = \{HH, HT, TH, TT\}$$

*We always use capital letters, like X, to represent the random variable, and lowercase letters, such as x, to represent the values of the random variable.*

This experiment does not involve a random variable, since the possible outcomes are not numerical. It is possible, however, to describe the outcomes of the experiment numerically by defining a random variable for the experiment. Let $X =$ the number of heads that appear on the two coins. The possible values of $X$ are $x = 0, 1,$ and 2. Every time you perform the experiment you do not know what will happen, but the laws of probability give you some insight into what is likely to occur. ∎

### 6.3.1 Probability Distribution of a Discrete Random Variable

The rules of probability that describe the way a random variable behaves are known as the **probability distribution** of the random variable. The probability distribution of a discrete random variable assigns a probability to each of the possible values that can happen.

> The *probability distribution* of a random variable, $X$, written as **p(x)**, gives the probability that the random variable will take on each of its possible values.

*We read p(x) as "p of x."*

The notation that we use for the probability distribution of a random variable is similar to the notation for the probability of an event. For a random variable,

$$p(x) = P(X = x) \quad \text{for all possible values of } X$$

Most often the probability distribution is written in the form of a table.

## EXAMPLE 6.13   Tossing Two Coins

*Writing the Probability Distribution*

In the experiment of tossing two coins we saw that $X =$ the number of heads is a random variable that can take on three values: 0, 1, and 2. Using the rules of probability that we learned, we can find the probability distribution of $X$, the number of heads in two tosses of a coin:

$p(0) = P(X = 0)$ corresponds to only one outcome, TT, so $p(0) = \frac{1}{4}$
$p(1) = P(X = 1)$ corresponds to two outcomes, HT and TH, so $p(1) = \frac{2}{4}$
$p(2) = P(X = 2)$ corresponds to one outcome, HH, so $p(2) = \frac{1}{4}$

In table form we would write this as

| $x$ | 0 | 1 | 2 |
|---|---|---|---|
| **p(x)** | $\frac{1}{4}$ | $\frac{2}{4}$ | $\frac{1}{4}$ |

The rules for probability distributions are the same as the rules for probabilities. For each value of the random variable, $X$,

1. $0 \le p(x) \le 1$ (probabilities are numbers between 0 and 1 inclusive)
2. $\Sigma p(x) = 1$ for all values of $x$ (the probabilities must sum to 1)

Notice that the outcomes of a random variable are *mutually exclusive*. This means that whenever we are interested in finding the probability that the random variable will take on one of its values OR another of its values we can use the simple addition rule.

## EXAMPLE 6.14   Newspaper Sales

*Finding Probabilities from a Probability Distribution*

The number of copies of *The Wall Street Journal (WSJ)* that are sold daily by a convenience store is a random variable $X$, which is described by the probability distribution

| $x$ | 0 | 1 | 2 | 3 | 4 | 5 |
|---|---|---|---|---|---|---|
| **p(x)** | 0.10 | 0.12 | 0.25 | 0.30 | 0.20 | 0.03 |

What is the probability that on any given day the convenience store sells exactly two copies of the *WSJ*?

We can obtain this answer directly from the table:

$$p(2) = 0.25$$

What is the probability that two or three copies are sold on any given day?

It is first important to recognize that we are trying to find $P(X = 2 \text{ OR } X = 3)$. Since the values of a random variable are mutually exclusive, we can simply add the probabilities to obtain the answer:

$$P(X = 2 \text{ OR } X = 3) = p(2) + p(3) = 0.25 + 0.30 = 0.55 \qquad \blacksquare$$

## Notation and Probabilities

Very often we are interested in the probability that a random variable takes on any one of a set of its possible values. In particular, we might be interested in finding the probability that the random variable takes on a value that is **"at least x,"** **"more than x,"** **"at most x,"** **"less than x,"** **"between $x_1$ and $x_2$,"** or **"between $x_1$ and $x_2$ inclusive."** There is nothing new that you need to know to find the probabilities. It just takes a little practice to be able to recognize each of the different problems and to write down what you are looking for. Since the words can get cumbersome, a standard notation is used.

As an example, we will use a random variable $X$ that can take on values of $x = 0, 1, 2, 3, \ldots, n$. Table 6.1 provides a summary of each problem, what it means, and the correct notation.

**TABLE 6.1   Probability Notation Summary**

| Find the probability that X takes on a value that is . . . | What it means | Notation |
|---|---|---|
| **at least x** | All of the values of the random variable that are the value $x$ or larger (up to $n$) | $P(X \geq x)$ |
| **more than x** | All of the values of the random variable that are larger than the value $x$ (up to $n$) | $P(X > x)$ |
| **at most x** | All of the values of the random variable that are the value $x$ or less | $P(X \leq x)$ |
| **less than x** | All of the values of the random variable that are smaller than the value $x$ | $P(X < x)$ |
| **between $x_1$ and $x_2$ inclusive** | All of the values of the random variable that start with the value $x_1$ and go up to and include the value $x_2$ | $P(x_1 \leq X \leq x_2)$ |
| **between $x_1$ and $x_2$** | All of the values of the random variable that are larger than the value $x_1$ and smaller than the value $x_2$ | $P(x_1 < X < x_2)$ |

### EXAMPLE 6.15   Newspaper Sales

*Finding Interval Probabilities*

The convenience store that sells the *WSJ* wants to know a bit more about the probability that it will sell the newspaper. The probability distribution of $X$, the number of copies of the *WSJ* sold per day, is given by

| $x$ | 0 | 1 | 2 | 3 | 4 | 5 |
|---|---|---|---|---|---|---|
| $p(x)$ | 0.10 | 0.12 | 0.25 | 0.30 | 0.20 | 0.03 |

In particular, the store needs to sell at least 3 copies per day to make a profit from the sales. What is the probability that the store will make a profit on any given day?

We are looking for the probability that the store sells at least 3 copies, or $P(X \geq 3)$. This means $P(X = 3 \text{ OR } X = 4 \text{ OR } X = 5)$, which is calculated as

$$p(3) + p(4) + p(5) = 0.30 + 0.20 + 0.03 = 0.53$$

What is the probability that the store will not make a profit?

We can answer this question directly by determining that the store will not make a profit if it sells less than 3 copies of the newspaper and finding $P(X < 3)$. We can also do this problem by making use of the complement of an event. The events "make a profit" or "$X \geq 3$" and "do not make a profit" or "$X < 3$" are complements of each other. This means that their probabilities must sum to 1. Since we already know that $P(X \geq 3) = 0.53$, we can find $P(X < 3)$ by taking

$$1 - P(X \leq 3) = 1 - 0.53 = 0.47$$ ■

 **TRY IT NOW!**

### Defective Diskettes
*Finding Interval Probabilities*

A company that sells computer diskettes in bulk packages for a warehouse club outlet knows that the number of defective diskettes in a package is a random variable with the probability distribution given below:

| $x$ | 0 | 1 | 2 | 3 | 4 | 5 | 6 |
|---|---|---|---|---|---|---|---|
| $p(x)$ | 0.30 | 0.21 | 0.12 | 0.10 | 0.10 | 0.09 | 0.08 |

Find the probability that a package of the diskettes will contain at least 3 defective disks.

Find the probability that the package will contain between 2 and 5 defective diskettes.

*(continued)*

Find the probability that the number of defective diskettes will be at most 2.

## 6.3.2 Probability Histograms

Random variables and their probability distributions are the models for the populations from which our sample data are taken. You learned in Chapter 3 that you can display quantitative data using a relative frequency table or a relative frequency histogram. In much the same way, a random variable can be displayed with a probability distribution table or a probability distribution histogram.

### EXAMPLE 6.16   Newspaper Sales

*Creating a Probability Histogram*

The convenience store that sells the *WSJ* would like to see what the distribution of newspaper sales looks like. The store creates a probability histogram for the random variable:

**Probability Distribution of Number of WSJs Sold**

The histogram shows that the probability distribution of sales is not very variable, approximately symmetric, and centered at about 3 newspapers per day.  ■

In a probability histogram the area of each rectangle is equal to the probability that the random variable will take on the given value. This may not seem important right now, but we use this fact later in the chapter when we move from discrete to continuous random variables.

# *TRY IT NOW!*

### Defective Diskettes

*Creating a Probability Histogram*

The company that sells computer diskettes in bulk packages for a warehouse club would like to have a picture of how the number of defective diskettes in a package behaves. The probability distribution is given below:

| $x$ | 0 | 1 | 2 | 3 | 4 | 5 | 6 |
|-----|------|------|------|------|------|------|------|
| $p(x)$ | 0.30 | 0.21 | 0.12 | 0.10 | 0.10 | 0.09 | 0.08 |

Create a probability histogram for the number of defective diskettes.

Use the probability histogram to describe the distribution of the number of defective diskettes in a package.

## 6.3.3 Exercises—Learning It!

6.6   The number of employees who call in sick on any given day in a small business is a random variable with probability distribution:

| $x$ | 0 | 1 | 2 | 3 | 4 | 5 | 6 |
|-----|------|------|------|------|------|------|------|
| $p(x)$ | 0.10 | 0.23 | 0.18 | 0.16 | 0.13 | 0.10 | 0.10 |

(a)  What is the probability that on any given day at most 4 employees call in sick?

(b)  What is the probability that between 2 and 4 employees call in sick?

(c)  What is the probability that more than 4 employees call in sick?

ANS.  Probability Distribution of Number of Defective Diskettes SKEWED RIGHT, MUCH MORE LIKELY TO FIND A SMALL NUMBER OF DEFECTIVES

**6.7**    The number of members who cannot get a tee time at a local country club is a random variable with probability distribution given by

| x | 0 | 1 | 2 | 3 | 4 | 5 | 6 | 7 | 8 |
|---|---|---|---|---|---|---|---|---|---|
| p(x) | 0.11 | 0.12 | 0.13 | 0.19 | 0.12 | 0.09 | 0.09 | 0.08 | 0.07 |

(a) What is the probability that between 2 and 5 members inclusive cannot get tee times?

(b) What is the probability that less than 3 cannot get tee times?

(c) What is the probability that at least 6 cannot get tee times?

(d) What is the probability that at most 4 cannot get tee times?

**6.8**    The number of times that a person gets a busy signal when calling the local cable television office is a random variable with the following probability distribution:

| x | 0 | 1 | 2 | 3 | 4 |
|---|---|---|---|---|---|
| p(x) | 0.23 | 0.34 | 0.17 | 0.15 | ? |

(a) What is p(4)?

(b) What is the probability that a person will not get a busy signal?

(c) What is the probability that a person gets at least 1 busy signal?

(d) What is the probability that a person gets more than 2 busy signals?

**6.9**    A company that packages small items for resale is looking at the problems of incorrect packaging. The company finds that in a box that is supposed to contain two dozen items, the number of missing items is a random variable with probability distribution:

| x | 0 | 1 | 2 | 3 | 4 | 5 | 6 |
|---|---|---|---|---|---|---|---|
| p(x) | 0.13 | 0.17 | 0.26 | 0.30 | 0.07 | 0.05 | 0.02 |

(a) What is the probability that in a box of 2 dozen, exactly 3 are missing?

(b) What is the probability that the number of missing items is less than 4?

(c) What is the probability that there are exactly 20 items in the box?

(d) What is the probability that there are more than 20 items in the box?

**6.10**   A large airline keeps track of the number of no-shows for one of its most important commuter flights. Over time the airline has found that the number of ticketed passengers who do not show up is a random variable with the following probability distribution:

| x | 0 | 1 | 2 | 3 | 4 | 5 | 6 | 7 | 8 |
|---|---|---|---|---|---|---|---|---|---|
| p(x) | 0.05 | 0.08 | 0.13 | 0.23 | 0.18 | 0.13 | 0.08 | 0.06 | 0.06 |

(a) What is the probability that at least 3 ticketed passengers do not show up for the flight?

(b) What is the probability that between 2 and 5 passengers do not show up for the flight?

(c) What is the probability that not more than 6 passengers do not show up for the flight?

(d) The aircraft used for the flight has 35 seats. If the airline routinely overbooks the flight by 4 passengers, what is the probability that on any given day every ticketed passenger who shows up will get a seat?

# 6.4   THE BINOMIAL PROBABILITY DISTRIBUTION

In the previous section you learned the definition of a random variable and a probability distribution. Each of the random variables that you looked at was described by a probability distribution that was given in a table. The random variables represented many different types of data that might be collected in a statistical study. In fact, most of the random variables that we see in the real world fall into specific categories and can be described by a set of special models or probability distributions. We now look at one of these models, the binomial probability distribution, in detail.

## 6.4.1 The Binomial Model

One of the most common types of data that people collect is data on the number of times that some phenomenon occurs in a sample of given size. For example, you may be interested in the number of people in a sample who are in favor of certain legislation or in the number of people who like a new flavor of ice cream. The sample does not have to consist of people; it could be the number of defective diskettes in a box of 10, or the number of times a coin turns up heads in a certain number of tosses. The random variable in each case is the *number of times the phenomenon occurs* in the sample. All of these types of data are examples of a **binomial random variable.**

> A *binomial random variable* is the number of successes in $n$ trials or in a sample of size $n$.

Certain characteristics define binomial random variables:

- There are a fixed number of identical trials of an experiment. *This is the same as taking a sample of size n. Just think of each trial as selecting the next item for the sample.*
- The outcome for each trial of the experiment can be classified in one of two ways: a *success,* S (when the phenomenon of interest happens) or a *failure,* F (when the phenomenon of interest does not happen).
- The probability that a success occurs in any sample element or on any trial of the experiment, $\pi$, is the same for each element or trial. This also means that the probability of a failure, which is $1 - \pi$, is also constant. *This means that nothing happens over the course of the experiment to change the probability of a success, such as a change in population.*
- The trials of the experiment are independent. *This means that the outcome from one trial does not affect the outcomes of subsequent trials.*
- The random variable is the number of successes that occur in the $n$ trials of the experiment.

In finding real-world situations that exactly fit the characteristics of a binomial random variable when we are sampling from a population it is difficult to ensure that the trials are independent of each other and that $\pi$, the probability of a success on any trial or the proportion of successes in the population, remains constant. Keep in mind that the reason we want the trials to be independent is that if they are not, then the probability of a success will change from trial to trial. This is a violation of the third characteristic. In fact, the only way to ensure independence when sampling, is to sample from a finite population with replacement or from an infinite population.

The first problem, the idea of sampling with replacement, is not very appealing from a practical point of view. Since the purpose of statistics is to obtain information, sampling the same item or person repeatedly will not add to the information contained in the sample. The second problem, the idea of an infinite population, is more philosophical. Although there are no truly infinite populations, many populations are infinite for all practical purposes, such as the number of people in the world (or even in a large country for that matter) or all of the production, past and present, of a machine or factory.

What happens when the population from which we sample is finite and we do not sample with replacement?

### EXAMPLE 6.17   CD Jewel Cases

***Sampling Without Replacement***

Suppose that you are inspecting a shipment of CD jewel cases for cracks in the cover. The box from which you are sampling contains 30 cases, of which 5 have cracked covers. The first time that you select an item from the population, the probability of getting a case with a cracked cover (the proportion of successes in the population) is $\frac{5}{30} = 0.167$. Now, you certainly are not going to sample with replacement in this situation. What would you gain by throwing a case you have already inspected back in the box, particularly if it is defective?

The next time you sample an item from the box, what is the probability of a success? If you are thinking. "It depends on what happened the first time," then you are absolutely correct! Look at the diagram below. You see that as you continue to sample from the box, the probability of obtaining a case with a cracked cover on *that trial* is different from the previous trial; that is, $\pi$ changes from trial to trial. The trials are not independent because the outcome of one trial directly affects the outcome of the next!

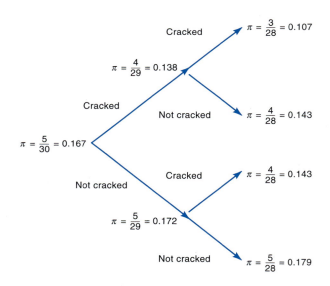

What could we do to fix the problem? The real issue is that when the denominator in the proportion is small and changing, the value of the fraction changes considerably. What happens if the denominator is much larger? Will the change still be noticeable?

Let's look at the same problem, but with everything increased proportionally. Suppose that the box contains 3000 jewel cases, of which 500 have cracked covers. In this case $\pi$, the probability or proportion of cracked cases, is still 0.167 when we begin sampling. Now what happens to $\pi$ as items are sampled and not replaced in the population? The next figure illustrates the results:

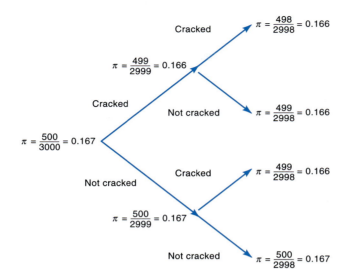

You see that although $\pi$ changes, the changes are much less noticeable. In fact, if you were writing $\pi$ to the nearest percent you would not see the change at all. ∎

Clearly, as the population gets larger the problem gets smaller. The magnitude of the change in $\pi$ depends on two things: the size of the population and the size of the sample that is drawn from the population. In general, if the population size is large and the sample size is small relative to the population size, then the random variable represented by the number of successes in the sample can be assumed to be binomial.

*Usually the population must be at least 1000 to ensure that the changes do not occur before the third decimal place and the sample size should be no more than 10% of the population.*

## EXAMPLE 6.18   Credit Problems

### Identifying a Binomial Random Variable

In addition to identifying variables, the Chamber of Commerce that is interested in the credit problems of small businesses asked the small businesses a series of questions about the problems they might be experiencing. One of these questions, to which the businesses responded Yes or No, was "Are you experiencing credit problems?" In previous applications of this questionnaire, the Chamber of Commerce found that approximately 10% of the businesses have reported having such problems.

Is the number of businesses who reported having credit problems in the sample of 166 a binomial random variable?

To answer this question we need to consider the five characteristics of the binomial random variable.

1. Yes. There are a fixed number of trials, in this case 166, and the selection was random.

2. Yes. Each trial can result in only one of 2 outcomes: a success (the company is having credit problems) or a failure (the company is not having credit problems).

3. Yes. If we assume that the time to return the survey was short enough, there would be no economic situations that would change the proportion from 10%.

4. Yes. Here we have to be careful. The sample of 166 is taken from a population of 1536 small businesses in the city. If 10% of the businesses do have credit problems, then that would correspond to 154 successes in the population. In

the worst case, the probability of a success could change to 0 (if all 154 successes in the population wound up in the sample of 166), but that is nearly impossible. (If it did happen, then either the sample was not random or else the estimate of 10% is not at all realistic.) Most likely it would change from 154/1536 (10.03%) at the beginning of the sampling to about 137/1370 (10%) when the entire sample of 166 was completed.

5. Yes. The data collected are the number of businesses in the sample that are experiencing credit-related problems. ■

In general, if the sample size is about 10% of the population then the assumption that the probability of a success does not change is a reasonable one.

### EXAMPLE 6.19   Wearing Seat Belts

*Identifying a Binomial Random Variable*

The state of Connecticut has seatbelt laws for drivers and passengers. It is known that 70% of the drivers in Connecticut wear their seatbelts when they drive. A police trap selects a random sample of five drivers and counts the number that are wearing seatbelts. Is the number of drivers wearing seatbelts in the sample of five a binomial random variable?

To answer this question we need to consider the five characteristics of a binomial random variable.

1. Yes. There are a fixed number of identical trials if we assume that each driver is selected randomly.

2. Yes. Each trial can result in only one of two outcomes: a success (the driver is wearing a seatbelt) or a failure (the driver is not wearing a seatbelt).

3. Yes. If we assume that the sample is taken over a short enough period of time that driving habits will not have changed drastically, then the 70% will remain constant.

4. Yes. If we assume that there is no way that the driver of a car that is stopped can influence the outcome for the next driver then the trials are independent. The population of the drivers in Connecticut is very large and the sample size is definitely small relative to the population so the trials will be independent and the proportion of successes in the population will remain essentially constant.

5. Yes. The data that are collected are the number of people wearing seatbelts in the sample of five selected at random. ■

It is important that you learn to recognize problems to which the binomial distribution applies. It is easy enough to do in a textbook (after some practice) but when you are actually practicing statistics you will need to know what to look for. Keep in mind the five characteristics of a binomial random variable when you are looking at both problems in this book and at data that have been collected in real statistical studies.

 **TRY IT NOW!**

### Loan Defaults

*Recognizing a Binomial Random Variable*

While the Chamber of Commerce is concerned about the problems of small businesses, it must also be sensitive to the problems that the lending institutions have when issuing credit. One of the problems that banks have with small businesses is default on loan payments. It is estimated that approximately 20% of all small businesses with less than 50 employees are at least six months behind in loan payments. The Chamber of Commerce that surveyed the small businesses of a city wants to look at this problem in more detail. It finds that of the 1536 small businesses in the city, 965 have less than 50 employees. It randomly selects 25 of these small businesses, check their credit histories, and count the number of companies in the sample of 25 that are at least six months behind in loan payments.

Does this qualify as a binomial probability distribution?

## 6.4.2 The Binomial Probability Distribution

Now that we understand binomial random variables, we need to know how to find their probability distribution. The probability distribution for the binomial random variable will depend on two quantities or parameters: $n$, the number of trials or the sample size, and $p$, the proportion of successes in the population. The random variable $X$ is the number of successes in $n$ trials of the experiment.

*Remember, we use parameters for populations, and a probability distribution is a model for a population.*

The probability distribution of $X$ is determined by this formula:

$$p(x) = \frac{n!}{(x!)(n-x)!} \pi^x (1-\pi)^{n-x} \quad \text{for } x = 0, 1, 2, \ldots, n$$

*Binomial probability distribution*

The formula considers the sample size, $n$, the probability of a success, $\pi$, and the value of $X$ that you are interested in, $x$. The number of trials of the experiment defines the possible values that the random variable can have. If there are $n$ trials of the experiment, then the smallest number of successes that can occur is 0 and the largest number of successes is $n$. Thus, a binomial random variable, $X$, can have values $x = 0, 1, 2, \ldots, n$. The probability of a success on any trial, $\pi$, will determine how the probabilities are distributed over the values of $X$. When $\pi$ is large, then you will expect to see a lot of successes in the sample, so the higher values of $X$ will have the larger probabilities. When $\pi$ is small, the lower values of $X$ will have the larger probabilities.

*The terms that contain the ! are known as factorials. $n!$ is equal to $n \times (n-1) \times \cdots \times 1$. The other factorial terms are calculated in a similar manner.*

You do not really need to know how to use the formula to find the probability distribution of a binomial random variable. There are tables for this purpose for many different combinations of $n$ and $\pi$. These tables are identical in use to the probability distribution tables that you used in the previous section. A good set of these tables is found in Appendix A at the end of this book.

### Binomial Probability Tables

The probability distribution tables for the binomial distribution are classified according to the value of $n$, the number of trials of the experiment. There are tables for values of $n$ from 5 to 30. Each table covers a range of values for $\pi$ from 0.05 to 0.95. A sample of some of the table for $n = 5$ is shown in Figure 6.1 on page 258.

*Most statistical software packages, such as Minitab and SPSS, and many spreadsheets, such as Excel, find binomial probabilities.*

| n = 5 | | | | | | π | | | | | |
|---|---|---|---|---|---|---|---|---|---|---|---|
| **x** | **0.05** | **0.10** | **0.20** | **0.25** | **. . .** | **0.50** | **0.60** | **0.70** | **0.75** | **. . .** | **0.90** | **0.95** |
| **0** | 0.774 | 0.590 | 0.328 | 0.237 | . . . | 0.031 | 0.010 | 0.002 | 0.001 | . . . | 0.000 | 0.000 |
| **1** | 0.204 | 0.328 | 0.410 | 0.396 | . . . | 0.156 | 0.077 | 0.028 | 0.015 | . . . | 0.000 | 0.000 |
| **2** | 0.021 | 0.073 | 0.205 | 0.264 | . . . | 0.313 | 0.230 | 0.132 | 0.088 | . . . | 0.008 | 0.001 |
| **3** | 0.001 | 0.008 | 0.051 | 0.088 | . . . | 0.313 | 0.346 | 0.309 | 0.264 | . . . | 0.073 | 0.021 |
| **4** | 0.000 | 0.000 | 0.006 | 0.015 | . . . | 0.156 | 0.259 | 0.360 | 0.396 | . . . | 0.328 | 0.204 |
| **5** | 0.000 | 0.000 | 0.000 | 0.001 | . . . | 0.031 | 0.078 | 0.168 | 0.237 | . . . | 0.590 | 0.774 |

**FIGURE 6.1**   Portion of binomial table for $n = 5$

*Do not forget that decimals and percentages are equivalent!*

The two columns that are shaded are the probability distribution table for $n = 5$ and $\pi = 0.70$ (or 70%). If you just take those columns and transpose as shown in Figure 6.2, then the result will look like every probability distribution table you have seen before!

Except for learning to recognize binomial random variables and using the tables, there is nothing new to learn about answering questions involving binomial random variables.

| x | 0 | 1 | 2 | 3 | 4 | 5 |
|---|---|---|---|---|---|---|
| **p(x)** | 0.002 | 0.028 | 0.132 | 0.309 | 0.360 | 0.168 |

**FIGURE 6.2**   Binomial probability distribution for $n = 5$ and $\pi = 0.70$

## EXAMPLE 6.20   Wearing Seatbelts

### Using the Binomial Probability Tables

The example about the Connecticut drivers identified the random variable for the problem as a binomial random variable with $n = 5$ and $\pi = 0.70$. What is the probability that, in the sample of 5 drivers, exactly 3 are wearing seatbelts?

Looking at the table in Figure 6.1 or 6.2, p(3) = 0.309.

What is the probability that in a sample of 5 drivers taken at most 2 are wearing seatbelts? Again, from the table in Figure 6.1 or 6.2,

$$P(X \le 2) = p(0) + p(1) + p(2) = 0.002 + 0.028 + 0.132 = 0.162$$   ■

You can see that there is really nothing new involved in actually solving problems that involve the binomial distribution. The parts that require some thinking and work are recognizing the problem as binomial, identifying the parameters $n$ and $\pi$ for the particular problem, and finding the correct table.

## TRY IT NOW!

### Loan Defaults

#### Solving Binomial Probability Problems

The Chamber of Commerce that is checking credit problems of small businesses estimated that 20% of all small businesses were at least six months behind in loan payments. The Chamber of Commerce took a random sample of 25 small businesses and counted the number of the businesses that were at least six months behind in loan payments.

Define a success for this problem.

Describe the random variable, $X$, in words.

Find the parameters of the binomial distribution for this problem.

Find the probability that in the sample of 25 businesses less than 6 were at least six months behind in loan payments.

Find the probability that between 4 and 9 inclusive were at least 6 months behind in loan payments.

It is important to recognize that the definition of a success is critical to problem solving with the binomial distribution. A success is not necessarily always something good, nor does it have to stay the same in a given problem. To determine what the success is, you need to look at the question that needs to be answered, define a success, and use the appropriate value of $\pi$.

### EXAMPLE 6.21   Parking Tickets

*Calculating Binomial Probabilities*

A local watchdog agency has been looking at parking problems outside the city court building. It estimates that 40% of all of the cars parked in the metered lot receive parking tickets for meter violations. The agency decides to take a random sample of 10 cars from the lot and check to see if they have a parking ticket on the windshield.

(a) Find the probability that, of the 10 cars sampled, exactly 6 have parking tickets.

For this problem a success is having a parking ticket and so we use the table with $n = 10$ and $\pi = 0.40$ and find that $P(X = 6) = 0.111$.

(b) Find the probability that between 4 and 7 inclusive do not have parking tickets.

*Caution! No matter what you may think, it is NOT easier to keep using $\pi = 0.40$ and try to change the question to be in terms of cars having parking tickets!*

For this problem a success is *not* having a parking ticket. Since 40% of the cars have parking tickets, we can use the definition of the complement to find that 100% − 40% = 60% do not have parking tickets.

Use $n = 10$ and $\pi = 0.60$ to get $P(4 \le X \le 7) = 0.111 + 0.201 + 0.251 + 0.215 = 0.778$.

If the watchdog group found in its sample of 10 cars that none of the cars had parking tickets, would you think that the estimate of 40% was reasonable?

With $n = 10$ and $\pi = 0.40$ the probability that no cars have tickets, $P(X = 0)$, is 0.006, which is very small. Thus, it is highly unlikely that if 40% of the cars get tickets that the sample of 10 would find no cars with tickets. The estimate seems to be high. ∎

The previous example is a preview of what the next part of statistics that you will study is all about. In the first part of this book you learned how to describe data that are collected and how to calculate different sample statistics. Now you are learning about probability models and how they can be used to determine how likely it is that different events will occur when we assume some parameters for our population data. In the previous example we looked at how well what we observed (the data) fit the probability model (binomial with $\pi = 40\%$ and $n = 10$). We used the probability that such an event would happen to come to the conclusion that the model did not seem appropriate. In the remainder of this book you will learn the more formal methods of hypothesis testing to accomplish this same thing.

## 6.4.3 The Mean and Standard Deviation of the Binomial Distribution

Since probability distributions are models of populations, and random variables are numerical, it makes sense that just like quantitative sample data, they have means and standard deviations. There are some general formulas for calculating the mean and standard deviation of a random variable, but we will concentrate on the probability distributions that we study.

Since probability distributions are population models, their means and standard deviations are represented by the Greek letters $\mu$ and $\sigma$. In particular, for a binomial random variable, $X$, the **mean, $\mu$,** and the **standard deviation, $\sigma$**, are found using the following formulas:

$$\mu = n\pi \quad \text{and} \quad \sigma = \sqrt{n\pi(1 - \pi)}$$

*Binomial mean and standard deviation*

You see that the mean and standard deviation depend on the parameters of the probability distribution, $n$ and $\pi$. The formula for the mean is actually quite intuitive if you think about it. If you knew, for example, that 40% of a certain population wore eyeglasses, and you took a sample of 10 people from that population, how many of the 10 would you *expect* to wear glasses? Instinctively you would take 40% of the 10, to get 4 people who wear glasses. From the formula for $\mu$ we would get

$$\mu = n\pi = (10)(0.40) = 4 \text{ people}$$

The formula for the standard deviation is really not at all intuitive and deriving it is beyond the scope of this text. To see how it is used, we can calculate the standard deviation of our example and find that

$$\sigma = \sqrt{n\pi(1 - \pi)} = \sqrt{(10)(0.40)(0.60)} = 1.55 \text{ people}$$

Thus, we know that if $n = 10$ and $\pi = 0.40$, then the number of people who wear glasses is a random variable with a mean of 4 and a standard deviation of 1.55.

## EXAMPLE 6.22   Credit Problems

### Calculating the Mean and Standard Deviation of a Binomial Random Variable

With a sample of $n = 166$ we could not use the tables to determine probabilities for the number of small businesses that reported having credit problems. Most statistical software packages will calculate probability distributions for binomial random variables for any values of $n$ and $\pi$, but we can find the mean and standard deviation of this random variable without any problems.

The mean represents the expected number of businesses that are experiencing credit problems in the sample of 166. That is, with $n = 166$ and $\pi = 0.10$ the mean is:

$$\mu = n\pi = (166)(0.10) = 16.6 \text{ businesses}$$

The standard deviation of the number of small businesses in 166 that report credit problems is

$$\sigma = \sqrt{n\pi(1 - \pi)} = \sqrt{(166)(0.10)(0.90)} = 3.87 \text{ businesses} \qquad \blacksquare$$

## EXAMPLE 6.23   Wearing Seatbelts

### Finding the Mean and Standard Deviation of a Binomial Random Variable

The Department of Transportation in Connecticut would like to know how many drivers the state troopers should expect to find wearing their seatbelts at the checkpoint. It decides to calculate the mean of the binomial random variable.

In this case, $n = 5$ and $\pi = 0.70$, so it calculates that

$$\mu = n\pi = (5)(0.70) = 3.5 \text{ people}$$

The Department would also like to know the standard deviation of the number that would be wearing seatbelts. In this case

$$\sigma = \sqrt{n\pi(1-\pi)} = \sqrt{(5)(0.70)(1-0.70)} = \sqrt{(5)(0.70)(0.30)} = \sqrt{1.05} = 1.02 \text{ people}$$ ■

Remember that the mean and standard deviation tell us something about how random variables behave. In particular, they tell us where the center of the probability distribution is located and how much the random variable will vary around that center.

## Try It Now!

### Loan Defaults
*Calculating the Mean and Standard Deviation of a Binomial Random Variable*

The Chamber of Commerce that was looking at the loan defaults for small businesses wants to know the mean and standard deviation for the binomial random variable with $n = 25$ and $\pi = 0.20$.

Find the mean and standard deviation of the number of small businesses in 25 that will default on their loans.

How can we use our knowledge of the mean and standard deviation? One thing we can do is to compare our knowledge of what *should* happen, to the reality of what *did* happen, to get some idea of how well the reality fits the theory. This is the basis for the work you will do in the remainder of this book, *inferential statistics*. Right now, we will look at this work in an informal way.

### EXAMPLE 6.24   Credit Problems

*Using the Mean and the Standard Deviation*

The Chamber of Commerce wonders whether or not the current random sample of 166 businesses supports its belief that 10% of the small businesses in the city are experiencing credit problems. It already has the mean and standard deviation for its model, so it decides to compare the model to the actual data.

The mean of the binomial random variable with $n = 166$ and $\pi = 0.10$ is 16.6 and the standard deviation is 3.87. From the data, the Chamber of Commerce analysts find that the number of businesses that answered yes to the question of credit problems was 25. Clearly, 25 is not equal to 16.6, but how different is it?

ANS. $\mu = 5$, $\sigma = 2$

If they consider the standard deviation they can calculate the number of standard deviations that their data value is from the mean:

$$z = \frac{25 - 16.6}{3.87} = 2.17$$

They know that the empirical rule holds only for symmetric distributions and they have no idea whether that applies here, but they do know that 2.17 standard deviations away from the mean is at least possibly unusual.

As a second check the analysts use their data to estimate the probability that a small business will report having credit problems. They found 25 such businesses in their sample of 166, so they estimate the probability that a business will report having credit problems to be $25/166 = 0.1506$ or approximately 15%. This is different from the 10% they were expecting, but the difference may just be a result of sampling.

As a third check they use a statistical software package to calculate the probability of having 25 successes in a sample of 166 when $\pi = 0.10$. They find that the probability is 0.0108, which is only slightly more than 1%. It would seem that perhaps this sample does not fit their earlier beliefs. ∎

## EXAMPLE 6.25   Parking Tickets

### Using the Mean and Standard Deviation

The watchdog agency that is looking at the problem of parking tickets would like to know what it should expect to see in the data it collects. The analysts in the agency decide to calculate the mean and standard deviation of the binomial random variable.

The mean is

$$\mu = (10)(0.40) = 4 \text{ cars}$$

and the standard deviation is

$$\sigma = \sqrt{n\pi(1 - \pi)} = \sqrt{(10)(0.40)(0.60)} = \sqrt{2.4} = 1.55 \text{ cars}$$

This tells the analysts that they should expect to find 4 cars with parking tickets. Wondering how much the number of cars might vary, they decide to find the probability that the number of tickets will be within 2 standard deviations of the mean. To do this they calculate $\mu \pm 2\sigma = 4 \pm (2)(1.55) = 4 \pm 3.1 = (0.9, 7.1)$. They want to find the probability that the random variable is between these two values, or $P(0.90 < X < 7.1)$.

How will they do this? The binomial tables certainly do not include numbers like 0.90 and 7.1! The analysts realize that they will have to convert the problem to the nearest integer values that satisfy the probability expression. That means they need to find $P(1 \leq X \leq 7)$. From the tables for $n = 10$ and $\pi = 0.40$ they find that the answer is 0.981. They interpret this to mean that 98.1% of the time the number of tickets they find will be between 1 and 7 inclusive. This gives them a good idea of what to expect in their data. ∎

The idea of comparing what should happen to what does happen is a critical idea in statistics. In the next few chapters we develop a formal method for doing this called *hypothesis testing*.

# Discovery Exercise 6.1
## EXPLORING THE BINOMIAL DISTRIBUTION

*Dear Mom and Dad: Send cash*
*According to* USA Today, *70% of college students receive spending money from their parents when at school.*

For this exercise, you will need to simulate selecting 30 samples of 5 students from this population of college students and observe whether they receive spending money from their parents. Consider the successful outcome to be "receives money" with $\pi = 0.70$ and the failure outcome to be "does not receive money." If your instructor does not provide you with a method, you can take ten (small) pieces of paper and write an S on 7 of them and an F on 3 of them. Put the papers in a bag or other container and select one at random to simulate an observation. *Note:* Be sure to replace the paper each time or $\pi$ will not always be 0.70.

1. Record an S when you select a student who receives money from his/her parents and an F when you select a student who does not receive spending money from his/her parents. For each sample, record the number of successes you sampled.

   In the last column compute a running estimate of $\pi$. Remember that $\pi$ is the probability of a successful outcome. In this case, $\pi$ is known to be 0.70. Let's see how close the estimate gets to 0.70 as the sample size increases. So, after the first sample is selected your estimate of $\pi$ is simply the number of successes divided by 5. After the second sample is selected, your estimate of $\pi$ is the number of successes in both samples divided by 10, and so forth.

| Sample number | Observation number 1 | 2 | 3 | 4 | 5 | X = Total number of successes in n = 5 trials | Running total number of successes | Running total of number sampled | Estimate of $\pi$ |
|---|---|---|---|---|---|---|---|---|---|
| 1 | | | | | | | | 5 | |
| 2 | | | | | | | | 10 | |
| 3 | | | | | | | | 15 | |
| 4 | | | | | | | | 20 | |
| 5 | | | | | | | | 25 | |
| 6 | | | | | | | | 30 | |
| 7 | | | | | | | | 35 | |
| 8 | | | | | | | | 40 | |
| 9 | | | | | | | | 45 | |
| 10 | | | | | | | | 50 | |
| 11 | | | | | | | | 55 | |
| 12 | | | | | | | | 60 | |
| 13 | | | | | | | | 65 | |
| 14 | | | | | | | | 70 | |
| 15 | | | | | | | | 75 | |
| 16 | | | | | | | | 80 | |
| 17 | | | | | | | | 85 | |
| 18 | | | | | | | | 90 | |
| 19 | | | | | | | | 95 | |
| 20 | | | | | | | | 100 | |
| 21 | | | | | | | | 105 | |
| 22 | | | | | | | | 110 | |
| 23 | | | | | | | | 115 | |
| 24 | | | | | | | | 120 | |

| Sample number | Observation number | | | | | $X =$ Total number of successes in $n = 5$ trials | Running total number of successes | Running total of number sampled | Estimate of $\pi$ |
|---|---|---|---|---|---|---|---|---|---|
| | 1 | 2 | 3 | 4 | 5 | | | | |
| 25 | | | | | | | | 125 | |
| 26 | | | | | | | | 130 | |
| 27 | | | | | | | | 135 | |
| 28 | | | | | | | | 140 | |
| 29 | | | | | | | | 145 | |
| 30 | | | | | | | | 150 | |

2. Using the results of the 7th column of the table, construct a relative frequency distribution for the number of successes in 5 trials.

| $X =$ number successes | Number of samples that had $X$ successes | Observed relative frequency = $X/30$ | Theoretical relative frequency |
|---|---|---|---|
| 0 | | | |
| 1 | | | |
| 2 | | | |
| 3 | | | |
| 4 | | | |
| 5 | | | |
| **Totals:** | **30** | **1.00** | **1.00** |

3. Display the relative frequency as a bar chart.

4. Complete the theoretical relative frequency column in the table above by using the binomial table with $n = 5$ and $\pi = 0.70$.

(continued)

5. Display the binomial distribution in a bar chart.

6. Compare the bar chart from step 3 to the binomial distribution displayed in step 5. How do they compare? Why are they different?

7. The graph below has a line at the theoretical value of $\pi$, 0.70. Graph your estimates of $\pi$ for each sample on the same graph. What happens to your estimate as the sample size increases?

## 6.4.4 Exploring the Binomial Distribution

You have started to become familiar with using the binomial probability tables and solving binomial probability problems, but you still may not have an understanding what role the parameters $n$ and $\pi$ play in determining what the probability distribution of a binomial random variable looks like.

In Figure 6.3 you see the effects of changing the value of the parameter $\pi$ for a fixed value of $n$. What you see agrees with our earlier thoughts that when $\pi$ is small it

**FIGURE 6.3**    Effects of changing $\pi$ when $n$ is fixed

is more likely that the number of successes will be small and the distribution skews to the right. When $\pi$ is large it is more likely that the number of successes will be large and the distribution is skewed to the left. You can also see that when $\pi$ is equal to 0.50 the distribution is symmetric.

If you look at the binomial table for $n = 10$ you will notice, not surprisingly, that the probability distributions for $\pi = 0.20$ and $\pi = 0.80$ are mirror images of each other.

Where does the mean fit into the picture? If you remember, the mean is the value that you expect to occur, that is, it is the most likely outcome. If you look at Figure 6.3, you see that in the first probability histogram the highest bar is for $X = 2$ and the mean of a binomial random variable with $n = 10$ and $\pi = 0.20$ is $\mu = (10)(0.20) = 2$.

Figure 6.4 shows the effects of changing the value of the parameter $n$ for a fixed value of $\pi$. The $y$ scale for each histogram is approximately the same. From the pic-

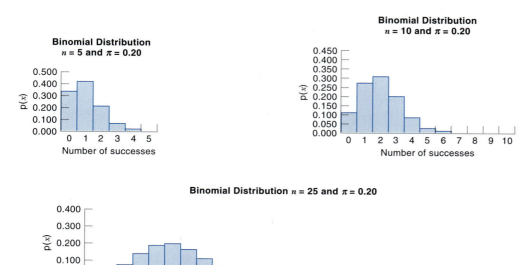

**FIGURE 6.4**    Effects of changing $n$ for a fixed value of $\pi$

ture you can see that, for the most part, the shape of the distribution remains the same as $n$ varies, but the number of possible values and the probability of each value of $X$ (bars in the histogram) changes. In fact, as $n$ gets larger, the individual probabilities get smaller. This makes sense, since we know that the probabilities must sum to 1 and if there are more values of $X$, then each value will get a smaller share of the total.

## 6.4.5 Exercises—Learning It!

**6.11**   The Board of Realtors of a small city reports that 80% of the homes that are sold have been on the market for more than 6 months. The Board takes a random sample of 15 homes that have recently been sold and counts the number that were on the market for more than six months. What is the probability that of the 15 houses in the sample

    (a) less than 12 have been on the market for more than six months?

    (b) between 8 and 13 have been on the market for more than six months?

    (c) at least 10 have been on the market for more than six months?

    (d) at most 4 have been on the market for less than six months?

**6.12**   The Department of Transportation for a city has found that 25% of all parking tickets that have been issued have not been paid within one month of issue. The Department takes a random sample of 20 parking tickets that were issued one month ago and counts the number that have not been paid. What is the probability that

    (a) at most 5 have not been paid?

    (b) between 4 and 8 inclusive have not been paid?

    (c) more than 7 have not been paid?

    (d) at least 6 have been paid?

**6.13**   A large construction firm estimates that 10% of the jobs that it manages are finished within the contracted time period. It looks at a random sample of 5 jobs that it has managed.

    (a) What is the probability that 4 of the five jobs were not completed within the contracted time period?

    (b) What is the mean number of jobs that are completed within the contracted time period?

    (c) What is the standard deviation of the number of jobs in 5 that are completed within the contracted time period?

**6.14**   A large bank that issues many loans for General Motors estimates that 70% of the loans are approved within 24 hours of application. If a consumer group takes a random sample of 25 recent GM loan applications, what is the probability that

    (a) at most 5 were not approved within 24 hours?

    (b) between 10 and 17 are approved within 24 hours?

    (c) What is the mean number of loans in 25 that will be approved within 24 hours?

    (d) What is the standard deviation of the number of loans in 25 that will be approved within 24 hours?

    (e) What is the probability that the number of loans in 25 that are approved within 24 hours is within 3 standard deviations of the mean?

**6.15**   Companies have been having problems with employees playing computer games at work for a long time. As the size and complexity of such games increases, computer system administrators find that network resources are being drained and that the games are using more and more hard disk space. A recent survey across various industries revealed that 30% of workers said that the last computer game they played had been played at work. If a random sample of 15 employees is taken, what is the probability that

    (a) less than 6 said they played their last computer game at work?

    (b) at least 4 said they played their last computer game at work?

    (c) at most 10 said they did not play their last computer game at work?

(d) What is the expected number of employees in a sample of size 15 who played their last computer game at work? What is the standard deviation of the number of employees in 15 who played their last computer game at work?

(e) What is the probability that the number of employees in 15 who played their last computer game at work is within 2 standard deviations of the mean?

# 6.5 CONTINUOUS RANDOM VARIABLES

In Chapter 2 you learned about the different types of data. In particular, you learned that there are two types of quantitative data: discrete and continuous. Discrete data are integer and are often a count of the number of times that something happens. The binomial distribution that you just studied is one good probability model for discrete data.

Continuous data occur when the variable of interest can take on any one of an infinite number of values over some *interval* on the real number line. Much of the time, continuous data result from taking data on a *measurement*. Examples of this would be height of students or grade point averages (GPA). The actual number of values that you can obtain is limited by the measuring instrument (we choose to measure height to the nearest inch or report GPA to two decimal places) but any value in the interval is valid.

When you learned about discrete random variables you learned that to find probabilities associated with the random variable, you simply added the relevant individual probabilities. With continuous random variables you will have to shift your thinking a little bit to see how probabilities are calculated.

When we looked at the effects of changing the value of $n$ in the binomial distribution, we saw that as $n$ increased, the number of possible values of $X$ increased and the individual probabilities got smaller. When $n = 10$ it is not that tedious to calculate $P(X \geq 4)$, but when $n = 20$ or $n = 25$ it is considerably more tedious to do it directly. You are adding many more terms and they are all very small numbers. If we let $n$ get very large, say $n = 100$, it would be tiring indeed! Figure 6.5 looks at the binomial distribution for $\pi = 0.50$ and $n = 10, 25, 50,$ and $100$.

**FIGURE 6.5**  Binomial distribution for large values of $n$

You can see from the graphs that if we connect the tops of all of the histogram bars, as $n$ gets larger, the curve becomes smooth. For continuous random variables, which can take on many more than 100 possible values, the probability distribution is called a **probability density function, $f(x)$,** and is represented by a smooth curve such as the one in Figure 6.6.

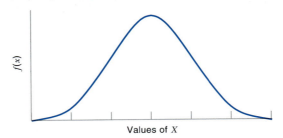

**Probability Density Function for a Continous Random Variable**

**FIGURE 6.6**   Probability distribution for a continuous random variable

> A *probability density function, $f(x)$,* is a smooth curve that represents the probability distribution of a continuous random variable.

You might be wondering how this curve relates to probabilities, since there is no corresponding probability table with numbers involved. How do you find the probabilities that you are interested in? These are good questions.

When we increased the value of $n$ for the binomial distribution, you saw that the individual probabilities got much smaller. In fact, if you look at the scales on the $y$ axes for Figure 6.5 you will see that for $n = 100$ the *largest* number on the graph is 0.08! Again, this makes sense because the total probability is always equal to 1 and when you have to divide it up over more and more possible values, each individual probability gets closer and closer to 0.

*For continuous random variables*
$P(x_1 < X < x_2)$ *is exactly the same as* $P(x_1 \leq X \leq x_2)$, *since* $P(X = x) = 0$.

In fact, for continuous random variables we can no longer talk about the probability that the random variable will assume one particular value, $P(X = x)$, because this is equal to 0. Instead, we can talk about the probability that the random variable will take on any one of the values over an *interval* of interest; that is, we can find $P(x_1 < X < x_2)$. This probability is represented by the area under the probability density curve as shown in Figure 6.7. Finding probabilities for a particular probability density involves the mathematics of calculus, but for certain distributions the probabilities have already been calculated and tabulated.

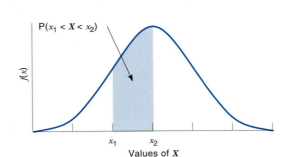

$P(x_1 < X < x_2)$

**FIGURE 6.7**   Probability represented by an area under the curve

In the next section we look at one particular continuous probability distribution that serves as a model for many natural phenomena. This is known as the *normal distribution.*

# 6.6 THE NORMAL DISTRIBUTION

## 6.6.1 The Normal Curve

The normal probability distribution is the symmetric, bell-shaped curve shown in Figure 6.8. It is usually referred to as the *normal curve*. This is the same curve we saw when we learned about the empirical rule.

In reality, most measurement data are modeled very well by this distribution. The actual formula for the probability density is

$$f(x) = \frac{1}{\sigma\sqrt{2\pi}}e^{-(x-\mu)^2/2\sigma^2} \qquad \text{for } -\infty < x < \infty$$

*Normal probability density*

where $e = 2.71828$ . . . and $\pi = 3.14159$. . . . If you look at the equation you can see that the probability density for a value, $x$, relies on two **parameters,** $\mu$ and $\sigma$.

*Here, $\pi$ is the familiar constant, 3.141 . . . , not related to the binomial distribution.*

> For a normal random variable, the parameter $\mu$ is the mean of the normal random variable, $X$, and $\sigma$ is the standard deviation.

When we refer to a normally distributed random variable we often use a special notation

$$X \sim N(\mu, \sigma)$$

This shorthand is equivalent to saying that the random variable $X$ is normally distributed with a mean of $\mu$ and a standard deviation of $\sigma$.

When we studied the binomial probability distribution we learned that the parameters, in that case $n$ and $\pi$, defined how the probability was distributed over the set of possible values of $X$. That is, they defined the shape of the probability histogram. In a similar way, $\mu$ and $\sigma$ define the way the normal distribution looks.

You have come across the mean in two other situations. When you learned about descriptive statistics, you learned that the mean was a measure of central tendency, or an estimate of the most typical data value. Then, when you studied the binomial distribution, we said that the mean is the most likely value, or the one that you expect to happen most often. For a normally distributed random variable, the mean is the center of the distribution. It is the value that determines the location of the probability density curve on the number line.

You also learned about the standard deviation. You know that the standard deviation measures how far the data are spread out around the mean. For a normally distributed random variable, $\sigma$ determines how spread out the probability density is around its center. You may remember that in Chapter 4 we used the mean and the standard deviation to obtain a mental image of the distribution of the data. The empirical rule told us that for a symmetric distribution, virtually all of the data fall within three standard deviations of the mean. Since the normal distribution is the fundamental symmetric curve, we can use this same idea to get a picture of a normal probability distribution.

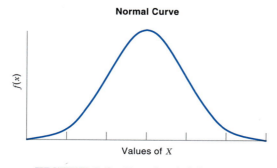

**Normal Curve**

$f(x)$

Values of $X$

**FIGURE 6.8** Normal probability curve

**Normal Curves with Equal Means and Different Standard Deviations**

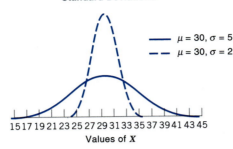

$\mu = 30, \sigma = 5$
$\mu = 30, \sigma = 2$

15 17 19 21 23 25 27 29 31 33 35 37 39 41 43 45
Values of $X$

**Normal Curves with Different Means and Equal Standard Deviations**

$\mu = 30, \sigma = 2$
$\mu = 30, \sigma = 2$

15 17 19 21 23 25 27 29 31 33 35 37 39 41 43 45
Values of $X$

**FIGURE 6.9**   Effects of changing $\mu$ and $\sigma$ on the normal curve

In Figure 6.9 you can see the effect that changes in $\mu$ and $\sigma$ have on the way the normal curve looks.

As the standard deviation increases, the distribution gets more spread out around its center. For the curve with a mean of 30 and a standard deviation of 5 you see that the ends of the curve are around 15 and 45, three standard deviations away from the mean! As the mean increases, the distribution moves to the right (in the direction of increasing values of $X$) and when the mean decreases the distribution moves to the left. Notice that although the appearance of the normal curve does change when $\mu$ and $\sigma$ are varied, the basic shape does not change. The curve is always bell-shaped and symmetric. This is an important feature of the normal probability distribution.

 **TRY IT NOW!**

## Food Expenditures

### *Looking at the Normal Curve*

The amount of money that a person working in a large city spends each week for lunch is a normally distributed random variable. For professional and management personnel the random variable has a mean of $35 and a standard deviation of $5. For hourly employees the mean is $30 with a standard deviation of $2.

Sketch the normal curves for each of the two random variables on the same graph.

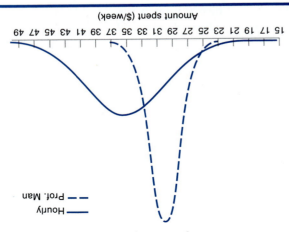

Amount spent ($/week)

15 17 19 21 23 25 27 29 31 33 35 37 39 41 43 45 47 49

Prof. Man.
Hourly

Weekly Lunch Expenditures                    ANS:

## 6.6.2 The Standard Normal Curve

When we learned about continuous random variables we said that the probability that the random variable takes on any value in an interval is equal to the area under the probability density curve. We also said that finding these probabilities requires the use of integral calculus, which is beyond the scope of this book. How will we find the probabilities we are interested in?

Remember that when we studied the binomial probability distribution, the formula to find probabilities was also complicated, and we solved the problem by using tables for different sets of values for $n$ and $\pi$. Since $\pi$ is constrained to be between 0 and 1, it was not hard to get tables for many different values of $n$.

The idea of using probability tables for the normal probability distribution seems like a good one, but there is a practical limitation. Unlike $n$ and $\pi$, the values of $\mu$ and $\sigma$ are not at all constrained. Depending on the variable of interest, there are an infinite number of different pairs of $\mu$ and $\sigma$ that we might be interested in. It is not feasible to have a table for every different problem we can think of.

The problem is solved by finding a way to transform every normally distributed random variable to a single, standard normal random variable. This standard normal random variable is known as **Z** and has a mean of 0 and a standard deviation of 1. We can use the shorthand notation to write that as **Z ~ N(0, 1).** Once we transform the problem we can use a single normal probability table, a **standard normal table,** to solve it.

> A **Z random variable** is normally distributed with a mean of 0 and a standard deviation of 1, $Z \sim N(0, 1)$.

> A **standard normal table** is a table of probabilities for a $Z$ random variable.

This is not mysterious, if you think about it for a minute. If the class average on an exam were a 70 and you wanted to make it a 75, you would ADD 5 points to everyone's grade. Similarly, to move the mean from $\mu$ to 0, you subtract $\mu$ from the values.

How does the transform work? Remember that the mean of a normally distributed random variable locates the normal curve on the number line. What we want to do is move any arbitrary normal random variable that is centered at $\mu$ down the number line to center at 0. It is not hard to see that this can be accomplished by subtracting the value of $\mu$ from the random variable, $X$.

The change from a standard deviation of $\sigma$ to a standard deviation of 1 is not difficult either. Essentially you need to scale the probability distribution so that the total area under the curve stays equal to 1 (one of the rules of probability). A value of $X$ that is two standard deviations away from $\mu$ for the original random variable must stay two standard deviations away from 0 when it is transformed to $Z$. To accomplish this we describe the distance that $X$ is from $\mu$, $X - \mu$, in terms of the number of standard deviations it is away from $\mu$. The transform from $X$ to $Z$ is then given by

$$Z = \frac{X - \mu}{\sigma}$$

*Formula for Z-score*

where $X \sim N(\mu, \sigma)$ and $Z \sim N(0, 1)$. At this point, an example will help you see what is happening.

*This is the population version of the z-scores you calculated in Chapter 4.*

## EXAMPLE 6.26 Aptitude Scores

*Transforming from X to Z*

Scores on an aptitude test given by the training department of a large company are normally distributed with a mean of 75 points and a standard deviation of 5 points.

The managers of the company are interested in knowing what proportion of the people that take the test score between 65 and 85 points.

The first thing we need to do is *draw a picture* representing the problem we are trying to solve.

**Aptitude Score**

55 57 59 61 63 65 67 69 71 73 75 77 79 81 83 85 87 89 91 93 95
Score

*Note: It may seem to be a waste of time to you to draw the pictures. A little time spent here will avoid a LOT of time wasted later on.*

The picture shows the distribution of the scores and the area under the curve that corresponds to the problem that management wants to solve. We know that to solve the problem we must transform it to a standard normal problem. To do this we must transform each of the values of $X$ that we are interested in to corresponding values of $Z$. In this case,

$$\text{for } X = 65: \quad Z = \frac{65 - 75}{5} = -2$$

$$\text{for } X = 85: \quad Z = \frac{85 - 75}{5} = 2$$

*Caution: It is easy to mix up X and μ when you use the formula for Z. Use the sign of the answer you get to check for errors. Make sure that the sign of your Z agrees with the location of X relative to μ.*

Neither of these values of $Z$ should be a surprise since you know that a $Z$-score measures the number of standard deviations that a value is from its mean. The score of 65 is 10 points, or exactly 2 standard deviations, *below* the mean of 75 and the score of 85 is exactly 2 standard deviations *above* the mean of 75. The sign on the value of $Z$ indicates whether the value of $X$ was above (+) or below (−) the mean.

If you look at a picture of the transformed problem you see that it does not look any different from the original picture except for the scale on the $x$ axis. It is not necessary to draw both $X$ and $Z$ pictures.

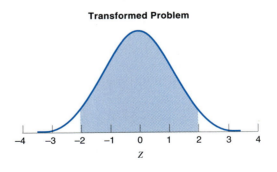

**Transformed Problem**

−4   −3   −2   −1   0   1   2   3   4
Z

Transforming problems about normally distributed random variables to problems about standard normals is not difficult. Just as in all probability problems the hardest part is being able to understand what problems you need to solve.

**TRY IT NOW!**

### Speed Reading
*Translating from* X *to* Z

The number of pages of a statistics textbook that a student can read in a given hour is a normally distributed random variable with a mean of 7 pages and a standard deviation of 1.5 pages. One of the professors who uses the book wants to know the probability that a randomly selected student can read more than 8.5 pages of the textbook in an hour.

Draw a picture that depicts the problem to be solved and find the *Z* values necessary to solve the problem.

The importance of drawing the picture that relates to the problem will be clear in the next section when we look at how to solve the probability problems.

## 6.6.3 The Standard Normal Tables

The tables for the binomial distribution that you used contained $P(X = x)$ for different combinations of $n$ and $\pi$. To find quantities like $P(x_1 \leq X \leq x_2)$ you summed the probabilities for all of the values of $X$ that were included in the interval. We know that $P(X = x) = 0$ for continuous random variables, so the probability tables for the standard normal probability distribution must contain some kind of **interval** probability.

> An **interval probability** gives the probability that a random variable will take on a value *between* two given values $P(x_1 \leq X \leq x_2)$.

ANS. $Z = 1$

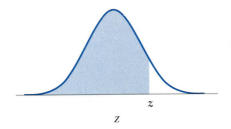

**FIGURE 6.10**     Probability given by the standard normal table

There are several different types of standard normal tables that can be used. They each give probabilities for different types of intervals. The tables that we use in this book give $P(Z < z)$; that is, the entries in the table are the probability that the random variable $Z$ will take on a value that is less than some value $z$. A picture that corresponds to this is shown in Figure 6.10.

The table that is included in Appendix A has two halves, for both negative and positive values of $Z$. This makes solving problems involving the table less cumbersome.

All tables for the standard normal probabilities are set up the same way. The values of $Z$ are defined to two decimal places, and the corresponding probability is located at the intersection of the appropriate row and column of the table. If you think of the $Z$ value as always being in the form $X.XX$, the rows contain the first two digits $X.X$ while the columns contain the last digit $0.0X$. It sounds complicated but it is easy once you are familiar with it. A portion of the table is shown in Figure 6.11.

To look up a value of $Z$, say $Z = 1.14$ you first locate the row that corresponds to the first two digits, 1.1. This is the shaded row in Figure 6.11. The column you need is the one that corresponds to the last digit, 0.04, which is also shaded. The entry in the table at the intersection of the row and column, 0.8729, gives the probability of obtaining a value of the random variable $Z$ that is less than 1.14, or the area under the curve to the left of $Z = 1.14$.

*Caution: When looking up z values such as 0.08, the first two digits are 0.0 not 0.8.*

The half of the table for negative values of $Z$ is used exactly the same way. It is important to be consistent about the way you calculate and look up $Z$ values. Since the tables give the values of $Z$ to two decimal places, you need to calculate $Z$ values to two decimal places, rounding correctly, and look them up that way.

|  |  |  |  | **Second decimal place** |  |  |  |  |  |  |
|---|---|---|---|---|---|---|---|---|---|---|
| $z$ | 0.00 | 0.01 | 0.02 | 0.03 | 0.04 | 0.05 | 0.06 | 0.07 | 0.08 | 0.09 |
| 0.0 | 0.5000 | 0.5040 | 0.5080 | 0.5120 | 0.5160 | 0.5199 | 0.5239 | 0.5279 | 0.5319 | 0.5359 |
| 0.1 | 0.5398 | 0.5438 | 0.5478 | 0.5517 | 0.5557 | 0.5596 | 0.5636 | 0.5675 | 0.5714 | 0.5753 |
| 0.2 | 0.5793 | 0.5832 | 0.5871 | 0.5910 | 0.5948 | 0.5987 | 0.6026 | 0.6064 | 0.6103 | 0.6141 |
| 0.3 | 0.6179 | 0.6217 | 0.6255 | 0.6293 | 0.6331 | 0.6368 | 0.6406 | 0.6443 | 0.6480 | 0.6517 |
| 0.4 | 0.6554 | 0.6591 | 0.6628 | 0.6664 | 0.6700 | 0.6736 | 0.6772 | 0.6808 | 0.6844 | 0.6879 |
| 0.5 | 0.6915 | 0.6950 | 0.6985 | 0.7019 | 0.7054 | 0.7088 | 0.7123 | 0.7157 | 0.7190 | 0.7224 |
| 0.6 | 0.7257 | 0.7291 | 0.7324 | 0.7357 | 0.7389 | 0.7422 | 0.7454 | 0.7486 | 0.7517 | 0.7549 |
| 0.7 | 0.7580 | 0.7611 | 0.7642 | 0.7673 | 0.7704 | 0.7734 | 0.7764 | 0.7794 | 0.7823 | 0.7852 |
| 0.8 | 0.7881 | 0.7910 | 0.7939 | 0.7967 | 0.7995 | 0.8023 | 0.8051 | 0.8078 | 0.8106 | 0.8133 |
| 0.9 | 0.8159 | 0.8186 | 0.8212 | 0.8238 | 0.8264 | 0.8289 | 0.8315 | 0.8340 | 0.8365 | 0.8389 |
| 1.0 | 0.8413 | 0.8438 | 0.8461 | 0.8485 | 0.8508 | 0.8531 | 0.8554 | 0.8577 | 0.8599 | 0.8621 |
| 1.1 | 0.8643 | 0.8665 | 0.8686 | 0.8708 | 0.8729 | 0.8749 | 0.8770 | 0.8790 | 0.8810 | 0.8830 |
| 1.2 | 0.8849 | 0.8869 | 0.8888 | 0.8907 | 0.8925 | 0.8944 | 0.8962 | 0.8980 | 0.8997 | 0.9015 |

**FIGURE 6.11**     The standard normal table

You now know how to use the table to find $P(Z < z)$. How does the procedure change if you need $P(Z > z)$ or $P(z_1 < Z < z_2)$? The first problem is not difficult to figure out. Since the tables give $P(Z < z)$ and since the total area under the curve must always equal 1, we can obtain $P(Z > z)$ by using the definition of the complement of an event. That is,

*For continuous random variables, probability is often referred to as the area under the curve.*

$$P(Z > z) = 1 - P(Z < z)$$

So, to find the area under the curve above a given value of $Z$ you look up the $Z$ value, find the probability below $Z$, and subtract it from 1. Thus, using our earlier example,

$$P(Z > 1.14) = 1 - 0.8729 = 0.1271$$

Once you are comfortable using the normal probability table you can find upper area probabilities easily by relying on the symmetry of the normal distribution. Since the normal probability distribution is symmetric about the mean, it must be true that the area to the right of a $Z$ value must be equal to the area to the left of the negative of that $Z$ value, or

$$P(Z > z) = P(Z < -z)$$

This is illustrated in Figure 6.12.

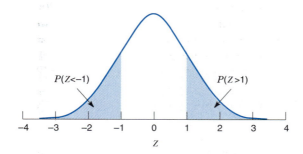

**FIGURE 6.12** Comparison of upper and lower probabilities

Using the tables to find $P(z_1 < Z < z_2)$ is not difficult either. Figure 6.13 illustrates the process.

When you look up the larger $Z$ value, in this case $Z = +2$, the table gives the entire area under the curve to the left of 2, which is 0.9772. Looking at the picture, you see that this includes the piece of the curve to the left of the lower value, in this case $Z = -2$, which is 0.0228. Since we do not want to include this lower portion we must subtract it from the first value. Thus, in general,

$$P(z_1 < Z < z_2) = P(Z < z_2) - P(Z < z_1)$$

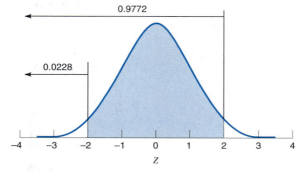

**FIGURE 6.13** Finding the area between two $Z$ values

and for this example,

$$P(-2 < Z < +2) = P(Z < +2) - P(Z < -2) = 0.9772 - 0.0228 - 0.9544$$

*Remember that Z can also be thought of as the number of standard deviations that a value is from its mean.*

If you think about it, you already knew this! Remember that the *Z* random variable is $N(0, 1)$, so finding $P(-2 < Z < +2)$ is the same as finding the probability that the random variable will be within two standard deviations of the mean. The answer, not surprisingly, is 0.9544 or 95.44%. Remember the empirical rule? It told us that for a symmetric distribution, approximately 95% of the data will fall within two standard deviations of the mean. This is where the empirical rule comes from.

Table 6.2 summarizes the rules for using the standard normal probability tables.

**TABLE 6.2   Rules for solving normal probability problems**

| To Find... | Area Under the Curve... | Look Up... |
|---|---|---|
| $P(Z < z)$ | below a value of $Z$ | the $Z$ value and use the table directly |
| $P(Z > z)$ | above a value of $Z$ | 1. the $Z$ value and subtract the value in the table from 1 |
|  |  | OR |
|  |  | 2. the negative of the $Z$ value and use the table directly |
| $P(z_1 < Z < z_2)$ | between two values of $Z$ | both $Z$ values and subtract the lower value from the higher value |

 **TRY IT NOW!**

## The Standard Normal Table
### *Using the Table to Find Probabilities*

For each of the following questions, *draw a picture* of what you are trying to find BEFORE you use the table to find it.

Find the probability that a *Z* random variable takes on a value that is less than 2.74.

Find the probability that a *Z* random variable is greater than 0.85.

*Drawing a picture will help you use the table correctly.*

Find the probability that $Z$ is between $-1.36$ and $1.87$.

## 6.6.4 Solving Normal Probability Problems

Just as you found with the binomial distribution, once you understand how the random variable and the probability distribution work, solving problems is really not hard. The hardest part is understanding what the problem really means.

The following steps are used to solve problems that involve normally distributed random variables with mean $\mu$ and standard deviation $\sigma$:

**Step 1:** Write down the information about the random variable, i.e., $X \sim N(\mu, \sigma)$.

**Step 2:** Draw a picture that represents the problem and write down the probability statement, e.g., $P(X > 30)$.

**Step 3:** Transform the values of $X$ that are involved in the problem into $Z$ values.

**Step 4:** Look up the $Z$ values on the standard normal table.

**Step 5:** Perform any additional calculations that need to be done, as described in Table 6.2.

### EXAMPLE 6.27 Aptitude Scores

*Solving Normal Probability Problems*

The company that administered the aptitude test to its employees wants to know what proportion of the people who take the test score between 65 and 85 points.

We have essentially answered this question in the course of this section, but we can put it all together here.

We know that $X \sim N(75, 5)$ and that we are looking for $P(65 < X < 85)$ or

**Aptitude Score**

55 57 59 61 63 65 67 69 71 73 75 77 79 81 83 85 87 89 91 93 95

Score

We found the $Z$ values of interest to be $-2$ and $+2$. Looking these values up in the normal tables, we find that the probability for $Z = -2$ is 0.0228 and for $+2$ it is 0.9772. From the picture we see that we are trying to find the area *between* these two values and so we subtract the probability values to find

$$0.9772 - 0.0228 = 0.9544$$

The company is also interested in knowing what proportion of the people who take the test score in what is considered to be the superior range, above 87 points.

We know that $X \sim N(75, 5)$. To answer their question we need to find $P(X > 87)$.

This is represented by the graph below:

**Aptitude Scores**

55 57 59 61 63 65 67 69 71 73 75 77 79 81 83 85 87 89 91 93 95
Score

Transforming the $X$ value of 87 to $Z$, we get

$$Z = \frac{87 - 75}{5} = \frac{12}{5} = 2.40$$

Notice that this answer makes sense, since it is positive and the $X$ value of interest is to the right of the mean.

Looking the $Z$ value of 2.40 up in the table, we get a probability of 0.9918.

Now we must decide whether there is anything left to do. The answer is most certainly YES! Common sense also tells you that 0.9918 cannot possibly be the answer. Look at how much of the area in the picture is shaded—it is very small, hardly over 99%! We are looking for the area *above* a value and the table gives the opposite area. We must subtract the value in the table from 1 to obtain

$$1 - 0.9918 = 0.0082 \text{ or } 0.82\%$$

Thus, the company sees that less than 1% of the people who take the test score in the superior range.                                                    ■

When doing problems, keep in mind that probabilities, proportions, and percentages are all ways of expressing the same quantity. Do not be misled by the way a question is asked. It is best to answer the question using the quantity type specified, but you are not wrong in using either decimals or percentages.

 **TRY IT NOW!**

## Speed Reading
### *Solving Normal Probability Problems*

The instructor who is interested in how many pages of the statistics text that students can read in an hour knows that the random variable is $N(7, 1.5)$.

Find the probability that a student could read more than 11.5 pages in an hour.

The instructor was worried about the percentage of students who could not finish reading a 5-page section in the given hour. What percentage of the students is this?

## 6.6.5 Finding Values That Correspond to Known Probabilities

In the previous section you learned how to use the standard normal probability tables to solve problems involving any normally distributed random variable. In these problems you were interested in finding the probability or the area under the curve that relates to a specific value of the random variable, $X$.

Another interesting type of problem which we can solve uses the normal distribution and the standard normal tables. Many people use the probabilities of the normal distribution to make decisions about what values of a variable define certain outcomes.

Suppose that a company wants to reward with an incentive bonus those salespeople whose sales are in the top 10%, or suppose that a city decides that only those people who score in the top 5% on a civil service exam will get job interviews. In many cases it does not make sense to simply take the top 5% of each group that takes the exam or the top 10% of each month's sales. If a particularly ill-suited group of people take the exam, the city could wind up interviewing people who are in no way qualified for the positions available. Similarly, the company could reward people in one month and deny people in the next who have the same sales figures.

ANS. 0.0013; 0.0918

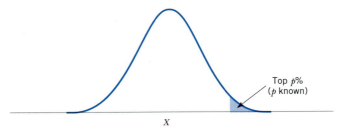

**FIGURE 6.14**   The "inverse" normal probability problem

Instead, these decision makers rely on the normal probability model to define the cutoff points for the variables of interest. How do they find the numbers that they are looking for?

You know how to use the normal probability tables to find the probability that corresponds to a given value of $X$ or $Z$. The problem we just described is really the exact opposite of that problem. The decision makers know the percentage or area under the curve that they want to define. They do not know the value of $Z$ (and subsequently $X$) that it corresponds to. This is illustrated in Figure 6.14.

Solving this type of problem involves using the normal probability tables "inside out." Previously you have located the $Z$ value on the *outside* of the table and determined the answer to the question by looking on the *inside* of the table. For this type of problem you will start by looking on the *inside* of the table for the known probability and move to the *outside* to find the answer. It sounds more complicated than it really is. Again, you simply need to be systematic in your approach.

To get started, we will concentrate on the technique. Suppose you want to find the value of $Z$ that has 10% of the area below it. This is illustrated in Figure 6.15.

Since we know that the normal tables give the area below the value of $Z$, and that is what we are looking for, we can look the probability up directly. At first glance it may seem like you are looking for a needle in a haystack, but with a little bit of common sense and insight you will see that it is really not difficult.

If you look at the picture we drew (and this is why we draw them), you can see immediately that the value of $Z$ that we want must be negative. We have already halved our search! Also, if you look at the tables, you will see that the numbers in the table are in numerical order. Once you find a suitable starting place for your search, you simply have to search systematically for the number you are looking for. The section of the normal table that contains the value we are looking for is shown in Figure 6.16.

This is probably a good place to point out that since the tables give probabilities to four decimal places, we are not likely to find the value we are looking for exactly. However, if the value you want is not there, you will always be able to locate two adjacent values, one smaller and one larger than the value you are looking for. Examining the table you see that the numbers from top to bottom are increasing and that at the row for $z = -1.3$ they are too small and at the $z = -1.2$ row they are too large. The value you want must be in there somewhere!. A more careful search will lead you to the two cells that are highlighted in the table, 0.0985 and 0.1003.

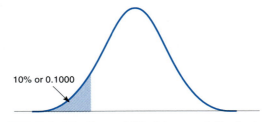

**FIGURE 6.15**   Bottom 10% of the normal distribution

| $z$ | 0.00 | 0.01 | 0.02 | 0.03 | 0.04 | 0.05 | 0.06 | 0.07 | 0.08 | 0.09 |
|---|---|---|---|---|---|---|---|---|---|---|
| −1.9 | 0.0287 | 0.0281 | 0.0274 | 0.0268 | 0.0262 | 0.0256 | 0.0250 | 0.0244 | 0.0239 | 0.0233 |
| −1.8 | 0.0359 | 0.0351 | 0.0344 | 0.0336 | 0.0329 | 0.0322 | 0.0314 | 0.0307 | 0.0301 | 0.0294 |
| −1.7 | 0.0446 | 0.0436 | 0.0427 | 0.0418 | 0.0409 | 0.0401 | 0.0392 | 0.0384 | 0.0375 | 0.0367 |
| −1.6 | 0.0548 | 0.0537 | 0.0526 | 0.0516 | 0.0505 | 0.0495 | 0.0485 | 0.0475 | 0.0465 | 0.0455 |
| −1.5 | 0.0668 | 0.0655 | 0.0643 | 0.0630 | 0.0618 | 0.0606 | 0.0594 | 0.0582 | 0.0571 | 0.0559 |
| −1.4 | 0.0808 | 0.0793 | 0.0778 | 0.0764 | 0.0749 | 0.0735 | 0.0721 | 0.0708 | 0.0694 | 0.0681 |
| −1.3 | 0.0968 | 0.0951 | 0.0934 | 0.0918 | 0.0901 | 0.0885 | 0.0869 | 0.0853 | 0.0838 | 0.0823 |
| −1.2 | 0.1151 | 0.1131 | 0.1112 | 0.1093 | 0.1075 | 0.1056 | 0.1038 | 0.1020 | 0.1003 | 0.0985 |
| −1.1 | 0.1357 | 0.1335 | 0.1314 | 0.1292 | 0.1271 | 0.1251 | 0.1230 | 0.1210 | 0.1190 | 0.1170 |
| −1.0 | 0.1587 | 0.1562 | 0.1539 | 0.1515 | 0.1492 | 0.1469 | 0.1446 | 0.1423 | 0.1401 | 0.1379 |
| −0.9 | 0.1841 | 0.1814 | 0.1788 | 0.1762 | 0.1736 | 0.1711 | 0.1685 | 0.1660 | 0.1635 | 0.1611 |
| −0.8 | 0.2119 | 0.2090 | 0.2061 | 0.2033 | 0.2005 | 0.1977 | 0.1949 | 0.1922 | 0.1894 | 0.1867 |

**FIGURE 6.16**   Normal probability table containing the probability 0.1000

As we guessed, one of these is a bit too small and the other is too large. The value of 0.0985 corresponds to a value of $Z$ of −1.29, while the 0.1003 corresponds to $z = -1.28$. Thus, the value of $Z$ that we are looking for must be in between these two.

At this point we need to discuss how we will determine the $Z$ values that we will use. Some people would suggest performing a linear interpolation between the two values to get an estimate of the correct value of $Z$. Even though the normal curve is certainly not linear, for small increments in $Z$ this is not a bad approximation. The only problem is, it is not worth the trouble.

For our purposes we will use the value of the probability that is closest to the value we are looking for unless there is something in the problem that tells us to use the one that is smaller or larger. The something would be some directional description of the percentage we are looking for, like "at least 10%," in which case we would choose 0.1003, or "at most 10%," which would lead us to the 0.0985. When we look at an actual application of this technique you will see how little difference the exact answer makes.

Now on to solving real problems. We start with a simple example that uses what we have already figured out.

## EXAMPLE 6.28   Aptitude Scores

### Solving the Inverse Problem

As hard as it may be to take, downsizing and cost cutting are a reality of business life. The company that gives employees the aptitude test has decided that people who score in the bottom 10% of the test scores will not receive any additional job training. If there are to be layoffs, these people will be among the first to be cut. The question is, what cutoff score on the test should the company use?

We already know that the bottom 10% corresponds to a $Z$-score of −1.28. But clearly this cannot be the answer to the question. How do we translate the $Z$-score back into the realm of the variable of interest, the test scores?

You know that the $Z$ value represents the number of standard deviations away from the mean that a value is. The negative sign on the $Z$ value tells you to move down from the mean (subtract the number of standard deviations). Thus, we know that the cutoff score must be 1.28 standard deviations *below* the mean. That is,

$$X = 75 - (1.28)(5) = 68.6 \text{ points}$$

The company will have to decide whether to use 68 points (and affect less than 10% of the people) or 69 points (and affect more than 10%). ■

From this example you see why it is not necessary to interpolate to get the exact value of Z. If the answer is to be used to make a decision, the extra precision gained by getting the exact Z value does not usually translate into any practical information. The test is not scored to the nearest tenth of a point. Even if it were, the change in the Z value affects the hundredths place of the score.

We used our understanding of the Z value to determine the cutoff score, but the formula we used really comes from algebraically solving the definition of the Z value for X:

$$Z = \frac{X - \mu}{\sigma} \longrightarrow X = \mu + Z\sigma$$

When Z is positive you will add to $\mu$ and when Z is negative you will subtract from $\mu$.

## TRY IT NOW!

### Speed Reading

*Solving the Inverse Problem*

The instructor who is interested in how fast students can read the statistics textbook would like to identify the bottom 25% of the class, in terms of the number of pages that they can read in an hour.

Find the number of pages per hour that defines the bottom 25% of the students.

When solving problems that define the probability and looking for the value of the random variable, it is important to know when you are specifying an area above the value of interest or an area below the value. When the area is below the value you are looking for, you can look up the specified probability directly. When the area is above the value you are looking for you will either have to subtract the specified probability from 1 and look that up, or look up the given probability and rely on the symmetry of the table.

### EXAMPLE 6.29   Aptitude Scores

*Specifying Upper Area Probabilities*

Although cost cutting and downsizing are negative aspects of the business world, the company with the aptitude test is also planning to give extra training to employees who score in the top 2% of those taking the test. The company would like to identify the score to use as the cutoff point.

If we draw a picture that represents the problem we see that the Z value we are looking for will most certainly be positive.

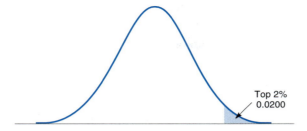

If you want to look up the value directly you will have to look up the bottom area or 0.9800. It is just as easy to use common sense and rely on the symmetry of the normal distribution. If you look up 0.0200 in the table you will find that it is not there, but that you can choose from 0.0197 and 0.0202. Since 0.0202 is closer to 2% (and slightly more generous to the employees) we will use the corresponding $Z$ value of $-2.05$.

But we just said that the answer has to be positive. This is where common sense, symmetry, and the picture come together. We know the answer should be positive from the picture and common sense, but we used the symmetry feature to make the look-up easier. We will use $+2.05$ to find the correct answer.

The cutoff score for the top 2% is then

$$75 + (2.05)(5) = 85.25 \text{ points}$$

The company can use 85 (and train more than 2%) or 86 (and train less than 2%). ∎

Inverse normal probability problems are an important application of the normal distribution. These types of problems are encountered many times in real life when standards or cutoff points need to be determined.

## 6.6.6 Exercises—Learning It!

**6.16** The amount of money spent by students for textbooks in a semester is a normally distributed random variable with a mean of $235 and a standard deviation of $15.

    (a) Sketch the normal distribution that describes the amount of money spent on textbooks in a semester.

    (b) What is the probability that a student spends between $220 and $250 in any semester?

    (c) What percentage of students spend more than $270 on textbooks in any semester?

    (d) What percentage of students spend less than $225 in a semester?

**6.17** On any given day, the number of leasable square feet of office space available in a small city is a normally distributed random variable with a mean of 850,000 square feet and a standard deviation of 25,000 square feet. The number of leasable square feet available in another small city is normally distributed with a mean of 900,000 square feet and a standard deviation of 25,000 square feet.

    (a) Sketch the distribution of leasable office space for both cities on the same graph.

    (b) What is the probability that the number of leasable square feet in the first city is less than 925,000 square feet?

    (c) What is the probability that the amount available in the second city is less than 925,000?

**6.18** The actual amount of a certain brand of orange juice in a container marked half gallon is a normally distributed random variable with a mean of 65 oz and a standard deviation of 0.35 oz.

    (a) What percentage of the containers contain more than 64.5 oz?

    (b) What percentage of the containers contain between 64 and 66 oz?

(c)  If federal law says that 98% of all containers must be at or above the labeled weight, does this brand of orange juice meet the requirement?

**6.19**  The amount of money per month earned by an auditor with 10 years experience is a normally distributed random variable with mean $3500 and standard deviation $240.

(a)  What percentage of auditors with 10 years experience earn more than $4000 per month?

(b)  What percentage of auditors with 10 years experience earn less than $3200 per month?

(c)  What is the probability that a randomly selected auditor earns between $3250 and $3800 per month?

(d)  What monthly income defines the top 10% of all auditors with 10 years experience?

# 6.7   USING EXCEL TO GENERATE PROBABILITY DISTRIBUTIONS

Excel has built-in functions that allow you to calculate probabilities and generate random data from many different probability distributions. In this section, we will look at using Excel with the binomial and normal probability distributions.

## 6.7.1 Calculating Binomial Probabilities in Excel

So far, you have relied on tables to find probabilities for the binomial probability distribution. What would you do if you needed to solve a problem for values of $n$ and $\pi$ that were not in the tables? Excel has built-in functions that calculate binomial probabilities for any values of $n$ and $\pi$.

Suppose that you want to calculate the probability that a random sample of $n = 50$ with $\pi = 0.45$ will have exactly 25 successes. You can use the Excel function BINOMDIST to calculate this probability. We will look at this function using the Function Wizard so that you understand how Excel functions work and how you can find out about Excel functions you have never used. We will also look at a macro included on the disk that accomplishes the same thing. Follow these steps:

1. To use a function, position the cursor in an empty cell in the worksheet.

2. From the main toolbar, click on the Function Wizard icon. The **Paste Function** dialog box opens. This box lists all of the different categories of functions available in Excel. Highlight **Statistical** for the function category and **BINOMDIST** for the function, as shown in Figure 6.17.

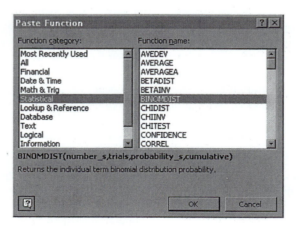

**FIGURE 6.17**   Choosing an Excel function

At the bottom you see the format for the function you are choosing and a description of what it does. Click **OK** and the dialog box for the BINOMDIST function opens.

As you place the cursor in each textbox, a description of the input for that box is described at the bottom of the dialog box. It is important that you read these descriptions so that you know what you are supposed to input.

3. Place the cursor in the textbox labeled **Number_s.** At the bottom you can see that this is where you enter the number of successes you are interested in. Type in the number "25."

4. Now, place the cursor in the textbox labeled **Trials** and enter "50," which is the number of trials that you have.

5. Place the cursor in the textbox labeled **Probability_s.** Notice that the bottom of the dialog box tells you that Excel is looking for the probability of a success on any trial, $\pi$. Type in "0.45."

6. The last textbox lets you indicate what kind of probability you want. Excel will calculate either $P(X = x)$ or $P(X \leq x)$, the cumulative probability. In this case we want $P(X = 25)$ so we will set this value to False. The completed dialog box should look like the one in Figure 6.18.

**FIGURE 6.18**   Completed BINOMDIST dialog box

7. Click on **OK** and the probability—in this case, 0.087330027—will appear in the cell in which you started.

You might have noticed that while you were filling in the dialog box, the status line in Excel was recording exactly what gets entered in the cell where you place the function, as shown in Figure 6.19. When you become familiar with particular functions, you can type the command in directly and bypass the Function Wizard. We will do that for the other functions in this chapter.

**FIGURE 6.19**   The BINOMDIST command

There is also a macro available for calculating binomial probabilities. To use this macro, choose **Macro** from the **Tools > Macro** menu and select **Binomial Probabilities** from the list of macros and click **Run.** The dialog box shown in Figure 6.20

*Remember, to use a macro you must open the MacDoIt.xls worksheet on the disk.*

**FIGURE 6.20**    Binomial Probabilities macro

opens. You will need to input the same values that you did using the Excel function. The textbox labeled **p** is where the value of $\pi$ goes. The macro gives both P($X = x$) and P($X \leq x$) as output.

You can choose to have the output placed into the current worksheet if you want. If you do, then when you click on **OK,** the output table is placed in the location you specify, as shown in Figure 6.21. If you do not, then the output just appears in the dialog box.

| Binomial Probabilities | | | |
|---|---|---|---|
| n | 50 | P(X=x) | 0.08733 |
| p | 0.45 | P(X<=x) | 0.803369 |
| x | 25 | | |

**FIGURE 6.21**    Binomial probability output

### 6.7.2 Calculating Normal Probabilities in Excel

Excel has functions either for calculating probabilities from a normal distribution with a given mean and standard deviation, or from a standard normal. There are also functions for solving inverse normal probability problems—that is, finding the value of $Z$ that corresponds to a given probability. Table 6.3 lists all of the functions, describes what they do, and explains what values are expected as input.

**TABLE 6.3**

| Function format: | What it does | Input |
|---|---|---|
| **NORMDIST**(x, mean, standard_deviation, cumulative) | Returns P($X \leq x$) for a normally distributed random variable | x = value of interest; mean = mean of the random variable; standard_deviation = standard deviation of the random variable; cumulative = true for P($X \leq x$) |
| **NORMINV**(probability, mean, standard_deviation) | Returns the value of $x$ that has the specified probability below it | probability = the area under the curve below the $X$ value of interest; mean = mean of the random variable; standard_deviation = standard deviation of the random variable |
| **NORMSDIST**(x) | Returns P($Z \leq x$) for a standard normal random variable | x = value of interest for $Z$ |
| **NORMSINV**(probability) | Returns the value of $Z$ that has the specified probability below it | probability = area under the standard normal curve below the $Z$ value of interest |

To calculate probabilities like $P(z_1 < Z < z_2)$ or $P(x_1 < X < x_2)$, you will need to calculate two probabilities and subtract them, just as you do with the tables. This is also true for $P(Z > z)$ and $P(X < x)$.

There is a macro available on the disk that will calculate any or all of these probabilities. To use it, select **Normal Probability** from the list of macros and click **Run.** The dialog box shown in Figure 6.22 opens. The input for the macro is very simple.

**FIGURE 6.22**    Dialog box for Normal Probability macro

To use the macro, follow these steps:

1. Fill in the mean, $\mu$, and the standard deviation, $\sigma$, of the random variable in the appropriate textboxes.

2. Look at the pictures on the right of the box. If you want only a lower-tail probability, fill in the textbox for **a**. If you want only an upper-tail probability, fill in the textbox for **b**. If you want the probability between two values, fill in the textboxes for both **a** and **b**.

3. If you want to solve the inverse normal problem, type the area under the normal curve *below* the value of interest in the textbox labeled **Percentile Input.**

4. Click **OK** to obtain the output.

You can choose to have the output placed in the worksheet, by checking the box labeled **Store the Model on Current Worksheet.** If you do not do this, then the answers will just appear in the dialog box when you click **OK.**

## 6.7.3 Generating Random Data in Excel

In addition to calculating probabilities in Excel, you can also generate random data that come from populations with specific probability distributions. Such data are useful in simulations and other exercises. To generate data from different probability distributions, you use the **Data Analysis** tools. From the **Data Analysis** tools list, select **Random Number Generation.** The dialog box shown in Figure 6.23 opens.

**FIGURE 6.23**   Random Number Generation dialog box

Excel will generate random data from several probability distributions, including the binomial and the normal. To see the list of available distributions, click on the drop down arrow in the textbox labeled **Distribution:** and the list appears as shown in Figure 6.24.

**FIGURE 6.24**   List of probability distributions

The top portion of the dialog box is the same for every probability distribution. In the textbox labeled **Number of Variables:** you enter the number of different random variables you want to generate. In most cases, this is 1. In the second textbox, labeled **Number of Random Numbers:** you input the number of observations from the probability distribution that you want to generate—usually either the sample size or the population size, depending on what you are generating.

We will generate 50 observations from a binomial distribution with $n = 20$ and $\pi = 0.70$. To do this, follow these steps.

1. Select **Binomial** from the list of distributions and the dialog box will change to allow input of the appropriate parameters.

2. In the textbox for **Number of Variables:** type "1" and in the textbox for **Number of Random Numbers:** type "50."

3. In the textbox labeled **p Value** enter the value for $\mu$, 0.70. Finally, in the textbox for **Number of Trials:** enter 20.

4. Specify where you want the output to appear and click **OK.** The completed dialog box should look like the one in Figure 6.25.

**FIGURE 6.25** The Binomial Random Variable dialog box

The data will be output to the location you specify. Figure 6.26 shows a portion of the output. Notice that the values are all less than 20 (which you would expect) and that most of the values are higher than 10, since $\pi$ is larger than 0.50.

| | A |
|---|---|
| 1 | 16 |
| 2 | 10 |
| 3 | 15 |
| 4 | 12 |
| 5 | 14 |
| 6 | 11 |
| 7 | 10 |
| 8 | 12 |
| 9 | 14 |
| 10 | 16 |
| 11 | 14 |
| 12 | 14 |
| 13 | 13 |
| 14 | 11 |
| 15 | 14 |

**FIGURE 6.26** Random data from binomial distribution

The steps for generating data from a normal distribution are identical to those for the binomial distribution. The only difference is that the parameters you enter for the normal distribution are the mean and standard deviation, $\pi$ and $\sigma$.

# CHAPTER 6 SUMMARY

When many people think of probability they think of flipping coins, spinning spinners, or gambling. While these are certainly aspects of probability, and are useful for illustrating some of the basic principles of probability theory, they are by no means all there is to the subject. Probability is an important and interesting subject in its own right, but the study of random variables and probability distributions is important in the study and development of statistics.

Probability is the bridge between what we studied in the first five chapters of this book, *descriptive statistics,* and what is to follow in subsequent chapters, *inferential statistics.* In descriptive statistics we use different techniques to describe sample data. In inferential statistics we will test hypotheses about the populations from which these samples came. Probability is the tool that allows us to reconcile what happened (descriptive) with what we think is true by determining how likely the outcomes of the experiment we perform are.

## Key Terms

| Term | Definition | Page reference |
|---|---|---|
| **A AND B** | The event **A AND B** is the event that A and B both occur. | 236 |
| **A OR B** | The event **A OR B** describes the event when either A happens or B happens or they both happen. | 236 |
| **Binomial random variable** | A **binomial random variable** is the number of successes in $n$ trials or in a sample of size $n$. | 253 |
| **Complement** | The **complement** of an event **A**, denoted $\mathbf{A}'$, is the set of all outcomes in the sample space, **S**, that do not correspond to the event **A**. | 234 |
| **Empirical probability** | An **empirical** probability is one that is calculated from sample data and is an estimate for the true probability. | 243 |
| **Event** | An **event, A,** is an outcome or a set of outcomes that are of interest to the experimenter. | 232 |
| **Experiment** | An **experiment** is any action whose outcomes are recordable data. | 230 |
| **Interval probability** | An **interval probability** gives the probability that a random variable will take on a value *between* two given values $\mathrm{P}(x_1 \le X \le x_2)$. | 275 |
| **Mutually exclusive** | Two events, A and B, are said to be **mutually exclusive** if they have no outcomes in common. | 237 |
| **Probability** | **Probability** is a measure of how likely it is that something will occur. | 230 |
| **Probability density function, $f(x)$** | A **probability density function, $f(x)$,** is a smooth curve that represents the probability distribution of a continuous random variable. | 270 |

| Term | Definition | Page reference |
|---|---|---|
| Probability distribution, p(x) | The **probability distribution** of a random variable $X$, written as **p(x)**, gives the probability that the random variable will take on each of its possible values. | 247 |
| Probability of an event, P(A) | The **probability of an event A, P(A),** is a measure of the likelihood that an event A will occur. | 232 |
| Random variable | A **random variable, X,** is a quantitative variable whose value varies according to the rules of probability. | 246 |
| Sample space, S | The **sample space, S,** is the set of all possible outcomes of an experiment. | 230 |
| Standard normal random variable, Z | A **standard normal random variable** is normally distributed with a mean of 0 and a standard deviation of 1, $Z \sim N(0, 1)$. | 273 |
| Standard normal table | A **standard normal table** is a table of probabilities for a $Z$ random variable. | 273 |

## Key Formulas

| Term | Formula | Page reference |
|---|---|---|
| Binomial mean and standard deviation | $\mu = n\pi$ and $\sigma = \sqrt{n\pi(1-\pi)}$ | 261 |
| Binomial probability formula | $p(x) = \dfrac{n!}{(x!)(n-x)!}\,\pi^x(1-\pi)^{n-x}$ for $x = 0, 1, \ldots, n$ | 257 |
| General addition rule | $P(A \text{ OR } B) = P(A) + P(B) - P(A \text{ AND } B)$ | 240 |
| Normal probability density | $f(x) = \dfrac{1}{\sigma\sqrt{2\pi}}\,e^{-(x-\mu)^2/2\sigma^2}$ for $-\infty < x < \infty$ | 271 |
| P(A) | $P(A) = \dfrac{\text{Number of ways that A can occur}}{\text{Total number of possible outcomes}}$ | 232 |
| | $P(A) = \dfrac{n_A}{N}$ | 232 |
| Probability of the complement, P(A′) | $P(A) + P(A') = 1$ | 234 |
| Simple addition rule | $P(A \text{ OR } B) = P(A) + P(B)$ | 237 |
| Z-score | $Z = \dfrac{X - \mu}{\sigma}$ | 273 |

# CHAPTER 6 EXERCISES

## Learning It!

**6.20** In the survey about satisfaction with local phone service, those respondents who rated their current service as excellent and those who rated it Poor–Very Poor were asked what type of company their current local service provider was. The results are given in the table below:

| Current Service Source | Excellent | Poor–Very Poor |
|---|---|---|
| Long distance company | 264 | 1394 |
| Local phone company | 444 | 1318 |
| Power company | 131 | 485 |
| Cable TV company | 215 | 431 |
| Cellular phone company | 198 | 572 |

(a) What is the probability that a person selected from this group will rate his service as excellent and will have a long distance company as his current provider?

(b) What is the probability that a person selected from this group will use a power company or a cable TV company as his current provider?

(c) What is the probability that a person selected from this group will rate his current service as Poor–Very Poor or use a local phone company?

(d) What is the probability that a person selected from this group will rate his service as Poor–Very Poor and will use a cellular phone company?

**6.21**  In a study on college students and binge drinking, researchers were interested in looking at binge drinking and gender. Questions were asked about the number of times that a student had five or more drinks in the last two weeks and gender. The results are shown in the table below:

|  | *Number of Times Had Five or More Drinks in a Row in Last Two Weeks* | | | | | |
|---|---|---|---|---|---|---|
| **Gender** | **0** | **1** | **2** | **3** | **4** | **5** |
| Male | 55 | 20 | 18 | 19 | 8 | 4 |
| Female | 75 | 14 | 12 | 24 | 2 | 0 |

(a) What is the probability that a student is a male and had five or more drinks in a row 2 times in the last two weeks?

(b) What is the probability that a student is female or had five or more drinks in a row 3 times in the last two weeks?

(c) What is the probability that a student had five or more drinks in a row in the last two weeks 3 or 4 times?

(d) What is the probability that a student did not have five or more drinks in a row in the last two weeks?

**6.22**  In the same study of binge drinking and college students, the researchers were also interested in the number of times that student experienced hangovers in the semester. The data they collected are found below:

|  | *Hangover Since Beginning of Semester* | | |
|---|---|---|---|
| **Gender** | **Not at All** | **Once** | **Twice or More** |
| Male | 61 | 23 | 40 |
| Female | 66 | 25 | 36 |

(a) What is the probability that a student is female and had a hangover twice or more during the semester?

(b) What is the probability that a student is male and has not had a hangover during the semester?

(c) What is the probability that a student had a hangover once or more during the semester?

**6.23**  Surveys indicate that 40% of all small businesses do not provide any kind of health insurance options for their employees. The Chamber of Commerce in a large city takes a random sample of 20 small businesses. What is the probability that in the sample of 20 small businesses

(a) at least 12 do not provide health insurance options for employees?

(b) at most 4 do not provide health insurance options for employees?

(c) between 8 and 13 do not provide health insurance options for employees?

(d) at least 5 provide health insurance options for employees?

**6.24**  Recent studies have shown that 30% of employees in the insurance industry telecommute four days a week. If a random sample of 15 insurance industry employees is taken, what is the probability that

(a) between 4 and 7 inclusive telecommute four days a week?

(b) less than 6 telecommute four days a week?

(c) at least 9 telecommute four days a week?

(d) between 5 and 10 do not telecommute four days a week?

**6.25** A survey done at a state university in New England found that 40% of all seniors have encountered academic problems related to binge drinking. If a random sample of 25 seniors at the university is taken, what is the probability that of the 25

(a) at most 6 have encountered academic problems related to binge drinking?

(b) between 4 and 9 have encountered academic problems related to binge drinking?

(c) more than 15 have encountered academic problems related to binge drinking?

(d) Find the mean and standard deviation of the number of seniors in a sample of 25 who have encountered problems related to binge drinking.

**6.26** The amount of office space allocated to production planners in consumer products companies is a normally distributed random variable with a mean of 120 square feet (sq ft) and a standard deviation of 6 sq ft.

(a) What percentage of production planners' offices have more than 135 sq ft?

(b) What percentage of production planners' offices have between 110 and 125 sq ft?

(c) What percentage of production planners' offices have less than 130 sq ft?

**6.27** The size of a gift/specialty store in a regional super mall is a normally distributed random variable with a mean of 8500 sq ft and a standard deviation of 260 sq ft. What is the probability that a randomly selected gift/specialty store in a regional super mall is

(a) more than 8000 sq ft?

(b) between 8300 and 9000 sq ft?

(c) less than 9500 sq ft?

**6.28** A recent study done at a university in New England found that 60% of all students have missed class in the semester because of drinking. A random sample of 20 students is taken. What is the probability that

(a) at least 15 have missed class because of drinking?

(b) between 12 and 17 inclusive have missed class because of drinking?

(c) less than 5 have not missed class because of drinking?

(d) at most 13 have not missed class because of drinking?

## Thinking About It!

**6.29** The company that is looking at the way employees took floating holidays collected data from 300 employees. The data are shown again below:

*Requires Exercise 6.2*

|  | *How Days Were Used* | | |
| --- | --- | --- | --- |
| **Type of Job** | **Took Actual Holiday** | **Added to Vacation** | **Took Random Days** |
| Professional | 5 | 17 | 51 |
| Clerical | 13 | 46 | 32 |
| Hourly | 53 | 78 | 5 |

(a) What is the probability that an employee selected at random is a clerical worker?

(b) Suppose you know that the person selected used the floating holidays as random days off. Now what is the probability that the person is a clerical worker?

(c) How do the probabilities in parts (a) and (b) compare? What is different about them? When you are given some information about the outcome of the experiment it reduces the size of the sample space. This type of probability is called a *conditional probability*. It is often described as the probability that an event occurs, given that you know another event has occurred.

(d) What is the probability that the person took the floating holidays on the actual holiday given that you know the person is a professional employee?

(e) If you know that the person selected is either a clerical worker or an hourly worker, what is the probability that the person took the days as added vacation days?

*Requires Exercise 6.3*

**6.30** Consider the computer magazine survey about consumers' plans to buy computers. The data are given below:

| Type of Computer | When Purchase Will Be Made | | |
| --- | --- | --- | --- |
| | **0–3 Months** | **3–6 Months** | **6–12 Months** |
| Notebook/Portable | 34 | 156 | 258 |
| Desktop | 56 | 346 | 128 |

(a) What is the probability that a person selected at random is planning to make a purchase in 0–3 months, given that the person plans to buy a notebook computer?

(b) What is the probability that a person selected is planning to buy a desktop computer given that the person plans to make the purchase in the next 6 months?

(c) If you know that the person selected is planning to buy a desktop computer, what is the probability that the person will make the purchase in the next 3–6 months?

(d) What is the probability that a person will buy a notebook computer, given that the person is not planning to buy it within the next 0–3 months?

**6.31** Consider the study of binge drinking and college students. The data are found below:

| Gender | Hangover Since Beginning of Semester | | |
| --- | --- | --- | --- |
| | **Not at All** | **Once** | **Twice or More** |
| Male | 61 | 23 | 40 |
| Female | 66 | 25 | 36 |

(a) Given that the student is female, what is the probability that she had a hangover twice or more during the semester?

(b) What is the probability that a student is male given that the student had a hangover once or less during the semester?

(c) What is the probability that a student has had two or more hangovers in a semester given that the student is male?

(d) Compare your answers to parts (a) and (c) and interpret the results for the researchers.

*Requires Exercise 6.23*

**6.32** The Chamber of Commerce that is interested in whether small businesses provide health care options for employees wants a few more questions answered. The surveys indicated that 40% of employers did not provide health insurance options for employees.

(a) In the sample of 20, how many should the Chamber of Commerce expect to provide health insurance options for employees?

(b) What is the standard deviation of the number of small businesses in the 20 that provide health insurance options?

(c) Find the probability that the number of small businesses in a sample of 20 will be within 2 standard deviations of the mean.

(d) How does this probability compare to the percentage predicted by the empirical rule? If it does not agree, why not?

*Requires Exercises 6.23, 6.32*

**6.33** The Chamber of Commerce looking at health insurance options provided by small businesses finds 4 that provide health insurance options for employees and thinks this is unusual.

(a) Do you agree that the findings are unusual? Why or why not?

(b) What would you tell the Chamber of Commerce this might mean?

**6.34** The manager of a regional super mall wants to compare the gift/specialty stores in her mall to the size of the smallest 20% of the gift/specialty shops in similar malls. She knows that the size of such stores is normally distributed with a mean of 8500 sq ft and a standard deviation of 260 sq ft.

(a) What square footage defines the smallest 20% of such stores?

(b) Harriet's Gift Boutique has complained that it is much smaller than any similar stores in other malls. If the size of this store is 7800 sq ft, is the complaint reasonable?

**6.35** The Dean of Students at another university in New England read the report on binge drinking and decided to conduct a small survey on her own campus. She took a random sample of 20 students and found that 15 of them had missed class because of drinking. She decides that this indicates that her university is within the norm for this problem. Do you agree with her conclusion? Why or why not?

*Requires Exercise 6.28*

## Doing It!

**6.36** The Chamber of Commerce that is studying the credit problems of small businesses asked them 3 questions to classify their business and 7 questions related to the issue of credit problems. A portion of the datafile and an explanation of each variable are given below:

*CHAMBER.XXX*

| Size | Employees | Nature | Problem | Understands | Concerned | Call | Loan | Collateral | Access |
|---|---|---|---|---|---|---|---|---|---|
| 2 | 2 | 1 | 1 | 2 | 1 | 2 | 2 | 0 | 2 |
| 1 | 2 | 3 | 1 | 2 | 2 | 0 | 2 | 2 | 1 |
| 4 | 3 | 1 | 2 | 1 | 2 | 2 | 2 | 2 | 0 |
| 1 | 1 | 1 | 2 | 1 | 2 | 2 | 2 | 2 | 2 |
| 1 | 2 | 1 | 2 | 0 | 0 | 0 | 0 | 0 | 0 |
| 3 | 1 | 5 | 2 | 0 | 2 | 2 | 2 | 2 | 1 |
| 3 | 3 | 2 | 2 | 1 | 1 | 2 | 2 | 2 | 1 |
| 2 | 2 | 3 | 2 | 1 | 2 | 2 | 2 | 2 | 2 |
| 3 | 5 | 2 | 2 | 1 | 2 | 2 | 2 | 1 | 2 |

- The variable *Size* refers to the annual sales of the company and is coded as follows:

  1 under $1 million
  2 $1–5 million
  3 $6–10 million
  4 $11–20 million
  5 over $20 million

- The variable *Employees* refers to the number of employees that the company currently employs. This variable was coded as

  1 0–5 employees
  2 6–10 employees
  3 11–50 employees
  4 51–150 employees
  5 151–250 employees
  6 Over 250 employees

- The variable Nature refers to the type of business and is coded as

  1 Manufacturing
  2 Retail
  3 Service
  4 Real Estate
  5 Other

- The next seven variables contain the response to the questions or statements indicated and are coded as follows:

  1 Yes
  2 No

- *Problem* "Are you experiencing credit related problems?"
- *Understd* "The bank understands my problems."
- *Concern* "I am concerned that my note might be recalled."
- *Call* "The bank is planning to recall my loan."
- *Loan* "The bank has called my loan."
- *Collateral* "The bank has demanded more collateral."
- *Access* "Access to credit is affecting my business."

(a) Look at the variable *Problem* that was discussed in Example 6.24. Separate the respondents according to the type of company that they have. For each type of industry, generate a binomial probability distribution for $\pi = 0.10$ and $n = $ the number of that type of company. For each type of company find the probability that the number of people reporting credit problems is the actual number found in the sample.

(b) For each type of company estimate the value of $\pi$. Compare these to each other, to the overall estimated value of $\pi$ and to the assumed value of 0.10. Report your conclusions.

(c) Repeat parts (a) and (b) for the different sizes of companies (both annual sales and number of employees).

(d) Now look at the variables related to loan recall, *Concern, Call,* and *Loan.* Use the data to estimate the percentage of small businesses that responded Yes to each of these questions. How does the number of nonresponses to the question affect your estimate of the percentage? Do you think that it is better to estimate the percentage of Yes respondents using the total of 166 companies or just the ones who responded to the question? Why?

(e) Investigate the variables *Concern, Call,* and *Loan* the same way you investigated *Problem.*

(f) Investigate the variable *Understd* the same way you investigated *Problem.*

(g) Investigate the variable *Access.*

(h) Prepare a report for the Chamber of Commerce about the credit problems and concerns of small businesses.

*TISSUES.XXX*     **6.37** In Chapter 3 you learned about some of the customer complaints that a company that manufactures tissues can get. One of the categories of complaints that made up a large percentage of the total was Dispensing. In that category, Sheets Tear on Removal was a significant factor.

The managers of the tissue company have decided to address this problem. They know that tensile strength is the factor that determines when a tissue will tear, and have decided that to solve the problem they will have to investigate the tensile strength of the tissues.

As part of the Quality Control program at the company, facial tissue has certain product specifications, that is, criteria that must be met, for the product to be acceptable to consumers. One of the characteristics that is specified is tensile strength.

The managers have decided to look at the current levels of tissue strength. They know the target values for the process and the parameters that should be met, and have decided to check to see if the process is meeting the current specifications. If it is not, then changes will need to be made to see that is does. If it is, then perhaps the process specifications will need to be changed. The managers will collect data on two variables:

• **Machine Direction (MD) Strength:** This is the strength in the direction that the machine pulls on the tissue during manufacture. It has to be high enough that the tissues do not break, causing machine down time.

• **Cross-Direction (CD) Strength:** This is the strength in the direction that the tissue is pulled out of the box. It is the variable that determines whether the sheets tear when you remove them from the box.

Samples were taken from tissue produced on a single tissue machine. The samples were taken over three different days and the results were recorded.

A portion of the datafile and an explanation of the variables are shown below:

| Day | MDStrength | CDStrength |
|-----|-----------|-----------|
| 1 | 1006 | 422 |
| 1 | 994 | 448 |
| 1 | 1032 | 423 |
| 1 | 875 | 435 |
| 1 | 1043 | 445 |
| 1 | 962 | 464 |
| 1 | 973 | 472 |

• *Day* keeps track of the day on which the sample was taken and goes from 1 to 3.
• *MDStrength* measures machine-directional strength and is measured in lb/ream.
• *CDStrength* measures cross-directional strength and is measured in lb/ream.

According to the specifications, *MDStrength* is supposed to be normally distributed with a mean of 1000 and a standard deviation of 50 lb/ream. *CDStrength* should be normally distributed with a mean of 400 and a standard deviation of 25 lb/ream.

(a) Use a computer software package to create normal probability tables for the specified distributions of *MDStrength*. Use increments of 50 and go from 800 to 1200.

(b) Use the table to determine the probability that a tissue manufactured according to specifications will have an *MDStrength* of less than 850 lb/ream.

(c) Create a relative frequency histogram for the variable *MDStrength* that goes from 800 to 1200 in class intervals of 50 units.

(d) Do the data in the histogram appear to have the shape of a normal distribution? What is the center?

(e) Look at the relative frequencies on the histogram you just created, and calculate what percentage of the data are within one standard deviation of the mean value of 1000. According to the empirical rule what percentage should be within one standard deviation? How do the actual data compare with the empirical rule prediction?

(f) Use the procedure above to determine the percentage of *MDStrength* data that are within two and three standard deviations of the mean.

(g) The critical specifications for *MDStrength* are 850 on the low side and 1075 on the high side. Use the frequency distribution to determine the percentage of the tissues that actually do not meet these specifications.

(h) How does this compare to the percent defective expected by the product specifications?

(i) Do you think that the company should be concerned about the difference? Why or why not?

(j) Create a set of normal probabilities for the theoretical distribution of *CDStrength*.

(k) The critical values for *CDStrength* are 480 on the high side and 390 on the low side. *CDStrength* that is too high creates a stiff tissue. Since the cross direction is the one in which tissues are pulled from the box, a value of *CDStrength* that is too low can cause sheets to tear on removal from the carton. According to the specifications, what percentage of the tissues should have CD strengths that are too high? too low?

(l) Create a graph of the theoretical distribution of *CDStrength*.

(m) Generate a set of descriptive statistics for the variable *CDStrength*. Compare the mean and the median. Do you think that the distribution of *CDStrength* is symmetric?

(n) Create a relative frequency histogram for the variable *CDStrength*. Does it support the assumption of normality?

(o) Compare the percentage of CD strength measurements that are within 1, 2, and 3 standard deviations of the mean to the predictions of the empirical rule. Is the assumption of normality still reasonable?

(p) Prepare a report to management that indicates whether the process appears to be running to the product specifications. In this report include any changes that need to be made (in terms of mean and standard deviation) to bring the process back to target values.

# SAMPLING DISTRIBUTIONS AND CONFIDENCE INTERVALS

## THE DIAPER COMPANY

Most large manufacturing companies use some form of Statistical Quality Control (SQC) in the manufacture of their products. One form of SQC that is often used is a control chart. A control chart looks at variation in data from samples of products taken over time.

A large company that manufactures disposable diapers collects data from its machines at random times during the workday. One of the variables that is measured is diaper weight. Diaper weight is an important factor in the manufacturing process for two reasons. First, the material that contributes most to diaper weight is the most expensive component of the diaper. Thus, it is reasonable to want to provide enough of this material, but not an excessive amount. The second reason is that the weight of a diaper relates to the consumer's perception of how well that diaper will absorb liquid. The target values, previously identified by consumer research, for diaper weight are a mean of 55 g and a standard deviation of 0.55 g.

When diapers are collected they are grouped in samples of size 5. The sample number and the individual diaper weights and bulks are recorded. The sample averages are then calculated and plotted on charts. Machine operators use these charts to tell them whether the machine is behaving as expected or whether the machine needs adjustment. The first few lines of the data set are shown below.

| Sample | Diaper | Weight | Bulk |
|--------|--------|--------|------|
| 1 | 1 | 55.87 | 0.419 |
| 1 | 2 | 55.35 | 0.380 |
| 1 | 3 | 54.50 | 0.365 |
| 1 | 4 | 53.97 | 0.406 |
| 1 | 5 | 54.29 | 0.360 |
| 2 | 1 | 54.85 | 0.397 |
| 2 | 2 | 54.79 | 0.405 |
| 2 | 3 | 54.65 | 0.393 |

## 7.1   CHAPTER OBJECTIVES

To properly use control charts and to make inferences about the population mean, you must understand the behavior of the sample mean, $\overline{X}$. Gaining an understanding of $\overline{X}$ is the focus of this chapter. We will look at control charts in more detail in Chapter 16. In Chapter 6, we saw that the behavior of a quantitative variable can be described by its probability distribution, and this distribution in turn can be described by parameters such as the mean and the variance. In this chapter we make the link from the probability concepts you learned in Chapter 6 to the inferential statistics techniques covered in the remaining chapters in this book.

Figure 7.1 is a drawing you first saw in Chapter 2. The larger circle on the left represents the population you are studying. The smaller circle on the right represents the sample you have taken from this population. This figure is being used to help you see the relationship between probability and inferential statistics. In reality, the sample circle should sit inside the population circle since it is a piece of it.

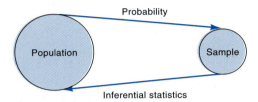

**FIGURE 7.1**     Relationship between probability and inferential statistics

Chapters 3, 4, and 5 taught you the techniques of descriptive statistics which help you describe the sample. These techniques work on the smaller circle and are very helpful in gaining an understanding of the sample data, but they do not let you use the information in the sample to draw conclusions or inferences about the population. In other words, you do not yet have the ability to link that smaller circle to the larger one. In the last chapter you learned some tools of probability. In particular, you learned how to calculate the likelihood of a particular sample being selected from a population. You were thus making the trip across the top arc in Figure 7.1, from the population to the sample.

But this is not really our desired goal. We wish ultimately to make the trip from the sample to the population along the bottom arc in Figure 7.1. We wish to use the information in the sample to make probabilistic statements about the behavior of the population. This is the goal of inferential statistics. The key that unlocks this trip is found in this chapter. The rest of the chapters in this book can be opened only with the key.

More specifically, this chapter covers:

- Motivation for Point Estimators
- Common Point Estimators
- Desirable Properties of Point Estimators
- Distribution of the Sample Mean: Large Sample or Known $\sigma$
- The Central Limit Theorem—A More Detailed Look
- Drawing Inferences by Using the Central Limit Theorem
- Large-Sample Confidence Intervals for the Mean
- Distribution of the Sample Mean: Small Sample and Unknown $\sigma$
- Small-Sample Confidence Intervals for the Mean
- Confidence Intervals for Qualitative Data
- Sample Size Calculations

# 7.2 MOTIVATION FOR POINT ESTIMATORS

Since we most likely will not know the value of the parameters that describe the population, we must resort to using the information contained in the sample. It seems logical that if we can identify a numerical descriptor for the sample, then this statistic, called a point estimate, might be used to estimate the corresponding measure for the population. This is the right idea, which leads us to the definition of a **point estimate** and **point estimator.**

> *Remember: A parameter is a numerical descriptor of the population. Parameter values are typically unknown.*

> A *point estimate* is a single number calculated from sample data. It is used to estimate a parameter of the population. A *point estimator* is the formula or rule that is used to calculate the point estimate for a particular set of data.

Actually you have been working with point estimates without really knowing it. Up to this point, values calculated from the sample have been called sample statistics. Now we establish the link between sample statistics and the corresponding population parameters. In doing so we will have created point estimates out of these sample statistics. Let's take a look at the diaper company introduced at the beginning of the chapter.

> *Sample statistics become point estimates.*

## EXAMPLE 7.1   The Diaper Company

*Some Possible Point Estimates*

Suppose a sample of 30 diapers was taken and the diaper weight was recorded in grams for each diaper. You decide to use the tools of descriptive statistics to describe this sample. The results of your analysis in Excel are shown below. Each of these values fits the definition of a point estimate. Each one is a single number calculated from the sample and it could be used to estimate an unknown population parameter. In the case of the sample mean, median, and mode, any of these values could be used as estimates of the true unknown mean weight of the population of all diapers manufactured.

| Weight | |
| --- | --- |
| Mean | 54.9833333 |
| Median | 54.86 |
| Mode | #N/A |
| Standard deviation | 0.63690162 |
| Range | 2.47 |
| Minimum | 53.97 |
| Maximum | 56.44 |
| Count | 30 |

*Note:* Excel uses #N/A to indicate that the Mode was unavailable. In most cases this means that there were no values that occurred more than once. ■

Before we go any further we must consider what parameters of the population we wish to estimate.

## 7.3   COMMON POINT ESTIMATORS

*Remember that a quantitative variable is numerical and a qualitative variable is descriptive.*

In Chapter 2 you learned that there are two major categories that can be used to classify variables: *qualitative* and *quantitative*. You also learned that the type of tools that can be used to analyze data depends on the type of data. Statisticians must know what kind of data they are analyzing before the analysis tools are selected and the inferences made.

### 7.3.1 Point Estimators for Quantitative Variables

If you are studying a *single quantitative variable,* then you typically wish to know the value of the *population mean* and the value of the *population standard deviation*. That is, you wish to know the center and the variability in the population. Remember that what you would like to know are the values of those parameters that describe the behavior of the population.

### EXAMPLE 7.2   The Diaper Company

*Parameters of Interest*

For the diaper company, the population consists of all diapers made by this company. The characteristic that is being studied is the weight of the diapers. This is a quantitative variable and so we are interested in knowing the mean weight of all diapers made by this company and the standard deviation of the weights of all diapers made by this company. Knowing the mean and standard deviation of the diaper weights will allow us to set up a control chart as described in the chapter introduction. These are the population parameters of interest.

We know that we cannot weigh all the diapers in the company's current inventory but rather must base our estimates of the population mean and the population standard deviation on information contained in a sample of the product taken from the manufacturing line.  ■

*Recall: This notation was introduced in Chapter 4.*

The population mean is always labeled with the Greek letter $\mu$ (pronounced mu) and the population standard deviation is labeled with the Greek letter $\sigma$ (pronounced sigma). These values are typically unknown. We will use point estimates to estimate these values. In particular, we will use the sample mean, $\overline{X}$, to estimate $\mu$ and we will use the sample standard deviation, $s$, to estimate $\sigma$. It makes logical sense to use the sample mean to estimate the population mean and the sample standard deviation to estimate the population standard deviation. In the next section we discuss in more detail why these are the point estimates of choice.

*Comparing two populations*

Now suppose you wish to compare two quantitative variables. For example, you may wish to compare:

- the weight of diapers produced by two different machines
- the sales of a product at two different locations
- the time it takes to get your burger at McDonalds and Burger King
- the salaries for men and women in the same occupation

There are many such situations when you are studying two quantitative variables and wish to compare them. In these cases you often want to know how the true population means compare and how the amount of variation in one population compares with the amount of variation in the other population. For the purposes of keeping track of the different populations we will arbitrarily label one population as population 1 and the other as population 2.

*Use subtraction to compare two population means.*

To compare two numbers, what computation might you do? For example, if you wanted to compare your salary to your friend's salary, what would you do with the two salary figures? If you said *subtract* one number from the other, then you are correct. By convention we always subtract the second mean from the first and thus we need to estimate the *true difference* between $\mu_1$ and $\mu_2$, or $\mu_1 - \mu_2$. It makes sense to estimate this true difference by using the actual difference in the two sample means or $\overline{X}_1 - \overline{X}_2$. If the two population means are the same, then the difference in the sample means should be close to zero. But remember that even if the two population means are the same and you select a sample from each population you will most likely get two slightly different sample means. This is due to the fact that you are looking at only a piece of each of the populations. This is what we have called sampling error. So just because the sample means are different, this does not at all imply that the population means are different.

Another way to compare two numbers is to find the *ratio* of one to the other. If the numbers are the same, then the ratio will be 1. We will use ratios to compare the amount of variation in the first population to the amount of variation in the second population. So far in this book we have been using the standard deviation as a measure of variation. Unfortunately, for mathematical reasons, to compare the variation in two populations you must compare the variances. Recall that the variance is just the standard deviation squared. So if the standard deviation is 3 grams then the variance is 9 grams squared. The variance is not usually quoted because the units of measure are so odd—no one thinks in terms of grams squared. In this case what we wish to estimate is $\sigma_1^2 / \sigma_2^2$. It makes sense to estimate this by calculating the ratio of the sample variances or $s_1^2 / s_2^2$.

*Use a ratio to compare variability.*

## TRY IT NOW!

### Sales of Gizmos

*Comparing Point Estimators Calculated From Samples Selected From Different Populations*

Use a software package such as Excel (Excel instructions can be found at the end of this chapter) or Minitab to simulate picking a sample of size 10 from two different populations with mean 100 and standard deviation 3, or use the samples below.

| Store 1 | | Store 2 | |
|---|---|---|---|
| 95 | 99 | 97 | 97 |
| 95 | 99 | 103 | 103 |
| 101 | 95 | 101 | 98 |
| 102 | 102 | 98 | 106 |
| 92 | 99 | 100 | 96 |

Suppose the first variable is daily sales of a new gizmo at store 1. The second variable is daily sales of the new gizmo at store 2. Select a sample of size 10 days from both stores. Assume that the daily sales at both stores are normally distributed with a mean of 100 and a standard deviation of 3.

Find $\overline{X}_1$, $\overline{X}_2$, and $\overline{X}_1 - \overline{X}_2$.

*(continued)*

What do you notice about the difference in sample means even though the population means are the same?

Find $s_1^2$ (store 1 standard deviation squared), $s_2^2$, and $s_1^2/s_2^2$.

What do you notice about the ratio of the two sample variances?

### 7.3.2 Point Estimators for Qualitative Variables

If the variable you are studying is a *qualitative variable* then you typically wish to know what *proportion* or *percentage* of the population has a particular characteristic. For example, you may wish to know the percentage of defective items in the population, the percentage of people who will purchase a new product, or the percentage of people who are in favor of a particular political candidate or issue. The true unknown *population percentage* is labeled $\pi$. In this case we will use the sample proportion, $p$, to estimate $\pi$.

*The use of $\pi$ as the notation for the population proportion*

Since Greek letters ($\mu$ and $\sigma$) have been used to represent population parameters, the same convention will be followed with respect to population proportions. Some books use $p$ for the population parameter and $\hat{p}$ for the sample proportion. This is done to avoid the use of $\pi$, which, of course, is traditionally reserved for the transcendental number 3.1415. The authors of this text feel that it makes sense to maintain the consistency of using Greek letters to denote population parameters.

### EXAMPLE 7.3   New Product Design

***Sample Proportion Used to Estimate True Proportion***

The company you work for is thinking about introducing a new product. You have been assigned the task of finding the percentage of consumers who would buy this new product if it were priced at $12.49. You cannot possibly survey all potential consumers to find the true percentage, but you know that taking a sample will provide you with the data to estimate this true percentage or proportion. Of the 100 people you survey, 62 indicate a willingness to buy this new product at this price. You find $p = 62/100 = 0.62$ or 62% as the estimate for the true proportion of consumers who would buy this new product.      ■

As we saw with quantitative variables, we often wish to compare two qualitative variables. For example, you may wish to compare

- the proportion of men who would buy a new product with the proportion of women who would buy it
- the proportion of defectives produced by the second shift with the proportion of defectives produced by the third shift
- the proportion of young people who like the new packaging with the proportion of older people who like it
- the proportion of bridges that are unsafe in Massachusetts compared with the corresponding proportion in Wyoming

When you are comparing two population proportions be sure that the proportions correspond to the same variable for the two different populations. Look at the first example in the list above. There are two populations here, men and women, and for both populations you are studying the proportion of the population that would buy the new product. You would not, however, compare the proportion of men who would buy the new product with the proportion of men who would not buy the product. This would be silly. These proportions would have to add up to one and so there is not anything to be learned by doing this.

As we did with the population means, we would like to compare the true population proportions $\pi_1$ to $\pi_2$ and we accomplish the comparison with a subtraction. So we wish to estimate the true difference in the population proportions, $\pi_1 - \pi_2$. It makes sense to estimate this true difference by using the difference in the two sample proportions or $p_1 - p_2$. If the two population proportions are the same, then the difference in the sample proportions should be close to zero. Remember that even if the two population proportions are both 0.20 and you select a sample from each population you will most likely get two slightly different sample proportions. This is again due to the fact that you are looking at only a piece of each of the populations. This is what we have called sampling error.

*Use subtraction to compare two population proportions.*

## 7.3.3 Summary of Commonly Used Point Estimators

The following table summarizes the most commonly used point estimators:

### TABLE 7.1   Summary of Common Point Estimators

| Type of Variable(s) | Population Parameter (unknown, what you wish to estimate) | Point Estimator or Sample Statistic (calculated from sample data) |
|---|---|---|
| A single quantitative variable | Population mean, $\mu$ | Sample mean, $\overline{X}$ |
|  | Population standard deviation, $\sigma$ | Sample standard deviation, $s$ |
| Two quantitative variables | Difference in population means, $\mu_1 - \mu_2$ | Difference in sample means, $\overline{X}_1 - \overline{X}_2$ |
|  | Ratio of population variances, $\sigma_1^2/\sigma_2^2$ | Ratio of sample variances, $s_1^2/s_2^2$ |
| A single qualitative variable | Population proportion, $\pi$ | Sample proportion, $p$ |
| Two qualitative variables | Difference in population proportions, $\pi_1 - \pi_2$ | Difference in sample proportions, $p_1 - p_2$ |

In the next section you will see that the estimators listed in column three of Table 7.1 are not the only estimators for the corresponding population parameters, but these are the most commonly used ones. The reason they are most commonly used is that, in a sense, they are the "best" estimators. In the next section we look at what makes them best.

## 7.3.4 Exercises—Learning It!

**7.1** Rocky's Hardware has 10 stores located in western Massachusetts. To plan inventory and staffing, the owner has recorded the daily sales for 15 days for two of the stores. The data are shown below:

> Store 1: $ 1981, 1802, 2053, 1877, 1647, 1931, 2065,
>    $ 1988, 1694, 1848, 1736, 1951, 2111, 1957, 2120
> Store 2: $ 2161, 2021, 1985, 2212, 2121, 2445, 1761, 2222,
>    $ 2025, 1917, 2038, 1956, 1911, 1991, 2072

(a) What is your estimate of the average daily sales for store 1?

(b) What is your estimate of the average daily sales for store 2?

(c) What is the difference in the two estimates you found in parts (a) and (b)?

(d) These samples were both selected from populations with a population mean of $\mu = \$2000$. Explain why they have different sample averages.

**7.2** Using the data for the hardware stores shown in Exercise 7.1, estimate the variability in the daily sales figures for both stores.

(a) What is your estimate of the variability of sales at store 1?

(b) What is your estimate of the variability of sales at store 2?

(c) What is the ratio of the two estimates you found in parts (a) and (b)?

(d) These samples were both selected from populations with a population variance of $\sigma^2 = (\$150)^2$. Explain why they have different sample variances.

**7.3** Jiffy Burger does not want its customers to wait more than 3 minutes on the average for their food. A sample of the wait time in minutes for 20 customers is collected and shown below:

| | | | | | | | | | |
|---|---|---|---|---|---|---|---|---|---|
| 3.4 | 3.3 | 3.3 | 3.3 | 3.3 | 3.4 | 3.4 | 3.3 | 3.2 | 3.4 |
| 3.3 | 3.3 | 3.3 | 3.3 | 3.4 | 3.2 | 3.5 | 3.2 | 3.4 | 3.1 |

(a) What is your estimate of the average customer wait at Jiffy Burger?

(b) How much difference is there between your estimate and the target of 3 minutes?

**7.4** A company is comparing the proportion of defective items produced by the second shift with the proportion of defective items produced by the third shift. In a sample of 100 items made by the second shift, 2 were found to be defective. In a sample of 100 items made by the third shift, 4 were found to be defective.

(a) What is your estimate of the proportion of defective items produced by the second shift?

(b) What is your estimate of the proportion of defective items produced by the third shift?

(c) What is the difference in the two estimates that you found in parts (a) and (b)?

**7.5** A company that buys blank VHS tapes for video recording is concerned about the amount of time that the tapes are able to record. The tapes are rated at 120 minutes, but the company knows there is variation in the actual recording time. Data are collected on 25 randomly selected tapes and the actual recording times are listed below:

| | | | | |
|---|---|---|---|---|
| 116 | 118 | 119 | 119 | 120 |
| 117 | 118 | 119 | 119 | 121 |
| 117 | 118 | 119 | 120 | 121 |
| 117 | 119 | 119 | 120 | 121 |
| 117 | 119 | 119 | 120 | 121 |

(a) What is your estimate of the average time available on the tape?

(b) Is your estimate different from 120 minutes? If so, by how much?

# 7.4 DESIRABLE PROPERTIES OF POINT ESTIMATORS

In this section, we develop the properties of point estimators. We do this by focusing on estimators for the population mean, $\mu$. However, the resulting properties apply to any point estimator.

In Chapter 4 you learned how to calculate measures of the middle of the sample data. In particular, you learned about the sample mean, $\overline{X}$, the sample median, and the sample mode. The trimmed mean was also examined as a measure of the middle of the sample data. Based on what we have seen so far, it makes sense to consider using one or more of these statistics as our point estimator of the unknown value $\mu$. Let's see how these might work for our diaper company.

**EXAMPLE 7.4**   **The Diaper Company**

*Possible Point Estimates of* $\mu$

Suppose you took a sample of 5 diapers and recorded the diaper weight (in grams) of each of the 5 diapers. The data are shown below:

$$55.87 \qquad 55.35 \qquad 54.50 \qquad 53.97 \qquad 54.29$$

The sample mean is 54.80 and the sample median (the middle score) is 54.50. There is no mode for this data set since none of the values occur more than once. We could use either the value of 54.80 or 54.50 as our point estimate for the unknown true mean weight of the diapers being made. ∎

It appears that the mean, the median, and the mode all fit the definition of a point estimator. In fact, we could dream up many other formulas that would fit the definition of a point estimator. For example, we could decide to average the minimum and maximum value in the sample and call this a point estimator. This is a legitimate point estimator, since the calculation is based only on sample data and yields a single number. Clearly, we need some criteria to judge which of these point estimators is the best one to use to estimate $\mu$.

Let's think about the criteria that we could use to make this judgment. If the point estimate is to be used as our best guess of the value of the unknown population parameter, in this case $\mu$, then we would like the point estimate to be close to $\mu$. Ideally, we would like our point estimate to hit $\mu$ on the nose. But we know that the chance of that happening is pretty slim. In fact, we know from Chapter 2 that the sample is only a piece of the population and if we were to take a different sample, we would get a different sample mean, a different sample median, and a different sample mode.

Let's follow the diaper company an hour after our first sample of 5 diapers.

**EXAMPLE 7.5**   **The Diaper Company**

*Second Sample Yields Slightly Different Values for Point Estimates*

A second sample of 5 diapers is taken an hour after the first sample. The diaper weights, in grams, are as follows:

$$54.85 \qquad 54.79 \qquad 54.65 \qquad 55.56 \qquad 55.82$$

Some quick calculations yield

Sample mean = 55.13     Sample median = 54.85     Sample mode = N/A ∎

The particular point estimates are different now. We used the same point estimators (formulas to calculate the sample mean and sample median) but we got different numbers because the data in this second sample were different from those of the first sample. This tells us that our point estimate depends on the particular sample we happen to choose and it changes from sample to sample. Combining this fact with the criterion of getting close to the unknown $\mu$, we can see that we want to use the point estimator that comes closest to $\mu$ for most samples. That is, we would like a point estimator that does not yield radically different numbers from sample to sample.

Suppose for a minute that you dreamed up a formula for a point estimator for $\mu$ and you used it with two different samples from the same population. For the first sample your point estimate was calculated to be 24 and for the second sample it was calculated to be 245. How much faith would you have in your estimate? "Not much" should be your answer. Why? The feature of this point estimator that is troubling is that it yields wildly different estimates from sample to sample. The estimator has too much variability; that is, it jumps around too much from sample to sample.

In summary, what we really want is a point estimator with the following two properties:

*Properties of point estimators*

1. The point estimator should yield a number close to the unknown population parameter.
2. The point estimator should not have a great deal of variability.

Even though we have been considering point estimators for the unknown population mean, $\mu$, these properties are desirable for any point estimator.

These two properties are more precisely stated as follows:

1. The point estimator should be unbiased.
2. The point estimator should have a small standard deviation.

*This is basically the same definition you saw in Chapter 2 when we were learning about bias in a sample.*

The word *unbiased* in statistics has basically the same meaning that it has in ordinary language. When the word is used as a character trait of somebody, it generally implies that the person is fair and does not favor something or someone. We would like our point estimator to behave in this fashion as well. We do not want to use a point estimator that always overestimates $\mu$ or always underestimates $\mu$. In this case the estimator would be biased or an unfair estimate. Sometimes our estimator $\overline{X}$ will be smaller than $\mu$ and sometimes it will be larger than $\mu$, but we know that it will be close to $\mu$ most of the time.

*Unbiased estimators are fair. They do not systematically overestimate or underestimate the parameter.*

> An *unbiased estimator* yields an estimate that is fair. It neither systematically overestimates the parameter nor systematically underestimates the parameter.

Proving that an estimator is unbiased is the job of the mathematical statisticians. They have shown that $\overline{X}$ is an unbiased estimator and we will believe them! They have also shown that the median is an unbiased estimator only if the population has a normal distribution. If a proposed point estimator is wildly biased, it may be obvious simply from the formula. Let's look at one such point estimator for $\mu$, one that is clearly ridiculous.

### EXAMPLE 7.6   The Diaper Company

#### An Example of a Biased Estimator

Someone in the diaper company proposes using the smallest observation in the sample as the estimate of $\mu$. For the first sample (shown in Example 7.4) this would yield an estimate of 53.97 g, and for the second sample (shown in Example 7.5) this formula would yield an estimate of 54.65 g. Clearly, this rule of selecting the minimum value as the estimate of $\mu$ is silly. It will always yield a number that is too small and is clearly an unfair estimate of the middle. ■

With regard to showing that $\overline{X}$ has a small standard deviation, we will again leave that to the mathematical statisticians. They have shown that $\overline{X}$ has a smaller standard deviation than the median or the mode; that is, it jumps around less from sample to sample. If you think about the way the sample mean, the sample median, and the sample mode are found you can see that this makes sense. The sample mean incorporates all of the data in the sample and in doing so washes out the effect of any single number in the sample that happens to be a bit unusual. Both the median and the mode are based on only a portion of the sample and thus tend to change more from sample to sample. Thus, the sample mean, $\overline{X}$, is the best estimator of the unknown population mean, $\mu$. The following example helps to demonstrate this fact.

*The sample mean is the point estimator for $\mu$ with the smallest standard deviation.*

## EXAMPLE 7.7   The Diaper Company

### Three More Samples

Let's follow the diaper company for 3 more hours. Each hour a sample of 5 observations is collected on the diaper weights (in grams). The observations are shown below along with the first two samples from Examples 7.4 and 7.5.

| | Sample Mean | Sample Median | Sample Mode |
|---|---|---|---|
| Hour 1: 55.87 55.35 54.50 53.97 54.29 | 54.80 | 54.50 | N/A |
| Hour 2: 54.85 54.79 54.65 55.56 55.82 | 55.13 | 54.85 | N/A |
| Hour 3: 54.40 56.44 54.11 54.67 54.42 | 54.81 | 54.42 | N/A |
| Hour 4: 54.69 54.55 54.56 54.22 54.87 | 54.58 | 54.56 | N/A |
| Hour 5: 55.50 54.44 56.30 54.00 55.68 | 55.18 | 55.50 | N/A |

*Recall that the sample mode is that value in the data set that occurs most frequently. Since none of the values repeat, there is no mode.*

You can quickly see that the sample mode is not a very useful point estimator. In any given sample of 5 observations, no weight appears more than once.

As we examine the sample median and the sample mean let's think about the two properties of point estimators that we have been discussing. We must look at where the values of the point estimator are centered and how much the values jump around from sample to sample. In this example, which numbers have more variation: the sample means or the sample medians? How can we tell? We can calculate the average and the standard deviation of the sample means and we can calculate the average and the standard deviation of the column of sample medians.

The average of the sample means is 54.90 g, which is very close to the target mean of 55.00 g. This tells us that the sample means are clustered around the target population mean, $\mu$. The standard deviation of the sample means is 0.251 g. For now, simply notice that this is quite a bit lower than the target standard deviation of 0.55 g. This is actually good news and we will return to this point in the next section.

The average of the sample medians is 54.77 g, which is also close to $\mu$ but not as close as the average of the sample means. The standard deviation of the sample medians is 0.441 g, telling us that the sample medians do jump around more than the sample means. That is, they have more dispersion. ■

This small example should give you a sense of what we mean by saying that the point estimator yields values that are clustered around the true population parameter and do not vary much from sample to sample.

*TRY IT NOW!*

## The Diaper Company
### *Comparing the Variability of Two Point Estimators*

Weights in grams for the next 5 hourly samples taken at the diaper company are shown below:

| | | | | | |
|---|---|---|---|---|---|
| Hour  6: | 54.89 | 55.06 | 55.45 | 55.23 | 55.75 |
| Hour  7: | 54.32 | 55.72 | 54.91 | 54.40 | 55.78 |
| Hour  8: | 54.14 | 55.18 | 55.78 | 55.37 | 55.69 |
| Hour  9: | 54.11 | 54.05 | 53.60 | 55.97 | 55.86 |
| Hour 10: | 55.21 | 55.40 | 53.87 | 55.09 | 55.70 |

For each sample, calculate the sample mean and the sample median.

Find the average of the sample means and the average of the sample medians.

Find the standard deviation of the sample means and the standard deviation of the sample medians.

Which point estimator has less variability?

ANS.

| | MEAN | MEDIAN |
|---|---|---|
| HOUR 6: | 55.28 | 55.23 |
| HOUR 7: | 55.03 | 54.91 |
| HOUR 8: | 55.23 | 55.37 |
| HOUR 9: | 54.72 | 54.11 |
| HOUR 10: | 55.05 | 55.21 |
| AVERAGE | 55.06 | 54.97 |
| ST. DEV.: | 0.22 | 0.51 |

THE MEAN HAS LESS VARIABILITY.

# 7.5  DISTRIBUTION OF THE SAMPLE MEAN, $\overline{X}$

As a result of your work in the previous section you have decided to use $\overline{X}$ as your point estimator for the true unknown population mean, $\mu$. Your boss reminds you that the average diaper weight should be 55.00 g. You know that your sample mean $\overline{X}$ will not equal 55.00 g, but it should be close to 55.00 g if the manufacturing process is working properly. You take a sample at hour 11 and find a sample mean of 54.00 g. By looking at the last 10 values of $\overline{X}$ you realize that a sample mean of 54.00 g is a bit low, but you are not sure if this was caused simply by sampling error or if the machine is working improperly. How can you decide?

## 7.5.1 Putting Z-scores and the Empirical Rule to Use

In Chapter 4 you learned about the *empirical rule* and *Z-scores*. These tools allowed you to decide if an individual observation was an outlier or unusual. To calculate the Z-score you used the following formula:

*A Z-score measures the number of standard deviations that a data value is from the mean.*

$$Z = \frac{\text{Distance between the data value and the average}}{\text{Standard deviation}}$$

Since we are trying to decide if an $\overline{X}$ value of 54.00 g is unusual, it seems like it would be a good idea to calculate the Z-score for the sample mean of 54.00 g. If 54.00 g is more than 3 standard deviations away from the target mean of 55.00 g, then we are pretty sure that there is a problem. Why? Because the empirical rule tells us that z-scores like this are very unusual and occur less than 1% of the time. This means that it would be *extremely unlikely* (less than 1% chance) that we would see a sample mean this small *if* the manufacturing process was producing diapers with a true mean weight of $\mu = 55.00$ g.

Now our data value is the observed sample mean, $\overline{X}$. This is a bit different from our previous use of Z-scores, but the idea is exactly the same. We are simply trying to decide if a particular $\overline{X}$ value is too unusual, whereas in Chapter 4 we were trying to decide if a particular individual observation was too unusual. We must rewrite the formula for $Z$ to reflect this fact:

$$Z = \frac{\overline{X} - \mu_{\overline{X}}}{\sigma_{\overline{X}}}$$

*Z STATISTIC FOR $\overline{X}$*

Notice that the average is now the average of all of the possible $\overline{X}$'s and is therefore labeled $\mu_{\overline{X}}$. The standard deviation is the standard deviation of the $\overline{X}$'s and is labeled $\sigma_{\overline{X}}$. Although this calculation for the Z-score looks different, it is really the same beast we used in Chapter 4. It simply measures the number of standard deviations that a particular $\overline{X}$ is above the mean of all possible $\overline{X}$'s.

So, we should just calculate the Z-score for the sample mean of 54.00 g. But if we look at the formula we see that to calculate the Z-score for the sample mean of 54.00 g, we need to know the *average of the sample means* and *the standard deviation of the sample means*. In the Try It Now Exercise in Section 7.4, you calculated the average of 5 sample means and the standard deviation of 5 sample means. We need to expand these calculations beyond 5 sample means to all possible sample means. In other words, what we need now is the average of all possible sample means and the standard deviation of all possible sample means.

Clearly, we do not want to select all possible samples from the population. This would defeat the whole idea of taking one sample to learn about the population. Fortunately, we can once again rely on our friends the mathematical statisticians who have proved a theorem that will give us the results we need. This theorem is the Central Limit Theorem and it is the foundation for virtually all of inferential statistics. It is the key that unlocks the rest of the course.

### 7.5.2 The Central Limit Theorem

Let's look at the diaper company data again. In Section 7.4 we looked at 10 hourly samples taken from the manufacturing process of the diaper company. Each sample consisted of $n = 5$ observations of diaper weight in grams. We saw that the value of the sample mean, $\overline{X}$, varied from sample to sample. We called $\overline{X}$ a point estimator, but if we think about it we can see that it is also a *random variable*. In Chapter 6 you learned that a random variable has a mean, a standard deviation, and a probability distribution. Thus, the random variable $\overline{X}$ has a mean, a standard deviation, and a probability distribution.

*Remember: A random variable is a quantitative variable whose value varies according to the rules of probability.*

When the random variable is a point estimator, then the standard deviation and the distribution are labeled with slightly different words. These are detailed in the next two definitions.

> The *standard error* is the standard deviation of a point estimator. It measures how much the point estimator or sample statistic varies from sample to sample.

> The probability distribution of a point estimator or a sample statistic is called a *sampling distribution*.

We looked at the mean and standard deviation of $\overline{X}$ in the last section. This concept is recapped in the next example.

### EXAMPLE 7.8   The Diaper Company

*Illustration of $\overline{X}$ as a Random Variable with a Mean, a Standard Error, and a Sampling Distribution*

In Section 7.4 we looked at 10 hourly samples. Each sample had $n = 5$ observations (remember sample size is always labeled as $n$). The 10 sample means (in grams) are shown below:

54.80    55.13    54.81    54.58    55.18    55.28    55.03    55.23    54.72    55.05

These 10 numbers are some of the possible values for $\overline{X}$ but not all of them. To list all of the possible sample means we would have to take all possible samples of size 5 from the population of all diapers manufactured by this company. This is not possible and, we will see, it will not be necessary.

We can begin to get a handle on how $\overline{X}$ behaves by looking at just these 10 values:

(a) The average of the 10 $\overline{X}$'s is 54.98 g.
(b) The standard deviation of the 10 $\overline{X}$'s is 0.24 g.
(c) The histogram, done in Excel, of these 10 $\overline{X}$'s looks like this:

*Remember: Excel displays the upper value of the class interval as though it were the middle of the class interval.*

**Histogram of 10 Sample Means**

Before stating the theorem let's see if we can anticipate the results by looking at Example 7.8. It seems like the mean of the $\overline{X}$'s is pretty close to the population mean, $\mu = 55$ g. The standard deviation of the $\overline{X}$'s is about half of the population standard deviation, $\sigma = 0.55$ g. It makes sense that the sample means would be more similar to each other (have a smaller standard deviation) than the individual diaper weights since the averaging process washes out the individual values. Finally, the sampling distribution or the histogram of the $\overline{X}$'s doesn't seem to have much of a shape right at the moment, but remember that we have only 10 $\overline{X}$'s in this example.

---

### Central Limit Theorem (CLT)

In random sampling from a population with mean $\mu$ and standard deviation $\sigma$, when $n$ is large enough, the distribution of $\overline{X}$ is

- approximately normal with
- a mean, $\mu_{\overline{X}}$, equal to $\mu$ and
- a standard deviation, $\sigma_{\overline{X}}$, equal to $\sigma/\sqrt{n}$

---

The Central Limit Theorem applies when you have a large enough sample size. A "large enough" sample size depends on how much the population distribution deviates from a normal distribution. Typically, if the sample size is larger than 30 then it is considered large enough. The larger the sample size, the better the normal approximation will be. If you are reasonably sure that the population you are sampling from has a normal distribution, then the results of the CLT hold even for small samples ($n \leq 30$).

*The CLT holds for small samples when the population has a normal or symmetric distribution.*

# 7.6 THE CENTRAL LIMIT THEOREM— A MORE DETAILED LOOK

Let's restate the Central Limit Theorem and examine each of the three points in the definition in terms of the diaper data.

## 7.6.1 The Shape of the Sampling Distribution of $\overline{X}$

The first point of the Central Limit Theorem is that if we took all possible random samples from an arbitrary population and calculated all the possible sample means, then the distribution of the sample means would be approximately normal. Remember that this applies if the sample size is large enough or if the underlying population distribution is normally distributed.

*The first point of the CLT: The distribution of all possible sample means is approximately normal.*

The diaper company is taking samples of size $n = 5$ every hour. Because each individual sample size is small ($n = 5$), to apply the CLT we will need to assume that the underlying distribution of diaper weights is normal or close to a normal shape. We can get a sense of the shape of the population distribution by examining a histogram of sample observations. The histogram of 260 diaper weights (representing 52 hours of sampling) is shown in Figure 7.2 (page 316).

This histogram gives us a visual picture of the sample, not the population. But we remember that we can learn about the population by studying a representative sample like this one. If the sample looks reasonably bell shaped we can say that the population is probably also bell shaped. In Chapter 15 we will learn a more precise way to make an inference about the shape of the population.

We just examined the distribution of a sample of 260 diaper weights and, based on the histogram of the sample, we concluded that the population is reasonably close to a normal distribution. Thus, we expect the distribution of the sample means to be

**FIGURE 7.2**   **The Diaper Company**   Histogram of individual diaper weights

approximately normal even though the sample size is small ($n = 5$). This can be seen visually by examining the histogram of 52 sample means shown in Figure 7.3. The distribution of the sample means does indeed look approximately normal. The Central Limit Theorem tells us that the sampling distribution of $\overline{X}$ will be approximately normal if we are sampling from a normal distribution, as is the case for the diaper data.

**FIGURE 7.3**   **The Diaper Company**   Histogram of 52 sample means

If the sample size is sufficiently large, the Central Limit Theorem tells us that the sampling distribution of $\overline{X}$ will be approximately normal regardless of the shape of the underlying population distribution. This is great news because most of the time we will not know the shape of the population from which we are sampling.

### 7.6.2 The Mean of the Sampling Distribution of $\overline{X}$

*The second point of the CLT: The average of all possible sample means is $\mu$.*

The second point of the Central Limit Theorem is that the mean of the sampling distribution of $\overline{X}$ equals the mean of the population you are sampling from. This means the center of the histogram of the $\overline{X}$'s should be $\mu$. This is consistent with what we have seen with the sample means from the diaper company. Most of the $\overline{X}$'s are close to $\mu$. The theorem tells us that if we took all possible samples of size $n$, $n$ sufficiently large, and calculated the sample mean for each of these samples and averaged the sample means, then we would get $\mu$. In terms of our notation this means $\mu_{\overline{X}} = \mu$. For our diaper company this means that the average of all the $\overline{X}$'s should be close to the target mean of $\mu = 55.00$ g. In fact, the average of the 52 sample means is 54.994 g. This can also be seen visually by examining the histogram of the sample means shown in Figure 7.3. The sample means are all clustered around the target mean of $\mu = 55.00$ g.

Why is this important? Clearly, we are not going to take more than one sample. We will take one sample and have one sample mean, $\overline{X}$, which will be our estimate of $\mu$. The Central Limit Theorem tells us that we can be reasonably sure that the sample mean that we get will be close to the true mean because the sample means cluster around the true mean, $\mu$. Later in this chapter we see that we quantify the words "reasonably sure" and "close" by using what is known as a confidence interval.

## 7.6.3 The Standard Error of the Sampling Distribution of $\overline{X}$

The third point of the theorem says that the standard deviation of the $\overline{X}$'s (also called the standard error) depends on two things: the amount of variability you start with in the population, $\sigma$, and the sample size, $n$.

*The third point of the CLT: The standard error of $\overline{X}$ is $\sigma/\sqrt{n}$.*

Let's think about what this means. Suppose you had a population of students who were all the *same age* and you picked 10 different samples from this population and calculated the average for each of the 10 samples. How different would the sample means be from each other? They will all be the same since everyone in the population is the same age! There is no variability, $\sigma = 0$, in the population and so there will be no variability in the $\overline{X}$'s. Clearly, this is an extreme example.

Now consider one population that has a mean of 25 and a standard deviation of 1 and one that has a mean of 25 and a standard deviation of 5. Graphs of these distributions are shown in Figure 7.4.

**Two Populations with Normal Distributions**

— Std Dev = 1
-- Std Dev = 5

10  13.1  16.2  19.3  22.4  25.5  28.6  31.7  34.8  37.9
Sales

**FIGURE 7.4** Graphs of two populations with the same mean but different standard deviations

Again consider taking 10 samples of size $n = 30$ from each population and calculate the 10 sample means. How would you expect the 10 sample means from the first population to behave compared to the 10 sample means from the second population? Let's examine the dotplots of the sample means shown in Figure 7.5.

In both cases, we see that the sample means cluster around $\mu = 25$, but the sample means from the first population appear to be more similar to each other (less dispersion) simply because the values in the first population are more similar to each other. Let's capture this in a result:

> As the standard deviation in the population increases, the standard error of $\overline{X}$ also increases.

Std. Dev = 1

Std. Dev = 5

23.50        24.00        24.50        25.00        25.50        26.00

**FIGURE 7.5** Dotplots of sample means

*TRY IT NOW!*

## The Central Limit Theorem
*Exploring the Third Point*

Use a software package such as Excel (Excel instructions are found at the end of the chapter) or Minitab to simulate picking 10 samples each of size $n = 35$ from two different populations:

Population 1: normal distribution with a mean of $\mu = 25$ and a standard deviation of $\sigma = 1$
Population 2: normal distribution with a mean of $\mu = 25$ and a standard deviation of $\sigma = 5$

Alternatively you can use the samples shown below:

**Samples from population 1**

| | | | | Sample | | | | | |
|---|---|---|---|---|---|---|---|---|---|
| 1 | 2 | 3 | 4 | 5 | 6 | 7 | 8 | 9 | 10 |
| 22.8 | 23.7 | 24.3 | 25.3 | 27.4 | 24.7 | 25.6 | 25.5 | 24.6 | 23.7 |
| 24.8 | 25.7 | 24.5 | 24.0 | 23.8 | 24.6 | 24.7 | 23.6 | 26.3 | 24.8 |
| 23.7 | 23.6 | 26.1 | 25.7 | 25.5 | 26.4 | 26.4 | 25.6 | 24.0 | 25.2 |
| 25.5 | 26.7 | 26.0 | 24.7 | 25.6 | 26.5 | 24.5 | 23.5 | 24.9 | 25.0 |
| 26.9 | 25.5 | 24.9 | 25.6 | 25.1 | 25.8 | 23.7 | 24.5 | 24.3 | 24.1 |
| 25.5 | 24.4 | 25.7 | 24.7 | 24.7 | 23.7 | 25.3 | 25.0 | 25.7 | 25.8 |
| 25.3 | 25.4 | 24.2 | 24.7 | 25.6 | 24.6 | 25.4 | 25.1 | 24.1 | 25.0 |
| 24.6 | 26.0 | 25.5 | 24.0 | 24.7 | 23.5 | 24.8 | 24.4 | 27.6 | 26.6 |
| 25.3 | 27.5 | 24.5 | 24.9 | 24.2 | 24.1 | 24.5 | 25.9 | 24.1 | 25.8 |
| 26.1 | 24.1 | 23.4 | 25.0 | 25.6 | 24.4 | 26.8 | 23.7 | 24.5 | 23.4 |
| 24.0 | 25.5 | 24.9 | 25.7 | 24.6 | 25.9 | 25.0 | 24.8 | 25.5 | 26.1 |
| 24.2 | 25.4 | 25.4 | 25.3 | 24.7 | 25.1 | 25.0 | 26.7 | 25.1 | 24.9 |
| 23.5 | 24.4 | 24.5 | 25.5 | 25.4 | 26.1 | 24.6 | 25.3 | 25.4 | 25.9 |
| 26.8 | 23.9 | 24.5 | 22.6 | 24.5 | 24.9 | 24.7 | 26.3 | 24.9 | 28.3 |
| 25.8 | 22.5 | 27.1 | 25.4 | 23.9 | 24.4 | 25.0 | 24.2 | 25.5 | 26.1 |
| 24.2 | 23.9 | 25.0 | 24.8 | 26.4 | 24.4 | 25.2 | 24.7 | 25.6 | 24.4 |
| 26.1 | 24.8 | 24.2 | 26.0 | 24.9 | 24.5 | 24.6 | 25.3 | 25.2 | 26.2 |
| 25.4 | 26.6 | 23.7 | 25.9 | 24.9 | 24.8 | 24.1 | 25.5 | 23.9 | 24.2 |
| 23.7 | 25.6 | 24.9 | 26.6 | 24.9 | 24.9 | 26.3 | 24.6 | 24.8 | 24.7 |
| 24.3 | 25.7 | 26.2 | 25.9 | 26.2 | 23.9 | 23.4 | 25.2 | 24.4 | 25.4 |
| 24.9 | 23.3 | 24.7 | 25.1 | 25.6 | 25.7 | 25.1 | 25.0 | 24.0 | 25.2 |
| 23.7 | 24.0 | 24.5 | 25.6 | 25.9 | 25.0 | 25.6 | 24.0 | 24.4 | 26.3 |
| 24.2 | 24.6 | 24.7 | 24.2 | 24.2 | 25.3 | 24.1 | 24.7 | 27.1 | 26.9 |
| 25.0 | 25.9 | 26.4 | 24.6 | 24.6 | 26.1 | 24.7 | 25.3 | 25.3 | 23.1 |
| 23.6 | 25.2 | 25.4 | 26.2 | 22.9 | 25.4 | 25.3 | 24.6 | 24.2 | 25.6 |
| 25.2 | 25.7 | 24.0 | 24.9 | 24.4 | 23.4 | 24.8 | 24.9 | 24.5 | 25.0 |
| 23.7 | 25.3 | 24.3 | 25.9 | 25.1 | 24.5 | 25.0 | 24.9 | 26.0 | 25.0 |
| 23.7 | 22.5 | 25.2 | 25.4 | 23.4 | 25.2 | 25.2 | 24.2 | 25.0 | 25.1 |
| 24.6 | 23.8 | 26.0 | 24.9 | 26.8 | 24.1 | 25.3 | 26.7 | 25.5 | 27.5 |
| 26.8 | 26.4 | 24.8 | 23.9 | 24.2 | 26.5 | 25.0 | 25.7 | 25.2 | 23.1 |
| 25.3 | 25.0 | 26.0 | 24.9 | 23.5 | 25.4 | 26.3 | 24.4 | 27.1 | 25.6 |
| 24.7 | 26.0 | 24.9 | 24.1 | 25.1 | 24.0 | 25.7 | 24.3 | 26.0 | 24.9 |
| 24.8 | 24.9 | 25.0 | 25.4 | 25.1 | 24.9 | 25.6 | 25.7 | 24.9 | 25.6 |
| 25.4 | 24.8 | 26.7 | 25.7 | 24.6 | 24.5 | 24.7 | 23.0 | 24.5 | 24.3 |
| 24.4 | 26.0 | 24.9 | 26.4 | 24.1 | 24.7 | 25.7 | 26.3 | 24.4 | 24.2 |

**Samples from population 2:**

| | | | | Sample | | | | | |
|---|---|---|---|---|---|---|---|---|---|
| **1** | **2** | **3** | **4** | **5** | **6** | **7** | **8** | **9** | **10** |
| 26.6 | 30.6 | 28.9 | 25.1 | 20.6 | 39.8 | 23.8 | 32.9 | 17.8 | 15.7 |
| 24.2 | 25.7 | 27.3 | 25.5 | 17.6 | 29.7 | 21.2 | 27.5 | 23.2 | 21.2 |
| 29.1 | 8.4 | 31.7 | 28.4 | 22.2 | 28.3 | 24.7 | 16.1 | 29.5 | 29.9 |
| 19.5 | 27.6 | 34.6 | 24.3 | 21.3 | 22.6 | 21.6 | 38.4 | 20.4 | 18.5 |
| 23.3 | 26.0 | 19.2 | 22.3 | 23.7 | 24.5 | 29.7 | 26.5 | 20.3 | 20.9 |
| 20.6 | 16.4 | 17.9 | 29.6 | 20.5 | 23.5 | 22.8 | 24.7 | 32.2 | 29.3 |
| 27.6 | 35.2 | 23.8 | 28.3 | 23.1 | 32.7 | 26.7 | 32.7 | 25.7 | 27.3 |
| 24.7 | 23.2 | 33.6 | 37.1 | 24.8 | 22.4 | 26.2 | 17.3 | 20.1 | 29.6 |
| 28.2 | 27.4 | 19.4 | 26.2 | 23.7 | 23.9 | 21.8 | 29.6 | 27.2 | 24.4 |
| 27.0 | 28.4 | 27.6 | 24.7 | 17.7 | 24.4 | 24.0 | 21.0 | 19.3 | 30.6 |
| 28.0 | 25.0 | 23.5 | 26.6 | 18.6 | 24.3 | 21.1 | 17.0 | 28.5 | 29.1 |
| 16.2 | 20.5 | 20.9 | 15.5 | 26.6 | 34.6 | 22.0 | 19.8 | 17.3 | 29.4 |
| 31.7 | 28.1 | 20.6 | 30.0 | 27.3 | 22.2 | 24.3 | 19.3 | 22.2 | 28.1 |
| 24.6 | 24.9 | 19.0 | 28.9 | 19.9 | 34.2 | 24.6 | 19.4 | 31.0 | 23.3 |
| 36.1 | 26.3 | 29.8 | 14.8 | 32.0 | 34.8 | 35.5 | 23.9 | 14.6 | 26.5 |
| 29.5 | 21.7 | 22.7 | 31.3 | 22.3 | 25.3 | 33.6 | 30.0 | 17.7 | 21.9 |
| 23.3 | 18.9 | 29.9 | 20.6 | 22.6 | 31.0 | 23.1 | 25.1 | 26.7 | 34.0 |
| 19.9 | 22.2 | 22.9 | 28.5 | 27.8 | 29.9 | 30.4 | 32.6 | 17.5 | 26.8 |
| 26.8 | 23.5 | 31.1 | 30.5 | 19.6 | 23.7 | 20.9 | 18.3 | 27.9 | 29.3 |
| 21.5 | 29.2 | 27.4 | 25.0 | 22.7 | 19.3 | 28.3 | 21.5 | 23.9 | 20.7 |
| 23.8 | 28.5 | 30.2 | 26.0 | 35.2 | 23.7 | 22.3 | 29.5 | 16.0 | 23.7 |
| 28.2 | 33.6 | 38.1 | 27.1 | 23.6 | 20.6 | 20.4 | 22.4 | 24.5 | 26.6 |
| 21.4 | 23.6 | 23.0 | 24.5 | 22.9 | 25.3 | 23.8 | 19.7 | 24.2 | 32.7 |
| 30.0 | 18.2 | 20.2 | 28.6 | 20.4 | 30.2 | 34.5 | 19.7 | 27.9 | 23.7 |
| 32.2 | 17.7 | 26.9 | 24.7 | 29.7 | 27.2 | 26.8 | 31.0 | 16.4 | 29.0 |
| 24.1 | 24.1 | 25.8 | 17.7 | 34.9 | 23.5 | 35.6 | 27.6 | 34.8 | 26.5 |
| 18.9 | 22.0 | 22.4 | 21.6 | 29.2 | 23.1 | 28.6 | 24.7 | 33.6 | 25.1 |
| 23.3 | 26.9 | 31.7 | 31.4 | 27.1 | 16.1 | 33.7 | 28.5 | 30.0 | 18.3 |
| 31.6 | 23.6 | 27.9 | 29.3 | 27.6 | 23.8 | 23.3 | 24.8 | 28.7 | 18.8 |
| 21.3 | 20.9 | 28.0 | 20.9 | 19.0 | 18.7 | 26.0 | 29.9 | 32.8 | 27.3 |
| 29.9 | 26.1 | 23.8 | 29.1 | 24.3 | 26.0 | 37.8 | 21.3 | 21.9 | 25.1 |
| 20.0 | 30.6 | 25.0 | 31.5 | 16.2 | 24.0 | 23.5 | 21.3 | 22.7 | 19.3 |
| 27.5 | 26.1 | 30.9 | 18.2 | 26.6 | 32.2 | 16.2 | 25.7 | 22.9 | 22.7 |
| 22.0 | 20.9 | 26.7 | 28.2 | 27.3 | 20.1 | 18.3 | 16.6 | 33.5 | 26.0 |
| 21.7 | 28.0 | 33.2 | 29.3 | 26.6 | 15.7 | 21.8 | 30.5 | 18.3 | 8.7 |

For each sample, calculate a sample mean, $\overline{X}$.

(*continued*)

Find the average and standard deviation of the $10\,\overline{X}$'s from population 1 samples.

Find the average and standard deviation of the $10\,\overline{X}$'s from population 2 samples.

The second number that influences the size of the standard error of $\overline{X}$ is the sample size, $n$. This should also make sense. Let's start by looking at the smallest possible sample size: $n = 1$. Clearly, the sample mean of one number is just the single value you have selected and these sample means will have the same variation as the population since $\sigma/\sqrt{n} = \sigma/\sqrt{1} = \sigma$. Typically, your sample is more than a single observation and thus $n$ is usually 2 or more. This means that no matter what the sample size is, the *amount of variation in the $\overline{X}$'s will be less than the amount of variation in the population you are sampling from.* This can be seen visually in the next example.

### EXAMPLE 7. 9 / The Diaper Company

***Frequency Histogram of Diaper Weights and Average Weights, $\overline{X}$***

The diaper company has taken 52 samples of size $n = 5$ over the last 52 hours. For each sample a sample mean, $\overline{X}$, has been calculated. The accompanying frequency histogram displays the distribution of the diaper weights ($5 \times 52 = 260$ observations) and the distribution of the 52 $\overline{X}$ values. Clearly, the distribution of the sample means has less variability.

ANS. THE AVERAGE OF THE $\overline{X}$'s SHOULD BE CLOSE TO $\mu = 25$ IN BOTH CASES. THE STANDARD DEVIATION OF THE $\overline{X}$'s FROM POPULATION 1 SAMPLES SHOULD BE SMALLER THAN THE STANDARD DEVIATION OF THE $\overline{X}$'s FROM POPULATION 2 SAMPLES.

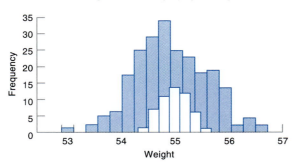

**Histogram of Average Diaper Weights**

Now let's see what the impact of increasing the sample size would be. Suppose we return to the population with a mean of 25 and a standard deviation of 1 and we take 10 samples of size 5 and calculate the 10 sample means. Now we take 10 samples of size 35 from this same population and calculate 10 more sample means. How would you expect the 10 sample means based on the samples with $n = 5$ to compare to the 10 sample means based on the samples with $n = 35$ to compare? We should expect that the sample means calculated from the bigger ($n = 35$) samples would be more similar (and thus have a smaller standard deviation) to each other since each sample mean contains more information and is thus a better estimate of $\mu$.

The Central Limit Theorem says that the standard error of $\overline{X}$, $\sigma_{\overline{X}}$, equals $\sigma/\sqrt{n}$. So as $n$ increases from 5 to 35 the denominator of the standard error gets bigger, which means we are dividing by a bigger number, yielding a smaller standard error. Let's follow this example and see what the numbers should be:

$\overline{X}$'s based on samples of size 5: The standard error should be $1/\sqrt{5} = 1/2.24 = 0.45$.

$\overline{X}$'s based on samples of size 30: The standard error should be $1/\sqrt{35} = 1/5.92 = 0.17$.

This is the second result:

As the sample size increases, the standard error of $\overline{X}$ decreases.

 **TRY IT NOW!**

## Central Limit Theorem

### *Impact of Sample Size on Standard Error*

Using the random number table in Appendix A and the 350 values shown in the previous **Try It Now!**, select 10 samples of size 5 from population 1. Review Section 2.4 if you need a refresher on how to use the random number table.

For each sample, calculate a sample mean, $\overline{X}$.

*(continued)*

Find the average and standard deviation of the 10 $\overline{X}$'s. Compare these values to the corresponding values that you found in the previous **Try It Now!** for population 1. In that case the sample size was $n = 35$.

### 7.6.4 Summary of Central Limit Theorem

Combining all three of the points of the Central Limit Theorem, we get Figure 7.6A, which displays the sampling distribution of $\overline{X}$, when $n$ is sufficiently large.

*Remember that the standard deviation of $\overline{X}$ is called the standard error. It is equal to $\sigma/\sqrt{n}$.*

      We know from our work on the normal distribution that 68% of values will fall within one standard deviation of the mean, 95% will fall within two standard deviations of the mean, and 99.7% will fall within three standard deviations of the mean. With regard to the random variable $\overline{X}$ this means that 68% of time we will observe a sample mean that falls within one standard error of the unknown, population mean, $\mu$. Similarly, 95% of the time we will observe a sample mean that falls within two standard errors of $\mu$, and 99.7% of the time we will observe a sample mean that falls within three standard errors of $\mu$. This idea leads to a concept known as a confidence interval or an interval estimate, which is developed in a later section of this chapter.

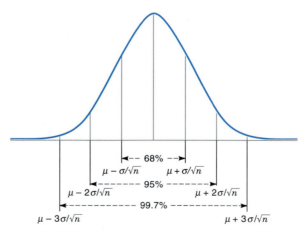

**FIGURE 7.6A**    Sampling distribution of $\overline{X}$

## Discovery Exercise 7.1
## THE CENTRAL LIMIT THEOREM IN ACTION

**Part I: Draw a picture of a normal distribution with mean of 10 and standard deviation of 5. This is the population we will sample from.**

**Part II: Generate and examine 100 random samples.**

(a) For this exercise you will need to generate 100 samples each consisting of 30 values selected from a normal distribution with a mean of 10 and a standard deviation of 5. If you do not know how to generate this data, your instructor will provide you with data.

   *Question:* If you plotted the values from one of the samples, what shape would the histogram have? _____ Try it and see.

   *Question:* What value would you expect the average of the 30 values from sample 1 to be close to? _____ Find the average of the first sample of 30 and see if it is close to the number you thought it would be.

   *Question:* Do you expect the average of the 30 values in sample 2 to be close to the same number? _____ Why or why not? _____

   *Question:* What value should the standard deviation of the first sample of 30 values be close to? _____ Find the standard deviation of the first sample of 30 values and check to see if you were right.

   *Question:* Do you expect the standard deviation of the 30 values in sample 2 to be close to the same number? _____ How about sample 3? _____

**Part III: Create a distribution of $\overline{X}$ for samples of size $n = 30$**

(a) For each of the 100 samples find the sample mean, $\overline{X}_1$, $\overline{X}_2$, etc. You should have a column of 100 $\overline{X}$ values.

   *Question:* Examine these $\overline{X}$ values. Are they all the same? _____ Why not?

   *Question:* Are they all equal to 10? Why not? What is this called? _____

(b) Construct a histogram of the 100 $\overline{X}$ values.

   *Question:* What shape does the distribution of $\overline{X}$ have? _____

   *Question:* Will the shape of the $\overline{X}$'s always be normal or is it just normal because we sample from a normal population? _____

(c) Find the average of the 100 $\overline{X}$ values and the standard deviation of the 100 $\overline{X}$ values.

The CLT tells us that the average of the $\overline{X}$'s should be close to the mean of the distribution of the population.

   *Question:* What is the mean of the population we sampled from? _____

   *Question:* What is the average of the 100 $\overline{X}$'s? _____

(*continued*)

*Question:* Are they close? _____

*Question:* How could you get them closer? _____

(d) The CLT tells us that the standard deviation of the $\overline{X}$'s should be about $\sigma/\sqrt{n}$.

*Question:* What is $\sigma$? _____

*Question:* What is $n$? _____

*Question:* So, what is the value that the standard deviation of the $\overline{X}$'s should be close to? _____

*Question:* What is the standard deviation of the $\overline{X}$'s? _____ Is it close to the value it was supposed to be? _____

*Question:* How could you get them closer?_____

# 7.7   DRAWING INFERENCES BY USING THE CENTRAL LIMIT THEOREM

Now we can return to the diaper company problem with which we started Section 7.5. We are ready to decide if an $\overline{X}$ value of 54.00 g is indeed unusual or if it is reasonable variation caused by sampling error when the process is running properly. Recall that the true mean diaper weight, $\mu$, is 55.00 g and the true population standard deviation $\sigma$ is 0.55 g when the manufacturing process is running properly.

## 7.7.1 Using the Central Limit Theorem

*Create a histogram of the diaper weights to check the assumption of normality.*

Let's create a picture similar to Figure 7.6A for the diaper manufacturing process. Recall that the samples each have 5 observations. Since the sample size is less than 30 we must assume that the underlying distribution of diaper weights is normally distributed. The $\overline{X}$'s should be centered at $\mu_{\overline{X}} = 55$ g and the standard error should be $\sigma_{\overline{X}} = \sigma/\sqrt{n} = 0.55/\sqrt{5} = 0.25$ g. Combining these results yields Figure 7.6B.

A sample mean of 54.00 g clearly falls in the tails of this distribution, telling us that if the process is running properly, the probability of seeing a sample mean this small is tiny. This conclusion has taken into account the fact the diaper weights themselves have some variability measured by $\sigma$ and the sample size is $n = 5$. Thus, we

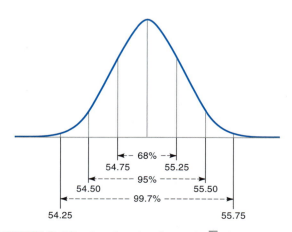

**FIGURE 7.6B**   Sampling distribution for $\overline{X}$ when $\mu = 55.00$

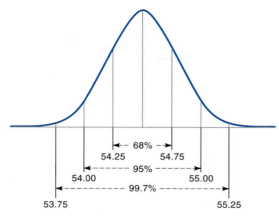

**FIGURE 7.6C**  Sampling distribution of $\overline{X}$ when $\mu = 54.50$

must conclude that something is not quite right with the process. We could also reach this conclusion by calculating the $Z$-score for our observed $\overline{X}$ value of 54 g:

$$Z = \frac{\overline{X} - \mu_{\overline{X}}}{\sigma_{\overline{X}}}$$

$$Z = \frac{54 - 55}{0.25} = -4.0$$

We know that a $Z$-score below $-3$ is extremely unusual. Thus, we conclude that we should check the machine.

Suppose for a moment that the process mean had shifted to 54.5 g. In this case the sampling distribution would look like Figure 7.6C.

Now a sample mean of $\overline{X} = 54$ g is not quite so unlikely. The Central Limit Theorem has given us the basis on which to judge whether the observed sample mean value of 54 g is too unusual and if so what could possibly have happened to the process that would result in such a low sample mean.

Let us examine some other problems that we can begin to tackle now that we have learned about the Central Limit Theorem.

## EXAMPLE 7. 10   Car Batteries

### *Using the Central Limit Theorem*

A company that produces car batteries warranties the batteries for 36 months. The battery lives have a standard deviation of 2 months. You and 8 of your friends purchased these batteries and you have collected data on the length of time each of them lasted. The data (in months) are shown below:

| 35.38 | 35.87 | 35.49 | 34.38 | 34.55 | 39.99 | 36.18 | 35.09 | 34.57 |

You calculate the average length of life of these batteries and find it to be 35.72 months. Do you have reason to be suspicious of the company claim on the basis of these data?

Let's calculate the $Z$-value for this sample mean value:

$$Z = \frac{\overline{X} - \mu_{\overline{X}}}{\sigma_{\overline{X}}}$$

$$Z = \frac{35.72 - 36}{2/\sqrt{9}} = -0.42$$

This $Z$-score tells us that the sample mean value of 35.72 months is within 1 standard deviation ($Z$-score between $-1$ and 1) of the warranty value of 36 months. We know that 68% of $Z$-scores are between $-1$ and 1. Thus, it is *not* an unusually low sample average and we have no basis to be suspicious of this company. ■

In this example we used the value of the population standard deviation claimed by the company. What happens if we do not know the population standard deviation? This is not unlikely in many cases. If we need to estimate the mean, why would we *know* the standard deviation? If the sample is sufficiently large, $n > 30$, then we can use the point estimate $s$ instead of $\sigma$ in the calculation of the z-score. The next example demonstrates this situation.

### EXAMPLE 7. 11   Manufacturer of Breakfast Cereals

*Using the Central Limit Theorem*

A manufacturer of breakfast cereals packs the cereal in 15-ounce boxes. Periodic quality checks must be made to ensure that the boxes are being filled adequately. A random sample of 35 boxes of the cereal are selected and weighed. The data are shown below:

| | | | | | | |
|---|---|---|---|---|---|---|
| 14.91 | 15.23 | 15.34 | 15.40 | 15.48 | 15.59 | 15.62 |
| 14.96 | 15.24 | 15.34 | 15.42 | 15.49 | 15.60 | 15.62 |
| 15.11 | 15.28 | 15.36 | 15.45 | 15.50 | 15.60 | 15.63 |
| 15.21 | 15.30 | 15.38 | 15.46 | 15.52 | 15.61 | 15.64 |
| 15.22 | 15.30 | 15.39 | 15.48 | 15.58 | 15.61 | 15.67 |

You find the average and standard deviation to be $\overline{X} = 15.42$ oz and $s = 0.19$ oz. Is there any reason to believe that the boxes are not being filled properly?

Calculating the Z-score for this sample value, we find

$$Z = \frac{\overline{X} - \mu_{\overline{X}}}{\sigma_{\overline{X}}}$$

$$Z = \frac{15.42 - 15}{0.19/\sqrt{35}} = \frac{0.42}{0.03} = 14.00$$

This Z-score is extremely large, telling us that it would be extremely unusual to see an average of 15.42 oz (based on a sample of 35 boxes) if the boxes were indeed being filled to an average of 15 oz. Therefore, we have rather strong evidence to indicate that the boxes are not being filled to an average of 15 oz. ∎

In the preceding example you should have noticed that we used $s$, the sample standard deviation, as an estimate of $\sigma$ in finding the Z-score, since $\sigma$ was not given. This isn't precisely the correct procedure but it is close enough given the large sample size. In this case, the formula to compute the approximate Z-score is given as follows:

*SCORE WHEN $\sigma$ IS UN-KNOWN AND $n$ IS LARGE*

$$Z = \frac{\overline{X} - \mu_{\overline{X}}}{s_{\overline{X}}} = \frac{\overline{X} - \mu_{\overline{X}}}{s/\sqrt{n}}$$

A more thorough discussion of this issue is addressed later in this chapter along with what to do if the standard deviation is unknown but the sample size is small ($n \le 30$).

 **TRY IT NOW!**

## Cost of Books

*Comparing the Sample Mean to the Claimed Population Mean*

A university states that students spend an average of $225 per semester on books. Based on your own experience you feel that this is an underestimate of the true expenditure. You ask 30 of your friends how much they spent on textbooks this past semester and you obtain the following data:

| | | | | | | | | | |
|---|---|---|---|---|---|---|---|---|---|
| 214 | 236 | 241 | 247 | 248 | 253 | 258 | 262 | 269 | 277 |
| 233 | 239 | 244 | 248 | 249 | 254 | 260 | 263 | 274 | 277 |
| 234 | 241 | 245 | 248 | 250 | 254 | 262 | 265 | 276 | 281 |

Based on these data, do you have reason to tell the university that its statement is inaccurate?

## 7.7.2 Exercises—Learning It!

**7.6**    A manufacturer of pain relievers claims that it takes an average of 12.75 min for a person to be relieved of headache pain after taking its pain reliever. The time it takes to get relief is normally distributed with a standard deviation of 0.5 min. A sample of 12 people is taken and the data are shown below:

   12.9    13.2    12.7    13.1    13.0    13.1    13.0    12.6    13.1    13.0    13.1    12.8

(a) Find the sample mean.

(b) Find the standard error of $\overline{X}$.

(c) If the manufacturer claims that the mean is 12.75 min, find the Z-score of the sample mean.

(d) What do you think of the manufacturer's claim based on the Z-score?

**7.7**    A company that sells personal computers via mail order claims that its customers wait an average of 3 minutes on hold before reaching a technical support staff member. The on-hold time is normally distributed with a standard deviation of 0.5 min. A sample of the on-hold times for 20 different customers is shown below:

| | | | |
|---|---|---|---|
| 3.21 | 3.13 | 3.72 | 2.88 |
| 2.84 | 3.49 | 3.04 | 2.98 |
| 2.56 | 2.91 | 3.34 | 2.84 |
| 2.73 | 3.16 | 3.26 | 3.41 |
| 3.29 | 2.67 | 2.98 | 2.31 |

(a) Find the sample mean.

(b) Find the standard error of $\overline{X}$.

(c) If the claimed mean is 3 min, find the Z-score for the sample mean.

(d) What do you think of the company's claim based on the Z-score?

**7.8**    The U-Make-It Hardware chain has 10 different stores within a certain geographical region. The dollar value of customer sales is normally distributed with an average sale of $35.25 and a standard deviation of $2.50. You have recently been hired to manage one of these stores and under your leadership the average sales based on a sample of 100 customers is $36.50. You are very proud of the increased average sale and point this out to senior management.

(a) Find the Z-score for $36.50.

(b) Based on this Z-score, is your pride justified?

ANS.
$\overline{X} = \$253.40$
$s = 15.51$
$Z = 10.03$
THE UNIVERSITY'S PUBLISHED AVERAGE IS INACCURATE.

**7.9**   Scores on a national test are normally distributed with a mean of 75 and a standard deviation of 5 points. At a certain high school 50 students took this test and their average was 80. On the basis of these data, can this high school be considered unusually strong?

**7.10**  Starting salaries nationally among business school graduates follow a normal distribution with a mean of $25,000 and a standard deviation of $1500. The Career and Human Resource Director at College XYZ has taken a sample of 25 of the most recent graduates of the business school and found that their average salary is $24,500.

(a) Between what two values will 99.7% of the salary averages fall?

(b) Based on your answer to part (a), what can you conclude about the average starting salary of the most recent graduates of College XYZ?

# 7.8  LARGE-SAMPLE CONFIDENCE INTERVALS FOR THE MEAN

In this chapter, you have learned about point estimators, which give you a single number to be used as an estimate of the population parameter. A confidence interval takes the point estimate a step further and gives you a range of values and a probability. The probability value tells you the likelihood that you have an interval that actually includes the value of the unknown population parameter.

## 7.8.1 The Basics of Confidence Intervals

Let's return to the problem that was presented at the beginning of this chapter. The diaper company must decide if the manufacturing process is running to specification. In particular, the diaper company must decide if the population mean, $\mu$, is 55.00 g. The company will decide this based on a sample of 30 observations that have a sample mean, $\overline{X} = 54.98$. We know that the sample mean, $\overline{X}$, is a reliable point estimate of $\mu$. But we also know that it is not likely to hit $\mu$ on the nose. Initially, it looks as if $\overline{X}$ is not very helpful. It is an estimate of $\mu$ that is pretty likely to be wrong! Ideally, we would like to use $\overline{X}$ to create a range of values that would definitely contain $\mu$. However, because of sampling error this is not possible. Instead, we can specify a high probability, say 0.90 or 0.95, that a particular range or interval covers the true mean, $\mu$. For example, we would like to be able to say that based on the sample data, the interval from 54.50 to 55.50 g covers $\mu$ with a probability of 0.95. Rewriting this in a more mathematical form gives us

$$\text{Probability}(54.5 \leq \mu \leq 55.5) = 0.95$$

This is an example of a confidence interval. Let's examine the components of the confidence interval. First of all, it has a lower bound for $\mu$. Call this $L$. In the example, $L$ is 54.50. The confidence interval also has an upper bound for $\mu$. Call this $U$. In the example, $U$ is 55.50.

Finally, it has a probability value, which is called the confidence level and is labeled $1 - \alpha$. In the example, $\alpha$ is 0.05 and $1 - \alpha$ is 0.95. In this case we are constructing a $100(1 - 0.05)\% = 95\%$ confidence interval. This means that if we were to construct 20 intervals based on 20 different sample means, we would expect 95% of them or 19 intervals to actually contain $\mu$. For any *individual* interval, $\mu$ is either in the interval or it is not. The problem is that, due to sampling error, we do not know for sure whether we have an interval which does include the value of $\mu$. Thus the probability refers to the chance that we have one of the intervals that does indeed contain $\mu$.

In general, a confidence interval for the population mean has the following form:

$$P(L \le \mu \le U) = 1 - \alpha$$

> A *confidence interval* or an *interval estimate* is a range of values with an associated probability or *confidence level,* $1 - \alpha$. The probability quantifies the chance that the interval contains the true population parameter.

Think about the information such a confidence interval gives you. Instead of just being able to say that the true $\mu$ is somewhere close to $\overline{X} = 54.98$ now you can say the probability that the interval from 54.50 to 55.50 g covers $\mu$ is 0.95. This gives you a way to quantify the error you know is inherent in $\overline{X}$.

The next few sections provide the details for calculating the lower and upper bounds for confidence intervals for $\mu$. These sections are followed by a section that focuses on the interpretation of confidence intervals.

## 7.8.2 Confidence Interval for $\mu$: Normally Distributed Population and Known Standard Deviation

We start by developing a confidence interval for the mean of a population that is normally distributed and for which the population standard deviation, $\sigma$, is known. We will see what happens as we relax both the assumption of normality and the assumption of a known standard deviation in the next section.

*Recall that an unbiased estimator of $\mu$ is a fair estimator. It does not systematically overestimate or underestimate $\mu$.*

As we start to look at how to calculate the lower and upper bounds for the confidence interval, think about where $\overline{X}$ should fall in the interval. Remember that $\overline{X}$ is an *unbiased* estimate of $\mu$ and so it makes sense to put $\overline{X}$ right in the middle of the interval as shown below:

*The value of $\overline{X}$ should always go in the center of a confidence interval for $\mu$.*

Notice that the distance from the lower bound to the middle of the interval has been labeled *e*. The distance from the middle of the interval to the upper bound has also been labeled *e*. Thus, the width of the interval is 2*e*. To find the lower bound we take $\overline{X}$ and subtract *e*, and to find the upper bound we take $\overline{X}$ and add the value of *e*. Now, all we need is the value of *e*, which stands for error.

Suppose we decide we want to construct a 95% confidence interval ($\alpha = 0.05$) for the population mean, $\mu$. In a sense, we wish to "cover" the value of $\mu$, 95% of the time. We know that $\overline{X}$ is the best point estimate we can use for $\mu$ and that $\overline{X}$ has a normal distribution when *n* is large enough. Remember that to cover 95% of the values that follow a normal distribution we need to go out about two standard deviations from the mean in both directions. Finally, we recall that the standard deviation of the sample mean, also known as the standard error, is equal to $\sigma/\sqrt{n}$. Bringing these three pieces of information together tells us that *e* should equal $2\sigma/\sqrt{n}$. This is just about right.

*The Central Limit Theorem tells us that the sampling distribution of $\overline{X}$ is a normal distribution.*

You may recall that in Chapter 6 you learned how to find *Z*-values given tail area probabilities. To get the correct value for *Z*, you must use that procedure with a tail area probability equal to $\alpha/2$. Label this value as $Z_{\alpha/2}$. For a 95% confidence interval we divide $\alpha = 0.05$ by 2 and find that the area in one of the tails is 0.025. This means we are given a probability and need the corresponding *Z*-value. Shown on page 330 is a portion of the *Z* table:

| $z$ | 0.00 | 0.01 | 0.02 | 0.03 | 0.04 | 0.05 | 0.06 | 0.07 | 0.08 | 0.09 |
|---|---|---|---|---|---|---|---|---|---|---|
| $-2.1$ | 0.0179 | 0.0174 | 0.0170 | 0.0166 | 0.0162 | 0.0158 | 0.0154 | 0.0150 | 0.0146 | 0.0143 |
| $-2.0$ | 0.0228 | 0.0222 | 0.0217 | 0.0212 | 0.0207 | 0.0202 | 0.0197 | 0.0192 | 0.0188 | 0.0183 |
| $-1.9$ | 0.0287 | 0.0281 | 0.0274 | 0.0268 | 0.0262 | 0.0256 | 0.0250 | 0.0244 | 0.0239 | 0.0233 |
| $-1.8$ | 0.0359 | 0.0351 | 0.0344 | 0.0336 | 0.0329 | 0.0322 | 0.0314 | 0.0307 | 0.0301 | 0.0294 |
| $-1.7$ | 0.0446 | 0.0436 | 0.0427 | 0.0418 | 0.0409 | 0.0401 | 0.0392 | 0.0384 | 0.0375 | 0.0367 |

We want to find the $Z$-value that has 0.025 in the lower tail. Look in the inside of the table for the number closest to 0.025 and see that it corresponds to a $Z$-value of $-1.96$. This gives us the value of $-Z_{\alpha/2}$. To get the positive cutoff value we can simply drop the negative sign because of the symmetry in the normal distribution. So $Z_{\alpha/2} = 1.96$. Figure 7.7 shows the normal distribution with 5% in the tail area.

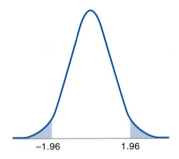

**FIGURE 7.7**   Normal distribution with 0.05 in the tails

So, the value of $e$ is calculated as $1.96\, \sigma/\sqrt{n}$. This gives us the following formulas for the lower and upper bounds for a 95% confidence interval for $\mu$:

*Lower and upper bounds for a 95% confidence interval for $\mu$*

$$L = \overline{X} - e = \overline{X} - \frac{1.96\,\sigma}{\sqrt{n}}$$

$$U = \overline{X} + e = \overline{X} + \frac{1.96\,\sigma}{\sqrt{n}}$$

Let's use these to find a 95% confidence interval for the diaper manufacturer.

**EXAMPLE 7.12**   **The Diaper Company**

*A 95% Confidence Interval for the Mean Weight*

A sample of 30 diapers yielded a sample average weight of 54.98. Based on this sample, find a 95% confidence interval for $\mu$.

Given information:   $\overline{X} = 54.98$

$n = 30$

$\sigma = 0.55$ g   (from the specifications)

Calculations:   Standard error $= \dfrac{\sigma}{\sqrt{n}} = \dfrac{0.55}{\sqrt{30}} = 0.10$

Error $= e = \dfrac{1.96\,\sigma}{\sqrt{n}} = (1.96)(0.10) = 0.20$

Lower bound $= \overline{X} - e = 54.98 - 0.20 = 54.78$

Upper bound $= \overline{X} + e = 54.98 + 0.20 = 55.18$

We can state that we are 95% confident that the true mean weight is between 54.78

and 55.18 g. This means that there is a 95% chance that we have constructed an interval that does indeed contain the true mean weight. Restating this mathematically gives us

$$P(54.78 \le \mu \le 55.18) = 0.95$$

Notice that the target value of 55.00 g is in between the lower and upper bound and thus a likely value for the population mean weight. ■

The formulas that we have used in this example are for a 95% confidence interval. Suppose instead that we wish to find a 90% confidence interval or a 99% confidence interval. The only adjustment that needs to be made to the formula is the value of Z that we multiply the standard error by to get e. For the 95% confidence level we used the value of Z that cut off a probability of 0.05 in the two tails of the normal distribution. The procedure for figuring out the correct Z-value for any confidence level is as follows:

*STEPS FOR FINDING THE VALUE OF Z FOR A CONFIDENCE INTERVAL*

**Step 1.** Take the value of $\alpha$ and divide it by 2.

**Step 2.** Look that value up in the body of the Z table (find the one closest to it).

**Step 3.** Read off the corresponding Z-value. This gives you the value of $-Z_{\alpha/2}$.

**Step 4.** Drop the negative sign to get the value of $Z_{\alpha/2}$.

Remember that Z tells you how many standard errors away from the mean that you need to go to get the coverage you desire.

In general, the formulas for a $100(1 - \alpha)\%$ confidence interval for $\mu$ when the population standard deviation, $\sigma$, is known are shown below:

*CONFIDENCE INTERVALS FOR $\mu$ WHEN $\sigma$ IS KNOWN*

Error: $\quad e = \dfrac{Z_{\alpha/2}}{\sqrt{n}}$

Width of the interval: $\quad w = 2e$

Lower bound: $\quad L = \overline{X} - e$

Upper bound: $\quad U = \overline{X} + e$

where $Z_{\alpha/2}$ is the value that cuts off $\alpha/2$ in the upper tail of the standard normal distribution and $\alpha/2$ in the lower tail of the standard normal distribution.

Some common values of $\alpha$ and the corresponding Z-values are shown in Table 7.2.

Notice that as the level of confidence increases so does the value of Z used in the formula. Since the value of Z affects the size of e, the interval gets wider as the confidence level increases. Are wider intervals better? Suppose you had a choice between 2 intervals with the same level of confidence: one that stated the average age, $\mu$, of the target population was between 2 and 95 years and one that stated the average age, $\mu$, of the target population was between 25 and 35 years. Which interval gives you better information? Clearly, the interval that is narrower, the second one. This small example shows us that the narrower the interval, the more precise our estimate. Thus, as you *increase* the level of confidence you *decrease* the precision of the estimate.

*As you increase the level of confidence, you lose precision.*

**TABLE 7.2 Common Values of $\alpha$ and the Corresponding Z-values**

| | Confidence Level $(1 - \alpha)$ | | | | |
|---|---|---|---|---|---|
| | 0.90 | 0.95 | 0.98 | 0.99 | 0.995 |
| $\alpha$ | 0.10 | 0.05 | 0.02 | 0.01 | 0.005 |
| $Z_{\alpha/2}$ | 1.645 | 1.96 | 2.33 | 2.58 | 2.81 |

## TRY IT NOW!

### The Bottle-Filling Problem
*Finding a Confidence Interval for $\mu$*

A sample of 36 bottles had a sample mean of $\overline{X} = 32.10$ oz. The population standard deviation, $\sigma$, was assumed to be 0.1 oz. Find a 95% confidence interval for $\mu$. How wide is the interval?

Now find a 98% confidence interval for $\mu$.

Which interval is wider?

---

*Assumptions of confidence interval formulae*

The procedure just described depends on the knowledge that $\overline{X}$ has a normal distribution and the population standard deviation is known. The Central Limit Theorem tells us that $\overline{X}$ has a normal distribution when $n > 30$ or when the underlying population distribution is normally distributed (or close to normal). Therefore, if you have a *large sample* ($n > 30$), then you can use this procedure to find confidence intervals for $\mu$ regardless of the shape of the underlying distribution.

If the *sample size is small* ($n < 30$), we should visually check the shape of the distribution of the sample data to get a rough idea of the population shape. If the histogram or other plot of the data looks reasonably symmetric in shape, then you can use the formulas we just developed.

### 7.8.3 What Happens if $\sigma$ Is Unknown or the Distribution Is Not Normal?

In many cases you do not know the population standard deviation, $\sigma$, and/or you do not have an underlying normally distributed population. First, consider what happens to the calculation of the confidence interval for $\mu$ if we do not know $\sigma$ but we still have a normally distributed population. There are two cases: large sample size and small sample size.

If you have a large sample size then you can use $s$, the sample standard deviation, to estimate $\sigma$. Using $s$ to approximate $\sigma$ in the formulas for the upper and lower limits of the confidence interval does not appreciably affect the confidence level. This is the same approach we took in computing the $Z$-score when we did not know $\sigma$ but the sample size was sufficiently large.

The formulas for a $100(1 - \alpha)\%$ confidence interval for the mean, $\mu$, of a normally distributed population when the population standard deviation, $\sigma$, is unknown and $n$ is sufficiently large are shown below:

| | | |
|---|---|---|
| Error: | $e = \dfrac{Zs}{\sqrt{n}}$ | CONFIDENCE INTERVAL FOR $\mu$ WHEN $\sigma$ IS UNKNOWN |
| Width of the interval: | $w = 2e$ | |
| Lower bound: | $L = \overline{X} - e$ | |
| Upper bound: | $U = \overline{X} + e$ | |

If you have a small sample size and the standard deviation is unknown then the sampling distribution of $\overline{X}$ is not normally distributed. The sampling distribution is called a $t$-distribution and is covered in the next section.

Finally, if you have a small sample size and the histogram of the data does not look roughly normal in shape and you do not know the population standard deviation, then you should not use this procedure for finding a confidence interval for $\mu$. It may be possible to use a transformation to get normally distributed data or it may be necessary to use a different distribution or a nonparametric procedure. Nonparametric procedures are typically not as powerful but do not require the assumption of an underlying normal population. Transformations and nonparametric procedures are out of the scope of this text but can be found in the next level of statistics courses/books. Table 7.3 summarizes the different scenarios.

### TABLE 7.3   Summary of Confidence Intervals for $\mu$

| Population | Standard deviation | Sample size | Confidence interval for $\mu$ |
|---|---|---|---|
| Normal | Known | $n \geq 1$ | $L = \overline{X} - Z\sigma/\sqrt{n}$ <br> $U = \overline{X} + Z\sigma/\sqrt{n}$ |
| Normal | Unknown | $n > 30$ | $L = \overline{X} - Zs/\sqrt{n}$ <br> $U = \overline{X} + Zs/\sqrt{n}$ |
| Normal | Unknown | $n \leq 30$ | Use $t$-distribution covered in Section 7.9 |
| Non-normal | Known | $n > 30$ | $L = \overline{X} - Z\sigma/\sqrt{n}$ <br> $U = \overline{X} + Z\sigma/\sqrt{n}$ |
| Non-normal | Unknown | $n \leq 30$ | Use transformation or nonparametric |

### 7.8.4 Interpreting the Confidence Interval

The last major point of discussion for confidence intervals has to do with what the words "I am 95% confident" imply. Many people feel that this means that there is a 95% chance that the population mean is in the interval you have constructed. This is wrong! After all, $\mu$ is a number even if it is unknown to us. Therefore, there is no probability associated with it being between two other numbers; it is either in the interval or it is not.

So, then, what is the proper interpretation of those words? The correct interpretation of the 95% confidence level has to do with the chance that you have an interval that does in fact contain the population parameter. Remember that if you take 100 different samples from the same population, you will get 100 different sample means and therefore 100 different confidence intervals. If each of them is a 95% confidence interval then theoretically 95 out of the 100 intervals will contain $\mu$ and 5 of them will not.

Suppose that 10 samples of size $n = 30$ were selected from the population of diapers. For each sample, an average diaper weight was calculated and a 90% confidence interval was calculated based on this value of $\overline{X}$. These 10 confidence intervals are shown in Table 7.4. Let's assume that the manufacturing process was running properly so the population mean is 55 g. Remember that the population standard deviation is 0.55 g. According to the theory, 9 out of 10 of these intervals should include the value of $\mu = 55$ g.

The last row of Table 7.4 indicates whether the interval includes the value of 55 g. You can see in Figure 7.8 that there is only 1 interval that does not include the value of 55 g and 9 that do. This is just what we expected, since 90% of 10 intervals is 9 intervals that cover $\mu$. The problem, of course, is that we don't know the actual value of $\mu$ and therefore we don't know if we have an interval that actually contains $\mu$ or not. The confidence level gives us the probability of our having an interval that does in fact contain the population parameter.

### TABLE 7.4  Ten 90% Confidence Intervals

| $\overline{X}$ | 55.21 | 54.94 | 54.99 | 55.05 | 55.06 | 54.97 | 55.11 | 54.92 | 55.03 | 55.10 |
|---|---|---|---|---|---|---|---|---|---|---|
| Error | 0.17 | 0.17 | 0.17 | 0.17 | 0.17 | 0.17 | 0.17 | 0.17 | 0.17 | 0.17 |
| Lower | 55.04 | 54.78 | 54.82 | 54.89 | 54.89 | 54.81 | 54.95 | 54.76 | 54.87 | 54.93 |
| Upper | 55.37 | 55.11 | 55.16 | 55.22 | 55.23 | 55.14 | 55.28 | 55.09 | 55.20 | 55.27 |
| Covers $\mu$? | No | Yes | Yes | Yes | Yes | Yes | Yes | Yes | Yes | Yes |

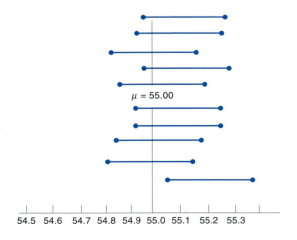

$\mu = 55.00$

54.5  54.6  54.7  54.8  54.9  55.0  55.1  55.2  55.3

**FIGURE 7.8**   Comparison of confidence intervals and $\mu$

To increase the probability that we have a "good" interval, we could widen the intervals. This is precisely what happens when you increase the confidence level. When you increase the confidence level from 90 to 95%, you are increasing the chance that you have a good interval, but you are losing precision by widening the interval. It is a tradeoff.

## 7.8.5 Exercises—Learning It!

**7.11** A university wants to estimate the average amount of money that students spend on textbooks in a semester. It takes a random sample of 45 students and find that the average amount of money spent was $282 with a standard deviation of $21. Find a 98% confidence interval estimate for the true mean amount of money spent on textbooks in a semester.

**7.12** The weights of jars of peanut butter are normally distributed with a standard deviation of 0.03 oz. A random sample of 12 jars has an average weight of 14.9 oz.

(a) Find a 95% confidence interval for the true mean weight of the jars.

(b) If the jars are labeled 15 oz, do you think that there is a problem?

**7.13** Quality Foods' regional manager wants to determine the average fat content per pound of steak sold in Portland, Maine. She's considering quitting her job and starting a cattle ranch after reading the article "Queen of the Range" in the January 1989 issue of *Successful Farming*. Her ranch would produce low-fat beef for sale to upscale consumers. She purchased one steak from each of the 18 stores in the area and determined the fat content per pound. The data (in ounces) are shown below:

| | |
|---|---|
| 1.2 | 1.3 |
| 1.6 | 1.3 |
| 1.3 | 1.6 |
| 0.7 | 1.1 |
| 1.3 | 1.2 |
| 1.7 | 1.3 |
| 0.9 | 0.9 |
| 1.8 | 1.4 |
| 1.3 | 1.4 |

The standard deviation of fat content is known to be 0.30 oz.

(a) Assuming that the fat content is normally distributed, calculate a 95% confidence interval for the average fat content per pound in steaks sold in the Portland, Maine, area.

(b) Her research shows that the steaks from her ranch will average 1.0 oz of fat per pound. Are her steaks significantly lower in fat content than the ones available in the Portland area? Explain why or why not.

**7.14** Is it a lottery jackpot or a bonus? A salary or a life's savings? No, it's the annual compensation paid out to top executives of public stock corporations. The following table shows the 1996 salary and bonus for the first few of the 43 executives, the firm they work for, and the type of firm (manufacturing or nonmanufacturing). *Data file: SALARY.XXX*

| Name | Firm | Type | 1996 Salary and Bonus |
|---|---|---|---|
| Eugene Freedman | Enesco Giftware Group/Stanhome Inc. | M | $4,505,500 |
| Thomas Wheeler | Massachusetts Mutual | NM | 3,569,998 |
| Terrence Murray | Fleet Financial Group | NM | 3,909,200 |
| Robert Shapiro | Monsanto | M | 2,920,000 |
| David E. Sams, Jr. | Massachusetts Mutual | NM | 2,760,183 |
| Charles Gifford | BankBoston | NM | 2,600,000 |
| George David | United Technologies | M | 2,375,000 |

SOURCE: *Springfield Sunday Republican* (20 July 1997)

Assuming that these 43 executives represent a random sample of executives, find a 90% confidence interval for the mean salary of executives.

**7.15** The United States Department of Labor publishes a great deal of data on a monthly basis. The following data on the average price of electricity were extracted from the World Wide Web:

**[Bureau of Labor Statistics Data]**
**Data extracted on: January 4, 1997**

Average   Price Data
Series    Catalog:
Series    ID          APU0000702111
Area:     U.S. City Average
Item:     Average price per kwh of electricity

| Jan-94 | 0.090 | Jan-95 | 0.091 | Jan-96 | 0.091 |
|--------|-------|--------|-------|--------|-------|
| Feb | 0.090 | Feb | 0.091 | Feb | 0.091 |
| Mar | 0.089 | Mar | 0.091 | Mar | 0.092 |
| Apr | 0.088 | Apr | 0.090 | Apr | 0.092 |
| May | 0.090 | May | 0.092 | May | 0.092 |
| Jun | 0.095 | Jun | 0.098 | Jun | 0.096 |
| Jul | 0.095 | Jul | 0.098 | Jul | 0.099 |
| Aug | 0.096 | Aug | 0.098 | Aug | 0.099 |
| Sep | 0.096 | Sep | 0.097 | Sep | 0.099 |
| Oct | 0.093 | Oct | 0.094 | Oct | 0.095 |
| Nov | 0.091 | Nov | 0.092 | Nov | 0.092 |
| Dec | 0.091 | Dec | 0.091 | Dec | 0.094 |

Treating these three years of data as a sample, find a 90% confidence interval for the average monthly price of electricity. Assume that the standard deviation is $.006 per kwh.

**7.16** A national grocery chain is considering opening a store at a particular location. To be sure that enough traffic goes by that location, the grocery chain took a sample of vehicles crossing the intersection on 40 days. The results are shown in the table below:

**Number of Cars Crossing Location/Day**

| 1431 | 1540 | 1293 | 1340 |
|------|------|------|------|
| 1302 | 1700 | 1533 | 1402 |
| 1255 | 1840 | 1272 | 1467 |
| 1377 | 1642 | 1572 | 1220 |
| 1450 | 1139 | 1520 | 1477 |
| 1483 | 1227 | 1227 | 1515 |
| 1529 | 1684 | 1257 | 1242 |
| 1588 | 1782 | 1238 | 1350 |
| 1535 | 1491 | 1276 | 1367 |
| 1533 | 1513 | 1420 | 1375 |

(a) Find a 95% confidence interval for the average number of cars that pass this location on a daily basis. The standard deviation is assumed to be 165 cars.

(b) The company has decided to open a store at this location only if there is a daily average of at least 1400 cars passing this location. Based on your confidence interval, would you advise the company to open a store at this location? Explain why or why not.

**7.17** The amount of time that it takes a student to complete an assignment in a statistics class has a standard deviation of 5 minutes. A group of 35 students are observed and it is found that the average time to complete the assignment was 55 minutes.

(a) Find a 99% confidence interval for the true mean time to complete the assignment.

(b) If the instructor believes that the students should be spending an average of 50 minutes on the assignment, what can you say about his belief?

7.18 Most companies have increased their dependence on computers and software. As a result, more employee time is spent on the telephone with technical support for the software. A sample of 22 times spent on hold for technical support is shown below:

**Time on Hold (minutes)**

| | |
|------|------|
| 8.6 | 12.9 |
| 12.7 | 11.4 |
| 8.7 | 10.3 |
| 12.2 | 7.9 |
| 11.7 | 9.2 |
| 7.3 | 10.5 |
| 9.8 | 11.8 |
| 14.5 | 10.9 |
| 13.0 | 11.5 |
| 12.9 | 10.6 |
| 11.4 | 11.7 |

Assume that the standard deviation is 2 minutes.

(a) Since the sample size is less than 30 and you are not told that the population of time on hold is normally distributed, display the data in a histogram, and comment on the shape of the data. Is it reasonable to assume that the data come from a population that has a normal distribution? Why or why not?

(b) Find a 99% confidence interval for the average amount of time spent on hold per call.

(c) Find a 95% confidence interval for the average amount of time spent on hold per call.

(d) If you are the manager of these employees and you are trying to argue for additional staff, which of the two confidence intervals would you use and why?

7.19 The symphony in a medium-size New England city is surveying the community to determine the average number of times during a year that a person would attend a concert at a particular price. The responses from 50 adults are shown below:

**Number of Times
Symphony Would Be
Attended in One Year**

| | | | | |
|---|---|---|---|---|
| 1 | 3 | 2 | 2 | 2 |
| 4 | 4 | 4 | 4 | 4 |
| 3 | 2 | 4 | 4 | 4 |
| 2 | 1 | 4 | 4 | 4 |
| 2 | 3 | 3 | 3 | 3 |
| 2 | 3 | 3 | 3 | 3 |
| 3 | 5 | 5 | 5 | 5 |
| 2 | 4 | 4 | 4 | 4 |
| 4 | 3 | 2 | 2 | 2 |
| 3 | 3 | 2 | 2 | 2 |

(a) Find a 90% confidence interval for the average number of times a year a person would attend a concert at the price studied.

(b) How might this confidence interval be useful to the management of the symphony?

## *Discovery Exercise 7.2*
## EXPLORING CONFIDENCE INTERVALS FOR $\mu$

From a population of college students across the United States, a sample was selected to find out how many hours per week a typical student spends playing sports.

**Part I: A random sample of 2500 students was selected. The sample mean, $\overline{X}$, was found to be 12.5 hours. The population standard deviation, $\sigma$, is known to be 1.05 hours. Given this information, find**

(a)  a 90% confidence interval for $\mu$

(b)  a 92% confidence interval for $\mu$

(c)  a 94% confidence interval for $\mu$

(d)  a 96% confidence interval for $\mu$

(e)  a 98% confidence interval for $\mu$

(f)  Discuss what happens to the size of the interval as the level of confidence increases.

**Part II: A random sample of 2500 students was selected. The sample mean, $\overline{X}$, was found to be 10.5 hours. The population standard deviation, $\sigma$, is known to be 1.05 hours. Given this information, find**

(a)  a 90% confidence interval for $\mu$

(b)  a 92% confidence interval for $\mu$

(*continued*)

(c) a 94% confidence interval for $\mu$

(d) a 96% confidence interval for $\mu$

(e) a 98% confidence interval for $\mu$

(f) Compare the intervals found in Part I with those found in Part II. Discuss what happened to the confidence interval due to the change in the value of the sample mean, $\overline{X}$.

**Part III: A random sample of 2500 students was selected. The sample mean, $\overline{X}$, was found to be 12.5 hours. Suppose you learn that the population standard deviation, $\sigma$, is actually 2.05 hours. Given this information, find**

(a) a 90% confidence interval for $\mu$

(b) a 92% confidence interval for $\mu$

(c) a 94% confidence interval for $\mu$

(d) a 96% confidence interval for $\mu$

(*continued*)

(e) a 98% confidence interval for $\mu$

(f) Compare the intervals found in Part I with those found in Part III. Discuss what happened to the confidence intervals due to the change in the value of the population standard deviation, $\sigma$.

**Part IV: A random sample of 2000 students was selected. The sample mean, $\overline{X}$, was found to be 12.5 hours. The population standard deviation, $\sigma$, is known to be 1.05 hours. Given this information, find**

(a) a 90% confidence interval for $\mu$

(b) a 92% confidence interval for $\mu$

(c) a 94% confidence interval for $\mu$

(d) a 96% confidence interval for $\mu$

(e) a 98% confidence interval for $\mu$

(f) Compare the intervals found in Part I with those found in Part IV. Discuss what happened to the confidence intervals due to the change in the value of the sample size, $n$.

# 7.9 DISTRIBUTION OF THE SAMPLE MEAN: SMALL SAMPLE AND UNKNOWN $\sigma$

In the previous section we noted that if the standard deviation of a normally distributed population is unknown and the sample size is small ($n \leq 30$), then the sampling distribution of $\overline{X}$ does not follow a normal distribution. As we did with the large sample size situation, logic tells us to try using $s$ in the formula for the $Z$-score instead of $\sigma$. When we do this a $t$-score is created instead of a $Z$-score. That is, the sampling distribution of $\overline{X}$ for small samples, when $\sigma$ is unknown and the population is normally distributed, follows what is called *Student's t distribution*. The $t$-score is calculated as follows:

*t-score*

$$t = \frac{\overline{X} - \mu}{s/\sqrt{n}}$$

Notice that the calculation for $t$ is just the same as for $Z$, with $\sigma$ replaced with $s$.

*This distribution is not named for you, the student!*

The $t$ distribution was first developed by a man named William Gosset who was working for Guinness Brewery in Ireland. The company did not allow research results to be published so he published his results under a pen name; Student was his pen name!

The graph of the $t$ distribution looks very much like the standard normal distribution. It is symmetric, it has a bell shape, and it is centered at zero just like the $Z$ distribution. The major difference between the $Z$ and the $t$ distributions has to do with the spread or variability of the distributions. The standard deviation of the $t$ distribution is not 1 like it is for the $Z$ distribution. Instead, the variability of the $t$ distribution is related to a number that is called the degrees of freedom. Thus, there are many different $t$ distributions, but they all have generally the same shape. A $t$ distribution with 5 degrees of freedom is shown in Figure 7.9.

*Each sample yields a different sample mean, $\overline{X}$, and a different sample standard deviation, s.*

Let's examine the $Z$ and $t$ calculations. The two calculations are shown side by side below:

$$Z = \frac{\overline{X} - \mu}{\sigma/\sqrt{n}} \qquad t = \frac{\overline{X} - \mu}{s/\sqrt{n}}$$

Think about what causes the variability in the $Z$ statistic. For a given sample size, say $n = 26$, the only thing that causes the $Z$ values to change from sample to sample is $\overline{X}$. The other elements of the calculation stay the same from sample to sample. Now look at the $t$ distribution. Both $\overline{X}$ and $s$ are changing from sample to sample. This causes $t$ to have more variability than $Z$. Based on this discussion we would correctly

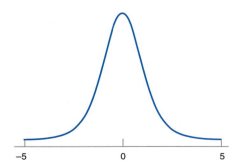

**FIGURE 7.9**   *t*-distribution with 5 degrees of freedom

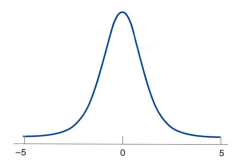

**FIGURE 7.10**  $t$ distribution with 25 degrees of freedom

conclude that $t$ will have "fatter" tails than $Z$ and indeed it does. The $t$ distribution for a sample size of $n = 26$ is shown in Figure 7.10.

You should ask, "Why do I need to know the value of $n$ to draw the graph?" You might remember that as we take larger and larger samples ($n$ gets larger) our estimate of the unknown population standard deviation gets better and better. Why is this so? Well, just think about the whole population as a big pie. When you take a small sample it is like taking a small piece of the pie and tasting it and trying to decide how the whole pie tastes based on that piece. As the piece you taste gets bigger you have a much better idea of how the whole pie tastes. Eventually, if you take a big enough sample you will eat the whole pie and you will know exactly how the whole pie tastes! This is the same as sampling the whole population. If the whole population were sampled you wouldn't need to estimate $\sigma$ because you would know it.

So as you take a larger and larger sample you get a better and better estimate of $\sigma$, which, in turn, causes $t$ to be less and less variable. All of this means that the graph of $t$ depends on the size of your sample. In particular, the degrees of freedom associated with the $t$ distribution is related to the sample size. The value for the degrees of freedom is $n - 1$, one less than the sample size. If you have taken a sample of size 26, then the $t$ distribution will have 25 degrees of freedom. This is the graph shown in Figure 7.10. Examine the $t$ distributions for various different degrees of freedom shown in Figure 7.11.

What do you notice about the graph of $t$ as the number of degrees of freedom gets larger (i.e., as the sample size gets larger)? You should see that the $t$ distribution gets tighter, meaning the variability of the distribution gets smaller. This is precisely what you should have expected to happen based on our discussion of what happens to our estimate $s$ as we sample bigger pieces of the pie. In fact, eventually, if we use a large enough value for $n$, the graph of $t$ will be indistinguishable from the standard normal distribution, $Z$. Typically, the value of $n$ that is considered large enough is $n > 30$.

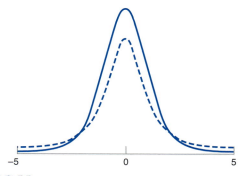

**FIGURE 7.11**  $t$ distributions with 1 and 50 degrees of freedom

# 7.10 SMALL-SAMPLE CONFIDENCE INTERVALS FOR THE MEAN

Now that we have an understanding of the $t$ distribution, the only thing that remains is to see how the confidence interval for $\mu$ changes for small samples. Replacing the $Z$ table value with a $t$ value in the formulas for the upper and lower limits of the confidence interval for $\mu$ gives us the following:

*CONFIDENCE INTERVAL FOR $\mu$, SMALL SAMPLE, AND $\sigma$ UNKNOWN*

| | | |
|---|---|---|
| Error: | $e = t_{\alpha/2,n-1}s/\sqrt{n}$ | |
| Width of the interval: | $w = 2e$ | |
| Lower bound: | $L = \overline{X} - e$ | |
| Upper bound: | $U = \overline{X} + e$ | |

In addition, we must now assume that the underlying population of the variable we are estimating is normally distributed.

## 7.10.1 Using the $t$ Table

The procedure for figuring out the correct $t$ value for any confidence level is similar to that used to find the correct $Z$ value. We need to find the $t$ value so that there is a total probability of $\alpha$ in the tails of the distribution. Since the $t$ distribution is symmetric, this means there will be $\alpha/2$ in each of the tail areas. Unlike with the $Z$ table, there is no need to work the table backwards. Typically the $t$ table is used to find the $t$ value that corresponds to a certain probability in one of the tails of the distribution rather than to find probabilities under the $t$ distribution. A portion of the $t$ table is shown in Figure 7.12.

The complete $t$ table is found in Appendix Table A.4. Notice that the table has several columns and several rows. The columns correspond to the tail area probability and the rows correspond to the number of degrees of freedom, which is one less than the sample size. The procedure for using the table is as follows:

*STEPS FOR FINDING THE VALUE OF $t$ FOR A CONFIDENCE INTERVAL*

**Step 1:** Take the value of $\alpha$ and divide it by 2.

**Step 2:** Use that column of the $t$ table.

**Step 3:** Find the number of degrees of freedom by calculating $n - 1$. Use that row of the $t$ table.

**Step 4:** Read off the corresponding $t$ value at the intersection of the row and column you have identified. Label this value $t_{\alpha/2,n-1}$ to indicate that the tail area probability is $\alpha/2$ and there are $n - 1$ degrees of freedom.

Let's look at an example to see how to use the $t$ table.

| | **Upper-Tail Areas** | | | | | |
|---|---|---|---|---|---|---|
| **Degrees of Freedom** | **0.25** | **0.1** | **0.05** | **0.025** | **0.01** | **0.005** |
| 20 | 0.687 | 1.325 | 1.725 | 2.086 | 2.528 | 2.845 |
| 21 | 0.686 | 1.323 | 1.721 | 2.080 | 2.518 | 2.831 |
| 22 | 0.686 | 1.321 | 1.717 | 2.074 | 2.508 | 2.819 |
| 23 | 0.685 | 1.320 | 1.714 | 2.069 | 2.500 | 2.807 |
| 24 | 0.685 | 1.318 | 1.711 | 2.064 | 2.492 | 2.797 |
| 25 | 0.684 | 1.316 | 1.708 | 2.060 | 2.485 | 2.787 |
| 26 | 0.684 | 1.315 | 1.706 | 2.056 | 2.479 | 2.779 |

**FIGURE 7.12** A portion of the $t$ table

## EXAMPLE 7.13

*Using the t Table*

Find the *t* value for a 95% confidence interval for a sample size of 25:

1. Since a 95% confidence interval is required, $\alpha$ is 0.05 and so $\alpha/2$ is 0.025.
2. Use the 0.025 column of the *t* table.
3. Since the sample size is $n = 25$, use the row corresponding to $n - 1 = 25 - 1 = 24$ degrees of freedom.
4. The *t* value is 2.064 in the table shown in Figure 7.12. Label it $t_{0.025,24}$. ■

Now let's use this *t* value to find a confidence interval for $\mu$ in the following example.

## EXAMPLE 7.14   Assembly Time

*Small-Sample Confidence Interval for $\mu$*

In pricing a new product, the manufacturer needs to determine the average number of minutes it takes to manufacture one item. A sample of 25 items is tracked and the number of minutes to make each unit is recorded below. The assembly time of a unit is known to have a normal distribution.

**Time to Assemble Item**

| | | | | | | | | | |
|---|---|---|---|---|---|---|---|---|---|
| 5.2 | 4.4 | 5.0 | 5.2 | 4.7 | 4.6 | 4.8 | 4.8 | 4.0 | 4.5 |
| 5.7 | 4.8 | 5.2 | 4.4 | 4.3 | 4.5 | 4.0 | 4.6 | 4.9 | 5.8 |
| 5.2 | 6.0 | 5.6 | 5.7 | 4.5 | | | | | |

To find the confidence interval for $\mu$, we need to calculate $\overline{X}$ and $s$ and find the appropriate *t* value from the table shown in Figure 7.12.

$$\overline{X} = 4.90 \text{ min}$$
$$s = 0.55 \text{ min}$$

The *t* value with 24 degrees of freedom and a tail area of 0.025 is $t_{0.025,24} = 2.064$.

The upper and lower 95% confidence limits for $\mu$ are calculated using the following formulas:

| | |
|---|---|
| Error: | $e = t_{\alpha/2,n-1} s/\sqrt{n} = (2.064)(0.55)/\sqrt{25} = 0.23$ |
| Width of the interval: | $w = 2e = (2)(0.23) = 0.46$ |
| Lower bound: | $L = \overline{X} - e = 4.90 - 0.23 = 4.67 \text{ min}$ |
| Upper bound: | $U = \overline{X} + e = 4.90 + 0.23 = 5.13 \text{ min}$ |

Notice that the *t* value for the 95% confidence interval with 24 degrees of freedom is 2.064. If we had known the value of $\sigma$, then we would have used a *Z* value of 1.96 instead of the *t* value. This would have made the value of *e* smaller, making the interval narrower. Confidence intervals constructed using *t* values are always wider than the corresponding interval would be if you could have used *Z*. This is a direct result of the increased variability due to having to estimate $\sigma$. ■

The *t* distribution is the correct sampling distribution to use whenever $\sigma$ is unknown. However, when the sample size is large, $n > 30$, it is common practice to use the *Z* distribution instead of the *t* distribution because the graphs are virtually identical. The original reason for using $Z$ instead of $t$ for large samples has all but vanished but the practice continues. Prior to the use of statistical software, analysts had to rely on printed $t$ tables. Since the $t$ distribution depends on the sample size, very large $t$ tables were needed to allow for many different sample sizes. Thus, it was generally agreed that $Z$ could be used for sample sizes greater than 30, eliminating the need for $t$ tables beyond 30 degrees of freedom. Software packages have eliminated the need for printed tables. Nevertheless, the convention of using $Z$ instead of $t$ for large samples has become entrenched, and in this book, when the sample size is larger than 30 we will use the $Z$ distribution. Note that you would certainly be correct to use $t$ whenever the standard deviation is unknown and that may be the approach taken by some software packages.

### 7.10.2 Exercises—Learning It!

**7.20**  The police department is concerned about the ability of officers to identify drunk drivers on the road. Before instituting a new training program they take a sample of 28 arrests and record the level of alcohol in the blood at the time of the arrest. Assume that the level of alcohol in the blood is normally distributed. The data are shown below.

**Alcohol Level**

| | |
|---|---|
| 92 | 204 |
| 93 | 182 |
| 108 | 173 |
| 173 | 105 |
| 194 | 153 |
| 133 | 150 |
| 207 | 180 |
| 127 | 209 |
| 256 | 141 |
| 184 | 151 |
| 253 | 133 |
| 159 | 147 |
| 101 | 209 |
| 133 | 252 |

(a)  Find a 90% confidence interval for the average alcohol level in the blood at the time of arrest.

(b)  Find a 95% confidence interval for the average alcohol level in the blood at the time of arrest.

**7.21**  A large amusement park has recently added 5 new rides, including a large roller coaster called the Mind Eraser. Management is concerned about the waiting times on the new roller coaster. A random sample of 10 people is selected and the time (in minutes) that each person waits to ride the Mind Eraser is recorded and shown below:

**Mind Eraser**

| | |
|---|---|
| 43 | 66 |
| 80 | 54 |
| 48 | 72 |
| 61 | 58 |
| 74 | 68 |

(a)  Find a 95% confidence interval for the average waiting time for the Mind Eraser, assuming that the waiting time is normally distributed.

(b)  The park management thinks that if customers have to wait more than 60 minutes for a ride, then the park should increase the staff to reduce the waiting time. Based on your confidence interval, does the park need to increase the staff? Explain why or why not.

**7.22** Hospital administrators are paying increasing attention to length of stay of patients. A sample of 14 patient stays (in days) are shown below:

**Patient Length of Stay**

| | |
|---|---|
| 7 | 6 |
| 2 | 5 |
| 6 | 4 |
| 7 | 4 |
| 8 | 2 |
| 8 | 3 |
| 3 | 7 |

(a) Find a 95% confidence interval for the length of patient stay, assuming that the length of stay is normally distributed.

(b) Find a 90% confidence interval for the length of patient stay, assuming that the length of stay is normally distributed.

(c) Which of these intervals should the administrator use in negotiating with insurance companies and why?

**7.23** The number of workers per vehicle at the top 10 car assembly plants is shown below:

| | |
|---|---|
| Nissan Smyrna | 2.22 |
| Toyota Cambridge | 2.35 |
| Honda East Liberty | 2.38 |
| Toyota Georgetown #1 | 2.50 |
| Chrysler Bramalea | 2.54 |
| Honda Marysville | 2.57 |
| Ford Atlanta | 2.63 |
| Ford Chicago | 2.66 |
| Chrysler Belvidere | 2.68 |
| GM Oshawa #1 | 2.68 |

SOURCE: *Manufacturing Engineering*, August 1997

Assuming that the number of workers per vehicle has a normal distribution, find a 95% confidence interval for the average number of workers per vehicle.

**7.24** The percentage of foster children adopted into families varies widely from state to state according to an 8 August 1997 article in the *Springfield Union News*. Congress is preparing bills to encourage the process in some states. The number of children adopted in 1996, as a percentage of all children available for adoption that year, is shown for a sample of 14 states:

| | |
|---|---|
| Alabama | 18.46 |
| Alaska | 31.31 |
| California | 34.78 |
| Kansas | 29.67 |
| Kentucky | 30.60 |
| Massachusetts | 46.27 |
| Minnesota | 20.66 |
| New Hampshire | 51.61 |
| North Dakota | 96.73 |
| Ohio | 25.08 |
| Rhode Island | 44.83 |
| Texas | 28.93 |
| Utah | 53.45 |
| Wyoming | 50.00 |

Find a 95% confidence interval for the average adoption rate, assuming that the adoption rates are normally distributed.

# 7.11  CONFIDENCE INTERVAL FOR QUALITATIVE DATA

Often we are interested in estimating what proportion of the population has a particular characteristic or has a particular opinion. In fact, most of the data printed in newspapers and magazines are the result of surveys and often proportions or percentages are reported. This is particularly true at election times. For example, the candidate for mayor wishes to know what proportion or percentage of voters in the city favor him.

*Remember: Qualitative data describe a particular characteristic of a sample item. They are most often nonnumerical in nature.*

The increasing focus on quality and meeting the needs of the customer has led to increased data collection, much of which is qualitative data. A manufacturing company is clearly interested in the proportion of products manufactured that are defective. All businesses, both manufacturing and service industries, are interested in knowing if their products/services are meeting the needs of the customer. As a result you are often asked to fill out a questionnaire about how you liked the product and/or service you received. The data from questionnaires such as these are often qualitative data.

If the data we are analyzing are qualitative data, then we are most likely interested in estimating the proportion, $\pi$, of population members that have a certain characteristic (one of the categories of the nominal variable). Confidence intervals for the population proportion, $\pi$, have the same basic structure as those for $\mu$. Remember that the sample proportion, $p$, is the best point estimate for $\pi$ and so we should center the confidence interval at the value of $p$.

### 7.11.1 Finding the Confidence Interval for $\pi$

*The point estimate, $p$, should be at the center of any confidence interval for $\pi$.*

To develop the corresponding formulas for the lower and upper bounds of a $100(1 - \alpha)\%$ confidence interval for $\pi$, we need to know that the sampling distribution of $p$ is a normal distribution. Again drawing on the properties of the normal distribution, we must extend the interval a certain number of standard errors to get the coverage we desire. The standard error of the point estimator, $p$, is $\sqrt{\pi(1 - \pi)/n}$. Notice that to calculate the standard error of $p$ you must know the value of $\pi$. But you are trying to estimate $\pi$, so, clearly, you do not know $\pi$. It makes the most sense to use $p$ as an estimate of $\pi$ in the formula for the standard error.

Combining these pieces of information with the knowledge that we wish to place $p$ at the center of the confidence interval, we arrive at the following formulas for the lower and upper bounds:

*FORMULA FOR CONFIDENCE INTERVAL FOR $\pi$*

Error: $\quad e = Z_{\alpha/2}\sqrt{\dfrac{\pi(1 - \pi)}{n}} \cong Z_{\alpha/2}\sqrt{\dfrac{p(1 - p)}{n}}$

Width of the interval: $\quad w = 2e$

Lower bound: $\quad L = p - e$

Upper bound: $\quad U = p + e$

where $Z_{\alpha/2}$ is the $Z$ value that cuts off $\alpha/2$ in the upper tail of the standard normal distribution and $\alpha/2$ in the lower tail of the standard normal distribution.

Let's look at an example.

### EXAMPLE 7.15   Beverly Hills, 90210

***Confidence Interval for $\pi$***

Hard times have hit the *Beverly Hills, 90210* crowd. Nearly a third of teens polled by Teenage Research Limited said they have been personally affected by the recession.

Where do teens get their money? A survey of 2000 teens showed that 47% of them get some money from their parents. Find a 95% confidence interval for the proportion of teens who get some money from their parents.

**Given information:**   $p = 0.47$

$n = 2000$

**Calculations:**   Standard error $= \sqrt{\dfrac{p(1-p)}{n}} = \sqrt{\dfrac{0.47(1-0.47)}{2000}} = 0.011$

Error $= e = Z_{\alpha/2} \sqrt{\dfrac{p(1-p)}{n}} = (1.96)(0.011) = 0.022$

Lower bound $= p - e = 0.47 - 0.02 = 0.45$

Upper bound $= p + e = 0.47 + 0.02 = 0.49$

So we can state that we are 95% confident that the percentage of all teens who get money from their parents is between 45 and 49%.   ■

## TRY IT NOW!

### Retirement Years
*Confidence Interval for $\pi$*

A survey shows that a growing number of Americans are willing to make sacrifices to become home owners despite increasing job and financial worries,. The Federal National Mortgage Association surveyed 1857 Americans and found that 67% would put off retirement for 10 years to own a home.

Find a 90% confidence interval for the proportion of all Americans who would put off retirement for 10 years to own a home.

## 7.11.2 Exercises—Learning It!

**7.25**  I asked 100 imaginary friends (only to avoid the time and cost of data collection) the following question: "Do you regularly watch MTV's *Beavis and Butthead*?" Of the 100 friends, 35 of them answered yes.

(a) Calculate a 95% confidence interval for the "viewership" of this show.

(b) MTV is considering canceling the show if less than one-third of the population regularly watches the show. Based on this information, what will MTV do?

ANS. $L = 0.652$, $U = 0.688$

**7.26** Many companies have been experiencing downsizing. There is some feeling that when rumors of a downsize begin to float through an organization, employees begin to use their sick days at a faster rate than normal for fear of losing them. Fifty employees were surveyed and asked if they would use their sick days freely in the face of a potential downsizing. Of the 50 employees in the sample, 11 said they would.

(a) Find a 95% confidence interval for the proportion of employees who would begin to use their sick days freely when faced with a potential downsizing.

(b) Based on your confidence interval, what are some recommendations you might suggest to managers?

**7.27** A poll of 450 registered Massachusetts voters found 46% of the Bay State voters opposed to allowing casino gambling in the state.

(a) Find a 95% confidence interval for the proportion of all Bay State voters opposed to allowing casino gambling in the state.

(b) Is 50% in the confidence interval? If so, what does this tell you? If not, what does this tell you?

(c) What would you recommend to the governor of Massachusetts, who is pushing for expanded gambling?

**7.28** Thousands of years ago people hung the yellow blossoms of Saint-John's-wort over their doorways, hoping to ward off evil spirits. Today, German physicians write nearly 3 million prescriptions a year for pills made from extracts of the plant, meant to relieve depression. But doctors on the other side of the Atlantic haven't followed suit. Scientists in the United States say that there is no reliable evidence that the herb can help.

To find out more, researchers at the University of Munich and the University of Texas in San Antonio studied 1500 people with mild to moderate depression. One-third were treated with Saint-John's-wort, one-third were treated with antidepressant drugs, and one-third were treated with a placebo. The researchers found that after 6 weeks of treatment 64% of the people taking Saint-John's-wort pills felt markedly better, compared to 59% of those receiving synthetic antidepressant drugs.

(a) Construct a 95% confidence interval for the proportion of people who would feel markedly better using Saint-John's-wort pills.

(b) Construct a 95% confidence interval for the proportion of people who would feel markedly better using synthetic antidepressant drugs.

(c) Compare the two confidence intervals. Do they overlap at all? What is your recommendation to doctors?

**7.29** In designing a new dormitory, a progressive university wishes to determine where students prefer to study in order to provide appropriate space. A survey of 100 randomly selected undergraduate students shows that 33% prefer to study in their room.

(a) Construct a 95% confidence interval for the proportion of students who prefer to study in their room.

(b) If the university has the option of making various sized rooms, what proportion of the rooms should be made larger to accommodate those who wish to study in their room?

**7.30** A medium-size city hospital is concerned about the number of ventilator-acquired respiratory infections in each of the past three years. The random sample of infections were studied and the number of infections that were vent-related infections is shown in the table below:

| Year | Number of Infections | Number of Vent-Related Infections |
|------|----------------------|-----------------------------------|
| 1995 | 13 | 10 |
| 1996 | 19 | 13 |
| 1997 | 14 | 11 |

(a) Find the sample proportion of vent-related infections for each of the three years.

(b) Construct a confidence interval for the proportion of vent-related infections for each of the three years.

(c) Should the hospital be concerned that the proportion of vent-related infections is increasing? Justify your answer using the confidence intervals found in part (b).

# 7.12  SAMPLE SIZE CALCULATIONS

Confidence intervals are easy to calculate and are commonly used to provide interval estimates for either the population mean or the population proportion. In fact, often a newspaper article will report the sample mean or sample proportion and a number called the sampling error or margin of error.

### EXAMPLE 7.16   Political Polls

*Illustration of Sampling Error*

According to an article written by the Associated Press and published in the *Springfield Union News* on 23 October 1996, a CNN–USA Today–Gallup tracking poll put the national polling gap at 23 points, with Clinton favored by 55%, Dole by 32%, and Perot by 8%. The survey of 754 likely voters had a 4-point margin of error.  ■

This sampling error or margin of error is what we have labeled error. Thus, if you wanted to construct the confidence interval from such a report you simply need to add and subtract the sampling error to the point estimate.

### EXAMPLE 7.17   Political Polls

*Constructing the Confidence Interval*

Continuing with the results of the CNN–USA Today–Gallup tracking poll, the proportion of voters who favor Clinton was 55% with a 4% margin of error. Thus, the percentage is between 51 and 59%. The article in the newspaper did not report the confidence level, but typically 95% confidence intervals are calculated.  ■

## 7.12.1 Understanding the Error

We have not really examined why this distance between the center of the confidence interval and either endpoint is called an *error*. As we have seen many times before, the term sampling error does not imply that you made an error. It does indicate that you have imperfect information about the population, since you studied only a sample of that population. There is another reason why this value is called an error. Consider the confidence interval calculated in Example 7.12 for the diaper manufacturer. In that example we found a 95% confidence interval for the mean diaper weight in grams. It is shown below:

*Sampling error is due to the fact that only a piece of the population has been studied.*

54.78          54.98          55.18

Remember that the value of 54.98 is the sample mean, $\overline{X}$. Suppose for the moment that we have found an interval that does contain $\mu$ but you don't know where in the interval $\mu$ is located. If you used $\overline{X}$ to estimate $\mu$, what is the largest amount by which you could miss the value of $\mu$? This *worst-case scenario* happens if $\mu$ is at either endpoint of the interval. If $\mu$ is at either of the endpoints, then our point estimate $\overline{X}$ is in error by the amount of $e = 0.20$ g. If $\mu$ is anywhere else in the interval, then our point estimate $\overline{X}$ is off by some amount less than 0.20 g. So we are 95% confident that the error is at most 0.20 g. Remember that a 95% confidence level tells you that

19 out of 20 intervals of this width contain the true population parameter. This is why we have labeled the distance from the middle of the interval to either endpoint as $e$ standing for error. It is precisely this value that is often quoted in newspaper or research articles, although the newspaper rarely tells you what the corresponding level of confidence is. Typically, a 95% level of confidence has been used.

You might wonder if it is possible to specify the size of this error to achieve a certain level of accuracy. This is indeed often desirable. By specifying the amount of error that you can tolerate in a particular situation you are determining the sample size needed. Often you cannot achieve a particular level of accuracy because to do so would require a sample size that you cannot afford. This is the subject of the next subsection.

## 7.12.2 Determining the Sample Size

In Chapter 2, we identified the following factors that are important in determining the size of the sample needed:

- The amount of variation in the population
- The amount of error that can be tolerated
- The extent of resources available
- The size of the population

We now know enough statistics to develop a formula for the sample size that incorporates the first two of these factors. You should remember that these factors were identified as most important to the sample size determination.

Recall that the expression for $e$ is one of the following, depending on whether you are estimating the mean, $\mu$, or the proportion, $\pi$:

$$\text{Estimating } \mu: \qquad e = Z_{\alpha/2}\sigma/\sqrt{n}$$

$$\text{Estimating } \pi: \qquad e = Z_{\alpha/2}\sqrt{\frac{p(1-p)}{n}}$$

Each of these equations can be rewritten and solved for $n$, the sample size, by using some basic algebra. When this is done you get two equations for $n$ that are algebraically equivalent to the two equations shown above:

$$\text{Estimating } \mu: \qquad n = \frac{Z_{\alpha/2}^2\,\sigma^2}{e^2}$$

$$\text{Estimating } \pi: \qquad n = \frac{Z_{\alpha/2}^2\,p(1-p)}{e^2}$$

Let's examine the equation for $n$ if we are trying to estimate the population mean, $\mu$. If we specify a certain level of confidence and a value for the maximum difference between $\overline{X}$ and $\mu$, then we can calculate how large a sample we must select. Let's see how this formula could be used by the diaper manufacturer.

## EXAMPLE 7.18   The Diaper Company

***Sample Size Determination***

The diaper manufacturer wants to be 95% confident that the estimate $\overline{X}$ is not in error by more than 0.10 g. That is, the diaper manufacturer wants to be sure that 19 out of 20 possible confidence intervals of width 0.20 g contain the true mean, $\mu$. Recall that the manufacturing process has a standard deviation of 0.55 g. Thus, we have

$$e = 0.10 \text{ g}$$
$$Z_{\alpha/2} = 1.96$$
$$\sigma = 0.55 \text{ g}$$

Using this information to calculate $n$ yields

$$n = \frac{Z_{\alpha/2}^2 \sigma^2}{e^2} = \frac{(1.96^2)(0.55^2)}{0.10^2} = 116.2$$

Since we can't sample a fractional diaper, we must round this value up to 117 diapers. ■

For sample size calculations you should always round up to guarantee the level of confidence and error you have specified.

*Always round up in sample size calculations.*

 **TRY IT NOW!**

### Bottle Filling

*Finding the sample size*

How many bottles does the bottle manufacturer need to sample to be 98% confident that the error is at most 0.05 oz? Remember that the population standard deviation is 0.1 oz.

Notice that we needed a value for the population standard deviation to calculate $n$. If this is unknown then we must estimate it. Unfortunately, often we need to take a sample to estimate $\sigma$ but we need $\sigma$ to figure out how large our sample needs to be. This is a classic "catch 22" situation. There appears to be no way out.

Fortunately, there are a couple of ways to handle this dilemma. First of all, we could take a small sample to get a rough estimate of the standard deviation and use this in the formula for $n$. Suppose that we took a sample of size 10 and we used this sample to calculate the sample standard deviation, $s$. This value of $s$ can then be used as an estimate for $\sigma$ and plugged into the formula for $n$. Suppose when we do this the formula tells us that we need a sample of size $n = 55$. Well, we already have data on 10 observations so we need an additional 45 to complete the sample. Another way is to use information about the variability of a similar product or process. This may not be perfect but it will give a rough idea of the value of $\sigma$ which can, in turn, be used in the sample size calculation.

*Ways to handle situations in which the standard deviation is unknown*

Very often the sample size determined from the formula cannot be collected because of limited resources (i.e., cost). If this happens then you must be willing to accept a lower level of confidence or tolerate a higher value of $e$ or both. Let's exam-

*For you to be more confident, the sample size must be larger.*

ine the impact on $n$ of varying the level of confidence. Before we do the calculations, let's see what we might anticipate. We know that as the level of confidence increases, the $Z$ value also increases. Since $Z$ is in the numerator of the formula for $n$, we would expect the sample size needed to increase as the level of confidence increases. This should make intuitive sense as well. For us to be more confident, the sample size must be larger.

### EXAMPLE 7.19   The Diaper Company

*Sample Sizes for Various Increasing Levels of Confidence*
*Require Increasing Sample Sizes*

Suppose the diaper manufacturer considers several different levels of confidence. The sample sizes needed to achieve an error of at most 0.10 g for various levels of confidence are calculated below:

90% confidence level:   $n = \dfrac{Z_{\alpha/2}^2\,\sigma^2}{e^2} = \dfrac{(1.645^2)\,(0.55^2)}{0.10^2} = 81.9$ rounded up to 82

95% confidence level:   $n = \dfrac{Z_{\alpha/2}^2\,\sigma^2}{e^2} = \dfrac{(1.96^2)\,(0.55^2)}{0.10^2} = 116.2$ rounded up to 117

98% confidence level:   $n = \dfrac{Z_{\alpha/2}^2\,\sigma^2}{e^2} = \dfrac{(2.33^2)\,(0.55^2)}{0.10^2} = 164.2$ rounded up to 165

99% confidence level:   $n = \dfrac{Z_{\alpha/2}^2\,\sigma^2}{e^2} = \dfrac{(2.58^2)\,(0.55^2)}{0.10^2} = 201.4$ rounded up to 202 ■

*As the maximum error increases, the sample size decreases.*

If the sample size you calculate is too expensive, the other way you can cut costs is to increase the size of the error you can tolerate. Let's see what happens to the sample size as the maximum error is increased. Again, a quick look at the formula for $n$ shows us that the value for $e$ is in the denominator. So by increasing $e$ we will be dividing by a larger number and hence the sample size needed will be smaller. Intuitively, this, too, makes sense. If you can tolerate a larger error, then you can take a smaller sample. Let's look at the diaper manufacturer from this perspective.

### EXAMPLE 7.20   The Diaper Company

*Impact of Larger Errors*

Consider the following values for the maximum error that the diaper company is willing to tolerate in its estimate of $\mu$. In all cases they wish to be 95% confident that the error is at most the specified value.

Error = 0.10 g   $n = \dfrac{Z_{\alpha/2}^2\,\sigma^2}{e^2} = \dfrac{(1.96^2)\,(0.55^2)}{0.10^2} = 116.2$ rounded up to 117

Error = 0.15 g   $n = \dfrac{Z_{\alpha/2}^2\,\sigma^2}{e^2} = \dfrac{(1.96^2)\,(0.55^2)}{0.15^2} = 51.6$ rounded up to 52

Error = 0.20 g   $n = \dfrac{Z_{\alpha/2}^2\,\sigma^2}{e^2} = \dfrac{(1.96^2)\,(0.55^2)}{0.20^2} = 29.05$ rounded up to 30

Error = 0.25 g   $n = \dfrac{Z_{\alpha/2}^2\,\sigma^2}{e^2} = \dfrac{(1.96^2)\,(0.55^2)}{0.25^2} = 18.6$ rounded up to 19 ■

Calculations of the sample size for situations when you wish to estimate the population proportion, $\pi$, are done in a similar manner. The only difference is the particular formula to be used. Consider the teens who are receiving money from their parents. We looked at this situation in Example 7.15.

**EXAMPLE 7.21** *Beverly Hills, 90210*

***Calculation of Sample Size for Proportions***

How many teens need to be sampled to be 95% confident that our estimate of $\pi$ is off by at most 5%? Remember that the point estimate is $p$, the sample proportion or percentage. If we want the sample proportion to be in error by at most 5%, then $e = 0.05$.

The formula for $n$ is

$$n = \frac{Z_{\alpha/2}^2 \, p(1 - p)}{e^2}$$

Plugging in the values we know gives us

$$n = \frac{(1.96^2) p(1 - p)}{0.05^2}$$

At this point we realize that we can't continue without a value for $p$ but we need a sample to get a sample proportion. ■

This is again a circular problem. We need $p$ to find $n$ but we need $n$ to get $p$! There are two approaches you can take here. First of all, if you have any information about the value of $p$ from previous samples or experience, then you should use that information as an estimate of $p$. If you have no information at all, then you should use a value of $p = 0.50$ in the formula for the sample size calculation. When you do this the sample size that is calculated is as large as it can be for the specified confidence level and error. It is the most conservative approach you can take and will often lead to a sample size larger than you really need. Let's finish Example 7.21 by using a value of $p = 0.50$.

**EXAMPLE 7.22** *Beverly Hills, 90210 (cont'd)*

***Sample Size Calculation for Proportion***

Using a value of $p = 0.50$ in the formula for $n$ gives us

$$n = \frac{(1.96^2) p(1 - p)}{0.05^2} = \frac{(1.96^2)(0.5)(1 - 0.5)}{0.05^2} = 384.16$$

which is rounded up to 385 teenagers. ■

 **TRY IT NOW!**

### Retirement Years

***Sample Size Calculation for $\pi$***

How many Americans must be sampled to determine the percentage who would put off retirement for 10 years to own a home? The estimate should not differ from the actual population proportion by more than 3% with a confidence of 90%.

The conclusions that we reached about what happens to the required sample size as the confidence level and the error vary are the same for proportions as for means. As the confidence level increases, the sample size needed to estimate the population proportion also increases. As the error that can be tolerated increases, the sample size needed decreases.

### 7.12.3 Exercises—Learning It!

*Requires Exercise 7.13*   **7.31**   How many stores must be sampled for the woman who wants to buy a ranch to be 95% confident that the error in estimating the average fat content per pound in steaks sold in the Portland, Maine, area is at most 0.05 oz?

*Requires Exercise 7.15*   **7.32**   How many months must be sampled for analysts to be 99% confident that the error in estimating the average monthly price per kwh of electricity is at most $.02?

*Requires Exercise 7.16*   **7.33**   How many days must be observed for the grocery store chain to be 90% confident that the error in estimating the average number of vehicles passing a certain location is at most 10 vehicles?

*Requires Exercise 7.25*   **7.34**   How many viewers must be surveyed to be 98% confident that the estimate of the viewership is in error by at most 3%?

*Requires Exercise 7.26*   **7.35**   How many employees must be sampled to be 95% confident that the error in estimating the proportion of employees who would begin to use sick days at a faster rate when rumors of downsizing are circulating is at most 0.02?

## 7.13   USING EXCEL TO FIND CONFIDENCE INTERVALS

Excel does not have a tool that automatically calculates confidence intervals. To do this, you will have to use some of the functions you already learned about in a formula. In this section, we will explain how to do so for a large-sample confidence interval for the mean. We will also explain how to use the Excel functions for the $t$ distribution.

### 7.13.1 Large-Sample Confidence Intervals for the Mean

To find a confidence interval for a population parameter, you need a point estimate and a formula for the error of the estimate. For the population mean, when $\sigma$ is known, or when $n$ is large ($n > 30$), the formula for the confidence interval is

$$\overline{X} \pm Z_{\alpha/2}\frac{\sigma}{\sqrt{n}}$$

To find a confidence interval for the mean using Excel, you must have a point estimate, $\overline{X}$. You can obtain this from the Descriptive Statistics in the Data Analysis Tools or by using the AVERAGE function. You will also need a value for $\sigma$, or else a sample estimate, $s$, based on a large sample. The last thing that you need is the $Z$ value, which you will obtain using the NORMSINV function discussed in Chapter 6.

Figure 7.13 shows a portion of an Excel worksheet that contains a sample of 50 diaper weights, and a set of summary statistics for the data.

To calculate a 95% confidence interval for the true weight of a diaper, follow these steps:

1. Position the cursor in the cell that you want to contain the upper limit of the confidence interval. In this case, we will use E1, which is just to the right of the summary data.

| | A | B | C | D |
|---|---|---|---|---|
| 1 | **Weight** | | *Weight* | |
| 2 | 55.87 | | | |
| 3 | 55.35 | | Mean | 54.993 |
| 4 | 54.50 | | Standard Error | 0.096962964 |
| 5 | 53.97 | | Median | 54.915 |
| 6 | 54.29 | | Mode | 54.4 |
| 7 | 54.85 | | Standard Deviation | 0.685631691 |
| 8 | 54.79 | | Sample Variance | 0.470090816 |
| 9 | 54.65 | | Kurtosis | -0.848226678 |
| 10 | 55.56 | | Skewness | 0.078374807 |
| 11 | 55.82 | | Range | 2.84 |
| 12 | 54.40 | | Minimum | 53.6 |
| 13 | 56.44 | | Maximum | 56.44 |
| 14 | 54.11 | | Sum | 2749.65 |
| 15 | 54.67 | | Count | 50 |
| 16 | 54.56 | | | |
| 17 | 54.69 | | | |

**FIGURE 7.13**   Diaper data and summary statistics

2. Since you want to let Excel know that you are entering a formula, start the formula by typing an equal sign "=". All formulas in Excel must start with an equal sign.

What you want to calculate is $\overline{X} + Z_{.025}\dfrac{s}{\sqrt{n}}$. We will use the sample standard deviation because our sample size is large enough.

3. After you type the equal sign, position the cursor on the cell that contains the value of $\overline{X}$, in this case D3. If you look at the status line or the cell itself, you see that the cell D3 is entered in the formula. Since in our confidence interval we want to use that *specific* cell all the time for the sample mean, you need to change the cell reference to an absolute reference.

4. With the cursor still in the cell, press F4. The cell reference D3 will change to $D$3. The $ tells Excel to look *always* in column D, *always* in row 3 for this value. This is important to know when you copy formulas.

5. Now, we want to add to $\overline{X}$ the quantity $Z_{.025}\dfrac{s}{\sqrt{n}}$. Type in + and then type NORMSINV(0.975).

*Remember: The NORM-SINV function returns the Z value for the lower tail probability.*

Remember that the quantity $\dfrac{s}{\sqrt{n}}$ is referred to as the standard error of the mean. When you look at the summary statistics that Excel calculates, you see that the second value is indeed the standard error of the mean. This was not meaningful to us when we first looked at numerical descriptors, so we did not talk about it then. Now we will use it to finish our confidence interval.

6. Type in a multiplication sign, "*", and position the cursor in the cell that contains the standard error, in this case, D4. Again, press F4 to make the cell an absolute reference.

7. The finished formula is shown in Figure 7.14.

| E | F | G | |
|---|---|---|---|
| =$D$3+NORMSINV(0.975)*$D$4 | | | |

**FIGURE 7.14**   Formula for upper limit of confidence interval

8. Hit Enter and the result, 55.18304, will be displayed.

You can calculate the lower confidence limit by typing the same equation in the cell next to the upper limit. Be sure to change the $Z$ value to a negative by using **NORMSINV(0.025).** To avoid retyping the whole equation, you can simply copy the one you just did and paste it into a new cell. Then, move to the display of the equation in the status bar and edit the **NORMSINV** parameter.

When you calculated the summary statistics, you might have noticed that there was an option available for Confidence Level for the Mean. Checking that box will add a new line to the end of the summary statistics output, which is supposed to be

$Z_{.025} \dfrac{s}{\sqrt{n}}$. We have not used this option because the calculation is wrong (not drastically, but it is definitely not correct).

## 7.13.2 The *t* Distribution in Excel

The difference between large- and small-sample confidence intervals for the mean is in the sampling distribution of $\overline{X}$. You know that when $\sigma$ is unknown and the sample size is not greater than 30, the Student-$t$ distribution applies. The formula for the confidence interval is

$$\overline{X} \pm t_{\alpha/2,\,n-1} \frac{s}{\sqrt{n}}$$

To calculate small-sample confidence intervals for the population mean with Excel, you will use the **TINV** function to get the appropriate $t$ values.

Just like other functions in Excel, the **TINV** function needs some input parameters. The format of the **TINV** function is **TINV(probability,deg_freedom).** The value that you input for **probability** is the value of $\alpha$, the total tail area probability. The value for **deg_freedom** is $n - 1$, the degrees of freedom. As an example, for a 95% confidence interval for $\mu$, based on a sample of size 25, the function would look like **TINV(0.05,24)** and would return the value 2.063898. There is no lower $t$ value in Excel. To obtain the lower confidence limit, you simply subtract the quantity $t_{\alpha/2,\,n-1} \dfrac{s}{\sqrt{n}}$ from $\overline{X}$.

# *CHAPTER 7 SUMMARY*

This chapter has covered the basics of estimating population parameters. In particular, you learned how to estimate the average of a numeric characteristic of a population, $\mu$, and the proportion of a population that has a certain characteristic, $\pi$. The estimates are calculated from a sample selected from the population. Each sample would therefore yield a slightly different estimate of the population parameter. Thus, the estimators are themselves random variables. Just like the random variables you studied in Chapter 6, the estimators have a distribution. When the random variable is an estimator, the distribution is called a sampling distribution. The sampling distribution has a mean and a standard deviation, which is called the standard error.

You have learned how to use the sampling distribution of $\overline{X}$ to calculate probabilities and make inferences about $\mu$. You have also learned how to create confidence intervals for $\mu$ and for $\pi$. Finally, you have learned how to calculate the required sample size to achieve a certain level of precision with a specified confidence.

## Key Terms

| Key Term | Definition | Page Reference |
|---|---|---|
| Point estimate | A **point estimate** is a single number calculated from sample data. It is used to estimate a parameter of the population. | 303 |
| Point estimator | A **point estimator** is the formula or rule that is used to calculate the point estimate for a particular set of data. | 303 |
| Unbiased estimator | An **unbiased estimator** yields an estimate that is fair. It neither systematically overestimates the parameter nor systematically underestimates the parameter. | 310 |
| Standard error | The **standard error** is the standard deviation of a point estimator. It measures how much the point estimator or sample statistic varies from sample to sample. | 314 |
| Sampling distribution | The distribution of a point estimator or a sample statistic is called a **sampling distribution.** | 314 |
| Central Limit Theorem | The **Central Limit Theorem** states that in random sampling from a population with mean $\mu$ and standard deviation $\sigma$, when $n$ is large enough, the distribution of $\overline{X}$ is approximately normal with a mean, $\mu_{\overline{X}}$, equal to $\mu$ and a standard deviation, $\sigma_{\overline{X}}$, equal to $\sigma/\sqrt{n}$. | 315 |
| Confidence interval | A **confidence interval** or an **interval estimate** is a range of values with an associated probability or confidence level, $1 - \alpha$. The probability quantifies the chance that the interval contains the true population parameter. | 329 |

## Key Formulas

| Description | Formula | Page Reference |
|---|---|---|
| **Z-score for $\overline{X}$** | $Z = \dfrac{\overline{X} - \mu_{\overline{X}}}{\sigma_{\overline{X}}}$ | 313 |
| **Mean of $\overline{X}$** | $\mu_{\overline{X}} = \mu$ | 315 |
| **Standard error of $\overline{X}$** | $\sigma_{\overline{X}} = \sigma/\sqrt{n}$ | 315 |
| **Confidence interval for $\mu$ when $\sigma$ is known** | Error: $e = \dfrac{Z_{\alpha/2}\sigma}{\sqrt{n}}$<br>Width of the interval: $w = 2e$<br>Lower bound: $L = \overline{X} - e$<br>Upper bound: $U = \overline{X} + e$ | 331 |
| **Large-sample confidence interval for $\mu$ when $\sigma$ is unknown** | Error: $e = \dfrac{Z_{\alpha/2}s}{\sqrt{n}}$<br>Width of the interval: $w = 2e$<br>Lower bound: $L = \overline{X} - e$<br>Upper bound: $U = \overline{X} + e$ | 333 |
| **Small-sample confidence interval for $\mu$ when $\sigma$ is unknown, normal population** | Error: $e = \dfrac{t_{\alpha/2,n-1}s}{\sqrt{n}}$<br>Width of the interval: $w = 2e$<br>Lower bound: $L = \overline{X} - e$<br>Upper bound: $U = \overline{X} + e$ | 346 |
| **Confidence interval for $\pi$** | Error: $e = Z_{\alpha/2}\sqrt{\dfrac{\pi(1 - \pi)}{n}} \cong Z_{\alpha/2}\sqrt{\dfrac{p(1 - p)}{n}}$<br>Width of the interval: $w = 2e$<br>Lower bound: $L = p - e$<br>Upper bound: $U = p + e$ | 350 |
| **Sample size for estimating $\mu$** | $n = \dfrac{Z_{\alpha/2}^2\,\sigma^2}{e^2}$ | 354 |
| **Sample size for estimating $\pi$** | $n = \dfrac{Z_{\alpha/2}^2\,p(1 - p)}{e^2}$ | 354 |

# CHAPTER 7 EXERCISES

## Learning It!

*Requires Example 7.11*

**7.36** The manufacturer of 15-oz breakfast cereal boxes samples and weighs 5 boxes every hour. The observations for seven hours are shown below:

| | | | | | | |
|---|---|---|---|---|---|---|
| 14.91 | 15.23 | 15.34 | 15.40 | 15.48 | 15.59 | 15.62 |
| 14.96 | 15.24 | 15.34 | 15.42 | 15.49 | 15.60 | 15.62 |
| 15.11 | 15.28 | 15.36 | 15.45 | 15.50 | 15.60 | 15.63 |
| 15.21 | 15.30 | 15.38 | 15.46 | 15.52 | 15.61 | 15.64 |
| 15.22 | 15.30 | 15.39 | 15.48 | 15.58 | 15.61 | 15.67 |

(a) Find the upper and lower confidence limits for the average weight of the breakfast cereals. Use $\alpha = 0.05$. Is the value of 15.00 oz contained in the interval?

(b) What does that tell you about the filling process?

*Requires Exercise 7.5*

**7.37** A company that buys blank VHS tapes for video recording is concerned about the actual amount of time that the tapes are able to record. The tapes are rated at 120 minutes, but the company knows there is variation in the actual time that the tapes can be used. The company collects data on 25 randomly selected tapes and finds the recording times listed below:

| 116 | 118 | 119 | 119 | 120 |
|-----|-----|-----|-----|-----|
| 117 | 118 | 119 | 119 | 121 |
| 117 | 118 | 119 | 120 | 121 |
| 117 | 119 | 119 | 120 | 121 |
| 117 | 119 | 119 | 120 | 121 |

(a) Find the average and the standard deviation of these data.

(b) Find a 95% confidence interval for the average recording time, assuming that the recording times are normally distributed.

(c) What do you conclude about the claim that the tapes have an average of 120 minutes of usable recording time? Use the confidence interval to support your conclusion.

**7.38** A cross-country runner is evaluating his times against a team average of 18.5 minutes. The standard deviation of his times was 0.25 minute. This season the runner's average is 17.0 minutes for 10 races.

(a) Assuming that race times are normally distributed, calculate the Z-score for the runner's average time.

(b) Using the Z-score you found in part (a), decide if this runner is unusually fast compared to the team.

**7.39** An insurance company is concerned about the amount of time it takes for operators who are responding to calls to the 800 number to access the database. It calls this time the response time. The Management Information Systems (MIS) department claims that the average response time is 10 seconds. A sample is taken and 40 response times are recorded. The results of generating descriptive statistics in Excel are shown below:

SUMMARY STATISTICS
FOR RESPONSE TIME

| | |
|---|---|
| Mean | 9.95656397 |
| Standard Error | 0.02942547 |
| Median | 9.9616274 |
| Mode | #N/A |
| Standard Deviation | 0.18610301 |
| Sample Variance | 0.03463433 |
| Kurtosis | $-0.29568133$ |
| Skewness | $-0.31707278$ |
| Range | 0.79489018 |
| Minimum | 9.49771882 |
| Maximum | 10.292609 |
| Sum | 398.262559 |
| Count | 40 |

(a) What are the average and standard deviation of the response times?

(b) Calculate the standard error of $\overline{X}$ using the standard deviation and $n$.

(c) Do you see the number calculated in part (b) anywhere in the output?

(d) Calculate the Z-score for the average response time.

(e) Based on the Z-score, do you think the MIS department's claim is correct?

**7.40** A company concerned about the health of its employees has offered the employees free membership in a local health club. One year after this benefit was adopted, a survey of 50 employees was done to determine the average number of hours per week they exercise. The 50 employees surveyed exercised an average of 3.5 hours a week with a standard deviation of 0.5 hour. The national average number of hours of exercise per week is 3 hours a week.

(a) Find the Z-score for the average of 3.5 hours.

(b) Display the sampling distribution of $\overline{X}$.

(c) Based on this Z-score, do you think the employees of this company exercise more than the national average?

**7.41**  The Bureau of Labor Statistics publishes a great deal of data on a monthly basis. The following data on the average price of bread were extracted from the World Wide Web.

**[Bureau of Labor Statistics Data]**

**Data extracted on: January 9, 1997**

Average   Price Data
Series    Catalog:
Series    ID   APU0000702111
Area:     U.S. City Average
Item:     Bread, White, pan (cost per pound/453 grams)

| 1994 | | 1995 | | 1996 | |
|---|---|---|---|---|---|
| Jan | 0.768 | Jan | 0.767 | Jan | 0.860 |
| Feb | 0.756 | Feb | 0.767 | Feb | 0.858 |
| Mar | 0.755 | Mar | 0.775 | Mar | 0.852 |
| Apr | 0.765 | Apr | 0.776 | Apr | 0.865 |
| May | 0.765 | May | 0.768 | May | 0.869 |
| Jun | 0.761 | Jun | 0.781 | Jun | 0.886 |
| Jul | 0.759 | Jul | 0.789 | Jul | 0.889 |
| Aug | 0.753 | Aug | 0.797 | Aug | 0.915 |
| Sep | 0.776 | Sep | 0.808 | Sep | 0.886 |
| Oct | 0.756 | Oct | 0.809 | Oct | 0.873 |
| Nov | 0.766 | Nov | 0.821 | Nov | 0.880 |
| Dec | 0.748 | Dec | 0.837 | Dec | 0.883 |

Treating these three years of data as a sample, find a 90% confidence interval for the average price of bread. Assume that the standard deviation is $0.110 per pound.

**7.42**  In an effort to improve the quality of the CD players that your company makes, you have started to sample the component parts that you purchase from an outside supplier. You will accept the shipment of parts only if there is less than 1% defectives in the shipment. Recognizing that you cannot test the entire shipment (or population), you select a sample of 25 components to test. You find 3 defectives in the sample of 25.

(a) Find a 90% confidence interval for the proportion of components in the population that are defective.

(b) Based on your confidence interval, should you accept the shipment? Why or why not?

**7.43**  A nationwide survey of practicing physicians will be taken to estimate $\mu$, the true mean number of prescriptions written per day. The desired margin of sampling error is 0.75. A pilot study revealed that a reasonable planning value for the population standard deviation is 5.

(a) If the desired level of confidence was 99%, how many physicians should be contacted in the survey to estimate $\mu$?

(b) If the desired confidence level were lowered to 95%, what would be the sample size?

(c) Discuss the results of parts (a) and (b).

**7.44**  A hotel is studying the proportions of rooms that are not ready when customers check in to the hotel.

(a) How many rooms must be in the sample for the hotel to be 95% confident that the margin of error is at most 1%?

(b) How many rooms must be in the sample for the hotel to be 95% confident that the margin of error is at most 3%?

## Thinking About It!

**7.45**  Suppose you are studying the length of stay at a hospital and the population had a mean of 8 days with a standard deviation of 2.5 days. The length of hospital stay has a normal distribution.

(a) Display the population distribution of hospital stay.

(b) If you observed the hospital stay of 20 patients, find the standard error of $\overline{X}$. Display the sampling distribution of $\overline{X}$ on the same graph as part (a) using a different color pen or pencil.

(c) If you observed the hospital stay of 30 patients, find the standard error of $\overline{X}$. Display the sampling distribution of $\overline{X}$ on the same graph as (a) and (b) using a different color.

(d) If you observed the hospital stay of 80 patients, find the standard error of $\overline{X}$. Display the sampling distribution of $\overline{X}$ on the same graph as (a) and (b) using a different color.

(e) What happens to the sampling distribution of $\overline{X}$ as the sample size increases? What does this tell you about the accuracy of $\overline{X}$ as an estimator for $\mu$?

**7.46** Suppose you are studying the salaries of 3 different populations that all have the same average salary: $\mu = \$25,000$. The 3 populations have different standard deviations: accounting major, $\sigma = 1000$; computer information systems major, $\sigma = 2000$; and marketing major, $\sigma = 3000$. You select a sample of 30 from each population.

(a) Find the standard error of $\overline{X}$ for each of the three groups.

(b) What happens to the standard error as the population variability increases?

(c) What does this tell you about the accuracy of $\overline{X}$ as an estimator for $\mu$?

**7.47** The 43 executives listed in Exercise 7.14 all work for firms that are headquartered or have significant operations in western Massachusetts.  *Requires Exercise 7.14*

(a) How does this affect the usefulness of the confidence interval for $\mu$?

(b) These firms can be divided into 2 groups: manufacturing (M) and nonmanufacturing (NM). Compute a 90% confidence interval for the mean executive salary for each of the two groups assuming the salaries are normally distributed.

(c) Do these confidence intervals overlap? If so, what do you think that indicates? If not, what does that tell you?

**7.48** According to a *New York Times* article of 31 December 1993, women are no more fully represented in French politics today than they were after universal suffrage was finally introduced 48 years ago. The article provides the percentage of seats held by women in several different countries and is shown below:

| Country | % Seats Held by Women in Each Country's Main Legislative Assembly |
| --- | --- |
| France | 6.1 |
| Italy | 8.1 |
| Britain | 9.1 |
| U.S. House | 11.0 |
| U.S. Senate | 7.0 |
| Spain | 15.7 |
| Germany | 21.6 |
| Netherlands | 27.3 |
| Sweden | 32.6 |
| Denmark | 33.5 |
| Norway | 38.0 |

(a) Leaving France out of the data set, construct a 95% confidence interval for the average percentage of seats held by women in international government.

(b) Does France appear to be atypical? Explain your answer using the confidence interval calculated in part (a).

**7.49** The results of a survey conducted to see if a newly introduced product is meeting the customers' needs are available to the manager. The survey results indicate that 52% of those in the survey were very satisfied with the new product. The margin of error was 5%. The product manager argues that more than half of the customers are very satisfied. What is wrong with his argument?

*Requires Exercise 7.24*     **7.50**   The state of Hawaii had an adoption rate of 9.6% in 1996. Using the confidence interval you found in Exercise 7.24, what can you conclude about Hawaii's adoption rate in relation to the national average?

## Doing It!

*Datafile: FCOJ.XXX*     **7.51**   The data in the file FCOJ.XXX have been collected over the past 11 years of futures trading of Frozen Concentrate Orange Juice (FCOJ). The FCOJ is traded on the New York Cotton Exchange. The unit of trading is 15,000 lb (one contract = 15,000 lb). The price quoted is in cents per pound. The data show average price for the month. The following is a brief description of data presented on each of the four sheets identified by name.

> *Popdata:* Population data for ALL 139 months (Jun 83–Dec 94).
> *Sample30:* A sample of 30 randomly selected months from the population. Fifty (50) samples are presented on this sheet.
> *Sample40:* A sample of 40 randomly selected months from the population. One sample is presented on this sheet.
> *Sample50:* A sample of 50 randomly selected months from the population. One sample is presented on this sheet.

(a) Using the 50 samples of size $n = 30$ (sheet *sample30*), calculate a 95% confidence interval for each of the 50 samples.

(b) Count how many of the 50 confidence intervals calculated in part (a) actually include the true population mean, $\mu$. Is your count consistent with what you expected? Why or why not?

(c) Calculate a 90, 95, and 99% confidence interval for the single sample on the sheet *sample40*. How does the width of the confidence interval vary with the level of confidence?

(d) Calculate a 95% confidence interval for each of the following: (i) the first sample on the sheet *sample30*, (ii) the single sample on the sheet *sample40* (*note:* you did this one in part (c) above), and (iii) the single sample on the sheet *sample50*. How does the width of the confidence interval vary with the sample size?

*Datafile: DIAPER.XXX*     **7.52**   The diaper company has collected data on 5 diapers per hour for a total of 52 hours. Thus, they have 260 rows of data. They have hired you to help them analyze these data. The data are located on the data disk in a file named DIAPER.XXX. A portion of the data is shown below:

| Sample | Diaper | Weight | Bulk |
|--------|--------|--------|------|
| 1 | 1 | 55.87 | 0.419 |
| 1 | 2 | 55.35 | 0.380 |
| 1 | 3 | 54.50 | 0.365 |

| Variable | Description |
|----------|-------------|
| *Sample* | Keeps track of the sequence in which the samples were taken and goes from 1 to 52 |
| *Diaper* | Keeps track of the individual diapers within a sample and goes from 1 to 5 |
| *Weight* | The diaper weight in grams |
| *Bulk* | The diaper bulk in millimeters |

In the chapter, the sampling distribution for the average diaper weight (Figure 7.3) was created when each sample had $n = 5$ observations. If you combine data from 2 samples and assume that a sample of size $n = 10$ was taken, you can examine the effects of a change in the sample size on the sampling distribution. Similarly, if you combine the data from 3 samples you can see the impact if the sample size had been $n = 15$. Finally if you combine the data from 4 samples you can see the impact if the sample size had been $n = 20$.

(a) Do this and display the histograms of the average weight. You will have 3 histograms in addition to Figure 7.3.

(b) For each set of average weights, find the mean and standard deviation and fill out the first three columns of the table below:

| Sample Size | Average of the $\overline{X}$'s | Standard Deviation of the $\overline{X}$'s | Theoretical Standard Error $(\sigma/\sqrt{n})$ |
|---|---|---|---|
| 5 | | | |
| 10 | | | |
| 15 | | | |
| 20 | | | |

(c) What do you notice about the values for the average of the $\overline{X}$'s as $n$ increases?

(d) What do you notice about the values for the standard deviation of the $\overline{X}$'s as $n$ increases?

(e) How do the standard errors that you found (last column of the table) compare to what the Central Limit Theorem says they should be? Complete the last column in the table above.

**7.53** Construct a histogram of the variable *Bulk* and calculate a set of summary statistics and a histogram for the variable *Bulk*. *Requires Exercise 7.52*

(a) Do the data appear to be normally distributed?

(b) Find the sample average bulk for each of the 52 samples of size $n = 5$. Call these *AverageBulk*.

(c) Calculate summary statistics and make a histogram of these $\overline{X}$'s.

(d) Compare the histogram for the variable *Bulk* to the histogram of the variable *AverageBulk* found in part (b).

**7.54** The product specifications for the diapers include a variable called *Density*, which is found by dividing the *Weight* of a diaper by its *Bulk*. *Requires Exercise 7.52*

(a) Create the variable *Density*.

(b) Calculate a set of summary statistics for *Density* and make a histogram of *Density*.

(c) Does it appear to be normally distributed?

(d) For each sample of size $n = 5$, find the *Average Density*.

# Part IV
# INFERENTIAL STATISTICS

CHAPTER *8*

# HYPOTHESIS TESTING: AN INTRODUCTION

## THE TISSUE STRENGTH PROBLEM

A manufacturer of tissues has received numerous complaints about its products. One of the frequent complaints is about the strength of the tissue. As we have noted many times before in this book, all companies are paying increasingly more attention to the customer in an attempt to "totally delight the customer." Clearly, in this case, the customers are not totally delighted with the tissues made by this company. Management has asked the statisticians to investigate the source of this problem by taking a sample.

There are two possible explanations. One possibility is that manufacturing is producing tissues that are not as strong as the specifications. In this case, an adjustment may be needed to the manufacturing process. The second possibility is that despite the fact that manufacturing is making the tissues according to specifications, the customers prefer stronger tissues. In this case, an adjustment to the specifications may be in order.

Your boss has asked you to find out if the tissues are in fact as strong as they are designed to be. There are two measures of tissue strength: machine-directional strength *(MDStrength)* measured in lb/ream and cross-directional strength *(CD-Strength)* also measured in lb/ream. You decide to collect data on 3 different days. A portion of the data file is shown below.

| Day | MDStrength | CDStrength |
|-----|------------|------------|
| 1 | 1006 | 422 |
| 1 | 994 | 448 |
| 1 | 1032 | 423 |
| 1 | 875 | 435 |
| 1 | 1043 | 445 |
| 1 | 962 | 464 |
| 1 | 973 | 472 |
| 1 | 1036 | 489 |
| 1 | 1084 | 440 |

The product specifications for *MDStrength* state that the mean should be 1000 lb/ream with standard deviation of 50 lb/ream. *CDStrength* is supposed to have a mean of 450 lb/ream with a standard deviation of 25 lb/ream.

# 8.1 CHAPTER OBJECTIVES

*Remember: A point estimate is a single number, calculated from sample data, which is used to estimate a population parameter.*

From Chapter 7 we learned that the sample mean is a good point estimate of the population mean. Suppose we took a sample of tissues and measured the strength of each tissue. We can use the data to calculate a sample mean strength. If that sample mean doesn't agree with the target mean strength, then we should tell the boss that the process needs to be adjusted. Simple, right? Well, not quite.

After a bit more thought, you remember that although the sample mean, $\overline{X}$, is a single number for any particular sample, if you pick a different sample you will probably get a different sample mean. In fact, there are many different possible values you could get for the sample mean, and virtually none of them will actually equal the true mean, $\mu$.

### EXAMPLE 8.1    Tissue Strength Problem

*Impact of Sampling Error*

Suppose that you take a sample of tissues and find the sample average *MDStrength* to be 1010 lb. Clearly, the sample average, $\overline{X} = 1010$, does not exactly equal the specification of 1000. Does this mean that the process is not running correctly? Probably not, but we will need a more precise way of deciding. An hour later you might take another sample and get a sample *MDStrength* of 995 lb. Again, you need to decide if the process is running correctly on the basis of that sample average, $\overline{X} = 995$ lb. Clearly, the sample average will be different each time you take a sample and will almost never be equal to 1000 even if the process is running correctly!

Remember that you are examining only a sample of the population of tissues, and each tissue manufactured has a slightly different MD strength. The variability in the tissue strengths combined with the fact that you have data on only a piece of the population leads to variability in the sample averages, the $\overline{X}$'s. Recall that the variability of $\overline{X}$ is called the standard error and it is based on the amount of variability in the population, $\sigma$, and the sample size, $n$. ∎

This example shows you that you should not simply compare the $\overline{X}$ value you get from the sample to the target mean strength, because even if the tissues were being made properly, the $\overline{X}$ that you observe will almost never be equal to that number. So, a simple point estimate will not do the job.

This chapter introduces the major concepts and philosophy of a technique called *hypothesis testing*, which will do the job. You might remember that when we talked about the need for inferential statistics we said that most of the techniques of inferential statistics could be classified as either estimation tools or hypothesis testing tools. This chapter lays the groundwork for hypothesis testing. The remaining chapters of the text rely heavily on the ideas developed in this chapter.

Specifically, in this chapter we will look at:

- What Is a Hypothesis Test?
- Overview of Hypotheses to Be Tested
- The Pieces of a Hypothesis Test
- Two-Tail Tests of the Mean: Large Sample
- Which Theory Should Go Into the Null Hypothesis?
- One-Tail Tests of the Mean: Large Sample
- What Error Could You Be Making?

## 8.2   WHAT IS A HYPOTHESIS TEST?

The word **hypothesis** has the same meaning in statistics as it does in everyday use. What does this word mean to you? Some possibilities are

- an idea
- an assumption
- a guess
- a theory

You actually work with many hypotheses on a daily basis without even realizing it. For example, you might think that exercising on a regular basis improves your overall ability to study. Your hypothesis, in this case, is that there is a relationship between exercising and effective studying. To decide if your hypothesis is correct you might use your recollection of past experiences.

> In statistics, a *hypothesis* is an idea, an assumption, or a theory about the behavior of one or more variables in one or more populations.

Once a hypothesis is formed, you must *test* it. You must decide whether to believe the hypothesis. The only information you have to help you decide is contained in your sample. Thus, to do a **hypothesis test,** you use the information in your sample data to help you decide if you should believe the hypothesis. Basically, you are trying to decide if the sample is consistent with the hypothesis (in which case you believe the hypothesis) or if the sample is inconsistent with the hypothesis (in which case you choose not to believe it or to reject it).

*The sample data are your evidence and on the basis of these data you must make a decision.*

> A *hypothesis test* is a statistical procedure that involves formulating a hypothesis and using sample data to decide on the validity of the hypothesis.

There are many different types of hypotheses that you could test and these are the subject of the next section. However, regardless of the specific hypothesis that you are testing, the basic procedure is the same. The purpose of this chapter is to explain the framework of the hypothesis testing procedure. The details of the test depend on the particular hypothesis that you are testing, but the purpose and general approach are the same for all tests. In fact, virtually all of the remaining chapters in this book are devoted to spelling out the details for the various tests you are likely to need for analyzing sample data and making informed business decisions. It is important that you not view these as separate and isolated chapters. They add flesh to the skeleton we will develop in this chapter. Think of each chapter as a variation on the same theme.

## 8.3   DESIGNING HYPOTHESES TO BE TESTED—AN OVERVIEW

Let's start by thinking about what type of variables we might be examining. In Chapter 2 you learned that two major categories can be used to classify variables, *qualitative* and *quantitative*. The distinction was made because different techniques are

*Different types of data require different analysis tools.*

used depending on the type of data you have. You have seen (Chapters 3–5 and 7) that different *descriptive* tools and point estimators are used, depending on what kind of data you wish to evaluate. Now we will see that there are different *inferential* tools for different kinds of data.

## 8.3.1 Hypotheses About Quantitative Variables

Sample data that are inherently numerical in form are called quantitative data. Recall from Chapter 6 that if we are analyzing a quantitative variable then we know that it can be described by its distribution. The distribution in turn can be described by parameters. Thus, if we are constructing a theory or hypothesis about a quantitative variable it might be a statement about

*Hypotheses about a quantitative variable*

- The shape of the distribution of the variable in one population
- The mean value, $\mu$, of the variable in one population
- How the mean value of the variable in one population, $\mu_1$, compares to the mean value of the variable in a second population, $\mu_2$
- The equality of the mean values of the variable in more than two populations
- The amount of variability, $\sigma^2$, of the variable in one population
- How the amount of variability of the variable in one population, $\sigma_1^2$, compares to the amount of variability of a variable in a second population, $\sigma_2^2$

All but the first bulleted item are generally referred to as tests on means and tests on variances. The first item requires a "goodness of fit" test. Goodness of fit tests are the subject of Chapter 15.

### EXAMPLE 8.2   Tissue Strength Problem

*Possible Hypotheses*

For the problem that you are investigating for your boss, tissue strength is a quantitative variable. An example of each of the hypotheses just discussed is shown for this variable:

- The variable *MDStrength* has a normal distribution.
- The population mean *MDStrength* is equal to 1000 lb/ream.
- The mean *MDStrength* of tissues made by machine 1 is greater than the mean *MDStrength* of tissues made by machine 2.
- The mean *MDStrength* of tissues made by machines 1, 2, and 3 are equal.
- The variability of the tissue *MDStrength* is less than 625 (lb/ream)$^2$. (Recall that the variance is the standard deviation squared.)
- The tissues made by machine 1 have more variability in the *MDStrength* characteristic than those tissues made by machine 2.  ■

*Remember: Qualitative data describe a particular characteristic of a sample item.*

Notice that a qualitative variable, in this case the number of the machine that made the tissue, was used to divide the data into two or more populations. This is often the major use of a qualitative variable, particularly nominal data. Recall that nominal data are one type of qualitative data. They are data that are created by assigning numbers to different categories when the numbers have no real meaning.

*TRY IT NOW!*

### 7-11 Stores
***Possible Hypotheses***

7-11 stores are located all over the Northeast. Management is studying sales data. Develop a specific hypothesis for each of the different types of hypotheses that we have discussed. Here's one hypothesis to get you started:

> *Daily Sales are normally distributed.*

## 8.3.2 Hypotheses About Nominal Variables

In addition to using nominal variables as a way to divide the data into two groups, we are often interested in what percentage or proportion of a population has a particular characteristic. For example, we might be interested in what proportion of the product we are manufacturing is defective. In this case, the nominal variable is the quality status of the product, nondefective or defective, and we are interested in the percentage or proportion of the population that has the quality status "defective." If the data we are analyzing are *nominal data,* the hypothesis might be a statement about

- The value of the proportion, $\pi$, of population members that have a certain characteristic (one of the categories of the nominal variable)
- How the proportion who have a certain characteristic in one population, $\pi_1$, compares with the corresponding proportion in a second population, $\pi_2$

### EXAMPLE 8.3   Cereal Manufacturer

***Possible Hypotheses***

A cereal company is considering a new package design. The company would like to know what percentage or proportion of the consumers like this new design. An example of each of the hypotheses just discussed is shown for this variable:

- At least 50% of the customers like this new design.
- A greater proportion of women than men like this new design.   ■

---

Notice that, in this example, we are interested in the proportion of the population that prefers the new design. In this case, the nominal variable is whether or not a consumer prefers the new design. We have used a second nominal variable (gender) to divide the consumers into two populations and have then constructed a hypothesis about the proportion of women who prefer the new design compared to the proportion of men who prefer the new design.

 *TRY IT NOW!*

### The Sports Complex
*Possible Hypotheses*

A particular university is considering building a new sports complex. It wishes to know if the sports complex would be widely used by students. Develop a specific hypothesis for each of the different types of hypotheses that we have discussed.

### 8.3.3 Hypotheses About Ordinal Variables

Another way in which qualitative data can be expressed numerically is when the order in which the numbers are assigned has some meaning. When numbers are used to name ordered categories the data are called *ordinal*. If we are analyzing ordinal data, the hypothesis might be a statement about

- The mean value, $\mu$, of the variable in one population
- How the mean value of the variable in one population, $\mu_1$, compares to the mean value of the variable in a second population, $\mu_2$
- The equality of the mean values of the variable in more than two populations

### EXAMPLE 8.4   The Soft Drink Manufacturer

*Hypotheses for Ordinal Data*

Suppose you asked a group of people to rank five different versions of a new soft drink. The resulting data for one person might look like this:

| Version | 3 | 2 | 5 | 1 | 4 |
|---------|---|---|---|---|---|
| Ranking | 1 | 2 | 3 | 4 | 5 |

In this case the numbers are not entirely meaningless since they indicate a relative position for each version of the product on a scale. But the variable ranking is not a quantitative variable because you cannot tell from this scale whether the person doing the ranking liked version 3 a lot and really hated the other 4 or whether versions 3 and 2 were similar and much superior to the remaining three versions. That is, the distances between the numbers assigned are not necessarily equal.

Possible hypotheses for this situation are:

- The average value for the ranking of version 3 is 1.5.
- Versions 2 and 5 have the same average ranking.
- There is no difference in the average rankings of the 5 different versions. ■

You can see that these hypotheses are very similar to the hypotheses we might use for a quantitative variable. However, the details of how to test these hypotheses when you have ordinal data are different. These tests are referred to as nonparametric tests. Although these tests are not presented in this text, the general hypothesis testing concepts that you will learn in this chapter and the next few chapters are also the basis for nonparametric tests.

### 8.3.4 Hypotheses About Two Variables

We started Section 8.2 with an example about exercise and effective studying. In this example we have two variables. How we investigate the relationship between these two variables depends on how the data are collected.

*Remember that you learned some techniques for displaying bivariate data in Chapter 5.*

Suppose we have number of hours per day that a person exercises and a rating of the effectiveness of the person's studying ranging from $1 = $ highly effective to $5 = $ highly ineffective. In this case we could treat the number of hours per day that a person exercises as a quantitative variable and use the effectiveness rating to divide the data into 5 populations. We could then construct the hypothesis that the average number of hours exercised is the same for students with an effectiveness rating of 1, 2, 3, 4, and 5. This type of hypothesis was discussed in Section 8.3.1.

However, if the data are collected in such a way that both variables are qualitative variables, then we need to construct the hypothesis that says that the variable exercise rating and study effectiveness rating are independent. This is typically called a test for independence and is the subject of Chapter 15.

### 8.3.5 Summary of Kinds of Hypotheses

Table 8.1 summarizes the various kinds of hypotheses you are likely to need to analyze sample data and make informed business decisions.

### TABLE 8.1  Summary of Various Hypotheses to Be Tested

|  | One Population (Chapters 8, 9, and 15) | Two Populations (Often a Qualitative Variable is Used to Divide the Data into 2 Populations) (Chapter 10) | More Than Two Populations (Chapters 14 and 15) |
|---|---|---|---|
| Quantitative data | 1. The shape of the distribution. 2. The mean value, $\mu$, of the variable. 3. The amount of variability, $\sigma^2$, of the variable. | 1. Compare the mean value of the variable in one population, $\mu_1$, to the mean value of the variable in a second population, $\mu_2$. 2. Compare the amount of variability of the variable in one population, $\sigma_1^2$, to the amount of variability of the variable in a second population, $\sigma_2^2$. | The equality of the mean values of the variable. |
| Nominal data | The value of the proportion, $\pi$, of population members that have a certain characteristic (one of the categories of the nominal variable). | Compare the proportion who have a certain characteristic in one population, $\pi_1$, with the corresponding proportion in a second population, $\pi_2$. | Test for independence. |

## 8.4   THE PIECES OF A HYPOTHESIS TEST

To carry out any hypothesis test, you need to set it up according to some general guidelines.

### 8.4.1 The Null and Alternative Hypotheses

The first step is to take your idea or hypothesis and construct two opposing views. One of these is called the **null hypothesis** and the other is called the **alternative hypothesis.**

### EXAMPLE 8.5   The Tissue Strength Problem

*Setting Up the Null and Alternative Hypotheses*

For the tissue strength problem that you are looking at you may choose to set it up this way:

| | |
|---|---|
| **Null hypothesis:** | The true mean *MDStrength* is equal to 1000 lb/ream. |
| **Alternative hypothesis:** | The true mean *MDStrength* is not equal to 1000 lb/ream. |

Clearly, if you believe the null hypothesis then you cannot believe the alternative hypothesis. You must choose between them on the basis of the evidence in the sample.

*Remember: You do not know μ! You are checking to see if μ is 1000. You are using the sample mean as your evidence.*

Using the notation of hypothesis testing, you would rewrite this as

$$H_0: \quad \mu = 1000 \text{ lb/ream}$$
$$H_A: \quad \mu \neq 1000 \text{ lb/ream}$$

> The *null hypothesis* is a statement about the population(s). It is referred to as $H_0$.
>
> The *alternative hypothesis* is a statement about the population(s) that is opposite to the null hypothesis. It is referred to as $H_A$.

Sometimes the alternative hypothesis is labeled $H_1$. It doesn't matter whether you use $H_1$ or $H_A$ as long as you are consistent. In this book we will always use $H_A$. You can see that the words have been rewritten in terms of the standard symbol for the population mean, $\mu$.

You should notice that no value of $\mu$ is part of both the null and the alternative hypotheses. The null and alternative hypotheses cannot overlap. That is, the two hypotheses are mutually exclusive. Not only is that true but, in addition, when considered together, the null and the alternative hypotheses must cover all the possibilities. This means the two hypotheses are exhaustive. For this example, this means that all the possible values of $\mu$ are included in either the null or the alternative hypothesis but not both. Finally, note that the "=" sign is in the null hypothesis. This will always be the case.

### EXAMPLE 8.6   The Bottle-Filling Problem

*Reinforces the Two Opposing Views Concept*

In any bottling process, a manufacturer will lose money if the bottles contain either more or less than the amount claimed on the label. Therefore, manufacturers pay close attention to the amount of their product that is dispensed by the bottle-filling machines. Suppose you are work      for a soda company and each of the soda bottles is supposed to contain 32 oz of s      You wish to set up a hypothesis test to decide if the bottle-filling machines are p      an average of 32 oz of soda into the bottles.

The two opposing views are      efore,

$$H_0: \quad \mu = 32 \text{ oz}$$
$$H_A: \quad \mu \neq 32 \text{ oz}$$    ■

Let's look at a situation in which the data to be analyzed are nominal data.

## EXAMPLE 8.7    U.S. Unemployment

### Null and Alternative Hypotheses for Test of Proportion

The U.S. government wishes to know what proportion of the workforce is unemployed. Last month the estimated proportion was 6% and the government wishes to set up a hypothesis test to decide if the proportion for this month has changed from 6%.

The two opposing views are, therefore,

$$H_0: \quad \pi = 0.06$$
$$H_A: \quad \pi \neq 0.06$$    ■

Now it is time for you to try your hand at setting up the null and alternative hypotheses.

 **TRY IT NOW!**

### The Potato Chip Manufacturer

#### Setting Up the Null and Alternative Hypotheses

Many people eat chips with their soda. Suppose a potato chip manufacturer is concerned that the bagging equipment may not be functioning properly when filling 10-oz bags. You have been asked to set up a hypothesis test that will help determine if there is a problem with the bagging equipment. What null and alternative hypotheses would you use?

## *Discovery Exercise 8.1*
## FORMULATING HYPOTHESES

Consider the population of all M&M packages like the one you have in your hand.

**Step 1:**   Identify as many different variables as you can. Be sure you have some quantitative and some qualitative variables. Record the values of these variables for your package. (*Hint:* You should carefully examine the package before you rip it open.)

**Step 2:**   Select one of the quantitative variables and set up a null and alternative hypothesis for a parameter of this variable.

**Step 3:**   Select one of the qualitative variables and set up a null and alternative hypothesis for a parameter of this variable.

**Step 4:**   As a class, agree on several quantitative and qualitative variables that you feel are important. Record the data for all of the teams on each of these variables.

**Step 5:**  Use the tools of descriptive statistics to display these data.




**Step 6:**  Enjoy the M&M's !


## 8.4.2 The Rejection Region

The next step is to determine how you will decide between the null and the alternative hypotheses. To be more specific, you must decide to reject or not to reject the null hypothesis. Your decision is always phrased with regard to the null hypothesis. If you choose not to reject the null hypothesis this means that the sample data are consistent with the null hypothesis. If you choose to reject the null hypothesis this means that the sample data are sufficiently inconsistent with the null hypothesis.

To make this decision you must set up what is known as the **rejection region.**

> The *rejection region* is the range of values of the test statistic that will lead you to reject the null hypothesis.

The test statistic is discussed in more detail in the next section. For example, the rejection region might be all values of $Z$ larger than 1.96. This is shown as the shaded region in Figure 8.1.

If this is the rejection region, we would reject the null hypothesis if the value of $Z$ calculated from the data is greater than 1.96. If the value of $Z$ calculated is less than or equal to 1.96, we would not reject the null hypothesis. The $Z$ statistic is the same

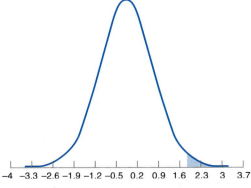

| | | | | | | | | | | |
|---|---|---|---|---|---|---|---|---|---|---|
| −4 | −3.3 | −2.6 | −1.9 | −1.2 | −0.5 | 0.2 | 0.9 | 1.6 | 2.3 | 3 | 3.7 |

**FIGURE 8.1**   Possible rejection region

one you learned about in Chapter 7. Often, $Z$ is the test statistic we need to use. This is discussed more in the next section.

The procedure for actually finding the rejection region depends on several things. The specifics are provided in the last section of this chapter and in upcoming chapters. For now, it is important that you understand the general concepts.

### 8.4.3 The Test Statistic

We have said that the information contained in your sample is the evidence that you will use to decide between $H_0$ and $H_A$. However, as you can see, the sample consists of $n$ observations. Somehow you must capture the information in the sample into a single number. This number will be called the **test statistic.**

> A *test statistic* is a number that captures the information in the sample. It will be used to choose between the null and alternative hypotheses.

If you think about it, you will see that this term is entirely consistent with what we have already defined. In general, a statistic is any number calculated from the sample data. In this case, you will calculate a number from the sample data and use this number to decide between the null and the alternative hypotheses. In other words, you will use this statistic to test the hypothesis. Hence, it is called a test statistic.

## 8.5   TWO-TAIL TESTS OF THE MEAN: LARGE SAMPLE

Now we will put these pieces together to form a *5-step hypothesis testing procedure:*

*Steps for any hypothesis test*

**Step 1:**   *Set up the null and alternative hypotheses.*

**Step 2:**   *Pick the value of α and find the rejection region.*

**Step 3:**   *Calculate the test statistic.*

**Step 4:**   *Decide whether or not to reject the null hypothesis.*

**Step 5:**   *Interpret the statistical decision in terms of the stated problem.*

There are two different cases to consider when testing the population mean: tests of the mean when you know something about the *population* standard deviation and tests of the mean when all you have is the *sample* standard deviation. The first case is covered in this chapter. This case also includes the situation where you do not know the population standard deviation, $\sigma$, but you have a sufficiently large sample size, $n > 30$. When this happens you use the sample standard deviation, $s$, as an estimate of $\sigma$. For this reason, the tests in this chapter are referred to as "large-sample tests" but they are used whenever you know the population standard deviation regardless of the sample size.

### 8.5.1 Two-Tail Test of the Mean: A Detailed Example

The first step is to take your idea or hypothesis about the mean and construct the null and the alternative hypotheses.

## EXAMPLE 8.8   The Tissue Company

### Step 1: Setting Up the Null and Alternative Hypotheses

Returning to the tissue strength example, suppose the manufacturer is interested in seeing if the MD strength is different from the target value of 1000 lb/ream. This means a two-tail test must be used. Remember that if a two-tail test is used, the "=" theory must be the null hypothesis. The first step of the procedure is then completed when we write the null and the alternative hypotheses:

$$H_0: \quad \mu = 1000 \text{ lb/ream}$$
$$H_A: \quad \mu \neq 1000 \text{ lb/ream} \qquad \blacksquare$$

Since we are talking about a hypothesis test of $\mu$, it makes sense to use the value of $\overline{X}$ as our evidence from the sample. Remember that the Central Limit Theorem (Chapter 7) tells you that the sample mean, $\overline{X}$, tends to have a normal distribution with a mean equal to the true population mean, $\mu$. It also tells you that the sample mean has a standard deviation or standard error equal to $\sigma/\sqrt{n}$. We can therefore standardize $\overline{X}$ by calculating

*Remember that $\overline{X}$ is the best point estimator for $\mu$.*

$$Z = \frac{\overline{X} - \mu}{\sigma/\sqrt{n}}$$

*Z test statistic*

which we know will have a standard normal distribution. So, it seems that the *test statistic* should be a $Z$-statistic.

Let's think about what the $Z$-score really tells us and what size $Z$-score should lead us to reject $H_0$. We are interested in detecting if the true mean has shifted from 1000. So, values of $\overline{X}$ that are far from 1000 would be the ones that lead us to be suspicious of $H_0$: $\mu = 1000$.

*Remember that the Z distribution is bell shaped and centered at the value of Z = 0.*

If the $\overline{X}$ value is considerably smaller than 1000, what will happen to the numerator of the $Z$-score? Let's see: The numerator is $(\overline{X} - \mu)$. If $\overline{X}$ is a lot smaller than $\mu$, then that difference will be a big negative number. For example, if $\overline{X} = 950$, then the difference is $-50$. As a result of the large negative value in the numerator, $Z$ will be a large negative number. Keep in mind that the difference of $-50$ is large relative to the size of the standard error. Similarly, if $\overline{X}$ is a number much larger than $\mu$ (say 1050), then the difference $(\overline{X} - \mu)$ will be a big positive number and $Z$ will be a large positive number. It seems then that *large negative values of Z* (resulting from $\overline{X}$ values a lot lower than 1000) and *large positive values of Z* (resulting from $\overline{X}$ values a lot higher than 1000) are the ones that will lead us to reject the null hypothesis. These $Z$ values are found in the tails of the normal distribution.

*The standard error of $\overline{X}$ is $\sigma/\sqrt{n}$.*

The specification of these cutoff points for $Z$ is completely determined by the value of $\alpha$. Remembering (from Chapter 6) that *probabilities* correspond to *areas* under a curve when you are working with the normal distribution, you want the total area in the two tails of the normal distribution to be equal to $\alpha$. Since the normal distribution is symmetric, the area in each tail must be $\alpha/2$.

Combining the idea that we will reject $H_0$ when the $Z$-score is too large in either the negative or positive direction with the requirement that the area in each tail is $\alpha/2$, gives us Figure 8.2 (page 384). The shaded region is called the *rejection region*. The rejection region consists of those values of the test statistic (in this case $Z$) that lead you to reject the null hypothesis. Thus, a $Z$ value larger than $Z_{\text{cutoff}}$ or a $Z$ value smaller than $-Z_{\text{cutoff}}$ will tell us to reject the null hypothesis.

You may recall that in Chapter 6 you learned how to find $Z$ values given tail area probabilities. To get the value for $Z_{\text{cutoff}}$ you must use that procedure with a tail area probability equal to $\alpha/2$. You might remember that the normal distribution is symmetric, so the lower cutoff value is simply the negative of this number and hence la-

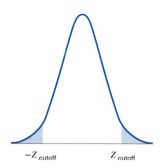

**FIGURE 8.2**   Rejection region for two-tail test of $\mu$

beled $-Z_{\text{cutoff}}$. Let's select a value for $\alpha$ and set up the rejection region for the tissue company.

### EXAMPLE 8.9   The Tissue Company

#### Step 2: Setting $\alpha$ and Finding the Rejection Region

The tissue company sets $\alpha$ to be 0.05. In this case, the area of the shaded region must be 0.05. Since the normal distribution is symmetrical, we can divide the 0.05 by 2 and find that the area in one of the tails is 0.025. What we want then is the $Z$ value that will guarantee that the tail area is 0.025. This means we are given a probability and need the corresponding $Z$ value. Shown below is a portion of the $Z$ table:

| $z$ | 0.00 | 0.01 | 0.02 | 0.03 | 0.04 | 0.05 | 0.06 | 0.07 | 0.08 | 0.09 |
|------|--------|--------|--------|--------|--------|--------|--------|--------|--------|--------|
| $-2.1$ | 0.0179 | 0.0174 | 0.0170 | 0.0166 | 0.0162 | 0.0158 | 0.0154 | 0.0150 | 0.0146 | 0.0143 |
| $-2.0$ | 0.0228 | 0.0222 | 0.0217 | 0.0212 | 0.0207 | 0.0202 | 0.0197 | 0.0192 | 0.0188 | 0.0183 |
| $-1.9$ | 0.0287 | 0.0281 | 0.0274 | 0.0268 | 0.0262 | 0.0256 | 0.0250 | 0.0244 | 0.0239 | 0.0233 |
| $-1.8$ | 0.0359 | 0.0351 | 0.0344 | 0.0336 | 0.0329 | 0.0322 | 0.0314 | 0.0307 | 0.0301 | 0.0294 |
| $-1.7$ | 0.0446 | 0.0436 | 0.0427 | 0.0418 | 0.0409 | 0.0401 | 0.0392 | 0.0384 | 0.0375 | 0.0367 |

We want to find the $Z_{\text{cutoff}}$ value that has 0.025 in the lower tail. Look in the inside of the table for the number closest to 0.025 and see that it corresponds to a $Z$ value of $-1.96$. This gives us the value of $-Z_{\text{cutoff}}$. To get the positive cutoff value we can simply drop the negative sign because of the symmetry in the normal distribution. So $Z_{\text{cutoff}} = 1.96$. Thus, we will reject $H_0$ if $Z$ is larger than 1.96 or smaller than $-1.96$, as shown in the shaded area below:

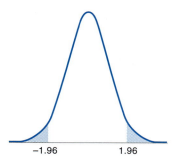

Actually, you should have been able to guess that the cutoff $Z$ value would be close to 2. Why? When you studied the normal distribution in Chapter 6 you saw that about 95% of the values fall within two standard deviations of the mean. For the standard normal distribution the mean is zero and the standard deviation is 1, so 95% of the values will fall between $+2$ and $-2$. We just found a $Z$ value of 1.96, which is very close to 2!

In general, the procedure for finding the rejection region for a two-tail test is as follows:

**Step 1:** *Take the value of α and divide it by 2.*

**Step 2:** *Look that value up in the body of the Z table (find the one closest to it).*

**Step 3:** *Read off the corresponding Z value. This gives you the value of* $-Z_{cutoff}$.

**Step 4:** *Drop the negative sign to get the value of* $Z_{cutoff}$.

**Step 5:** *The rejection region consists of all values of Z greater than* $Z_{cutoff}$ *or less than* $-Z_{cutoff}$.

*STEPS FOR FINDING THE REJECTION REGION FOR A TWO-TAIL TEST*

 **TRY IT NOW!**

### The Tissue Company

*Finding the Rejection Region*

Suppose the tissue company decided to set α at 0.10. Find the rejection region.

The third step is to capture the information in the sample into a single number called the test statistic. It is the sample mean, $\overline{X}$, that is particularly relevant, since you are testing an idea about the true population mean, $\mu$. Suppose the sample mean *MDStrength* of 36 tissue samples is found to be 980 lb/ream. On the surface it looks like maybe the process needs to be adjusted. After all, 980 is 20 units less than the target value of 1000. You must remember that this $\overline{X}$ value of 980 is based on a sample. If you took another 36 tissue samples you might get a sample mean of 1020. You cannot decide if a difference of 20 is really big or not until you compare it to the standard error! This is precisely what the Z-score calculation does.

We can calculate $Z$ using the following formula for standardizing $\overline{X}$:

$$Z = \frac{\overline{X} - \mu}{\sigma/\sqrt{n}}$$

It seems like we have a problem since we need to know $\mu$ to calculate the Z-statistic but we don't know $\mu$. This is easily resolved at this point because $\mu$ is the *target* specification value. In general, the value of $\mu$ that you use is the number that you are testing your sample evidence against, the value in your null hypothesis.

Let's see how this test statistic would be calculated for the tissue strength problem that we have been following.

*Remember that $\overline{X}$ is the best point estimator for $\mu$.*

**EXAMPLE 8.10**   **The Tissue Company**

*Step 3: Calculating the Test Statistic*

Recall that the population standard deviation, $\sigma$, for the tissue manufacturer is 50. The $Z$-statistic is then

$$Z = \frac{\overline{X} - \mu}{\sigma/\sqrt{n}}$$

$$Z = \frac{980 - 1000}{50/\sqrt{36}} = \frac{-20}{8.333} = -2.40$$

Notice that you use the target value of 1000 as the value for $\mu$ in this calculation.  ■

The fourth step is to decide whether or not to reject the null hypothesis. If the $Z$ value that was calculated at step 3 falls in the rejection region, then we reject $H_0$. If the $Z$ value does not fall in the rejection region, then we fail to reject $H_0$. Let's do the fourth step for the tissue company.

**EXAMPLE 8.11**   **The Tissue Company**

*Step 4: Using the Test Statistic to Decide Whether or Not to Reject the Null Hypothesis*

The rejection region for this two-tail test is shown below:

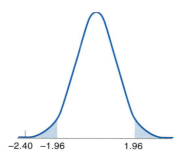

The calculated $Z$ value of $-2.40$ is marked on this line and clearly falls in the rejection region. Thus, we reject $H_0$ and believe the alternative hypothesis.  ■

The fifth step is to interpret the statistical decision in terms of the stated problem. At this step you must "translate" the result of the statistical test into a recommended action, or decision, or reach a conclusion depending on the nature of the problem. Let's complete the hypothesis test for the tissue company

**EXAMPLE 8.12**   **The Tissue Company**

*Step 5: Interpreting the Statistical Decision*

At step 4 we rejected the null hypothesis. Therefore, we believe the alternative hypothesis. The alternative hypothesis was that the true mean was not equal to 1000 lb/ream. Therefore, on the basis of the evidence in the sample that is captured in the $Z$ value of $-2.4$, you decide that the process needs adjustment.

Your decision is based on the fact that it is unlikely that you would get such a large $Z$ value ($Z = -2.4$) if the process is really OK. But is it possible that the process really does not need adjustment? You bet. Maybe you got a sample with a large $\overline{X}$, which gave you the large $Z$ value. How likely is this? The chance of this happening is precisely $\alpha$. In this case, that chance is at most 5%, since $\alpha$ was set to 0.05.  ■

## *Discovery Exercise 8.2*

## EXPLORING THE IMPACT OF
## VARYING THE VALUE OF $\alpha$

You may remember that in Chapter 4 we examined the data collected by a company that was concerned about the length of time its customers were on hold. The data shown below were first presented to you in Example 4.14. There are 50 observations on customer hold times:

| 0.6 | 4.6 | 5.6 | 6.3 | 6.8 | 7.5 | 7.8 | 8.3 | 8.9 | 9.6 |
| 2.9 | 4.7 | 6.0 | 6.3 | 6.9 | 7.5 | 7.9 | 8.4 | 9.2 | 10.1 |
| 3.4 | 5.2 | 6.0 | 6.6 | 6.9 | 7.6 | 8.0 | 8.4 | 9.2 | 10.2 |
| 3.8 | 5.5 | 6.1 | 6.6 | 7.0 | 7.6 | 8.1 | 8.6 | 9.4 | 10.7 |
| 4.5 | 5.5 | 6.1 | 6.7 | 7.2 | 7.8 | 8.2 | 8.6 | 9.4 | 11.1 |

The company wishes to test the hypothesis that the true mean hold time is 7 minutes. The standard deviation of hold times is known to be 2 minutes.

**(a)** Test this hypothesis using a value of $\alpha = 0.10$.

**Step 1:**

**Step 2:**

**Step 3:**

*(continued)*

**Step 4:**




**Step 5:**




**(b)** Now vary the value of $\alpha$ and complete the following table.

| $\alpha$ | Rejection Region | Decision |
|---|---|---|
| 0.10 (from part a above) | | |
| 0.05 | | |
| 0.02 | | |
| 0.01 | | |

**(c)** What happened to your decision as the value of $\alpha$ changed?


## 8.5.2 Understanding, Calculating and Using *p* values

You may have noticed that the rejection region is tied to the value of $\alpha$ that we selected. You may therefore be wondering if selecting a different value of $\alpha$ would have
led to a different conclusion. This is indeed a good point and worthy of consideration. Let's follow the details of the hypothesis-testing procedure through another example and then examine the impact of selecting a different value for $\alpha$.

### EXAMPLE 8.13    The Bottle-Filling Problem

*Step 1: Setting Up the Null and Alternative Hypotheses*

In Example 8.6 we looked at the hypothesis set up to test whether a soda machine was
dispensing 32 oz of soda. Recall that we wished to know if the machine was underfill-

ing or overfilling the cups. Since we are interested in detecting differences from 32 oz in either direction, we used a two-tail test. The null and alternative hypotheses are shown below:

$$H_0: \quad \mu = 32 \text{ oz}$$
$$H_A: \quad \mu \neq 32 \text{ oz}$$

Thus, step 1 has been done. ∎

Continuing with this example, we will complete step 2 by selecting the value of $\alpha$ and finding the rejection region.

### EXAMPLE 8.14   The Bottle-Filling Problem

*Step 2: Selecting $\alpha$ and Finding the Rejection Region*

Set $\alpha = 0.05$. The rejection region is the same as the one we found for the tissue company. Reject $H_0$ if $Z < -1.96$ or $Z > 1.96$. ∎

Suppose you observe the machine filling 30 bottles and observe the following data:

| | | | | | |
|---|---|---|---|---|---|
| 32.12 | 32.04 | 32.04 | 32.00 | 32.01 | 32.12 |
| 32.13 | 32.05 | 32.05 | 31.72 | 32.08 | 32.03 |
| 32.02 | 32.22 | 32.22 | 32.22 | 31.89 | 32.20 |
| 31.73 | 31.95 | 31.95 | 32.07 | 32.13 | 32.13 |
| 31.98 | 32.21 | 32.21 | 31.97 | 32.09 | 31.97 |

You also know from the machine specifications that the population standard deviation is supposed to be 0.1 oz. What do you do with the sample information? Remember, step 3 is to capture the information in the sample into a single number called the test statistic. Let's complete step 3 for the bottle-filling problem.

### EXAMPLE 8.15   The Bottle-Filling Problem

*Step 3: Calculating the Test Statistic*

To calculate the $Z$ test statistic:

$$Z = \frac{\overline{X} - \mu}{\sigma/\sqrt{n}}$$

we clearly need $\overline{X}$. Calculating $\overline{X}$, we get 32.05 oz . The $Z$ test statistic is then

$$Z = \frac{32.05 - 32}{0.1/\sqrt{30}} = \frac{0.05}{0.018} = 2.78$$ ∎

Now we are ready to use the evidence in the sample, which has been captured in the $Z$ test statistic, to decide if we should reject the null hypothesis or not. Step 4 for the bottle-filling problem is completed in the next example.

### EXAMPLE 8.16   The Bottle-Filling Problem

*Step 4: Using the Test Statistic*

The value of $Z = 2.78$ clearly falls in the rejection region. So our statistical decision is to reject the null hypothesis and conclude that the true mean is not equal to 32 oz. ∎

Step 5 requires that we interpret this statistical decision in terms of the problem statement.

**EXAMPLE 8.17**   The Bottle-Filling Problem

***Step 5: Interpreting the Statistical Decision***

We concluded that the true mean of the bottle-filling process was not 32 oz. This tells us that there is something wrong with the filling process. Possibly the true mean has shifted to a value other than 32 oz.   ■

Let's consider a variety of different $\alpha$ values for the bottle-filling example and see what the impact is on the rejection region and ultimately on the decision.

|              |                   | Decision |
| --- | --- | --- |

| $\alpha$ | $Z_{\text{cutoff}}$ | $-Z_{\text{cutoff}}$  0  $Z_{\text{cutoff}}$ |
| --- | --- | --- |
| 0.10  | 1.64 or 1.65 | $2.78 > 1.64$, so reject $H_0$ |
| 0.05  | 1.96         | $2.78 > 1.96$, so reject $H_0$ |
| 0.01  | 2.58         | $2.78 > 2.58$, so reject $H_0$ |
| 0.005 | 2.81         | $2.78 < 2.81$, so fail to reject $H_0$ |
| 0.001 | 3.29         | $2.78 < 3.29$, so fail to reject $H_0$ |

*Remember: To get the cutoff Z value you divide $\alpha$ by 2 and look up that value in the body of the Z table.*

Notice that when you change $\alpha$ from 0.01 to 0.005 your decision switches from rejecting $H_0$ to failing to reject $H_0$. Initially, this should make you a bit uncomfortable. After all, you need to tell your boss whether this machine is working correctly and you get a different answer depending on how you set $\alpha$! There will always be some value for $\alpha$ where your decision switches. This is called the ***p* value.**

> The ***p* value** is defined to be the smallest value of $\alpha$ for which you can reject $H_0$.

For some value of $\alpha$, the value of the test statistic (based on sample data) will be the same as the $Z_{\text{cutoff}}$ that defines the rejection region. When that happens you have found the $p$ value.

*Calculating the p value is just like the calculations done in Chapter 7.*

Remember that $\alpha$ is a probability and therefore so is the *p value*. In fact you have actually calculated $p$ values in Chapter 7 without knowing it. In that chapter you calculated the probability of observing a particular $Z$ value. This is exactly what needs to be done to get the $p$ value. You must calculate the chance of observing the particular sample mean, $\overline{X}$, if the null hypothesis is true. Let's calculate the $p$ value for the bottle-filling problem.

**EXAMPLE 8.18**   The Bottle-Filling Problem

***Calculating the p Value***

For the bottle-filling problem, we observed data and calculated a sample mean of $\overline{X} = 32.05$ oz and a $Z$ statistic of 2.78. To calculate the $p$ value we must determine the chance that we would have observed this $\overline{X}$ value if the null hypothesis was true. The Central Limit Theorem tells us that $\overline{X}$ has a normal distribution which we know is a continuous distribution. We also remember that there is no probability (zero area under the curve) of observing a specific value of a continuous random variable. Thus, we need to calculate the probability of observing the particular $\overline{X}$ value or one even more extreme. Since the test is a two-tail test we need to double the tail area probability. The $p$ value is calculated by finding

$$p \text{ value} = 2P(\overline{X} > 32.05)$$

This calculation is done by converting the random variable $\overline{X}$ to a $Z$ value and using the procedure developed in Chapter 7. But we already have the $Z$ value from the test statistic so we know we can simply calculate

$$p\,\text{value} = 2P(Z > 2.78)$$

Using the symmetry of the normal curve, we find that this is just two times the tail area probability for $Z = -2.78$. Using the standard normal table, we get

$$p\,\text{value} = 2(0.0027) = 0.0054$$

This result is consistent with our previous analysis. We expected a $p$ value less than 0.01. ∎

In this example the $p$ value is very small. In particular, the $p$ value is less than 0.01. This tells us that the smallest value of $\alpha$ for which we can reject $H_0$ is 0.0054. It also tells us that there is a chance of 54 in 10,000 that we would have observed a sample mean of $\overline{X} = 32.05$ oz, if the population mean was really 32 oz. Clearly, the data are very inconsistent with the null hypothesis. The data are telling you to reject the null hypothesis for virtually every possible value of $\alpha$.

Consider another two-tail test where the calculated $Z$ value is 1.80. The same type of analysis can be done to get an idea of what the $p$ value is for this case.

| | | Decision |
|---|---|---|
| $\alpha$ | $Z_{cutoff}$ | $-Z_{cutoff} \quad 0 \quad Z_{cutoff}$ |
| 0.10 | 1.64 or 1.65 | $1.8 > 1.64$, so reject $H_0$ |
| 0.05 | 1.96 | $1.8 < 1.96$, so do not reject $H_0$ |
| 0.01 | 2.58 | $1.8 < 2.58$, so do not reject $H_0$ |
| 0.005 | 2.81 | $1.8 < 2.81$, so do not reject $H_0$ |
| 0.001 | 3.29 | $1.8 < 3.29$, so do not reject $H_0$ |

In this case, the decision switches from rejecting $H_0$ to not rejecting $H_0$ somewhere between $\alpha = 0.10$ and $\alpha = 0.05$. Thus, the $p$ value is somewhere between 0.10 and 0.05. To calculate the $p$ value for this situation we would find

$$p\,\text{value} = 2P(Z > 1.8) = (2)(0.0359) = 0.0718$$

A $p$ value between 0.05 and 0.10 tells you that the data are not strongly consistent with the null hypothesis, but neither are they strongly inconsistent with the null hypothesis. In such cases the data are "on the fence."

Finally, consider an example where the data are extremely consistent with $H_0$. Suppose you are doing a two-tail test of $\mu$ and the calculated $Z$ statistic is 0.95. The same analysis is shown below.

| | | Decision |
|---|---|---|
| $\alpha$ | $Z_{cutoff}$ | $-Z_{cutoff} \quad 0 \quad Z_{cutoff}$ |
| 0.10 | 1.64 or 1.65 | $0.95 < 1.64$, so do not reject $H_0$ |
| 0.05 | 1.96 | $0.95 < 1.96$, so do not reject $H_0$ |
| 0.01 | 2.58 | $0.95 < 2.58$, so do not reject $H_0$ |
| 0.005 | 2.81 | $0.95 < 2.81$, so do not reject $H_0$ |
| 0.001 | 3.29 | $0.95 < 3.29$, so do not reject $H_0$ |

As you can see for all values of $\alpha$ less than or equal to 0.10, we would not reject $H_0$. See if you can find the value for $\alpha$ where the decision would switch to a rejection. It is clearly larger than 0.10. The calculation is shown below:

$$p\text{ value} = 2P(Z > 0.95) = (2)(0.1711) = 0.3422$$

You should also have realized that a $Z$ value of less than 1 means that the observed sample mean is less than 1 standard error away from the hypothesized value and we know that such $Z$ values are likely. This also points to the same conclusion. So in this case our $p$ value is large ($>0.10$), which means we will almost always fail to reject the null hypothesis. The data are *highly consistent* with the null hypothesis.

In summary, we have the following table:

| $p$ Value | Data are . . . | Decision |
|---|---|---|
| $<0.05$ | Highly inconsistent with $H_0$ | Strong rejection of $H_0$ |
| Between 0.05 and 0.10 | "On the fence" | Collect more data (if possible) |
| $>0.10$ | Highly consistent with $H_0$ | Strong failure to reject $H_0$ |

 **TRY IT NOW!**

**The Tissue Company**

*Finding the p Value*

Find the $p$ value for the tissue company's two-tail test of $\mu$. Recall that the average *MDStrength* was found to be 980 lb/ream and the $Z$ statistic was calculated to be $-2.40$.

## 8.5.3 Exercises—Learning It!

8. 1   The School Committee members of a midsize New England city agreed that a strict discipline code has caused an increase in the number of student suspensions. The number of suspensions for a sample of schools in this city for the period September 1992–February 1993 are shown below:

| | | | |
|---|---|---|---|
| Central | 245 | Putnam | 1024 |
| MCDI | 1 | Kiley | 56 |
| Chestnut | 65 | Central Academy | 254 |
| Duggan | 133 | Commerce | 114 |
| Kennedy | 97 | Bridge | 7 |
| Forest Park | 149 | | |

The average number of suspensions for the previous year was 130.5 with a population standard deviation of 158.2.

(a) Set up the null and the alternative hypotheses to test if the average number of suspensions has changed.

(b) Test your hypothesis using $\alpha = 0.05$.

(c) Find the $p$ value.

(d) Display the data to see if it is reasonable to assume that the underlying population distribution is normal.

(e) Based on the $p$ value, what can you conclude about the average number of suspensions?

**8.2**   The Educational Testing Service (ETS) designs and administers the SAT exams. Recently the format of the exam changed and the claim has been made that the new exam can be completed in an average time of 120 minutes. A sample of 50 new exam times yielded an average time of 115 min. The standard deviation is assumed to be 2 minutes.

(a) Set up the null and the alternative hypotheses to test if average time to complete the exam has changed from 120 minutes.

(b) Test your hypothesis using $\alpha = 0.05$.

(c) Find the $p$ value.

(d) Based on the $p$ value, what can you conclude about the average time to complete the new exam?

**8.3**   A vending machine that dispenses coffee into cups must fill the cups with 7.8 oz of liquid. Before selling the vending machine to a college or business, the company tests the machine to be sure it is dispensing an average amount of 7.8 oz of coffee. A sample of 20 amounts is shown below. The amount of coffee dispensed is assumed to be normally distributed with a standard deviation of 0.05 oz.

**Ounces Dispensed**

| | |
|------|------|
| 7.78 | 7.79 |
| 7.82 | 7.82 |
| 7.87 | 7.84 |
| 7.80 | 7.82 |
| 7.80 | 7.78 |
| 7.83 | 7.75 |
| 7.85 | 7.83 |
| 7.84 | 7.73 |
| 7.82 | 7.87 |
| 7.81 | 7.88 |

(a) Set up the null and the alternative hypotheses to test if average amount of coffee dispensed is different from 7.8 ounces.

(b) Test your hypothesis using $\alpha = 0.05$.

(c) Find the $p$ value.

(d) Based on the $p$ value, what can you conclude about the average amount of fluid dispensed by the machine?

**8.4**   According the data released by the United States Bureau of Labor, the average price of a half-gallon of ice cream in 1996 was $2.86. The monthly averages for the price of a half-gallon of ice cream at 12 stores in your town in 1996 are shown below.

**Price for Half-Gallon of Ice Cream**

| | | | | | | | | | | | |
|------|------|------|------|------|------|------|------|------|------|------|------|
| $2.87 | $2.93 | $3.13 | $2.92 | $2.99 | $3.01 | $3.02 | $3.01 | $2.93 | $3.08 | $2.67 | $2.89 |

Assume that the ice cream prices are normally distributed with a standard deviation of $0.10.

(a) Set up the null and the alternative hypotheses to test if the average price of a half-gallon of ice cream in your town is different from the 1996 national average.

(b) Test your hypothesis using $\alpha = 0.05$.

(c) Find the $p$ value.

(d) What can you conclude about the average price of a half-gallon of ice cream in your town?

**8.5** Your manager has just read an article stating that, on the average, employees spend one hour a day playing video games at work. She would like you to see if her employees are really spending this amount of work time playing games. A random sample of 35 employees played games an average of 55 minutes per day with a standard deviation of 5 minutes. You report this to your manager and she is relieved to learn that her employees are playing less than the national average.

You quickly remind her that she has information on a sample of employees and suggest that she test the hypothesis that the population mean is 60 minutes. She is impressed with your knowledge of statistics and asks you to do the hypothesis test and report back to her.

(a) Set up the null and the alternative hypotheses to test if the average amount of time spent playing games at work is different from 60 minutes.

(b) Test your hypothesis using $\alpha = 0.05$.

(c) Find the $p$ value.

(d) Based on the $p$ value, are the employees at this company any different than the average employee nationwide?

**8.6** A cereal manufacturer is concerned that the boxes of cereal not be underfilled or overfilled. Each box of cereal is supposed to contain 13 oz of cereal. A test is done to see if the machines are really putting an average of 13 oz of cereal into the boxes. A sample of 30 boxes is tested. The average weight is 12.58 oz and the standard deviation is 0.25 oz.

(a) Set up the null and the alternative hypotheses to test if average number of ounces in the cereal boxes is different from 12.58 oz.

(b) Test your hypothesis using $\alpha = 0.10$.

(c) Find the $p$ value.

(d) Based on the $p$ value, what can you conclude about the average number of ounces in the cereal boxes?

# 8.6 WHICH THEORY SHOULD GO INTO THE NULL HYPOTHESIS?

In Section 8.4.1 we said that the first step is to construct two opposing views. We called one of them the null hypothesis and one of them the alternative hypothesis. Because of the way the hypothesis testing procedure works, it is important to carefully consider which of the views you are going to call the null hypothesis and which one you will call the alternative hypothesis.

Several different approaches may be taken to determine how the null and alternative hypotheses should be set up. These different approaches are the subject of the next two subsections. The first subsection introduces the concept of two-tail and one-tail tests, and the second subsection illustrates the conservative nature of the hypothesis testing procedure. Depending on the specifics of the problem you are trying to solve, one of these approaches will shed some light on resolving the question, Which theory should go into the null hypothesis?

## 8.6.1 Two-Tail Tests and One-Tail Tests

Much of our discussion so far has been about null and alternative hypotheses, which are called **two-tail tests.** The format is familiar to you by now and is shown in the following definition.

> A *two-tail test* of the population mean has the following null and alternative hypotheses:
>
> $$H_0: \quad \mu = \text{[a specific number]}$$
> $$H_A: \quad \mu \neq \text{[a specific number]}$$

The null hypothesis of a two-tail test claims that the mean (or whatever parameter you are testing) is actually *equal* to the particular number stated. The opposing view is clearly that the mean is *not equal* to that particular value, and this is the alternative hypothesis.

*Remember: You could be testing a variance or a proportion instead.*

When you use a two-tail test you are interested in seeing if the true mean is *different* from the number specified. You wish to know if the true mean is *higher than* the number or *lower than* the number. In other words, you want to test for deviations from the number in either direction—on the high side or the low side.

If you are doing a two-tail test, there is no decision to be made about how the null and the alternative hypotheses should be set up. The hypothesis testing procedure requires that the view with the equal sign be in the null hypothesis.

### EXAMPLE 8.19  The Tissue Strength Problem

***Setup as a Two-Tail Test of the Mean***

For the tissue strength example, we could set up the following two-tail test:

$$H_0: \quad \mu = 1000 \text{ lb/ream}$$
$$H_A: \quad \mu \neq 1000 \text{ lb/ream}$$

Here the specific number is 1000.  ∎

Both the soda bottle-filling problem and the potato chip-packaging problem discussed in the previous section are also two-tail tests. In each of these cases the view that the true mean, $\mu$, was equal to 32 oz (for the soda) or 10 oz (for the chips) was the null hypothesis.

Consider another example.

### EXAMPLE 8.20  Testing Supplies

***Illustrating the Setup of a Two-Tail Test of a Proportion***

Suppose you are purchasing 3.5-in. disks from Disks Are Us. The company claims that only 1% of the disks it manufactures are defective. As the purchasing agent you wish to check this claim. The null and alternative hypotheses are shown below:

$$H_0: \quad \pi = 0.01$$
$$H_A: \quad \pi \neq 0.01$$  ∎

Notice that, in this example, we used a test on proportions, since we were interested in the percentage of the population of disks that were defective. The quality status of the defective can be considered a nominal variable.

 ***TRY IT NOW!***

### The Chapperel Steel Company
***Setting Up the Null and Alternative Hypotheses for a Two-Tail Test of the Mean***

Another recent management approach is to have employees become actual partners of the business. Chapperel Steel Company has done exactly this and the company feels that one of the benefits of this concept is that the average number of sick days will decrease. Prior to implementing this program, Chapperel had an average of 7.2 sick days per employee.

*(continued)*

Set up the null and alternative hypotheses to test if the average number of sick days per employee is different from 7.2.

---

Sometimes you really wish to see only if the population mean (or whatever parameter you are testing) is lower than the stated value. In this case, you are interested only in testing if the true mean, proportion, or variance is *less than* some number. Then you should use what is called a **lower-tail test.** A lower-tail test is one of two types of one-tail tests.

> A *lower-tail test* of a population mean has the following null and alternative hypotheses:
>
> $$H_0: \quad \mu \geq [\text{a specific number}]$$
> $$H_A: \quad \mu < [\text{a specific number}]$$

As with the two-tail test you could just as easily be testing a proportion, a variance, or the difference between two means or proportions.

### EXAMPLE 8.21   The Tissue Strength Problem

***Set Up as a Lower-Tail Test***

Suppose the manufacturer of the tissues is worried only about the tissue strength being less than the 1000-lb/ream specified value. Then the test would be set up as follows:

$$H_0: \quad \mu \geq 1000 \text{ lb/ream}$$
$$H_A: \quad \mu < 1000 \text{ lb/ream} \qquad \blacksquare$$

Notice that the viewpoint that the manufacturer is interested in testing became the alternative hypothesis. You should also notice that the inequality sign with the equal sign attached to it (in this case $\geq$) became part of the null hypothesis.

### EXAMPLE 8.22   The Chapperel Steel Company

***Illustrating a Lower-Tail Hypothesis Test***

Reconsider the Chapperel Steel Company example. The company is particularly interested in seeing if their partner idea has in fact reduced the average number of sick days from 7.2 per year. It would make more sense to set this test up as a lower-tail test:

$$H_0: \quad \mu \geq 7.2 \text{ days}$$
$$H_A: \quad \mu < 7.2 \text{ days} \qquad \blacksquare$$

You can see that we have placed the theory that management wishes to test into the alternative hypothesis and again the inequality with the equal sign (in this case $\geq$) is placed into the null hypothesis.

---

ANS. $H_0: \ \mu = 7.2$ DAYS; $H_A: \ \mu \neq 7.2$ DAYS

*TRY IT NOW!*

### The Bank Example

*Lower-Tail Test*

Suppose a bank knows that its customers are waiting in line an average of 10.2 minutes during the lunch hour. The branch manager has decided to add an additional teller during the 12–2 P.M. period and wishes to test the hypothesis that the average wait has decreased due to the additional teller. Set up the null and alternative hypotheses for the bank manager.

You may also wish to see if the true mean is greater than the stated value. In this case, you are interested only in testing if the true mean is *greater than* some number. This is also called an **upper-tail test.** An upper-tail test is the other kind of one-tail test. In general, such a test would look like this:

> An *upper-tail test* of a population mean has the following null and alternative hypotheses:
>
> $$H_0: \quad \mu \leq \text{[a specific number]}$$
> $$H_A: \quad \mu > \text{[a specific number]}$$

### EXAMPLE 8.23   The Tissue Strength Problem

*Set Up as an Upper-Tail Test*

For example, suppose the manufacturer of the tissues is worried only about the tissue strength being greater than the 1000-lb/ream specified value. Then, the test would be set up as follows:

$$H_0: \quad \mu \leq 1000 \text{ lb/ream}$$
$$H_A: \quad \mu > 1000 \text{ lb/ream}$$ ∎

Here again, notice that the manufacturer's contention becomes the alternative hypothesis. The inequality with the equal sign (in this case ≤) is again placed into the null hypothesis.

### EXAMPLE 8.24   Quality Tip Corp.

*Set Up as an Upper-Tail Test of Proportions*

Reconsider the cereal manufacturer from Example 8.3. One of the hypotheses that the company might be interested in studying was the hypothesis that a greater proportion of women (population 1) like the new packaging design than men (population 2). In this case we have two populations but it is still an upper-tail test. The appropriate null and alternative hypotheses are shown below:

$$H_0: \quad \pi_1 \leq \pi_2$$
$$H_A: \quad \pi_1 > \pi_2$$ ∎

In Example 8.24 we see again that the theory management wishes to test is placed into the alternative hypothesis and the inequality with the equal sign (in this case $\leq$) is again placed into the null hypothesis.

 **TRY IT NOW!**

### New Advertising Program
*Setting Up an Upper-Tail Test*

Suppose a company has implemented a new advertising program in the hopes of increasing sales from last year's annual average of $4.3 million. Set up the null and alternative hypotheses to test the theory that sales have increased.

### *Summary*

A Two-Tail Test

- Is used to test if the parameter has shifted away from a certain number in either direction, increased or decreased.
- Must always be set up so the "=" theory is the null hypothesis.
- Is used when the problem statement has the key words *changed* or *different* in the problem statement.

One-Tail Tests
 A Lower-Tail Test

- Is used to test if the parameter has shifted to a number less than a certain number.
- Must always be set up with the "=" as part of the null hypothesis.
- Is used when the problem statement has the key words *decreased, reduced,* or *less than.*
- The theory that you wish to "prove" is placed into the alternative hypothesis.

ANS. $H_0$: $\mu \leq \$4.3$ MILLION; $H_A$: $\mu > \$4.3$ MILLION

An Upper-Tail Test

- Is used to test if the parameter has shifted to a number more than a certain number.
- Must always be set up with the "=" as part of the null hypothesis.
- Is used when the problem statement has the key words *increased, greater than.*
- The theory that you wish to "prove" is placed into the alternative hypothesis.

## 8.6.2 What View Requires No Action?

The next approach is particularly useful if you have decided to use a one-tail test. This is because, as we saw in the preceding section, there is really no choice about how to set up the null and alternative hypotheses once you decide to use a two-tail test.

   This approach considers the question, What view requires that I take no action? Typically, this is the view that the population is behaving as it should be or as someone claims it should be. This view becomes the null hypothesis. Sometimes people call this the status quo.

### EXAMPLE 8.25   The Tissue Strength Problem

*Manufacturing Specification Becomes the Null Hypothesis*

For the tissue strength problem that we have been examining, the view that requires no action (or adjustments) is to believe that the manufacturing specification is being met. In this case, the specification is that the mean *MDStrength* is greater than or equal to 1000 lb/ream. This becomes our null hypothesis. Thus, $H_0$: $\mu \geq 1000$ lb/ream. Once we have the null hypothesis, it is easy to construct the alternative hypothesis since it has to cover all the other cases. So we have $H_A$: $\mu < 1000$ lb/ream. This hypothesis test is shown below:

$$H_0: \quad \mu \geq 1000 \text{ lb/ream}$$
$$H_A: \quad \mu < 1000 \text{ lb/ream}$$ ■

   Now you should ask, why do it this way? The reason has to do with the fact that the hypothesis testing procedure is a conservative procedure. As such it behaves like a conservative person. A conservative person will take action only when he/she is very sure that action needs to be taken.

   One way to help you understand what this means is to think about the belief in Santa Claus. Many young children believe in Santa Claus until they are about 7 or 8 years old. In this case, the null hypothesis would be $H_0$: Santa Claus exists and the alternative hypothesis would be $H_A$: Santa Claus does not exist. Children start off believing in Santa Claus (it is never proven to them) and they continue to believe that Santa Claus exists until there is overwhelming evidence to the contrary.

   In a similar manner, the hypothesis testing procedure will tell us to believe the null hypothesis unless the evidence in the sample data overwhelmingly contradicts the null hypothesis. In other words, the status quo, or the view that implies that no action be taken, should be placed in the null hypothesis. It will be assumed to be correct until the data in the sample are *really* incompatible with it.

   This approach is just like our judicial system. A person is assumed innocent until the evidence is so strong that it is impossible to continue to believe that the person is innocent. Remember that the instructions to the jurors are always to prove "beyond a reasonable doubt" that the person is guilty. This means that even if most of the evidence indicates that the person is guilty, if there is still some reasonable doubt, the jurors must find the person not guilty.

# TRY IT NOW!

## Judicial System

### *Setting Up the Null and Alternative Hypotheses*

If you think about the judicial system in terms of a hypothesis test, how would you set up the null and the alternative hypotheses?

Consider another example.

### EXAMPLE 8.26   The Soda Machine Problem

### *Demonstrating the Approach, Which View Requires*
### *That No Action Be Taken? for a Two-Tail Test*

A soda machine should dispense, on average, 8 oz of liquid into cups. Having purchased soda from this machine in the past, you recall that sometimes you get a small amount of soda and sometimes the cup overflows. Using the approach described in this section, you would ask, What view requires that no action be taken? Since the view that requires us to take no action is that the true mean fill is 8 oz, you should put this statement in the null hypothesis. Thus, you would get

$$H_0: \quad \mu = 8 \text{ oz}$$
$$H_A: \quad \mu \neq 8 \text{ oz}$$
■

# TRY IT NOW!

## VCR Manufacturer

### *Setting Up the Hypotheses so the Status Quo Is in the Null Hypothesis*

Suppose a manufacturer of VCRs claims that the average life of his VCRs is at least three years. You have a VCR made by this company and have had problems with it, and so you question this claim. Set up the hypothesis test to investigate the manufacturer's claim.

ANS. $H_0: \quad \mu \geq 3; H_A: \quad \mu < 3$

ANS. $H_0:$ PERSON IS INNOCENT; $H_A:$ PERSON IS GUILTY

## 8.6.3 Exercises—Learning It!

**8.7** Administrators at a small college are concerned that part-time evening students may not be familiar with all the services of the College. They wish to offer an orientation program to these students, but recognize that most of the part-time students work during the day and are generally very busy. They do not want to prepare an elaborate presentation if only a handful of part-time students will attend. They will conduct the orientation if more than 25% of the part-time students are interested in attending. Set up the null and alternative hypotheses to be used to decide if the administrators should offer the session.

**8.8** A company is thinking about setting up an on-site day-care program for its employees. The CEO has stated that she will do so only if more than 80% of the employees favor such a decision. Set up the null and alternative hypotheses to be tested.

**8.9** A human resource manager feels that men use e-mail less than women. Set up the null and alternative hypotheses that should be tested to determine if the human resource manager is correct.

**8.10** In an attempt to improve quality many manufacturers are developing partnerships with their suppliers. A local fast-food burger outfit has partnered with its supplier of potatoes. The burger outfit buys potatoes in bags that weigh 20 lb. It wishes to set up the null and alternative hypotheses to test if the bags do weigh on the average 20 lb.

**8.11** You are a connoisseur of chocolate chip cookies and you do not think that Nabisco's claim that every bag of Chips Ahoy cookies has 1000 chocolate morsels is correct. Set up the null and alternative hypotheses to test this claim.

**8.12** Anti-lock brake systems (ABS) have been hailed as a revolutionary safety feature. A study by the National Highway Traffic Safety Administration looked at fatal accidents. The claim is that cars with ABS are in fewer fatal crashes than those without. Set up the null and alternative hypotheses to test this claim.

**8.13** A college placement office wonders if the average entry level salary is different this year than last year's value of $25,000. Set up the null and alternative hypotheses to test this question.

**8.14** The college placement office in Exercise 8.13 also wonders if there is a difference between the average salary of engineering graduates and business school graduates. Set up the null and alternative hypotheses to see if these averages are different. *Requires Exercise 8.13*

**8.15** Your new television has a 1-year warranty. You are given the option to buy a 3-year warranty and wonder if it is worth it. You wish to test the hypothesis that the average time before a problem occurs is more than 3 years. Set up the null and alternative hypotheses to test this belief.

**8.16** It seems like you spend more money on groceries during the summer months when you eat more ice cream and drink more fluids. You know that you spend an average of $25/week on groceries during the winter months. Set up the null and alternative hypotheses to decide if, on the average, you spend more than this amount per week during the summer.

**8.17** M&M/Mars claims that at least 20% of the M&M's in each package are the new blue color. Set up the null and alternative hypotheses to test this claim.

**8.18** The computer center at a university claims that the average amount of time that students spend on-line has increased from last year's average of 1 hour per day. Set up the null and alternative hypotheses to test this claim.

# 8.7   ONE-TAIL TESTS OF THE MEAN: LARGE SAMPLE

In Section 8.5 you learned how to do a two-tail hypothesis test of the mean. In this section, we adapt that procedure to do a one-tail hypothesis test of the mean, $\mu$. The procedure applies when you know the standard deviation of the population or if you have a sufficiently large sample. In the latter case, the only change in the procedure is to use the sample standard deviation, $s$, instead of $\sigma$. Therefore, we again use the label of "large sample" to describe these tests.

## 8.7.1 Lower-Tail Tests of the Mean

In Section 8.6.1 we saw that often we are interested in whether the mean has shifted in *one* direction. Reconsider the tissue strength problem. Most likely the company is interested in checking to see if the true average *MDStrength* has decreased from the specification of 1000 lb/ream. Thus, a lower-tail test would make more sense in this case. Let's use this example to illustrate the procedure for a one-tail hypothesis test of $\mu$ when the population standard deviation is known.

Fortunately, the steps for completing a one-tail test of the mean when the population standard deviation is known are quite similar to those we used in the previous section. In fact, the only thing that will change is the procedure for finding the rejection region in step 2. Recall that we have been using the following 5-step hypothesis testing procedure:

*Steps for any hypothesis test*

**Step 1:**   *Set up the null and alternative hypotheses.*

**Step 2:**   *Pick the value of $\alpha$ and find the rejection region.*

**Step 3:**   *Calculate the test statistic.*

**Step 4:**   *Decide whether to reject the null hypothesis.*

**Step 5:**   *Interpret the statistical decision in terms of the stated problem.*

Reconsider the problem that the tissue company faces as a one-tail test. The first step is to set up the null and alternative hypotheses.

### EXAMPLE 8.27   The Tissue Company

*Step 1: Setting Up Null and Alternative Hypotheses*

Since the tissue manufacturer is checking to see if the true mean is *less than* 1000 lb/ream, this must go into the alternative hypothesis. So, we have the following setup.

$$H_0: \quad \mu \geq 1000 \text{ lb/ream}$$
$$H_A: \quad \mu < 1000 \text{ lb/ream} \qquad \blacksquare$$

We called this type of test a *lower-tail test.* The fact that the $<$ sign is in the alternative hypothesis makes it a lower-tail test. Next, you will see why it is called a lower-tail test as we find the rejection region.

*Again, when we talk about "a lot lower" we must consider the difference in terms of the size of the standard error.*

The second step is to select a value for $\alpha$ and find the rejection region. Again, let's think about what size $Z$ values should lead us to reject the null hypothesis. We are interested in detecting if the true mean has decreased below 1000 lb/ream. So, values of $\overline{X}$ that are much lower than 1000 would lead us to be suspicious of the null hypothesis. Following the same logic that we used in the previous section, we can see that if $\overline{X}$ is a lot lower than 1000 then $\overline{X} - \mu$ will be negative and $Z$ will be a negative number. Clearly, large negative $Z$ values will then lead us to reject the null hypothesis.

How about large positive $Z$ values? To get a positive $Z$ value, $\overline{X}$ would have to be bigger than 1000. If this is the case, would we ever want to reject the null hypothesis? No! An $\overline{X}$ value of say, 1150, would certainly give us a large positive $Z$ value but a sample mean of 1150 is consistent with the null hypothesis. Remember that we reject the null hypothesis only when the sample evidence is inconsistent with the null hypothesis. Thus, any $\overline{X}$ value larger than 1000 would lead us to continue to believe the null hypothesis. In fact, we wouldn't even have to do the hypothesis test in such cases!

So we have decided that only large negative $Z$ values will lead us to reject the null hypothesis. We need only decide how large in the negative direction $Z$ has to be before we reject $H_0$. Keep in mind that we arrived at this rejection region by simple logic! The particular value of the cutoff point for $Z$ is again completely determined by the value of $\alpha$.

Combining the idea that we will reject $H_0$ when the $Z$ value is too large in the negative direction with the requirement that we incorrectly reject $H_0$ only a certain percentage ($\alpha$) of the time, we arrive at the following picture:

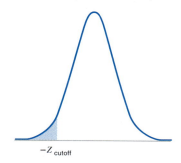

$-Z_{\text{cutoff}}$

*Remember: The rejection region consists of those values of Z that lead you to reject $H_0$.*

The shaded region is the rejection region. The rejection region is the *lower tail* of the normal distribution and hence the name *lower-tail test*. Remember that the rejection region consists of those values of $Z$, the test statistic, that lead you to reject the null hypothesis. Thus, a $Z$ value smaller than $-Z_{\text{cutoff}}$ (or a large negative $Z$ value) will lead you to reject the null hypothesis.

By using the standard normal table we can find this negative cutoff $Z$ value so that the probability that we are in the rejection region is $\alpha$. The rejection region for the tissue company can be found using this procedure.

### EXAMPLE 8.28   The Tissue Company

**Step 2: Selecting $\alpha$ and Finding the Rejection Region**

Suppose $\alpha$ is set at 0.05. In this case, the shaded area must be 0.05. Since we are specifying the probability of 0.05, we look for 0.05 in the body of the standard $Z$ table and see that the closest values are 0.0495 and 0.0505 corresponding to $Z$ values of $-1.65$ and $-1.64$, respectively. It is OK to use either of these values or the average of these two numbers, which would be $-1.645$. Thus, we will reject $H_0$ if $Z$ is smaller than $-1.645$ as shown in the shaded area below:

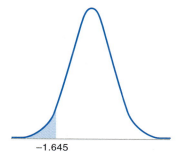

$-1.645$

You may remember that for the two-tail test when we set $\alpha = 0.05$, we found the rejection region to be

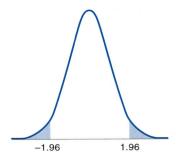

-1.96        1.96

Can you explain why the negative $Z_{cutoff}$ is different for the two-tail test? Remember that for the two-tail test, we split the value of $\alpha = 0.05$ in half before looking it up in the table. Thus, the cutoff value of $-1.96$ has an area of only $0.025$ compared to the full value of $0.05$ we used to get the cutoff value of $-1.645$.

 **TRY IT NOW!**

### Finding the Rejection Region

Try to predict what will happen to the $-Z_{cutoff}$ value for the one-tail test if $\alpha = 0.025$. Now find it to confirm your guess.

In general, the procedure for finding the rejection region for a lower-tail test is as follows:

*Steps for finding the rejection region for a lower-tail test*

**Step 1:**   *Look up the value of $\alpha$ in the body of the Z table (find the one closest to it).*

**Step 2:**   *Read off the corresponding value: call it $-Z_{cutoff}$.*

**Step 3:**   *The rejection region consists of all values of Z that are less than $-Z_{cutoff}$ as shown below:*

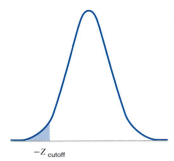

$-Z_{cutoff}$

The third step is to capture the information in the sample into a single number called the test statistic. It is the same calculation done in the previous section and is repeated here.

## EXAMPLE 8.29   The Tissue Company

### Step 3: Calculating the Test Statistic

Recall that the population standard deviation, $\sigma$, for the tissue manufacturer is 50 lb/ream and that the sample mean, $\overline{X}$, was observed to be 980 lb/ream. The $Z$ statistic is then

$$Z = \frac{\overline{X} - \mu}{\sigma/\sqrt{n}}$$

$$Z = \frac{980 - 1000}{50/\sqrt{36}} = \frac{-20}{8.333} = -2.4 \qquad \blacksquare$$

Now we can finish the tissue strength problem. At step 4 the value of the test statistic is examined to see if it falls in the rejection region. Finally, step 5 requires that you interpret the statistical decision in terms of the problem statement.

## EXAMPLE 8.30   The Tissue Company

### Step 4: Deciding Whether the Null Hypothesis Should Be Rejected

### Step 5: Interpreting the Statistical Decision

The rejection region for this one-tail, lower-tail alternative hypothesis test is shown below:

−1.645

Clearly the calculated $Z$ value of $-2.40$ falls in the rejection region. Thus, we reject $H_0$ and conclude that the true mean has decreased below 1000 lb/ream.

Remember that there is less than a 5% chance of observing a sample mean as low as 980 lb/ream if in fact the true mean really has not decreased below 1000! The precise calculation can be found based on the $p$ value.

Let's look at another example.   $\blacksquare$

## EXAMPLE 8.31   The Chapperel Steel Company

### Lower-Tail Hypothesis Test

In Example 8.22 we set up the hypothesis test for the company that had adopted the partnering approach with its employees. Remember that the company was interested in seeing if this approach did in fact reduce the average number of sick days per year from 7.2. Suppose a sample of 20 employees of this company used an average of 6.9 sick days the first year after the partner approach was implemented. National personnel data indicate that the standard deviation is 0.5 day.

Is there enough evidence to conclude that the partnering approach did decrease the average number of sick days? Use $\alpha = 0.05$.

Following the hypothesis test procedure, step 1 requires you to set up the null and alternative hypotheses. Since the company is trying to show that the average number of sick days has decreased, that must be the alternative hypothesis. So, we have the following null and alternative hypotheses:

$$H_0: \quad \mu \geq 7.2$$
$$H_A: \quad \mu < 7.2$$

Step 2 requires you to set $\alpha$ and find the rejection region. Lacking any specific direction from the company, set $\alpha = 0.05$. Since it is a one-tail, lower-tail alternative test, you look up 0.05 in the body of the standard normal table and find $-Z_{cutoff}$ to be $-1.645$. The rejection region is shown below:

−1.645

Step 3 requires the calculation of the $Z$ statistic:

$$Z = \frac{\overline{X} - \mu}{\sigma/\sqrt{n}} = \frac{6.9 - 7.2}{0.5/\sqrt{20}} = -2.68$$

Step 4 asks you to check to see if the value of the test statistic is in the rejection region. Clearly, $Z = -2.68$ is in the rejection region and we reject $H_0$ and conclude that $\mu < 7.2$.

Step 5 requires that you restate the statistical conclusion in terms of the problem. In this case, the data indicate that the average number of sick days used has been reduced. ∎

 **TRY IT NOW!**

## Frozen Foods

### *Lower-Tail Test of the Mean*

Jake Bramhall can identify the make, model, and number of cylinders of any passing car but he can't tell the difference between stewed tomatoes and tomato paste. While more men are pushing shopping carts these days, many like Mr. Bramhall show little aptitude in the supermarket and display markedly different purchasing behavior than women. A study done by Consumer Network Inc. shows that the average amount of money spent by 100 single men on facial tissues was $7.38. On the basis of these data can you conclude that men spend less money on facial tissues than the average $8.19 spent by women on facial tissues? Use a population standard deviation of $3.50 and an $\alpha$ value of 0.05.

Step 1:

Step 2:

Step 3:

Step 4:

Step 5:

Are the results different if you use $\alpha = 0.01$?

---

## 8.7.2 Upper-Tail Test of $\mu$

So far all of the one-tail tests that we have performed have been lower-tail tests. That is, the $<$ symbol was used in the alternative hypothesis. We must now consider what will be different in our procedure if we need to do an upper-tail test, that is, one with the $>$ symbol in the alternative hypothesis. Let's look at an example.

### EXAMPLE 8.32    Frozen Foods

**Step 1: Setting Up the Null and Alternative Hypotheses for an Upper-Tail Test**

In the same study of supermarket behavior that you just looked at in the *Try It Now!* exercise, the average amount of money spent of frozen dinner/entrees by men was found to be $51.48. There were 100 men in the study, and the population standard deviation can be assumed to be $10. Is there any evidence to indicate that the men spend more than the national average for women of $40.71?

The major steps of the hypothesis testing procedure are the same. Here again, the only difference is in the rules for finding the rejection region used in step 2.

Since we are interested in showing that the men spend *more* on frozen dinner/entrees, we know it is a one-tail test with the > symbol in the alternative hypothesis. That gives us the following null and alternative hypotheses:

$$H_0: \quad \mu \leq \$40.71$$
$$H_A: \quad \mu > \$40.71$$

That completes step 1.                                                                   ■

Step 2 asks us to select a value for $\alpha$ and find the rejection region. If you look back at the logic that led us to the rejection region for the lower-tail test, you will see that in this case it is values of $\overline{X}$ that are much larger than 40.71 that will lead us to reject $H_0$. These values of $\overline{X}$ will yield large positive $Z$ statistics. This observation coupled with what we have learned about the behavior of the normal distribution gives the following picture:

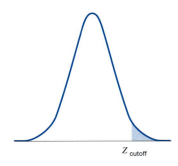

$Z_{\text{cutoff}}$

Here again the shaded region has an area of $\alpha$ and we will reject $H_0$ only when $Z$ is bigger than $Z_{\text{cutoff}}$, thus falling into the rejection region.

### EXAMPLE 8.33   Frozen Foods

#### Step 2: Selecting the Value for $\alpha$ and Finding the Rejection Region

Suppose we set $\alpha = 0.05$. In this case the shaded area must equal 0.05. Remember that the normal distribution table gives us only lower-tail areas as the probabilities. But, since we know that the distribution is symmetric we can look up 0.05 and simply drop the negative sign in order to get $Z_{\text{cutoff}}$. The value for $Z_{\text{cutoff}}$ is arrived at by dropping the negative sign in front of either $-1.64$ or $-1.65$. Hence, use either 1.64 or 1.65 or the midpoint, 1.645.

The rejection region is shown below:

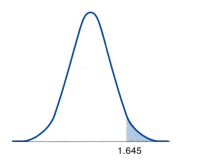

1.645                                    ■

This example has shown us the steps for the rejection region for an upper-tail test:

*Steps for finding the rejection region for an upper-tail test*

**Step 1:**   *Look up the value of $\alpha$ in the body of the Z table (find the one closest to it).*

**Step 2:** *Read off the corresponding value and drop the negative sign. Call it $Z_{cutoff}$.*

**Step 3:** *The rejection region consists of all values of $Z$ that are greater than $Z_{cutoff}$ as shown below:*

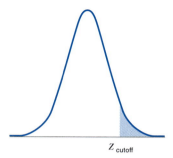

$Z_{cutoff}$

We are now ready to complete this example.

## EXAMPLE 8.34   Frozen Foods

*Steps 3,4, and 5*

For step 3 we calculate the $Z$ test statistic as follows:

$$Z = \frac{\overline{X} - \mu}{\sigma/\sqrt{n}} = \frac{51.48 - 40.71}{10/\sqrt{100}} = 10.77$$

Finishing this example, we do steps 4 and 5. Clearly, our $Z$ value of 10.77 falls in the rejection region and we should reject the null hypothesis and conclude that men spend more than $40.71 on average for frozen dinners/entrees!  ∎

# TRY IT NOW!

### Supermarket Survey
*Upper-Tail Test of the Mean*

From the same supermarket survey it is found that the 100 men spent, on the average, $19.98 on low-calorie soft drinks. Is there enough evidence to conclude that men spend more than women, who, on the average, spend $18.86?  Assume that the population standard deviation is $10 and use $\alpha = 0.05$.

## 8.7.3 Adjustments to $p$ Value Calculation for One-Tail Tests

In Section 8.5.2 we calculated the $p$ value for a two-tail hypothesis test of $\mu$ when $\sigma$ was known. The procedure for one-tail tests is the same except that you do not double the tail area probability.

ANS. $Z = 1.12$; NO

**EXAMPLE 8.35**   Frozen Foods

*Calculating the p Value*

Consider the frozen foods example (Example 8.34). This problem used an *upper-tail test of μ* with $\sigma$ known. A sample of 100 men spent $51.48 on the average on frozen foods. The $Z$ statistic was calculated to be 10.77. A quick look at this $Z$ statistic tells us that the $p$ value will be extremely small.

The $p$ value is calculated as follows:

$$p \text{ value} = P(Z > 10.77) = \text{virtually } 0$$    ■

## 8.7.4 Exercises—Learning It!

8.19   A major manufacturer of glue products thinks it has found a way to make the glue adhere longer than the current average of 90 days. The manufacturer wishes to see if the glue products made this way have an average time to failure greater than 90 days. A sample of 30 tubes of the new glue yields an average of 93 days before failing. The failure time is normally distributed with a standard deviation of 3 days.

(a) Set up the null and the alternative hypotheses to test if average time to failure is greater than 90 days.

(b) Test your hypothesis using $\alpha = 0.05$.

(c) Find the $p$ value.

(d) Based on the $p$ value, what can you conclude about the average time to failure for the new product?

8.20   Recent medical research indicates that skin cancer patients who receive a new medication for skin cancer live longer than those who do not. The average length of life prior to the development of this medication was 18 months. The medical community wishes to test the claim made by the developers of this drug. A sample of 35 patients who received the medication lived an average of 21 months. The standard deviation is 5 months.

(a) Set up the null and the alternative hypotheses to test if average length of life has increased from 18 months.

(b) Test your hypothesis using $\alpha = 0.05$.

(c) Find the $p$ value.

(d) Based on the $p$ value, what can you conclude about the average length of life for patients who receive the vaccine?

8.21   An automobile company thinks that with new designs, its cars will last longer before having a problem. For this reason, the company wishes to extend the warranty that comes with the vehicle in hopes of attracting more customers. Before making this change, the idea is tested. Prior to the design changes, the cars lasted on the average 43 months before having a major problem. A sample consisting of 50 cars was tested. The cars lasted an average of 44 months before having a major problem. The standard deviation is 2 months.

(a) Set up the null and the alternative hypotheses to test if average time before having a major problem is longer than 43 months.

(b) Test your hypothesis using $\alpha = 0.05$.

(c) Find the $p$ value.

(d) Based on the $p$ value, what can you conclude about the average time before having a major problem?

8.22   A manufacturer of top-of-the line tennis rackets claims that its Smack Em tennis racket will change a player's game. A tennis pro currently serves the ball at an average speed of 115 mph with a standard deviation of 2.5 mph. The speeds are normally distributed. The tennis pro decides to test the company's claim and records the speed of his serve for 15 balls using the Smack Em racket. The data are shown in the following table:

| Speed (mph) | |
| --- | --- |
| 117.3 | 115.9 |
| 115.1 | 115.2 |
| 116.0 | 115.0 |
| 116.2 | 113.0 |
| 112.9 | 120.8 |
| 115.4 | 116.9 |
| 113.8 | 114.4 |
| 114.2 | |

(a) Set up the null and the alternative hypotheses to test if the average service speed has increased using the new racket.

(b) Test your hypothesis using $\alpha = 0.05$.

(c) Find the $p$ value.

(d) Based on the $p$ value, should the tennis pro invest in the new racket?

8.23 In an attempt to improve quality, many manufacturers are developing partnerships with their suppliers. A local fast-food burger outfit has partnered with its supplier of potatoes. The burger outfit buys potatoes in bags that weigh 20 lb. It does not wish to accept underweight bags of potatoes. A sample of 40 bags shows an average weight of 19.95 lb with a standard deviation of 0.1 lb.

(a) Set up the null and the alternative hypotheses to test if the average bag weighs at least 20 lb.

(b) Test your hypothesis using $\alpha = 0.05$.

(c) Find the $p$ value.

(d) Based on the $p$ value, should the burger outfit accept the shipment of potatoes?

# 8.8 WHAT ERROR COULD YOU BE MAKING?

Since you will be deciding between the null and the alternative hypotheses using only sample information, there is always a chance that you will be wrong. That is, you may choose to believe the null hypothesis when, in fact, it is not true. Alternatively, you may choose to believe the alternative hypothesis (rejecting the null hypothesis) when in fact it is not true. These are the two ways that you could be wrong when you perform a hypothesis test.

In Chapter 6 you learned the basics of probability. You saw that probabilities are a way to measure the chance that something occurs. In hypothesis testing, we use probabilities to measure the chance of being wrong. If we can state that there is a 5% chance that we have made an error, then we have a sense of how often we will be wrong.

Let's look in more detail at the two ways that you could be wrong.

## 8.8.1 Two Types of Errors

Four outcomes could result from the decisions you make in each hypothesis test situation. Look at Table 8.2 (page 412) and think carefully about how each of these outcomes could occur.

As you can see from Table 8.2, there are two ways that you could make the right choice and two ways that you could make a mistake. Unfortunately, you will not know whether you made the right decision or you made an error. This is because to know which category you are in you would have to know what the population actually looked like and thus if $H_0$ or $H_A$ were true. If you knew that, you wouldn't be taking a sample!

**TABLE 8.2   Possible Outcomes in a Hypothesis Test**

| Sample Indicates You Should | In Actuality: | |
| --- | --- | --- |
| | $H_0$ is True | $H_A$ is True |
| Believe $H_0$ | You made the right choice—no error | Error |
| Believe $H_A$ | Error | You made the right choice—no error |

Let's look at these four possible outcomes from the perspective of the tissue manufacturer. They are presented in Table 8.3.

**TABLE 8.3   Four Possible Outcomes for the Tissue Strength Problem**

| Sample Indicates You Should | In Actuality: | |
| --- | --- | --- |
| | $H_0$ is True<br>$\mu = 1000$ | $H_A$ is True<br>$\mu \neq 1000$ |
| Believe $H_0$:<br>You believe that $\mu = 1000$ and so the product specs are being met. | You correctly decide that the product specs are being met. | You **incorrectly** decide that the product specs are being met when they are not. |
| Believe $H_A$:<br>You believe that $\mu \neq 1000$ and so the product specs are not being met. | You **incorrectly** decide that the product specs are not being met when they are. | You correctly decide that the product specs are not being met. |

To be clear which of the two possible errors we are talking about, we need to give them names. Look at Table 8.4, which shows the same four situations with each error named.

**TABLE 8.4   Name Those Errors: Type I and Type II**

| Sample Indicates You Should | In Actuality: | |
| --- | --- | --- |
| | $H_0$ is True | $H_A$ is True |
| Believe $H_0$ | You made the right choice—no error | Type II error |
| Believe $H_A$ | Type I error | You made the right choice—no error |

You can see that statisticians are uncreative at naming errors and they are simply called **Type I** and **Type II errors.** These errors are stated in the following definitions:

A *Type I error* is made when you reject the null hypothesis and the null hypothesis is actually true. In other words, you incorrectly reject a true null hypothesis.

A *Type II error* is made when you fail to reject the null hypothesis and the null hypothesis is false. In other words, you continue to believe a false null hypothesis.

**EXAMPLE 8.36**    **The Bottle-Filling Problem**

*Type I and Type II Errors*

Returning to the bottle-filling problem in Example 8.6, look at the Type I and Type II errors for this hypothesis test. We had set up the following null and alternative hypotheses:

$$H_0: \quad \mu = 32 \text{ oz}$$
$$H_A: \quad \mu \neq 32 \text{ oz}$$

Remember that a Type I error is made if you believe $H_A$ but $H_0$ is true. For this problem that would mean that you believe the true mean is not equal to 32 oz ($H_A$) and therefore you believe the bottles are not being filled properly. You adjust the bottle filling machine when in fact the true mean is really 32 oz ($H_0$) and the machine did not need to be adjusted.

   A Type II error is made if you believe $H_0$ is true when $H_A$ is true. In this case, you would believe that the soda machine is properly filling the bottles when, in fact, it is either overfilling them or underfilling them. So, you do not adjust a machine that needs to be adjusted.  ∎

 **TRY IT NOW!**

**The Potato Chip Manufacturer**

*Examining the Type I and Type II Errors*

Find the Type I and Type II errors for the hypothesis test that you set up for the potato chip manufacturer of this chapter.

## 8.8.2 Probability of Making an Error

Since we can't eliminate sampling error unless we sample the entire population, the next best thing we can do is try to control the chances that we are in error. The standard method of labeling the chance of making these errors is given in the following definition.

ANS. A TYPE I ERROR IS MADE IF YOU CONCLUDE THAT THE BAGS ARE NOT BEING FILLED PROPERLY WHEN THEY ARE BEING FILLED PROPERLY. A TYPE II ERROR IS MADE IF YOU CONCLUDE THAT THE BAGS ARE BEING FILLED TO 10 OUNCES BUT THEY ARE REALLY BEING OVER OR UNDER FILLED.

> The probability of making a Type I error is called $\alpha$ **(alpha)**.
>
> The probability of making a Type II error is called $\beta$ **(beta)**.

Clearly, $\alpha$ and $\beta$ must be numbers between 0 and 1 since they are probabilities. As the investigator, you will get to decide the value of $\alpha$. This means that you can specify the chances of making a Type I error to be anything you wish. Once you set $\alpha$, the value of $\beta$ is completely determined. Your first thought might be to set $\alpha$ to be as small as possible so there is hardly any chance of making a Type I error. Well, there is a price to pay for making $\alpha$ really small. As $\alpha$ gets smaller, the size of $\beta$ gets larger. So, just like most things in life, there is a trade-off. You can force the chance of making a Type I error to be really small but then you have to live with a greater chance of making a Type II error. This trade-off is pictured in Figure 8.3.

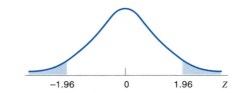

**FIGURE 8.3**   Two-sided rejection region with $\alpha = 0.05$

The area of the shaded region is equal to $\alpha$. This is the chance that we reject $H_0$ when we should not have done so. If we look at Figure 8.4 we can see that the value of $\alpha$ has been reduced. This means we will be less likely to reject $H_0$, when we should not reject it. If there is less chance of rejecting the null hypothesis, then there must be a greater chance of continuing to believe the null hypothesis when we should not. This is the definition of $\beta$.

**FIGURE 8.4**   Two-sided rejection region with $\alpha = 0.02$

Thus, as $\alpha \uparrow \beta \downarrow$ and as $\alpha \downarrow \beta \uparrow$

Let's examine this trade-off in terms of the tissue strength problem.

### EXAMPLE 8.37   The Tissue Strength Problem

***Possible Value for $\alpha$***

In terms of the tissue strengths, suppose you decide to set the chance of making a Type I error, $\alpha$, really small, say 0.01. This means that only 1% of the time will you incorrectly decide that the product specs are not being met, when really they are OK. This is the good news.

The trade-off is that the chance of making a Type II error increases. In this case, it means that the likelihood of deciding that the product specs are OK when they are not gets larger. How large the likelihood gets will depend on the real value of $\mu$.   ■

The standard value of $\alpha$ used to test hypotheses is 0.05. However, if making a Type I error is much more costly than making a Type II error, then you should consider testing at a value of $\alpha$ equal to 0.01. This will, of course, cause the value of $\beta$ to increase. If, on the other hand, you feel that making a Type II error is particularly costly, then you should consider testing at a value of $\alpha = 0.10$. If both errors are equally costly then you should probably use the standard value of $\alpha = 0.05$.

Consider another example.

### EXAMPLE 8.38   The Airplane Bolts Problem

***Considering the Costs of the Errors***

Suppose you are inspecting bolts that are used to fasten engines to airplanes. The mean strength of these bolts must be at least 100 pounds per square inch (psi). You need to decide if a shipment of these bolts is OK—that is, if they are strong enough. You realize that you cannot test them all because once you test a bolt it cannot be used in the airplane. This is what we call destructive sampling. If you test them all you will have none to use. You decide to take a sample of bolts, and on the basis of the mean strength of the sample you will decide if the shipment is OK.

One possible way to set up the hypothesis test is as follows:

$$H_0: \quad \mu \geq 100 \text{ psi} \qquad \text{(The shipment is OK.)}$$
$$H_A: \quad \mu < 100 \text{ psi} \qquad \text{(The shipment is not OK.)}$$

*Remember: A Type I error is that you believe the alternative hypothesis when, in fact, the null hypothesis is true.*

For this hypothesis test, what is a Type I error? In this case, you would believe that the shipment is not OK when in fact it is OK. So, you would send the bolts back for rework when they did not need to be reworked.

Now let's look at a Type II error for this problem. In this case, you would keep the bolts even though they are not strong enough.

Which of these errors is more costly?

*Recall: A Type II error is that you believe the null hypothesis when, in fact, the alternative hypothesis is true.*

• Send bolts back for rework when they did not need to be reworked

• Keep bolts that are not strong enough.

We would probably all agree that keeping bolts that are not strong enough is the more costly error (in terms of human life). This is a Type II error. Thus, we should try to minimize the value of $\beta$ by testing at $\alpha$ equal to 0.10. ∎

## TRY IT NOW!

### New Package Design:

***Setting the Value of $\alpha$***

We have seen that a one-sided test is often used to investigate whether a new method of advertising or producing something is better than the existing method. Consider a company that is trying a new package design for its product. The average sales for this product are currently $1500/month. The null and alternative hypotheses would be

$$H_0: \quad \mu \leq \$1500$$
$$H_A: \quad \mu > \$1500$$

*(continued)*

In terms of the company's decision to adopt or not adopt this new design, what are the Type I and Type II errors?

What value of $\alpha$ would you suggest be used to conduct the test?

## 8.8.3 Exercises—Learning It!

**8.24** Administrators at a small college are concerned that part-time evening students may not be familiar with all the services of the College. They wish to offer an orientation program to these students but recognize that most of the part-time students work during the day and are generally very busy. The administrators do not want to prepare an elaborate presentation if only a handful of part-time students will attend. Hence, they will conduct the orientation if more than 25% of the part-time students are interested in attending.

(a) State the consequences of a Type I error.

(b) State the consequence of a Type II error.

(c) Suggest a value for $\alpha$, and justify your choice.

**8.25** A company is thinking about setting up an on-site day-care program for its employees. The CEO has stated that she will do so only if more than 80% of the employees favor such a decision.

(a) State the consequences of a Type I error.

(b) State the consequence of a Type II error.

(c) Suggest a value for $\alpha$, and justify your choice.

**8.26** In an attempt to improve quality, many manufacturers are developing partnerships with their suppliers. A local fast-food burger outfit has partnered with its supplier of potatoes. The burger outfit buys potatoes in bags that weigh 20 lb. It does not wish to accept underweight bags of potatoes.

(a) State the consequences of a Type I error.

(b) State the consequence of a Type II error.

(c) Suggest a value for $\alpha$, and justify your choice.

ANS. TYPE I ERROR IS TO CONCLUDE THE NEW DESIGN IS BETTER WHEN IT IS NOT. TYPE II ERROR IS TO CONCLUDE THE NEW DESIGN IS NOT BETTER WHEN IT IS. THESE ARE EQUALLY COSTLY ERRORS SO USE $\alpha = 0.05$.

**8.27** You are a connoisseur of chocolate chip cookies and you do not think that Nabisco's claim that every bag of Chips Ahoy cookies has 1000 chocolate morsels is correct.

(a) State the consequences of a Type I error.

(b) State the consequence of a Type II error.

(c) Suggest a value for $\alpha$, and justify your choice.

**8.28** Anti-lock brake systems (ABS) have been hailed as a revolutionary safety feature. A study by the National Highway Traffic Safety Administration looked at fatal accidents. The claim is that cars with ABS are in fewer fatal crashes than those without.

*Requires
Exercise 8.12*

(a) State the consequences of a Type I error.

(b) State the consequence of a Type II error.

(c) Suggest a value for $\alpha$, and justify your choice.

**8.29** A College Placement Office wonders if there is a difference between the average salary of engineering graduates and business school graduates.

*Requires
Exercise 8.13*

(a) State the consequences of a Type I error.

(b) State the consequence of a Type II error.

(c) Suggest a value for $\alpha$, and justify your choice.

**8.30** Your new television has a 1-year warranty. You are given the option to buy a 3-year warranty and wonder if it is worth it. You wish to test the hypothesis that the average time before a problem occurs is more than 3 years.

(a) State the consequences of a Type I error.

(b) State the consequence of a Type II error.

(c) Suggest a value for $\alpha$, and justify your choice.

**8.31** M&M/Mars claims that at least 20% of the M&M's in each package are the new blue color.

(a) State the consequences of a Type I error.

(b) State the consequence of a Type II error.

(c) Suggest a value for $\alpha$, and justify your choice.

**8.32** A computer center is arguing for more computers in the lab for students at a mid-size college. The computer center at a university claims that the average amount of time that students spend on-line has increase from last year's average of 1 hour per day.

(a) State the consequences of a Type I error.

(b) State the consequence of a Type II error.

(c) Suggest a value for $\alpha$, and justify your choice.

# 8.9   HYPOTHESIS TESTING IN EXCEL

Hypothesis tests from real data are sometimes inconvenient to do using a pencil and paper. While Excel has many statistical analysis tools already available, it does not have a tool for large-sample hypothesis tests for the mean. In this section we will look at a macro that allows you to do large-sample hypothesis tests for the population mean.

The macro provides output for both two-sided and one-sided hypothesis tests. You will use whatever output is pertinent to your test. We will test the tissue strength data to see if the population mean is really 1000 lb/ream, which is a two-sided test. The macro assumes that you have the data located in an Excel worksheet. A portion of the data and the summary statistics are shown in Figure 8.5 (page 418).

| | A | B | C | D |
|---|---|---|---|---|
| 1 | Day | **MDStrength** | MDStrength | |
| 2 | 1 | 1006 | | |
| 3 | 1 | 994 | Mean | 990.8266667 |
| 4 | 1 | 1032 | Standard Error | 6.66764065 |
| 5 | 1 | 875 | Median | 991 |
| 6 | 1 | 1043 | Mode | 992 |
| 7 | 1 | 962 | Standard Deviation | 57.74346186 |
| 8 | 1 | 973 | Sample Variance | 3334.307387 |
| 9 | 1 | 1036 | Kurtosis | -0.180965938 |
| 10 | 1 | 1084 | Skewness | -0.264673668 |
| 11 | 1 | 1067 | Range | 261 |
| 12 | 1 | 939 | Minimum | 830 |
| 13 | 1 | 984 | Maximum | 1091 |
| 14 | 1 | 1001 | Sum | 74312 |
| 15 | 1 | 1076 | Count | 75 |

**FIGURE 8.5**   Tissue data and summary statistics

To perform a large sample hypothesis test for a mean, follow these steps:

1. Make sure that you have the MacDoIt.xls worksheet open and select **Tools** > **Macro** > **Macros** to obtain the list of macros. From the list, select **Z test** and click **Run.** A new worksheet is added to the current workbook and the table shown in Figure 8.6 appears. The table opens with values already in place. You will use Excel functions with your data to fill in the Input section of the table.

| | A | B | C |
|---|---|---|---|
| 1 | **Hypothesis Test for Mean - Z-test** | | |
| 2 | **Input** | | |
| 3 | **Sample Mean** | m | 990.8266667 |
| 4 | **Hypothesized Mean** | $\mu_0$ | 1000 |
| 5 | **Standard Deviation** | $\sigma$ | 57.74346186 |
| 6 | **Sample Size** | n | 75 |
| 7 | **Significance Level** | $\alpha$ | 0.05 |
| 8 | **Output** | | |
| 9 | **Standard Error** | StdErr | 6.6676 |
| 10 | **Z-Statistic** | z | -1.3758 |
| 11 | **Lower-Tail Test for** $H_A$: $\mu < \mu_0$ | | |
| 12 | **Lower-Tail Critical Z** | $Z_L$ | -1.6449 |
| 13 | **Lower-Tail p-value** | p-value$_L$ | 0.0844 |
| 14 | **Two-Tail Test for HA:** $\mu \neq \mu_0$ | | |
| 15 | **Two-Tail Absolute Critical Z** | $Z_T$ | 1.9600 |
| 16 | **Two-Tail p-value** | p-value$_T$ | 0.1689 |
| 17 | **Upper-Tail Test for HA:** $\mu > \mu_0$ | | |
| 18 | **Upper-Tail Critical Z** | $Z_U$ | 1.6449 |
| 19 | **Upper-Tail p-value** | p-value$_U$ | 0.9156 |

**FIGURE 8.6**   Table for Z test

*If you get a #DIV/0! error, you switched to the z test sheet before typing the last parenthesis. It might be easier to copy the data to the sheet with the table to avoid this problem.*

2. The first value that should be input is the sample mean. Position the cursor in the cell next to the one labeled m. You are going to type in the Excel function **AVERAGE,** which expects a data range as its input. Type **=average(** and switch to the worksheet that contains your data. Highlight the data and move the cursor to the status line of the worksheet that contains the data. Type in a right parenthesis ")" and hit Enter to finish the function. You will see the value of $\overline{X}$ appear in the table and the cursor will move to the next input cell.

3. Enter the hypothesized value for the population mean. In this case we will enter 1000 because that was the target value for *MDStrength*. Hit Enter to move the cursor to the next input cell.

4. Input the value of $\sigma$. Since you are doing a $Z$ test, you are expected to know the population standard deviation. If you do not, but you have a large sample, then you can use the sample standard deviation, $s$. To do this, use the Excel function **STDEV** in this cell. This function works exactly like the one for average.

5. Hit Enter and input the sample size—in this case, 75.

6. Hit Enter again and enter the level of significance for the test.

When you hit Enter the last time, the table will update and give the test output as shown in Figure 8.7.

| | A | B | C |
|---|---|---|---|
| 1 | **Hypothesis Test for Mean - Z-test** | | |
| 2 | **Input** | | |
| 3 | *Sample Mean* | m | 98 |
| 4 | *Hypothesized Mean* | $\mu_0$ | 100 |
| 5 | *Standard Deviation* | $\sigma$ | 10 |
| 6 | *Sample Size* | n | 30 |
| 7 | *Significance Level* | $\alpha$ | 0.05 |
| 8 | **Output** | | |
| 9 | *Standard Error* | StdErr | 1.8257 |
| 10 | *Z-Statistic* | z | -1.0954 |
| 11 | **Lower-Tail Test for $H_A: \mu < \mu_0$** | | |
| 12 | *Lower-Tail Critical Z* | $Z_L$ | -1.6449 |
| 13 | *Lower-Tail p-value* | $p\text{-value}_L$ | 0.1367 |
| 14 | **Two-Tail Test for HA: $\mu \neq \mu_0$** | | |
| 15 | *Two-Tail Absolute Critical Z* | $Z_T$ | 1.9600 |
| 16 | *Two-Tail p-value* | $p\text{-value}_T$ | 0.2733 |
| 17 | **Upper-Tail Test for HA: $\mu > \mu_0$** | | |
| 18 | *Upper-Tail Critical Z* | $Z_U$ | 1.6449 |
| 19 | *Upper-Tail p-value* | $p\text{-value}_U$ | 0.8633 |

**FIGURE 8.7**    $Z$ test table with input and output

The output section of the $Z$ test includes the standard error of the mean and the value of the $Z$ test statistic. In addition, it includes information for all three different ways that the null and alternative hypotheses can be set up. In this case you want to look at the section labeled Two-Tail Test. You see that the critical value for the test is $\pm 1.96$. Comparing the test statistic to this value, you would fail to reject $H_0$. The $p$ value for the test is also provided in the output.

# *CHAPTER 8 SUMMARY*

In this chapter you have learned the key steps involved in doing any hypothesis test. You first formulate two opposing viewpoints called the null and alternative hypotheses. These hypotheses are typically theories or ideas about the value of one or more population parameters. The technique of hypothesis testing helps you decide between these opposing hypotheses using the sample data as the evidence upon which to base your decision. In doing any hypothesis test there are two possible errors you can make. These are called Type I and Type II errors. The probabilities of making these errors are labeled $\alpha$ and $\beta$, respectively. It is desirable to make both of these probabilities small, but there is a tradeoff.

In this chapter you have also learned the procedure for doing large-sample tests of the mean. This procedure applies whenever you know the population standard deviation or you have a sufficiently large sample size, $n > 30$. Thus, the tests are called large-sample tests.

## Key Terms

| Term | Definition | Page Reference |
|---|---|---|
| **Hypothesis** | A **hypothesis** is an idea, an assumption, or a theory about the behavior of one or more variables in one or more populations. | 373 |
| **Hypothesis test** | A **hypothesis test** is a statistical procedure that involves formulating a hypothesis and using sample data to decide on the validity of the hypothesis. | 373 |
| **Null hypothesis** | The **null hypothesis** is a statement about the population(s). It is referred to as $H_0$. | 378 |
| **Alternative hypothesis** | The **alternative hypothesis** is a statement about the population(s) that is opposite to the null hypothesis. It is referred to as $H_A$. | 378 |
| **Rejection region** | The **rejection region** is the range of values of the test statistic that will lead you to reject the null hypothesis. | 381 |
| **Test statistic** | A **test statistic** is a number that captures the information in the sample. It will be used to choose between the null and alternative hypotheses. | 382 |
| **$p$ value** | The **$p$ value** is defined to be the smallest value of $\alpha$ for which you can reject $H_0$. This is also called the level of significance of the test. | 390 |
| **Two-tail hypothesis test** | A **two-tail test** of the population mean has the following null and alternative hypotheses: $H_0: \mu =$ [a specific number] $H_A: \mu \neq$ [a specific number] | 394 |

| Term | Definition | Page Reference |
|------|-----------|----------------|
| Lower-tail hypothesis test | A **lower-tail hypothesis test** of the population mean has the following null and alternative hypotheses: $H_0: \mu \geq$ [a specific number] $H_A: \mu <$ [a specific number] | 396 |
| Upper-tail hypothesis test | An **upper-tail hypothesis test** of the population mean has the following null and alternative hypotheses: $H_0: \mu \leq$ [a specific number] $H_A: \mu >$ [a specific number] | 397 |
| Type I error | A **Type I error** is made when you reject the null hypothesis and the null hypothesis is actually true. In other words, you incorrectly reject a true null hypothesis. | 412 |
| Type II error | A **Type II error** is made when you fail to reject a false null hypothesis. In other words, you fail to reject $H_0$ when you should have rejected it. | 413 |
| Alpha ($\alpha$) | The probability of making a Type I error is called **$\alpha$**. | 414 |
| Beta ($\beta$) | The probability of making a Type II error is called **$\beta$**. | 414 |

## Key Formulas

| Term | Formula | Page Reference |
|------|---------|----------------|
| $Z$ test statistic | $Z = \dfrac{\overline{X} - \mu}{\sigma/\sqrt{n}}$ | 385 |

# CHAPTER 8 EXERCISES

## Learning It!

**8.33** The manufacturer of an over-the-counter pain reliever claims that its product brings pain relief to headache sufferers in an average of 3.5 minutes. To be able to make this claim in its television advertisements, the manufacturer was required by a particular television network to present statistical evidence in support of the claim.

    (a) Is this a one-tail test or a two-tail test?

    (b) Set up the null and the alternative hypotheses.

    (c) What is a Type I error?

    (d) What is a Type II error?

    (e) A sample of 40 headache sufferers is used. They report that it took an average of 3.3 minutes to get some relief. If the standard deviation is 0.5 minute, perform the hypothesis test using $\alpha = 0.02$.

**8.34** Pharmaceutical companies spend millions of dollars annually on research and development of new drugs. After a new drug is formulated, the pharmaceutical company must subject it to lengthy and involved testing before receiving the necessary permission from the Food and Drug Administration (FDA) to market the drug. The pharmaceutical company must provide substantial evidence that a new drug is safe prior to receiving FDA approval, so that the FDA can confidently certify the safety of the drug.

    (a) Set up the null and alternative hypotheses.

    (b) What is a Type I error?

    (c) What is a Type II error?

**8.35** Suppose a quality manager for a catsup company is interested in testing whether the mean number of ounces of catsup per family-size bottle differs from the labeled amount of 20 oz.

    (a) Is this a one-tail or a two-tail test?

    (b) Set up the null and the alternative hypotheses.

(c) What is a Type I error?

(d) What is a Type II error?

(e) A sample of 30 catsup bottles is checked. The average number of ounces in the sampled bottles is 19.97 oz. If the standard deviation is 0.1 oz, perform the hypothesis test using $\alpha = 0.05$.

(f) Find the $p$ value. What conclusion can you draw about the mean number of ounces in catsup bottles?

8.36   The LEGO Group is an international company which makes the LEGO blocks that many of us have played with at some time. Many of its products require that the production process perform according to specifications.  One of the products is Little People and the diameter of the neck of each of the Little People must be 0.5 inch so that it can be attached to the head properly. LEGO is interested in testing to see if this process is performing according to specifications.

(a) Is this a one-tail test or a two-tail test?

(b) Set up the null and the alternative hypotheses.

(c) What is a Type I error?

(d) What is a Type II error?

(e) A sample of 30 blocks is tested. The average diameter of the sampled necks was 0.48 inch. If the standard deviation is .05 inch, perform the hypothesis test using $\alpha = 0.02$.

(f) Find the $p$ value and make a recommendation to LEGO.

8.37   Another concept from the TQM (total quality management) movement is the idea of building better relationships with your vendor.  In doing so the quality of the incoming raw material is improved. If you buy raw materials from me you may wish to sample some of the material I sell you to test the quality claims that I have made. Suppose I claim that the weight of the material I sold you was on the average 10 lb.  You wish to test this claim.

(a) Is this a one-tail test or a two-tail test?

(b) Set up the null and alternative hypotheses.

(c) What is a Type I error?

(d) What is a Type II error?

(e) A sample of 30 yields an average weight of 10.03 lb. If the standard deviation is 0.03 lb, perform the hypothesis test using $\alpha = 0.05$.

(f) Find the $p$ value and decide if you would like to continue to purchase material from one supplier.

8.38   An up-and-coming restaurant chain is trying to decide whether to locate in the town of Longmortgage. It will locate there only if the average number of days of the week that people eat out is three or more.  The company wishes to investigate the town of Longmortgage as a potential location.

(a) Is this a one-tail test or a two-tail test?

(b) Set up the null and the alternative hypotheses.

(c) What is a Type I error?

(d) What is a Type II error?

## Thinking About It!

*Requires Exercise 8.2*

8.39   Would it be possible to switch the null and the alternative hypotheses for the ETS problem? Explain why or why not.

*Requires Exercise 8.19*

8.40   For the glue manufacturer, which of the possible errors is the more costly error? Consider the consequences of making a Type I error and the consequences of making a Type II error.

*Requires Exercise 8.20*

8.41   Consider the case of the medical research described in Exercise 8.20.

(a) What position would you be taking if you had made the null hypothesis: $H_0: \mu \geq 18$?

(b) What factors other than the medication might influence the length of time a cancer patient lives?

**8.42** What are the implications of the car company setting the warranty too long? *Requires Exercise 8.21*

**8.43** Consider the cereal manufacturer. *Requires Exercise 8.6*
   (a) From the cereal manufacturer's perspective, what is the more costly error?
   (b) From the consumer's perspective, what is the more costly error?

**8.44** For the television network, what are the implications and consequences of a Type I error? A Type II error? *Requires Exercise 8.33*

**8.45** Rewrite the hypothesis test for the FDA problem by switching the null and alternative hypotheses. *Requires Exercise 8.34*
   (a) What is a Type I error for this new setup?
   (b) What is a Type II error for this new setup?
   (c) Compare your answers with those you found in Exercise 8.34. What has happened to the errors?
   (d) Which setup do you feel is better and why?

**8.46** If you took a sample of catsup bottles and observed a sample mean, $\overline{X} = 19.7$ oz, does this automatically mean that the bottle-filling machine is not working properly? Explain why or why not. *Requires Exercise 8.35*

**8.47** What are the implications for LEGO if the process that manufactures the necks for the Little People is not producing necks with the specified diameter? *Requires Exercise 8.36*

**8.48** As the receiver of raw material, explain why you would not want to test every component of raw material (i.e., why not test the population instead of just examining a sample). *Requires Exercise 8.37*

**8.49** From the perspective of the restaurant chain, what are the consequences of a Type I error? a Type II error? Which error is more costly? *Requires Exercise 8.38*

## Exercises—Doing It!

**8.50** Let us return to the tissue manufacturer presented at the beginning of this chapter. Recall that the manufacturer was concerned about customer complaints involving sheets tearing on removal. It decided to look at the manufacturing process to see how it compared to the product specification and to see if any changes needed to be made. Making changes to the manufacturing process is a big job and before proceeding the manufacturer would like to be a little more certain that the changes need to be made. A sample of 225 was taken from a single tissue machine. The samples were taken over three different days and the results are stored in the data file TISSUES.XXX. A portion of the data file is shown below: *Datafile: TISSUES.XXX*

| Day | MDStrength | CDStrength |
|-----|-----------|-----------|
| 1 | 1006 | 422 |
| 1 | 994 | 440 |
| 1 | 1032 | 423 |
| 1 | 875 | 435 |
| 1 | 1043 | 445 |

   The variable *Day* keeps track of the day on which the tissue was produced and ranges from 1 to 3.
   The variable *MDStrength* measures the machine-directional strength and is measured in lb/ream. The product specifications for *MDStrength* state that the measurement is normally distributed with $\mu = 1000$ lb/ream and $\sigma = 50$ lb/ream.
   The variable *CDStrength* measures the cross-directional strength and is measured in lb/ream. The product specifications for *CDStrength* state that the measurement is normally distributed with a mean of 450 lb/ream and a standard deviation of 25 lb/ream.
   After giving the matter some thought the engineers looking at the tissue manufacturing process wondered if the fact that the measurements were taken on three different days might bias the analysis. Since operating conditions are not always the

same on any given day, they wondered if the process was off target on all three days, or if it was off target on one or two days. They decided to check each day individually.

(a) Perform a hypothesis test at the 0.05 level of significance to determine whether *MD-Strength* was 1000 on each of the three days separately. Use the target standard deviation of 50 and write your results in the table below:

| Day | Z Statistic | Decision (reject $H_0$/fail to reject $H_0$) |
|---|---|---|
| 1 | | |
| 2 | | |
| 3 | | |
| All 3 days together | | |

(b) Using these results, do you think the company can assume that *MDStrength* is running on target? Why or why not?

(c) Using a 0.05 level of significance, do the sample data indicate that the variable *CD-Strength* was running according to target specifications on each of the three days? Do you conclude the same for the three-day period as a whole?

| Day | Z Statistic | Decision (reject $H_0$/fail to reject $H_0$) |
|---|---|---|
| 1 | | |
| 2 | | |
| 3 | | |
| All 3 days together | | |

(d) Using these results, do you think the company can assume that *CDStrength* is running on target? Why or why not?

(e) Since the consumer problem that prompted the study was sheets tearing on removal, the company decides that it is really interested in knowing only whether the average *CDStrength* is less than the target of 450. If it is not, the company will not make any adjustments. Redo the tests in part (c) as a one-tail test to reflect this change. What is the result?

(f) One of the other variables that needs to be considered is some measure of total strength, which is related to both the dispensing defects and another important tissue variable, softness. Often strength and softness are tradeoffs. Using both *MD-Strength* and *CDStrength,* the company can calculate the Geometric Mean Tensile (GMT) strength, which is equal to the square root of the product of the two variables. This variable is also subject to process specifications.

Unfortunately, the group is divided on the subject of GMT. The operation's specialists think that since the company has critical specifications for the measurement, they need to check if those specifications are being met. The machine operators say that since GMT is calculated from two other variables that have been tested, they do not need to perform a hypothesis test concerning these data.

(g) What do you think? Is it possible for both *MDStrength* and *CDStrength* to be running according to specs but yet have GMT out of specs?  Experiment with the data and see.

(h) Using the results of your analysis prepare a report to management about the current tissue manufacturing process. Make a recommendation on whether the process needs to be adjusted or whether it is running on target. Remember, if it is not running on target, the management will be considering making changes to the specifications to reduce customer complaints about dispensing. Address this point in your report.

**8.51** A manufacturer of electronic telecommunications equipment was receiving complaints from the field about low volume on long-distance connections. Aunt Sue in California couldn't hear Cousin Fred in Florida.

A string of amplifiers manufactured by the company was being used to boost the signal at various points along the way. The boosting ability of the amplifiers (the "gain") was naturally the prime suspect in the case.

The design of the amplifiers calls for a gain of 10 decibels (dB). This means that the output from the amplifier should be about 10 times stronger than the input signal. This amplification makes up for the natural fading of the signal over long-distance connections. Because it is difficult to make every amplifier with a gain of exactly 10 dB, the design allows amplifiers to be considered acceptable if the gain falls between 7.75 and 12.2 dB. These permissible minimum and maximum values are sometimes called the specification (or spec) limits. The average value is to be 10 dB. Because there are literally hundreds of amplifiers boosting the signal on a long connection, low-gain amplifiers should be balanced by high-gain amplifiers to give an acceptable volume level.

The quality improvement team investigating the "couldn't hear" condition arranged to test the gain of 120 amplifiers. The results of the tests are listed below and are found on the disk file named AMPLIF.xxx.

**Gain of 120 Tested Amplifiers**       *Filename: AMPLIF.XXX*

| | | | | | | | | | |
|---|---|---|---|---|---|---|---|---|---|
| 8.1 | 10.4 | 8.8 | 9.7 | 7.8 | 9.9 | 11.7 | 8.0 | 9.3 | 9.0 |
| 8.2 | 8.9 | 10.1 | 9.4 | 9.2 | 7.9 | 9.5 | 10.9 | 7.8 | 8.3 |
| 9.1 | 8.4 | 9.6 | 11.1 | 7.9 | 8.5 | 8.7 | 7.8 | 10.5 | 8.5 |
| 11.5 | 8.0 | 7.9 | 8.3 | 10.0 | 9.4 | 9.2 | 10.7 | 9.0 | 8.7 |
| 9.3 | 9.7 | 8.7 | 8.9 | 8.6 | 9.5 | 9.4 | 8.8 | 8.3 | 8.2 |
| 8.4 | 9.1 | 10.1 | 7.8 | 8.1 | 8.8 | 9.2 | 8.4 | 7.8 | 8.0 |
| 7.9 | 8.5 | 9.2 | 8.7 | 10.2 | 7.9 | 9.8 | 8.3 | 9.0 | 9.6 |
| 9.9 | 10.6 | 8.6 | 9.4 | 8.8 | 8.2 | 10.5 | 9.7 | 9.1 | 8.0 |
| 8.7 | 9.8 | 8.5 | 8.9 | 9.1 | 8.4 | 8.1 | 9.5 | 8.7 | 9.3 |
| 8.1 | 10.1 | 9.6 | 8.3 | 8.0 | 9.8 | 9.0 | 8.9 | 8.1 | 9.7 |
| 8.6 | 9.2 | 8.5 | 9.6 | 9.0 | 10.7 | 8.6 | 10.0 | 8.8 | 8.6 |
| 8.5 | 8.2 | 9.0 | 10.2 | 9.5 | 8.3 | 8.9 | 9.1 | 10.3 | 8.4 |

SOURCE: *Quality Progress,* September 1990

(a) How many of the amplifiers fell within the specification limits?

(b) Does this mean that the data were of little value?

(c) Generate a histogram of the data to get a better "picture" of the data. What shape does it have and what does that tell you about the problem?

(d) Calculate the standard summary statistics for these data: mean, median, trimmed mean, standard deviation, variance, and range.

(e) Although the amplifiers were designed to have 10-dB gain, very few of them actually had a measured gain of 10 dB. Furthermore, very few amplifiers had exactly the same gain. Speculate about what could be causing that variation.

(f) Test the hypothesis that the true average gain is 10 dB.

(g) Find a 95% confidence interval for the true average gain, $\mu$.

(h) Based on your analysis in parts (a)–(f), what are your recommendations to the manufacturer of the amplifiers?

# INFERENCES: ONE POPULATION

## THE HOSPITAL

The health-care industry is as concerned with quality as manufacturing companies. With the rising costs of health care, many hospitals are looking for ways to contain costs without sacrificing quality of care. Studies have shown that a simple way to cut down on the spread of bacteria is to be sure that all employees wash their hands adequately. The literature indicates that a minimum of 5 seconds is needed.

To be sure that the medical staff was washing their hands long enough, a large hospital observed a sample of employees washing their hands. The employees did not know they were being observed. In total, 28 employees were observed. The average hand-washing time was found to be 2.556 seconds with a sample standard deviation of 3.755 seconds. On the basis of these data, the hospital must decide if the employees should participate in a training session to learn about the consequences of inadequate hand washing.

A portion of the data set is shown below:

| Observation | Unit | Time 1 |
|---|---|---|
| 1 | CCU | 3 |
| 2 | CCU | 2 |
| 3 | CCU | 0 |
| 4 | CCU | 5 |
| 5 | CCU | 2 |
| 6 | CCU | 0 |

## 9.1   CHAPTER OBJECTIVES

The hospital described above does not know the standard deviation of hand-washing times and the sample size is small ($n \leq 30$). Therefore, the hypothesis testing procedure introduced in Chapter 8 cannot be used. Now it is time to examine what to do if the population standard deviation is unknown and the sample size is small. In addition to handling small-sample tests of the mean, we will also see how to perform a hypothesis test for other population parameters.

If we are analyzing a *quantitative variable,* such as time an employee washes his/her hands, then we know that it can be described by its distribution. The distribution, in turn, can be described by parameters. Thus, if we are constructing a theory or hypothesis about a *quantitative variable* it might be a statement about

- The mean value, $\mu$, of the variable in one population
- The amount of variability, $\sigma^2$, of the variable in one population

If the data we are analyzing are *nominal data,* such as whether or not an employee is satisfied with the job, the hypothesis might be a statement about

- The value of the proportion, $\pi$, of population members that have a certain characteristic (one of the categories of the nominal variable)

In this chapter we look at the numerical details of the test statistics and rejection regions for hypothesis tests about a single population parameter. We will be studying only one population. For example, we will look at the average sales of a particular product or the proportion of people who favor legalizing marijuana. In Chapter 10 we extend these results to compare the behavior of two populations. This might entail comparing the average sales of two products or comparing the views of men and women on the issue of legalizing marijuana.

*Remember: Both $\mu$ and $\pi$ describe the behavior of the population not the sample. They are typically unknown.*

This chapter covers the following material:

- Hypothesis Test of the Population Mean, $\mu$: Small Sample
- Hypothesis Test of the Population Variance, $\sigma^2$
- Hypothesis Test of the Population Proportion, $\pi$
- The Relationship Between Hypothesis Testing and Confidence Intervals

## 9.2   HYPOTHESIS TEST OF THE MEAN: SMALL SAMPLE

*Remember: To find the sample variance you use:*
$$s^2 = \frac{n \sum x^2 - (\sum x)^2}{n(n-1)}$$

In the previous chapter you learned to perform a large-sample hypothesis test of the mean. In our examples, we have seen that sometimes there is information about the population standard deviation from either the product specs or from some previous study. However, more often than not, the population standard deviation is not known. In this case the sample standard deviation, $s$, must be used to calculate an estimate of the unknown population standard deviation, $\sigma$. If the sample is sufficiently large, $n > 30$, then you can use the $Z$ test statistic from Chapter 8 to do a hypothesis test on the mean. However, very often you have a small sample.

Now let us think about the implications of having a small sample on our hypothesis test. The steps of any hypothesis test are listed below:

*Steps for any hypothesis test*

**Step 1:**  *Set up the null and alternative hypotheses.*

**Step 2:**  *Pick the value of $\alpha$ and find the rejection region.*

**Step 3:**   *Calculate the test statistic.*

**Step 4:**   *Decide whether or not to reject the null hypothesis.*

**Step 5:**   *Interpret the statistical decision in terms of the stated problem.*

Step 1 will not be affected since we are just setting up the null and the alternative hypotheses. If you guessed that steps 2 and 3 are affected then you were right. In Chapter 8, we calculated a $Z$ statistic as our test statistic in step 3. But the formula for $Z$ uses $\sigma$, which is now unknown, and our estimate is based on a small sample. Logic tells us simply to use $s$ in the formula instead of $\sigma$. However, when we do this the test statistic no longer follows a normal distribution. Instead, it has a *t distribution*.

You used the $t$ distribution in Chapter 7 when you learned how to find a confidence interval for the mean of a normally distributed population when $\sigma$ was unknown and the sample size was small. We will use our knowledge of the $t$ distribution to do a hypothesis test of $\mu$ for precisely this situation.

*Remember that a test statistic is a single number based on the sample which allows you to decide between the null and the alternative hypotheses.*

t *test statistic*

The test statistic is calculated as follows:

$$t = \frac{\overline{X} - \mu}{s/\sqrt{n}}$$

Notice that the calculation for the $t$ statistic is just the same as $Z$ with $\sigma$ replaced with $s$.

*Remember: The population must have a normal distribution in order to use the* t *statistic.*

### 9.2.1 Two-Tail Test of the Mean: Small Sample

Let's first look at two-tail tests of the mean when $\sigma$ is unknown.

### EXAMPLE 9.1   The Hospital

***Two-Tail Test of $\mu$: Small Sample***

The table shown below gives the hand-washing times (in seconds) for the 28 employees who were observed:

| | | | | | | | | | |
|---|---|---|---|---|---|---|---|---|---|
| 3 | 3 | 0 | 6 | 2 | 0 | 1 | 0 | 5 | 2 |
| 2 | 0 | 0 | 0 | 0 | 6 | 15 | 0 | 5 | 0 |
| 0 | 1 | 0 | 10 | 2 | 0 | 8 | 0 | | |

The hospital wishes to test if the population average washing time is different from 5 seconds. Use $\alpha = 0.05$.

**Step 1:**   *Set up the null and alternative hypotheses.* Since the company is interested in detecting shifts from the value of 5 seconds in either direction, we know that this is a two-tail test. So we have

$$H_0:\quad \mu = 5 \text{ s}$$
$$H_A:\quad \mu \neq 5 \text{ s}$$

**Step 2:**   *Select a value for $\alpha$ and find the rejection region.* The value of $\alpha$ has been specified as 0.05. From our work in the previous section, we know that we will reject $H_0$ if the calculated $t$ statistic is too large or too small.

We need to use the $t$ table to find the value of $t_{\text{cutoff}}$. Since this is a two-tail test we know that we must divide $\alpha$ in half. Thus, we want $0.05/2 = 0.025$ in the upper-tail area and 0.025 in the lower-tail area. Using the column labeled 0.025 and the row for $n - 1 = 28 - 1 = 27$ degrees of freedom, we find the $t_{\text{cutoff}}$ value to be 2.052.

Since the $t$ distribution is symmetric, the value for $-t_{cutoff}$ is $-2.052$ and the rejection region is shown as the shaded area below:

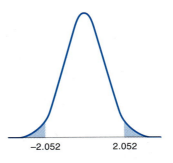

-2.052          2.052

*The $t_{cutoff}$ value will always be a bit larger than the corresponding $z_{cutoff}$ value.*

Do you remember what the value for the $z_{cutoff}$ was when $\alpha = 0.05$? It was 1.96. So you can see that the $t_{cutoff}$ value is just a bit larger than the corresponding $z_{cutoff}$ value. This will always be the case, and as the degrees of freedom get larger, the $t_{cutoff}$ value will get closer and closer to 1.96. Look down the column for $\alpha = 0.025$ and see that, as the degrees of freedom increase (due to larger sample sizes) the values for $t_{cutoff}$ get closer and closer to 1.96.

**Step 3:**  *Calculate the test statistic and the p value.* From the data we can calculate $\overline{X}$ and $s$ and find them to be $\overline{X} = 2.536$ s and $s = 3.687$ s and we know $n = 28$. With this information we can calculate the $t$ test statistic:

$$t = \frac{\overline{X} - \mu}{s/\sqrt{n}} = \frac{2.536 - 5}{3.687/\sqrt{28}} = -3.54$$

Find the $p$ value. To get this probability you will most likely need to use some software support such as Minitab or Excel. This is because the $t$ table is not designed to be used to find probabilities but rather to find the cutoff values for the rejection region.

$$p \text{ value} = 2P(t < -3.54) = (2)(0.00075) = 0.0015$$

The output shown below is from MINITAB. You can see the $t$ statistic labeled T. In addition, the $p$ value is calculated by Minitab and shown in the last column labeled P.

```
T-TEST OF THE MEAN
Test of mu = 5.000 vs mu not = 5.000

Variable    N    Mean    StDev    SE Mean       T         P
Hand was    28   2.536   3.687    0.697      -3.54    0.0015
```

**Step 4:**  Here we check to see if the $t$ statistic calculated at step 3 falls in the rejection region. Since our $t$ statistic from step 3 falls in the rejection region, we reject $H_0$. We can also reach this conclusion by examining the $p$ value from Minitab. Since the $p$ value is considerably smaller than 0.05, we would reject the null hypothesis.

**Step 5:**  We translate this decision in terms of the problem statement and conclude that the average employee hand-washing time is different from 5 seconds.  ■

Let's look at another example.

## EXAMPLE 9.2  The Cereal Company

*Two-Tail Test of μ When σ Is Unknown*

We have seen that packaging the correct amount is critical to most companies. Underfilling the package will clearly yield customer complaints and overfilling the packages will cost the company money. Suppose you are working for a cereal manufacturer and each box of cereal is supposed to contain 13 oz of cereal. The weight of the cereal boxes is assumed to be normally distributed. The company has taken a sample of 25 cereal boxes and found the following weights:

| | | | | |
|---|---|---|---|---|
| 12.985 | 12.976 | 13.107 | 13.006 | 12.910 |
| 12.755 | 12.938 | 13.139 | 13.015 | 13.033 |
| 13.029 | 12.887 | 13.049 | 12.823 | 12.910 |
| 13.073 | 13.024 | 13.088 | 13.061 | 13.111 |
| 13.050 | 13.141 | 13.008 | 13.149 | 12.907 |

Is there evidence that the true mean is not equal to 13 oz? Use $\alpha = 0.05$.

**Step 1:** *Set up the null and alternative hypotheses.* Since the company is interested in detecting differences from the value of 13 ounces in either direction, we know that this is a two-tail test. So we have

$$H_0: \quad \mu = 13 \text{ oz}$$
$$H_A: \quad \mu \neq 13 \text{ oz}$$

**Step 2:** *Select a value for α and find the rejection region.* The value of $\alpha$ has been specified as 0.05. From our work in the previous section, we know that we will reject $H_0$ if the calculated $t$ statistic is too large or too small.

We need to use the $t$ table to find the value of $t_{\text{cutoff}}$. Since this is a two-tail test we know that we must divide $\alpha$ in half. Thus, we want $0.05/2 = 0.025$ in the upper-tail area and 0.025 in the lower-tail area. Using the column labeled 0.025 and the row for $n - 1 = 25 - 1 = 24$ degrees of freedom, we find the $t_{\text{cutoff}}$ value to be 2.064. Since the $t$ distribution is symmetric, the value for $-t_{\text{cutoff}}$ is $-2.064$ and the rejection region is shown as the shaded area below:

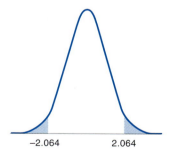

$$-2.064 \qquad 2.064$$

**Step 3:** *Calculate the test statistic and the* p *value.* From the data we can calculate $\overline{X}$ and $s$ and find them to be $\overline{X} = 13.007$ oz and $s = 0.1004$, and we know $n = 25$. With this information we can calculate the $t$ test statistic:

$$t = \frac{\overline{X} - \mu}{s/\sqrt{n}} = \frac{13.007 - 13}{0.1004/\sqrt{25}} = 0.35$$

Alternatively, we can use a software package such as Minitab to find the test statistic and the $p$ value. The output from Minitab is shown at the top of page 432.

```
T-TEST OF THE MEAN
Test of mu = 13.0000 vs. mu ≠ 13.0000

Variable    n      Mean     StDev     SEMean      T       P
Cereal     25    13.0070   0.1004    0.0201    0.35    0.73
```

**Step 4:** Here we check to see if the *t* statistic calculated at step 3 falls in the rejection region. Since our *t* statistic from step 3 does not fall in the rejection region, we do not reject $H_0$. Using the *p* value to make this same decision, we see that the *p* value is larger than 0.05, leading us to fail to reject $H_0$.

**Step 5:** We translate this decision in terms of the problem statement and conclude that the cereal boxes are being filled properly.  ■

Let's summarize the procedure for finding the rejection region for a two-tail test of $\mu$ when $\sigma$ is unknown:

*Steps for finding the rejection region for a two-tail test of $\mu$ when $\sigma$ is unknown*

**Step 1:** *Divide the value of $\alpha$ in half and use that column of the* t *table.*

**Step 2:** *Find the number of degrees of freedom by calculating* n − 1 *and use that row of the* t *table.*

**Step 3:** *Find the value of* t$_{cutoff}$ *in that row and column.*

**Step 4:** *The value of* −t$_{cutoff}$ *is the same as* t$_{cutoff}$ *with a negative sign.*

 **TRY IT NOW!**

## The Soda Machine

### *Test of Population Mean When $\sigma$ Is Unknown*

In Chapter 8 we looked at the hypothesis test to see if a soda machine was correctly dispensing 32 oz of soda. The amount dispensed is assumed to be normally distributed. The machine was not working properly if the bottles were overfilled or underfilled. You observed the machine filling 30 bottles and collected the following data:

|       |       |       |       |       |       |
|-------|-------|-------|-------|-------|-------|
| 32.12 | 32.04 | 32.04 | 32.00 | 32.01 | 32.12 |
| 32.13 | 32.05 | 32.05 | 31.72 | 32.08 | 32.03 |
| 32.02 | 32.22 | 32.22 | 32.22 | 31.89 | 32.20 |
| 31.73 | 31.95 | 31.95 | 32.07 | 32.13 | 32.13 |
| 31.98 | 32.21 | 32.21 | 31.97 | 32.09 | 31.97 |

This time assume that you do not have any information about the population standard deviation and so you must calculate the sample standard deviation.

Is there any evidence to indicate that the machine is not filling the bottles properly? Use $\alpha = 0.05$.

**Step 1:**

**Step 2:**

**Step 3:**

**Step 4:**

**Step 5:**

## 9.2.2 One-Tail Test of the Mean: Small Sample

In the previous section we learned that the two-tail hypothesis testing procedure for $\mu$ is affected in two major ways by the lack of knowledge about $\sigma$. The test statistic becomes a $t$ test instead of a $Z$ test statistic, and the rejection region cutoff values must be found from the $t$ table rather than the $Z$ table. The same can be said about one-tail tests of the mean when $\sigma$ is unknown.

The only step in the procedure that we need to update is finding the rejection region using the $t$ table for one-tail tests. The form of the rejection region is the same as when $\sigma$ is known. The only difference is in finding the cutoff values. Remember that we constructed the rejection regions by following a series of logical arguments as to what values of $\overline{X}$ would lead us to reject the null hypothesis. These arguments still apply.

For a one-tail test we want to reject $H_0$ if the calculated $t$ statistic is too small. Thus, we have the rejection region shaded in Figure 9.1.

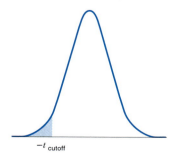

$-t_{\text{cutoff}}$

**FIGURE 9.1**    Rejection region for a lower-tail test of $\mu$

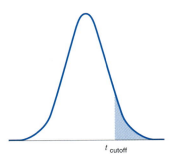

$t$ cutoff

**FIGURE 9.2**    Rejection region for an upper-tail test of $\mu$

For an upper-tail test we want to reject $H_0$ if the calculated $t$ statistic is too large. The rejection region for this type of test is shaded in Figure 9.2.

The cutoff values are found in a manner similar to that used for the two-tail test except that there is no reason to split the value of $\alpha$ in half. These steps are summarized here:

*Steps for finding the rejection region for one-tail tests of $\mu$: small sample*

**Step 1:**  *Use the column of the* t *table that corresponds to the value of $\alpha$ you have selected.*

**Step 2:**  *Find the number of degrees of freedom by calculating* n $-$ 1 *and use that row of the* t *table.*

**Step 3:**  *For upper-tail tests the desired* t$_{cutoff}$ *value is found at the intersection of that row and column. For lower-tail tests, place a negative sign in front of the value to get the value of* $-$t$_{cutoff}$.

Let's look at some examples.

### EXAMPLE 9.3   The Hospital

***One-Tail Test of $\mu$: Small Sample***

The hospital is most likely interested in testing whether the mean hand-washing time is less than 5 seconds. This is a one-tail test of $\mu$.

**Step 1:**  *Set up the null and alternative hypotheses.* Since the hospital is interested in seeing if average hand-washing times are *less than 5 seconds*, this is a lower-tail test. The null and alternative hypotheses are shown below:

$$H_0: \quad \mu \geq 5 \text{ s}$$
$$H_A: \quad \mu < 5 \text{ s}$$

**Step 2:**  *Select a value for $\alpha$ and find the rejection region.* The value of $\alpha$ has been set at 0.05. The specific value of $t_{cutoff}$ is found by looking in the $t$ table. Use the column labeled 0.05 and the row corresponding to 27 degrees of freedom ($28 - 1 = 27$). The table value at the intersection of this column and row is $-1.703$. This is the value of $-t_{cutoff}$.

**Step 3:**  *Calculate the test statistic and the* p *value.* Since the value of $\sigma$ is not known we must calculate both $\overline{X}$ and $s$ from the sample data. The output from Minitab is shown below.

```
T-TEST OF THE MEAN
Test of mu = 5.000 vs mu < 5.000
```

| Variable | N | Mean | StDev | SE Mean | T | P |
|----------|-----|-------|-------|---------|-------|--------|
| Hand was | 28 | 2.536 | 3.687 | 0.697 | $-3.54$ | 0.0007 |

**Step 4:** *Decide whether to reject the null hypothesis.* Since the calculated test statistic of $t = -3.54$ is in the rejection region and the $p$ value is less than 0.05, we reject $H_0$.

**Step 5:** *State the statistical decision in terms of the problem.* Based on these data, we can conclude that the average hand-washing time is less than 5 seconds.

■

Let's look at one more example.

## EXAMPLE 9.4   New Marketing Plan

*Upper-Tail Test of $\mu$ When $\sigma$ Is Unknown*

A company is trying out a new marketing plan and wishes to evaluate its success. Prior to the new advertising scheme the average store sales per week was $4000. The new method is tried on a random sample of 15 stores and the following sales data were collected:

| | | |
|---|---|---|
| $4128 | $4132 | $4163 |
| $4148 | $4157 | $4039 |
| $4028 | $4146 | $4174 |
| $4190 | $4054 | $4181 |
| $4088 | $4069 | $4099 |

On the basis of these data, can the company conclude that the new marketing plan worked? The weekly sales are assumed to be normally distributed. Use $\alpha = 0.05$.

**Step 1:** *Set up the null and alternative hypotheses.* Since the company is interested in seeing if average sales have *increased,* this is an upper-tail test. The null and alternative hypotheses are shown below:

$$H_0: \quad \mu \leq \$4000$$
$$H_A: \quad \mu > \$4000$$

**Step 2:** *Select a value for $\alpha$ and find the rejection region.* The value of $\alpha$ has been set at 0.05. The specific value of $t_{cutoff}$ is found by looking in the $t$ table. Use the column labeled 0.05 and the row corresponding to 14 degrees of freedom ($15 - 1 = 14$). The table value at the intersection of this column and row is 1.761. This is the value of $t_{cutoff}$.

**Step 3:** *Calculate the test statistic and the* p *value.* Since the value of $\sigma$ is not known we must calculate both $\overline{X}$ and $s$ from the sample data:

$$\overline{X} = \$4119.7 \qquad s = \$53.4 \qquad t = \frac{4119.7 - 4000}{53.4/\sqrt{15}} = 8.69$$

```
T-TEST OF THE MEAN
Test of mu = 4000.0 vs. mu > 4000.0

Variable    n     Mean    StDev    SE Mean     T       P
sales       15   4119.7    53.4      13.8     8.69   0.0000
```

**Step 4:** *Decide whether to reject the null hypothesis.* Since the calculated test statistic of $t = 8.69$ is in the rejection region, we reject $H_0$. The $p$ value of 0.0000 also tells us to reject $H_0$.

**Step 5:** *State the statistical decision in terms of the problem.* Based on these data, we can conclude that the advertising scheme has indeed increased sales. ■

**TRY IT NOW!**

### Diameter of Washers

*Lower-Tail Test of μ: Small Sample*

Your company purchases washers. It is important that the diameter of the hole not be more than 0.5 inch. If the hole in the center of the washer is too large then your company will not be able to use the washer. The diameters are assumed to be normally distributed. You have just received a shipment of 10,000 of these washers. You decide to sample 20 to check to be sure that the diameters are not more than 0.5 inch, on the average. Your measurements (in inches) are

|        |        |        |        |
|--------|--------|--------|--------|
| 0.5053 | 0.5098 | 0.4606 | 0.4606 |
| 0.4711 | 0.4627 | 0.4800 | 0.4800 |
| 0.4672 | 0.5642 | 0.5495 | 0.5495 |
| 0.4672 | 0.5346 | 0.5745 | 0.5745 |
| 0.5340 | 0.3767 | 0.3933 | 0.3933 |

Should you accept the shipment? Use $\alpha = 0.05$.

Step 1:   *Set up the null and the alternative hypotheses.*

Step 2:   *Select a value for $\alpha$ and find the rejection region.*

Step 3:   *Calculate the test statistic and the* p *value.*

Step 4:   *Decide whether to reject the null hypothesis.*

Step 5:   *State the statistical decision in terms of the problem.*

## 9.2.3 Summary of Tests of the Mean: Small Sample

In the last two sections we have seen several more examples of the five-step hypothesis testing procedure. In each case, regardless of whether it was a two-tail or one-tail test, the same five steps were utilized. There were two major differences from the tests of $\mu$ when $\sigma$ is known: (1) a $t$ test statistic was used instead of $Z$ and (2) the cutoff values for the rejection region were found by using the $t$ table instead of the $Z$ table.

The rejection regions are summarized below:

| Type of Test | Rejection Region |
|---|---|
| Two-tail test of $\mu$ | Reject $H_0$ if $t < -t_{\text{cutoff } \alpha/2}$ or if $t > t_{\text{cutoff } \alpha/2}$ |
| $\quad H_0: \mu = $ [a specific number] | |
| $\quad H_A: \mu \neq $ [a specific number] | |
| Lower-tail test of $\mu$ | Reject $H_0$ if $t < -t_{\text{cutoff } \alpha}$ |
| $\quad H_0: \mu \geq $ [a specific number] | |
| $\quad H_A: \mu < $ [a specific number] | |
| Upper-tail test of $\mu$ | Reject $H_0$ if $t > t_{\text{cutoff } \alpha}$ |
| $\quad H_0: \mu \leq $ [a specific number] | |
| $\quad H_A: \mu > $ [a specific number] | |

## 9.2.4 Exercises—Learning It!

**9.1**  The cost of common goods and service in 5 cities is shown in the table below *(USA Today):*

| City | Aspirin (100) | Fast food (hamburger, fries, soft drink) | Woman's haircut/blow dry | Toothpaste (6.4 oz) |
|---|---|---|---|---|
| Los Angeles | $7.69 | $4.15 | $20.11 | $2.42 |
| Tokyo | $35.93 | $7.62 | $76.24 | $4.24 |
| London | $9.69 | $5.80 | $44.35 | $3.63 |
| Sydney | $7.43 | $4.53 | $29.93 | $2.08 |
| Mexico City | $1.16 | $3.63 | $17.94 | $1.08 |

(a) You have just returned from a business trip and you lost your receipt for the aspirin you purchased but would like to be reimbursed by your company (since you had take the aspirin after a stressful business meeting!). You guesstimate a cost of $10.00. Your boss claims that the average cost of aspirin is less than $10.00. Using these data, can you "prove" your boss wrong? Conduct the necessary hypothesis test. Assume that all costs are normally distributed.

(b) Based on these data, is there enough evidence to support your submitting a cost of $10.00 for the fast-food meal on your trip?

(c) If you remove Tokyo from the data set do your answers to parts (a) and (b) change? What does this tell you about the effect of outliers on the hypothesis test of $\mu$ when you have a small sample?

**9.2**  The marketing material for a New England ski resort advertises that they can make snow whenever the temperature is 32°F or below. To demonstrate how often this happens their material includes the following line graph of the weekly average temperatures.

**Weekly Average Temperatures**
**15 December–15 March**

The data that generated this graph are shown below:

| Week | 1 | 2 | 3 | 4 | 5 | 6 | 7 | 8 | 9 | 10 | 11 | 12 | 13 | 14 |
|---|---|---|---|---|---|---|---|---|---|---|---|---|---|---|
| Temperature | 18 | 19 | 24 | 35 | 33 | 14 | 22 | 20 | 23 | 33 | 27 | 23 | 30 | 35 |

Is there enough evidence for the ski resort to claim that the average weekly temperature is less than 32°F? Assume that the average weekly temperature is normally distributed.

**9.3**  While job prospects for nurses were once good, many now face an uncertain future as hospitals cut staff and train unlicensed workers to do some of their jobs. The change is reflected in enrollment at several area nursing schools. The following table shows the enrollment at nursing schools in Western Massachusetts:

| School | 1995 | 1996 |
|---|---|---|
| Holyoke Community College | 60 | 45 |
| Berkshire Community College | 96 | 96 |
| Greenfield Community College | 90 | 79 |
| Springfield Technical Community College | 146 | 135 |
| American International College | 189 | 180 |
| Elms College | 298 | 300 |
| Umass–Amherst | 411 | 429 |
| Baystate Medical Center | 150 | 130 |

Using the 1996 data, test the hypothesis that the average enrollment in nursing programs has decreased from the average of 180 in 1995. Assume that enrollment in nursing programs is normally distributed.

**9.4**  If you like shopping for the best deal on long-distance phone service, then you'll enjoy sorting through offers from 10 different marketers vying to be your energy supplier. Residents of 16 communities will be the first in Massachusetts to wade into the coming nationwide experiment in deregulation of the natural gas industry. The average customer uses 1232 therms of natural gas, for which the average cost has been $520.24. The table below shows the proposed costs to deliver 1232 therms of natural gas from 10 competitors:

| Company | Cost ($) |
| --- | --- |
| All Energy Marketing Co. | 478.66 |
| Broad Street/Energy One | 450.24 |
| Global Energy Services | 468.16 |
| Green Mountain Energy Partners | 471.24 |
| KBC Energy Services | 435.53 |
| Louis Dreyfus Energy Services | 472.24 |
| National Fuel Resources | 468.22 |
| NorAm Energy | 442.20 |
| WEPCO Gas | 443.52 |
| Western Gas Resources | 457.81 |

Is there enough evidence to conclude that the average cost to the customer from the competitors is less than $520.24? Assume that the costs are normally distributed.

**9.5**  Computer centers at universities and colleges are certainly aware of the increased number of web surfers. To begin to understand the demands that will be made on the computer center resources, one school studied 25 children in grades 7 to 12. The number of hours that these children spent on the Internet in one week is shown below:

| | | | | |
| --- | --- | --- | --- | --- |
| 5.0 | 4.4 | 5.7 | 5.6 | 5.5 |
| 5.2 | 5.0 | 4.8 | 3.6 | 4.1 |
| 4.6 | 4.9 | 4.0 | 6.7 | 5.5 |
| 5.4 | 6.7 | 5.8 | 5.4 | 4.8 |
| 5.9 | 5.1 | 3.8 | 4.1 | 6.7 |

Is there enough evidence to indicate that children spend more than an average of 5 hours per week web surfing? Assume that the time spent web surfing is normally distributed.

## 9.3  $\chi^2$ TEST OF A SINGLE VARIANCE

Hypothesis tests of the population variance, $\sigma^2$, follow the same basic steps that were used to do a hypothesis test of the population mean and the population proportion.

*Remember: The variance is the standard deviation squared.*

There are two types of situations for which you might be interested in doing a hypothesis test of the population variance. First, you may wish to see if a manufacturing process is running to the specified standard deviation. You cannot test the standard deviation but must instead test the population variance. This is the case for the tissue manufacturer. Second, to perform some other statistical analysis of the data, you may need to know the population variance. In this case you would use sample data to test to see if the population variance is, in fact, a particular value.

Just like with tests of the mean, you can do either two-sided or one-sided tests of the variance. To decide which type of test you need, you must determine what you are trying to learn from your study. The key words and guidelines are found in Chapter 8.

### 9.3.1 Two-Tail Hypothesis Test of the Variance

To learn how to do a two-sided test of the variance we will complete the five steps of the hypothesis testing procedure for the tissue manufacturer looked at in Chapter 8.

### EXAMPLE 9.5  The Tissue Manufacturer

***Step 1: Setting Up the Null and Alternative Hypotheses for a Test of the Population Variance***

The variable *MDStrength* should have a standard deviation of 50 lb/ream. Also, the population variance should be $50^2 = 2500$ (lb/ream)$^2$. This gives us the following null and alternative hypotheses:

$$H_0: \quad \sigma^2 = 2500 \text{ (lb/ream)}^2$$
$$H_A: \quad \sigma^2 \neq 2500 \text{ (lb/ream)}^2 \qquad \blacksquare$$

The next step in the hypothesis testing procedure is to select a value for $\alpha$ and find the rejection region. To do this we need to know what test statistic will be used in step 3 of the procedure. This is the main difference between a hypothesis test of the population mean, $\mu$, and the population variance, $\sigma^2$. In testing means, we used the sample mean as the basis for our decision to reject or fail to reject the null hypothesis. Because the Central Limit Theorem told us that $\overline{X}$ has an approximately normal distribution for sufficiently large sample sizes, the appropriate test statistic for a large-sample test of the mean is a $Z$ statistic and thus the rejection region is determined by $Z$ values. As we have seen, in the small-sample case, $\overline{X}$ follows a $t$ distribution for normally distributed populations.

If we are testing the population variance then the sample variance, $s^2$, will be used as the basis for deciding between $H_0$ and $H_A$. Relying once again on the mathematical statisticians to do the theoretical work, we learn that the sampling distribution associated with the sample variance, $s^2$, is called the chi-square ($\chi^2$) distribution. The rejection region will be determined by critical values from the chi-square distribution. An example of a chi-square distribution is shown in Figure 9.3.

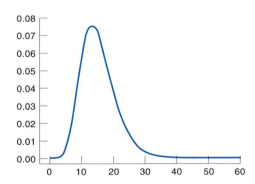

**FIGURE 9.3**  A chi-square distribution

The chi-square distribution, just like any distribution, describes how the random variable behaves. Recall from Chapter 6 that a distribution tells you the most likely values of the random variable (where there is the most area under the curve) and the least likely values of the random variable (where there is little to no area under the curve). You will see the chi-square distribution again in Chapter 15.

If the variable being studied is assumed to be normally distributed, then the statistic to test whether or not the population variance is equal to a particular value is calculated as follows:

$$\chi^2 = \frac{(n-1)s^2}{\sigma^2}$$

*Chi-square test statistic*

where

$n$ = sample size

$s^2$ = the sample variance

$\sigma^2$ = the hypothesized value of the population variance under the null hypothesis

Notice that this test statistic is basically comparing the variability contained in the sample and reflected in the sample variance to the value of the population variance that is being tested. In this case, the comparison takes the form of a ratio. In testing means, the comparison of the sample evidence, reflected in $\overline{X}$, to the value of the population mean being tested is accomplished by a subtraction. In both cases the idea is the same: Is the sample evidence consistent with the null hypothesis?

*The chi-square test assumes that the underlying population distribution is normal.*

This test statistic has been labeled with the Greek letter $\chi$ (chi) and a squared symbol, hence the name chi-square test statistic. Like the $t$ distribution, the shape of the chi-square distribution is determined by the number of degrees of freedom. This test statistic has $n-1$ degrees of freedom. Unlike the $Z$ and $t$ distributions, the $\chi^2$ distribution is not symmetric.

Recall that the rejection region is the set of those values of the test statistic that would lead you to reject the null hypothesis. Let's think a minute about what values of the chi-square test statistic would lead us to reject the null hypothesis. Clearly, if the sample variance exactly equals the population variance then the ratio of $s^2$ to $\sigma^2$ will be one and the test statistic will be equal to $n-1$. In the case of the tissue manufacturer, this would be $36 - 1 = 35$. In this situation the sample evidence is clearly consistent with the null hypothesis and we would fail to reject $H_0$.

If the sample variance is quite different from the population variance being tested then it will be either larger or smaller than $\sigma^2$. If the sample variance is a great deal lower than the population variance being tested then the ratio of $s^2$ to $\sigma^2$ will be a fraction and the test statistic will be some value less than $n-1$. A similar analysis tells us that if the sample variance is a great deal larger than the population variance being tested then the ratio of $s^2$ to $\sigma^2$ will be greater than one and the test statistic will be larger than $n-1$. These are the situations that would lead us to reject $H_0$. This gives us the basic structure for the rejection region. We will reject $H_0$ if the value of the test statistic is much less than $n-1$ or much greater than $n-1$. Since this is a two-sided rejection region, the area in each tail must be $\alpha/2$. A typical rejection region is shown as the shaded region in Figure 9.4.

***Remember:*** *The total area of the rejection region is $\alpha$.*

**Chi-square Distribution with 5 Degrees of Freedom**

**FIGURE 9.4** Rejection region for a two-sided test of the variance

**TABLE 9.1    A Portion of the Chi-square Table**

| Degrees of Freedom | Upper-Tail Areas | | | | | | | | | | | |
|---|---|---|---|---|---|---|---|---|---|---|---|---|
| | **0.995** | **0.99** | **0.975** | **0.95** | **0.9** | **0.75** | **0.25** | **0.1** | **0.05** | **0.025** | **0.01** | **0.005** |
| 32 | 15.134 | 16.362 | 18.291 | 20.072 | 22.271 | 26.304 | 36.973 | 42.585 | 46.194 | 49.480 | 53.486 | 56.328 |
| 33 | 15.815 | 17.073 | 19.047 | 20.867 | 23.110 | 27.219 | 38.058 | 43.745 | 47.400 | 50.725 | 54.775 | 57.648 |
| 34 | 16.501 | 17.789 | 19.806 | 21.664 | 23.952 | 28.136 | 39.141 | 44.903 | 48.602 | 51.966 | 56.061 | 58.964 |
| 35 | 17.192 | 18.509 | 20.569 | 22.465 | 24.797 | 29.054 | 40.223 | 46.059 | 49.802 | 53.203 | 57.342 | 60.275 |

The specific values for $\chi^2_{\text{lower}}$ and $\chi^2_{\text{upper}}$ must be found from the chi-square table. A portion of this table is shown in Table 9.1. The complete table is found in Appendix A.5.

Let's use this table to complete step 2 of the hypothesis testing procedure for the tissue company.

### EXAMPLE 9.6    The Tissue Company

***Rejection Region for Variance Test***

Since the sample size was $n = 36$, we must use the row with $n - 1 = 35$ degrees of freedom. To get the value for $\chi^2_{\text{upper}}$, use the column labeled 0.025 in the upper tail.

The table value at the intersection of the correct row and column is 53.203. You may have noticed that the column values correspond to upper-tail areas. There is no need to have another table for lower-tail areas. To have the area of the lower tail be 0.025, the upper tail must be 0.975.

To get the value for $\chi^2_{\text{lower}}$ use the column labeled 0.975 in the upper tail. The table value at the intersection of the row with 35 degrees of freedom and 0.975 in the upper tail is 20.569. ∎

To get a sense of how different the sample variance must be from the population variance in order to reject $H_0$, let's examine this rejection region for a moment. The upper cutoff value is 53.203. If we divide this by 35, the number of degrees of freedom, we get 1.52. Thus, even if the sample variance is 1.5 times larger than the population variance you are testing it against, you would not yet be in the rejection region. Likewise if we take the lower cutoff value of 20.569 and divide it by 35 we get 0.588. So, if the sample variance is, for example, 60% (0.60) of the population variance you are testing it against, you would not reject the null hypothesis. This tells us that the sample evidence must be "pretty different" from the population variance you are testing it against in order to reject the null hypothesis. The chi-square table quantifies "pretty different" for us.

Now we can complete step 3 of the hypothesis testing procedure. Remember that at this step the test statistic is calculated.

### EXAMPLE 9.7    The Tissue Company

***Calculating the Chi-square Test Statistic***

A sample of 36 tissues had a sample standard deviation of 60.1 lb/ream. The test statistic is

$$\chi^2 = \frac{(n-1)s^2}{\sigma^2}$$

$$= \frac{(36-1)(60.1)^2}{2500} = 50.57$$

∎

The fourth step of the hypothesis testing procedure is to decide whether to reject the null hypothesis on the basis of the test statistic. The fifth step is to interpret this decision in terms of the original problem statement. Let's complete the test of the variance for the tissue manufacturer.

## EXAMPLE 9.8   The Tissue Company

### Steps 4 and 5 of the Hypothesis Testing Procedure

Since the value of 50.57 does not fall in the rejection region, we fail to reject the null hypothesis. This tells us that the manufacturing process is running according to specifications for the variability of the variable *MDStrength*. ∎

 **TRY IT NOW!**

## The Cereal Company

### Testing the Variance

A cereal manufacturer wishes to test if the population variance of the weight of the boxes is equal to $0.0500$ oz². A random sample of 20 boxes has a standard deviation of $s = 0.25$ oz. Use $\alpha = 0.05$.

Step 1:

Step 2:

Step 3:

*(continued)*

Step 4:

Step 5:

## 9.3.2 One-Sided Tests of the Variance

If you are testing the population variance, most of the time you are interested in doing a two-sided test. However, it is possible to do a one-sided test of the variance. The only change in the procedure needed to complete a one-sided test of the variance is in step 2. A one-sided rejection region is used in this case. The rejection regions for the two possibilities are shown in Figure 9.5.

$$H_0: \quad \sigma^2 \geq \text{[a specific number]} \qquad H_0: \quad \sigma^2 \leq \text{[a specific number]}$$
$$H_A: \quad \sigma^2 < \text{[a specific number]} \qquad H_A: \quad \sigma^2 > \text{[a specific number]}$$

**Lower-tail test of the variance**

0    5    10    15    20    25

**Upper-tail test of the variance**

0    5    10    15    20    25

**FIGURE 9.5**  Rejection regions for one-sided tests of the variance

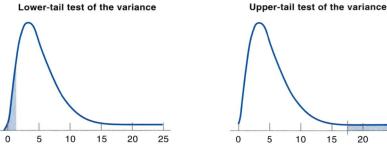

ANS. $H_0: \sigma^2 = 0.0500$, $H_A: \sigma^2 \neq 0.0500$; $X_{upper} = 32.852$, $X_{lower} = 8.907$; TEST STATISTIC = 23.75; DO NOT REJECT $H_0$.

**EXAMPLE 9.9** **The Tissue Company**

*One-Sided Test of the Variance*

Let's reconsider the variability for *MDStrength* for the tissue manufacturer. Suppose that the company wishes to be sure that the variance is at most 2500 lb/ream. The key words "at most" tell us this is an upper-tail test of the variance.

**Step 1:**

$$H_0: \quad \sigma^2 \le 2500 \text{ lb/ream}$$
$$H_A: \quad \sigma^2 > 2500 \text{ lb/ream}$$

**Step 2:** Set $\alpha = 0.05$. Use the chi-square table with 35 degrees of freedom to find the rejection region. The table value at the intersection of the row for 35 degrees of freedom and the column with 0.05 in the upper-tail area is 49.802. Note that we did not divide $\alpha$ in half because this is a one-sided test with a one-sided rejection region.

**Step 3:** The test statistic is calculated below:

$$\chi^2 = \frac{(n-1)s^2}{\sigma^2}$$
$$= \frac{(36-1)(60.1)^2}{2500} = 50.57$$

**Step 4:** This test statistic falls in the rejection region, so we reject the null hypothesis.

**Step 5:** We conclude that the population variance is greater than 2500 lb/ream and the process should be adjusted. ∎

*A note on ethics in data analysis*

You should notice that we got two different results based on the same sample of data. When we used a two-sided test of the variance we failed to reject the null hypothesis and concluded that the process was running properly. But when we used an upper-tail test of the variance we concluded that the population variance was greater than 2500 lb/ream and that the process was not running properly. You must decide what you are interested in detecting before you do the test. This is an example of how to "lie with statistics." You might say that the data do not lie. You are right about that in one sense. However, by manipulating the type of test that is done and the value of $\alpha$ that is used, you can change the conclusions of the analysis. If you are using the results of someone else's data analysis you must ask the right questions to be sure that the data have not been unfairly manipulated to support a particular position.

## 9.3.3 Summary of Tests on the Variance

In this section all of the tests used the same test statistic in step 3. The purpose of the test statistic is always the same: It gives you a single number that summarizes the sample information and helps you to decide between the null and alternative hypotheses. The test statistic for tests on variances is

$$\chi^2 = \frac{(n-1)s^2}{\sigma^2}$$

At step 2 we did see that the rejection region is different depending on what type of test we are doing. These differences are summarized in the next table.

| Type of Test | Rejection Region |
|---|---|
| Two-tail test of $\sigma^2$ | Reject $H_0$ if $\chi^2 < \chi^2_{lower}$ or if $\chi^2 > \chi^2_{upper}$ |
| $\quad H_0: \sigma^2 =$ [a specific number] | |
| $\quad H_A: \sigma^2 \neq$ [a specific number] | |
| Lower-tail test of $\sigma^2$ | Reject $H_0$ if $\chi^2 < \chi^2_{lower}$ |
| $\quad H_0: \sigma^2 \geq$ [a specific number] | |
| $\quad H_A: \sigma^2 <$ [a specific number] | |
| Upper-tail test of $\sigma^2$ | Reject $H_0$ if $\chi^2 > \chi^2_{upper}$ |
| $\quad H_0: \sigma^2 \leq$ [a specific number] | |
| $\quad H_A: \sigma^2 >$ [a specific number] | |

## 9.3.4 Exercises—Learning It!

**9.6**  A company that sells mail-order computer systems has been planning inventory and staffing based on an assumption that the variance of their weekly sales is 180 ($\$1000^2$). The weekly sales are normally distributed. The company selects 15 weeks at random from the past year and obtains the data (in thousands of dollars) shown below:

Weekly sales  191  222  222  223  223  225  227  228  229  232  234  234  236  244  253

(a) What is the sample variance for these data?

(b) Set up the hypotheses to test whether the population variance is different from 180.

(c) At the 0.05 level of significance, what can you conclude about the company's assumption?

**9.7**  In manufacturing, the amount of material that is wasted or lost during a process is very important. In preparing financial estimates, a company assumes that the percent material lost for its new process has a variance $10\%^2$. After the new process has been running for a month and appears to be stable, the cost analyst looks at the percent material lost and finds the following data:

Daily Loss     10  12    12  13    14  14    18  19    19  20

(a) What is the sample variance for these data?

(b) Set up the hypotheses to test whether the actual variance is greater than the value the company has been assuming. Assume that the daily loss is normally distributed.

(c) At the 0.05 level of significance, what can you conclude?

**9.8**  Lead time for an order is a critical factor in inventory planning. A company has noticed that inventory for assembling its new product line has been consistently short in the past few months. It has been assumed that lead time is normally distributed with a standard deviation 2 days. The company decides to look at the last 20 orders placed, and finds that the lead times (in days) are

| | | | |
|---|---|---|---|
| 1 | 4 | 6 | 7 |
| 4 | 5 | 6 | 8 |
| 4 | 5 | 7 | 8 |
| 4 | 6 | 7 | 10 |
| 4 | 6 | 7 | 11 |

(a) What is the sample variance for these data?

(b) Set up the hypotheses to test whether or not the actual variance is greater than the value the company has been assuming. Be sure you use the variance here.

(c) At the 0.05 level of significance, what can you conclude?

**9.9**  In an effort to understand the cost overruns by a particular department in a company, data were collected on the amount of the overrun ($) for 24 different days:

| | | | | | |
|---|---|---|---|---|---|
| 87.3 | 93.7 | 96.8 | 98.4 | 100.9 | 107.8 |
| 89.9 | 94.9 | 97.0 | 99.6 | 101.3 | 109.7 |
| 91.5 | 96.5 | 97.1 | 100.3 | 105.7 | 111.5 |
| 93.6 | 96.7 | 97.3 | 100.4 | 107.7 | 114.2 |

(a) Calculate the sample variance for the daily cost overrun.

(b) At the 0.05 level of significance, do the data agree with the assumption that the variance of the daily overrun is 50 (dollars)$^2$. Assume that the cost overruns are normally distributed.

# 9.4 TEST OF A SINGLE PROPORTION

All of the tests that we have done thus far in the chapter have been tests on $\mu$. However, very often it is not the average value that we are interested in but rather some proportion of the population that behaves in a certain manner. If the data we are analyzing are *nominal data,* the hypothesis might be a statement about

- The value of the proportion, $\pi$, of population members that have a certain characteristic (one of the categories of the nominal variable)

For example, the hypothesis might be a statement about the proportion of

- Residents of Swingfield who are in favor of a casino
- Students who are interested in graduate school
- Vaccinated patients who remain cancer free
- CEOs who use computers as a major tool
- People who are unemployed

In these cases, the parameter of interest is a proportion. In Chapter 7 you learned that an estimator for the true unknown proportion, $\pi$, is the corresponding sample proportion, $p$. For example, if 100 people are surveyed and 58 are in favor of a casino, then the sample proportion, $p$, is 58/100 or 0.58.

Like $\bar{X}$, the sample proportion, $p$, rarely if ever actually equals the true population proportion, $\pi$. Remember that this is not because you did something wrong but rather because you are looking at only a piece of the population and not the entire population. This is what we have called sampling error. It is a fact of life in statistics. You cannot eliminate sampling error unless you examine the entire population.

Since the sample proportion is not likely to equal the population proportion, we again see that the estimate alone is inadequate for making decisions. Suppose, for example, that more than 60% of the voters must be in favor of a casino and the sample proportion comes out to be 0.58. Does this automatically mean that the casino should be killed? Is it possible that more than 60% of the population favors the casino but only 58% of those sampled favor the casino? Possibly. How possible again depends on the standard error. So we need to develop the hypothesis testing procedure for tests on proportions. We start by looking at two-tail tests of proportions.

## 9.4.1 Two-Tail Tests of Proportions

Fortunately much of what we have already learned about hypothesis testing can be easily adapted to tests of proportions. In fact, we will use precisely the same five steps that we have been following for any hypothesis test:

**Step 1:** *Set up the null and alternative hypotheses.*

**Step 2:** *Pick the value of $\alpha$ and find the rejection region.*

**Step 3:** *Calculate the test statistic.*

**Step 4:** *Decide whether or not to reject the null hypothesis.*

**Step 5:** *Interpret the statistical decision in terms of the stated problem.*

*Steps for any hypothesis test*

Let's use an example to see how these steps work for a two-tail test of proportions.

### EXAMPLE 9.10  Beverly Hills, 90210

***Two-Tail Test of Proportions***

Hard times have hit the *Beverly Hills, 90210* crowd. Nearly a third of teens polled by Teenage Research Limited said they have been personally affected by the recession. Where do teens get their money? A survey of 2000 teens showed that 47% of them get some money from their parents. Suppose you wish to do a hypothesis test to see if half or 50% of all teens get some money from their parents.

> **Step 1:** *Set up the null and alternative hypotheses.* The same general guidelines from Chapter 8 apply to setting up the null and the alternative hypotheses for a test of proportions. Since we are interested in seeing if the true proportion *differs* from 50% in either direction, we use a two-tail test. For this example, the null and alternative hypotheses are

$$H_0: \quad \pi = 0.50$$
$$H_A: \quad \pi \neq 0.50$$

> **Step 2:** *Pick the value of α and find the rejection region.* In Chapter 7 you learned that the sampling distribution for *p* is the normal distribution. Therefore, the test statistic for proportions is a *Z* statistic. Thus, the form of the rejection region is exactly the same as the rejection region for a two-tail test on the mean when the standard deviation is known. If we set $\alpha = 0.05$, then the rejection region is as follows: reject $H_0$ if $Z > 1.96$ or $Z < -1.96$.

> **Step 3:** *Calculate the test statistic and the* p *value.* Now we must consider what information in the sample will help us make a decision between the null and the alternative hypotheses. Clearly, we must use the sample proportion, *p*. But we know that we cannot simply use the value of *p* alone. We must take into account the fact that *p* will vary from sample to sample. As we have repeatedly seen, this is taken into account by dividing by the standard error of the estimator. The appropriate test statistic is

*__Z__ test statistic for proportions*

$$Z = \frac{p - \pi}{\sqrt{\pi(1 - \pi)/n}}$$

where $\pi$ is the hypothesized value of the population proportion.

This test statistic has the same basic form as the *Z* statistic we used in testing $\mu$. In that case we used

$$Z = \frac{\bar{X} - \mu}{\sigma/\sqrt{n}}$$

In the numerator we are calculating the difference between the value of the population proportion, $\pi$, and the sample proportion, *p*. This is equivalent to the calculation that we did in the numerator of the *Z* calculation for $\mu$. In the denominator of both *Z* statistics we use the standard error of the estimator. In the case of proportions, this is $\sqrt{\pi(1 - \pi)/n}$.

For this example the *Z* statistic is

$$Z = \frac{0.47 - 0.50}{\sqrt{0.50(1 - 0.50)/2000}} = -2.68$$

Remember that the $p$ value is the smallest value of $\alpha$ for which you can reject $H_0$. To complete the hypothesis test you should calculate the $p$ value. The $p$ value for a test of proportions is calculated the same way as it was for the test of means. Since this is a two-tail test we must double the tail area probability. The $Z$ statistic for this example was found to be $-2.68$. So, the $p$ value is found as follows:

$$p\text{ value} = 2P(Z > 2.68) = (2)(0.0037) = 0.0074$$

If you select an $\alpha$ of 0.05 as we have done in this example, then you will reject $H_0$.

**Step 4:** *Decide whether or not to reject $H_0$.* Since the value of $Z = -2.68$ is in the rejection region, we reject $H_0$.

**Step 5:** *Interpret the statistical decision in terms of the problem.* We conclude that the true proportion of teenagers who receive spending money from their parents is different from 0.50. ■

 **TRY IT NOW!**

### Poll of Americans

*Test of Proportion*

"People believe the country is still very, very screwed up," said Charles Cook of 10 political experts who analyzed the polling data upon its release at a Capitol Hill luncheon. The poll was conducted by sampling 2001 adults between 16 March and 21 March 1994. In the poll only 24% of Americans are satisfied with the country's economic course. Is there evidence that 25% of Americans are satisfied with the country's economic course? Use $\alpha = 0.05$.

Step 1:

Step 2:

*(continued)*

Step 3:

Step 4:

Step 5:

Find and interpret the $p$ value.

## 9.4.2 One-Tail Test of Proportions

Finally, we move to one-tail tests of proportions. Here again the five-step hypothesis testing procedure is identical to the one we have been using. The test statistic is the same as that used for a two-tail test of proportions and the rejection regions are the same as those used for one-tail tests of the mean.

**Step 1:**   *Set up the null and alternative hypotheses.* There are two possible ways to set up a one-tail test of proportions. We have already looked at the is-sues that should be addressed in deciding how to set up the hypothesis test. Here we will simply present the two forms again using the terms upper-tail and lower-tail test.

| Upper-Tail Test | Lower-Tail Test |
|---|---|
| $H_0$: $\pi \leq$ [a specific number] | $H_0$: $\pi \geq$ [a specific number] |
| $H_A$: $\pi >$ [a specific number] | $H_A$: $\pi <$ [a specific number] |

*Remember that the specific number must be a number from 0 to 1.*

**Step 2:** *Select the value of $\alpha$ and find the rejection region.* Since the test statistic is a $Z$ test, the rejection regions for the one-tail tests are precisely the same as the ones we used in testing means.

**Step 3:** *Calculate the test statistic.* We have seen in the previous section that the appropriate test statistic is

$$Z = \frac{p - \pi}{\sqrt{\pi(1 - \pi)/n}}$$

**Steps 4** and **5** remain the same.

Consider an example.

## EXAMPLE 9.11   Testing Raw Materials

### An Upper-Tail Test of Proportions

Your company has been building a long-term relationship with one of its suppliers in the spirit of TQM. To validate the supplier's claim that at most 5% of its products are defective, you take a sample of 100 items and find six defectives. Should you believe the supplier's claim? Test using $\alpha = 0.05$.

**Step 1:** *Set up the null and alternative hypotheses.* Since you are interested in seeing if there is at most 5% defectives, this is a one-tail test. You will believe the supplier's claim unless the data indicate otherwise. This is an upper-tail test with the following setup:

$$H_0: \quad \pi \leq 0.05 \quad \text{(Supplier's claim is correct.)}$$
$$H_A: \quad \pi > 0.05 \quad \text{(Supplier's claim is incorrect.)}$$

**Step 2:** *Select a value for $\alpha$ and find the rejection region.* Since $\alpha = 0.05$, the rejection region is as shown below:

1.645

**Step 3:** *Calculate the test statistic.* The sample proportion was found to be $p = 6/100 = 0.06$ and the sample size was 100. Using this information we can calculate the test statistic:

$$Z = \frac{p - \pi}{\sqrt{\pi(1 - \pi)/n}} = \frac{0.06 - 0.05}{\sqrt{0.05(1 - 0.05)/100}} = 0.46$$

Notice that 0.05 is used as the value for $\pi$ since that is the hypothesized value of the population proportion under $H_0$.

**Step 4:** *Decide whether or not to reject $H_0$.* Since $Z = 0.46$ does not fall in the rejection region, we fail to reject $H_0$.

> **Step 5:** *Interpret the statistical decision in terms of the problem.* You cannot conclude with confidence that the vendor has more than 5% defectives in the entire shipment.

To finish the example we should calculate the $p$ value. This is an upper-tail test so we do not double the tail area probability. The $p$ value is found as follows:

$$p\text{ value} = P(Z > 0.46) = 0.3228$$

The large $p$ value tells us that the data are highly consistent with the null hypothesis. This gives us greater confidence in accepting the shipment. ■

 **TRY IT NOW!**

### The Soft Drink Company
*One-Tail Test of Proportion*

The Coca-Cola Company is interested in entering the fruit drink market. Before bringing its new product, Fruitopia, to the market the company wishes to be sure that it will capture more than 20% of the fruit drink market. A survey of 1000 people shows that 210 respondents prefer Fruitopia to other fruit drinks. Is there enough evidence to allow Coca-Cola to proceed with the new product? Use $\alpha = 0.05$.

Step 1:

Step 2:

Step 3:

Step 4:

Step 5:

## 9.4.3 Summary of Tests on Proportions

In this section all of the tests use the same test statistic in step 3. The purpose of the test statistic is always the same: It gives you a single number that summarizes the sample information and helps you to decide between the null and alternative hypotheses. The test statistic for tests on proportions is

$$z = \frac{p - \pi}{\sqrt{\pi(1 - \pi)/n}}$$

At step 2 we did see that the rejection region is different depending on what type of test we are performing. This is summarized below:

| Type of Test | Rejection Region |
|---|---|
| Two-tail test of $\pi$ | Reject $H_0$ if $Z < -Z_{\text{cutoff } \alpha/2}$ or if $Z > Z_{\text{cutoff } \alpha/2}$ |
| $\quad H_0: \pi = $ [a specific number between 0 and 1] | |
| $\quad H_A: \pi \neq $ [a specific number between 0 and 1] | |
| Lower-tail test of $\pi$ | Reject $H_0$ if $Z < -Z_{\text{cutoff } \alpha}$ |
| $\quad H_0: \pi \geq $ [a specific number between 0 and 1] | |
| $\quad H_A: \pi < $ [a specific number between 0 and 1] | |
| Upper-tail test of $\pi$ | Reject $H_0$ if $Z > Z_{\text{cutoff } \alpha}$ |
| $\quad H_0: \pi \leq $ [a specific number between 0 and 1] | |
| $\quad H_A: \pi > $ [a specific number between 0 and 1] | |

ANS: $H_0: \pi \leq 0.20$, $H_A: \pi > 0.20$; CRITICAL VALUE = 1.645; Z = 0.79; FAIL TO REJECT $H_0$; THEY SHOULD NOT PROCEED WITH FRUITOPIA.

## 9.4.4 Exercises - Learning It!

**9.10**   Companies are increasingly concerned about employees playing video games at work. In addition to reducing productivity, this habit slows down networks and uses valuable storage space. A recent article stated that 80% of all employees play video games at work at least once a week. A large company that employs many engineers wonders if its employees are as bad as the article claims. If they are, the company will install software that detects and removes video games from the network. The company surveys (anonymously) 100 employees and finds that 85 of the employees surveyed had played video games at work in the past week.

   (a) Set up the null and alternative hypotheses to test whether the proportion of the company's employees that play video games is greater than the proportion stated in the article.
   (b) At the 0.05 level of significance, test the hypotheses.
   (c) What do you recommend that the company do?

**9.11**   An alumni office is interested in serving their alumni better in order to encourage more donations to the college. A survey of 200 alumni was conducted to determine whether half-day training sessions offered on the campus were of interest. If more than 75% of the alumni were interested, the college would start a program. The survey showed that 160 of the alumni surveyed were interested in such a program.

   (a) Set up the null and alternative hypotheses to test whether the college should implement the program.
   (b) At the 0.05 level of significance, test the hypotheses.
   (c) What do you recommend that the college do?

**9.12**   A company that makes computer keyboards has specifications that allow it to produce a product that has a maximum of 3% defective. The company has been receiving more customer complaints than usual. A sample of 50 keyboards has 2 defectives.

   (a) Set up the null and alternative hypotheses to test whether the proportion defective keyboards has exceeded the amount allowed by the specifications.
   (b) At the 0.05 level of significance, test the hypotheses.
   (c) What do you recommend that the company do?

**9.13**   A university in the Northeast claims in its brochures that it has an acceptance rate of 60%. A sample of 300 high school seniors who applied to this university shows that 148 of them were accepted.

   (a) Set up the null and alternative hypotheses to test whether the acceptance rate is what the university claims.
   (b) At the 0.05 level of significance, test the hypotheses.
   (c) Is there a need for the university to change its literature?

**9.14**   "Computer Jobs Increase by 6%" is the headline on an article in a national newspaper. A study of the classified job ads in that same newspaper indicates that out of 2202 advertisements in a typical Sunday edition, 502 were for computer related jobs. Statistics released by the Bureau of Labor Statistics for the last 12 months show that 15% of available jobs were computer related.

   (a) Set up the null and alternative hypotheses to test whether the headline is correct.
   (b) At the 0.05 level of significance, test the hypotheses.
   (c) What conclusion can you reach?

# *Discovery Exercise 9.1*
## EXPLORING THE CONNECTION BETWEEN CONFIDENCE INTERVALS AND HYPOTHESIS TESTING

### Part I

A recent survey is offering the first evidence that PCs are replacing TVs as the primary source of home recreation, information, and entertainment. The survey was conducted among 1200 homes nation-wide. The average computer user spends 9.5 hours per week in front of the PC but only 8 hours per week watching prime-time TV. Assume that the standard deviation of hours spent in front of a PC is 3 hours/week.

(a) Test the hypothesis that the average number of hours per week a computer user spends in front of a PC is different from 8 hours. Use $\alpha = 0.05$.

(b) Using the same data, construct a 95% confidence interval for $\mu$, the population average time spent in front of a PC.

(c) Is the value of 8 in the confidence interval you constructed in part (b)?

(d) Did you reject the null hypothesis in part (a)?

### Part II

The Casual Businesswear Employee Survey was conducted to assess the attitudes and behavior of white-collar employees whose companies allow casual dress on some basis. The study was national in scope and the sample size was 752 people. Of those in the sample, 609 agree that allowing casual dress improves morale.

(a) Test the hypothesis that the proportion of white-collar employees who agree that allowing casual dress improves morale is different from 80%. Use $\alpha = 0.05$.

(b) Construct a 95% confidence interval for $\pi$.

*(continued)*

(c) Is the value of 0.80 in the confidence interval?

(d) Did you fail to reject the null hypothesis?

**Part III**

Summarize.

Based on these two situations, speculate how you would complete the following statements:

If the value of the parameter being tested (the one in the null hypothesis) is not in the confidence interval then I will _____ the null hypothesis.

If the value of the parameter being tested (the one in the null hypothesis) is in the confidence interval then I will _____ the null hypothesis.

## 9.5  CONNECTION BETWEEN HYPOTHESIS TESTING AND CONFIDENCE INTERVALS

Now you have learned about both confidence intervals and hypothesis testing. These two techniques are actually closely related even though they are used for different purposes. You can see the relationship between these techniques easily if we reconsider one of the hypothesis testing examples from this chapter and calculate the corresponding confidence interval. Let's look at the tissue company.

### EXAMPLE 9.12   The Tissue Company

*Relationship of Confidence Interval to Hypothesis Test*

A sample of 36 tissues was taken and the average *MDStrength* was 980 lb/ream. The standard deviation of the process is 50 lb/ream. In Example 8.11, we tested the hypothesis that the population mean *MDStrength* was 1000 lb/ream. The following hypothesis test was done:

$$H_0: \quad \mu = 1000 \text{ lb/ream}$$
$$H_A: \quad \mu \neq 1000 \text{ lb/ream}$$

The value of the test statistic was $Z = -2.4$. Based on this $Z$ value and an $\alpha$ value of 0.05, the null hypothesis was rejected, leading us to conclude that the population mean, $\mu$, was not 1000 lb/ream.

Now construct a 95% confidence interval for $\mu$ using these data:

$$\text{Upper} = \overline{X} + Z(\sigma/\sqrt{n}) = 980 + (1.96)(50/\sqrt{36}) = 980 + 16.33 = 996.33 \text{ lb/ream}$$
$$\text{Lower} = \overline{X} - Z(\sigma/\sqrt{n}) = 980 - (1.96)(50/\sqrt{36}) = 980 - 16.33 = 963.67 \text{ lb/ream}$$

Notice that the value of $\mu$ of 1000 lb/ream is not in the confidence interval. This means that 1000 is not a likely value of $\mu$. Using the same data we rejected the null hypothesis, concluding that $\mu$ was not 1000 lb/ream. You can see that the results from the confidence interval are consistent with the conclusions drawn from the hypothesis test.

This result can be stated in general for any hypothesis test, whether it be a test of $\mu$, $\sigma^2$, or $\pi$. If the value of the parameter being tested (the one in the null hypothesis) is in the confidence interval, then you will fail to reject the null hypothesis of a two-tail test. If the value of the parameter being tested is not in the confidence interval, then you will reject the null hypothesis. Of course, the same value of $\alpha$ must be used for both the confidence interval construction and the hypothesis test, and the hypothesis test must be a two-tail test.

## 9.6 HYPOTHESIS TESTING IN EXCEL

One of the things you learned in this chapter is that all hypothesis tests have the same basic form. What changes from test to test is how to calculate the test statistic and which sampling distribution applies. It is most important to understand when each test is appropriate.

This is also true about using Excel to perform hypothesis tests. All that changes for the different tests is the input.

### 9.6.1 Small-Sample Tests for the Population Mean

Since Excel does not have any analysis tools for single population hypothesis tests, we have written a macro that will do this analysis. Suppose that we were looking at the hand washing data, and that we want to perform the test to see if the mean time spent washing hands by the 28 employees is different from 5 seconds. Since we do not have the population standard deviation and the sample size is small ($n < 30$), the $t$ test is appropriate.

You should have the data you want to analyze in an Excel worksheet and have the MacDoIt.xls worksheet open so that the macros are available. From the list of macros, choose **t Test**. The table shown in Figure 9.6 opens in a new worksheet. You can see that it is very similar to the one for the $Z$ test.

| | A | B | C |
|---|---|---|---|
| 1 | Hypothesis Test for Mean - t-Test | | |
| 2 | Input | | |
| 3 | Sample Mean | m | 95 |
| 4 | Hypothesized Mean | $\mu_0$ | 100 |
| 5 | Standard Deviation | s | 10 |
| 6 | Sample Size | n | 16 |
| 7 | Significance Level | $\alpha$ | 0.05 |
| 8 | Output | | |
| 9 | Standard Error | StdErr | 2.5 |
| 10 | Degrees of Freedom | df | 15 |
| 11 | t-Statistic | t | -2 |
| 12 | Lower-Tail Test for $H_A$: $\mu < \mu_0$ | | |
| 13 | Lower-Tail Critical t | $t_L$ | -1.7531 |
| 14 | Lower-Tail p-value | p-value$_L$ | 0.0320 |
| 15 | Two-Tail Test for HA: $\mu \neq \mu_0$ | | |
| 16 | Two-Tail Absolute Critical t | $t_T$ | -2.1315 |
| 17 | Two-Tail p-value | p-value$_T$ | 0.0639 |
| 18 | Upper-Tail Test for HA: $\mu > \mu_0$ | | |
| 19 | Upper-Tail Critical t | $t_U$ | 1.7531 |
| 20 | Upper-Tail p-value | p-value$_U$ | 0.9680 |

**FIGURE 9.6** Table for $t$ test macro

Follow these steps to carry out the test:

1. Position the cursor in the Input section in the row for the Sample Mean. Type **=AVERAGE(** and highlight the range that contains the data. Remember to type in the right parenthesis *before* you switch back to the sheet with the table.

2. After you hit Enter, you must type in the value for the hypothesized mean. In this case, we are testing whether the mean washing time is different from 5 seconds, so type in "5" and hit Enter.

3. Now you must enter the value for the sample standard deviation using the **STDEV** function.

4. Enter the sample size, in this case 28, and hit Enter again.

5. Now enter the level of significance for the test. Let's test the hypothesis at the 0.05 level of significance.

After you hit Enter the last time, the table will update. The results are shown in Figure 9.7.

| | A | B | C |
|---|---|---|---|
| 1 | Hypothesis Test for Mean - *t*-Test | | |
| 2 | Input | | |
| 3 | Sample Mean | m | 2.428571429 |
| 4 | Hypothesized Mean | $\mu_0$ | 5 |
| 5 | Standard Deviation | s | 3.655828137 |
| 6 | Sample Size | n | 28 |
| 7 | Significance Level | $\alpha$ | 0.05 |
| 8 | Output | | |
| 9 | Standard Error | StdErr | 0.690886578 |
| 10 | Degrees of Freedom | df | 27 |
| 11 | *t*-Statistic | t | -3.721925791 |
| 12 | Lower-Tail Test for $H_A$: $\mu < \mu_0$ | | |
| 13 | Lower-Tail Critical *t* | $t_L$ | -1.7033 |
| 14 | Lower-Tail p-value | p-value$_L$ | 0.0005 |
| 15 | Two-Tail Test for HA: $\mu \neq \mu_0$ | | |
| 16 | Two-Tail Absolute Critical *t* | $t_T$ | -2.0518 |
| 17 | Two-Tail p-value | p-value$_T$ | 0.0009 |
| 18 | Upper-Tail Test for HA: $\mu > \mu_0$ | | |
| 19 | Upper-Tail Critical *t* | $t_U$ | 1.7033 |
| 20 | Upper-Tail p-value | p-value$_U$ | 0.9995 |

**FIGURE 9.7**   Output from *t* test macro

From the output section, we see that the value of the *t* statistic is $-3.722$. Since this is a two-tailed test, look at that section of the output and compare the test statistic to the critical value of $-2.0518$. The decision is to reject $H_0$ and conclude that the mean hand washing time is not 5 seconds. We also see that the *p* value of the test is 0.0009, which is much smaller than the specified level of significance.

Recall that one of the assumptions of the *t* test for the mean is that the data are normally distributed. Although you cannot test this assumption formally yet, you can (and most certainly should) create a graphical display of the data to see whether the assumption of normality is valid. Using the boxplot macro in Excel produces the plot shown in Figure 9.8.

From the plot you can see that the data are really rather skewed to the right and that the assumption might not be a good one for these data.

**FIGURE 9.8**   Boxplot of hand washing data

## 9.6.2 $\chi^2$ Tests for the Variance in Excel

There is no built-in tool or macro for the $\chi^2$ test for the variance, but the test can be performed by using some built-in Excel functions. Most of these functions should be familiar to you.

We will look at the problem of the tissue manufacturer. When we did the test on the mean tissue strength in Chapter 8, we used the target value for the standard deviation, 50 lb/ream. Now we will test to see whether it was reasonable to assume that the population variance is 2500 (lb/ream)$^2$, using a $\chi^2$ test. The data are the same data that we used in Chapter 8.

Since we want to know whether the true population variance is different from the target value of 2500 (lb/ream)$^2$, we will do a two-tailed test. To do the test, we will need to calculate the value of the test statistic and the critical value for the test. We will also calculate the $p$ value for the test.

The test statistic for a $\chi^2$ test for the variance is

$$\chi^2 = \frac{(n-1)s^2}{\sigma^2}$$

To calculate this using Excel, the following steps are used:

1. To organize the output, put labels in worksheet cells to identify all of the output values. Then when you are done, you will know which value is which. In this case, label cells for $s^2$, $\sigma^2$, the sample size, the test statistic, the critical value of the test, and the $p$ value of the test. An example of this is shown in Figure 9.9.

| E | F |
|---|---|
| $s^2$ | |
| $\sigma^2$ | |
| n | |
| Test Statistic | |
| Critical Value | |
| p value | |
| | |
| | |

**FIGURE 9.9**   Setup for $\chi^2$ test

2. In the cell adjacent to the one labeled $s^2$, use the **STDEV** function to calculate the sample standard deviation, and then square it. The formula will look like the one shown in Figure 9.10 (page 460).

$$\boxed{=}\quad \text{=STDEV(B1:B76)^2}$$

**FIGURE 9.10**   Formula for variance

As an alternative to this, you can use the **VAR** function, which calculates the sample variance. The input for this function is the data range, just as in the **STDEV** function.

3.  In the cell adjacent to the one labeled $\sigma^2$, type in 2500, and type 75 in the one adjacent to $n$.

4.  To calculate the test statistic, we will need the three values just input. In the cell adjacent to Test Statistic, type in the following formula:

$$\boxed{=}\quad \text{=(F3-1)*F1/F2}$$

*Note: You need parentheses for $n - 1$.*

These are the cell locations from the worksheet we are using as an example. Be sure to use the cell locations from *your* worksheet if they are different. You could enter the numbers directly in the formula, but it really is easier to use the cell locations.

5.  We will use the **CHIINV** function to find the two critical values for the test. The **CHIINV** function works the same way as the **TINV** function. The input values are the upper tail probability and the degrees of freedom. Since $\alpha$ for this test is 0.05 and it is a two-tailed test, the probability for the lower critical value will be 0.975, and 0.025 for the upper value. The degrees of freedom are $75 - 1 = 74$. The upper tail formula is shown below:

$$\boxed{=}\quad \text{=CHIINV(0.025,74)}$$

The output value is 99.67838. Place the formulas for the upper and lower critical values in the two cells adjacent to the one labeled Critical Value.

6.  Remember that the $p$ value of the test is associated with the test statistic. To find the $p$ value of the test, we will use the **CHIDIST** function. This function accepts as input the value of the $\chi^2$ variable and the degrees of freedom, and outputs the tail area probability. In the cell adjacent to the one labeled $p$ value, type

$$\boxed{=}\quad \text{=CHIDIST(F4,74)*2}$$

The cell F4 is the reference of the cell that contains the test statistic value. We multiply the $p$ value by 2 because this is a two-tailed test.

The complete output is shown in Figure 9.11.

| E | F | G |
|---|---|---|
| $s^2$ | 3334.307 | 3334.307 |
| $\sigma^2$ | 2500 | |
| n | 75 | |
| Test Statistic | 98.6955 | |
| Critical Value | 99.67838 | 52.10282 |
| p value | 0.058305 | |

**FIGURE 9.11**   Output for $\chi^2$ test

Comparing the test statistic to the critical value, we find that we cannot reject $H_0$. That is, it is reasonable to assume that the population variance is 2500 $(lb/ream)^2$. The decision is a rather close one, as you can see from the $p$ value.

Remember that the $\chi^2$ test, just like the $t$ test, assumes that the population is normally distributed. Before you report any results, you should check this assumption with some type of graphical display.

Any hypothesis test can be done in Excel using the same type of procedure outlined here, although it is easier if there are built-in tools or macros available.

### 9.6.3 Hypothesis Tests for a Single Proportion in Excel

Although Excel does not have a built-in analysis tool for the single population proportion test, we have a macro that will accomplish this task. The macro is very similar to the ones for the $Z$ and $t$ tests.

Suppose we want to look at the data on the proportion of teenagers who get money from their parents. We want to determine whether the true proportion of teenagers who get money from their parents is different from 0.50 or 50%. The data for this problem have been partially analyzed, and we know that in the sample of 2000 teenagers, the proportion who receive money from their parents is $p = 0.47$ or 47%.

To carry out the test, follow this procedure:

1. Select **Tools > Macro > Macros** from the menu and run the **SinglePropTest** macro from the list. The table shown in Figure 9.12 opens in a new worksheet.

| | A | B | C | D | E | F |
|---|---|---|---|---|---|---|
| 1 | | | | | | |
| 2 | | Test Statistic for Proportion | | | | |
| 3 | | | | | | |
| 4 | | Sample Proportion | $p_s$ | 0.5 | | |
| 5 | | Hypothesized Proportion | $\pi$ | 0.5 | | |
| 6 | | Sample Size | $n$ | 100 | | |
| 7 | | | | | | |
| 8 | | Test Statistic | $Z$ | 0.0000 | | |
| 9 | | Significance Level | $\alpha$ | 0.05 | | |
| 10 | | Critical Values | | | | |
| 11 | | Lower Tail Test | Z1 | -1.6449 | p-value1 | 0.5000 |
| 12 | | Upper Tail Test | Z2 | 1.6449 | p-value2 | 0.5000 |
| 13 | | Two Tailed Test | Z3 | -1.9600 | | |

**FIGURE 9.12** Table for a single proportion test

2. Fill in each of the cells in the top portion of the table with the relevant data: $p = 0.47$, $\pi = 0.50$, and $n = 2000$.

3. Fill in the significance level of the test; in this case, we will use $\alpha = 0.05$.

The table will update and the result is shown in Figure 9.13 (page 462).

From the output, we see that the two-tailed critical value is $-1.96$ and that the test statistic is $-2.6833$. The decision is to reject $H_0$ and conclude that the true proportion is different from 50%. Also, we see that the $p$ value of the test is much smaller than the level of significance.

| | A | B | C | D | E | F |
|---|---|---|---|---|---|---|
| 1 | | | | | | |
| 2 | | Test Statistic for Proportion | | | | |
| 3 | | | | | | |
| 4 | | Sample Proportion | p$_s$ | 0.47 | | |
| 5 | | Hypothesized Proportion | $\pi$ | 0.5 | | |
| 6 | | Sample Size | n | 2000 | | |
| 7 | | | | | | |
| 8 | | Test Statistic | Z | -2.6833 | | |
| 9 | | Significance Level | $\alpha$ | 0.05 | | |
| 10 | | *Critical Values* | | | | |
| 11 | | Lower Tail Test | Z1 | -1.6449 | p-value1 | 0.0036 |
| 12 | | Upper Tail Test | Z2 | 1.6449 | p-value2 | 0.0036 |
| 13 | | Two Tailed Test | Z3 | -1.9600 | p-value 3 | 0.00729 |

**FIGURE 9.13** Output for single proportion test

You might wonder how you would proceed with the test if you had raw data and not the value of $p$ available. In this case you would use the Pivot Table tool to obtain a frequency table. The steps to do this were explained in Chapter 3. As part of the pivot table, you can either display the output as a count or as a percentage of the total.

# *CHAPTER 9 SUMMARY*

In this chapter you have learned how to do a hypothesis test of the population mean, population variance, and population proportion. The five-step hypothesis test procedure is the same for any hypothesis test. The differences in the tests are in the rejection regions and the calculation of the test statistic. For a test of the population mean, you use a $Z$ test if the population standard deviation is known; otherwise, you use a $t$ test. For a hypothesis test of the population variance use a chi-square test, and for a test of proportions use a $Z$ test.

You have seen that in any hypothesis test the rejection region and thus the final decision depend on the value of $\alpha$. You can alter the outcome of the test by adjusting the value of $\alpha$. To handle the potential unethical use of hypothesis testing and to provide management with more information than simply whether or not the null hypothesis was rejected, you learned about $p$ values. You have seen that the $p$ value frees you from specifying a value of $\alpha$ for the test. By reporting the $p$ value for the test, you put the decision to reject or fail to reject the null hypothesis in the hands of management. The decision maker must weigh the costs of Type I and Type II errors and make a decision using these costs and the $p$ value. The final section of this chapter ties the results of a hypothesis test to the corresponding confidence interval calculation.

## Key Formulas

| Term | Formula | Page Reference |
|------|---------|----------------|
| $t$ test statistic | $t = \dfrac{\overline{X} - \mu}{s/\sqrt{n}}$ | 430 |
| Chi-square statistic | $\chi^2 = \dfrac{(n-1)s^2}{\sigma^2}$ | 441 |
| $Z$ test statistic for proportions | $Z = \dfrac{p - \pi}{\sqrt{\pi(1-\pi)/n}}$ | 448 |

# CHAPTER 9 EXERCISES

## Learning It!

**9.15** Most traffic lights are set so that there is enough time for pedestrians to cross the road safely. A recent study indicates that a large number of elderly cannot get across the road in the usual 15 seconds allowed. To determine the average amount of time it takes senior citizens to cross the street, a study was taken of 25 seniors. On the average, it took them 19.5 seconds to cross the street, with a sample standard deviation of five seconds. Assume the time to cross the road has a normal distribution.

(a) Set up the null and alternative hypotheses to see if the data show that it does indeed take seniors longer than 15 seconds to cross the street.

(b) In terms of traffic flow, what are the implications of a Type I error?

(c) What are the implications of a Type II error?

(d) Find the value of the test statistic.

(e) If $\alpha = 0.05$, what is the rejection region?

(f) What is your recommendation to the city?

**9.16** Kids think Moms are doing a great job! According to a recent poll conducted by Massachusetts Mutual Insurance Company, 90% of children feel their mothers spend enough time with them. A similar survey was conducted in New York to see if children there felt the same way. Of 1000 children surveyed in New York 880 felt that their mothers spend enough time. Do these data indicate that less than 90% of the New York children feel their mothers spend enough time with them?

(a) Set up the null and alternative hypotheses.

(b) What is the value of the test statistic?

(c) If $\alpha = 0.10$, what is the rejection region?

(d) What do you conclude about the children of New York?

(e) What might be causing this difference?

**9.17** In a new advertising campaign, Coca-Cola pokes fun at the decision to revive its famous contoured bottle. Atlanta-based Coke faced significant technical hurdles in bringing out the well-known bottle in a new material: plastic. Coke also conducted extensive marketing tests to make sure that one of the world's best-known packages, seldom seen since the 1970s, would be a hit in the 1990s.

Coca-Cola wishes to be sure that at least 60% of the consumers prefer the contoured bottle. Of 3,000 consumers nationwide, 1900 prefer the new bottle. Is this sufficient evidence to give the new bottle the go-ahead?

(a) Set up the null and alternative hypotheses for Coca-Cola.

(b) In terms of Coke's decision to introduce the new bottle or not, what is a Type I error?

(c) In terms of Coke's decision to introduce the new bottle or not, what is a Type II error?

(d) What is the value of the test statistic?

(e) If $\alpha = 0.05$, what is the rejection region? What is your decision?

**9.18** You purchased your home in 1987 for $165,000, which was the average price at that time. Now you are thinking of selling your home. You take a sample of homes sold in your neighborhood during the past 2 months and find the following sale prices in $:

| 130,000 | 135,500 | 136,000 | 140,000 | 160,000 |
|---------|---------|---------|---------|---------|
| 167,000 | 168,000 | 174,500 | 177,400 | 180,000 |

(a) Based on these data, conduct the hypothesis test to see if the average selling price has increased from $165,000. Assume that the selling price of homes in this town is normally distributed.

(b) What advice would you give this homeowner regarding selling his home at this time? Explain your answer in terms of the results of the hypothesis test.

**9.19** According to the *Springfield Union News,* the New England Blizzard may not yet be ready to take a place among the American Basketball League's elite, but the Hartford–Springfield franchise is in a whole other dimension when it comes to getting people through the turnstiles. The next table shows the 1997 ABL attendance figures for the other ABL teams:

| Team | Dates | Attendance |
|------|-------|-----------|
| Portland | 4 | 19,856 |
| San Jose | 5 | 23,775 |
| Seattle | 3 | 9,233 |
| Philadelphia | 5 | 15,308 |
| Colorado | 3 | 9,176 |
| Columbus | 4 | 11,239 |
| Atlanta | 4 | 10,776 |
| Long Beach | 3 | 5,372 |

(a) Calculate the average attendance for the ABL league without the Blizzard.

(b) The Blizzard had an attendance figure of 49,879 in 6 games. Find the average attendance for the sample.

(c) Test the hypothesis that the Blizzard had a higher average attendance than the ABL league. Assume that attendance figures are normally distributed.

(d) What conclusions can you share with the owners of the Blizzard franchise?

(e) Upon further investigation you discover that the Blizzard plays at both the Hartford Civic Center and the Springfield Civic Center. What additional information would you like to know about the attendance figures to help the Blizzard increase attendance?

9.20 In the wake of a large number of repeat crime offenders, many states are considering the death penalty. A sample of 100 citizens of a New England state was taken and 56 of the 100 people favored the death penalty.

(a) Legislatures in favor of the death penalty are using the results of this survey to say that the majority of the people in the state are in favor of the death penalty. Conduct the appropriate hypothesis test to determine if these legislatures are justified in their remark.

(b) What conclusion can be reached from this survey about the preference for citizens of this state toward the death penalty?

## Thinking About It!

9.21 Coca-Cola has asked you your recommendation about their new bottle. Basically, the company wants to know if you are secure in your decision before it makes any drastic changes. You decide to test how much the procedure you chose affected your decision. *Requires Exercise 9.17*

(a) Suppose you had used $\alpha = 0.10$. What is your decision?

(b) Now use $\alpha = 0.01$. What is your decision?

(c) Find the $p$ value for this problem.

(d) Based on this new information, what would you tell the Coca-Cola company to do?

9.22 Look again at your recommendation to the city about the length of its walk light signal. In writing your report, you start to think about whether or not the study you did was adequate. *Requires Exercise 9.15*

(a) What other factors might be useful in determining the length of time to allow people to cross the street safely?

(b) What are the drawbacks of making decisions such as this based on the mean?

(c) What might be a better statistic to use in this case?

9.23 Many states are considering allowing casino gambling in an attempt to revitalize urban areas. Opponents say gambling would be a magnet for crime and would shift spending away from entertainment and other industries. A recent poll conducted by the Boston *Sunday Herald* found that 46% of the Bay State voters are opposed to gambling, while 36% support the proposal. Eighteen percent of those surveyed had no opinion or were neutral. The poll of 450 registered voters across the state was conducted between a Tuesday and a Friday. Legislatures will endorse the casinos if less than 50% of the voters oppose it. Do the data indicate that this is the case?

(a) Set up the null and the alternative hypotheses.

(b) What should you do with the 18% who had no opinion or were neutral?

(c) Find the value of the test statistic.

(d) If the members of the legislature wish to be conservative, how should they set $\alpha$?

(e) If the members of the legislature wish to be more aggressive on this issue, how should they set $\alpha$?

(f) Using the value of $\alpha$ set in part (d), find the rejection region.

(g) What is your recommendation to your local representative?

(h) Using the value of $\alpha$ set in part (e), find the rejection region.

(i) What is your recommendation now to your local representative?

*Requires Exercise 9.1*

**9.24** Remember the problem where you were trying to convince your boss that your expense report for the aspirin purchase was justified? Your boss just asked you for the data you used to justify the expense. He points out that the price of aspirin in Tokyo seems a bit high and asks you to reduce your estimate. The data are repeated below:

| City | Aspirin (100) | Fast food (hamburger, fries, soft drink) | Woman's haircut/blow dry | Toothpaste (6.4 oz) |
|---|---|---|---|---|
| Los Angeles | $7.69 | $4.15 | $20.11 | $2.42 |
| Tokyo | $35.93 | $7.62 | $76.24 | $4.24 |
| London | $9.69 | $5.80 | $44.35 | $3.63 |
| Sydney | $7.43 | $4.53 | $29.93 | $2.08 |
| Mexico City | $1.16 | $3.63 | $17.94 | $1.08 |

(a) If you remove Tokyo from the data set should you change your expense report?

(b) What does this tell you about the effect of outliers on the hypothesis test of $\mu$ when you have a small sample?

## Doing It!

*Data file:*
*HOSPITAL.XXX*

**9.25** A portion of the data file for the hospital that collected the data on hand washing is shown below:

| Observation | Unit | Time 1 | Time 2 |
|---|---|---|---|
| 1 | CCU | 3 | 16 |
| 2 | CCU | 2 | 7 |
| 3 | CCU | 0 | 5 |
| 4 | CCU | 5 | 8 |
| 5 | CCU | 2 | 15 |
| 6 | CCU | 0 | 15 |
| 7 | CCU | 2 | 20 |
| 8 | CCU | 3 | 16 |
| 9 | CCU | 0 | 18 |
| 10 | IMCU | 1 | 16 |
| 11 | IMCU | 2 | 8 |

You can see that information was also collected on the unit of the employee and a second time, labeled Time 2, was also recorded. Assume that the hand-washing times are normally distributed and are recorded in seconds.

(a) Find the sample means and sample standard deviations for the first hand-washing time, Time 1, for each unit.

(b) Test to see if any of the units had an average hand-washing time, Time 1, less than 5 seconds.

(c) The second hand-washing time, Time 2, was recorded after the employees received some training on the effects of not washing their hands long enough. Find the sample means and sample standard deviations for the second hand-washing time, Time 2, for each unit.

(d) Test to see if any of the units had an average hand-washing time after training, Time 2, of less than 5 seconds.

(e) Display all the initial hand-washing times, Time 1, in a histogram. Is the assumption of normality reasonable?

(f) Display all the second hand-washing times, Time 2, in a histogram. Is the assumption of normality reasonable?

(g) Test the hypothesis that the variance of the initial hand-washing time is 14 seconds.

(h) Test the hypothesis that the variance of the hand-washing times after training is greater than 14 seconds.

(i) Find the sample proportion of employees who increased their hand-washing time after training. Test the hypothesis that the proportion of employees who increased their time after training is greater than 0.80.

(j) Based on your analysis, write a report to the manager of the hospital.

CHAPTER **10**

# COMPARING TWO
# POPULATIONS

## TRAINING

As the number of management consultant firms rapidly grows, more and more companies are relying on company outsiders to provide different types of training. While using company personnel might be more cost effective, management feels that employees attach a stigma to in-house trainers (often their peers or even subordinates) and feels that outside consultants provide a more credible service.

In an attempt to determine whether there are any real differences in the results achieved by in-house and external trainers, a large corporation decided to try an experiment with its quality inspectors. The corporation took two groups of employees at a large manufacturing location who were training for the American Society for Quality (ASQ) certification exam as Certified Mechanical Inspector (CMI). One group used the on-site statistician to provide training, while the other group used an outside consultant.

The company measured several variables, including a "pretraining" score for the exam, an evaluation survey for the training course, and whether the engineer passed the actual certification examination. Gender and age of the employees were also noted. A portion of the data is shown below:

| Employee | Training | Gender | Age | Pre Test | Post Test | Certification Exam |
|----------|----------|--------|-----|----------|-----------|--------------------|
| 1 | I | M | 43 | 27 | 64 | F |
| 2 | I | M | 40 | 33 | 71 | P |
| 3 | I | M | 36 | 41 | 60 | F |
| 4 | I | M | 34 | 45 | 51 | F |
| 5 | I | M | 43 | 38 | 65 | F |
| 6 | I | M | 29 | 48 | 63 | F |
| 7 | I | M | 38 | 55 | 58 | P |

## 10.1   CHAPTER OBJECTIVES

In the previous chapter you learned how to do hypothesis tests for many different population parameters. In each case you tested hypotheses about whether the sample data could have come from a population whose parameter was equal to a specific value. While these tests provide valuable information to decision makers, they are not the only questions that you might want answered. For example, suppose you wanted to know the answer to the following questions:

- Who uses electronic mail more frequently, senior managers or midlevel managers?
- Is it better to promote a person from within the organization or hire from outside the organization?
- Do men and women behave the same way in the supermarket?
- Does one design of a golf ball travel farther than another?
- Do college graduates actually earn more money than high school graduates?

*In Chapter 9 we looked at hypothesis tests of a mean, proportion, or variance from a single population.*

Although the questions listed above cover widely different topics, they do have a common theme. In each case a *comparison* is being made. In this chapter we will extend the hypothesis testing procedure that we learned in Chapter 9 to handle the *comparison* of two population means, two population proportions, or two population variances. The basic five-step hypothesis testing procedure is the same. The only difference is the test statistic to be used.

This chapter covers the following material:

- Collecting Data From Two Populations
- Large-Sample Test of the Difference in Two Population Means
- Small-Sample Test of the Difference in Two Population Means
- Test of the Difference in Two Population Means—Paired Data
- Hypothesis Test of the Difference in Two Population Proportions
- Hypothesis Test of the Difference in Two Population Variances

## 10.2   COLLECTING DATA FROM TWO POPULATIONS

When you are comparing characteristics of two different populations, you must have a sample from each of the populations. These samples are usually selected independently of each other. In other words, the selection of one sample should not have any effect on the selection of the second sample. We will label all of the parameters of one population with a subscript 1 and all the parameters of the second population with a subscript 2. It does not matter which population you label 1 or 2. The populations and samples are shown in Figure 10.1.

**FIGURE 10.1**   Two populations and two samples

Consider the question posed about whether men or women spend more money on frozen foods. We could label the population of males as population 1 and the population of females as population 2. If we do this then the parameters and statistics corresponding to the male population will be identified with a subscript 1 and those describing the female population will carry a subscript 2. These are shown in the following lists.

| | Population 1: Males | Population 2: Females |
|---|---|---|
| Size of population | $N_1$ | $N_2$ |
| Population mean | $\mu_1$ | $\mu_2$ |
| Population standard deviation | $\sigma_1$ | $\sigma_2$ |
| Sample size | $n_1$ | $n_2$ |
| Sample mean | $\overline{X}_1$ | $\overline{X}_2$ |
| Sample standard deviation | $s_1$ | $s_2$ |

For this example, we would select a sample of men shoppers and a separate sample of women shoppers. We would ask all members of the sample how much money they spent on frozen foods in the past week. It is not necessary that the sample sizes be equal, but if possible it is desirable to have both sample sizes ($n_1$, $n_2$) greater than or equal to 30. The reason for this stems from the fact that the Central Limit Theorem generally applies when the sample size is 30 or greater. Remember that we developed the $Z$ test statistic based on the knowledge that the sample mean, $\overline{X}$, has a normal distribution.

However, often a single sample is selected and a qualitative variable is used to identify two populations for comparison. For the food shoppers example, we might select one sample of shoppers and then record the gender of the respondent as part of the data. This means that the data can then be divided into the two comparison populations after the data have been collected. If at the same time you collect data on the age of the person as "under 40" or "40 and over" then you can also compare the average frozen food expenditure for younger buyers to that of older buyers. Clearly, spending differences that are identified by gender or age could be of great assistance in developing a marketing strategy.

*Qualitative variables are often used to identify two populations for comparison.*

## 10.3 HYPOTHESIS TEST OF THE DIFFERENCE IN TWO POPULATION MEANS—OVERVIEW

There are several different cases to consider if you are testing a hypothesis of the difference in two population means. To determine into which case your data fit, you must first see if you have information on the population standard deviation of both populations. As you saw in Chapter 9, sometimes you know the value of the population standard deviation but most of the time you do not. If you know the value of the standard deviation for each of the two populations that you are comparing, then you will use a $Z$ test statistic. In this case, it does not matter how large the samples are. The underlying populations must be normally distributed. This is covered in Section 10.4.

*Population standard deviations are known or the sample sizes are large.*

If you do not know the value of the standard deviation for each of the two populations, then you must check the size of the sample. If each of the samples has 30 or more observations then you will be in what is commonly referred to as the "large-sample" case. This case is very similar to the situation when you know the standard deviations.

*Remember that when $n \geq 30$ the Central Limit Theorem generally applies.*

*Population standard deviations are unknown and sample sizes are small.*

If you do not know the population standard deviations and one or both of your samples has less than 30 observations, then you will be in the "small-sample" case. You must then decide if the variances in each of the two populations are the same, even though they are unknown, or different. Section 10.5 covers the case when you do not know the value of the standard deviations but you do know that the populations are normally distributed.

Finally, sometimes the data are collected in such a way that the samples are not independent. This happens quite often when you are comparing two different medical procedures or treatments. You wish to be sure that the differences that you observe are not simply due to the fact that you have different patients with different medical histories. In this case you might use the same patients in both samples. If your samples are dependent then you must use what is known as a paired $t$ test to compare the population means. This is the subject of Section 10.6.

*Paired data must be analyzed differently.*

In summary, three sections will be devoted to hypothesis testing of the difference in means. Section 10.4 will cover tests when the standard deviation is known, Section 10.5 will cover the cases when the standard deviation is not known, and Section 10.6 will cover the test for paired data. Regardless of the particular case, each hypothesis test will follow the five-step procedure outlined in Chapters 8 and 9. It is repeated here for you:

*Steps for any hypothesis test*

**Step 1:**   Set up the null and alternative hypotheses.

**Step 2:**   Pick the value of $\alpha$ and find the rejection region.

**Step 3:**   Calculate the test statistic and the $p$ value.

**Step 4:**   Decide whether or not to reject the null hypothesis.

**Step 5:**   Interpret the statistical decision in terms of the stated problem.

Since you are already familiar with hypothesis testing, in this chapter we often combine the first two steps, which define the test procedure, into a single step.

# 10.4   LARGE-SAMPLE TESTS OF THE DIFFERENCE IN TWO POPULATION MEANS

## 10.4.1 Large-Sample Tests of Two Means With Known Standard Deviations

The basic test concerning two population means occurs when we want to know whether the two samples come from populations with equal means and we *assume that the population standard deviations are known*. Although this may not seem reasonable, there are cases where a historical or specified standard deviation is appropriate.

### EXAMPLE 10.1   The Tissue Manufacturer

*Identifying the Two Populations*

The company that is looking at tissue strength wants to look at the problem again. Since tissue strength seems to vary quite a bit, your boss has asked you to find out if there is any difference in the strength of the tissues made on one day compared to another day. The company uses two measures of tissue strength: machine-directional strength *(MDStrength)* measured in lb/ream and cross-directional strength *(CD-Strength)* also measured in lb/ream. Right now he wants you to look at the *MD-Strength*.

The product specifications for *MDStrength* state that the mean should be 1000 lb/ream with standard deviation of 50 lb/ream. You have collected data on 75 sheets for each of three different days. For each sheet you recorded the *MDStrength* and the *CDStrength*.

The first thing to do is to identify the two populations to be compared. The tissue manufacturer wants to compare the strength of tissues made on day 1 to the strength of tissues made on day 2. Population 1 is all tissue sheets made on day 1 and we have a sample of size $n_1 = 75$ from this population. All tissue sheets made on day 2 comprise population 2 and we have a sample of $n_2 = 75$ from this population. ■

In this case, we will use the specified values of the population variance to perform the test. Of course, this is not something you want to do blindly. Just because you have specifications does not mean that you have compliance. Still, for our example we already established that the specified values for the population standard deviation are valid.

*Use the techniques of Chapter 9 to test the variance for each of the populations.*

Once you can establish that the values of the population standard deviations for both populations are known, you know what case we are dealing with and we are ready to proceed with the hypothesis test.

- *The first step of the procedure is to construct the null and alternative hypotheses.*

As with tests of a single mean, there are three different ways to set up the hypothesis test. The guidelines for determining which of these tests you should use are the same as presented in Chapter 9. The three possible setups are shown below:

**Two-Sided Test**

*Remember the key words different, less than, and greater than tell you which test to use.*

$H_0$: $\mu_1 = \mu_2$ or $H_0$: $\mu_1 - \mu_2 = 0$      Use this test if you wish to test if the mean of
$H_A$: $\mu_1 \neq \mu_2$      $H_A$: $\mu_1 - \mu_2 \neq 0$      population 1 is *different* from the mean of population 2.

**Lower-Tail Test**

$H_0$: $\mu_1 \geq \mu_2$ or $H_0$: $\mu_1 - \mu_2 \geq 0$      Use this test if you wish to test if the mean of
$H_A$: $\mu_1 < \mu_2$      $H_A$: $\mu_1 - \mu_2 < 0$      population 1 is *less than* the mean of population 2.

**Upper-Tail Test**

$H_0$: $\mu_1 \leq \mu_2$ or $H_0$: $\mu_1 - \mu_2 \leq 0$      Use this test if you wish to test if the mean of
$H_A$: $\mu_1 > \mu_2$      $H_A$: $\mu_1 - \mu_2 > 0$      population 1 is *greater than* the mean of population 2.

*Note: You should never have $\overline{X}$'s in the statement of the hypotheses. There is no need to formulate a hypothesis about the sample means because you know the values. You do not know the population means.*

Notice that the equals sign is always part of the null hypothesis. Also notice that the hypotheses are statements about the *relationship* between the size of the mean of population 1 and the mean of population 2. The tests do not give you information about the value of the means, only about how the value of $\mu_1$ compares to the value of $\mu_2$.

You can see that there are two ways to state each of the hypotheses. For each setup shown above the first way of writing the test makes a statement about the relative value of $\mu_1$ to $\mu_2$. The second way of writing the same test makes a statement about the value of the difference, $\mu_1 - \mu_2$. They are equivalent to each other. Look at the two-sided test. Clearly, if $\mu_1 = \mu_2$ then the difference between them must be zero. It is also possible to test for differences of values other than zero. For example, you could test that the difference between the two means is 10 or any other value you like. The procedure is basically the same with only a slight difference in the test statistic formula.

### EXAMPLE 10.2   The Tissue Manufacturer

*Setting Up the Test*

The tissue manufacturer wishes to see if the mean tissue strengths are the same for day 1 and day 2 tissues. The variable to be examined is *MDStrength*. The null and alternative hypotheses are

$$H_0: \quad \mu_1 = \mu_2$$
$$H_A: \quad \mu_1 \neq \mu_2$$

In addition to setting up the hypotheses, the test is defined by choosing a level of significance, $\alpha$.

*The rejection regions are found using the procedures from Chapter 8.*

Suppose for this test we choose $\alpha = 0.05$. The test statistic to be calculated in step 3 is a $Z$ statistic so we know how to find the rejection region for the two-sided test. Since we know that the total area of the rejection region must be equal to $\alpha$ and $\alpha = 0.05$, we split it in half and require the area of each tail of the normal distribution to be 0.025. From the standard normal tables we find the critical values for the test to be $\pm 1.96$. ■

Once the test has been defined, the data are collected and processed. This allows us to proceed with the third step of the procedure, calculating the value of the test statistic.

Remember that the test statistic is calculated from sample data and is used to decide between the null and the alternative hypotheses. If we are trying to decide if $\mu_1 = \mu_2$, then it makes sense to look at the size of the difference between $\overline{X}_1$ and $\overline{X}_2$. We know that even if the two population means are exactly the same, we will almost never get two equal sample means. So we don't expect the difference $\overline{X}_1 - \overline{X}_2$ to be zero even if the null hypothesis is true. However, if the null hypothesis is true, then the difference $\overline{X}_1 - \overline{X}_2$ should be small relative to the size of the standard error. Clearly, large differences in the sample means would lead us to be suspicious of the null hypothesis. The appropriate test statistic is a $Z$ statistic and is calculated as follows:

*Formula for test statistic for difference between two population means*

$$Z = \frac{(\overline{X}_1 - \overline{X}_2) - 0}{\sqrt{\sigma_1^2/n_1 + \sigma_2^2/n_2}}$$

*Remember: A point estimator is calculated from sample information only and is used to estimate an unknown population parameter.*

The numerator of the test statistic is simply comparing the evidence, $\overline{X}_1 - \overline{X}_2$, to the difference in the population means if $H_0$ is true. For the hypotheses we are using, if $H_0$ is true then the true difference is zero, so you see that the comparison is between $\overline{X}_1 - \overline{X}_2$ and zero. As we saw in Chapter 9, the size of the difference is always measured in terms of the standard error. Thus, the denominator is the standard error of the point estimator. In this case we are trying to estimate $\mu_1 - \mu_2$ and the natural point estimator is $\overline{X}_1 - \overline{X}_2$. The standard error of $\overline{X}_1 - \overline{X}_2$ is a natural extension of the standard error of $\overline{X}$, $\sqrt{\sigma^2/n}$, which we often wrote as $\sigma/\sqrt{n}$.

When we are testing for a difference of zero, the last part of the numerator is often left out entirely. This makes sense, but it is important that you know what you are really testing. When you need to test for a difference equal to some value other than zero, then you need to put that value in the formula for the test statistic in place of the zero. For example, if you wished to test to see if the difference in the populations means was equal to ten, then you would change the null and alternative hypotheses to the following:

$$H_0: \quad \mu_1 - \mu_2 = 10$$
$$H_A: \quad \mu_1 - \mu_2 \neq 10$$

You would then use the value of 10 in the formula for the test statistic instead of zero. This is called the hypothesized difference and is labeled $d$.

### EXAMPLE 10.3   The Tissue Manufacturer

*Calculation of Test Statistic*

Now we are ready to calculate the test statistic for the tissue manufacturer. The relevant data are shown below:

| Population 1: Day 1 Tissues | Population 2: Day 2 Tissues |
|---|---|
| $n_1 = 75$ | $n_2 = 75$ |
| $\overline{X}_1 = 990.8$ | $\overline{X}_2 = 977.0$ |
| $\sigma_1 = 50$ lb/ream | $\sigma_2 = 50$ lb/ream |

We can calculate the value of the test statistic $Z$ as follows:

$$Z = \frac{(990.8 - 977.0) - 0}{\sqrt{50^2/75 + 50^2/75}} = 1.69$$

Using the standard normal table, we can find the $p$ value of the test by looking up the test statistic of 1.69. We find that the area to the right of $Z = 1.69$ is 0.0455. Since this is a two-sided test, we double that and find that the $p$ value is 0.09. ∎

Although much of the work is done at this point, the test procedure is not. The fourth step of the hypothesis testing procedure requires you to reject or fail to reject the null hypothesis on the basis of the test statistic. At the fifth step of the procedure you interpret this decision in terms of the problem. Let's finish the tissue manufacturer problem now.

### EXAMPLE 10.4   Tissue Manufacturer

*Finishing the Test*

The test statistic for the comparison of tissue strengths was 1.69. The rejection region for the test had critical values of ±1.96. Since 1.69 is not outside the critical values, that is, it is not in the rejection region of the test, we fail to reject the null hypothesis. Our conclusion is that there is no evidence that the mean *MDStrength* of tissues made on day 1 is different than that of those made on day 2.

From previous experience in hypothesis testing, you know that you should also examine the $p$ value for these data. The $p$ value is less than 0.10. This means that if you set $\alpha$ larger than the $p$ value, say at 0.10, then you would reject the null hypothesis of equal means. The decision, in this case, depends very much on the choice of $\alpha$. We might say that the decision is "on the fence" with regard to these two hypotheses. Perhaps we should compare day 2 to day 3 to get a better understanding of what is happening. ∎

*TRY IT NOW!*

The tissue manufacturer also recorded values of *MDStrength* for 75 tissues made on day 3. The sample mean *MDStrength* for that day was found to be 1000.32 lb/ream. Is there any evidence that the average *MDStrength* is different on day 2 than day 3? Use $\alpha = 0.05$.

Identify

population 1 :
population 2 :

Step 1:

Step 2:

Step 3:

Step 4:

Step 5:

Once the conclusion of the test is reached, the real work in terms of decision making begins. The statistical analysis simply confirms or fails to confirm a hypothesis. It does not help you make decisions about the problem you are trying to solve.

## EXAMPLE 10.5  Tissue Strength

### Interpreting and Using the Results of a Test

As a result of our tests, we have concluded that the average *MDStrength* of the tissues made on day 1 and day 2 are equal but the average *MDStrength* of the tissues made on day 2 and day 3 are different. Clearly, something is not quite right with this process. How can we interpret this? What should we do next?

What we know as a result of this test is that the tissue strength is not behaving entirely as it is supposed to. It is not clear *how* the mean strength differs, or whether it is *always* different or just *sometimes different*. It is also not clear on which, if any, of the three days the *MDStrength* is correct.

This would be a good time to graph the data and see what is happening. This would help us look for any patterns in the data. ■

To answer the question we started with, we would like to compare all 3 days. That is, we would like to do a hypothesis test that looks like this:

$$H_0: \quad \mu_1 = \mu_2 = \mu_3$$
$$H_A: \quad \text{At least one of them is different.}$$

This is a test of more than 2 means and you cannot do such a test using the techniques of this chapter. You need to use a technique called analysis of variance, which is covered in Chapter 12.

## 10.4.2 Large-Sample Tests of Two Means With Unknown Standard Deviations

The large-sample test for the difference between two population means requires that *both* of the sample sizes be greater than 30. Remember that the only differences in the various hypothesis tests are in the calculation of the test statistic.

The test statistic for this case is very similar to the case when you know the standard deviations. Since the sample sizes are large, each individual sample standard deviation is a good estimate of the corresponding unknown population standard deviation. So, we simply use each of the sample standard deviations in the formula instead of the corresponding values of $\sigma$.

*Large-sample test of the difference between two means*

The test statistic becomes

$$Z = \frac{(\overline{X}_1 - \overline{X}_2) - 0}{\sqrt{s_1^2/n_1 + s_2^2/n_2}}$$

Finding the rejection region is the same as it is for any hypothesis test involving the $Z$ distribution—it depends on $\alpha$ and whether the test is one-sided or two-sided.

### EXAMPLE 10.6 Training Issues

***Setting Up the Hypotheses***

Recall the corporation that was interested in whether there was a difference in the results of in-house vs. outside training programs. The employees involved in the training completed a sample copy of the CMI certification examination at the end of each of the respective training programs. the company wants to know if the outside trainers achieve better results. In this case the two populations are:

Population 1: Employees trained by an outside consultant
Population 2: Employees trained in-house

Since the company is interested in whether one group is *better* than the other, it will use a one-sided test and the hypotheses are

$$H_0: \quad \mu_1 \leq \mu_2$$
$$H_A: \quad \mu_1 > \mu_2$$

Once the hypotheses are set up and you know the population parameters you are testing, you need to make a decision about which specific type test you will use. To do this you need to assemble the data you have and look at standard deviation and sample size. With tests about the mean, the first question is whether the population standard deviation is known. If it is, then you would perform the $Z$ test defined in the previous section. If it is not, then you need to look at the sample sizes to make the next decision. When both sample sizes $n_1$ and $n_2$ are large ($\geq 30$), then the test is also a $Z$ test with the test statistic just described.

### EXAMPLE 10.7 Training Issues

***Setting Up the Test***

The company looking at training issues assembled and processed its data as shown below:

|  | Population 1 (Outside Consultant) | Population 2 (In-house Training) |
|---|---|---|
| Sample size | $n_1 = 50$ | $n_2 = 47$ |
| Sample mean | $\overline{X}_1 = 75.46$ | $\overline{X}_2 = 63.55$ |
| Sample standard deviation | $s_1 = 4.12$ | $s_2 = 7.69$ |

You see that the company had 50 employees in the group receiving outside training and 47 people in the group being trained in-house. Since the study was a new one, the company had only sample measures of variation. It will be able to use a $Z$ test, because of the large sample sizes.

Once the company knows it will use a $Z$ test, it can find the rejection region. In this case there is no real reason to choose a level of significance other than $\alpha = 0.05$, so that is what is used.

The one-sided rejection region is found using the techniques of Chapter 9. The value of $Z_{cutoff}$ is found to be 1.645. That is, the rejection region is to the right of the value 1.645.

Steps 1 and 2 of the hypothesis test define the test procedure to be used. The only other computational portion of the test is to calculate the test statistic. This is completely determined by the test procedure and so it is a matter of using the correct formula.

### EXAMPLE 10.8  Training Issues

*Performing the Test*

The corporation looking at training issues knows that it is using a $Z$ test for its data. The test statistic is calculated as follows:

$$Z = \frac{(75.46 - 63.55) - 0}{\sqrt{4.12^2/50 + 7.69^2/47}}$$

$$Z = \frac{11.91}{1.2640} = 9.42$$

To find the $p$ value of the test, we look up the $Z$ score of 9.42 on the standard normal table and find that the $p$ value is 0.0000. ■

The remainder of the hypothesis testing procedure is the same for this test as it has been for any of the tests we have done. So let's complete the test by doing steps 4 and 5.

### EXAMPLE 10.9  Training Issues

*Conclusions and Interpretation*

Step 4: *Decide whether to reject the null hypothesis or fail to reject it based on the test statistic.* Since $Z = 9.42$ is most definitely in the rejection region, we reject the null hypothesis.

Step 5: *Interpret the decision in terms of the original problem statement.* Since we reject the null hypothesis, we conclude that the mean post-test score for the group trained by an outside consultant is higher than the group trained in-house. ■

You are probably not surprised at this conclusion since the average for the outside group was higher than the one for the inside group. However, this is not sufficient reason to conclude that they are different. You cannot simply look at the two $\overline{X}$ values and conclude that employees trained by an outside consultant did better on the post-test than those trained by an in-house person. Yes, $\overline{X}_1$ is higher than $\overline{X}_2$ if you look only at the observed difference. It is possible that such a difference might occur due to the random nature of the sampling and testing procedure. The only way you can tell if there is a "significant" difference is to compare the observed difference in the sample means to the size of the standard error. This is precisely what the test statistic does for you.

*Remember that the standard error is the standard deviation of the point estimator. It is the yardstick by which we judge all differences.*

## 📖 TRY IT NOW!

The management of the corporation looking at training issues are not convinced that they should switch to outside consultants for all of their training programs. They want to look at how the two groups compared on the pretest.

(*continued*)

The relevant data are given below:

|  | Population 1 (Outside Consultant) | Population 2 (In-house Trainer) |
|---|---|---|
| Sample size | $n_1 = 50$ | $n_2 = 47$ |
| Mean score on pretest | $\overline{X}_1 = 43.52$ | $\overline{X}_2 = 47.66$ |
| Standard deviation of pretest score | $s_1 = 14.80$ | $s_2 = 12.80$ |

Test to see if there is a difference in the mean pretest scores. Use $\alpha = 0.05$.

Step 1:

Step 2:

Step 3:

Step 4:

Step 5:

## 10.4.3 Exercises—Learning It!

**10.1** Many studies have been done comparing consumer behavior of men and women. One such on-going study concerns take-out food. In particular, the study focuses on whether there is a difference in the mean number of times per month that men and women buy take-out food for dinner. The most recent results of the study are shown below:

| | Population | |
| | Men | Women |
|---|---|---|
| Sample size | $n_1 = 34$ | $n_2 = 28$ |
| Sample mean | $\overline{X}_1 = 25.6$ | $\overline{X}_2 = 21.2$ |
| Population standard deviation | $\sigma_1 = 4.2$ | $\sigma_2 = 3.8$ |

Because the study has so much historical data, information is known about the population standard deviations.

(a) Set up the hypotheses to test whether there is a difference in the mean number of times per month that a person buys take-out food for dinner for men and women.

(b) Use the $Z$ test with known population variances to set up and perform the test. Use a level of significance of 0.05.

(c) Find the $p$ value for the test.

(d) Do the data provide evidence that the mean number of times per month for men differs from that for women?

(e) Does the choice of $\alpha$ in this case affect the decision?

**10.2** Professional employees who work for large corporations often contend that the mean salary paid by a company differs by location in the United States. To test that claim, data were collected on financial analysts working for a large corporation at locations in New England and in the upper Midwest. Because there is an extensive history of salary data, the population standard deviations are available. The study found the following results:

| | Population | |
| | New England | Upper Midwest |
|---|---|---|
| Sample size | $n_1 = 25$ | $n_2 = 18$ |
| Sample mean | $\overline{X}_1 = \$38,348$ | $\overline{X}_2 = \$36,782$ |
| Sample standard deviation | $s_1 = \$2336$ | $s_2 = \$2258$ |

(a) Set up the appropriate hypotheses to test whether the company's analysts in New England were paid more, on the average, than those working in the upper Midwest.

    (b) Use the $Z$ test with known population variances to set up and perform the test. Use a level of significance of 0.05.

    (c) Find the $p$ value for the test.

    (d) Do the data support the contention that the mean pay for analysts in New England is higher than that of analysts in the upper Midwest?

**10.3** A study at a university in New England focuses on binge drinking by students. The people who administered the study wanted to determine whether students who lived on campus had more episodes of binge drinking per semester on the average than those who lived at home. They surveyed all students who were enrolled in a required health course and obtained the following data:

|  | *Population* | |
|---|---|---|
|  | **On Campus** | **At Home** |
| Sample size | $n_1 = 220$ | $n_2 = 196$ |
| Sample mean | $\overline{X}_1 = 37.3$ | $\overline{X}_2 = 35.6$ |
| Population standard deviation | $\sigma_1 = 8.6$ | $\sigma_2 = 10.1$ |

The study used the same survey instrument used in many other similar studies and so they used the population standard deviation.

    (a) Set up the appropriate hypotheses to test whether students who live at home have fewer episodes of binge drinking per semester, on the average, than those who live on campus.

    (b) Use the $Z$ test with known population variances to set up and perform the test. Use a level of significance of 0.05.

    (c) Find the $p$ value for the test.

    (d) Do the data indicate that students who live at home binge less on the average?

**10.4** A company uses two suppliers to provide paper for the copy machines. The company has experienced an excessive number of paper jams and wonders if there is a difference in the paper provided by each supplier. The company collects data on the number of jams per ream of paper (500 sheets) over a period of 2 months. The summary data are

|  | *Population* | |
|---|---|---|
|  | **Supplier 1** | **Supplier 2** |
| Sample size | $n_1 = 45$ | $n_2 = 52$ |
| Sample mean | $\overline{X}_1 = 8.1$ | $\overline{X}_2 = 9.3$ |
| Population standard deviation | $\sigma_1 = 1.1$ | $\sigma_2 = 1.2$ |

    (a) Set up the appropriate hypotheses to test whether paper from the two suppliers results in the same average number of jams per ream.

    (b) Use the $Z$ test with known population variances to set up and perform the test. Use a level of significance of 0.05.

    (c) Find the $p$ value for the test.

    (d) Do the data indicate that there is a difference between suppliers on average?

**10.5** You are considering selling your house and need to choose between 2 real estate agencies. You collect data on the number of weeks that a house is on the market for both agencies. The summary statistics are shown below:

|  | *Population* | |
|---|---|---|
|  | **Agency A** | **Agency B** |
| Sample size | $n_1 = 25$ | $n_2 = 20$ |
| Sample mean | $\overline{X}_1 = 22.3$ | $\overline{X}_2 = 18.5$ |
| Sample standard deviation | $s_1 = 1.5$ | $s_2 = 2.2$ |

    (a) At the 0.05 level of significance, test to see if houses listed with Agency A are on the market on the average longer than those listed with Agency B.

    (b) Find the $p$ value for the test.

    (c) Which agency would you use and why?

# 10.5 SMALL-SAMPLE TESTS OF THE DIFFERENCE IN TWO POPULATION MEANS

When the population standard deviation is unknown and the sample size is large, the $Z$ test can be used. If the sample size is not large, then it is necessary to use a $t$ test, just as we did in the one-population tests.

## 10.5.1 Small-Sample Tests of Two Means With Unknown But Equal Standard Deviations

Unfortunately, we do not always have the luxury of large samples. Remember that data collection is expensive, so often the sample size for one or both of the samples is less than 30. This is called the "small-sample" case.

*The reason that 30 is used as the cutoff point has to do with the fact that the Central Limit Theorem applies for sample sizes of 30 or more.*

Just as in the one-population case, the small-sample tests for two population means will involve the $t$ distribution. The $t$ test carries with it the same assumption that the populations involved in the test are normally distributed. In addition, though, the two-population case requires that you know whether or not the two population variances can be considered equal. In this section, we look at both cases.

### EXAMPLE 10.10   Training and Gender

*Setting Up the Hypotheses*

In addition to looking at the results of training in-house vs. using an outside consultant, the company decided to look at gender issues in training. It took the group of people that had been trained in-house and differentiated them according to gender. The company wanted to know if there was any difference in the post-test scores for males vs. females.

We can call the females population 1 and the males population 2. Since the company is interested only in whether the groups are different, it can use a two-sided test. The hypotheses are

$$H_0: \quad \mu_1 = \mu_2$$
$$H_A: \quad \mu_1 \neq \mu_2$$ ∎

The first approach is to assume that even though you don't know the variability in the two populations, you know enough to believe that there is the same amount of variability in each population. At first this may seem a bit odd to you but actually it is not such a ridiculous assumption. Remember that you are comparing the means of two populations. Since you are measuring the same variable in both populations, it is quite possible that the two populations have a normal distribution with the same shape but with different means. Figure 10.2 illustrates such a situation.

There is, of course, a way to tell by doing a hypothesis test to compare two variances. This is the subject of Section 10.8.

**FIGURE 10.2**   Two normally distributed populations with the same variance but different means

For now, we will assume that equal variance has been confirmed through a statistical procedure. We have two estimates of this variance, one from each sample. We have $s_1^2$ and $s_2^2$, which are two different estimates of the same number. Instead of using two different estimates of the unknown but common population variance, it makes more sense to combine the data and find one better estimate of the variance. That is, instead of using $s_1^2$ and $s_2^2$ in the formula for the test statistic, we will "pool" the data from the samples and get one estimate of the common variance. This is called the **pooled variance.**

One way to estimate the pooled variance is simply to put all the data together from both samples and calculate the sample variance of this "big" sample. There are two reasons for not doing that. One reason is that if, indeed, the two population means are different, then the consolidated data will have a mean that is not from *either* population. This will result in a distorted measure of variance. In addition, at this point most people have already calculated the sample variances for each sample individually. Thus, we will find the pooled variance by building from the individual sample variances. Your first thought might be to average the two sample variances. This is the right idea, but instead we will use a weighted average. If, for example, one of the samples sizes is 10 and the other one is 25, then the sample variance from the sample based on 25 observations is more accurate and should carry a greater weight in calculating the pooled variance. The formula for the pooled variance, $s_p^2$, is shown below:

*Formula for the pooled variance*

$$s_p^2 = \frac{(n_1 - 1)s_1^2 + (n_2 - 1)s_2^2}{n_1 + n_2 - 2}$$

This formula weights the sample variances by one less than the sample size. If the sample sizes are the same, this formula is the same as a simple average of the two sample variances.

## EXAMPLE 10.11   Training and Gender

### Calculating the Pooled Variance

The data for the post-tests on the in-house group are given below:

|  | Population 1<br>Females Trained In-House | Population 2<br>Males Trained In-House |
|---|---|---|
| Sample size | $n_1 = 20$ | $n_2 = 27$ |
| Sample average | $\overline{X}_1 = 63.95$ | $\overline{X}_2 = 63.26$ |
| Sample standard deviation | $s_1 = 8.86$ | $s_2 = 6.87$ |

*As a double check, be sure the number you end up with for the pooled standard deviation is in between the two sample standard deviations. If it is not, recheck your work, as you have made a mistake.*

The calculation of the pooled variance of the in-house post test scores is then

$$s_p^2 = \frac{(20 - 1)(8.86^2) + (27 - 1)(6.87^2)}{20 + 27 - 2}$$

$$s_p^2 = \frac{1491.4924 + 1227.1194}{45} = \frac{2718.6118}{45} = 60.4136$$

$$s_p = \sqrt{60.4136} = 7.77$$ ■

Now we are ready to consider the test statistic for the small-sample case when the population variances are equal. The test statistic in this case has a $t$ distribution with $(n_1 + n_2 - 2)$ degrees of freedom. Remember that for simpler notation we can write the $t$ value as $t_{\alpha, n-1}$. The test statistic is calculated as follows:

*Test statistic using pooled variance*

$$t = \frac{\overline{X}_1 - \overline{X}_2}{\sqrt{s_p^2/n_1 + s_p^2/n_2}}$$

or equivalently

$$t = \frac{\overline{X}_1 - \overline{X}_2}{s_p\sqrt{1/n_1 + 1/n_2}}$$

Knowing the degrees of freedom and the formula for the test statistic, we can complete the hypothesis test.

### EXAMPLE 10.12    Gender and Training

*Conducting the Hypothesis Test*

Now the company looking at training issues can find the rejection region for the test and calculate the test statistic. Since it is doing a two-tail test, the cutoff values will be the $t$ statistic with 0.025 in the tail area and 45 degrees of freedom, $t_{0.025,\,45}$. From the $t$ table the critical values are found to be $\pm 2.014$. The test statistic is calculated as

$$t = \frac{63.95 - 63.26}{7.77\sqrt{\frac{1}{20} + \frac{1}{27}}}$$

$$t = \frac{0.69}{7.77\sqrt{0.08704}} = \frac{0.69}{(7.77)(0.2950)} = 0.3010$$

The company uses a statistical software package to look up the $t$ statistic and finds that the area outside the test statistic is 0.3824. Since this is a two-sided test, it doubles that to find that $p$ value $= 0.7648$.

   The results of the test show that since 0.3010 is not in the rejection region of the test, the company cannot reject $H_0$. There is not enough evidence to say that the mean test score for the male in-house group is different than the mean post-test score for the female in-house group. ■

## TRY IT NOW!

The company also wants to look at gender differences in the test scores for the group of employees trained by the outside consultant. It assembles the relevant data and finds the following:

| | Population 1 Females Trained by Outside Consultant | Population 2 Males Trained by Outside Consultant |
|---|---|---|
| Sample size | $n_1 = 21$ | $n_2 = 29$ |
| Sample mean | $\overline{X}_1 = 75.9$ | $\overline{X}_2 = 75.2$ |
| Sample standard deviation | $s_1 = 3.9$ | $s_2 = 4.4$ |

Find the pooled variance for the data.

(*continued*)

Test to see if there is evidence that the mean post-test score for the females trained by the outside consultant is different than the mean post-test score for the males trained by the outside consultant. Use a level of significance of 0.05.

Step 1:

Step 2:

Step 3:

Step 4:

Step 5:

## 10.5.2 Small-Sample Tests of Two Means With Unknown and Unequal Standard Deviations

If you test for equality of variances and find that the population variances are not equal, then you cannot pool the data. This is known as the Behrens–Fisher problem. Figure 10.3 depicts this situation. In this case both of the populations have a normal distribution with different variances.

*Second way to handle small samples*

**FIGURE 10.3**   Two normally distributed populations with unequal variances

The test statistic is called a separate-variance *t* test. As the name indicates, you must use the individual sample variances in the test statistic. In fact, the formula for calculating the test statistic looks exactly like the *Z* statistic for the large-sample case when the variances are unknown. It is calculated as follows:

$$t = \frac{(\overline{X}_1 - \overline{X}_2) - d}{\sqrt{s_1^2/n_1 + s_2^2/n_2}}$$

*Small-sample test with unknown standard deviations*

It does not follow a *Z* distribution because the estimates of the unknown variances are based on small samples. We know from our work in Chapter 9 that the appropriate distribution for small-sample situations is the *t* distribution. This is the case here as well, but the number of degrees of freedom is found according to a rather more complicated formula than you have seen thus far. This test statistic can be approximated by a $t$ distribution with $\nu$ (Greek letter pronounced "new") degrees of freedom. The formula for $\nu$ is given as follows:

*Formula for degrees of freedom for small samples and unequal variances*

$$\nu = \frac{\left(\dfrac{s_1^2}{n_1} + \dfrac{s_2^2}{n_2}\right)^2}{\dfrac{(s_1^2/n_1)^2}{n_1 + 1} + \dfrac{(s_2^2/n_2)^2}{n_2 + 1}} - 2$$

*Note: There are several equivalent formulas for calculating the degrees of freedom in this case.*

When you calculate $\nu$ you will most likely get a fractional result. But we know that degrees of freedom are integer values. Simply drop the fractional portion of the result to get just an integer value for the degrees of freedom. You may wonder why we do this. The reason is that the number that results for the degrees of freedom in this case will be smaller than the one used in the pooled variance case. The number of degrees of freedom in the test defines the test's ability to detect significant differences in the population means. That is, when you assume that the population variances are *not* equal, your hypothesis test becomes less *powerful* and the chances that you will make a Type II error for a given value of $\alpha$ increase. By not pooling variances you are more likely to miss true differences in the population means.

### EXAMPLE 10. 13   Training and Gender

***Small-Sample Test with Unequal Variances***

Suppose we repeat the test on gender in the in-house group, this time assuming that the population variances are not equal. How does this change the analysis? The relevant data are again:

|  | Population 1<br>Females Trained In-House | Population 2<br>Males Trained In-House |
|---|---|---|
| Sample size | $n_1 = 20$ | $n_2 = 27$ |
| Sample average | $\overline{X}_1 = 63.95$ | $\overline{X}_2 = 63.26$ |
| Sample standard deviation | $s_1 = 8.86$ | $s_2 = 6.87$ |

*Section 10.8 will cover the procedure for testing for equal variances.*

**Step 1:** Construct the null and the alternative hypotheses.

$$H_0: \quad \mu_1 = \mu_2$$
$$H_A: \quad \mu_1 \neq \mu_2$$

**Step 2:** Select $\alpha$ and find the rejection region.

To do this we need to find the degrees of freedom to use. Substituting in the formula we get

$$\nu = \frac{\left(\dfrac{8.86^2}{20} + \dfrac{6.87^2}{27}\right)^2}{\dfrac{(8.86^2/20)^2}{20 + 1} + \dfrac{(6.87^2/27)^2}{27 + 1}} - 2$$

$$= \frac{32.18308 \ldots}{0.84272 \ldots} - 2$$

$$= 38.18939 - 2 = 36.189$$

Taking the integer portion gives us 36 degrees of freedom. The value for $t_{\text{cutoff}}$ is $t_{0.025,\,36}$, found using a computer software package to be $\pm 2.0281$. ∎

Many statistical packages offer you the option of doing the $t$ test assuming equal variances or assuming unequal variances. To make an informed choice about whether you should pool the data, you must do a test for the equality of the variances. This is the subject of Section 10.8.

## 10.5.3 Exercises—Learning It

**10.6** The Board of Realtors for Greater Bridgeport, CT, wants to know whether the average price of a single family home in the area has increased in the past two years. They take a random sample of homes sold in 1995 and 1996 and calculate the following statistics:

|  | Population | |
|---|---|---|
|  | **1995** | **1996** |
| Sample size | $n_1 = 25$ | $n_2 = 25$ |
| Sample mean | $\overline{X}_1 = \$151{,}166$ | $\overline{X}_2 = \$160{,}669$ |
| Sample standard deviation | $s_1 = \$5332$ | $s_2 = \$6468$ |

(a) Set up the appropriate hypotheses to test whether there has been an increase in the mean selling price of a home in Greater Bridgeport.

(b) Calculate the pooled variance for the data.

(c) Assuming that the data are normally distributed, use the small-population test with equal variances to test the hypotheses. Use a level of significance of 0.10.

(d) Do the data provide evidence that the average price of a home in 1996 is higher than it was in 1995?

**10.7** The cost of shipping is of great concern to mail-order businesses. Traditionally, mail-order companies have used United Parcel Services (UPS) for their shipping. The U.S. Postal Service has been promoting their parcel post service as a competitor for this service. To determine whether parcel post is better than UPS ground, a mail-order company sent 20 packages via UPS ground and 20 packages via parcel post and tracked the number of working days for the packages to reach their intended destinations. A summary of the data is given below:

|  | Population | |
|---|---|---|
|  | **UPS Ground** | **U.S. Postal Service** |
| Sample size | $n_1 = 20$ | $n_2 = 20$ |
| Sample mean | $\overline{X}_1 = 7.3$ | $\overline{X}_2 = 7.1$ |
| Sample standard deviation | $s_1 = 0.9$ | $s_2 = 2.3$ |

(a) Set up the appropriate hypotheses to test whether the U.S. Postal Service is better than UPS Ground.

(b) Assuming that the data are normally distributed, use the small-population test with unequal variances to test the hypotheses. Use a level of significance of 0.05.

(c) Do the data provide evidence that the average delivery time for Parcel Post is less than UPS Ground?

**10.8** After months of working overtime, you have saved for some money for a set of new golf clubs and you want to make sure that you are buying the best. You can get a really good deal on Brand X clubs, but would make the sacrifice to buy Brand Z if they really improve your game. The salesperson allows you to take the number 3 wood from each brand and use them to hit balls on a driving range. The data for your efforts are summarized below:

|  | Population | |
|---|---|---|
|  | **Brand X** | **Brand Z** |
| Sample size | $n_X = 15$ | $n_Z = 15$ |
| Sample mean | $\overline{X}_X = 255$ | $\overline{X}_Z = 271$ |
| Sample standard deviation | $s_X = 8.7$ | $s_Z = 9.1$ |

(a) Set up the appropriate hypotheses to test whether Brand Z clubs are better than Brand X.

(b) Calculate the pooled variance for the data.

(c) Assuming that the data are normally distributed, use the small-population test with equal variances to test the hypotheses. Use a level of significance of 0.05.

(d) Should you spend the extra cash or save it for golf balls?

**10.9**  Few things are more frustrating than needing information and having to wait for it. The time spent on hold waiting for technical support to resolve software problems is one of the largest drains on productivity experienced by small businesses. Before a company will consider switching from its current word processing software (Microsoft Word for Windows 95) to a competitor (WordPerfect for Windows) the company will have to be convinced that it will spend less time on hold. The company decides to perform a test by placing 10 calls to each technical support line and recording the amount of time (in minutes) that the caller is on hold. The summary data are

|  | Population | |
| --- | --- | --- |
|  | **MS Word** | **WordPerfect** |
| Sample size | $n_1 = 10$ | $n_2 = 10$ |
| Sample mean | $\overline{X}_1 = 11.2$ | $\overline{X}_2 = 9.7$ |
| Sample standard deviation | $s_1 = 3.4$ | $s_2 = 1.6$ |

(a) Assuming that the data are normally distributed, use the small-sample test with unequal variances to test the hypotheses that the company will spend less time per call, on the average, with WordPerfect technical support.

(b) Should the company switch software?

**10.10**  Are all discount mail-order companies the same? In an attempt to answer this question, data were collected on the price of software for the top ten business software packages reported by *PC Magazine*, August 1997. Two well-known mail-order companies were asked for prices on each piece of software. The data are shown below:

| Top Ten Business Software Packages | Computability Price (\$) | PC Connection Price (\$) |
| --- | --- | --- |
| Windows 95 Upgrade | 88 | 95 |
| Norton Anti-Virus | 59 | 70 |
| McAfee ViruScan | 49 | 60 |
| First Aid 97 Deluxe | 54 | 58 |
| Clean Sweep III | 37 | 37 |
| Norton Utilities | 68 | 75 |
| Netscape Navigator | 45 | 40 |
| MS Office Pro 97 Upgrade | 300 | 310 |
| First Aid 97 | 32 | 35 |
| Win Fax Pro | 95 | 95 |

(a) For each vendor, calculate the average price and the standard deviation of the prices.

(b) Based on these statistics, do you think that the companies are the same or different? Why?

(c) Set up the hypothesis test to determine whether the average price for software is different at the two mail-order companies.

(d) Assume that the variances are equal and calculate the pooled variance.

(e) Assuming that the data are normally distributed, use the $t$ test with equal variances to perform the test at the 0.05 level of significance.

(f) What can you conclude about software prices at the two companies?

(g) What assumption was violated in this example?

# 10.6  SUMMARY OF TESTS OF THE DIFFERENCE IN TWO POPULATION MEANS—INDEPENDENT SAMPLES

The previous two sections have presented several different situations in which you might find yourself when testing the difference in two population means. This is a good place to summarize what you have learned so far. Table 10.1 lists the criteria for each of the tests.

**TABLE 10.1  Summary of Tests**

| Populations | Variances | Sample Sizes | Test |
|---|---|---|---|
| Independent and normal | Known | No restrictions | $Z$ test |
| Independent and non-normal | Unknown | $n_1$ and $n_2 \geq 30$ | $Z$ test |
| Independent and normal | Unknown but assumed equal | $n_1$ and $n_2 < 30$ | $t$ test with a pooled variance |
| Independent and normal | Unknown and assumed not equal | $n_1$ and $n_2 < 30$ | Modified $t$ test (Behrens–Fisher problem) |

For all but one of these situations the populations must be normally distributed. It is unlikely that you will know information about the distribution of the populations, so you must use the sample data to decide if this assumption is reasonable. To determine whether the data are normally distributed you should use a statistical test such as a chi-square goodness of fit test or a normal probability plot. The chi-square test is presented in Chapter 15 and normal probability plots are discussed in Chapter 11. Right now, you can consider the assumption of normality by using the descriptive techniques you learned in Chapters 3 and 4. You can look at boxplots, histograms, or dotplots of the data. Since the assumption of normality is an issue in small-sample tests, histograms often do not work well as visual tools. While these graphs are certainly not rigorous tests, unless there are large departures from normality (very skewed or non-mound-shaped distributions) the test will be valid. You could also apply the empirical rule for normality, which was covered in Chapter 4.

## EXAMPLE 10.14  Training and Gender

### Checking for Normality of Data

The company looking at gender issues in training knows that it needs to check the data to see if it is reasonable to assume that the population scores (males and females) are normally distributed. The company decides to make boxplots of the data and finds the following:

**Post-Test Scores for In-house Training**

From the boxplots the data appear to be reasonably symmetric—the medians are located near the centers of the boxes and the whiskers are approximately the same length. It would appear that normality is a reasonable assumption. ■

A quick look at the summary table tells you that all of these tests are for *independent samples*. Recall that this means that you select a sample from population 1 and it has no impact on the selection of the sample from population 2. This is typically what is done. However, sometimes it is necessary or desirable to use dependent samples instead. This is the subject of the next section.

## 10.7 TEST OF TWO POPULATION MEANS—DEPENDENT SAMPLES

### 10.7.1 Tests With Dependent Populations

In many cases, such as the example on training programs, data are collected *before* and *after* some experiment is performed. This type of data is known as **pretest** and **post-test** data. When data of this type are collected, the samples come not from *independent* populations, but from *dependent* populations. Often, in this type of situation, we are interested in knowing whether the experimental procedure has an effect on the variable being studied.

> **Pretest** and **post-test** conditions exist when data are collected on the same sample elements before and after some experiment is performed.

This is not the only way samples can be **dependent.** Sometimes each member of the sample from population 1 is matched or paired with a member of the sample from population 2. For example, in studying the effect on sales of a new product display, you would want to compare stores that are either similar in size or in similar locations. Otherwise, the apparent increase in sales that is observed might simply be due to the fact that a larger, busier store was randomly chosen to try out the new display. In this case, it is possible that the sales did not really increase due to the new display.

> **Dependent samples** are related to each other. The members of one sample are identical to or matched or paired with the members in the other sample according to some characteristic.

### *Discovery Exercise 10.1*
### INTRODUCTION TO EXPERIMENTAL DESIGN

Did you ever wonder why the restrooms in restaurants have a sign instructing all employees to wash their hands before returning to work? The owner of a large restaurant chain wanted to increase the amount of time that employees washed their hands after reading about the health implications of washing for only a few seconds. A random sample of 10 employees was selected. The amount of time each employee washed was recorded. The data were collected in such a way that the employees did not know they were being observed.

These employees were then educated on the benefits of hand washing. They watched a health video that detailed the benefits of increasing the amount of time they washed. One week after the training, employee hand washing was timed again. The times (in seconds) are shown in the next table:

| Employee | Before Training | After Training |
|:---:|:---:|:---:|
| 1 | 3 | 3 |
| 2 | 3 | 3 |
| 3 | 2 | 4 |
| 4 | 4 | 4 |
| 5 | 3 | 5 |
| 6 | 3 | 4 |
| 7 | 4 | 4 |
| 8 | 3 | 4 |
| 9 | 5 | 6 |
| 10 | 2 | 3 |

(a) Use a two-population paired $t$ test to test the hypothesis that no learning has occurred. Use a pooled variance.

$$H_0: \quad \mu_1 = \mu_2$$
$$H_A: \quad \mu_1 \neq \mu_2$$

(b) What is your conclusion?

(c) Why is this surprising? (*Hint:* Of the 10 employees, how many of them washed at least as long after the training?)

(d) Explain how these data are different from most of the two-sample data sets you have looked at up to this point in the chapter. Why might this be important?

(e) Suppose 10 different employees were used for the second timing. Would this make a difference? Explain why or why not.

In the situation described above, one of the characteristics that could be used to pair or match the sample members is the size of the store. A second is the location of the store. This is the first departure we have seen from random sampling. In fact, you are *trying* to match the items that are sampled in an attempt to control other variables that influence the data. Clearly, the size of the store will influence sales. In a sense, you match sample elements to keep the comparison "fair." Similarly, it would not be fair to compare the average miles per gallon of two cars driven by two people with very different driving styles. You would want either the same person testing both cars or people matched on the basis of their driving styles. There may be more than one characteristic that you wish to use to match the members of the two samples. This discussion is really the beginning of a topic known as experimental design. In designing the statistical experiment, you are trying to get the most information for your money. Paired data are the simplest case of a designed experiment. It fits naturally here because often paired data are collected to compare two population means and yet the data are incorrectly analyzed using the methods discussed in the last section.

You should *not* analyze these paired or matched data using the small-sample *t* test that was described in the previous section. If you do, you might miss differences that exist but are masked by improper analysis. In this section we see how to analyze dependent or paired data.

## 10.7.2 The Paired Sample Test

The problem with using the tests for independent populations is that if there is a large amount of variation among the sample elements, then the standard errors of the estimates will be high. Since we divide by the standard error to obtain our test statistic, this value is often smaller than it should be. Real differences in the populations are *hidden* by the amount of variation among the sample elements.

**TABLE 10.2** **Data set up for paired difference test**

| | Variable | | |
|---|---|---|---|
| **Observation** | **Pretest** $X_1$ | **Post-test** $X_2$ | **Difference** $d = (X_2 - X_1)$ |
| 1 | $x_{11}$ | $x_{21}$ | $d_1$ |
| 2 | $x_{12}$ | $x_{22}$ | $d_2$ |
| 3 | $x_{13}$ | $x_{23}$ | $d_3$ |
| $\vdots$ | $\vdots$ | $\vdots$ | $\vdots$ |
| $n$ | $x_{1n}$ | $x_{2n}$ | $d_n$ |
| Average | | | $\bar{d}$ |

How can we fix this problem? Well, if you think about it, the real question we are asking is whether the difference between the pretest and post-test measurements is significantly different from zero. In that case, we can look at the differences as our measurements and perform a test about them. The hypotheses for this test are then

$$H_0: \ \mu_d = 0 \qquad H_0: \ \mu_d \leq 0 \qquad H_0: \ \mu_d \geq 0$$
$$H_A: \ \mu_d \neq 0 \qquad H_A: \ \mu_d > 0 \qquad H_A: \ \mu_d < 0$$

Table 10.2 shows how the data for such a test are set up. Each observation in the sample is a row in the table. The second and third columns are the values of the pretest (population 1) and post-test (population 2). For example, in the second column, the entry $x_{13}$ is the pretest value for the third sample element. The fourth column, labeled Difference, is calculated by subtracting the pretest score from the post-test score. These differences are labeled $d_1, d_2, \ldots, d_n$. Finally, the average of these difference is found and labeled $\bar{d}$. That is,

$$\bar{d} = \frac{\sum\limits_{i=1}^{n} d_i}{n}$$

*Average difference*

## EXAMPLE 10.15 Training Issues

### Setting Up the Hypotheses for a Paired Difference Test

The company that was looking at the results of two different types of training programs wonders if there is really an improvement realized from any type of training program. It decides to analyze the in-house training program it has and look at the difference in the pretest and post-test scores for the people who have gone through the training. The company wants to know if the training increases the scores. The company knows it must use a paired difference test because the two populations are dependent. Since it wants to know if the program *increases* scores, it will use a one-sided test. The question is, in what direction do each of the hypotheses go?

If the training program does increase scores, then the differences (post-test − pretest) will be positive. In this case, the test should be an upper-tail test and the appropriate hypotheses are

$$H_0: \ \mu_d \leq 0$$
$$H_A: \ \mu_d > 0$$ ∎

You see that the direction of the test depends on what the experimenter thinks the effect of the experiment will be and the order in which you subtract the pretest and post-test values. Most paired difference tests are one-sided.

### EXAMPLE 10.16   Sleep Apnea

*Setting up Hypotheses*

Because of skyrocketing health-care costs, many hospital administrators are working to contain costs. Studies are being done to see if diagnosis and treatment of some conditions can be done at home at a significantly reduced cost. One such area of study is in the diagnosis and treatment of obstructive sleep apnea syndrome (OSAS), which affects 2 to 5% of the adult male population and is characterized by extremely heavy snoring. As a result of OSAS, breathing is suspended either partially or entirely, which can lead to suffocation.

Once OSAS is diagnosed, the traditional treatment calls for hospitalization to begin nasal continuous positive airway pressure (NCPAP). In an effort to reduce costs, a study was done where NCPAP was initiated at home.

One of the variables observed for each patient before and after treatment was the number of obstructions. If the treatment worked, then the average number of obstructions post-treatment should be less than the pretreatment average. The hospital running the study wants to know if home treatment of OSAS was effective.

Since the number of obstructions should decrease after NCPAP if treatment is effective, the analyst knows that the test should be one-tailed, and the differences (post-test − pretest) should be negative. This means that the test should be a lower-tail test and the appropriate hypotheses are

$$H_0: \quad \mu_d \geq 0$$
$$H_A: \quad \mu_d < 0$$ ∎

The hypothesis test is now essentially a test of a single mean. We know how to do this test from Chapter 9. The paired difference test is considered to be a small-sample test, so the test statistic will be a $t$ statistic. The only difference is that the mean we are testing is the mean difference between two populations rather than the mean of a single population. Otherwise, the procedure is the same.

We need to define the test statistic for the test. For a paired difference test, the test statistic has a $t$ distribution with $n - 1$ degrees of freedom. Remember that in this test $n$ is the *number of pairs* and not the number of data points:

*Paired difference test*

$$t = \frac{\bar{d}}{s_d/\sqrt{n}}$$

where $s_d$ is the standard deviation of the differences and is found in the usual way. Since the test is a $t$ test, we must, as always, have normally distributed populations.

Figure 10.4 shows typical data for a paired sample test. Since the data are plotted by observation, you can clearly see that the values after treatment are higher than the values before treatment for every observation in the sample. The statistical test will see if the differences are statistically significant.

**FIGURE 10.4**   Pre- and post-test data for similar observations

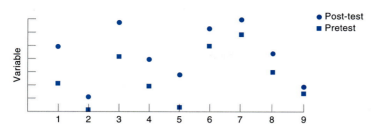

**FIGURE 10.5**    Pre- and post-test data for highly variable observations

If we used the small, independent sample $t$ test we would average all of the values in the first sample and all of the values in the second sample. In doing so we would lose some of the information contained in the sample. To see this effect, let's look at some different data to make a point. When you look at Figure 10.5, it is not obvious that the post-test data have a higher average if you do not view the data in pairs. The variation from sample element to sample element hides the differences between the pretest and post-test data. This is what happens when you simply average all of the pretreatment values and all of the post-treatment values and compare the sample means using a $t$ test.

### EXAMPLE 10.17    Training Issues

***Performing the Paired Sample Test***

The company that wants to know the value of training programs has 47 people who participated in the in-house training program. It analyzes the data and finds the following summary statistics:

$$n = 47$$
$$\bar{d} = 15.89$$
$$s_d = 8.45$$

The company decides to test with $\alpha = 0.05$. Since there are 47 pairs of data, the $t$ test will have $n - 1 = 46$ degrees of freedom and the critical values will be $t_{0.05,\ 46} = 1.679$, obtained from statistical software.

The test statistic is calculated to be

$$t = \frac{15.89}{8.45/\sqrt{47}} = 12.89$$

Clearly, this has a $p$ value of 0.0000. ■

### EXAMPLE 10.18    OSAS Data

***Doing the Paired Sample Test***

Let's complete the OSAS example using a level of significance of 0.05. If the treatment was effective, then the average difference should be significantly less than zero. If the treatment was not effective, then the pretest and post-test values would be close and the differences small, resulting in an average difference close to zero. The actual data for the $n = 9$ patients are shown in the table on page 498.

| | Number of Obstructions | | |
|---|---|---|---|
| Patient | Pretreatment | Post-Treatment | Difference |
| 1 | 365 | 22 | − 343 |
| 2 | 107 | 1 | − 106 |
| 3 | 28 | 12 | − 16 |
| 4 | 40 | 8 | − 32 |
| 5 | 48 | 3 | − 45 |
| 6 | 64 | 8 | − 56 |
| 7 | 109 | 0 | − 109 |
| 8 | 55 | 4 | − 51 |
| 9 | 20 | 0 | − 20 |
| $\overline{d}$ | | | − 86.44 |
| $s_d$ | | | 101.83 |

The test statistic follows a $t$ distribution with $n − 1$ degrees of freedom. The value of $n$ is equal to the number of pairs or the number of differences. In this case, $n = 9$ so there are 8 degrees of freedom. We need to find $t_{0.05, 8}$. From the $t$ table, the critical value is −1.860.

The test statistic looks just like the $t$ statistic for a single population mean. It is calculated as

$$t = \frac{-86.44}{101.83/\sqrt{9}}$$
$$= -2.546$$

The $p$ value of the test is found by looking up the test statistic in the $t$ distribution with 8 degrees of freedom. We find that the $p$ value is 0.0172. ∎

Once the testing procedure is done we require a decision and then an interpretation.

### EXAMPLE 10.19   Does Training Work?

*Deciding What the Test Results Mean*

The company looking at whether training programs are effective looks at the results of the test. It sees that the test statistic is in the rejection region, so they will reject $H_0$.

The company concludes that the average difference in score before and after the training program is greater than zero. This means that the training program is effective. ∎

### EXAMPLE 10.20   OSAS Data

*Conclusions*

In the study looking at the effectiveness of home treatment of sleep apnea, we see that the test statistic of −2.546 is in the rejection region, so we reject $H_0$.

The hospital administration can conclude that the average difference between the number of obstructions after and before home treatment is less than zero—that is, it has gone down. Therefore, the home treatment is effective. ∎

*TRY IT NOW!*

In addition to measuring the number of obstructions before and after home treatment in the OSAS study, the number of hypopnea both pre- and post-treatment for each patient is shown below.

| Patient | Pretreatment | Post-Treatment | Difference |
|---------|--------------|----------------|------------|
| | | *Hypopnea* | |
| 1 | 208 | 22 | |
| 2 | 297 | 0 | |
| 3 | 126 | 17 | |
| 4 | 150 | 13 | |
| 5 | 133 | 0 | |
| 6 | 201 | 10 | |
| 7 | 272 | 7 | |
| 8 | 137 | 1 | |
| 9 | 310 | 2 | |
| $\bar{d}$ | | | |
| $s_d$ | | | |

Complete the table above.

Is there evidence that the treatment has reduced the number of hypopnea at the 0.05 level of significance?

Step 1:

Step 2:

Step 3:

(*continued*)

Step 4:




Step 5:



## 10.7.3 Exercises—Learning It!

**10.11** Having learned about the paired *t* test you realize that you really should have used the test for the data on software price comparison. The data are repeated below:

| Top Ten Business Software Packages | Computability Price ($) | PC Connection Price ($) |
|---|---|---|
| Windows 95 Upgrade | 88 | 95 |
| Norton Anti-Virus | 59 | 70 |
| McAfee ViruScan | 49 | 60 |
| First Aid 97 Deluxe | 54 | 58 |
| Clean Sweep III | 37 | 37 |
| Norton Utilities | 68 | 75 |
| Netscape Navigator | 45 | 40 |
| MS Office Pro 97 Upgrade | 300 | 310 |
| First Aid 97 | 32 | 35 |
| Win Fax Pro | 95 | 95 |

    (a) Calculate the differences between the prices for each type of software package. Just looking at the differences, do you think that one company charges more than the other? Why or why not?

    (b) Calculate the average difference and the standard deviation of the differences.

    (c) Set up the hypotheses to test whether the mean difference in price between the two companies is zero.

    (d) Assuming that the data are normally distributed, at the 0.05 level of significance, is there a difference in the mean price of software for the two companies?

    (e) Did these results differ from the last time you analyzed the data? Why do you think this happened?

**10.12** A hospital administrator is concerned about the length of time that the nursing staff washes their hands. A recent study in health care showed that longer washing greatly reduces the spread of germs. The hospital observed the amount time that a sample of nine nurses in the Cardiac Care Unit (CCU) washed their hands. The data were collected in such a way that the employees did not know that they were being observed. The hospital then showed the nurses an educational video on the negative effects of shortened time spent hand washing. After the video, the hospital again watched and timed the group of nurses washing their hands. The data are shown below:

| Observation | Unit | Time 1 (s) | Time 2 (s) |
|---|---|---|---|
| 1 | CCU | 3 | 16 |
| 2 | CCU | 2 | 7 |
| 3 | CCU | 0 | 5 |
| 4 | CCU | 5 | 8 |
| 5 | CCU | 2 | 15 |
| 6 | CCU | 0 | 15 |
| 7 | CCU | 2 | 20 |
| 8 | CCU | 3 | 16 |
| 9 | CCU | 0 | 18 |

(a) Calculate the differences between the times for each nurse. Just looking at the differences, do you think that, on the average, they washed their hands longer the second time? Why or why not?

(b) Calculate the average difference and the standard deviation of the differences.

(c) Set up the hypotheses to test whether there was an increase in the average amount of time spent washing hands.

(d) Assuming that the data are normally distributed, at the 0.05 level of significance, what can you conclude?

(e) Can you conclude that the video caused the nurses to wash their hands longer? Why or why not?

**10.13** While it cannot be denied that advertising impacts sales, it is also true that advertising is expensive and companies want to advertise in ways that have the greatest benefit for the amount of money spent. A company that sells snack food designs two different advertising strategies, one focusing on print media and the other on radio. It ran the campaigns in a total of 16 different cities, paired on population size and measured the sales ($1000) in the week directly following the beginning of the campaign. The data are given below:

| Print Sales | Radio Sales |
|---|---|
| 28.3 | 22.1 |
| 24.6 | 19.1 |
| 23.1 | 20.3 |
| 21.0 | 24.4 |
| 25.7 | 22.4 |
| 22.5 | 19.2 |
| 32.0 | 22.8 |
| 23.5 | 20.3 |
| 24.3 | 25.5 |
| 25.2 | 22.6 |
| 23.3 | 24.9 |
| 25.3 | 29.7 |
| 22.2 | 22.2 |
| 23.4 | 28.5 |
| 23.9 | 28.2 |
| 25.7 | 21.6 |

(a) Calculate the differences in sales for each pair of cities.

(b) Based on these differences, do you think there is a difference in mean sales for the two types of advertising campaigns? Why or why not?

(c) Calculate the average difference and the standard deviation of the differences.

(d) Set up the hypotheses to test whether there is a difference in mean sales due to type of advertising.

(e) Assuming that the data are normally distributed, at the 0.01 level of significance, what can you conclude?

**10.14** How does time spent using the computer impact the speed with which you work? A software company ran a study that looked at the effectiveness with which a person uses a mouse. It selected 10 people, matched on computer skills, and measured the speed with which they moved a mouse at the beginning of a long session of computer use and after two hours of use. The data (in hundredths of a second) are shown below:

| | |
|---|---|
| 67 | 57 |
| 64 | 53 |
| 69 | 71 |
| 88 | 61 |
| 72 | 73 |
| 80 | 50 |
| 85 | 53 |
| 116 | 80 |
| 77 | 63 |
| 78 | 41 |

(a) Calculate the differences between the times for each person. Just looking at the differences, do you think that there was a change in the average speed after two hours? Why or why not?

(b) Calculate the average difference and the standard deviation of the differences.

(c) Set up the hypotheses to test whether there was a change in the mean speed with which the people moved the mouse.

(d) Assuming that the data are normally distributed, at the 0.05 level of significance, what can you conclude?

**10.15** Does gender impact the use of electronic mail (e-mail)? An insurance company studied the use of e-mail in its organization by counting the number of business-related e-mails generated by 10 men and 10 women matched on job position in a day. The data are given below:

| Men | Women |
|---|---|
| 82 | 48 |
| 77 | 61 |
| 78 | 56 |
| 83 | 59 |
| 82 | 58 |
| 78 | 56 |
| 81 | 60 |
| 74 | 64 |
| 86 | 59 |
| 76 | 63 |

(a) Calculate the differences in the number of business-related e-mails for each male/female pair. Just looking at the differences, do you think that men use e-mail more? Why or why not?

(b) Calculate the average difference and the standard deviation of the differences.

(c) Set up the hypotheses to test whether the average number of business-related e-mails generated by men is greater than that generated by women.

(d) Assuming that the data are normally distributed, at the 0.05 level of significance, what can you conclude?

# 10.8  HYPOTHESIS TEST FOR THE DIFFERENCE IN TWO POPULATION PROPORTIONS

## 10.8.1 Two-Population Tests With Qualitative Data

One of the things you should realize by now is that different types of data require different statistical techniques. So far, all of the two-population tests that we have looked at involve quantitative data. In this section, we look at the two-population tests for a kind of qualitative data—population proportions or percentages.

A lot of data are available in the form of proportions or percentages. You see this type of data all the time in the newspaper. Here are some examples taken from the newspaper.

- A study of Americans in 1987 found that 71% said the government should take care of people who can't take care of themselves. In 1994 the percentage was 57%. Is there evidence that Americans have become more cynical and less compassionate?

- A recent nationwide poll of 1225 adults showed that blacks and whites differ in their opinions of the causes and solutions of society's problems. But do they really? Seventy percent of blacks feel progress has been made in easing racial tension in the past decade compared to 65% of whites. Based on these data, do blacks and whites really feel differently about this issue?

- Researchers in San Francisco using the diseased cells of melanoma patients have developed a vaccine that they say dramatically reduces the recurrence of the deadliest form of skin cancer. After three years, 70% of those vaccinated remained cancer-free, compared to 20% of patients treated with surgery alone.

- A study of 1049 men and women aged 18 to 65 shows that a greater percentage of women (86%) find it difficult to have sex without emotional involvement compared to men (71%).

- A study conducted by an on-line service found that 30% of respondents under age 45 drove sports cars compared to 17% of the 45 or over population.

Each of the studies cited above has a structure similar to those we have already looked at in this chapter. In each case two populations are being compared and we have taken a sample from each population. What has changed is that the parameter being analyzed is no longer the mean, but the population proportion. For each sample, the percentage of the sample that has a certain characteristic is found. These percentages then need to be compared.

## 10.8.2 The Test for Two-Population Proportions

We know that even if the percentage of two populations that have a certain characteristic were exactly the same, we would almost never get exactly the same percentage in two samples from the populations. This is due to sampling error. The question then becomes, how large a difference in the percentages is large enough to declare the difference statistically significant? That is, how large does the difference have to be for us to be reasonably sure that there really is a difference in the behavior of the two populations, that something more than just sampling error is causing the difference? This is the same question we addressed in comparing means. This time the population parameters being compared are proportions and the evidence from the sample is summarized by the sample proportions. Our notation will follow the pattern established for the tests of means.

|  | Population 1 | Population 2 |
|---|---|---|
| Size of population | $N_1$ | $N_2$ |
| Sample size | $n_1$ | $n_2$ |
| Population proportion | $\pi_1$ | $\pi_2$ |
| Number in sample that have the characteristic being studied | $x_1$ | $x_2$ |
| Sample proportion | $p_1 = x_1/n_1$ | $p_2 = x_2/n_2$ |

As in the tests for comparing two population means, there are two different but equivalent ways of stating the null and alternative hypotheses. You can also test for a difference in proportions other than zero. The null and alternative hypotheses for the upper-tail and lower-tail tests of differences in proportions are shown below.

**Lower-Tail Test**

$H_0: \ \pi_1 \geq \pi_2$ or $H_0: \ \pi_1 - \pi_2 \geq 0$    Use this test if you wish to test if the pro-
$H_A: \ \pi_1 < \pi_2$       $H_A: \ \pi_1 - \pi_2 < 0$    portion of population 1 is *less than* the proportion of population 2.

**Upper-Tail Test**

$H_0: \ \pi_1 \leq \pi_2$      $H_0: \ \pi_1 - \pi_2 \leq 0$    Use this test if you wish to test if the propor-
$H_A: \ \pi_1 > \pi_2$      $H_A: \ \pi_1 - \pi_2 > 0$    tion of population 1 is *greater than* the proportion of population 2.

### EXAMPLE 10.21    Is Training Really Better?

*Null and Alternative Hypotheses*

The company that is looking at training programs has learned that the mean score on the sample CMI certification exam is higher for the group of employees that was trained by the outside consultant than for the group that was trained in-house. Some upper managers are ready to abandon all in-house training and move to outside training immediately. Other managers, however, are not so sure. They reason that the end result is really whether or not the employees pass the actual certification exam. If the percentage of employees that pass is the same in both groups, then why bother with the more expensive route? They decide to look at the proportion of employees who pass the exam in each group. If the percentage that pass for the outside consultant is higher than the percentage for the in-house group, then they will consider switching.

The ASQ does not publish scores for people who take the exam; scores are simply classified as "Pass" or "Fail." The data for the two groups in question are found to be

| | Population 1 In-House | Population 2 Outside Consultant |
|---|---|---|
| Sample size | 47 | 50 |
| Population proportion | $\pi_1$ | $\pi_2$ |
| Number that passed CMI | 13 | 20 |
| Sample proportion | $p_1 = \dfrac{13}{47} = 0.277$ or 27.7% | $p_2 = \dfrac{20}{50} = 0.400$ or 40.0% |

Since the managers want to know if the proportion for the outside trained group is higher than that of the in-house group, they want to do a one-tail test. They set up their hypotheses as

$$H_0: \ \pi_1 \geq \pi_2 \quad \text{or} \quad H_0: \ \pi_1 - \pi_2 \geq 0$$
$$H_A: \ \pi_1 < \pi_2 \quad\quad\quad\; H_A: \ \pi_1 - \pi_2 < 0$$

This may seem strange to you, since the managers want to know if the outside group has a higher percentage of passes than the in-house group, but if you think about it, that is the same as saying the in-house group has a lower percentage of passes. The managers used the latter because of the way they labeled their populations. ∎

After you become proficient at hypothesis testing you might want to change from the standard subscripts of 1 and 2 to denote populations. It is often easier to use subscripts that reflect the two populations, such as M and F for comparisons of males and females or O and I for outside and in-house. In fact, we will do that in the next example.

## EXAMPLE 10.22   E-Mail Usage

### Null and Alternative Hypotheses

Although most companies have implemented some form of electronic mail (e-mail) to increase the frequency and timeliness of information, very little information is available on the impact that e-mail has had on organizational behavior. A study of 5 different companies was conducted to learn how e-mail is being used at different levels of the organization. A total of 70 people responded to the survey. The respondents provided information on their title (VP/Director, Manager, Staff/Administrator), their gender, and the degree to which they used e-mail for various purposes. All companies in this study had utilized e-mail for at least one year. In this case, the study focused on the differences in the proportion of males and females that used e-mail to communicate decisions.

|  | Population 1<br>Men | Population 2<br>Women |
|---|---|---|
| Sample size | 35 | 35 |
| Population proportion | $\pi_M$ | $\pi_W$ |
| Number that use e-mail | 13 | 27 |
| Sample proportion | $p_M = \dfrac{13}{35} = 0.371$ or | $p_W = \dfrac{27}{35} = 0.771$ or |
|  | 37.1% | 77.1% |

We wish to decide if there is evidence of different use of e-mail by men and women. Since we wish to detect differences in the two population proportions we know to use a two-sided test. The null and alternative hypotheses are shown below:

$$H_0: \quad \pi_M = \pi_W \quad \text{or} \quad H_0: \quad \pi_M - \pi_W = 0$$
$$H_A: \quad \pi_M \neq \pi_W \qquad\qquad H_A: \quad \pi_M - \pi_W \neq 0$$

*Remember that sample proportions must always be between 0 and 1 or between 0 and 100 if expressed as a percentage.*

Recall that the test statistic of a single-population proportion was a $Z$ statistic. The extension to comparing two-population proportions is also a $Z$ test statistic. Therefore, we already know how to find the rejection region.

The test statistic for this test is constructed using logic similar to what we have used before. The two sample proportions are compared by subtracting one from the other. To determine if this difference is "large," it is compared to the standard error of the estimate. In this case, the estimate of the true difference in the population proportions, $\pi_1 - \pi_2$, is the difference in the sample proportions, $p_1 - p_2$. The standard error of this estimate is similar to the standard error for a single-sample proportion. The test statistic is then given by the formula

$$Z = \frac{p_1 - p_2}{\sqrt{\bar{p}(1 - \bar{p})\left(\dfrac{1}{n_1} + \dfrac{1}{n_2}\right)}}$$

*Test statistic for comparing two population proportions*

*Remember that the test statistic is calculated under the assumption that the null hypothesis is true.*

Notice that $\bar{p}$ is used in the calculation of the standard error. The value of $\bar{p}$ is similar to the pooled variance in the sense that it combines all of the sample data. If the null hypothesis is true and the population proportions are equal, then the best estimate of the common but unknown population proportion is obtained by pooling the data. The formula for $\bar{p}$ is then

*Common population proportion*

$$\bar{p} = \frac{x_1 + x_2}{n_1 + n_2}$$

## EXAMPLE 10.23   Is Training Really Better?

### Performing the Test

After thinking about the problem, the company testing to see if the proportion of passes from the in-house training group is lower than that of the outside training group decides that a Type I error will be more costly. The managers reason that if they reject $H_0$ in error and conclude that the in-house group has a significantly lower percentage of passes, they will wind up making changes and implementing expensive programs in error. They decide to test at the 0.01 level of significance.

The critical value for the test is $Z = -2.33$.

They calculate the value of $\bar{p}$ to be

$$\frac{13 + 20}{47 + 50} = 0.340 \text{ or } 34\%$$

The value of the test statistic is then

$$Z = \frac{0.277 - 0.400}{\sqrt{(0.340)(0.660)\left(\frac{1}{47} + \frac{1}{50}\right)}}$$

$$= -1.28 \qquad\blacksquare$$

## EXAMPLE 10.24   E-Mail Usage

### Doing the Test

For the study of e-mail usage, there is no reason to test at any level other than 0.05. The rejection region for a two-tail $Z$ test is defined by the critical values $\pm 1.96$.

The value of $\bar{p}$ is calculated as follows:

$$\bar{p} = \frac{13 + 27}{35 + 35} = 0.571$$

and the test statistic is found to be

$$Z = \frac{0.371 - 0.771}{\sqrt{(0.571)(0.429)\left(\frac{1}{35} + \frac{1}{35}\right)}}$$

$$= -3.39 \qquad\blacksquare$$

As in any hypothesis test, we are not finished until we make a decision about $H_0$ and interpret that decision in terms of the problem we are trying to solve.

## EXAMPLE 10.25   Is Training Really Better?

### Conclusions

The group of managers who are looking at training programs found that the test statistic for their data was $-1.28$. Since this is not in the rejection region, they fail to reject $H_0$. Thus, they conclude that there is not enough evidence to say that the group trained in-house has a lower percentage of passes on the certification exam than the group trained by the outside consultant.

Does this end the issue? Not by a long shot. The managers who are promoting the idea of using outside consultants are still not convinced that it is not a better route. They argue that even if the percentage of passes *is* the same in both groups there are other factors to consider. They want to look at employee perceptions about the training programs. They also argue that the higher test scores on the post-test must mean something. Perhaps the true difference will be evident in subsequent job performance.

What you see here is that statistics *cannot* solve the problem. Both groups have legitimate arguments. What the group doing the study has to decide on is *what* measurable outcome they really want to see. If it is the percentage of their employees who pass the certification exam, then in-house training is probably sufficient. If they are interested in employee attitudes and job performance issues, then they have a lot more work to do before they can make a good decision. ■

### EXAMPLE 10.26  E-Mail Usage

*Conclusions*

For the question about using e-mail we see that the test statistic is in the rejection region, so we reject the null hypothesis. We conclude that a different proportion of men and women use e-mail to communicate decisions. Although this tells the company something, the results of the test do not indicate whether men or women use e-mail more. If this is what the company wanted to know, then it should have done a one-sided test. ■

As you can see, testing the difference in two population proportions is not very different from testing the difference in two population means. This test has been developed parallel to the procedure for testing the difference in two population means. It has also been developed as a natural extension of the test of a single-population proportion.

 **TRY IT NOW!**

### E-mail Usage

*Tests of Two Proportions*

The company looking at use of e-mail is also interested in the use of e-mail for personal messages. After seeing the results of the last test, it decides that what it really wants to know is if a higher proportion of women than men use e-mail to send personal messages. Use $\alpha = 0.05$. The data are shown below:

|  | Population 1 Men | Population 2 Women |
| --- | --- | --- |
| Sample size | 35 | 35 |
| Population proportion | $\pi_1$ | $\pi_2$ |
| Number that use e-mail | 15 | 18 |
| Sample proportion | | |

Finish the table and then complete the hypothesis test.

Step 1:

*(continued)*

Step 2:

Step 3:

Step 4:

Step 5:

## 10.8.3 Exercises—Learning It!

**10.16** People who watch major league baseball may have noticed a decrease in the percentage of successful stolen bases. The increase in slide-step pitching motion and players who focus on homers may be behind the decline, according to an article in *The New York Times*, 24 May 1994. In 1993 there were 957 stolen bases in 1446 attempts, while in 1994 there were 807 stolen bases in 1193 attempts.

ANS. $p_1 = 0.43$, $p_2 = 0.51$; $H_0$: $\pi_2 \leq \pi_1$; $H_A$: $\pi_2 > \pi_1$; $Z_{\text{cutoff}} = 1.645$; $\bar{p} = 0.47$; $Z = 0.67$; FAIL TO REJECT $H_0$; MEN

(a) Calculate the sample proportion of stolen bases for each year.

(b) Set up the hypotheses to test whether there has been a decrease in the percentage of successful stolen bases.

(c) At the 0.05 level of significance, has the percentage really decreased?

**10.17** Does television match reality? A recent study looked at the percentage of television characters who survived cardiopulmonary resuscitation (CPR) compared to those who survive in real life. The study looked at 34 instances where CPR was applied to a television character and found that 25 survived, while data provided by Emergency Medical Technicians found that in 40 similar cases, only 16 survived.

(a) Calculate the sample proportion of people who survive CPR for television and for reality.

(b) Set up the hypotheses to test whether there is a difference in the proportion of people who survive CPR in the two groups.

(c) At the 0.05 level of significance, does television reflect reality?

**10.18** Women who smoke suffer an increased risk of dying of breast cancer, according to a recently published study. In the study, out of 319,000 women who never smoked there were 468 deaths from breast cancer, while out of 120,000 smokers, there were 187 deaths.

(a) Calculate the sample proportion of women who died of breast cancer for smokers and nonsmokers.

(b) Set up the hypotheses to test whether the proportion of women who die of breast cancer is higher for smokers than for nonsmokers.

(c) At the 0.05 level of significance, can you conclude that smoking causes breast cancer? If not, what can you conclude?

**10.19** Selling personal computers is big business and consumers are becoming increasingly aware of vendor reputation. A recent study of two vendors of desktop personal computers reports on the units that need repair for Dell Computer and Gateway 2000. Of 1584 computers manufactured by Dell Computer, 427 needed repair, while for Gateway 2000, 825 of 2662 computers needed repair.

(a) Calculate the sample proportion of computers needing repair for each company.

(b) Set up the hypotheses to test whether the proportion of computers needing repairs is different for the two companies.

(c) At the 0.05 level of significance, what can you conclude?

**10.20** How punctual are Amtrak trains? One of the reasons that passenger trains are late is that almost all of the tracks are owned and maintained by freight companies who control the schedules for use. Amtrak's definition of on time is based on trip length. After customer complaints about late trains, Amtrak has tried to improve and claims that it has. Of 100 runs of the Metroliner between New York City and Washington, DC, in March 1993, 87 were on time, while in the same period in 1994, 90 were on time.

(a) Calculate the sample proportion of on-time trips for each time period.

(b) Can Amtrak really claim it has improved ($\alpha = 0.05$)? Why or why not?

# 10.9 HYPOTHESIS TEST OF THE DIFFERENCE IN TWO POPULATION VARIANCES

## 10.9.1 The F Test for Comparing Population Variances

Remember the test to compare population means when the samples are small and the standard deviations are unknown? In working with small samples, we had to decide whether the populations had a common but unknown variance. If they did then we pooled the data and calculated a pooled estimate of the variance, $s_p^2$. We did not, however, discuss how to decide if the populations had a common variance. Now it is time to learn how to determine this. A hypothesis test of variances follows the same five steps that we have repeatedly used. Once again, the only difference is the test statistic.

To decide if we should pool the data we need to test to see if two population variances are *equal*. Thus, we should use a two-sided test. The null and alternative hypotheses are shown below:

*Remember: the key words different, less than, and greater than tell you which test to use.*

**Two-Sided Test**

$H_0: \sigma_1^2 = \sigma_2^2$   or   $H_0: \sigma_1^2/\sigma_2^2 = 1$    Use this test if you wish to test if the variance
$H_A: \sigma_1^2 \neq \sigma_2^2$      $H_A: \sigma_1^2/\sigma_2^2 \neq 1$    of population 1 is *different* from the variance of population 2.

As with the tests to compare two population means or proportions, it is also possible to do one-sided tests. These are shown below.

**Lower-Tail Test**

$H_0: \sigma_1^2 \geq \sigma_2^2$   or   $H_0: \sigma_1^2/\sigma_2^2 \geq 1$    Use this test if you wish to test if the variance of
$H_A: \sigma_1^2 < \sigma_2^2$      $H_A: \sigma_1^2/\sigma_2^2 < 1$    population 1 is *less than* the variance of population 2.

*Because of the difficulty in finding the rejection region for a lower-tail test, most one-sided tests are conveniently done as upper-tail tests.*

**Upper-Tail Test**

$H_0: \sigma_1^2 \leq \sigma_2^2$    $H_0: \sigma_1^2/\sigma_2^2 \leq 1$    Use this test if you wish to test if the variance of
$H_A: \sigma_1^2 > \sigma_2^2$    $H_A: \sigma_1^2/\sigma_2^2 > 1$    population 1 is *greater than* the variance of population 2.

Since we are trying to decide how two population variances compare, it makes sense to compare the sample variances. We have seen two ways of comparing numbers. One way is to subtract one number from the other and see how close the difference is to zero. If the two numbers are estimates of the same unknown population parameter, then their difference will be close to zero. The other way to compare two numbers is to divide one number by the other and see how close the ratio is to one. If the two numbers are estimates of the same unknown population parameter, then their ratio will be close to one. Notice that the second way of writing each of the variance tests shown above uses a ratio and a comparison to one. When comparing variances, you should always use a ratio to compare them and not a difference.

*The test statistic for comparing variances is an F statistic.*

We know from our work in Chapter 7 that the best estimator of the population variance is the sample variance. Extending this idea to two populations, the point estimate for the ratio of the population variances is the ratio of the sample variances. This is also the test statistic. It is shown below:

*Formula for test statistic for comparing two variances*

$$F = \frac{s_1^2}{s_2^2}$$

*The F distribution is determined by two sets of degrees of freedom.*

Notice that this ratio is labeled $F$. This means that the test statistic follows an $F$ distribution, if the two original populations are normally distributed. The $F$ distribution was briefly introduced to you in Chapter 7. It is named after the famous statistician R. A. Fisher. Like the $\chi^2$ distribution, which we used to test a single-population variance, the specific shape of the $F$ distribution is determined by its degrees of freedom. But the $F$ distribution has not one, but two values that determine its shape. One of these is called the degrees of freedom in the numerator and it is equal to one less than the sample size on which $s_1^2$ is based, $n_1 - 1$. The other one is called the degrees of freedom in the denominator and is equal to one less than the sample size on which $s_2^2$ is based, $n_2 - 1$.

The procedure for finding the rejection regions is similar to that which we have used for other tests. If you are a doing a two-sided test, then the rejection region is two-sided; if you are doing a one-sided test, then the rejection region is one-sided. These are shown in Figure 10.6.

The critical values that define the rejection region are labeled $F_{\text{upper,df1,df2}}$ and $F_{\text{lower,df1,df2}}$ rather than $F_{\text{cutoff}}$ and $-F_{\text{cutoff}}$, the notation you might have expected. To find the values for $F_{\text{upper,df1,df2}}$ and $F_{\text{lower,df1,df2}}$, we need to notice that the shape of the $F$ distribution is not symmetric and the distribution is not centered at zero. Therefore, the absolute values of $F_{\text{upper}}$ and $F_{\text{lower}}$ are not the same and they will always be greater than zero. This is why they are not labeled $F_{\text{cutoff}}$ and $-F_{\text{cutoff}}$. It happens that there is a relationship between the upper and lower $F$ values. In particular, $F_{\text{lower,df1,df2}}$ can be found from an upper-tail value as follows:

*The symbols $df_1$ and $df_2$ stand for the number of degrees of freedom in the numerator and the denominator, respectively.*

$$F_{\text{lower,df1,df2}} = \frac{1}{F_{\text{upper,df2,df1}}}$$

*Formula for finding the lower critical value for an F test*

That is, the lower critical value is found by taking the reciprocal of the upper critical value *with the degrees of freedom reversed*. Therefore, we need table values only for $F_{\text{upper}}$.

Since the $F$ distribution is determined by two sets of degrees of freedom, labeled $df_1$ and $df_2$, we will need an entire table to specify the values for $F_{\text{upper}}$ that cut off a certain amount of probability in the upper tail of the distribution. Unlike the other tables that we have encountered so far, each upper-tail area probability requires a separate table. For instance, Table A.6a shows the values for $F_{\text{upper}}$ that cut off an area of 0.01 in the upper tail. The row and column we need to use depends on the number of degrees of freedom in the numerator and the denominator. However, if we need 0.05 in the upper-tail area, then we must use Table A.6b. This is the first time we have encountered a different table for each value of $\alpha$. When we used the $t$ or the $\chi^2$ distribution, each column corresponded to a different value of $\alpha$.

 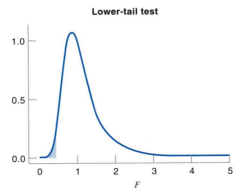

**FIGURE 10.6**    Rejection regions for test of variances

### EXAMPLE 10.27   Training and Gender

*Setting Up the Test*

When the company looking at training programs wanted to know if there was a difference in the post-test scores for males versus females, we made an assumption about whether or not the population variances were equal. The relevant data from the problem were

|  | Population 1<br>Females Trained In-House | Population 2<br>Males Trained In-House |
|---|---|---|
| Sample size | $n_1 = 20$ | $n_2 = 27$ |
| Sample average | $\overline{X}_1 = 63.95$ | $\overline{X}_2 = 63.26$ |
| Sample standard deviation | $s_1 = 8.86$ | $s_2 = 6.87$ |

To really determine whether this assumption is reasonable we need to test the hypotheses:

$$H_0 : \sigma_1^2 = \sigma_2^2$$
$$H_A : \sigma_1^2 \neq \sigma_2^2$$

With these hypotheses, if we reject $H_0$, then we will conclude that the variances *are* different and we will know that we cannot pool the variances. For the moment we will assume that the populations are normally distributed and we will do the test at the 0.10 level of significance, which means that we need to use the $F$ table with the upper-tail area of 0.05, Table A.6b. A portion of that table is shown below:

| | | **Numerator Degrees of Freedom (df$_1$)** | | | | | | | | |
|---|---|---|---|---|---|---|---|---|---|---|
| | | **18** | **19** | **20** | **21** | **22** | **23** | **24** | **25** | **26** | **27** |
| | **18** | 2.217 | 2.203 | 2.191 | 2.179 | 2.168 | 2.159 | 2.150 | 2.141 | 2.134 | 2.126 |
| | **19** | 2.182 | 2.168 | 2.155 | 2.144 | 2.133 | 2.123 | 2.114 | 2.106 | 2.098 | 2.090 |
| | **20** | 2.151 | 2.137 | 2.124 | 2.112 | 2.102 | 2.092 | 2.082 | 2.074 | 2.066 | 2.059 |
| **Denominator** | **21** | 2.123 | 2.109 | 2.096 | 2.084 | 2.073 | 2.063 | 2.054 | 2.045 | 2.037 | 2.030 |
| **degrees of** | **22** | 2.098 | 2.084 | 2.071 | 2.059 | 2.048 | 2.038 | 2.028 | 2.020 | 2.012 | 2.004 |
| **freedom** | **23** | 2.075 | 2.061 | 2.048 | 2.036 | 2.025 | 2.014 | 2.005 | 1.996 | 1.988 | 1.981 |
| **(df$_2$)** | **24** | 2.054 | 2.040 | 2.027 | 2.015 | 2.003 | 1.993 | 1.984 | 1.975 | 1.967 | 1.959 |
| | **25** | 2.035 | 2.021 | 2.007 | 1.995 | 1.984 | 1.974 | 1.964 | 1.955 | 1.947 | 1.939 |
| | **26** | 2.018 | 2.003 | 1.990 | 1.978 | 1.966 | 1.956 | 1.946 | 1.938 | 1.929 | 1.921 |
| | **27** | 2.002 | 1.987 | 1.974 | 1.961 | 1.950 | 1.940 | 1.930 | 1.921 | 1.913 | 1.905 |

*To find $F_{lower,df1,df2}$ you must reverse the degrees of freedom and find the corresponding $F_{upper}$ value and then take the reciprocal.*

To find the upper critical value we will use the numerator degrees of freedom as $20 - 1 = 19$ and the denominator degrees of freedom as $27 - 1 = 26$. The value of $F_{upper,19,26}$ is 2.003. Now, the lower critical value that we want is $F_{lower,19,26}$, so using the relationship that

$$F_{lower,19,26} = \frac{1}{F_{upper,26,19}}$$

we find that

$$F_{lower,19,26} = \frac{1}{2.098} = 0.3356$$ ■

This example has illustrated the procedure for finding the rejection region for a two-sided test. Let's summarize the steps for finding a two-sided rejection region and note the differences for a one-sided rejection region.

**Two-sided rejection region:**

**Step 1:** Divide the value of $\alpha$ in half.

**Step 2:** Find the $F$ table that corresponds to those values that cut off $\alpha/2$ in the upper tail of the distribution.

**Step 3:** Find $F_{\text{upper,df1,df2}}$ at the intersection of the column corresponding to $\text{df}_1 = n_1 - 1$ and the row corresponding to $\text{df}_2 = n_2 - 1$.

**Step 4:** Find $F_{\text{lower}}$ by first finding $F_{\text{upper,df2,df1}}$ with $n_2 - 1$ degrees of freedom in the numerator and $n_1 - 1$ degrees of freedom in the denominator. Then take the reciprocal of this number.

**Step 5:** Reject $H_0$ if the $F$ statistic is larger than $F_{\text{upper}}$ or less than $F_{\text{lower}}$.

If we are using a one-sided test of the variances, the main difference in finding the rejection region is that you do not split $\alpha$ in half. In addition, you need to find only one of the critical values: either $F_{\text{upper}}$ or $F_{\text{lower}}$.

**Upper-tail rejection region:**

**Step 1:** Find the $F$ table that corresponds to those values that cut off $\alpha$ in the upper tail of the distribution.

*Steps for finding the rejection regions for the F distribution*

**Step 2:** Find $F_{\text{upper,df1,df2}}$ at the intersection of the column corresponding to $\text{df}_1 = n_1 - 1$ and the row corresponding to $\text{df}_2 = n_2 - 1$.

**Step 3:** Reject $H_0$ if the $F$ statistic is larger than $F_{\text{upper}}$.

**Lower-tail rejection region:**

**Step 1:** Find the $F$ table that corresponds to those values that cut off $\alpha$ in the upper tail of the distribution.

*To avoid the tedious problem of finding the lower-tail rejection region for a one-tail test, it is easier to just set the test up as an upper-tail test by defining population 1 as the one with the larger of the two sample variances.*

**Step 2:** Find $F_{\text{lower}}$ by first finding $F_{\text{upper,df2,df1}}$ with $n_2 - 1$ degrees of freedom in the numerator and $n_1 - 1$ degrees of freedom in the denominator. Then take the reciprocal of this number.

**Step 3:** Reject $H_0$ if the $F$ statistic is less than $F_{\text{lower}}$.

The remainder of the hypothesis testing procedure involves calculating the test statistic and making a decision both in terms of the null and alternative hypotheses and the original problem. We can now finish the example.

## EXAMPLE 10.28 Training and Gender

*Concluding the Test*

For the training data we can calculate the test statistic as

$$F = \frac{8.86^2}{6.87^2} = 1.663$$

Since the critical values for the test were 0.3356 and 2.003, we see that we cannot reject $H_0$. This means that there is not enough evidence to say that the variances are different and so the assumption that they are equal is reasonable. ∎

 **TRY IT NOW!**

### Training and Gender

*Test for Equality of Variances*

When you looked at gender differences for the group of employees trained by the outside consultant, you did the test using the pooled variance. That is, you assumed the population variances were equal. The relevant data were as shown in the table on page 514.

(*continued*)

|  | Population 1<br>Females Trained by<br>Outside Consultant | Population 2<br>Males Trained by<br>Outside Consultant |
| --- | --- | --- |
| Sample size | $n_1 = 21$ | $n_2 = 29$ |
| Sample mean | $\overline{X}_1 = 75.9$ | $\overline{X}_2 = 75.2$ |
| Sample standard deviation | $s_1 = 3.9$ | $s_2 = 4.4$ |

Assume that the data are normally distributed and perform a test to see whether the assumption of equal variances was reasonable. Use a level of significance of 0.10.

Step 1:

Step 2:

Step 3:

Step 4:

Step 5:

## 10.9.2 Exercises—Learning It!

**10.21** Consider the problem in which The Board of Realtors for Greater Bridgeport, CT, was looking at the average selling prices of homes. The data are given again below:

| | Population | |
|---|---|---|
| | **1995** | **1996** |
| Sample size | $n_1 = 25$ | $n_2 = 25$ |
| Sample mean | $\overline{X}_1 = \$151,166$ | $\overline{X}_2 = \$160,669$ |
| Sample standard deviation | $s_M = \$5332$ | $s_W = \$6468$ |

(a) Assuming that the populations are normally distributed, set up the hypotheses to test whether the population variances are equal at the 0.10 level of significance.

(b) Was the decision to test using the pooled variance justified?

**10.22** In your quest for the perfect golf clubs you made an assumption about the population variances when you tested your hypotheses. The data you collected are given below:

| | Population | |
|---|---|---|
| | **Brand X** | **Brand Z** |
| Sample size | $n_X = 15$ | $n_Z = 15$ |
| Sample mean | $\overline{X}_X = 255$ | $\overline{X}_Z = 271$ |
| Sample standard deviation | $s_X = 8.7$ | $s_Z = 9.1$ |

(a) Set up the appropriate hypotheses to test whether the variance of Brand Z clubs is the same as the variance for Brand X.

(b) Assuming that the populations are normally distributed, at the 0.10 level of significance was your decision to pool the variances a good one?

(c) In general, would a difference in variation between the clubs be a factor in your purchase decision?

**10.23** The members of the Chamber of Commerce of a small city in Fairfield County, CT, are looking at the amount of vacant office space in their city compared to a similar city. The data are given below:

| | Population | |
|---|---|---|
| | **Our City** | **Their City** |
| Sample size | $n_1 = 12$ | $n_2 = 12$ |
| Sample mean | $\overline{X}_1 = 210,700$ | $\overline{X}_2 = 167,607$ |
| Sample standard deviation | $s_1 = 2200$ | $s_2 = 2100$ |

Assuming that the populations are normally distributed, at the 0.02 level of significance, should they have pooled the variances?

**10.24** A company has two different production lines that make the plastic cards used for credit cards and ATM cards. Both lines use $\overline{X}$ control charts to make sure that they run to the target specification, and both have been in control for the past six weeks. Recently, however, the quality manager has noticed that one of the machines (machine A) has many more items being rejected for the measurement on the width of the card. Since both machines are running to target, he decides the problem must be with the variability and decides to run a test. He samples 40 items from each production line and calculates the following summary statistics:

| | Population | |
|---|---|---|
| | Machine A | Machine B |
| Sample size | $n_A = 40$ | $n_B = 40$ |
| Sample standard deviation | $s_A = 1.1$ mm | $s_B = 0.62$ mm |

(a) Set up the hypotheses to test whether the variance of machine A is greater than the variance of machine B.

(b) Assuming that the populations are normally distributed, perform the test at the 0.10 level of significance.

(c) Is the quality manager correct in his perception that machine A is more variable?

**10.25** A large utility company is considering two sites for locating a large-scale wind energy conversion system (windmill). In selecting a site both the average speed and the variation in speed are important: the more consistent the wind speeds, the more efficient the energy conversion. The variability for the two sites is measured using a sample of 30 wind speed observations at each site. The summary data are given below:

| | Population | |
|---|---|---|
| | Site 1 | Site 2 |
| Sample size | $n_1 = 30$ | $n_2 = 30$ |
| Sample standard deviation | $s_1 = 1.8$ mph | $s_2 = 2.62$ mph |

(a) Set up the hypotheses to test whether the variability in wind speed at the two sites is the same.

(b) Assuming that the populations are normally distributed, perform the test at the 0.10 level of significance.

(c) What can you tell the utility company about the variability at the two sites?

# 10.10   TWO POPULATION HYPOTHESIS TESTS IN EXCEL

Excel has built-in data analysis tools for two population hypothesis tests. Once you know which test to use, the procedure is not very different from test to test.

### 10.10.1 Large-Sample Tests of Two Means

For a large-sample test of two population means, either the population standard deviations must be known or else the sample size from each population must be large—that is, greater than 30. Suppose that we want to consider the data from the company that was looking at in-house vs. outside consultant training. They want to know whether the results obtained by the outside consultant are better than those of the in-house trainers. This is a one-tailed test, and the hypotheses are

$$H_0: \mu_1 \le \mu_2$$
$$H_A: \mu_1 > \mu_2$$

To use Excel to perform the hypothesis test, the data must be in a spreadsheet. If you do not know the population standard deviations and are using the sample standard deviations, you must first calculate these using either the Descriptive Statistics tool or the **STDEV** or **VAR** functions. The value you will actually need is the variance, so if you calculate the standard deviation, you will have to square it to find the variance.

In our example, the population standard deviations are not known, but the sample sizes are $n_1 = 50$ and $n_2 = 47$, respectively. To perform the test, open the **Tools > Data Analysis** menu and from the list choose **z Test: Two Sample for Means.** This will open the dialog box for the test as shown in Figure 10.7.

**FIGURE 10.7**   Dialog box for $Z$ test for two means

Use the following procedure to perform the test:

1. Position the cursor in the textbox next to **Variable 1 Range:** and highlight the range that contains the post-test data for the first population—in this case, the group trained by the outside trainer.

2. With the cursor in the textbox for **Variable 2 Range:** highlight the location of the data for the second population, the in-house group.

3. Position the cursor in the textbox for **Hypothesized Mean Difference** and enter 0. Since we want to know whether $\mu_1 > \mu_2$, this is the same as $\mu_1 - \mu_2 > 0$.

4. Now, put the cursor in the textbox for **Variable 1 Variance (known)** and enter the *value* of the variance for population 1. For the training example, the variance in post-test scores for the outside group is $(4.12)^2 = 16.97$.

   *Note: You must enter the value of the variance itself. You cannot enter a formula or highlight a data range here.*

5. Repeat this for **Variable 2 Variance (known).**

6. If there were labels in any of the data ranges you highlighted, check **Labels** checkbox.

7. Enter the level of significance for the test. In this case, we will use the default value of $\alpha = 0.05$.

8. Finally, indicate where you want the output from the test to appear.

When you have entered all the data, the dialog box should look like the one in Figure 10.8.

**FIGURE 10.8**   *Z* test dialog box

**9.** Hit Enter to perform the test. The resulting output is shown in Figure 10.9.

| z-Test: Two Sample for Means | | |
|---|---|---|
| | *Variable 1* | *Variable 2* |
| Mean | 75.46 | 63.55319 |
| Known Variance | 16.97 | 59.14 |
| Observations | 50 | 47 |
| Hypothesized Mean Difference | 0 | |
| z | 9.419937922 | |
| P(Z<=z) one-tail | 0 | |
| z Critical one-tail | 1.644853 | |
| P(Z<=z) two-tail | 0 | |
| z Critical two-tail | 1.959961082 | |

**FIGURE 10.9**   Output from *z* Test for two sample means

From the output, we see that the value of the test statistic, labeled z in the output, is 9.419937922 (which is the same as the 9.42 we obtained by hand). Since the critical value for a one-tailed test is 1.644853, we can reject $H_0$ and conclude that the mean for the outside group is higher than the mean for the in-house group.

The procedure gives the $p$ value for the test, but does not label it as such. The value labeled P(Z <= z) one-tail is the $p$ value for the one-tailed test, and P(Z <= z) two-tail is the $p$ value for the two-tailed test.

## 10.10.2 Small-Sample Tests for Two Population Means

When you do not know the population standard deviations and the sample sizes are not large, the appropriate test is the $t$ test. To choose the correct $t$ test, you must decide whether the two population variances are the same or whether they differ from each other. Excel provides $t$ test procedures for both cases.

Let's look at the training data for the outside consultant. Suppose that now we want to know whether there is a difference in post-test scores for males and females. Since there were only 50 people in the original sample, it is not possible that both sample sizes are larger than 30. Therefore, we must use a $t$ test. We will use the $t$ test that assumes the two population variances are equal.

From the **Data Analysis Tools** menu, select **t-Test: Two Sample Assuming Equal Variances.** The dialog box that opens should look very similar to the one for the $Z$ test. The only difference is that you are not required to put in the population variances. The $t$ test assumes that they are unknown and the procedure will calculate them from the data. Fill in the dialog box, just as you did for the $Z$ test. An example of the completed box is shown in Figure 10.10.

**FIGURE 10.10**   $t$ Test for two means, variances assumed equal

Click **OK;** the output is placed in the location you specified. The output for this test is shown in Figure 10.11.

From the output, we see that the value of the $t$ statistic is $-0.575819805$. Since we wanted to determine whether there was a difference between males and females, it is a two-sided test and the critical values are $\pm 2.01063358$. Comparing the test statistic and the critical value, we fail to reject $H_0$ and conclude that there is no difference in the scores for males and females in the outside group. The output from the $t$ test does include the $p$ value of the test, although it is not labeled as such. The value labeled $P(T <= t)$ two-tail is the $p$-value for the two-tailed test. Notice that it is twice the value for the one-tailed test.

| t-Test: Two-Sample Assuming Equal Variances | | |
|---|---|---|
| | Variable 1 | Variable 2 |
| Mean | 75.17241379 | 75.85714 |
| Variance | 18.93349754 | 14.82857 |
| Observations | 29 | 21 |
| Pooled Variance | 17.22311166 | |
| Hypothesized Mean Difference | 0 | |
| df | 48 | |
| t Stat | -0.575819805 | |
| P(T<=t) one-tail | 0.283713515 | |
| t Critical one-tail | 1.677224191 | |
| P(T<=t) two-tail | 0.567427031 | |
| t Critical two-tail | 2.01063358 | |

**FIGURE 10.11**   Output for $t$ test assuming equal variances

What if we cannot assume that the variances are equal? In this case, the two variances are 18.93 and 14.83, which do not appear to be that different. However, we can do the same test, assuming that the variances are not equal, and see what changes. From the list of **Data Analysis** tools, select **t Test: Two Sample Assuming Unequal Variances.** There are no differences in the dialog box. This makes sense, because there are no differences in the test inputs, just the outputs. Entering the same data as in the previous test produces the output shown in Figure 10.12.

| t-Test: Two-Sample Assuming Unequal Variances | | |
|---|---|---|
| | Variable 1 | Variable 2 |
| Mean | 75.17241379 | 75.85714 |
| Variance | 18.93349754 | 14.82857 |
| Observations | 29 | 21 |
| Hypothesized Mean Difference | 0 | |
| df | 46 | |
| t Stat | -0.58736595 | |
| P(T<=t) one-tail | 0.2799151 | |
| t Critical one-tail | 1.678658919 | |
| P(T<=t) two-tail | 0.559830201 | |
| t Critical two-tail | 2.012893674 | |

**FIGURE 10.12**   Output from $t$ test assuming unequal variances

From the output, you see that the only real difference is in the value for degrees of freedom. The test assuming unequal variances is a more conservative test. Since the degrees of freedom have changed, the critical values will also change. The conclusion of the test in this example is the same.

Do not forget that all of the $t$ tests *assume the populations are normally distributed.* You need to check this assumption before reporting any results.

## 10.10.3 Paired Difference Test in Excel

When the two populations in a hypothesis test are related, we must use a paired difference test. There is an Excel data analysis tool for the paired difference test.

Suppose that we want to look at the training scores to determine whether training really changes test scores. In this case, the two populations are related; they are the same people, so we must use a paired difference test. We will look at the in-house group, using the same data as for the previous examples. The hypothesis test is a one-sided test, because we want to know whether training increases test scores. We will define the difference in scores to be post-test − pretest.

From the list of data analysis tools, select **t Test: Paired Two Sample** for Means. The dialog box is the same as the dialog boxes for the other $t$ tests that we have looked at. In this case, the first variable is the post-test scores for the in-house group, and the second variable is the pretest scores for the same group. The output from the test is shown in Figure 10.13. By now, the output should look familiar.

From the output, you can see that the value of the test statistic is 7.17031. Comparing this to the one-tail critical value of 1.678658919, we reject $H_0$ and conclude that the test scores have increased. The $p$ value of the test is 0.0000.

| t-Test: Paired Two Sample for Means | | |
|---|---|---|
| | *Variable 1* | *Variable 2* |
| Mean | 63.55319149 | 47.65957 |
| Variance | 59.20906568 | 163.7946 |
| Observations | 47 | 47 |
| Pearson Correlation | -0.040209148 | |
| Hypothesized Mean Difference | 0 | |
| df | 46 | |
| t Stat | 7.170308992 | |
| P(T<=t) one-tail | 2.53155E-09 | |
| t Critical one-tail | 1.678658919 | |
| P(T<=t) two-tail | 5.0631E-09 | |
| t Critical two-tail | 2.012893674 | |

**FIGURE 10.13**   Output from paired difference test

## 10.10.4 Two Population Tests of Proportions

Excel does not have a built-in test for comparing two population proportions, but we have supplied a macro that will perform the test. The macro assumes that you have the proportion of frequency of successes calculated for each population. You can use pivot tables to do this.

The company looking at training scores wants to determine whether the proportion of people who pass the test from the outside training group is greater than the proportion for the in-house group. An example of the pivot table is shown in Figure 10.14. The counts for each cell are displayed as percentage of row.

| Count of Certificaton Exam | Certificaton Exam | | |
|---|---|---|---|
| Training | F | P | Grand Total |
| I | 0.723 | 0.277 | 1.000 |
| O | 0.600 | 0.400 | 1.000 |
| Grand Total | 0.660 | 0.340 | 1.000 |

**FIGURE 10.14**   Pivot table for training data

From the list of MacDoIt macros, choose **TwoPropTest** and click **Run.** A table will open in a new worksheet, just as in the macros for the one-population tests. We will let the outside group be population 1 and the in-house group be population 2. In the top portion of the table, type in the sample proportions and sample sizes for each group. The table will update automatically. The results of the test are shown in Figure 10.15 (page 522).

| | A | B | C | D | E |
|---|---|---|---|---|---|
| 2 | | | | | |
| 3 | | **Proportion** | | **Sample Size** | |
| 4 | Sample 1 | p1 | 0.4000 | n1 | 50 |
| 5 | Sample 2 | p2 | 0.2770 | n2 | 47 |
| 6 | Pooled | pp | 0.3404 | | |
| 7 | | | | | |
| 8 | Test Statistic | Z | 1.2777 | | |
| 9 | Significance Level | α | 0.05 | | |
| 10 | *Critical Values* | | | | |
| 11 | Lower Tail | Z1 | -1.6449 | p-value1 | 0.100684 |
| 12 | Upper Tail | Z2 | 1.6449 | p-value2 | 0.100684 |
| 13 | Two Tail | Z3 | -1.9600 | p-value3 | 0.201367 |

**FIGURE 10.15** Table for two proportion test

The test statistic is 1.2777. The test is an upper-tail test so the critical value is 1.6449 with a *p* value of 0.1007. From the results, we conclude that we cannot reject $H_0$. There is not enough evidence to say that the proportion that passes from the outside group is higher.

### 10.10.5 *F* Test for Comparing Two Variances

When we looked at the small-sample tests for comparing population means, we had to make a decision about whether the population variances were equal. Excel has an analysis tool to test hypotheses about population variances.

We can use this test to determine which *t* test was correct. From the list of Data Analysis tools, select **F Test Two Sample for Variances.** The dialog box is the same one we have seen for the other two population tests.

The only difference is not obvious from the dialog box—the test is done as an upper-tail test. So, to test whether the variances are not equal, you must enter *half* the value of $\alpha$ that you want to use and you must enter the larger variance as population 1. Finish filling out the dialog box and click **OK.** The output is shown in Figure 10.16.

| F-Test Two-Sample for Variances | | |
|---|---|---|
| | *Variable 1* | *Variable 2* |
| Mean | 75.17241379 | 75.85714286 |
| Variance | 18.93349754 | 14.82857143 |
| Observations | 29 | 21 |
| df | 28 | 20 |
| F | 1.27682546 | |
| P(F<=f) one-tail | 0.289003794 | |
| F Critical one-tail | 2.365666774 | |

**FIGURE 10.16** Output from two sample variance test

From the output, the test statistic is labeled F and is 1.27682546. Comparing this to the critical value of 2.365666774, we conclude that we cannot reject $H_0$ and that it is reasonable to assume that the variances are equal. The *p* value of 0.289003794 given is the one-tailed *p* value.

# CHAPTER 10 SUMMARY

## Key Terms

| Term | Definition | Page Reference |
|------|-----------|---------------|
| Dependent samples | **Dependent samples** are related to each other. The members of one sample are identical to or matched or paired with the members in the other sample according to some characteristic. | 492 |
| Pretest and post-test | **Pretest** and **post-test** conditions exist when data are collected on the same sample elements before and after some experiment is performed. | 492 |

## Key Formulas

| Term | Formula | Page Reference |
|------|---------|---------------|
| Formula for average difference, $\bar{d}$ | $\bar{d} = \dfrac{\sum\limits_{i=1}^{n} d_i}{n}$ | 495 |
| Formula for common population proportion, $\bar{p}$ | $\bar{p} = \dfrac{x_1 + x_2}{n_1 + n_2}$ | 506 |
| Formula for degrees of freedom for small-sample, unequal variances test | $\nu = \dfrac{\left(\dfrac{s_1^2}{n_1} + \dfrac{s_2^2}{n_2}\right)^2}{\dfrac{(s_1^2/n_1)^2}{n_1 + 1} + \dfrac{(s_2^2/n_2)^2}{n_2 + 1}} - 2$ | 488 |
| Variances test formula for $F_{\text{lower}}$ | $F_{\text{lower,df1,df2}} = \dfrac{1}{F_{\text{upper,df2,df1}}}$ | 511 |
| Formula for pooled variance, $s_p^2$ | $s_p^2 = \dfrac{(n_1 - 1)s_1^2 + (n_2 - 1)s_2^2}{n_1 + n_2 - 2}$ | 484 |
| Test for paired difference | $t = \dfrac{\bar{d}}{s_d/\sqrt{n}}$ | 496 |
| Test for two population means: | | |
| • Variances known | $Z = \dfrac{(\overline{X}_1 - \overline{X}_2) - 0}{\sqrt{\sigma_1^2/n_1 + \sigma_2^2/n_2}}$ | 474 |
| • Variances unknown, $n_1, n_2 \geq 30$ | $Z = \dfrac{(\overline{X}_1 - \overline{X}_2) - 0}{\sqrt{s_1^2/n_1 + s_2^2/n_2}}$ | 478 |
| | $t = \dfrac{(\overline{X}_1 - \overline{X}_2) - d}{\sqrt{s_1^2/n_1 + s_2^2/n_2}}$ | 487 |
| • Variances unknown but equal, $n_1, n_2 < 30$ | $t = \dfrac{(\overline{X}_1 - \overline{X}_2)}{s_p\sqrt{1/n_1 + 1/n_2}}$ | 485 |

| Term | Formula | Page Reference |
|---|---|---|
| Test for two population proportions | $$Z = \dfrac{p_1 - p_2}{\sqrt{\bar{p}(1 - \bar{p})\left(\dfrac{1}{n_1} + \dfrac{1}{n_2}\right)}}$$ | 505 |
| Test for two population variances | $$F = \dfrac{s_1^2}{s_2^2}$$ | 510 |

# CHAPTER 10 EXERCISES

## Learning It

**10.26** The members of the Chamber of Commerce of a small city in Fairfield County, CT, are wondering whether they need to worry about the amount of vacant office space in the city. They would consider lobbying for tax incentives for businesses if they find that there is more vacant office space in their city than in a comparable area. They take weekly data on the number of square feet of vacant office space, for the second quarter of 1996 for each city and find the following:

|  | Population | |
|---|---|---|
|  | **Our City** | **Their City** |
| Sample size | $n_1 = 12$ | $n_2 = 12$ |
| Sample mean | $\overline{X}_1 = 210{,}700$ | $\overline{X}_2 = 167{,}607$ |
| Sample standard deviation | $s_1 = 2200$ | $s_2 = 2100$ |

(a) Calculate the pooled variance for the data.

(b) Assuming that the data are normally distributed, at the 0.01 level of significance, should they lobby for tax incentives?

**10.27** A recent study of consumer behavior focused on the amount of money spent monthly on frozen foods. The study wanted to determine whether there was a difference in the average amount of money spent for men and women. Data were collected on samples of 50 men and 50 women and the following information was found:

|  | Population | |
|---|---|---|
|  | **Men** | **Women** |
| Sample size | $n_M = 50$ | $n_W = 50$ |
| Sample mean | $\overline{X}_M = \$72.24$ | $\overline{X}_W = \$67.44$ |
| Sample standard deviation | $s_M = \$8.23$ | $s_W = \$8.12$ |

(a) Set up the appropriate hypotheses to test whether men spend more per month, on the average, for frozen foods than women.

(b) Use the large-population test with unknown variances to test the hypotheses. Use a level of significance of 0.10.

(c) Do the data provide evidence that the average amount spent per month on frozen food by men is greater than by women?

(d) What is the $p$ value of the test?

**10.28** The nurses who were part of the hand-washing experiment are still not convinced that the length of time spent washing hands makes that much difference. They design their own study and decide to have each nurse in the CCU wash his or her hands twice, once for 2.5 seconds and once for 15 seconds. After each washing they do a bacteria culture and measure the number of bacteria that remain on the person's hands. The data are shown in the next table.

| Observation | Culture 1 2.5 s | Culture 2 15 s |
|---|---|---|
| 1 | 66 | 78 |
| 2 | 132 | 115 |
| 3 | 120 | 93 |
| 4 | 187 | 48 |
| 5 | 190 | 77 |
| 6 | 17 | 3 |
| 7 | 33 | 12 |
| 8 | 92 | 12 |
| 9 | 1000 | 146 |

(a) Calculate the differences between the number of bacteria for each nurse. Just looking at the differences, do you think that washing longer decreased the amount of bacteria? Why or why not?

(b) Calculate the average difference and the standard deviation of the differences.

(c) Set up the hypotheses to test whether there was a decrease in the average amount of bacteria after washing longer.

(d) Assuming that the data are normally distributed, at the 0.05 level of significance, what can you conclude?

(e) Suppose that you were told that the second episode of hand washing was done right after the first. Would that change the way you interpret the results of the study?

## Thinking About It

**10.29** It has been a widely held belief that the switch to participative management would increase employees' buy-in to the company. One of the benefits that should be realized is a reduction in the number of sick days that employees use. A company that has made the switch in some departments wonders if this has been true. It decides to sample 25 employees from each of two manufacturing departments. The first has been using a participative management style for almost two years and the second is still using a traditional management style. The data on the number of sick days used by each employee in the past 12 months are found below:

| Participative | | | | | Traditional | | | | |
|---|---|---|---|---|---|---|---|---|---|
| 1 | 3 | 5 | 5 | 6 | 0 | 5 | 6 | 7 | 9 |
| 1 | 4 | 5 | 6 | 7 | 3 | 5 | 7 | 7 | 9 |
| 2 | 4 | 5 | 6 | 8 | 4 | 6 | 7 | 7 | 10 |
| 2 | 4 | 5 | 6 | 8 | 4 | 6 | 7 | 8 | 11 |
| 3 | 4 | 5 | 6 | 8 | 5 | 6 | 7 | 8 | 11 |

At the 0.05 level of significance, do the data provide enough evidence to say that employees who use participative management styles use, on the average, fewer sick days than those who use a traditional management style? Be sure to justify any assumptions you make in selecting the test procedure you use.

**10.30** The software company that is looking at the time to failure of the diskettes (hours) it uses decides to look at an alternative supplier of the product. The data for the current supplier and for the new supplier are shown below:

| Current Supplier | | | | Alternative Supplier | | | |
|---|---|---|---|---|---|---|---|
| 486 | 494 | 502 | 508 | 489 | 492 | 495 | 498 |
| 490 | 496 | 504 | 510 | 489 | 492 | 496 | 499 |
| 491 | 498 | 505 | 514 | 491 | 493 | 497 | 502 |
| 491 | 498 | 506 | 515 | 492 | 493 | 497 | 503 |
| 494 | 498 | 507 | 527 | 492 | 494 | 497 | 505 |

The company has decided that if the mean time to failure for the new supplier is longer than it is for the current supplier, it will switch suppliers.

(a) What level of significance would you suggest the company use? Justify your choice.

(b) Assuming that the data are normally distributed, should the company switch suppliers? Use the level of significance you chose in part (a).

(c) What impact did your choice of $\alpha$ have on the decision?

*Requires Exercise 10.11*   **10.31** You are still wondering about the results of the test on the difference in software prices. You wonder why the two tests came to different conclusions and figure that it must be the amount of variability in the software prices for the packages chosen.

| Top Ten Business Software Packages | Computability Price ($) | PC Connection Price ($) |
|---|---|---|
| Windows 95 Upgrade | 88 | 95 |
| Norton Anti-Virus | 59 | 70 |
| McAfee ViruScan | 49 | 60 |
| First Aid 97 Deluxe | 54 | 58 |
| Clean Sweep III | 37 | 37 |
| Norton Utilities | 68 | 75 |
| Netscape Navigator | 45 | 40 |
| MS Office Pro 97 Upgrade | 300 | 310 |
| First Aid 97 | 32 | 35 |
| Win Fax Pro | 95 | 95 |

(a) Look at the data again. Does any software have a price that seems to be very different from the others? If so, which one?

(b) Drop the data for the unusual observation and perform the hypothesis test again using the test for independent samples.

(c) Does anything change from the last time you did the test? If so, what?

(d) Does dropping the observation change the decision?

(e) Do you think this was the right test to use? Why or why not?

**10.32** A study was recently completed by an insurance company concerning a particular surgical procedure. The study looked at the hospital records of 40 patients at two different hospitals and compared the length of patient stay. The data were analyzed using Minitab. Output showing the descriptive statistics for the data is given below:

```
DESCRIPTIVE STATISTICS

Length of Stay    N      Mean    Median   Trim Mean   St.Dev.   SE Mean
Hospital 1        40     7.725   7.500    7.667       2.562     0.405
Hospital 2        40     10.350  10.000   10.222      3.340     0.528

Length of Stay    Min    Max      Q1       Q3
Hospital 1        2.000  14.000   6.000    9.000
Hospital 2        5.000  18.000   8.000    13.500
```

(a) Set up the hypotheses necessary to test whether the patients at hospital 1 had, on the average, a shorter stay than those at hospital 2.

(b) What type of test would you use to make this decision? Why?

(c) Perform the appropriate hypothesis test. Use a level of significance of 0.05.

(d) Do the data provide evidence that the mean stay at hospital 1 is shorter than at hospital 2?

*Requires Exercise 10.29*   **10.33** After looking at the results of the data analysis, the Director of Human Resources at the company looking at sick days and type of management writes a memo to the Vice President of Human Resources suggesting that the company change all units over to Participative Management. He cites the results of the test and states that the data "provide evidence that Participative Management causes people to take fewer sick days." Since you did the analysis, he gives you the memo to read before he sends it.

(a) Do you agree with the Director of Human Resources? Why or why not?

(b) Write a memo to the Director of Human Resources explaining your reaction. Include plans for further study if you think it is warranted.

**10.34** Reconsider the study of the amount of money spent monthly on frozen foods. Data were collected on samples of 50 men and 50 women:

*Requires Exercise 10.27*

| | Population | |
|---|---|---|
| | **Men** | **Women** |
| Sample size | $n_M = 50$ | $n_W = 50$ |
| Sample mean | $X_M = \$72.24$ | $X_W = \$67.44$ |
| Sample standard deviation | $s_M = \$8.23$ | $s_W = \$8.12$ |

(a) Do you think your decision was sensitive to the value chosen for $\alpha$? Why or why not?

(b) Suppose that you were interested only in whether the average spent by men was different than the average spent by women. How would this have changed the setup of the test? Would it have changed the conclusion?

**10.35** After looking at the software price data again, and based on the results of the paired test, you are considering buying your software from Computability. You decide to check the ads for each company one more time to see if there are any hidden catches and you notice that shipping charges for PC Connection are $5 while for Computability they are $16.95.

*Requires Exercises 10.11, 10.31*

At the 0.05 level of significance, who will you buy your software from?

**10.36** Since the data were available, the Nursing Supervisor was interested in knowing whether there was a difference in the average amount of time that nurses from two different departments spent washing their hands. She was not sure whether to pool the variances, and Minitab does not do an $F$ test on variances, so she decided to run the $t$ test both ways. The Minitab output is shown below:

```
Two Sample T-Test and Confidence Interval

Two sample T for C9
C8      N    Mean    StDev    SE Mean
IMCU    10   1.00    1.89     0.60
N4      8    4.87    5.79     2.0
95% CI for mu (IMCU) − mu (N4 ): (−8.79, 1.0)
T-Test mu (IMCU) = mu (N4 ) (vs not =): T = −1.82 P = 0.11 DF = 8


Two Sample T-Test and Confidence Interval

Two sample T for C9
C8      N    Mean    StDev    SE
IMCU    10   1.00    1.89     0.60
N4      8    4.87    5.79     2.0
95% CI for mu (IMCU) − mu (N4 ): (−7.98, 0.2)
T-Test mu (IMCU) = mu (N4 ) (vs not =): T = −2.00 P = 0.063 DF =
16
```

(a) Interpret the results of the output for the first test, without pooling the variances. How many degrees of freedom are there for this test? If you use a level of significance of 0.10 what would you conclude about the two departments?

(b) Interpret the results of the output for the second test, pooling the variances. How many degrees of freedom are there for this test? If you use a level of significance of 0.10 what would you conclude about the two departments?

(c) How does this example confirm what you learned about the effects of pooling the variances?

# Doing It!

**10.37** Every year the *Wall Street Journal* has a feature on compensation of CEOs of different companies. The data include type of company, amount of salary, amount of bonus, % change from the previous year, and several other compensation forms. A sample of the data is shown at the top of page 528.

*Datafile: BOSSSAL.XXX*

| Company | 1996 Salary (000) | 1996 Bonus (000) | % Change from 1995 | Options Gains | Other | Total Direct Compensation (000) | Present Value of Options Grants (000) |
|---|---|---|---|---|---|---|---|
| Basic materials | | | | | | | |
| Air Products | $738.10 | $473.00 | − 8.70% | $560.60 | $390.50 | $2162.10 | $887.00 |
| Alcoa | 750 | 810 | − 22 | 6113.60 | 0 | 7673.60 | 13,905.60 |
| Alumax | 800 | 797.1 | 27.3 | 0 | 670 | 2267.10 | 8,491.10 |
| Armco | 559.2 | 0 | − 42.2 | 0 | 0 | 559.2 | 422.7 |
| Asarco | 811.7 | 485 | − 20.5 | 0 | 539.8 | 1836.40 | 571.6 |

(a) Look at the 1996 salary data for cyclical and noncyclical companies. Create a plot of salaries for each group. Based on the graphs do you think the data are normally distributed? Why or why not?

(b) Perform the appropriate hypothesis test to determine whether the variances in salary for the two types of companies are equal. Based on this test, can you assume equal variances?

(c) Based on the results of your answers to parts (a) and (b) select the appropriate test procedure to test whether or not the mean salary for cyclical and noncyclical companies is the same.

(d) Perform the test at the 0.05 level of significance. What is your conclusion?

(e) Repeat the procedure you used to answer parts (a) to (d) to determine whether the mean bonus for the two types of companies is different.

(f) Look at the entire set of data. One variable that is reported is the change in compensation level from the previous year (%). Calculate the proportion of companies whose CEOs received increases in compensation. Do the same thing for the proportion who received decreases in compensation.

(g) At the 0.05 level of significance, is the proportion of those receiving positive increases greater than the proportion of those receiving decreases?

(h) Do the necessary analysis to compare the mean salaries of the technology companies to those of the industrial companies.

## *REALLY DOING IT!*
## *Who Spends Money—What Do They Buy?:*
## *Inferential Statistics*

A large marketing research firm has an international client that has business interests in many consumer industries such as food, apparel, alcohol, and automobile fuel. As a person with some statistical experience, you have been hired by this marketing research company. Before spending any of the client's money on surveying consumers, the client has asked you to report on general trends in consumer spending. Specifically, you are asked which business areas have experienced increases, and what are the characteristics of people who are spending money in these areas. You report will guide future survey design and eventually new product development and advertising expenditures.

Using your Internet surfing skills, you find the web site for the U.S. Census Bureau and find the results of the Consumer Expenditure Survey conducted periodically by the Census Bureau. The most recent survey available online is from 1993 and so you decide to use it. The Consumer Expenditure Survey includes information on many variables but you select for study only 49 variables that you think would be helpful. A portion of the data is shown below:

| AGE2 | AGE_REF | ALCBEVCQ | ALCBEVPQ | APPARCQ | APPARPQ | AS_COMP1 | AS_COMP2 |
|------|---------|----------|----------|---------|---------|----------|----------|
|      | 46      | 97       | 30       | 0       | 33      | 2        | 1        |
|      | 78      | 0        | 0        | 0       | 0       | 0        | 1        |
|      | 55      | 0        | 65       | 0       | 159     | 0        | 1        |
| 73   | 78      | 0        | 40       | 0       | 145.83  | 1        | 1        |
| 45   | 45      | 85       | 170      | 482     | 845.4   | 1        | 1        |
| 52   | 56      | 54       | 27       | 175     | 840     | 2        | 1        |
| 36   | 35      | 40       | 20       | 375.38  | 120     | 1        | 1        |
|      | 63      | 0        | 0        | 22      | 10      | 1        | 0        |

Each row in the data file (CONSEXP.XXX) represents one person's (the Reference Person) responses to the survey questions. As is the case with much publicly available data, the variable names are cryptic and the qualitative variables are coded. Clearly you need to know what each variable means and also what the codes mean. This information is detailed in Appendix B.

One portion of your client's business is in the food industry. The company owns restaurants and supermarkets. The first thing you decide to do is examine trends in the amount of money spent on food. You have information on the amount of money spent in the current quarter of the survey year and the amount spent in the previous quarter.

1. Did consumers spend significantly more money in the current quarter of the survey year than in the previous quarter on food? If so, is the difference meaningful?

2. Was there a change in how the money spent on food was distributed between the amount of money spent on eating out and the amount of money spent for food eaten at home?

3. Do households with teenagers spend more money on food eaten at home?

4. Do households where both adults are working spend more money on eating out?

Having done the analysis of food expenditures, you decide to take the same approach for the other business areas. In addition to looking for changes in expenditures between quarters, you wonder what other demographic factors affect spending habits. You set about doing the appropriate statistical tests to provide your client with a report by the end of the week.

# *REALLY DOING IT*
## *Advising the President: Inferential Statistics*

The President is pleased with your exploratory data analysis. While he agrees with your impressions, he is hesitant to propose new legislation based on subjective interpretation of the data. You are smarter now, too, and you recognize the need for the formal statistical tools of hypothesis testing.

Remember the data available to you are from the U.S. Census Bureau and are found in a data file named CENSUS.XXX:

| Region division, and state | Resident population\1 | | | Resident population\1 | | |
|---|---|---|---|---|---|---|
| | Total,\2 1994 (1,000) | Percent increase,\3 1990-94 (%) | 65 years old and over, 1994 (%) | Householder 65 years old and over, 1994\3 (%) | In metro areas, 1992\4 (%) | Net domestic immigration, 1990-94\1 |
| United States | 260,341 | 4.7 | 12.7 | 21.8 | 79.7 | 0 |
| Northeast | 51,396 | 1.2 | 14.1 | 23.7 | 89.4 | (1,439,371) |
| New England | 13,270 | 0.5 | 13.9 | 22.9 | 84.1 | (364,734) |
| Maine | 1,240 | 1.0 | 13.9 | 22.8 | 35.7 | (12,109) |
| New Hampshire | 1,137 | 2.5 | 11.9 | 19.5 | 59.4 | (9,912) |
| Vermont | 580 | 3.1 | 12.1 | 20.1 | 27.0 | 2,287 |
| Massachusetts | 6,041 | 0.4 | 14.1 | 23.3 | 96.2 | (183,995) |
| Rhode Island | 997 | −0.7 | 15.6 | 25.6 | 93.6 | (35,219) |
| Connecticut | 3,275 | −0.4 | 14.2 | 23.1 | 95.7 | (125,786) |

The attacks continue on the health care front. Specifically, legislators from the South Central region claim that they need additional federal funding to encourage doctors and nurses to practice there. Before the President supports such legislation he wants to know the following:

1. Is the availability of medical professionals in the South Central states really lower than the availability in the rest of the country?

2. Are the South Central states the only area of the country that need this type of support? If you had to make a recommendation, what area(s) would you support?

A group of Senators from the Midwest region says that they will support the health care initiatives for the South Central region only if they are assured that they have equity in the area of education, specifically teachers' salaries.

3. Can the President ignore their lobbying? Why or why not?

You now know that you need to be prepared to advise the President if legislators from other regions are going to try similar tactics in other categories. Do the necessary analyses and prepare your report.

# Part V
# MODEL BUILDING

# CHAPTER 11

# REGRESSION ANALYSIS

## IS TQM WORKING?

A company that manufactures computer storage media has had a Total Quality Management (TQM) program in place for the past two years. The program incorporates Quality Circles, Statistical Quality Control, and Team-Based Decision Making in the production of removable storage media, i.e., floppy diskettes. Before the company decides to expand the program to other departments it would like to assess whether the program has been effective. According to the literature, when TQM programs are used they should result in increased productivity and quality, and decreased waste and delay.

The company asked the production team to assemble some data so that it can decide whether the program is successful. The team decided to collect data on five different variables, **machine speed, waste, delay, rate of operation,** and **average outgoing quality** for the two-year period that the program has been in place. A sample of the data is shown below:

| Month | Speed | Waste (%) | RateOper | Delay (h) | Quality |
|-------|-------|-----------|----------|-----------|---------|
| 1 | 375 | 8.9 | 14.5 | 6.2642 | 95 |
| 2 | 334 | 9.9 | 12.8 | 6.4854 | 93 |
| 3 | 356 | 8.7 | 12.7 | 6.8372 | 94 |
| 4 | 378 | 9.5 | 13.9 | 5.7134 | 93 |
| 5 | 373 | 9.8 | 13.7 | 6.3136 | 94 |
| 6 | 381 | 8.8 | 14.7 | 6.2034 | 92 |

In particular, the company managers would like to know how production rate, waste, and delay are related to the speed at which a machine runs. They know that they have been able to run the production process at a faster speed in the past two years, but they are not sure that the increase in speed translates to an increase in productivity or a decrease in waste and delay.

## 11.1   CHAPTER OBJECTIVES

In Chapter 5 you learned a little bit about relationships between quantitative variables. In particular, you learned that there are different types of relationships between two variables and that when the relationship is linear you can use least squares to find the equation to describe the relationship.

For example, a company might want to predict sales of a particular product. The company knows that there are many different variables that might affect sales, but it does not know which variable(s) are most important or exactly how they relate to sales. The company needs to find a model that will enable it to predict sales as a function of the other variables. In *regression analysis*, this model is an *equation*.

In the first part of the chapter we look at *simple linear regression*, which predicts the value of $Y$ as a linear function of a single independent variable, $X$. Once we have established the basics of regression analysis in the simple linear case we will look at *multiple regression models*, which predict the value of $Y$ as a function of a set of independent variables.

In this chapter you will learn how to

- Find the linear regression equation for a dependent variable $Y$ as a function of a single independent variable $X$
- Determine whether a relationship between $X$ and $Y$ exists
- Analyze the results of a regression analysis to determine whether the simple linear model is appropriate

## 11.2   THE SIMPLE LINEAR REGRESSION MODEL

### 11.2.1 Deterministic and Statistical Relationships

In some cases where two variables, $x$ and $y$, are related, the relationship is *deterministic*, or *functional*. This means that when a value of $x$ is selected, the value of $y$ is uniquely determined. For example, if we were interested in the relationship between the total cost of an order, $y$, and the number of items ordered, $x$, we can describe this relationship by an equation such as

$$y = \$50 + \$1.20x$$

where the value of $50 might represent an ordering cost and the $1.20 is the cost per item ordered. Figure 11.1 illustrates this type of relationship.

If a person were to order $x = 100$ items, then the corresponding cost would be $y = \$50 + (\$1.20)(100) = \$170$. Every person who orders 100 items will incur the cost of $170. That is, the value of $y$ is *unique*, for a given value of $x$.

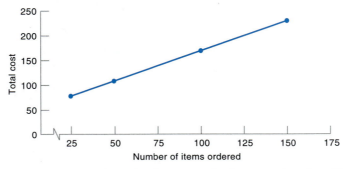

**FIGURE 11.1**   Deterministic relationship between total order cost and number of items ordered

While many real problems are described by this type of relationship, we are interested in a different situation when we study *linear regression*. When we look at the relationship between two variables, X and Y, we are interested in situations where the value of Y varies for a given value of X. That is, we are interested in the *statistical relationship* between two variables.

Suppose that we are looking at the relationship between dollars spent in advertising and the revenues from sales. Clearly, we expect the two variables to be related, but we do not expect that every time a company spends $x in advertising it will always have $y in revenues. We know that there are other *factors,* or *variables,* such as the type of product, location, and various economic factors that will affect the value of Y for a given value of X.

When we collect our data we are collecting *pairs* of observations on the two variables, X and Y. Thus, we will have a set of n data pairs:

$$(x_1, y_1), (x_2, y_2), \ldots, (x_n, y_n)$$

A plot of the data might look like the one in Figure 11.2. You can see that the two variables are related, but that a particular value of advertising expenditures can result in more than one value for revenue. This type of plot, a scatter plot, is of primary importance in exploring relationships between variables, and should be done *before* any type of statistical analysis is performed.

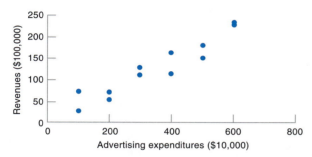

**FIGURE 11.2** Statistical relationship between revenue and advertising expenditures

## EXAMPLE 11.1  Is TQM Working?

### *Plotting the Data to Look for a Linear Relationship*

The company that is looking at its TQM program has decided that if the program is effective then the observed increase in machine speed should be accompanied by an increase in productivity and a decrease in waste and delay.

The company decides to look at productivity and machine speed to see if there is any relationship between them that can be investigated further. The machine speed is measured in items per minute that the machine is set to produce. The measure chosen for productivity is Rate of Operation, which is a measure of the usable throughput of the machine per hour (in thousands of units).

After looking at the plot it seems that a linear relationship is a reasonable model, and so the company decides to proceed with a simple linear regression analysis. ∎

In the case of simple linear regression we would like to find an equation, or model, that will allow us to predict a value for a variable, Y. Ideally, since the value of Y depends on many different factors, we would like to find some variable X that does a good job of predicting Y with a linear equation. By a "good" job of predicting, we mean that the prediction we obtain is *useful* for purposes of planning or problem solving. In regression analysis, the X variable is assumed to be controlled or at least controllable. This means that its values can be fixed by the person collecting the data.

### EXAMPLE 11.2   HMO Health

***Looking at the Relationship Between Variables***

As approaches to health-care coverage change, Health Maintenance Organizations (HMOs) are growing in popularity. Some business analysts wondered how the increase in popularity affected the financial health of the HMOs. They collected data on revenue and number of members for 10 different HMOs:

| HMO | Members (million) | Revenue ($ billion) |
| --- | --- | --- |
| United HealthCare | 4.24 | 5.49 |
| Humana | 3.19 | 4.63 |
| FHP International | 1.83 | 3.86 |
| PacifiCare Health systems | 1.62 | 3.60 |
| U.S. Healthcare | 2.07 | 3.43 |
| WellPoint Health Network | 2.30 | 2.91 |
| Health Systems International | 1.83 | 2.74 |
| Foundation Health | 2.15 | 2.40 |
| Oxford Health Plans | 0.97 | 1.71 |
| Physician Corp. of America | 0.89 | 1.20 |

*Source: The Economist, 6 April 1996.*

In this case the dependent variable is revenue, measured in billions of dollars and the independent variable is number of members, in millions. The analysts looked at a scatter plot of the data to see if a relationship existed:

From the plot it appeared that a linear relationship existed between revenue and number of members. This meant that a straight line should provide a good model for predicting revenue for an HMO as a function of the number of members of the HMO. ∎

As you learn additional and more sophisticated statistical tools, you should not lose sight of the descriptive tools you learned first. Plotting data is essential to any statistical analysis and should be one of the first things you do so that you can determine what analysis, if any, is appropriate.

## TRY IT NOW!

### Increasing Capacity

*Plotting Data to Look at the Relationship*

An oil company is trying to determine how the number of refining sites available for refining crude oil relates to the overall refining capacity. It would use this information to determine whether or not expansion will provide the increase in capacity that it wants or whether other steps to increase capacity will be necessary. The company collects data on other competitive companies and finds the following:

| Oil Company | Number of Sites | Refining Capacity (million tons per year) |
|---|---|---|
| Royal Dutch/Shell | 13 | 81.82 |
| Exxon | 10 | 81.82 |
| Agip | 13 | 58.18 |
| BP | 8 | 43.64 |
| Repsol | 5 | 40.00 |
| Total | 7 | 36.36 |
| Turkish Petroleum | 4 | 34.55 |
| Elf | 8 | 32.73 |
| Mobil | 7 | 29.09 |
| Petrofina | 3 | 25.45 |

Source: *The Economist,* 15 July 1995.

Use the grid below to create a scatter plot of the data.

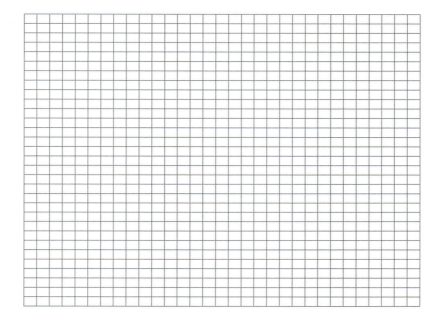

*(continued)*

Do you think that a linear model is a good one?

## 11.2.2 The Simple Linear Regression Model

The objective of simple linear regression is to find a linear equation that describes the existing relationship between two variables, $X$ and $Y$. The equation we find will be based on data taken from some population. Just as in any problems involving sampling from a population, we are trying to find an *estimate* for the true **regression model** between the two variables.

> The true relationship between the variables $X$ and $Y$, the **simple linear regression model,** can be described by the equation
>
> $$y = \beta_0 + \beta_1 x + \varepsilon$$

*At this point we are assuming that a plot of the data has determined that a **linear model** is appropriate.*

This equation says that for a given value of the variable $X = x$, the actual value of $Y$ will be determined by the expression $\beta_0 + \beta_1 x$, plus some random variation, $\varepsilon$, due to other, unmeasured, factors. Thus, if we knew the values of $\beta_0$, the true population intercept, and $\beta_1$, the true population slope, we could predict the value of $Y$ to within some random error, $\varepsilon$. Figure 11.3 shows the population model for a linear regression.

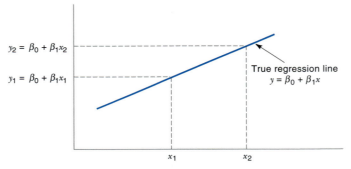

$y_2 = \beta_0 + \beta_1 x_2$

$y_1 = \beta_0 + \beta_1 x_1$

True regression line
$y = \beta_0 + \beta_1 x$

$x_1 \qquad x_2$

**FIGURE 11.3** The true regression model showing how $Y$ varies for a given value of $X$

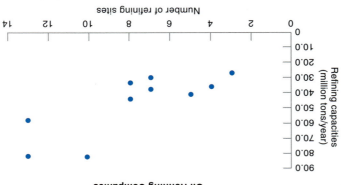

ANS. YES

You can see that for a given value of $X = x_1$, the values of $Y$ vary around the regression line. This variation is measured by the error term, $\varepsilon$. One of the assumptions of regression analysis is that the $\varepsilon$ terms are normally distributed with a mean of 0 and a standard deviation of $\sigma$. We discuss this assumption along with some others later in the chapter.

How can we find estimates for the population values $\beta_0$ and $\beta_1$? We would like to find values that do the best job of describing the relationship between the variables. If you look at the data on advertising and revenue in Figure 11.2 you can probably imagine a straight line that you might draw that captures the relationship. Figure 11.4 shows such a line along with the data.

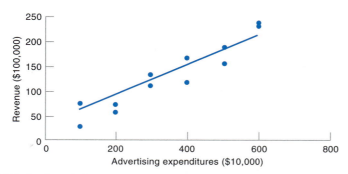

**FIGURE 11.4**  Straight line approximating the relationship between advertising and revenue

The equation of the line that we draw will be:

$$\hat{y} = b_0 + b_1 x$$

where $\hat{y}$ is the predicted value of $Y$ for a particular value of $X = x$. The quantities $b_0$ and $b_1$ are the *estimates* of the population values $\beta_0$ and $\beta_1$. This line is called the *regression line of y on x* or the estimate of the *simple regression model.*

The problem is that there are many different lines that we might draw depending on what our criterion for "best" is. A list of possible criteria might be

- Hit as many points as possible.
- Have an equal number of points above and below the line.
- Pick two representative points and connect them.
- Connect the first and the last points.

The trouble with these criteria is that they do not produce a unique line. That is, there are many lines that hit 3 or 4 or 5 points and many lines that have an equal number of points above and below the line. (The last criterion actually produces a unique line, but it is not a very good criterion.) Figure 11.5 (page 540) illustrates the problem. In Figure 11.5a both of the lines drawn have six points above and six points below. Figure 11.5b has several different lines, all of which go through exactly three of the data points.

You may wonder why it is important that we find a unique line to fit the data. Remember that we want to use the model (the equation) to predict values for $Y$, the dependent variable, for different possible values of $X$, the independent variable. One of the important features of a good model is that it be consistent. If the technique we use to find the line can produce many different models, then how will the user know

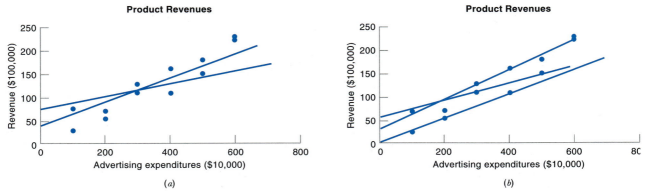

**FIGURE 11.5**   A single criterion can produce many different lines

which model or prediction to use? If everyone is allowed to choose the model he or she likes best, then the technique is, for all practical purposes, useless. You might as well just pick a number out of a hat.

## 11.2.3 The Least-Squares Line

At this point, we are sure about two things. We want the method we use for finding the equation of the line to produce one that is unique, and we want it to be a *good* representation of reality. While there are certainly many ideas about what is good, we can agree that the line should be close to as many of the data points as is possible. Figure 11.6 shows a set of data and a line drawn to represent the relationship between the variables. Although the line does not actually go through any of the data points, it is very close to most of them. The distance from each data point to the line is shown. These distances are called the ***deviations*** or ***errors*** of the line. We would like to find a line that somehow minimizes the overall deviation of the data points from the line.

> The distance between the predicted value of $Y$, $\hat{y}$, and the actual value of $Y$, $y$, is called the **deviation** or **error.**

When you learned about the variance in Chapter 4, you learned that when deviations are both positive and negative (in this case because the data points fall both above and below the line) you cannot simply look at the *sum* or *total* deviation, because it will be 0. For the variance, you solved the problem by squaring the deviations and then adding them together. We will use a similar approach to solving this problem.

The technique that we will use to find the line that best fits the data will be the **least-squares method.**

**FIGURE 11.6**   Deviations between the data points and the line

> The technique that finds the equation of the line that minimizes the total or sum of the squared deviations between the actual data points and the line is called the **least-squares method.**

You may wonder why we always square negative quantities to make them positive instead of just taking the absolute value. One reason is that mathematically the square is a nicer quantity to work with than the absolute value. The other reason, in the case of fitting a line to a set of data, is that using the square of the error makes the line try to fit *all* of the points. This is because the penalty for avoiding certain points is much larger. Suppose that the actual distance from a data point to the line is $-10$ units. If we simply wanted to minimize the *absolute* deviation, then the penalty attached to missing that point is $|-10| = 10$. When we minimize the square of the distance, the penalty for missing that point becomes $(-10)^2 = 100$ and so the line tries to get closer to that point.

The least-squares method finds the equation of the line

$$\hat{y} = b_0 + b_1 x$$

that minimizes

$$\sum_{i=1}^{n} (y_i - \hat{y}_i)^2$$

the total of squared deviations from the data points to the line. The values for $b_0$ (the intercept of the line) and $b_1$ (the slope of the line) are found by using the following equations:

$$b_1 = \frac{n \sum_{i=1}^{n} x_i y_i - \sum_{i=1}^{n} x_i \sum_{i=1}^{n} y_i}{n \sum_{i=1}^{n} x_i^2 - \left(\sum_{i=1}^{n} x_i\right)^2} \quad \text{or} \quad \frac{n \sum xy - \sum x \sum y}{n \sum x^2 - \left(\sum x\right)^2}$$

*Formulas for the slope and intercept of a least-squares line*

and

$$b_0 = \frac{\sum_{i=1}^{n} y_i}{n} - b_1 \frac{\sum_{i=1}^{n} x_i}{n} \quad \text{or} \quad b_0 = \bar{y} - b_1 \bar{x}$$

Although these equations may seem a bit complex, if you look at them carefully you will see that only five quantities have to be calculated. Four of the quantities involve the data values for $X$ and $Y$, and the fifth is simply the number of observations in the sample, $n$. For example, look at the quantity $\sum xy$. This is simply the sum of the products of the $x$ and $y$ values for each data point. The easiest way to look at what is involved in the calculations is to make a table with a column for each sum needed. The table will look like the one in Table 11.1.

**TABLE 11.1  Table for Calculating the Least-Squares Line**

| Observation Number | $x$ | $y$ | $xy$ | $x^2$ |
|---|---|---|---|---|
| 1 | $x_1$ | $y_1$ | $x_1 y_1$ | $x_1 x_1$ |
| 2 | $x_2$ | $y_2$ | $x_2 y_2$ | $x_2 x_2$ |
| . | . | . | . | . |
| . | . | . | . | . |
| . | . | . | . | . |
| $n$ | $x_n$ | $y_n$ | $x_n y_n$ | $x_n x_n$ |
| Totals | $\sum x$ | $\sum y$ | $\sum xy$ | $\sum x^2$ |

### EXAMPLE 11.3   Is TQM Working?

*Finding the Least-Squares Line*

*Do not round until you have done the calculations.*

The manufacturing company that is trying to assess the effectiveness of its TQM program is looking at the relationship between productivity as measured by rate of operation ($Y$) and machine speed ($X$). It has plotted the data and thinks that the plots indicates that the linear model is a good one. To find the equation of the least-squares line, the managers of the company have assembled the following information:

$$\Sigma x = 10,340 \qquad \Sigma y = 391.4 \qquad \Sigma xy = 162,951.9 \qquad \Sigma x^2 = 4,306,918 \qquad n = 25$$

They first calculate $b_1$, the estimate of the slope:

*Although we would never use the estimate of $b_1$ to this many decimal places in the final equation, again, you should not round until after you use the value in the equation for $b_0$. In the final equation, the values for $b_0$ and $b_1$ should be rounded to reflect the precision of the original data.*

$$b_1 = \frac{(25)(162,951.9) - (10,340)(391.4)}{(25)(4,306,918) - (10,340)^2} = \frac{26,721.5}{757,350} = 0.035282894$$

and then use that estimate to find $b_0$, the estimate of the $y$ intercept of the line:

$$b_0 = \frac{391.4}{25} - (0.035282894)\left(\frac{10,340}{25}\right) = 1.062995$$

The equation of the regression line relating rate of operation to machine speed is

$$\hat{y} = 1.06 + 0.035x$$

This means that there is a positive relationship between machine speed and productivity, that is, as machine speed increases, so does productivity. In fact, the equation tells the managers that for every unit increase in machine speed, the productivity goes up by 0.035. This is the definition of the slope of a line.

Although the positive relationship is encouraging, the people looking at the equation wonder if it really means anything. That is, they wonder if the increase in productivity realized by increasing speed is *significant*. The equation of the line alone cannot tell them that. They need to do further analysis. ∎

The equation of the least-squares regression line does give some information about the relationship between the independent variable $Y$ and the dependent variable, $X$. The *sign* of the slope estimate tells whether the relationship is positive or negative. The *value* of the slope gives the change that will occur in $Y$ when $X$ is changed by one unit.

It is more difficult to explain the interpretation of the intercept. By definition, the intercept of a line is the value of $Y$ when $X = 0$. In some situations, this number can be thought of as the value of the dependent variable that is related to *other factors* that are not considered in this model. For example, when we look at the relationship between sales and advertising, it is possible that for $0 spent on advertising, there will still be sales of a product.

In many cases, however, it does not make sense for the value of $Y$ to be nonzero when $X = 0$. In the TQM example, if the machine speed is 0 there *is no productivity*, that is, $Y$ must be 0. We discuss this problem in more detail as we proceed.

### EXAMPLE 11.4   HMO Health

*Calculating the Least-Squares Line*

After they plotted the data and looked at the graph, the analysts who were interested in the relationship between HMO revenues ($Y$), and number of members ($X$), decided to use least squares to find the equation of the regression line relating the two variables. They assembled the data shown in the next table.

| Observation Number | Members, $X$ (million) | Revenue, $Y$ ($ billion) | $XY$ | $X^2$ |
|---|---|---|---|---|
| 1 | 4.24 | 5.49 | 23.2776 | 17.9776 |
| 2 | 3.19 | 4.63 | 14.7697 | 10.1761 |
| 3 | 1.83 | 3.86 | 7.0638 | 3.3489 |
| 4 | 1.62 | 3.60 | 5.8320 | 2.6244 |
| 5 | 2.07 | 3.43 | 7.1001 | 4.2849 |
| 6 | 2.3 | 2.91 | 6.6930 | 5.29 |
| 7 | 1.83 | 2.74 | 5.0142 | 3.3489 |
| 8 | 2.15 | 2.40 | 5.1600 | 4.6225 |
| 9 | 0.97 | 1.71 | 1.6587 | 0.9409 |
| 10 | 0.89 | 1.20 | 1.0680 | 0.7921 |
| **Totals** | **21.09** | **31.97** | **77.6371** | **53.4063** |

*Although finding the equation of the regression line by hand is not difficult, it is tedious, even for small data sets. Many calculators have the capability to find the equation of the least-squares line. If you have such a calculator you might want to learn how to use it now.*

Substituting into the equations they found

$$b_1 = \frac{(10)(77.6371) - (21.09)(31.97)}{(10)(53.4063) - (21.09)^2} = \frac{102.1237}{89.2749} = 1.143924$$

$$b_0 = \frac{31.97}{10} - (1.143924)\left(\frac{21.09}{10}\right) = 0.7844643$$

Thus, the regression equation of HMO revenues on number of members is

$$\hat{y} = 0.78 + 1.14x$$

They realize that the equation means that for an increase in members of 1 million, the revenues will increase by $1.14 billion. The intercept of the line is 0.78, which would mean that when an HMO has no members, it will still generate $0.78 billion in revenues. This is a case where interpretation of the intercept does not make sense. ∎

Because least-squares analysis is mechanical, the equation alone cannot tell us whether the relationship is real or whether, in fact, the two variables are unrelated. You can find a least-squares line relating any two variables, but that certainly does not mean that the two variables are really related. To make that decision, further statistical tools are needed.

Most spreadsheet programs and all statistical software also perform regression analysis. It is not necessary to use the computer just to find the equation of the regression line, but further analysis can really only be done using a computer package. Although it is not necessary now, the remainder of this chapter assumes that you have access to some statistical software package. At the end of this section we look at regression output from different software packages. As the chapter progresses we will use the output extensively to make decisions about the problems we are trying to solve.

 **TRY IT NOW!**

### Increasing Capacity
*Finding the Equation of the Least-Squares Regression Line*

The oil company that is looking at increasing refining capacity has decided that a linear relationship is appropriate.

*(continued)*

Fill in the table below or use some other means to find the equation of the least-squares line:

| Observation Number | Number of Sites, $X$ | Capacity, $Y$ | $XY$ | $X^2$ |
|---|---|---|---|---|
| 1 | 13 | 81.82 | | |
| 2 | 10 | 81.82 | | |
| 3 | 13 | 58.18 | | |
| 4 | 8 | 43.64 | | |
| 5 | 5 | 40.00 | | |
| 6 | 7 | 36.36 | | |
| 7 | 4 | 34.55 | | |
| 8 | 8 | 32.73 | | |
| 9 | 7 | 29.09 | | |
| 10 | 3 | 25.45 | | |
| **Totals** | | | | |

Interpret the meaning of the estimate of the slope of the line. Does the $y$ intercept make sense for these data?

## 11.2.4 Using the Computer to Do Regression Analysis

Any statistical software that you might be using is capable of performing regression analysis. In addition, most spreadsheet packages also do regression. The output from the analyses will look different, but they have many common elements. We will start by identifying the estimates for the parameters of the regression equation in the output for several software packages. As we progress in the chapter we will continue to use the output in conjunction with other analyses.

In Figure 11.7 you see the regression output from two different computer packages for the TQM problem. The output from any software package contains three sections: (1) general information about the regression line, (2) information about the regression coefficients, and (3) ANOVA information. For each output, the coefficients of the regression equation are highlighted.

As you can see from the output, the equation is identified in a slightly different way for each package. You can also see that there is a lot more to the output of a regression analysis than just the equation. It is important that you understand the language of regression so that you can identify the output from whatever software package that you are using.

## 11.2.5 Using the Regression Equation to Make Predictions

Once we have the equation of the regression line, we can use the equation to predict values of the dependent variable, $Y$, for different values of the independent variable. We do this by substituting different values of $X$ into the regression equation and calculating the corresponding value of $\hat{y}$. Remember in Figure 11.3 we saw that the values of $Y$ vary around the true regression line.

**Regression Analysis**

The regression equation is
RateOper = 1.06 + 0.0353 Speed

```
Predictor        Coef     Stdev   t-ratio       p
Constant        1.063     1.084      0.98   0.337
Speed        0.035283  0.002611     13.51   0.000

s = 0.4544   R-sq = 88.8%   R-sq(adj) = 88.3%
```

Analysis of Variance

```
SOURCE        DF       SS       MS       F       p
Regression     1   37.712   37.712  182.64   0.000
Error         23    4.749    0.206
Total         24   42.462
```

Unusual Observations

| Obs. | Speed | RateOper | Fit | Stdev.Fit | Residual | St.Resid |
|------|-------|----------|---------|-----------|----------|----------|
| 2 | 334 | 12.8000 | 12.8475 | 0.2268 | −0.0475 | −0.12 X |
| 3 | 356 | 12.7000 | 13.6237 | 0.1757 | −0.9237 | −2.20R |

R denotes an obs. with a large st. resid.
X denotes an obs. whose X value gives it large influence.

Minitab

SUMMARY OUTPUT

*Regression Statistics*

| | |
|---|---|
| Multiple R | 0.942419659 |
| R Square | 0.888154813 |
| Adjusted R Square | 0.883291979 |
| Standard Error | 0.454404799 |
| Observations | 25 |

ANOVA

| | df | SS | MS | F | Significance F |
|---|-----|------------|------------|------------|----------------|
| Regression | 1 | 37.7124744 | 37.7124744 | 182.6413924 | 1.9909E-12 |
| Residual | 23 | 4.749125596 | 0.206483722 | | |
| Total | 24 | 42.4616 | | | |

| | Coefficients | Standard Error | t Stat | P-value | Lower 95.0% | Upper |
|---|--------------|----------------|--------------|-------------|--------------|-------|
| Intercept | 1.062994916 | 1.083622206 | 0.980964501 | 0.336817039 | −1.178645356 | 3.304 |
| Speed | 0.035282894 | 0.002610746 | 13.51448824 | 1.9909E-12 | 0.029882162 | 0.040 |

Excel

**FIGURE 11.7**   Regression output for Minitab and Excel

The value of $\hat{y}$ that we find is really a prediction of the mean value of $Y$ for a given value of $X$.

## EXAMPLE 11.5   Is TQM Working?

### Prediction Using the Regression Equation

The company that is looking at effectiveness of its TQM program found that the regression equation for its sample data was

$$\hat{y} = 1.06 + 0.035x$$

The managers would like to know what the productivity will be if the machine runs at 350 items per minute, since this is a speed that they have been able to run for quite some time. They substitute into the equation and find that

$$\hat{y} = 1.06 + 0.035(350) = 13.31 \text{ thousand units per hour}$$

They would also like to know what the productivity will be when they run at 450 items per minute, since this is a speed that they have achieved only since TQM was instituted. They find

$$\hat{y} = 1.06 + 0.035(450) = 16.81 \text{ thousand units per hour}$$

From this it would seem that productivity does increase with speed.   ■

When we first looked at the regression equation and talked about what the parameters mean, we noticed that the slope of the line does not always make sense. What happens if we use the regression equation to predict $Y$ when $X = 0$?

### EXAMPLE 11.6   Is TQM Working?

*Using the Regression Equation When X = 0*

One of the workers who is involved in the TQM study thinks that there is something wrong with the model. He uses the equation to predict productivity for a machine speed of 0 items per minute and finds

$$\hat{y} = 1.06 + 0.035(0) = 1.06 \text{ thousand parts per minute}$$

If TQM can accomplish this it must be pretty good, since productivity is impossible when the machine is not running! He is not so sure about using the model to predict productivity, if it makes errors like this.   ■

When using a regression model we must remember that the model is constructed only from sample data. It is really relevant only *over the range of observed values.* The equation tells you about the relationship between the dependent and independent variables, only in this range. There are two kinds of predictions that you can do, **interpolation** and **extrapolation.**

> Using the equation to predict values of $Y$ within the range of the $X$ data is called **interpolation.** Predicting values of $Y$ for values of $X$ outside the observed range is called **extrapolation.**

Extrapolation is risky and really should almost never be done. Without data you have no idea what the relationship beyond your "boundaries" is like. Over a larger range of the dependent variable, the relationship might change shape and be nonlinear. It might even change direction, making your predictions totally inappropriate.

## TRY IT NOW!

### Increasing Capacity

*Using the Regression Equation to Predict the Value of* Y

Use the equation of the regression line you found earlier to predict the refining capacity for each of the observed values of *X*, the number of sites.

| Observation Number | Number of Sites, $X$ | Capacity, $Y$ | $\hat{y} = 9.03 + 4.79x$ |
|---|---|---|---|
| 1 | 13 | 81.82 | |
| 2 | 10 | 81.82 | |
| 3 | 13 | 58.18 | |
| 4 | 8 | 43.64 | |
| 5 | 5 | 40.00 | |
| 6 | 7 | 36.36 | |
| 7 | 4 | 34.55 | |
| 8 | 8 | 32.73 | |
| 9 | 7 | 29.09 | |
| 10 | 3 | 25.45 | |

## 11.2.6 Calculating Residuals

We used least squares to find the line for regression analysis, because we wanted the line we obtained to be close to the actual data points. How can we assess how well the regression line accomplishes this?

If we plot the regression line on the same plot with the data we can get a *visual* or *graphical* idea of the connection between the model and the data. The proximity of the data points to the line will give an overall picture of how well the regression line describes the relationship between the variables.

### EXAMPLE 11.7   Is TQM Working?

*Plotting the Regression Line and Calculating Residuals*

The company assessing its TQM decided to plot its data and the regression equation together to try to get a visual idea of what the least-squares line did.

The graph shows that most of the data points are very close to the line. The company felt that the line it obtained did a pretty good job of describing the relationship between machine speed and productivity. ■

### EXAMPLE 11.8   HMO Health

*Plotting the Regression Line*

The analysts who are looking at the relationship between revenues and number of members of HMOs created a plot of the data and the regression equation that they calculated.

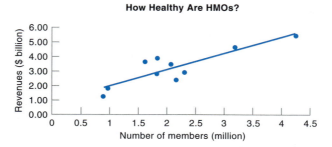

**How Healthy Are HMOs?**

They see that some of the points are very close to the line, but some others are farther away. Still, it looks like the line does a pretty good job of describing the relationship between revenues and number of members. The analysts wonder if they can get a numerical measure of the differences between the predicted values and the actual data. ■

A *numerical* measure of the agreement between the line and the data is obtained by looking at the differences between the observed and expected values of *Y* for each value of *X*. In regression analysis, these differences are known as the **residuals.**

> The difference between the observed value of *Y* ($y_i$), and the predicted value of *Y* from the regression equation ($\hat{y}_i$), for a value of $X = x_i$ is called the $i$th **residual,** $e_i$.

The *sign* of the residual, positive or negative, tells whether the prediction was lower or higher than the actual data value. The *size* of the residual gives an idea of how much the actual data vary around the line.

### EXAMPLE 11.9    Is TQM Working?

*Calculating Residuals*

To obtain a better idea of how well the regression line agreed with their data, the workers decided to calculate the residuals for two predicted values close to the ones that they were interested in. They looked at their data and found that they had data for machine speeds of 373 and 447 items per minute, which were close to the 375 and 450 speeds. They used the regression equation to find the predicted values for $X = 373$:

$$\hat{y} = 1.06 + 0.035(373) = 14.1$$

and for $X = 447$:

$$\hat{y} = 1.06 + 0.035(447) = 16.7$$

and calculated the residuals, $y - \hat{y}$, for $X = 373$:

$$y - \hat{y} = 13.7 - 14.1 = -0.4$$

and for $X = 447$:

$$y - \hat{y} = 16.4 - 16.7 = -0.3$$

In each case, the difference between the observed data and the line was negative. The magnitude of the deviation was about 0.3 or 0.4, which is about 300 or 400 items per hour. ■

The residuals play a very important role in regression analysis. As you just saw, they give a numerical value to the differences between the model and the actual data for each of the data points in the sample. We have already discussed that, for a statistical relationship, the actual values of $Y$ will vary for a fixed value of $X$. The residuals also provide a means to measure the overall variation in $Y$ for any value of $X$. In addition, as we will see later on in this chapter, the residuals are used to determine the appropriateness of a linear model.

## EXAMPLE 11.10   HMO Health

### Calculating Residuals

For the HMO problem, the analysts wanted to use the model to predict revenues for different size HMOs. They were interested in predicting $Y$ for $X = 1.0$ (1 million members) and 2.0 (2 million members). To have a way to compare the predicted values to the actual data, they chose the data points with $X = 0.97$ and 2.07, since these points were closest to the values of interest. The analysts calculated the predicted values using the equation $\hat{y} = 0.78 + 1.14x$ for each value and then calculated the residuals. The results were

| Observation Number | $x_i$ | $y_i$ | $\hat{y}_i = 0.78 + 1.14x_i$ | $e_i = y_i - \hat{y}_i$ |
|---|---|---|---|---|
| 9 | 0.97 | 1.71 | $\hat{y}_9 = 0.78 + 1.14(0.97) = 1.89$ | $e_9 = 1.71 - 1.89 = -0.18$ |
| 5 | 2.07 | 3.43 | $\hat{y}_5 = 0.78 + 1.14(2.07) = 3.14$ | $e_5 = 3.43 - 3.14 = 0.29$ |

The differences between the predicted revenues and the actual revenues were $180 million and $290 million or about 10%. ■

Although residuals are important, they cannot tell you everything about how good the model will be for predicting values of $Y$. Just because the residual for a certain data point is large, it does not mean that the regression equation does a poor job of predicting $Y$ for values of $X$ near that point. It may be that the data point used for the residual is unusually far from the regression line and that the prediction is actually more representative of the population. For this reason it helps to look at the residuals in context. Graphing the regression line with the data points provides additional insight about individual residuals.

## EXAMPLE 11.11   HMO Health

### Looking at the Residuals Graphically

When the HMO analysts looked at the plot of the data and the regression line, they saw that the data points that they used to calculate the residuals were actually very close to the regression line.

They felt that the errors for the number of members that they were interested in were not unusually large. ■

## TRY IT NOW!

### Increasing Capacity

#### *Calculating the Residuals*

The oil company that is looking at the relationship between refining capacity and the number of refining sites wants to get a better idea of how the regression line relates to the actual data. It decides to calculate
the residuals for each observed value of $X$, the number of sites. Find the residuals and fill in the table
below:

| Observation Number | Number of Sites, $X$ | Refining Capacity $Y$ | $\hat{y}_i = 9.03 + 4.79x$ | $e_i = y_i - \hat{y}_i$ |
|---|---|---|---|---|
| 1 | 13 | 81.82 | 71.3 | |
| 2 | 10 | 81.82 | 56.9 | |
| 3 | 13 | 58.18 | 71.3 | |
| 4 | 8 | 43.64 | 47.4 | |
| 5 | 5 | 40.00 | 33.0 | |
| 6 | 7 | 36.36 | 42.6 | |
| 7 | 4 | 34.55 | 28.2 | |
| 8 | 8 | 32.73 | 47.4 | |
| 9 | 7 | 29.09 | 42.6 | |
| 10 | 3 | 25.45 | 23.4 | |

Source: *The Economist*, 15 July 1995.

To get a picture of how the residuals and the regression line fit together, the company also decides to graph
the regression line on a plot of the data. Graph the regression line on the data plot. How well do you think
the line represents the data?

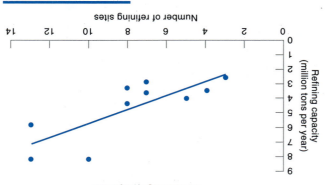

## 11.2.7 The Standard Error of the Estimate

Since each residual corresponds to a single data point we know that we cannot use any single residual to determine how much the data differ from the model. Still, we would like some way to measure how much the data points vary around the regression line. This measure is known as the **standard error of the estimate,** $s_{y|x}$.

> The **standard error of the estimate,** $s_{y|x}$, is a measure of how much the data vary around the regression line.

The quantity that we are trying to measure is the *overall* or *average* variation of the data around the line. Remember that the residuals measure the *individual* variation of each data point from the line. We can obtain an overall measure of this variability by using something similar to the formula we used to measure the variability of the data from its mean in Chapter 4. In fact,

$$s_{y|x} = \sqrt{\frac{\Sigma(y - \hat{y})^2}{n - 2}} = \sqrt{\frac{\Sigma e^2}{n - 2}}$$

There are several different computational formulas for finding the standard error of the estimates. The one we have shown makes it easiest to understand what the quantity really measures. It is most convenient if you have already calculated the residuals. We will not look at any other formulas, because, as we already said, beyond finding the least-squares equation, regression analysis should be done using a computer.

## EXAMPLE 11.12   HMO Health

### Calculating the Standard Error of the Estimate

Since they knew that any single residual might be unusually large or small, the analysts decided to calculate the standard error of the estimate for the regression line. They calculated the residuals and squared residuals for each data point as

| Observation Number | Members, $X$ | Revenue, $Y$ | $\hat{y}$ | $e = y - \hat{y}$ | $e^2$ |
|---|---|---|---|---|---|
| 1 | 4.24 | 5.49 | 5.61 | −0.12 | 0.01 |
| 2 | 3.19 | 4.63 | 4.42 | 0.21 | 0.04 |
| 3 | 1.83 | 3.86 | 2.87 | 0.99 | 0.98 |
| 4 | 1.62 | 3.60 | 2.63 | 0.97 | 0.94 |
| 5 | 2.07 | 3.43 | 3.14 | 0.29 | 0.08 |
| 6 | 2.30 | 2.91 | 3.40 | −0.49 | 0.24 |
| 7 | 1.83 | 2.74 | 2.87 | −0.13 | 0.02 |
| 8 | 2.15 | 2.40 | 3.23 | −0.83 | 0.69 |
| 9 | 0.97 | 1.71 | 1.89 | −0.18 | 0.03 |
| 10 | 0.89 | 1.20 | 1.79 | −0.59 | 0.35 |
| **Total** | | | | | **3.38** |

Then, the standard error of the estimate is

$$s_{y|x} = \sqrt{\frac{3.38}{8}} = \sqrt{0.423} = 0.65$$

This means that the typical deviation or distance of each data point from the line is 0.65 or $650 million. ∎

The standard error of the estimate is analogous to the standard deviation of data from the mean and to the standard error of the mean, which we have studied before. From our knowledge of what these quantities measure, and the Empirical Rule, we can get an intuitive sense of what the quantity $s_{y|x}$ means. We can get a *mental image* of how the data are spread out around the line. Later on in the chapter we develop some more formal ways to use the standard error of the estimate to look at predictions.

In Section 11.2.4 we looked at the computer output for the HMO problem for several different software packages. Figure 11.8 shows the portion of the output from each of the packages that contains the standard error of the estimate. You will notice that in the Excel output, it is called the "standard error" while in the Minitab output it is simply labeled "S".

SUMMARY OUTPUT

| Regression Statistics | |
| --- | --- |
| Multiple R | 0.879773847 |
| R Square | 0.774002022 |
| Adjusted R Square | 0.745752275 |
| Standard Error | 0.65297731 |
| Observations | 10 |

### Excel Output

```
Regression Analysis

The regression equation is
Rev $bn = 0.784 + 1.14 Members (m)

Predictor     Coef    StDev     T       P
Constant     0.7845   0.5050   1.55    0.159
Members      1.1439   0.2185   5.23    0.000


S = 0.6530    R-Sq = 77.4%    R-Sq(adj) = 74.6%
```

### Minitab Output

**FIGURE 11.8**   Computer output showing the standard error of the estimate

## 11.2.8 Exercises—Learning It!

**11.1**   In a study about postage tariffs and revenue, the European Economic Community wanted to look at the relationship between the number of postal employees in a country and the amount of domestic mail that was processed. They collected data for six different countries and found

| Country | Number of Staff | Domestic Traffic (billions of pieces) |
| --- | --- | --- |
| Germany | 342,413 | 18.32 |
| France | 289,156 | 23.87 |
| Italy | 221,534 | 6.62 |
| Britain | 189,000 | 16.75 |
| Spain | 65,355 | 4.06 |
| Sweden | 52,251 | 4.21 |

Source: *The Economist*, 15 June 1996.

(a) Create a scatter plot of the data and determine whether a linear relationship is appropriate.

(b) Find the regression equation for the data.

(c) The Netherlands employs 53,560 postal workers. Use the regression equation to predict the domestic mail traffic for the Netherlands.

(d) Calculate the values of the residuals for France and Germany.

**11.2** In trying to look at the effects of shopping center expansion, the Commerce Department decided to look at the relationship between the number of shopping centers and the retail sales for different states in the same region. It collected the data for the North Central states in the United States and found the following:

**North Central States
Retail Sales—1994**

| State | Number of Shopping Centers | Retail Sales ($ billion) |
|---|---|---|
| Ohio | 1559 | 34.8 |
| Indiana | 848 | 18.0 |
| Illinois | 1961 | 34.1 |
| Michigan | 967 | 21.0 |
| Wisconsin | 588 | 12.0 |
| Minnesota | 436 | 11.5 |
| Iowa | 277 | 6.3 |
| Missouri | 834 | 18.9 |
| North Dakota | 84 | 1.8 |
| South Dakota | 51 | 1.1 |
| Nebraska | 245 | 4.8 |
| Kansas | 458 | 9.7 |

Source: *Statistical Abstract of the United States 1995.*

(a) Create a scatter plot of the data.

(b) Find the regression equation relating retail sales and number of shopping centers.

(c) Plot the regression line on the same plot as the data. Do you think the line fits the data well? Why or why not?

(d) Use the regression line to predict retail sales for each state.

(e) Calculate the residuals for each state. Which state has the largest residual? Which state has the smallest? Do the residuals support your answer to part (d)?

(f) Find the standard error of the estimate.

**11.3** As part of an international study on energy consumption, data were collected on the number of cars in a country and the total travel in kilometers. The data for 12 of the countries are shown below:

| Country | Total Cars (millions) | Travel (billion km) |
|---|---|---|
| United States | 142.35 | 3140.29 |
| Finland | 1.82 | 34.66 |
| Denmark | 1.66 | 30.76 |
| Britain | 21.32 | 352.76 |
| Australia | 8.53 | 138.22 |
| Sweden | 3.32 | 53.21 |
| Netherlands | 5.53 | 83.69 |
| France | 23.27 | 348.20 |
| Norway | 1.59 | 23.54 |
| Italy | 26.12 | 367.85 |
| Germany | 43.75 | 608.52 |
| Japan | 40.25 | 439.30 |

Source: *The Economist,* 22 June 1996.

(a) Create a scatter plot of the data. Do you think that there is a linear relationship between number of kilometers traveled and the number of cars?

(b) Find the least-squares regression line for the data. Interpret the value of the slope.

(c) Does the intercept make sense for these data? Why or why not?

(d) Plot the regression line on the same plot with the data. Does the line make you feel confident about predicting travel as a function of the number of cars?

(e) Use the regression line to predict the number of kilometers traveled for Sweden and Japan. How well do the predictions agree with the original data?

11.4 Large companies are always looking for expanding markets. Media companies in particular are looking at foreign markets for expansion. To obtain some information about expansion possibilities in Central Europe, a major communications company collected data on revenues from television advertising and the percentage of households with cable or satellite television for seven countries.

| Country | % Cable | TV Advertising Revenues ($ million) |
|---|---|---|
| Hungary | 55 | 170 |
| Slovakia | 47 | 28 |
| Poland | 33 | 369 |
| Czech Republic | 28 | 105 |
| Romania | 27 | 40 |
| Russia | 15 | 250 |
| Bulgaria | 9 | 23 |

Source: *The Economist,* 6 July 1996.

(a) Create a scatter plot of the data. Does it appear that advertising revenues are related to the percentage of households that have cable or satellite TV?

(b) Find the equation of the regression line for the data.

(c) Plot the regression line on the same plot as the data. Do you think that the line does a good job of predicting TV advertising revenues? Why or why not?

(d) From the plot, which country do you think will have the largest residual? Which will have the smallest?

(e) In the Ukraine, 8% of the households have cable or satellite television. Use the line to predict TV advertising revenues for the Ukraine.

(f) Find the residuals for each country. Do the values support your answer to part (d)?

(g) Calculate the standard error of the estimate for the regression line.

11.5 As part of the anti-crime bill passed in 1995, the U.S. government granted money to different cities to hire new beat-patrol officers. Nine cities in New Jersey were given grants and hired officers. In trying to assess the program, government analysts wanted to look at the relationship between the number of officers hired and the amount of the grant. The data are

| Grant ($) | Number of Officers |
|---|---|
| 150,000 | 2 |
| 375,000 | 5 |
| 471,125 | 6 |
| 70,967 | 1 |
| 450,000 | 6 |
| 525,000 | 7 |
| 375,370 | 7 |
| 750,000 | 10 |
| 1,000,000 | 12 |

Source: *Justice Department.*

(a) Create a scatter plot of the data.

(b) Find the equation of the linear regression line.

(c) Plot the regression line on the same plot as the data.

(d) How well do you think the line fits the data?

(e) Use the regression equation to predict the number of officers hired for each city.

(f) Calculate the residuals and the standard error of the estimate.

**11.6** For many countries, tourism is an important part of revenues. In trying to predict tourism revenues, one of the independent variables that is considered important is the number of foreign visitors to the country. Data for six different countries were collected:

| Country | Number of Visitors (million) | Tourism Receipts ($ billion) |
|---|---|---|
| France | 60 | 27.3 |
| Spain | 48 | 25.1 |
| US | 45 | 58.4 |
| Italy | 30 | 27.1 |
| Britain | 23 | 17.5 |
| Germany | 15 | 11.9 |

Source: *The Economist,* July 1996.

(a) Create a scatter plot of the data and find the equation of the linear regression line.

(b) Use the regression line to predict tourism revenues for Italy and the United States.

(c) For which country does the regression line do a better job of predicting tourism receipts?

(d) Calculate the residuals and the standard error of the estimate.

**11.7** As communications and the media change and become more important in the economy, the radio industry has become an area of concern. One of the major communications companies wanted to look at the relationship between the number of radio stations a company owned and the revenues generated by radio. It collected the following data:

**American Radio Companies**

| Company | Number of Stations | Revenues ($ billion) |
|---|---|---|
| Westinghouse/Infinity | 83 | 1.05 |
| Jacor | 57 | 0.31 |
| Clear Channel | 104 | 0.31 |
| Evergreen | 35 | 0.30 |
| Disney/ABC | 21 | 0.29 |
| American Radio Systems | 63 | 0.23 |
| SFX | 67 | 0.22 |
| Chancellor | 41 | 0.21 |
| Cox | 38 | 0.21 |
| Bonneville | 20 | 0.12 |

Source: *The Economist,* 29 June 1996, *Duncan's American Radio.*

(a) Create a scatter plot of the data. Do you think that there is a relationship between the number of stations that a company owns and the revenues generated by radio?

(b) Find the linear regression line for the data.

(c) Use the regression line to predict the radio revenues for Chancellor and Westinghouse/Infinity. Which prediction is better?

(d) Leave out the data point for Westinghouse/Infinity and recalculate the regression line. Do you think that this line will do a better job of predicting radio revenues? Why or why not?

# 11.3 INFERENCES ABOUT THE LINEAR REGRESSION MODEL

The tools you have learned for finding the simple linear regression equation are descriptive. The method of least squares allows you to find the equation of the line that best describes the sample data. Looking at the residuals allows you to get some idea about how well the line fits the data. Residuals are useful to some degree but, as in most cases, we would really like to be able to make *inferences* about the population based on our sample data.

In particular, we would like to know

- if the relationship described by the regression equation is *meaningful*.
- how well the equation enables us to predict values of the dependent variable.
- how useful the predictions are for decision making.

To accomplish these tasks we need to learn some inferential methods related to regression analysis.

### 11.3.1 Hypothesis Testing About the Slope, $\beta_1$

Saying that there is a linear relationship between $X$ and $Y$ means that as $X$ changes, $Y$ changes in some predictable, corresponding way. The parameter that describes the magnitude of the relationship is the slope of the line. Remember that the definition of slope is the change in $Y$ for a unit change in $X$. Thus, if the variables $X$ and $Y$ are related, the slope of the line will be some number. If there is no relationship between $X$ and $Y$, then the slope of the line is zero. That is, we say that as $X$ changes, $Y$ *does not change* in a related way (Figure 11.9). Thus, one of the ways that we can determine if the relationship between the two variables is real is to decide whether the slope of the regression line is equal to zero.

**FIGURE 11.9**   (a) Line with nonzero slope; (b) line with zero slope

While looking at data gives you an idea about what might be true in the population, you know quite well by now that you cannot simply look at the numbers to make these decisions. If the data consist of small numbers, the slope and intercept values, $b_1$ and $b_0$, will naturally be small. In the same way, if the data are large numbers, the corresponding slope and intercepts might or might not be large. You need to consider the *standard error* of the statistic in question to make decisions involving the values. We will use hypothesis testing to decide whether the slope of the regression line is *significantly different from zero*.

*You know that if you took a different sample your calculated regression line would not have exactly the same slope and intercept values. Thus, these estimates are sample statistics and have sampling distributions.*

The first step in testing a hypothesis is to set up the appropriate hypotheses. In this case we want to test

$$H_0: \quad \beta_1 = 0$$
$$H_A: \quad \beta_1 \neq 0$$

If the test results in rejecting the null hypothesis, then we will conclude that the slope of the regression line is not equal to zero, and that the relationship between the $X$ and $Y$ variables is real.

Our estimate of $\beta_1$ is $b_1$, and to proceed with the steps of the hypothesis test we need to know about the sampling distribution of $b_1$. It turns out that the sampling distribution associated with the least-squares estimate of the slope is the Student $t$ distribution. The test statistic for our hypothesis test is therefore

*Formula for* t *statistic*

$$t = \frac{b_1 - \beta_1}{s_{b_1}}$$

which has a $t$ distribution with $n - 2$ degrees of freedom. In the formula, $s_{b_1}$ is the **standard error of the slope** $b_1$ and is calculated by

$$s_{b_1} = \frac{s_{y|x}}{\sqrt{\Sigma x^2 - (\Sigma x)^2/n}}$$

*Formula for standard error of* $b_1$

Although the equation may look complicated we have said that for any purposes other than finding the regression line, we will be using statistical software to find the values we need. Since the test is a two-sided test, once the significance level of the test, $\alpha$, is chosen, the critical values of the test are $\pm t_{\alpha/2, n-2}$. We now have the test set up. All that remains is to perform the test and make a decision.

*This does not mean that the values cannot be calculated by hand. Once you find the regression coefficients and the standard error you have all of the pieces you need to find the test statistic.*

## EXAMPLE 11.13   Is TQM Working?

### Hypothesis Test About the Slope Coefficient

The company looking at its TQM program has determined that the equation of the regression line relating productivity to machine speed is

$$\hat{y} = 1.06 + 0.035x$$

When the regression line and the data are graphed, it appears that the regression line does a good job of representing the data. Still, the company looks at the value of the slope, 0.035, and wonders if it really means anything. The company decides to perform a hypothesis test at the 0.05 level of significance, to see if the relationship is real.

The hypotheses are

$$H_0: \quad \beta_1 = 0$$
$$H_A: \quad \beta_1 \neq 0$$

Since the value of $\alpha$ is 0.05 and there are 25 data points, the critical values of the test are $\pm t_{0.025, 23}$ or $\pm 2.069$.

From previous analyses the company knows that $\Sigma x = 10{,}340$, $\Sigma x^2 = 4{,}306{,}918$, and $s_{y|x} = 0.4544$. The value of the test statistic is

$$t = \frac{0.035 - 0}{\dfrac{(0.4544)}{\sqrt{4{,}306{,}918 - (10{,}340)^2/25}}} = \frac{0.035}{0.00261} = 13.51$$

*Remember that the hypothesized value for* $\beta_1$ *is zero.*

Comparing the value of the test statistic to the critical values, the company sees that 13.51 is definitely beyond the critical values and so it will reject $H_0$ and conclude that the slope of the regression line is not equal to zero. That is, there is a significant linear relationship between productivity and machine speed. ■

When we first talked about computer output, we said that there are usually three sections in the output. We are interested in the section with information about the regression coefficients. Figure 11.10 (page 558) shows the relevant sections of the computer output for two different statistical packages.

In the output you are given all of the information needed to perform the hypothesis test about the slope coefficient. You can either look up the critical value for the level of significance that you have chosen and compare it to the $t$ statistic, or use the $p$ value of the test, which is also included in the output. In all cases, the $p$ value of the test is 0.000. Since this is less than our chosen $\alpha$, we reject $H_0$.

| Predictor | Coef | StDev | T | P |
|-----------|------|-------|---|---|
| Constant | 1.063 | 1.084 | 0.98 | 0.337 |
| Speed | 0.035283 | 0.002611 | 13.51 | 0.000 |

Minitab

| | Coefficients | Standard Error | t Stat | P-value |
|---|---|---|---|---|
| Intercept | 1.062994916 | 1.083622206 | 0.980964501 | 0.336817039 |
| Speed | 0.035282894 | 0.002610746 | 13.51448824 | 1.9909E-12 |

Excel

**FIGURE 11.10**    *t*-test portion of computer output

## EXAMPLE 11.14   HMO Health

### Hypothesis Test About the Slope of the Regression Line

The analysts who were looking at the relationship between HMO revenues and number of members used statistical software to perform the regression analysis and to determine whether the relationship is significant. They want to test, at the 0.05 level of significance, the hypotheses

$$H_0: \quad \beta_1 = 0$$
$$H_A: \quad \beta_1 \neq 0$$

At the 0.05 level of significance with $10 - 2 = 8$ degrees of freedom, the critical values for the test are $\pm 2.306$.

They obtained the following output:

```
Regression Analysis

The regression equation is
Rev $bn = 0.784 + 1.14 Members (m)

Predictor    Coef    StDev      T      P
Constant    0.7845   0.5050    1.55   0.159
Members     1.1439   0.2185    5.23   0.000
```

From the output line that corresponds to the slope estimate, they see that the coefficient is 1.1439 and the standard error of the slope is 0.2185. The corresponding $t$ statistic is 5.23. Since the critical value for the test is 2.306, they reject $H_0$ and conclude that the slope of the regression line is not zero. That is, there is a significant linear relationship between revenues and number of members.

They can also use the $p$ value of the test to make the decision. From the output, the $p$ value of the test is 0.000. Since that is less than the specified $\alpha$ of 0.05, they know that they should reject $H_0$. ∎

You might have noticed that in the computer output there is corresponding information about $b_0$, the intercept of the regression line. It is tempting to want to test hypotheses about $\beta_0$, the intercept, in particular whether it is equal to zero. *Do not do this.* Remember that the regression line is valid only in the range of $X$ that has been observed in the data. If you do not have $X$ data for $X = 0$, then you cannot make a determination about whether the intercept should remain in the equation. Removing the intercept from the equation when the slope is not equal to zero means that the regression line will move downward, parallel to itself, and will no longer be the line of best fit for the data.

*TRY IT NOW!*

## Increasing Capacity

### *Testing for Significance of the Regression Model*

The oil company that is looking at increasing capacity wants to determine whether the relationship between refining capacity and number of refining sites that it calculated is significant.

    Write down the hypotheses that the company needs to test.

The company decides to use a 0.01 level of significance for the test. Find the critical values for the test.

> *Remember that you need to find the degrees of freedom, in this case $n - 2$.*

It used a computer software package to run the analysis and obtained the following output:

|  | Coefficients | Standard Error | tStat | P-value | Lower 95% | Upper 95% |
|---|---|---|---|---|---|---|
| Intercept | 9.028295455 | 11.04587353 | 0.817345539 | 0.437393793 | −16.44355105 | 34.50014196 |
| Number of sites | 4.786628788 | 1.307226853 | 3.661666509 | 0.006385827 | 1.77215631 | 7.801101266 |

From the computer output, find the slope of the regression line, the standard error of the slope, and the value of the $t$ statistic.

Perform the hypothesis test and make a decision about the regression line.

Find the $p$ value of the test from the output and explain how you could use the $p$ value on the output to make the same decision.

---

    Once we have determined that the relationship between $X$ and $Y$ is significant, we can perform some additional analyses to see if the predictions we obtain are useful for the purposes of decision making and to determine the strength of the relationship.

---

ANS. $H_0: \beta_1 = 0$, $H_A: \beta_1 \neq 0$; $\mp 3.355$, $4.79$, $1.31$, $3.66$; REJECT $H_0$, THE REGRESSION IS SIGNIFICANT. SINCE $0.01 > 0.0064$ WE CAN REJECT $H_0$.

## 11.3.2 Partitioning the Variance in Linear Regression

In statistics, we are always interested in looking at the *variation* in data. In Chapter 4 you learned about the standard deviation and how that can be used to measure the variation of data around its mean value. When we are looking at the relationship between two variables, $Y$ (the dependent variable) and $X$ (the independent variable), we are interested in how $Y$ *varies*, considering the values of the variable, $X$.

When two variables, $X$ and $Y$, are linearly related, the variation of the $y$ values from the overall mean of $Y$ can be attributed to two different factors: its relationship with $X$ and pure chance. In fact, the variation in the $y$ values from the overall mean of $Y$ can be partitioned, or divided, into parts called **sums of squares:**

> **SST**—the total variation in the y values around the mean $\bar{y}$
> **SSR**—the variation in $Y$ that is caused by $Y$'s relationship with $X$
> **SSE**—the variation in $Y$ that remains unexplained

*Notice that SSE is related to the residuals that you learned about previously.*

The quantities **SST, SSR,** and **SSE** are known as the **sums of squares:** SST is the total sum of squares, SSR is the regression sum of squares, and SSE is the error sum of squares. Figure 11.11 illustrates how these quantities relate to the data and to the regression line. From the picture you see that SST represents the distance of the data value $y_i$ from the overall mean $\bar{y}$. SSR represents the distance of the predicted value $\hat{y}$ from the overall mean $\bar{y}$ and SSE represents the distance of a data value $y_i$ from the predicted value $\hat{y}$. The formulas for calculating these quantities are

$$SST = \sum (y - \bar{y})^2 \qquad SSR = \sum (\hat{y} - \bar{y})^2 \qquad SSE = \sum (y - \hat{y})^2$$

In addition you see that

$$SST = SSR + SSE$$

When $X$ and $Y$ are related, SSR is a large part of the total variation. This implies that a major reason that $Y$ varies so much is *because* it is related to $X$. When this is true, the SSE component of the variation is small and is the variation in $Y$ that happens "naturally" or entirely due to chance.

When $X$ and $Y$ are not related, the regression line is horizontal ($\beta_1 = 0$) and the SSR component of the variation disappears. The SSE part of the variation becomes dominant and we say that we cannot really explain the variation in $Y$ using the linear model with $X$.

We can use these ideas to create a formal test for the significance of the linear

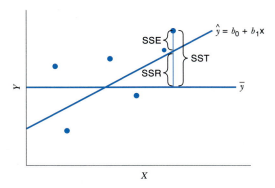

**FIGURE 11.11**   Components of the variation in y value

regression model. The test is referred to as an *analysis of variance (ANOVA)* test, because it is based on looking at the *variation* in the *Y* variable.

The hypotheses that we test are

$H_0$: The linear regression model is significant.

$H_A$: The linear regression model is not significant.

The test statistic for this test uses the **mean squares,** which are obtained by dividing the sums of squares, SSR and SSE, by their respective degrees of freedom:

$$\text{MSR} = \frac{\text{SSR}}{1} \quad \text{and} \quad \text{MSE} = \frac{\text{SSE}}{n-2}$$

The test looks at the ratio of the regression mean square (MSR) to the error mean square (MSE). As we just saw, when the regression model is really significant (that is, when *X* and *Y* are really linearly related), SSR (and therefore MSR) will be much larger than SSE (and therefore MSE). The ratio of the two mean squares has an *F* distribution with 1 and $n-2$ degrees of freedom. If the ratio is larger than the critical *F* value, then the model is significant. Thus, the ANOVA test is a one-sided or one-tail test.

*You saw the* F *distribution when you looked at the test for comparing the variances of two normal populations in Chapter 10.*

Calculation of SST, SSR, and SSE is not difficult, but for even small data sets it can be computationally exhausting. If you have any two of the components, you can obtain the third by subtraction. You can see from the formulas that SSE is really the sum of the squared residuals that you used to find $s_{y|x}$, and that SST is really the numerator portion of the variance of *Y*.

If you have been doing the analyses by hand or using a calculator, you probably have already calculated most of the quantities you need. However, if you are doing this much analysis you really should be using a statistical software package to do the job. You remember that we learned that output from a statistical analysis program includes a section known as the ANOVA section. Figure 11.12 shows the ANOVA output for Excel and Minitab.

You might wonder why you need another method for testing significance of a linear regression model. In reality you need only one method, since both tests (the *t* test and the *F* test) yield the same decision. The difference is, to some degree, a matter of preference. Other statistical analyses use the ANOVA approach and some people prefer that type of test. In addition, the ANOVA approach is used in multiple regression and model building as you will see in the next chapter.

ANOVA

| | df | SS | MS | F | Significance F |
|---|---|---|---|---|---|
| Regression | 1 | 37.7124744 | 37.7124744 | 182.6413924 | 1.9909E-12 |
| Residual | 23 | 4.749125596 | 0.206483722 | | |
| Total | 24 | 42.4616 | | | |

Excel Output

Analysis of Variance

| Source | DF | SS | MS | F | P |
|---|---|---|---|---|---|
| Regression | 1 | 37.712 | 37.712 | 182.64 | 0.000 |
| Error | 23 | 4.749 | 0.206 | | |
| Total | 24 | 42.462 | | | |

Minitab Output

**FIGURE 11.12** Computer ANOVA output for regression analysis

The quantities SST, SSR, and SSE generated during the ANOVA approach also yield additional information about how good the linear regression model is. One of the reasons that we would like to know whether two variables are related is so that we can use the information to predict one given the other. Another reason is so that we can control or reduce the variation in one quantity by controlling the variation in the other. The regression sum of squares (SSR) measures the amount of the variation in the $Y$ variable that can be accounted for or *explained by* $Y$'s relationship with $X$. If you look at SSR as a portion of SST, then you can determine the amount of the variability in $Y$ that can be explained, or accounted for. This value is called the **coefficient of determination, $R^2$**:

*Formula for coefficient of determination*

$$R^2 = \frac{\text{SSR}}{\text{SST}} \, 100\%$$

$R^2$ is usually part of the general information in the output from statistical packages.

The coefficient of determination gives you a measure of how much the variation in $Y$ could be reduced if $X$ were controlled to a single value. This is a way of measuring how useful a model is for planning purposes. Even if a linear regression model is significant, it may not explain enough of the variation in the $Y$ variable to be useful for planning. When this is the case, you may want to use a regression model with more than one independent or *explanatory* variables. These models are called multiple regression models and are discussed in Chapter 12.

## 11.3.3 Exercises—Learning It!

**11.8**   The data on the number of postal employees and the amount of mail processed were analyzed using Minitab. A portion of the output is given below:

```
Predictor        Coef        StDev       T      P
Constant        0.955        4.608      0.21   0.846
_ Staff    0.00005872   0.00002087    2.81   0.048

S = 5.461   R-Sq = 66.4%   R-Sq(adj) = 58.1%
```

(a) Set up the hypotheses to test whether the slope coefficient is equal to 0.

(b) There were 6 observations in the sample. At the 0.05 level of significance what are the critical values of the test?

(c) From the output, find the value of the coefficient, the standard error of the slope, and the value of the $t$ statistic.

(d) Compare the $t$ statistic to the critical value and determine whether the relationship between the number of postal employees and the amount of mail processed is significant.

**11.9**   Consider the data from Exercise 11.4 which looked at the relationship between the percentage of households with cable TV in Central Europe and TV Advertising revenues. The regression model is

$$\hat{y} = 127.2 + 0.442x \qquad s_{y|x} = 143.2 \qquad \Sigma x = 214 \qquad \Sigma x^2 = 8142$$

(a) Find the standard error of the slope.

(b) Set up the appropriate hypotheses and find the value of the test statistic.

(c) At the 0.10 level of significance, test to see if the slope coefficient is significantly different from 0.

**11.10**   In Exercise 11.6 you looked at the relationship between the number of visitors to a country, $X$, and the receipts from tourism, $Y$. The regression equation is

$$\hat{y} = 9.95 + 0.487x \qquad s_{y|x} = 15.50 \qquad \Sigma x = 221 \qquad \Sigma x^2 = 9583$$

(a) Calculate the standard error of the slope.

(b) Set up the hypotheses to determine whether the regression equation is significant at the 0.05 level.

(c) Find the critical values of the test and calculate the test statistic.

(d) Is the relationship between number of visitors and tourism receipts significant?

**11.11** Consider the data from Exercise 11.3 on the number of kilometers of travel, $Y$, and the number of cars, $X$, for different countries. The results of a regression analysis using Microsoft Excel are given below:

|  | Coefficients | Standard Error | t Stat | P-value |
|---|---|---|---|---|
| Intercept | −106.212984 | 55.14747208 | −1.92598102 | 0.082984278 |
| Total Cars | 2.15816E-05 | 1.194E-06 | 18.07496921 | 5.75839E-09 |

(a) From the output, find the slope coefficient, the standard error of the slope, the $t$ statistic, and the $p$ value of the test.

(b) At the 0.01 level of significance use the $p$ value to determine if the relationship between number of kilometers traveled and the number of cars is significant. (The $p$ value in Excel is a one-sided $p$ value.)

(c) There were 12 observations in the data set. Find the critical values for the test and verify your answer to part (b).

**11.12** In Exercise 11.5 you looked at the relationship between the amount of money, $X$, that a city was granted and the number of new police officers hired, $Y$, for nine different cities in New Jersey. The Minitab output from the regression analysis is shown below:

```
Predictor         Coef         StDev         T
Constant        0.7077        0.5360       1.32
Grant ($)    0.00001191    0.00000100      11.88
```

(a) Use the output to find the regression equation relating $X$ and $Y$.

(b) Use the output to test whether the relationship between the amount of the grant and the number of new officers is significant.

**11.13** Data on the number of radio stations owned by 10 different companies, $X$, and the revenues generated by radio, $Y$, were examined in Exercise 11.7. The data were analyzed using Microsoft Excel and the output is given below:

|  | Coefficients | Standard Error | t Stat | P-value |
|---|---|---|---|---|
| Intercept | 0.085444208 | 0.176860232 | 0.483117131 | 0.641956743 |
| # of Stations | 0.004546468 | 0.003006584 | 1.512170824 | 0.16894211 |

Use the output to determine whether there is a significant relationship between the number of radio stations owned and the revenues generated by radio for a company.

**11.14** Consider the data from Exercise 11.2 on the number of shopping centers and retail sales for the 12 North Central states. The data were analyzed with Minitab and the output is shown here

```
Predictor        Coef        StDev        T        P
Constant        1.331       1.006       1.32     0.215
No. Shop      0.019021    0.001129    16.84     0.000
```

At the 0.01 level of significance, is the relationship between the number of shopping centers and retail sales significant?

# 11.4 PREDICTION AND CONFIDENCE INTERVALS

We have discussed the fact that in a *statistical* relationship the values of the dependent variable will vary for any single value of the independent variable. No matter how strong the relationship between $X$ and $Y$ is, there is some amount of inherent variation that will exist in $Y$ for any given value of $X$. This natural variation is identified in the population regression model $y = \beta_0 + \beta_1 x + \varepsilon$ as the quantity $\varepsilon$.

When the relationship between $X$ and $Y$ is strong, the variation in $Y$ for a given value of $X$ is small and can be thought of simply as *natural* variation. If you look at a plot of the data and the regression line, the points will be very close to the line. That is, the variation that exists is entirely due to chance and we should not waste time trying to figure out why it happens. In this case, the predictions we obtain from our estimate of the regression line will be fairly precise and can be used for the purposes of making decisions.

When the relationship between $X$ and $Y$ is not so strong, the variation in $Y$ for a given value of $X$ will be variation present because our linear model is *not adequate*. In this case, the data points will not be close to the regression line, the relationship might not be linear, or there may be other factors that cause $Y$ to vary. In this case, even if $X$ and $Y$ have a statistically significant relationship, the predictions obtained from the regression model might not be precise enough for decision making purposes.

In Section 11.2 you learned that the standard error of the estimate, $s_{y|x}$, is a measure of the overall variation in $Y$ for any value of $X$. In this section we put that value in context and see how it can be used to evaluate the usefulness of a regression model.

## 11.4.1 Confidence Intervals for Regression Analysis

The value of $\hat{y}$ obtained from the regression model is an estimate of the *mean value of Y for a given value of X*. That is, the value $\hat{y}$ is an *estimate* of the true mean, $\mu_{y|x}$. If the variable $X$ represented advertising expenditures, then the regression model relating $X$ to sales of a product, $Y$, would predict the *average* sales for a given level of advertising. Because our estimate is based on sample data, we know that there is some error in that estimate.

In Chapter 9 you learned how to find confidence intervals for the mean of a population, $\mu$, based on a sample estimate, $\overline{X}$. We can use the same ideas to create a confidence interval for $\mu_{y|x}$ based on our sample estimate, $\hat{y}$.

To find a confidence interval for a population parameter we need to know the sampling distribution of the estimate and the standard error of the sample statistic. The estimate $\hat{y}$ has a Student $t$ distribution with $n - 2$ degrees of freedom. The standard error is related to the standard error of the estimate, $s_{y|x}$. The formula for a $(1 - \alpha)100\%$ **confidence interval** for the mean value of $Y$ for a given value of $X = x_i$, $\mu_{y|x_i}$, is

$$\hat{y}_i - t_{\alpha/2, n-2} s_{y|x} \sqrt{\frac{1}{n} + \frac{(x_i - \overline{x})^2}{\Sigma x^2 - (\Sigma x)^2 / n}} \leq \mu_{y|x_i} \leq \hat{y}_i + t_{\alpha/2, n-2} s_{y|x} \sqrt{\frac{1}{n} + \frac{(x_i - \overline{x})^2}{\Sigma x^2 - (\Sigma x)^2 / n}}$$

> A **confidence interval** provides an estimate for the mean value of $Y$ ($\mu_{y|x}$) at a particular value of $X$.

Again, the expression looks formidable, but it does not contain anything we have not already dealt with!

Previously, when you found confidence intervals, you found that the width of the interval was related to the level of confidence that you wanted. This is also true for the confidence interval for $\mu_{y|x}$, but the width of the interval also depends on how far the selected value of $X$ is from the average value of $X$. The farther the value of interest is from the average $X$ value, the wider (and less precise) the confidence interval becomes! Figure 11.13 illustrates how the confidence interval is related to the values of the $X$ variable.

**FIGURE 11.13**  Confidence interval for the mean estimate

If you look at the formula for the confidence interval again, you will see that when $X = \overline{X}$, the part of the formula in the square root reduces to $1/\sqrt{n}$ and the formula looks exactly like the confidence interval for the mean that you learned previously.

### EXAMPLE 11.15  Is TQM Working?

***Calculating a Confidence Interval for the Mean Response***

The manufacturing group that is looking at the effectiveness of TQM wants to know something about the accuracy of the predictions from the regression model. To accomplish this, it decides to find 95% confidence intervals for two machine speeds that it runs frequently, 400 and 450 items per minute. The regression model is

$$\hat{y} = 1.06 + 0.035x$$

The first step is to find the predicted value for the 400 speed:

$$\hat{y} = 1.06 + 0.035(400) = 15.06$$

*To actually calculate the interval it is probably easiest to calculate the $\pm$ part and then add and subtract that value from the prediction, $\hat{y}$.*

From previous work the manufacturers know the predicted value for 450 is 16.81 and $s_{y|x} = 0.4544$, $\Sigma x = 10{,}340$, and $\Sigma x^2 = 4{,}306{,}918$. They calculate that $\overline{X} = 413.6$.

For a 95% confidence interval the $t$ values are $\pm t_{0.025,23} = \pm 2.069$, so they calculate

$$(2.069)(0.4544)\sqrt{\frac{1}{25} + \frac{(400 - 413.6)^2}{4{,}306{,}918 - (10{,}340)^2/25}}$$

$$= (2.069)(0.4544)\sqrt{0.0461054\ldots} = 0.202$$

So, the confidence limits are

$$15.06 \pm 0.20 = (14.86, 15.26)$$

or

$$(14.86 \le \mu_{y|400} \le 15.26)$$

The manufacturers see that the estimate of the mean productivity is somewhere between 14.86 (or 14,860 units) per hour and 15.26 (or 15,260 units) per hour. The width of the interval is

$$15.26 - 14.86 = 0.40 \text{ or } 400 \text{ units/hr.}$$

which is fairly precise for their purposes.

*Notice that the only change in the calculation is substituting 450 for 400 in the square root part of the formula.*

They do a similar calculation for a speed of 450 items per minute and obtain:

$$(2.069)(0.4544)\sqrt{\frac{1}{25} + \frac{(450 - 413.6)^2}{4{,}306{,}918 - (10{,}340)^2/25}}$$

$$= (2.069)(0.4544)\sqrt{0.0837367\ldots} = 0.272$$

So the 95% confidence interval for a speed of 450 items per minutes is

$$16.81 \pm 0.27 = (16.54, \ 17.08)$$

or

$$(16.54 \le \mu_{y|450} \le 17.08)$$

That means that at a speed of 450 items per minute, average hourly production will be between 16,540 and 17,080 items. ∎

It is not really difficult to calculate confidence intervals for one or two estimates, but if you want intervals for more estimates, using a computer is much easier.

### EXAMPLE 11.16   HMO Health

*Finding Confidence Intervals for the Mean Estimate*

The analysts who are looking at the relationship between HMO revenues and number of members would like to know how accurate the estimates from the regression model are. They use Minitab to find 95% confidence intervals for the mean for each value of $X$ in the data set. The output (edited) from the analysis is shown below:

| X Value | Fit | 95.0% CI |
|---------|-------|------------------|
| 0.89 | 1.803 | (1.025, 2.580) |
| 0.97 | 1.894 | (1.148, 2.640) |
| 1.62 | 2.638 | (2.101, 3.174) |
| 1.83 | 2.878 | (2.381, 3.374) |
| 1.83 | 2.878 | (2.381, 3.374) |
| 2.07 | 3.152 | (2.676, 3.629) |
| 2.15 | 3.244 | (2.767, 3.721) |
| 2.30 | 3.415 | (2.930, 3.901) |
| 3.19 | 4.434 | (3.710, 5.157) |
| 4.24 | 5.635 | (4.460, 6.809) |

They see that the uncertainty in the estimated revenue (the width of the interval) goes from a low of $0.953 billion to a high of $2.349 billion. For the middle $X$ values (close to the mean) the width of the interval is about $1 billion. They feel that the confidence intervals are very wide, especially considering the size of the actual estimates. It might not be practical to use this model to predict mean revenues. ∎

Most statistical software packages will calculate the confidence intervals for either a single $X$ value or for an entire set of $X$ values. If you are using a spreadsheet program, entering the formula is not a difficult task.

*TRY IT NOW!*

### Increasing Capacity

*Finding Confidence Intervals for the Mean Predicted Value*

After calculating the regression model and deciding that the model is significant, the analysts at the oil company would like to know about the accuracy of the estimates from the model. They decide to calculate 95% confidence intervals for $X = 8$ and 13 sites. They know from previous work that for the set of 10 observations in the model, $s_{y|x} = 13.43$, $\Sigma x = 78$, and $\Sigma x^2 = 714$.

Find 95% confidence intervals for the mean estimates.

Do you think that these estimates would be useful for planning purposes? Why or why not?

## 11.4.2 Prediction Intervals for Regression Analysis

A confidence interval for the mean estimate from the regression model gives the decision maker an idea of just how accurate the estimate is *over a long period of time.* That is, it is an interval estimate for the mean. This is certainly of importance for decisions that involve long-time horizons, but what about models that are important for predictions in a single instance?

If a company is looking at the relationship between sales and advertising, a confidence interval will tell it that, for a specific level of advertising, the *average* sales will vary between these two amounts. This is certainly useful information, but some additional information might also be helpful. The company might be interested in the accuracy of a *single observation.* For example, it might like to have an interval estimate for next month's sales for some specific advertising outlay. In a problem related to cash flow, this information would be more important than an overall mean. Intervals for individual estimates are called **prediction intervals** in regression analysis.

> A **prediction interval** gives an estimate for an individual value of $Y$ at a particular value of $X$.

The formula used to calculate a $(1 - \alpha)100\%$ prediction interval for a regression model is almost identical to the one for the confidence interval:

$$\hat{y}_i - t_{\alpha/2,n-2}s_{y|x}\sqrt{1 + \frac{1}{n} + \frac{(x_i - \bar{x})^2}{\Sigma x^2 - (\Sigma x)^2/n}} \leq y \leq \hat{y}_i + t_{\alpha/2,n-2}s_{y|x}\sqrt{1 + \frac{1}{n} + \frac{(x_i - \bar{x})^2}{\Sigma x^2 - (\Sigma x)^2/n}}$$

You see that the only difference is the "1" in the square root part of the formula. The impact of this change is that the prediction intervals for a specific value of $X$ will be *wider* than the corresponding confidence interval. If you think about it, a wider interval for prediction makes sense. You know from the ideas of sampling and the Central Limit Theorem that the distribution of individual observations is always more variable than the distribution of the sample means.

### EXAMPLE 11.17   Is TQM Working?

***Finding a Prediction Interval for a Regression Model***

The company that is looking at the TQM program thinks that the estimates for the mean production are useful, but the analysts would like to know how much variability there is in the estimate for the hourly productivity. They decide to find 95% prediction intervals for the estimates for speeds of 400 and 450 units per min. The relevant values for the 25 observations are $\hat{y}_{400} = 15.06$, $\hat{y}_{450} = 16.81$, $s_{y|x} = 0.4544$, $\Sigma x = 10,340$, and $\Sigma x^2 = 4,306,918$. The $t$ value has not changed from the confidence interval calculations and so the analysts calculate the interval as follows:

$$(2.069)(0.4544)\sqrt{1 + \frac{1}{25} + \frac{(400 - 413.6)^2}{4,306,918 - (10,340)^2/25}}$$

$$= (2.069)(0.4544)\sqrt{1.0461055 \ldots} = 0.962$$

The 95% prediction interval is then

$$15.06 \pm 0.96$$

or

$$(14.10 \leq y \leq 16.02)$$

So, in a given hour, at 400 units per minute, production might vary from 14,100 to 16,020 units. Similarly for the 450 speed they find

$$(2.069)(0.4544)\sqrt{1 + \frac{1}{25} + \frac{(450 - 413.6)^2}{4,306,918 - (10,340)^2/25}}$$

$$= (2.069)(0.4544)\sqrt{1.0837367 \ldots} = 0.979$$

The prediction interval is

$$16.81 \pm 0.98$$

or

$$(15.83 \leq y \leq 17.79)$$

At 450 units per minute, hourly production will vary from 15,830 to 17,790 units.

They see that the variability for an individual month's productivity is greater than the variability for average production. The width of the intervals is about 0.97 or 970 items, which is still useful for planning purposes. ∎

Prediction intervals are important when the individual values for the $Y$ variable are critical to the decision process. If, for example, the company looking at TQM is interested in planning warehouse capacity based on the amount of usable product it has, then a long-term estimate might not make sense.

 **TRY IT NOW!**

### Increasing Capacity

*Calculating Prediction Intervals for Regression Estimates*

The oil company analysts decide to calculate 95% prediction intervals for the two $X$ values that they are interested in. The relevant values from the set of 10 observations are $s_{y|x} = 13.43$, $\Sigma x = 78$, and $\Sigma x^2 = 714$.

Find 95% prediction intervals for $X = 8$ and $X = 13$ refining sites.

Do you think that confidence intervals or prediction intervals would be more appropriate for the oil company's purposes?

## 11.4.3 Exercises — Learning It!

**11.15** For the data on federal grant money and police officers hired the relevant data are     *Requires Exercise 11.5*

$$\Sigma x = 4{,}167{,}462 \qquad \Sigma x^2 = 2{,}571{,}647{,}717{,}614 \qquad s_{y|x} = 0.8033 \qquad n = 9$$

(a) Find a 95% confidence interval for the mean number of officers hired when the amount of the grant is $350,000.

(b) Find a 95% prediction interval for the number of officers hired when the amount of the grant is $350,000.

**11.16** For the data on the number of postal workers in a country and the amount of mail     *Requires Exercise 11.1*
processed, the relevant data are:

$$\Sigma x = 1{,}159{,}709 \qquad \Sigma x^2 = 292{,}657{,}611{,}087 \qquad s_{y|x} = 5.4614 \qquad n = 6$$

(a) Find 95% confidence intervals for the mean amount of mail processed when the number of postal workers is 50,000 and 250,000.

(b) Find 95% prediction intervals for the same two values of $X$.

*Requires Exercise 11.2*     **11.17** For the data on the number of shopping centers and retail sales for the North Central States, the relevant data are:

$$\Sigma x = 8308 \qquad \Sigma x^2 = 9,517,766 \qquad s_{y|x} = 2.1918 \qquad n = 12$$

(a) Find 98% confidence intervals for the mean retail sales when the number of shopping centers is 1000 and 1500.

(b) Which interval is wider? Why?

(c) Find 98% prediction intervals for the same values of X.

(d) Do you think that prediction or confidence intervals are more appropriate for these data?

*Requires Exercise 11.3*     **11.18** For the data on the number of automobiles in a country and the total number of kilometers traveled, the relevant data are

$$\Sigma x = 319.51 \qquad \Sigma x^2 = 25,598.8975 \qquad s_{y|x} = 156.0981 \qquad n = 12$$

(a) Find 99% confidence intervals for the number of kilometers traveled for $X = 1.5$ and 10 (million) cars.

(b) Find 99% prediction intervals for the two values of X from part (a).

## 11.5   CORRELATION ANALYSIS

In many instances, when people talk about regression analysis they also talk about *correlation analysis*. The two topics are related, but not interchangeable. Both regression and correlation analyses deal with bivariate quantitative data and the relationship between the two variables. The main purpose of regression analysis is to find an equation or model that allows the decision maker to predict the value of the dependent variable. On the other hand, *correlation analysis* simply *measures the strength* of the relationship between two quantitative variables. The output of the analysis is a single number. In correlation analysis, there is no need to identify which variable is dependent and which is independent, since prediction is not the end result.

### 11.5.1 The Correlation Coefficient

We have talked about the types of linear relationships that can exist between two variables. In Figure 11.14 you see three types of relationships: perfect negative, none, and perfect positive.

If we simply want to measure the strength of a relationship between two variables, we use the **correlation coefficient.**

> The **correlation coefficient** is used as a measure of the strength of a linear relationship. A correlation of $-1$ corresponds to a perfect negative relationship, a correlation of 0 corresponds to no relationship, and a correlation of $+1$ corresponds to a perfect positive relationship.

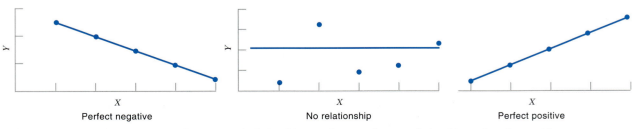

**FIGURE 11.14**   Three types of relationships: perfect negative, no relationship, and perfect positive

The **correlation coefficient, *r*,** is calculated using the formula

$$r = \frac{\Sigma xy - (\Sigma x)(\Sigma y)/n}{\sqrt{\Sigma x^2 - (\Sigma x)^2/n}\sqrt{\Sigma y^2 - (\Sigma y)^2/n}}$$

Again, we have a complicated formula, but it uses all of the familiar elements. Any scientific calculator that does regression will output the correlation coefficient. Software that performs regression analysis usually does not output the correlation coefficient directly because correlation and regression are not interchangeable. In the next section you will learn a statistic that is part of regression that can be used to obtain the correlation coefficient.

In many cases, the correlation coefficient is used along with regression analysis as another measure of how good the regression model is. It is important to recognize that the correlation coefficient can be used as a statistic in its own right when prediction of one variable as a function of the other is not appropriate or not necessary.

For example, suppose that a company is interested in knowing whether there is a relationship between the score on an aptitude test and the number of months that a person remains in an entry level position. The company does not necessarily want to *predict* the number of months in the entry level position; it simply wants to know if the test score is related to that variable. In this case the company could calculate the correlation between score on the test and number of months in the entry level job.

## EXAMPLE 11.18   Is TQM Working?

*Calculating the Correlation Coefficient*

The company looking at its TQM program wonders what the correlation coefficient is for the relationship between machine speed and productivity. It decides to calculate the correlation. The relevant numbers are

$$\Sigma x = 10{,}340 \qquad \Sigma y = 391.4 \qquad \Sigma xy = 162{,}951.9 \qquad \Sigma x^2 = 4{,}306{,}918$$
$$\Sigma y^2 = 6170.22 \qquad n = 25$$

The calculation is

$$r = \frac{162{,}951.9 - (10{,}340)(391.4)/25}{\sqrt{4{,}306{,}918 - (10{,}340)^2/25}\sqrt{6170.22 - (391.4)^2/25}}$$

$$= \frac{1068.86}{(174.051\ \ldots)(6.516\ \ldots)} = 0.9424$$

The correlation coefficient is very close to $+1$, which indicates a strong, positive relationship between machine speed and productivity.  ∎

You may know what values indicate good correlation and what values indicate that there is not really a relationship between two variables. The answer is "it depends." When data are collected under uncontrolled circumstances, there is usually a lot of variation in the data from other, uncontrolled, unmeasured variables. When this happens, the correlation between two variables might not appear to be strong because of the additional factors. This will also be reflected in the widths of the confidence and prediction intervals. Sometimes a correlation in the range of 0.6 to 0.7 (or

$-0.6$ to $-0.7$), which might not provide *useful* information, is an indication that further investigation is appropriate.

### EXAMPLE 11.19   HMO Health

*Calculating the Correlation Coefficient*

The analysts looking at the relationship between revenues and number of members for various HMOs are pretty certain that the relationship between the two variables will be a strong, positive one, so they decide to calculate the correlation coefficient to check their assumption. The numbers that they need to perform the calculation are

$$n = 10 \quad \Sigma x = 21.09 \quad \Sigma y = 31.97 \quad \Sigma xy = 77.6371$$
$$\Sigma x^2 = 53.4063 \quad \Sigma y^2 = 117.3013$$

The calculation is

$$\frac{77.6371 - (21.09)(31.97)/10}{\sqrt{53.4063 - (21.09)^2/10}\sqrt{117.3013 - (31.97)^2/10}} = \frac{10.2124}{11.60795 \ldots} = 0.8798$$

Looking at the value, the analysts feel that the number does indicate the strong positive correlation that they expected to find. ∎

 **TRY IT NOW!**

### Increasing Capacity
*Calculating the Correlation Coefficient*

The relevant data to calculate the correlation coefficient for the oil company problem are

$$n = 10 \quad \Sigma x = 78 \quad \Sigma y = 463.64 \quad \Sigma xy = 4121.86 \quad \Sigma x^2 = 714$$

$$\Sigma y^2 = 25{,}359.3224$$

Find the correlation coefficient for the data.

The correlation coefficient is also related to one of the quantities that we looked at in regression analysis, the coefficient of determination, $R^2$. The value of $r$ is equal to the positive square root of $R^2$.

$$r = \sqrt{R^2}$$

# 11.6 REGRESSION ASSUMPTIONS AND RESIDUAL ANALYSIS

While regression is certainly one of the most powerful and frequently used tools in statistical analysis, it is also one that is often misused. This sometimes happens because the model is built on certain *assumptions* and people are often unaware of the assumptions or choose to ignore them. As you have learned before in this book, when you ignore the *assumptions* behind a statistical tool, the results of the analysis might be totally incorrect. In this section we will look at the assumptions of the simple linear regression model and the ramifications of violating those assumptions. We will also look at some simple techniques for determining whether the assumptions of the model are appropriate for a particular set of data.

Another reason that the simple linear model is misused is that people do not realize that the model is inappropriate for the data being analyzed. This can be true even when the results of the analysis indicate that the simple linear model is appropriate. We will also look at ways to determine when the simple linear model is not appropriate.

## 11.6.1 Assumptions and Problems in the Regression Model

Remember that the simple linear model is given by

$$y = \beta_0 + \beta_1 x + \varepsilon$$

and that $\varepsilon$ represents the random error in the model. Most of the assumptions that are built into the simple linear regression model involve this error term.

The basic assumptions about the error term $\varepsilon$ are

1. It has a mean value of zero ($\mu_\varepsilon = 0$).
2. For every value of $X$, the standard deviation, $\sigma$, of $\varepsilon$ is the same.
3. The distribution of $\varepsilon$ is normal.
4. The error terms for the different observations are not correlated with each other.

What happens if these assumptions are not valid? Obviously, the simple linear model is not correct, but what does that really mean? It may mean several different things. One possible result of violating the assumptions is that you may decide that a model is significant when it is not, or vice versa. Another result is that although you may be correct in your decision about the significance of the model, the equation you obtain is completely wrong. That is, although the model may give reasonable predictions for $Y$, the coefficients in the model, $b_0$ and $b_1$, cannot be interpreted correctly. When this is the case, using the model for planning or control purposes is very risky.

In addition to violating the assumptions of the regression model, there are other factors that make the linear model inappropriate. Some of these factors are

- Fitting a linear model when the true relationship is nonlinear
- Influence of outliers on the regression equation

Sometimes data that appear to have a linear relationship in a scatter plot may actually have a nonlinear relationship. It is very often hard to see curves or nonlinear patterns when looking at scatter plots. The results of the analysis might indicate that the linear model is appropriate when in fact a nonlinear model would be better. Also, sometimes certain observations in the data set have a large influence on the regression equation. This often happens when the $X$ value for an observation is far away from the other values. When these data points are removed, the entire analysis may change.

How can we tell if we are violating the assumptions or using the linear model incorrectly? One tool that answers many of these questions is called *residual analysis*. You have already learned how to calculate residuals. Now we will look at them as a diagnostic tool.

According to the assumptions of linear regression, for all values of the independent variable, the residuals of the model should be normally distributed random variables with a mean of zero and a standard deviation of $\sigma$. That is, the $\varepsilon$ are $N(0, \sigma)$. If we plot the residuals from the regression model, the $e_i$'s, as a function of $X$, we should expect to see that they are randomly distributed around a value of 0. Furthermore, the amount of variation around 0 should be the same for all values of $X$. When the plot exhibits departures from this it indicates that certain of the model assumptions are not valid or that a linear model was not appropriate.

In Figure 11.15 we see several different plots of the residuals versus the values of the independent variable, $X$. The first plot shows what the residual plot should look like when the assumptions of the linear regression model are met and when the simple linear model is appropriate. You can see that the residuals are randomly distributed around the mean value of 0 and that there is no apparent pattern in the plot. The others plots show different patterns that can result when the assumptions are violated or when the model is not appropriate. Figure 11.15b shows a typical plot that results when the linear model is used on data that have a nonlinear relationship. The residuals have a systematic pattern to them. Figures 11.15c and d show plots that can result when the assumption of equal variances is violated.

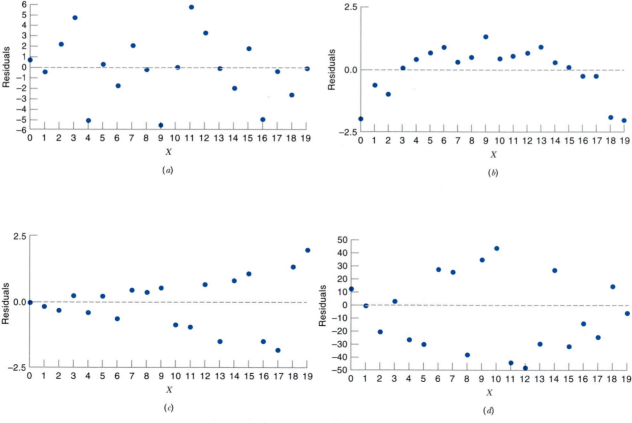

**FIGURE 11.15**   Examples of residual plots

Certain statistical packages plot the residuals against the fitted values ($\hat{y}$) instead of $X$. When the regression model is significant, $X$ and $Y$ are related and so the plots will not be different in shape. Also, some packages use "standardized residuals" (the residuals divided by the standard error of the residuals) instead of the residuals. Again, this does not produce a difference in the plots.

In addition to checking that the residuals have a mean of zero and that they are randomly dispersed around this value, we must check the shape of the distribution to see if it is normal. You remember that this is also an assumption for some of the hypothesis tests that you learned in Chapters 9 and 10. How can we check to see whether data come from a normal distribution? There are several ways to accomplish this. We will look at one *informal* method, making a histogram, and one *formal* method, a normal probability plot, that is available in many statistical software packages.

If you look at a data distribution to see if it might be normally distributed, you will look to see that it is approximately bell-shaped, that is, unimodal and symmetric. Figure 11.16 shows histograms of the residuals from the same plots that you looked at in Figure 11.15. In each case, a normal curve with a mean of 0 is overlaid on the histogram.

You may notice that even in the case where the residuals looked randomly distributed, it does not look like the residuals come from a normal distribution. It may be that the residuals are not normally distributed, or it may be that there is just not enough data to make a really good histogram. This is one of the problems with such an informal method.

Another, more formal method for determining whether data are normally distributed is a plot called a **normal probability plot.**

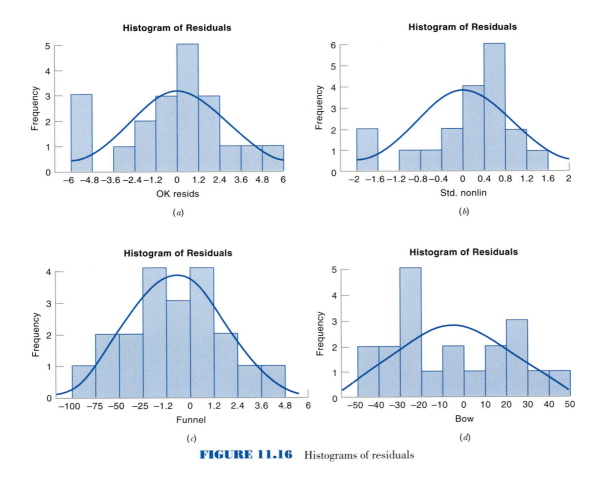

**FIGURE 11.16**   Histograms of residuals

A **normal probability plot** is a plot of the ordered data against their expected values under a normal distribution. When data are normally distributed, the plot will be a straight line.

The methods for creating a normal plot are well beyond the scope of this book, but they are easily created with statistical software packages. In fact, most statistical software make normal plots as part of their output from regression analysis. Figure 11.17 shows a set of residual plots, including a normal plot from Minitab, for each of the sets of residuals we have been looking at.

Another factor that can cause problems in using the linear regression model is the effects of observations that have excessive influence on the equation of the line. This can happen as a result of two things: data pairs that are outliers, and data points whose $X$ values are very far removed from the rest of the data. Both of these factors can cause the regression model to shift toward the suspect data point and skew the resulting regression line.

There are a few ways to determine whether a data set has such an observation. One visual method, as seen in the Minitab output in Figure 11.17, is to create a control chart for the residuals and see if any of the values indicate an "out of control" situation. Another visual method is to make a boxplot of the residuals or the $X$ data values and see if any are determined to be outliers.

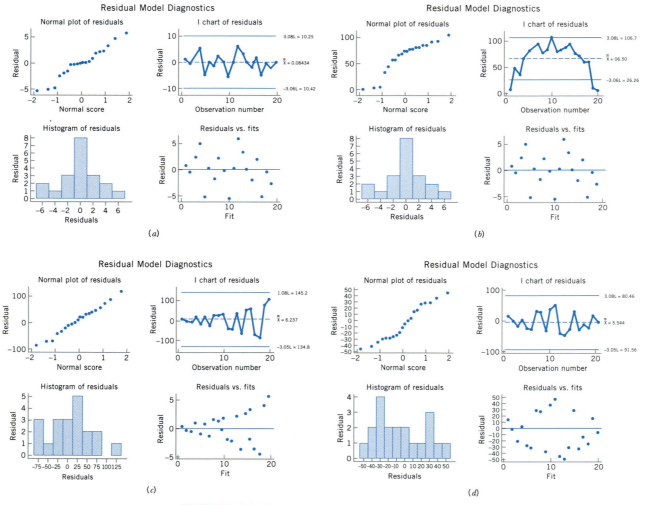

**FIGURE 11.17** Regression diagnostic plots (Minitab)

| Obs | Members | HMORev $ | Fit | StDev Fit | Residual | St Resid |
|---|---|---|---|---|---|---|
| 1 | 4.24 | 5.486 | 5.631 | 0.509 | −0.146 | −0.36 X |
| 2 | 3.19 | 4.629 | 4.432 | 0.313 | 0.197 | 0.34 |
| 3 | 1.83 | 3.857 | 2.878 | 0.215 | 0.979 | 1.59 |
| 4 | 1.62 | 3.600 | 2.639 | 0.232 | 0.961 | 1.58 |
| 5 | 2.07 | 3.429 | 3.153 | 0.206 | 0.276 | 0.45 |
| 6 | 2.30 | 2.914 | 3.415 | 0.210 | −0.501 | −0.81 |
| 7 | 1.83 | 2.743 | 2.878 | 0.215 | −0.136 | −0.22 |
| 8 | 2.15 | 2.400 | 3.244 | 0.206 | −0.844 | −1.37 |
| 9 | 0.97 | 1.714 | 1.896 | 0.323 | −0.182 | −0.32 |
| 10 | 0.89 | 1.200 | 1.805 | 0.336 | −0.605 | −1.08 |

X denotes an observation whose X value gives it large influence.

| Obs | of Sta | RadioRev | Fit | StDev Fit | Residual | St Resid |
|---|---|---|---|---|---|---|
| 1 | 83 | 1.0500 | 0.4628 | 0.1191 | 0.5872 | 2.75P |
| 2 | 57 | 0.3143 | 0.3446 | 0.0783 | −0.0303 | −0.13 |
| 3 | 104 | 0.3143 | 0.5583 | 0.1720 | −0.2440 | −1.40 |
| 4 | 35 | 0.3048 | 0.2446 | 0.0942 | 0.0602 | 0.27 |
| 5 | 21 | 0.2857 | 0.1809 | 0.1232 | 0.1048 | 0.50 |
| 6 | 63 | 0.2286 | 0.3719 | 0.0831 | −0.1433 | −0.62 |
| 7 | 67 | 0.2190 | 0.3901 | 0.0882 | −0.1710 | −0.75 |
| 8 | 41 | 0.2095 | 0.2718 | 0.0852 | −0.0623 | −0.27 |
| 9 | 38 | 0.2095 | 0.2582 | 0.0894 | −0.0487 | −0.21 |
| 10 | 20 | 0.1238 | 0.1764 | 0.1256 | −0.0526 | −0.25 |

**FIGURE 11.18**   Warning output from Minitab

It is also possible to determine whether a data point has a large influence on the model by using some statistical techniques. Many statistical software packages use these techniques and report the problem as part of the output. Figure 11.18 shows this part of the output for Minitab.

## EXAMPLE 11.20   Is TQM Working?

*Checking Regression Assumptions Using Residuals*

The company that is using linear regression to determine whether its TQM program is working decides to make sure that what it has done so far is correct. Since the linear regression model is significant, the analysts want to make sure that it is valid and that they can use it for decision-making purposes. They use Minitab to analyze the data and looked at the output of residual plots. The output is shown below:

From the histogram and normal probability plots it appears that there might be a problem with the assumption of normality. The plots of the residuals versus the $\hat{y}$'s appears to be random, although there is an impression that the data actually might have a nonlinear relationship. The control chart indicates that there are no problems with extreme values.

Before the analysts decide to use this model, they will look more carefully at the data and see if another, perhaps nonlinear model, would be better. ∎

## 11.6.2 Exercises—Learning It!

*Requires Exercises 11.5, 11.15*

**11.19** Look at the regression model you developed for the data on federal grant money and police officers hired.

(a) Make a plot of the residuals vs. the values of the independent variable.

(b) Look at the plot you obtained. Does it appear that a linear model is really appropriate for these data? Why or why not?

(c) Looking at the same residual plot, does the assumption of equal variances appear reasonable? Why or why not?

(d) Use an appropriate graphical technique to display the distribution of the residuals. Be sure to consider how many residuals you have in selecting the technique. Does the assumption of normality appear to be reasonable?

(e) Create a normal probability plot of the residuals. What does this make you think about the normality assumption?

(f) Considering your answers to parts (a)–(e), do you think that the linear regression model was a good one for these data. Be sure to cite specifics in your answer.

*Requires Exercises 11.1, 11.16*

**11.20** Consider the regression model you found for the data on the number of postal workers in a country and the amount of mail processed.

(a) Make a plot of the residuals versus the values of the independent variables.

(b) What does this lead you to believe about the appropriateness of the linear model?

(c) Now plot the residuals versus the predicted values of $y$.

(d) How do the two graphs compare? Why do you think this happened?

(e) Make a normal probability plot of the data. Do you think the assumption of normality is reasonable?

(f) Considering your answers to the parts (a)–(e), do you think that the linear regression model is appropriate for these data. Why or why not?

*Requires Exercises 11.2, 11.17*

**11.21** Look at the model you found for the data on the number of shopping centers and retail sales for the North Central states.

(a) Make a plot of the residuals versus the independent variable.

(b) From the plot, does it appear that a linear model is appropriate?

(c) Looking at the residual plot, do you think that the assumption of equality of variances is violated?

(d) Make a graph that shows the distribution of the residuals.

(e) Does it appear that the residuals are normally distributed?

(f) Considering your answers to parts (a)–(e), do you think that the linear regression model is appropriate for these data? Why or why not?

*Requires Exercises 11.3, 11.18*

**11.22** Look at the model you found for the data on the number of automobiles in a country and the total number of kilometers traveled.

(a) Make a residual plot of the data versus the independent variables.

(b) What does the residual plot tell you about the assumption of a linear relationship? The assumption of equal variances?

(c) Using the residual plot, the computer output, or a plot of the values of the independent variable, do you think that any observation had a lot of influence on the model? If so, which point(s) are they?

(d) Do you think that the assumption of normality is reasonable?

(e) Considering your answers to parts (a)–(d), do you think that the linear model is appropriate for these data? Why or why not?

# 11.7 SIMPLE LINEAR REGRESSION IN EXCEL

In Chapter 5, you learned how to find the equation of the least squares line in Excel by creating a scatter plot and inserting a linear trend line. However, there is more to understanding the relationship between two variables than just finding the equation of the line. You can do much more to analyze the simple linear regression model in Excel by using a procedure from the data analysis tools.

## 11.7.1 The Linear Regression Model in Analysis

Let's look at the problem of predicting HMO revenues based on the number of members. The data should be located in an Excel worksheet similar to the one shown in Figure 11.19.

| | A | B | C |
|---|---|---|---|
| 1 | HMOs | Members (m) | Rev $bn |
| 2 | United HealthCare | 4.24 | 5.49 |
| 3 | Humana | 3.19 | 4.63 |
| 4 | FHP International | 1.83 | 3.86 |
| 5 | PacifiCare Health systems | 1.62 | 3.60 |
| 6 | U.S. Healthcare | 2.07 | 3.43 |
| 7 | WellPoint Health Network | 2.3 | 2.91 |
| 8 | Health Systems International | 1.83 | 2.74 |
| 9 | Foundation Health | 2.15 | 2.40 |
| 10 | Oxford Health Plans | 0.97 | 1.71 |
| 11 | Physician Corp. of America | 0.89 | 1.20 |

**FIGURE 11.19**    The HMO data in Excel

To find the simple linear regression model for a set of data, follow these steps:

1. From the list of data analysis tools, select **Regression.** The dialog box shown in Figure 11.20 (page 580) opens.

   The dialog box has two sections, Input and Output. The input section gives Excel information about your data. The output section lets you make decisions about specific types of extra output that are available.

2. Position the cursor in the textbox labeled **Input Y Range:** and highlight the data range for the *Y* variable, in this case, Revenues.

3. Move the cursor to the textbox for **Input X Range:** and highlight the data range of the *X* variable, in this case, Members.

4. If the data ranges contain labels, click on the **Labels** checkbox. If you want confidence intervals for the regression estimates, click the checkbox for **Confidence Level.**

**FIGURE 11.20**     The regression dialog box

5. Specify the location where you want the output to appear, either in the current sheet, in a new worksheet, or in a new workbook.

In addition to standard regression output, Excel will create some of the diagnostic plots that were discussed in this chapter. You can choose to have the residuals and/or standardized residuals as part of the output. You can also have residual plots, fitted line plots, and normal probability plots. We recommend that you not create some of these plots since the values are calculated incorrectly.

6. Click the checkbox for **Residuals.** Do *not* check the **Standardized Residuals** checkbox. Excel does not calculate these values correctly.

7. Click the checkboxes for **Residual Plots** and **Line Fit Plots.** Do *not* click the checkbox for **Normal Probability Plot.** The plot is not created correctly.

The completed dialog box should resemble the one in Figure 11.21.

**FIGURE 11.21**     Completed regression dialog box

**8.** Click on **OK.** The output will appear in the location you specified.

The output from the regression analysis is quite extensive. While the whole output is still highlighted, you should select **Format > Column > AutoFit Selection** so that the columns adjust to fit the various output sections.

The first section of the output is the Summary Section. This section gives information about the regression model including the coefficients, the values of $R^2$, the standard error, and the ANOVA output. This output is shown in Figure 11.22.

| | | | | | | | |
|---|---|---|---|---|---|---|---|
| 13 | SUMMARY OUTPUT | | | | | | |
| 14 | | | | | | | |
| 15 | *Regression Statistics* | | | | | | |
| 16 | Multiple R | 0.879773847 | | | | | |
| 17 | R Square | 0.774002022 | | | | | |
| 18 | Adjusted R Square | 0.745752275 | | | | | |
| 19 | Standard Error | 0.65297731 | | | | | |
| 20 | Observations | 10 | | | | | |
| 21 | | | | | | | |
| 22 | ANOVA | | | | | | |
| 23 | | *df* | *SS* | *MS* | *F* | *Significance F* | |
| 24 | Regression | 1 | 11.68217506 | 11.68217506 | 27.3985468 | 0.000788678 | |
| 25 | Residual | 8 | 3.411034941 | 0.426379368 | | | |
| 26 | Total | 9 | 15.09321 | | | | |
| 27 | | | | | | | |
| 28 | | *Coefficients* | *Standard Error* | *t Stat* | *P-value* | *Lower 95%* | *Upper 95%* |
| 29 | Intercept | 0.784464301 | 0.505044505 | 1.55325777 | 0.158965828 | -0.380171168 | 1.94909977 |
| 30 | Members (m) | 1.143923992 | 0.218541241 | 5.23436212 | 0.000788678 | 0.639966661 | 1.647881323 |

**FIGURE 11.22**   Summary section of regression output

The coefficients for the regression equation are at the end of the section. You see that the equation is $\hat{y} = 0.78 + 1.1439x$. In the same section, you see the information for testing whether the regression model is significant. You can perform this test using the ANOVA output just above. In both cases, we see that the regression model is significant.

The top section of the output, Regression Statistics, includes the value for $R^2$ and the standard error. Excel does not include the correlation coefficient specifically, but you can get it by taking the square root of the $R^2$ value.

The second section of the output you created is the Residual section, as shown in Figure 11.23. This section contains the values of $\hat{y}$ and the residuals.

| | | | |
|---|---|---|---|
| 34 | RESIDUAL OUTPUT | | |
| 35 | | | |
| 36 | *Observation* | *Predicted Rev $bn* | *Residuals* |
| 37 | 1 | 5.634702027 | -0.144702027 |
| 38 | 2 | 4.433581835 | 0.196418165 |
| 39 | 3 | 2.877845206 | 0.982154794 |
| 40 | 4 | 2.637621168 | 0.962378832 |
| 41 | 5 | 3.152386964 | 0.277613036 |
| 42 | 6 | 3.415489482 | -0.505489482 |
| 43 | 7 | 2.877845206 | -0.137845206 |
| 44 | 8 | 3.243900884 | -0.843900884 |
| 45 | 9 | 1.894070573 | -0.184070573 |
| 46 | 10 | 1.802556654 | -0.602556654 |

**FIGURE 11.23**   Residual output

The plots that are created as part of the output are stacked on top of each other. Figure 11.24 shows the plots after they have been moved in the worksheet.

**FIGURE 11.24**   Plots from regression analysis

The residual plot is a plot of the residuals vs. the $X$ variable. From this plot, you can look at whether the assumption of equal variances is violated. You can also get information about whether a linear model is appropriate, or whether any observation exerts a lot of influence on the model.

The Line Fit plot shows the predicted and fitted values of $y$ on the same plot. This gives you information about how well the model fits the data. If the predictions were perfect, the two sets of symbols would coincide exactly.

## 11.7.2 Confidence and Prediction Intervals Using Excel

Although Excel does not provide an option to output confidence intervals or prediction intervals as part of the regression routine, it is not hard to calculate these intervals. All of the values that are needed are included as part of the regression output. You can use formulas to find the upper and lower values for the intervals.

Recall that the formulas for the $(1 - \alpha)100\%$ confidence interval for the mean value of $Y$ at a particular value of $X$ are given by

$$\hat{y}_i - t_{\alpha/2, n-2}s_{y|x}\sqrt{\frac{1}{n} + \frac{(x_i - \overline{x})^2}{\Sigma x^2 - (\Sigma x)^2/n}} \leq \mu_{y|x_i} \leq \hat{y}_i + t_{\alpha/2, n-2}s_{y|x}\sqrt{\frac{1}{n} + \frac{(x_i - \overline{x})^2}{\Sigma x^2 - (\Sigma x)^2/n}}$$

and the formulas for the corresponding prediction intervals are given by

$$\hat{y}_i - t_{\alpha/2, n-2}s_{y|x}\sqrt{1 + \frac{1}{n} + \frac{(x_i - \overline{x})^2}{\Sigma x^2 - (\Sigma x)^2/n}} \leq \mu_{y|x_i}$$

$$\leq \hat{y}_i + t_{\alpha/2, n-2}s_{y|x}\sqrt{1 + \frac{1}{n} + \frac{(x_i - \overline{x})^2}{\Sigma x^2 - (\Sigma x)^2/n}}$$

Excel allows you to obtain the predicted values $\hat{y}$ as part of the output; all other parts of the formula are either part of the output or can be calculated using other Excel functions. Unless you are adept at manipulating formulas, entering a formula this complicated in Excel is best accomplished by breaking it into smaller pieces and then combining the final results. For example, you should calculate the quantity under the square root separately and take the square root in a separate step. This minimizes possible problems with using parentheses.

We have provided a macro that will calculate 95% prediction intervals for an individual value $Y$ at a particular value of $X$. To use the macro, the $X$ and $Y$ data should be in columns of a worksheet with no titles at the top. The $X$ column must come first. From the list of **Tools > Macro > Macros,** choose **Prediction Interval Limits** and click **Run.** The predicted values, the residuals, and the upper and lower prediction intervals are output to the worksheet as shown in Figure 11.25.

| Prediction Interval Limits | | | | | |
|---|---|---|---|---|---|
| b_0 | 0.784464301 | xBar | 2.109 | | |
| b_1 | 1.143923992 | $t_\alpha$ | 2.306005626 | | |
| $\alpha$ | 5% | SumX$^2$ | 53.4063 | | |
| n | 10 | $s_{XY}$ | 0.65297731 | | |
| | | | | | |
| **X** | **Y** | **Prediction** | **Residual** | **LPL** | **UPL** |
| 4.24 | 5.49 | 5.6347 | -0.1447 | 3.7249 | 7.5445 |
| 3.19 | 4.63 | 4.4336 | 0.1964 | 2.7630 | 6.1042 |
| 1.83 | 3.86 | 2.8778 | 0.9822 | 1.2923 | 4.4634 |
| 1.62 | 3.60 | 2.6376 | 0.9624 | 1.0392 | 4.2360 |
| 2.07 | 3.43 | 3.1524 | 0.2776 | 1.5730 | 4.7318 |
| 2.3 | 2.91 | 3.4155 | -0.5055 | 1.8333 | 4.9977 |
| 1.83 | 2.74 | 2.8778 | -0.1378 | 1.2923 | 4.4634 |
| 2.15 | 2.40 | 3.2439 | -0.8439 | 1.6645 | 4.8233 |
| 0.97 | 1.71 | 1.8941 | -0.1841 | 0.2137 | 3.5744 |
| 0.89 | 1.20 | 1.8026 | -0.6026 | 0.1080 | 3.4971 |

**FIGURE 11.25**    Output from prediction interval macro

*Note: The data are originally located in the top left of the worksheet. The macro moves it down the sheet.*

You can use the formulas in the macro to help you set up formulas for the confidence intervals.

# *CHAPTER 11 SUMMARY*

Linear regression analysis is a powerful tool for determining how two variables are related. The regression equation can be used for *description, control,* and *prediction.* Description is important when the user is simply trying to understand the way that two variables are related. Control describes when the model is used to set standards or reduce variability. Prediction is when the model is used to determine what the resulting $Y$ value should be when $X$ takes on certain values. Although finding the simple linear model itself is not numerically difficult, performing a complete regression analysis requires the use of computer software to completely understand the model.

We have also seen that even though the simple linear model may be significant, it might not be correct. It is necessary to test the *assumptions* of the linear model to see whether the model you obtain is appropriate.

## Key Terms

| Term | Definition | Page Reference |
|---|---|---|
| **Simple linear regression model** | The true relationship between the variables $X$ and $Y$, the simple linear regression model, can be described by $$y = \beta_0 + \beta_1 x + \varepsilon$$ | 538 |
| **Deviation** | The distance between the predicted value of $Y, \hat{y},$ and the actual value of $Y, y,$ is called the deviation or error. | 540 |
| **Least-squares method** | The technique that finds the equation of the line that minimizes the total or sum of the squared deviations between the actual data points and the line is called the least-squares method. | 541 |
| **$\hat{y}$ (y hat)** | The value of $\hat{y}$ that we find is really a prediction of the mean value of $Y$ for a given value of $X$. | 545 |
| **Interpolation and extrapolation** | Using the equation to predict values of $Y$ within the range of the $X$ data is called interpolation. Predicting values of $Y$ for values of $X$ outside the observed range is called extrapolation. | 546 |
| **Residual ($e_i$)** | The difference between the observed value of $Y (y_i),$ and the predicted value of $Y$ from the regression equation, $(\hat{y}_i),$ for a value of $X = x_i$ is called the $i$th residual, $e_i$. | 548 |
| **Standard error of the estimate ($s_{y \mid x}$)** | The standard error of the estimate, $s_{y \mid x},$ is a measure of how much the data vary around the regression line. | 551 |

| Term | Definition | Page Reference |
|---|---|---|
| Sums of squares | **SST**—the total variation in the $y$ values around the mean $\bar{y}$.<br>**SST**—the variation in $Y$ that is caused by $Y$'s relationship with $X$.<br>**SSE**—the variation in $Y$ that remains unexplained. | 560 |
| Confidence interval | A confidence interval provides an estimate for the mean value of $Y$ ($\mu_{y\,|\,x}$) at a particular value of $X$. | 564 |
| Prediction interval | A prediction interval gives an estimate for an individual value of $Y$ at a particular value of $X$. | 567 |
| Correlation coefficient | The correlation coefficient is used as a measure of the strength of a linear relationship. A correlation of $-1$ corresponds to a perfect negative relationship, a correlation of 0 corresponds to no relationship, and a correlation of $+1$ corresponds to a perfect positive relationship. | 570 |
| Normal probability plot | A normal probability plot is a plot of the ordered data against their expected values under a normal distribution. When data are normally distributed, the plot will be a straight line. | 576 |

## Key Formulas

| Term | Formula | Page Reference |
|---|---|---|
| $b_1$ | $b_1 = \dfrac{n\sum\limits_{i=1}^{n} x_i y_i - \sum\limits_{i=1}^{n} x_i \sum\limits_{i=1}^{n} y_i}{n\sum\limits_{i=1}^{n} x_i^2 - \left(\sum\limits_{i=1}^{n} x_i\right)^2}$   or   $\dfrac{n\sum xy - \sum x \sum y}{n\sum x^2 - \left(\sum x\right)^2}$ | 541 |
| $b_0$ | $b_0 = \dfrac{\sum\limits_{i=1}^{n} y_i}{n} - b_1 \dfrac{\sum\limits_{i=1}^{n} x_i}{n}$   or   $b_0 = \bar{y} - b_1 \bar{x}$ | 541 |
| Standard error of the estimate $s_{y\,|\,x}$ | $s_{y\,|\,x} = \sqrt{\dfrac{\sum(y - \hat{y})^2}{n-2}} = \sqrt{\dfrac{\sum e^2}{n-2}}$ | 551 |
| Standard error of the slope | $s_{b_1} = \dfrac{s_{y\,|\,x}}{\sqrt{\sum x^2 - (\sum x)^2/n}}$ | 557 |
| Test statistic for slope | $t = \dfrac{b_1 - \beta_1}{s_{b_1}}$ | 556 |
| Sums of squares | $\text{SST} = \Sigma(y - \bar{y})^2 \qquad \text{SSR} = \Sigma(\hat{y} - \bar{y})^2 \qquad \text{SSE} = \Sigma(y - \hat{y})^2$ | 560 |
| Mean squares | $\text{MSR} = \dfrac{\text{SSR}}{1} \quad \text{and} \quad \text{MSE} = \dfrac{\text{SSE}}{n-2}$ | 561 |
| Coefficient of determination, $R^2$ | $R^2 = \dfrac{\text{SSR}}{\text{SST}}\,100\%$ | 562 |
| Confidence interval | $\hat{y}_i - t_{\alpha/2,\,n-2}\,s_{y\,|\,x}\sqrt{\dfrac{1}{n} + \dfrac{(x_i - \bar{x})^2}{\Sigma x^2 - (\Sigma x)^2/n}} \leq \mu_{y\,|\,x_i}$<br><br>$\leq \hat{y}_i + t_{\alpha/2,\,n-2}\,s_{y\,|\,x}\sqrt{\dfrac{1}{n} + \dfrac{(x_i - \bar{x})^2}{\Sigma x^2 - (\Sigma x)^2/n}}$ | 564 |
| Prediction interval | $\hat{y}_i - t_{\alpha/2,\,n-2}\,s_{y\,|\,x}\sqrt{1 + \dfrac{1}{n} + \dfrac{(x_i - \bar{x})^2}{\Sigma x^2 - (\Sigma x)^2/n}} \leq \mu_{y|x_i}$<br><br>$\leq \hat{y}_i + t_{\alpha/2,\,n-2}\,s_{y|x}\sqrt{1 + \dfrac{1}{n} + \dfrac{(x_i - \bar{x})^2}{\Sigma x^2 - (\Sigma x)^2/n}}$ | 568 |
| Correlation coefficient, $r$ | $r = \dfrac{\Sigma xy - (\Sigma x)(\Sigma y)/n}{\sqrt{\Sigma x^2 - (\Sigma x)^2/n}\ \sqrt{\Sigma y^2 - (\Sigma y)^2/n}}$ | 571 |

# CHAPTER 11 EXERCISES

## Learning It!

**11.23** How much does advertising impact market penetration? To assess the impact of advertising in the tobacco industry, a study looked at the amount of money spent on advertising a particular brand of cigarettes and brand preference among adolescents and adults. The data are shown below:

|  |  | Brand Preference | |
| --- | --- | --- | --- |
| **Brand** | **Advertising ($ million)** | **Adolescent (%)** | **Adult (%)** |
| Marlboro | 75 | 60 | 23.5 |
| Camel | 43 | 13.3 | 6.7 |
| Newport | 35 | 12.7 | 4.8 |
| Kool | 21 | 1.2 | 3.9 |
| Winston | 17 | 1.2 | 3.9 |
| Benson & Hedges | 4 | 1.0 | 3.0 |
| Salem | 3 | 0.3 | 2.5 |

Source: *Center for Disease Control Website.*

(a) Look at the data for brand preference for adolescents and amount spent on advertising. Which variable is the dependent variable? Which is the independent variable?

(b) Create a scatter plot of advertising and adolescent brand preference. Do you think that there is a linear relationship between the two variables? Why or why not?

(c) Now create another scatter plot using adult brand preference instead. How does this plot compare to the one for adolescent brand preference? From the plots, do you think that adolescent or adult brand preference is more strongly related to advertising expenditures? Why?

(d) Find the least-squares line for adolescent brand and advertising expenditures.

(e) Interpret the meaning of the slope and intercept for the model. Do they make sense?

(f) Use the model to predict adolescent brand preference for each brand studied. How well do the predicted values agree with the actual data?

(g) Using $\alpha = 0.05$, is the model significant?

**11.24** Retention of students is one of the largest problems facing colleges and universities today. Loss of students impacts revenues not only from tuition but from state and federal funding programs. If a university can understand what factors impact student retention it can form strategic plans aimed at keeping students enrolled. Academic performance is often cited as one reason that students leave. Data were collected from 20 colleges in the Midwest on the freshman retention rate (% of freshmen who stay for a second year) and the 25th percentile score on the American College Testing (ACT) examination. (The ACT test is a college entrance examination similar to the Scholastic Assessment Test, SAT.) The data are given below:

| College | ACT 25th Percentile | Freshman Retention Rate |
| --- | --- | --- |
| 1 | 21 | 0.82 |
| 2 | 25 | 0.86 |
| 3 | 22 | 0.78 |
| 4 | 22 | 0.82 |
| 5 | 22 | 0.82 |
| 6 | 23 | 0.87 |
| 7 | 23 | 0.90 |
| 8 | 21 | 0.85 |
| 9 | 23 | 0.84 |
| 10 | 20 | 0.80 |
| 11 | 21 | 0.84 |

*(continued)*

| College | ACT 25th Percentile | Freshman Retention Rate |
|---|---|---|
| 12 | 23 | 0.82 |
| 13 | 21 | 0.82 |
| 14 | 23 | 0.83 |
| 15 | 26 | 0.78 |
| 16 | 21 | 0.79 |
| 17 | 22 | 0.83 |
| 18 | 20 | 0.78 |
| 19 | 22 | 0.78 |
| 20 | 21 | 0.61 |

(a) Which variable is the independent variable? Which is the dependent variable?

(b) Create a scatter plot of the data. Does it appear that freshman retention rate is related to the 25th percentile on the ACT exam?

(c) Find the equation of the regression line for the data.

(d) Plot the regression line on the same plot as the data. Do you think that the line does a good job of predicting freshman retention at a college or university? Why or why not?

(e) Calculate the standard error of the estimate, $s_{y|x}$, for the regression line.

(f) At the 0.05 level, is the model significant?

**11.25** The British Bankers' Association wanted to look at the relationship between the amount of deposits made (in billions of £) and the number of customers that a bank had. Analysts collected data on six different large banks and found the following information:

| Bank Name | Deposits (£ billion) | Customers (million) |
|---|---|---|
| Abbey National | 101.7 | 13.6 |
| Barclays | 108.2 | 10.0 |
| Lloyds | 96.9 | 15.0 |
| National Westminster | 113.8 | 7.5 |
| Woolrich | 27.5 | 4.0 |
| Halifax | 77.1 | 7.6 |

(a) Which variable is the independent variable? Which is the dependent variable?

(b) Create a scatter plot of the data. Does it appear that the amount of deposits is related to the number of customers?

(c) Find the equation of the regression line for the data.

(d) Plot the regression line on the same plot as the data. Do you think that the line does a good job of predicting the amount of deposits? Why or why not?

(e) Calculate the standard error of the estimate, $s_{y|x}$, for the regression line.

(f) At the 0.05 level, is the model significant?

**11.26** An education task force looking at poverty levels in the United States has collected data for each state and the District of Columbia on the total number of people below the poverty level and the number of adults over the age of 25 who did not graduate from high school for the year 1993. The data are shown in the table below:

*Datafile:*
*POVERTY.XXX*

| State | Number (1000) Below Poverty Level | Number of Adults Over 25 Not HS Graduates | State | Number (1000) Below Poverty Level | Number of Adults Over 25 Not HS Graduates |
|---|---|---|---|---|---|
| Alabama | 725 | 843,638 | District of Columbia | 158 | 109,866 |
| Alaska | 52 | 43,244 | Florida | 2507 | 2,271,074 |
| Arizona | 615 | 491,080 | Georgia | 919 | 1,169,815 |
| Arkansas | 484 | 503,481 | Hawaii | 91 | 141,506 |
| California | 5803 | 4,450,528 | Idaho | 150 | 121,787 |
| Colorado | 354 | 328,056 | Illinois | 1600 | 1,735,789 |
| Connecticut | 277 | 457,208 | Indiana | 704 | 850,014 |
| Delaware | 73 | 96,472 | Iowa | 290 | 353,800 |

*(continued)*

| State | Number (1000) Below Poverty Level | Number of Adults Over 25 Not HS Graduates | State | Number (1000) Below Poverty Level | Number of Adults Over 25 Not HS Graduates |
|---|---|---|---|---|---|
| Kansas | 327 | 293,272 | North Dakota | 70 | 92,427 |
| Kentucky | 763 | 825,857 | Ohio | 1461 | 1,684,888 |
| Louisiana | 1119 | 803,872 | Oklahoma | 662 | 506,961 |
| Maine | 196 | 168,460 | Oregon | 363 | 343,609 |
| Maryland | 479 | 673,932 | Pennsylvania | 1598 | 1,994,278 |
| Massachusetts | 641 | 792,657 | Rhode Island | 108 | 184,344 |
| Michigan | 1475 | 1,356,759 | South Carolina | 678 | 687,260 |
| Minnesota | 506 | 488,765 | South Dakota | 102 | 98,720 |
| Mississippi | 639 | 549,685 | Tennessee | 998 | 1,033,914 |
| Missouri | 832 | 858,368 | Texas | 3177 | 2,872,559 |
| Montana | 127 | 96,469 | Utah | 203 | 133,315 |
| Nebraska | 169 | 181,072 | Vermont | 59 | 68,637 |
| Nevada | 141 | 167,628 | Virginia | 627 | 987,203 |
| New Hampshire | 112 | 127,423 | Washington | 634 | 505,783 |
| New Jersey | 866 | 1,205,206 | West Virginia | 400 | 398,527 |
| New Mexico | 282 | 229,974 | Wisconsin | 636 | 662,072 |
| New York | 2981 | 2,977,604 | Wyoming | 64 | 47,113 |
| North Carolina | 966 | 1,277,747 | | | |

(a) The task force would like to find a model that will predict the number of people living below the poverty level from the number of adults who are not high school graduates. Which is the dependent variable and which is the independent variable?

(b) Create a scatter plot of the data. Do you think that a linear relationship exists between the two variables?

(c) Find the linear regression model for the data.

(d) Interpret the meaning of the slope and the $y$ intercept of the model. Do you think that the $y$ intercept makes sense for these data?

(e) Use the model to predict the number of people who are below the poverty level for the states of Wyoming, Connecticut and North Carolina. What are the residuals for these three observations?

(f) At the 0.05 level of significance, is the model significant?

## Thinking About It!

*Requires Exercises 11.2, 11.17*

**11.27** The Commerce Department also has data available on number of shopping centers and retail sales for the South Central states. The data are given below:

| State | Number of Shopping Centers | Retail Sales ($ billiion) |
|---|---|---|
| Kentucky | 593 | 11.7 |
| Tennessee | 1137 | 19.1 |
| Alabama | 601 | 13.2 |
| Mississippi | 418 | 7.2 |
| Arkansas | 339 | 6.4 |
| Louisiana | 676 | 15.6 |
| Oklahoma | 556 | 11.4 |
| Texas | 2824 | 72.7 |

Source: *Statistical Abstract of the United States 1995.*

(a) Find the linear regression model for the South Central states.

(b) How does the equation for the South Central states compare to the one you found for the North Central states?

(c) Would you have expected them to be exactly the same? Why or why not?

(d) What similarities would you expect them to have? What differences?

(e) Combine the data from the North Central and South Central states and find the linear regression model for the combined data.

(f) Do you think that the individual models are better or worse than the combined model? On what criteria do you base your conclusion?

**11.28** Look at the data on advertising and brand preference for cigarettes again.

*Requires Exercise 11.23*

(a) Find the simple linear regression model that relates adult brand preference to advertising expenditures.

(b) At the 0.05 level, is the model significant?

(c) What is $R^2$ for this model? What does it mean? What is the correlation coefficient?

(d) Compare the model for adult brand preference to the one you found for adolescent brand preference. Which variable do you think is more strongly related to advertising expenditures? Why?

(e) Do you think that these data help to support the claims against the tobacco industry that its advertising targeted teenagers? Why or why not?

**11.29** Consider the model you developed relating poverty levels and education.

*Requires Exercise 11.26*

(a) Find 95% confidence intervals for the mean number of people who live below the poverty level when the number of adults who are not high school graduates is (i) 500,000, (ii) 1,000,000, and (iii) 2,000,000.

(b) Find 95% prediction intervals for the number of people who live below the poverty level when the number of adults who are not high school graduates is (i) 500,000, (ii) 1,000,000, and (iii) 2,000,000.

(c) Do you think that the model is useful for predicting the number of people below the poverty level? Why or why not?

(d) The task force contends that if it can institute programs to reduce the number of adults who have not finished high school by 10% in each state, it will have a significant impact on the number of people who are living below the poverty level. Do you agree or disagree with this statement? Why?

**11.30** Consider the data on the British banks.

*Requires Exercise 11.25*

(a) What is $R^2$ for this model? What does it mean? What is the correlation coefficient?

(b) Find 95% confidence intervals for the mean amount of deposits when the number of customers is (i) 12 million, (ii) 15 million, and (iii) 7 million.

(c) Find 95% prediction intervals for the amount of deposits when the number of customers is (i) 12 million, (ii) 15 million, and (iii) 7 million.

(d) Do you think that the model is useful for predicting the amount of deposits? Why or why not?

**11.31** The British Bankers' Association decided to look at another variable to predict deposits. It chose the number of branches for each bank. The data are given below:

*Requires Exercises 11.25, 11.30*

| Bank Name | Deposits (£ billion) | Branches 1996 |
|---|---|---|
| Abbey National | 101.7 | 867 |
| Barclays | 108.2 | 1997 |
| Lloyds | 96.9 | 2797 |
| National Westminster | 113.8 | 1920 |
| Woolrich | 27.5 | 430 |
| Halifax | 77.1 | 938 |

(a) Find the simple linear regression model that relates deposits to number of branches.

(b) At the 0.05 level, is the model significant?

(c) What is $R^2$ for this model? What does it mean? What is the correlation coefficient?

(d) Compare the model for predicting deposits using the number of branches to the one you found for number of customers. Which variable do you think is more strongly related to deposits? Why?

*Requires Exercises 11.23, 11.28*

**11.32** Look at the model that you developed for adolescent brand preference for cigarettes and the amount of money spent on advertising.

(a) Make a plot of the residuals versus the values of the independent variables.

(b) What does this lead you to believe about the appropriateness of the linear model?

(c) Now plot the residuals versus the predicted values of $y$.

(d) How do the two graphs compare? Why do you think this happened?

(e) Make a normal probability plot of the data. Do you think the assumption of normality is reasonable?

(f) Considering your answers to parts (a)–(e), do you think that the linear regression model is appropriate for these data? Why or why not?

*Requires Exercises 11.17, 11.27*

**11.33** Look at the model you developed for shopping centers and retail sales for the South Central states.

(a) Make a plot of the residuals versus the values of the independent variables.

(b) What does this lead you to believe about the appropriateness of the linear model?

(c) Now plot the residuals versus the predicted values of $y$.

(d) How do the two graphs compare? Why do you think this happened?

(e) Make a normal probability plot of the data. Do you think the assumption of normality is reasonable?

(f) Considering your answers to parts (a)–(e), do you think that the linear regression model is appropriate for these data? Why or why not?

*Requires Exercises 11.26, 11.29*

**11.34** Consider the model that you developed for poverty levels and education.

(a) Make a plot of the residuals versus the values of the independent variables.

(b) What does this lead you to believe about the appropriateness of the linear model?

(c) Now plot the residuals versus the predicted values of $y$.

(d) How do the two graphs compare? Why do you think this happened?

(e) Make a normal probability plot of the data. Do you think the assumption of normality is reasonable?

(f) Considering your answers to parts (a)–(e), do you think that the linear regression model is appropriate for these data? Why or why not?

## Doing It!

*Datafile: FACULTY.XXX*

**11.35** The provost at Aluacha Balaclava College wants to look at the faculty salary data in some other ways. She has collected data on salary, years of service, rank, school, gender, and tenure. A sample of the data is shown below:

| Salary ($) | Years of Service | Rank | Schools | Gender M/F | Tenure Y/N |
|---|---|---|---|---|---|
| 53,316 | 22 | ASST | BUSINESS | F | Y |
| 64,375 | 11 | PROF | BUSINESS | M | Y |
| 63,501 | 7 | ASSO | BUSINESS | M | Y |
| 59,426 | 6 | ASSO | BUSINESS | M | N |
| 49,058 | 20 | ASSO | BUSINESS | M | Y |
| 94,969 | 4 | PROF | BUSINESS | M | N |
| 54,762 | 21 | ASST | BUSINESS | M | Y |
| 55,516 | 9 | ASSO | BUSINESS | M | Y |

(a) Make scatter plots of salary versus years of service for the entire faculty. Do you think that there is a linear relationship between the two variables? Why or why not?

(b) Find the simple linear regression model with salary as the dependent variable and years of service as the independent variable.

(c) At the 0.05 level, is there a significant linear relationship between salary and years of service?

(d) What is the value of $R^2$ for this model? What does it mean?

(e) Separate the data by rank and create scatter plots of salary versus years of service for each rank. How do these plots compare to the one for all of the faculty?

(f) Find a simple linear regression model for salary versus years of service for each rank. How do the regression coefficients for each rank compare to each other? How do they compare to the overall model for all of the faculty?

(g) At the 0.05 level of significance, are any of the models significant?

(h) What are the $R^2$ values for each of the models?

(i) Calculate the residuals for all significant models. Create a plot of residuals versus predicted values for each model. Is the simple linear model appropriate for these data? Why or why not?

(j) For each significant model, investigate the assumption of normality. Do you think the assumption is valid for these data? Why or why not?

(k) Are there any data values that might be exerting a strong influence on the model? If so, which ones? Drop any unusual values from the data and rerun the models. How does this change the results?

(l) Perform any additional analyses that you feel would be helpful to determine whether or not the simple linear model is an appropriate one for these data. Write a report with your conclusions.

# CHAPTER *12*

# *MULTIPLE REGRESSION MODELS*

## WHERE SHOULD WE LIVE?

One of the problems that face city planners and real estate developers is the price of homes in an area. What factors influence the prices of homes? On a micro level, real estate developers can look at factors directly related to the individual dwelling, such as square footage, number of bedrooms, and number of bathrooms. For large-scale planning purposes, a micro model is not appropriate. Many other global factors affect the price of a home, such as location, school systems, and access to public transportation.

The cities of Dallas and Fort Worth, Texas, wanted to look at the various "commuter" suburbs of the city that have grown up over the years and find a model that would enable them to understand what factors play a role in the prices of homes in these areas. They decided to look at 41 different communities regarded as "commuter" suburbs of Dallas/Fort Worth, and collected data on 5 different variables that they thought would impact the average selling price of a home. The variables that they looked at were average household income, total instructional expenditures, mean SAT scores, population diversity, and the violent crime rate. A portion of the data is shown below:

| Suburb | 1994 Avg. Value of Home Sold | Avg. Household Income | Total Instructional Expenditures per Pupil | Mean SAT Score | Population Diversity (% non-white) | 1993/94 Violent Crime (per 1000) |
|---|---|---|---|---|---|---|
| Addison | $204,800 | $49,803 | $2,600 | 980 | 32.40% | 4.5 |
| Allen | $99,000 | $60,031 | $2,185 | 955 | 9.30% | 1.9 |
| Arlington | $97,613 | $50,188 | $2,154 | 929 | 14.70% | 8.3 |
| Balch Springs | $48,300 | $37,111 | $1,947 | 935 | 24.00% | 14.6 |
| Bedford | $173,073 | $58,982 | $2,473 | 951 | 10.10% | 2.3 |
| Carrollton | $108,700 | $58,551 | $2,600 | 980 | 22.10% | 2.5 |
| Cedar Hill | $84,500 | $55,768 | $2,174 | 901 | 24.10% | 1.1 |
| Cockrell Hill | $58,200 | $30,434 | $2,714 | 775 | 70.10% | 8.3 |
| Colleyville | $235,427 | $121,962 | $2,471 | 952 | 4.90% | 0.4 |
| Coppell | $149,500 | $78,917 | $2,472 | 925 | 13.80% | 0.5 |
| Desoto | $109,300 | $62,575 | $2,306 | 885 | 27.20% | 3.1 |

They wanted to find a model that would explain the average value of homes in a location based on these variables. They felt that this model would help them make decisions about current as well as future communities.

## 12.1   CHAPTER OBJECTIVES

The purposes of regression models are to *predict,* to *explain,* and to *control.* Many real-world phenomena are too complex for the simple linear model that we looked at in Chapter 10. Some variables that we are interested in depend on *more than one* independent variable. When this is the case, we use *multiple regression models* to describe the relationship between the independent variable, $Y$, and the set of dependent variables, $X_1, X_2, \ldots, X_k$.

In this chapter you will learn how to

- find the regression equation for a dependent variable $Y$ as a function of a set of independent variables, $X_1, X_2, \ldots, X_k$.
- determine whether the relationship is significant
- determine which variables contribute to the model and which do not
- analyze the results of a regression analysis to determine whether the model is appropriate

## 12.2   THE MULTIPLE REGRESSION MODEL

### 12.2.1 A Description of the Multiple Regression Model

When two quantitative variables are related in a linear manner, we can use a simple linear regression model to describe the relationship. We saw that, in many cases, the simple linear model does a good job of predicting or describing the way one variable behaves relative to another. In some cases, though, the simple linear model was not appropriate because the relationship between the variables was not linear, or because one or more of the assumptions of the simple linear model was violated. In other cases, the simple linear model may have been appropriate and even statistically significant, but the results might not have been *useful* for the purposes of making decisions.

One of the quantities that we looked at in simple linear regression was the value of $R^2$, the coefficient of determination. You remember that $R^2$ is a measure of the amount of variation in the dependent variable, $Y$, that can be explained by the linear model involving the independent variable $X$. Sometimes even when a linear model is significant, the independent variable, $X$, does not do a good enough job of explaining how $Y$ behaves. One possible solution to this problem is to find another, better, independent variable to work with, but there may not be a single variable that will do the job. Another solution, the one we will look at in this chapter, is to find a *set* of independent variables, known as **input variables,** that together provide a good model for predicting the dependent variable known as the **output variable.**

> The dependent variable, $Y$, is often referred to as the **output variable,** while the independent variables $X_1, X_2, \ldots, X_k$ are referred to as the **input variables.**

Models that describe this type of relationship are known as **multiple regression models.**

> The true relationship between the independent variable $Y$ and the set of independent variables, $X_1, X_2, \ldots, X_k$, the **multiple regression model,** can be described by the equation
>
> $$y = \beta_0 + \beta_1 x_1 + \beta_2 x_2 + \cdots + \beta_k x_k + \varepsilon$$

This equation says that the actual value of the variable $Y$ will be determined by the equation $\beta_0 + \beta_1 x_1 + \beta_2 x_2 + \cdots + \beta_k x_k$, plus some random variation, $\varepsilon$, due to other, unmeasured, factors.

The coefficients $\beta_1, \beta_2, \ldots, \beta_k$ are similar to the slope coefficient $\beta_1$ in the simple linear model, $y = \beta_0 + \beta_1 x_1 + \varepsilon$, with just a small difference. In the simple linear model, $\beta_1$ represents the slope of the line, or the change in the dependent variable, $Y$, for a unit change in the independent variable, $X$. In the multiple regression model, the parameters $\beta_1, \beta_2, \ldots, \beta_k$ are really not the slope, because we are not talking about a line, but they have a similar interpretation. Each of the $\beta_i$ coefficients represents the change that will occur in the independent variable, $Y$, if the variable associated with the coefficient of interest, $X_i$, is changed by one unit *and all other variables in the model are held constant*. In the case of only two independent variables, the multiple regression equation has a physical interpretation: The equation is the *plane* that describes the relationship, similar to the line in the one-variable model.

## 12.2.2 The Least-Squares Multiple Regression Model

Our objective in the multiple regression model is to find the estimates for the population values $\beta_1, \beta_2, \ldots, \beta_k$ that do the "best" job of describing the real relationship between $Y$ and the set of $X$ variables that we have chosen.

You remember that, in the simple linear model, we used the method of least squares to find the equation of the line that minimized the sum of the squared differences between the actual values of $Y$ and the values predicted by the model. In the simple linear model you were given equations to find the values of $b_0$ and $b_1$, although you found out that, except for finding the equation of the model, most of the analysis in simple linear regression is best done using an appropriate computer software package, such as Minitab or Excel.

Least squares can also be extended to models with more than one independent variable. That is, we want to find the equation

$$\hat{y} = b_0 + b_1 x_1 + b_2 x_2 + \cdots + b_k x_k$$

that minimizes the total squared deviation between the actual data and the values obtained from the regression model. Remember that $\hat{y}$ is the value of the dependent variable as predicted by the model *for a specific set of values of the independent variables,* $X_1 = x_1, X_2 = x_2, \ldots, X_k = x_k$. Finding the values of $b_0, b_1, b_2, \ldots, b_k$ involves solving a system of $k$ simultaneous equations and is entirely too tedious to do by hand. For such an analysis, an appropriate software package is essential for finding the regression model.

## EXAMPLE 12.1   Texas Real Estate

### Finding the Multiple Regression Model

The city planners that are looking at the prices of homes in the suburbs of Dallas/Fort Worth, Texas, are trying to find the relationship between the average price of a home ($Y$) and a set of five independent variables, *average household income* ($X_1$), *total instructional expenditures,* ($X_2$), *mean SAT scores* ($X_3$), *population diversity* ( % *nonwhite*) ($X_4$), *and the violent crime rate* (per 1000) ($X_5$). Their equation will be

$$\hat{y} = b_0 + b_1 x_1 + b_2 x_2 + b_3 x_3 + b_4 x_4 + b_5 x_5$$

The analysts use Minitab to find the equation of the multiple regression model that best fits the data. The results are shown on page 596.

```
Regression Analysis

The regression equation is
1994 Avg. Value _of Home Sold = - 140352 + 1.64 Avg. Household Income
    + 39.2 Total Instructional Expenditure + 90 Mean SAT Score
    - 47820 Diversity - 1132 1993/94 Avg. Violent Crime
```

The computer output gives the values of the coefficients as $b_0 = -140{,}352$, $b_1 = 1.64$, $b_2 = 39.2$, $b_3 = 90$, $b_4 = -47{,}820$, and $b_5 = -1132$.

   The analysts interpret the coefficients as the change in the average selling price of a home in dollars for a unit change in each of the variables. That is, if the average household income in a location increases by one dollar, the average selling price of a home in that location will increase by $1.64, *as long as all other variables remain constant.* Similarly, if the mean SAT score increases by one point, the average price of a home will increase by $90. Again, this assumes that nothing else in the model has changed.

   From the coefficients the analysts see that for the most part the model makes sense. That is, as income, total educational expenditures, and SAT scores go up, so do the selling prices of homes. Similarly, as the violent crime rate increases, the price of the homes goes down. The analysts are not pleased about the coefficient of $X_4$, but they know that it reflects a reality.   ■

   The output from statistical software for multiple regression models is almost identical to the output for the simple linear model. Figure 12.1 shows the output from Excel and Minitab. The section of the output that contains the coefficients of the multiple regression model is highlighted.

   You see that the basic sections of the output are the same, general information about the model, information about the coefficients, and the ANOVA information. We will look at the output in more detail as the chapter progresses.

## EXAMPLE 12.2   Phillips' Petroleum

### Finding the Regression Model

Financial analysts would like to understand the relationship between common stockholders' equity ($/share) and total revenue (billions of dollars) and total assets (billions of dollars). Data on these variables for Phillips' Petroleum from 1986 to 1996 are shown in the table below:

| Year | Equity, $Y$ | Revenue, $X_1$ | Assets, $X_2$ |
|------|------|------|------|
| 1986 | 7.55 | 10.018 | 9.186 |
| 1987 | 7.08 | 10.917 | 8.772 |
| 1988 | 8.69 | 11.490 | 8.417 |
| 1989 | 8.74 | 12.492 | 7.832 |
| 1990 | 10.51 | 13.975 | 8.301 |
| 1991 | 10.61 | 13.259 | 8.298 |
| 1992 | 10.37 | 12.140 | 8.489 |
| 1993 | 10.28 | 12.545 | 7.961 |
| 1994 | 11.29 | 12.367 | 8.042 |
| 1995 | 12.16 | 13.521 | 8.493 |
| 1996 | 16.15 | 15.807 | 9.120 |

The analysts decide to use a multiple regression model to find a model with equity as the dependent variable and revenue and assets as the independent variables. A portion of the computer output for their analysis is shown on page 598.

SUMMARY OUTPUT

| Regression Statistics | |
|---|---|
| Multiple R | 0.866778185 |
| R Square | 0.751304423 |
| Adjusted R Square | 0.715776483 |
| Standard Error | 37957.7302 |
| Observation | 41 |

ANOVA

| | df | SS | MS | F | Significance F |
|---|---|---|---|---|---|
| Regression | 5 | 1.52341E+11 | 30468171569 | 21.14686162 | 1.08463E-09 |
| Residual | 35 | 50427624859 | 1440789282 | | |
| Total | 40 | 2.02768E+11 | | | |

| | Coefficients | StandardError | tStat | P-value | Lower 95% | Upper 95% | Lower 95.0% |
|---|---|---|---|---|---|---|---|
| Intercept | −140351.6689 | 112957.8302 | −1.242513853 | 0.222307657 | −369668.5358 | 88965.198 | −369668.5358 |
| Avg. Household Income | 1.637671454 | 0.312118144 | 5.246960128 | 7.59702E-06 | 1.004037162 | 2.271305746 | 1.004037162 |
| Total Instructional Expenditures per pupil | 39.15871129 | 21.78140875 | 1.797804345 | 0.080839566 | −5.059953319 | 83.3773759 | −5.059953319 |
| Mean SAT | 89.63285763 | 113.8778516 | 0.787096493 | 0.436523067 | −141.5517541 | 320.8174694 | −141.5517541 |
| Population Diversity (% non-white) | −47820.09741 | 62143.05497 | −0.769516359 | 0.446749366 | −173977.3601 | 78337.1653 | −173977.3601 |
| 1993/94 Avg. Violent Crime (per 100) | −1132.378569 | 2146.787097 | −0.527475952 | 0.601191161 | −5490.5934 | 3225.836262 | −5490.5934 |

*Note: The software packages do not highlight these areas.*

Excel

**Regression Analysis**

The regression equation is
1994 Avg. Value _of Home Sold = - 140352 + 1.64 Avg. Household_ Income
         + 39.2 Total Instructional_ Expenditur + 90 Mean SAT_ Score
         - 47820 Population Diversity_ (% non-white
         - 1132 1993/94 Avg. Violent _Crime (per 1000)

```
Predictor       Coef        StDev          T          P
Constant     -140352       112958       -1.24      0.222
Avg. Hou      1.6377       0.3121        5.25      0.000
Total In       39.16        21.78        1.80      0.081
Mean SAT        89.6        113.9        0.79      0.437
Populati      -47820        62143       -0.77      0.447
1993/94        -1132         2147       -0.53      0.601

S = 37958      R-Sq = 75.1%      R-Sq(adj) = 71.6%

Analysis of Variance

Source       DF          SS            MS          F          P
Regression    5 1.52341E+11   30468171569      21.15      0.000
Error        35 50427624859    1440789282
Total        40 2.02768E+11

Source       DF       Seq SS
Avg. Hou      1 1.46103E+11
Total In      1  3261836377
Mean SAT      1  1202680560
Populati      1  1372121063
1993/94       1   400872070

Unusual Observations
Obs    Avg. Hou    1994 Avg         Fit   StDev Fit   Residual   St Resid
  1      49803       204800      110273       13150      94527      2.65R
  8      30434        58200       42311       28127      15889      0.62 X
 21     196697       361100      397689       30411     -36589     -1.61 X
 38      84007       269926      178710        9319      91216      2.48R
 39      87854       269926      178242       13518      91684      2.58R
```

Minitab

**FIGURE 12.1** Computer output from Excel and Minitab

SUMMARY OUTPUT

| Regression Statistics | |
| --- | --- |
| Multiple R | 0.923228385 |
| R Square | 0.852350651 |
| Adjusted R Square | 0.815438314 |
| Standard Error | 1.067662859 |
| Observations | 11 |

ANOVA

| | df | SS | MS | F | Significance F |
| --- | --- | --- | --- | --- | --- |
| Regression | 2 | 52.64353179 | 26.32176589 | 23.09121324 | 0.000475254 |
| Residual | 8 | 9.119231848 | 1.139903981 | | |
| Total | 10 | 61.76276364 | | | |

| | Coefficients | StandardError | tStat | P-value | Lower 95% |
| --- | --- | --- | --- | --- | --- |
| Intercept | −16.71847039 | 7.179010344 | −2.328798761 | 0.048252087 | −33.27330864 |
| Revenue | 1.452643014 | 0.216118353 | 6.721516208 | 0.00014936 | 0.954272876 |
| Assets | 1.034291795 | 0.769247372 | 1.344550312 | 0.215646057 | −0.739596974 |

Excel output for Example 12.2

The model is $\hat{y} = -16.72 + 1.45x_1 + 1.03x_2$.

From this model, they see that an increase in revenue of 1 billion dollars will increase equity of $1.45/share for a fixed value of assets, while an increase in assets of 1 billion dollars will increase equity by $1.03/share for a fixed revenue. The constant for the model is −16.72, which indicates that with revenue and assets equal to zero, shareholder equity is −16.72 billion dollars. This does not make much sense, but it would be extrapolating anyway. ∎

 **TRY IT NOW!**

## Order Filling

### *Finding the Multiple Regression Model*

A mail-order catalog company is looking at the time it takes to prepare an order for shipping. In particular, the company is looking for the amount of time that is spent collecting the items ordered and packing them. In this operation, an employee (a checker) is given an order to fill. Items are located in bins in one of six different sections of the warehouse. The checkers move around the warehouse retrieving the items and packing them into the shipping cartons. The company has looked at the operation in some detail and believes that three major variables are involved in the process: the number of items ordered, the number of different locations (sections of the warehouse) in which the items are located, and the experience level (in months) of the checker. Data are collected on 45 orders. A portion of the data is shown below:

| Time (min) | Items | Locations | Experience (months) |
| --- | --- | --- | --- |
| 9.3 | 1 | 6 | 8 |
| 4.4 | 1 | 3 | 13 |
| 4.4 | 5 | 2 | 3 |
| 5.6 | 3 | 3 | 5 |
| 4.9 | 11 | 1 | 4 |
| 8.8 | 14 | 3 | 5 |

The company wants to know how the time it takes to fill an order is related to the other three variables, so it decides to use a multiple regression model. The output from the computer software is given here.

SUMMARY OUTPUT

| Regression Statistics | |
| --- | --- |
| Multiple R | 0.915838993 |
| R Square | 0.838761061 |
| Adjusted R Square | 0.82696309 |
| Standard Error | 1.021396967 |
| Observations | 45 |

ANOVA

| | df | SS | MS | F | Significance F |
| --- | --- | --- | --- | --- | --- |
| Regression | 3 | 222.5057888 | 74.16859627 | 71.09366969 | 2.71039E-16 |
| Residual | 41 | 42.7733223 | 1.043251763 | | |
| Total | 44 | 265.2791111 | | | |

| | Coefficients | StandardError | tStat | P-value | Lower 95% |
| --- | --- | --- | --- | --- | --- |
| Intercept | 0.215224599 | 0.55856837 | 0.385314691 | 0.701996141 | −0.912827729 |
| Number of Items | 0.281391801 | 0.029191616 | 9.639473395 | 4.26275E-12 | 0.222438105 |
| Locations | 1.259340956 | 0.103175681 | 12.20579253 | 3.10059E-15 | 1.050973327 |
| Experience | 0.019087782 | 0.047667374 | 0.400437032 | 0.690913642 | −0.077178487 |

Write down the equation of the regression model.

Interpret the value of each of the coefficients of the model.

## 12.2.3 Using the Multiple Regression Model for Prediction

Just as in the case with the simple linear model, the multiple regression equation can be used to predict the value of the dependent variable for a specific set of values of the independent variables. We must ask the same question: How useful are these predictions?

You remember from the simple linear model that one way to determine how useful a model is, is to look at the *residuals,* the differences between the actual values of the dependent variable and the values predicted by the model for a specific set of input variables. The value of the residual for the $i$th observation is

$$e_i = y_i - \hat{y}_i$$

The size of the residual gives a measure of how good a particular model is at predicting the value of the dependent variable.

ANS. $\hat{y} = 0.21 + 0.28x_1 + 1.26x_2 + 0.02x_3$; ASSUMING THAT ALL OTHER FACTORS REMAIN CONSTANT, IF YOU INCREASE THE NUMBER OF ITEMS BY 1 THEN THE TIME THE TIME INCREASES BY 1.26 MINUTES; AND IF YOU INCREASE THE NUMBER OF LOCATIONS, THEN THE TIME THE TIME INCREASES BY 1.26 MINUTES; AND IF YOU INCREASE EXPERIENCE BY 1 MONTH, THEN THE TIME THE TIME INCREASES BY 0.02 MINUTE.

**EXAMPLE 12.3**   Texas Real Estate

*Using the Model to Make Predictions*

The group looking at average selling prices of homes in Dallas/Fort Worth suburbs decides to use the model to predict the selling price for two of the communities in the data set to see how well the values predicted by the model agree with the data. To decide which locations to pick, the analysts make a boxplot of the data:

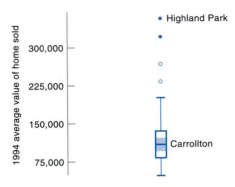

After looking at the boxplot, they decide to look at Highland Park because it appears to be an outlier and Carrollton because it is at the center of the data. Using the regression model

> 1994 avg. value of home sold = −140,352 + 1.64 avg. household income
> + 39.2 total instructional expenditure
> + 90 mean SAT score
> − 47,820 population diversity
> − 1132 1993/94 avg. violent crime

for Carrollton they calculate

$$\hat{y} = -140{,}352 + 1.64(58{,}551) + 39.2(2600) + 90(980) - 47{,}820(0.221)$$
$$- 1132(2.5) = \$132{,}393$$

and for Highland Park

$$\hat{y} = -140{,}352 + 1.64(196{,}697) + 39.2(3159) + 90(1070) - 47{,}820(0.037)$$
$$- 1132(1.7) = \$398{,}670$$

They compare this to the actual data of $361,100 for Highland Park and $108,700 for Carrollton and find the actual deviations to be

Carrollton:   $108,700 − $132,393 = −$23,693   % error = ($23,693/$108,700)(100)
= 21.8%

Highland Park:   $361,100 − $398,670 = −$37,570   % error = ($37,570/$361,100)(100)
= 10.4%

They observe that while the actual deviation is larger for Highland Park, the percent error is larger for Carrollton.   ■

As you saw in the previous example, it is not difficult to use the multiple regression model to predict the values of the independent variable. However, as models become more complex, it can get tedious. In fact, if you want to compare the predicted and actual values for every observation in the sample, it will certainly be time-consuming. For this reason, almost all software used for regression analysis allows you to find predicted values of $Y$, for either all of the actual sets of $X$ data or for some specific data of interest as well as the residuals.

## EXAMPLE 12.4   Phillips' Petroleum

### Using the Model for Prediction

The people looking at a model to predict equity using revenue and assets wonder how well the multiple regression model does the job. They use Minitab to analyze the data, and as part of the output, they have the software calculate for each set of input data, the predicted equity, and the residual. A portion of the output is shown below:

```
Residual Output

Observation   Predicted Equity      Residuals
         1         7.335111748    0.214888252
         2         8.212841015   -1.132841015
         3         8.678031875    0.011968125
         4         9.528519475    0.788519475
         5        12.16787192    -1.657871917
         6        11.12467664    -0.514676643
         7         9.696718843    0.673281157
         8         9.738933196    0.541066804
         9         9.564140375    1.725859625
        10        11.70695601     0.453043987
        11        15.6761989      0.473801102
```

They see that the largest deviation is about \$1.73/share, while the smallest is \$0.01/share. ■

 

Another way to look at how good a job the model does at predicting the values of the independent variable is to create a scatter plot of the predicted values. If the model is a perfect fit then you would expect the predicted values to be equal to the actual data, that is, $\hat{y} = y$. If you plot the predicted values versus the actual values, the results should form a straight line that intercepts the origin if the fit is good.

 

## EXAMPLE 12.5   Phillips' Petroleum

### Plotting the Predicted versus the Actual Values

The analysts in the group looking at predicting equity want to look at how well the model predicts equity graphically. They decide to plot the predicted values ($\hat{y}$'s) versus the actual values ($y$'s). They also plot the line $\hat{y} = y$ on the plot for reference. They see that the points fall very close to the line. It appears that the model provides a good fit.

■

## TRY IT NOW!

### Order Filling
*Finding Predicted Values*

The mail-order catalog company that is looking at the time it takes to prepare an order for shipping wants to see how well the model it has found predicts the time to fill an order. The data for six of the observations are shown below:

| Time (min) | Items | Locations | Experience (months) |
|---|---|---|---|
| 9.3 | 6 | 1 | 8 |
| 4.4 | 3 | 1 | 13 |
| 4.4 | 5 | 2 | 3 |
| 5.6 | 3 | 3 | 5 |
| 4.9 | 11 | 1 | 4 |
| 8.8 | 14 | 3 | 5 |

Use the model to find the predicted time to fill an order for each of the six sets of input data you have.

Compare the predicted results to the actual data. Do you think that this model does a good job of predicting the time to fill an order? Why or why not?

## 12.2.4 Exercises — Learning It!

*TRAFFAT.XXX*

**12.1** A group of legislators wanted to look at factors that affect the number of traffic fatalities. They collected some data for 1994 from the National Transportation Safety Board on the number of fatalities for 50 states and the District of Columbia (DC), the number of licensed drivers, the number of registered vehicles, and the number of vehicle miles traveled. A portion of the data is shown below:

| State | Fatalities | Population | Licensed Drivers, $X_1$ (thousands) | Registered Vehicles, $X_2$ (thousands) | Vehicle Miles $X_3$ (millions) |
|---|---|---|---|---|---|
| AL | 1083 | 4,219 | 3,043 | 3,422 | 48,956 |
| AK | 85 | 606 | 443 | 508 | 4,150 |
| AZ | 903 | 4,075 | 2,654 | 2,980 | 38,774 |
| AR | 610 | 2,453 | 1,770 | 1,560 | 24,948 |
| CA | 4226 | 31,431 | 20,359 | 23,518 | 271,943 |
| CO | 585 | 3,656 | 2,620 | 3,144 | 33,705 |

SOURCE: *Statistical Abstract of the United States 1996.*

As you can see from the data table, they also had information on population for each state, but they did not think that it would prove useful. They decided to find a multiple regression model that predicts the number of fatalities using the other variables as the set of independent variables. The output from the computer software is shown below:

SUMMARY OUTPUT

**Regression Statistics**

| | |
|---|---|
| Multiple R | 0.982548538 |
| R Square | 0.96540163 |
| Adjusted R Square | 0.963193224 |
| Standard Error | 154.5407481 |
| Observations | 51 |

ANOVA

| | df | SS | MS | F | Significance F |
|---|---|---|---|---|---|
| Regression | 3 | 31321046.9 | 10440348.97 | 437.1485021 | 2.54274E-34 |
| Residual | 47 | 1122493.613 | 23882.84282 | | |
| Total | 50 | 32443540.51 | | | |

| | Coefficients | Standard Error | tStat | P-value | Lower 95% |
|---|---|---|---|---|---|
| Intercept | 51.7481659 | 30.43306219 | 1.700392999 | 0.095666076 | −9.475200509 |
| X Variable 1 | 0.06294764 | 0.048829545 | 1.289130172 | 0.20366155 | −0.035284642 |
| X Variable 2 | −0.211896991 | 0.055989427 | −3.784589385 | 0.000435566 | −0.324533083 |
| X Variable 3 | 0.029349954 | 0.003525079 | 8.326041167 | 8.34285E-11 | 0.022258416 |

(a) How many independent variables are there in the model proposed? What are they?
(b) Use the computer output to write down the regression model.
(c) Interpret the coefficients of the model.
(d) Use the model to predict the number of traffic fatalities for the states shown in the data table.
(e) Compare the predicted values from the model to the actual values. Based on the plot, does the model do a good job of predicting the number of traffic fatalities?

**12.2** In a study on mandatory seat-belt laws, a group of lobbyists wanted to determine whether there was a relationship between the percentage of motorists who comply with the law, the size of the fine for violation of the law, and the amount of time that the law has been in effect. They decided to look at data on 48 locations with mandatory seat-belt laws. A portion of the data that they had available is shown below:   *SEATBELT.XXX*

| State | Percent Compliance | Fine | Years in Effect |
|---|---|---|---|
| AL | 55 | 25 | 5.0 |
| AK | 69 | 15 | 6.8 |
| AZ | 60 | 10 | 6.5 |
| AR | 51 | 30 | 6.0 |
| CA | 83 | 20 | 11.5 |
| CO | 54 | 15 | 10.0 |

*Note.* ME and NH do not have mandatory seat-belt laws. Data not available for Wyoming.
*Source:* Bureau of Transportation.

(a) Which variable do you think is the dependent variable and which are the independent variables? Why?
(b) The output from a multiple regression model is shown on page 604. Does this model agree with your answer to part (a)?

```
Regression Analysis

The regression equation
Percent Compliance = 0.416 + 0.00187 Fine + 0.0194 Years In Effect

47 cases used 4 cases contain missing values

Predictor       Coef        StDev          T         P
Constant       0.41558     0.05153       8.06     0.000
Fine           0.001874    0.001050      1.78     0.081
Years In       0.019397    0.004672      4.15     0.000


S = 0.09682      R-Sq = 30.8%      R-Sq(adj) = 27.7%

Analysis of Variance

Source          DF          SS           MS         F        P
Regression       2       0.183552     0.091776    9.79     0.000
Error           44       0.412431     0.009373
Total           46       0.595983


Source          DF        Seq SS
Fine             1       0.021952
Years In         1       0.161600
```

(c) Interpret the values of the coefficients in the model.

(d) Use the model to predict the percent compliance with the law for the six states shown in the data table.

(e) Make a plot of the values of $\hat{y}$ vs. $y$ for these states.

(f) From the plot, do you think that the model does a good job of predicting compliance with seat-belt laws? Why or why not?

*CRIME.XXX*   **12.3**  A committee studying the issue of crime in large cities wanted to know if it was possible to find a relationship between the number of serious crimes in the city and several other crime/sociological variables. The committee members decided to focus their study on the number of police officers in the city, the percentage of arrested males who tested positive for drugs, the number of families in the city below the poverty level, the number of people between 16 and 19 who were not enrolled in school and were not high school graduates, and the number of single female parent households. They were focusing on 23 large cities, but all of the data were not available for all of the cities in the study. A portion of the data is shown below:

| City | Police Officers | Non-Enrolled/ Not Grad | Families Below Poverty Level | Serious Crimes | Single Female Parent Households | Male Arrests Drugs (%) |
|---|---|---|---|---|---|---|
| Atlanta, GA | 1,533 | 3,247 | 21,686 | 76,398 | 36,577 | 69 |
| Birmingham, AL | 724 | 1,891 | 14,075 | 33,895 | 22,391 | 64 |
| Chicago, IL | 12,132 | 27,838 | 116,645 | NA | 197,631 | 69 |
| Cleveland, OH | 1,682 | 5,262 | 31,340 | 45,610 | 45,455 | 64 |
| Dallas, TX | 2,857 | 10,702 | 35,085 | 154,929 | 54,434 | 59 |
| Denver, CO | 1,361 | 3,312 | 14,417 | 36,558 | 24,215 | 60 |
| Detroit, MI | 3954 | 13,455 | 71,673 | 127,080 | 113,553 | 58 |
| Fort Lauderdale, FL | 463 | 1,212 | 4,545 | 24,334 | 7,346 | 64 |

SOURCE: *County and City Data Book 1994.*

(a) Which variable is the dependent variable? Which are the independent variables?

(b) Look at the set of input variables. Do you think that it is reasonable that there is a relationship between the number of serious crimes in a city and these variables? Why or why not? Can you think of other variables that the committee should consider?

(c) Use a computer software package to find the multiple regression model for the data. Interpret the coefficients in the model.

(d) Use the software package to find the predicted value of the number of serious crimes for each of the cities in the study and calculate the residuals. Which city has the best fit? Which has the worst?

(e) Use the model to predict the number of serious crimes for the three cities that were missing those data. How do these predictions fit into the overall data set?

(f) Make a plot of the predicted versus actual data. Overall, how well do you think the model predicts the number of serious crimes in a city?

12.4 A survey done by a local Internet provider in a medium-sized city focused on the amount of time per week that the computer was used for Internet access. Analysts surveyed 35 households and collected data on the number of hours of Internet use per week, the number of computers in the household, the number of children in the household, household income, and the education level of the head of the household (in years of education). *INTERNET.XXX*

(a) Decide which variable in this study is the dependent variable and which are the independent variables.

(b) Which two independent variables do you think are most likely related to the number of hours of Internet use per week?

(c) Use a computer software package to find the model that predicts hours of use using the two variables that you identified in part (b). Interpret the coefficients of the model.

(d) Use the computer software to find predicted values and residuals for each of the 35 households in the survey. Make a plot of the predicted values versus the actual data.

(e) Based on the plot do you think that the model you found does a good job of predicting the number of hours of Internet use per week? Why or why not?

# 12.3 ASSESSING THE MULTIPLE REGRESSION MODEL

In the previous section you learned about the multiple regression model and what it means. In addition to finding the multiple regression model you did some investigation of the *quality* of the model. This investigation, looking at the size of the residuals, was subjective in that there was no formal statistical test attached to it. In this section we look at formal, *statistical,* methods for assessing the multiple regression model we obtain.

## 12.3.1 Testing the Significance of the Model

Once we have the regression equation, we would like to know whether that relationship is *statistically significant.* That is, we want to know whether the model, defined by the set of input or independent variables, does a good job of explaining or accounting for the variation in the output or dependent variable.

In the simple linear model, you learned two ways to test the significance of the model. The *t* test tested whether the slope coefficient, $\beta_1$, was nonzero. If $\beta_1$ is not zero, then there is a significant relationship between the dependent and independent variables. The *F* test, the ANOVA test, tested whether the amount of variation in the dependent variable that was explained by the linear model was larger than the amount left unexplained. In multiple regression analysis we test the significance of the *model as a whole.* The hypotheses are

$$H_0: \quad \beta_1 = \beta_2 = \cdots = \beta_k = 0$$
$$H_A: \quad \text{At least one of the } \beta_i\text{'s} \neq 0.$$

To do this, we use the *F* test. Remember that the variation in the *y* values from the overall mean of *Y* can be partitioned, or divided, into parts called Sums of Squares:

SST—the total variation in the $y$ values around the mean $\bar{y}$

SSR—the variation in $Y$ that is caused by $Y$'s relationship with $X$

SSE—the variation in $Y$ that remains unexplained.

You remember from the simple linear model that SST = SSR + SSE and that SST is known as the total sum of squares and represents the distance of the data value $y_i$ from the overall mean $\bar{y}$. SSR is the regression sum of squares and represents the distance of the predicted value $\hat{y}$ from the overall mean $\bar{y}$. SSE is the error sum of squares and represents the distance of a particular data value $y_i$ from the predicted value $\hat{y}$. The formulas for calculating these quantities are

$$SST = \Sigma(y - \bar{y})^2 \qquad SSR = \Sigma(\hat{y} - \bar{y})^2 \qquad SSE = \Sigma(y - \hat{y})^2$$

When the dependent variable, $Y$, is related to the set of independent variables included in the model, then SSR is large and SSE is small. When $Y$ is not related to the set of $X$ variables, then SSE will dominate. The test statistic, $F$, uses the Mean Squares, MSR and MSE, which are calculated by dividing the sums of squares by their respective degrees of freedom.

$$F = \frac{MSR}{MSE}$$

*Formula for* F *statistic*

How many degrees of freedom does each sum of squares have? The entire model with $n$ observations has $n - 1$ degrees of freedom, so SST has $n - 1$ degrees of freedom. Since the model has $k$ independent variables, SSR has $k$ degrees of freedom, and by subtraction SSE has $(n - 1) - k$ or $n - k - 1$ degrees of freedom. Figure 12.2 shows a general ANOVA table for a multiple-regression model:

| Source of Variation | Degrees of Freedom | Sum of Squares | Mean Square | F Value |
|---|---|---|---|---|
| Regression | $k$ | SSR | $MSR = \dfrac{SSR}{k}$ | $\dfrac{MSR}{MSE}$ |
| Error | $n - k - 1$ | SSE | $MSE = \dfrac{SSE}{n - k - 1}$ | |
| Total | $n - 1$ | SST | | |

**FIGURE 12.2** ANOVA table for multiple regression model

## EXAMPLE 12.6  Texas Real Estate

### The ANOVA Table

The planners who are interested in predicting the average selling price of a home in the Dallas/Fort Worth area used Minitab to perform the regression analysis and get the multiple regression model. They looked at the residuals to see how well the model was able to predict the selling price, but now they need to see if the model is significant.

They want to test the hypotheses

$$H_0: \quad \beta_1 = \beta_2 = \beta_3 = \beta_4 = \beta_5 = 0$$
$$H_A: \quad \text{At least one of the coefficients is not equal to 0.}$$

To do this they look at the output from the Minitab analysis again:

```
Regression Analysis

The regression equation is
1994 Avg. Value of Home Sold = - 140352 + 1.64 Avg. Household Income
      + 39.2 Total Instructional Expenditure + 90 Mean SAT Score
      - 47820 Population Diversity - 1132 1993/94 Avg. Violent Crime

Predictor       Coef       StDev            T          P
Constant      -140352     112958        -1.24      0.222
Avg. Hou       1.6377      0.3121         5.25      0.000
Total In        39.16       21.78         1.80      0.081
Mean SAT         89.6       113.9         0.79      0.437
Populati       -47820       62143        -0.77      0.447
1993/94         -1132        2147        -0.53      0.601

S = 37958        R-Sq = 75.1%      R-Sq(adj) = 71.6%

Analysis of Variance

Source       DF          SS          MS          F         P
Regression    5 1.52341E+11 30468171569      21.15     0.000
Error        35 50427624859  1440789282
Total        40 2.02768E+11
```

They look at the section of the output labeled Analysis of Variance. From the table they see that the value of the test statistic, $F$, is:

$$\frac{\text{MSR}}{\text{MSE}} = \frac{30,468,171,569}{1,440,789,282} = 21.15$$

Now they must determine what this means. ∎

In an ANOVA test we are looking at the ratio of the explained variation to the unexplained variation. If it is high enough (if the explained variation is significantly larger than the unexplained part), then we can conclude that the model is significant. The $F$ test is then a one-sided, right-tail, test.

You remember that an $F$ statistic has two parameters, the numerator degrees of freedom and the denominator degrees of freedom. In this case the numerator, MSR, has $k$ (the number of variables in the model) degrees of freedom and the denominator, MSE, has $n - k - 1$ degrees of freedom. For a test at the $\alpha$ level of significance, the $F$ critical value is

$$F_{\alpha,k,n-k-1}$$

*Symbol for* F *critical value*

## EXAMPLE 12.7 Texas Real Estate

### Testing the Significance of the Multiple Regression Model

The planners looking at the average selling price of houses want to perform the test at the 0.05 level of significance. Since they are looking at 41 different locations and their model has 5 different independent variables, they know that they need an $F$ critical value with 5 degrees of freedom in the numerator and $41 - 5 - 1 = 35$ degrees of freedom in the denominator. From the tables of the $F$ distribution, they find that the critical value is $F_{0.05,5,35} = 2.4852$.

The test statistic is 21.15, which is definitely outside the critical value. They reject $H_0$ and conclude that the model is significant. ∎

In the previous example, we showed the calculation for the test statistic and found the critical value using the $F$ distribution tables. When you are using the computer to perform the analysis this is not always necessary. Most statistical software packages print out the value of the test statistic. Unless the software asks for the level of significance of the test it will not print out the critical value, but many packages do print out the $p$ value of the test. Remember that you can use the $p$ value to determine whether or not to reject $H_0$ by comparing it to the level of significance you choose. If the $p$ value is less than the chosen level of significance, then you reject $H_0$; if not, then you fail to reject.

### EXAMPLE 12.8    Phillips' Petroleum

*Testing Significance of the Multiple Regression*

The analysts who are looking at predicting shareholders' equity using total revenues and assets need to know whether their model is significant. They want to test the hypotheses

$$H_0: \quad \beta_1 = \beta_2 = 0$$
$$H_A: \quad \text{At least one coefficient is not equal to 0.}$$

at the 0.05 level of significance.

They look again at the output from the Excel computer analysis:

SUMMARY OUTPUT

| Regression Statistics | |
| --- | --- |
| Multiple R | 0.923228385 |
| R Square | 0.852350651 |
| Adjusted R Square | 0.815438314 |
| Standard Error | 1.067662859 |
| Observations | 11 |

ANOVA

| | df | SS | MS | F | Significance F |
| --- | --- | --- | --- | --- | --- |
| Regression | 2 | 52.64353179 | 26.32176589 | 23.09121324 | 0.000475254 |
| Residual | 8 | 9.119231848 | 1.139903981 | | |
| Total | 10 | 61.76276364 | | | |

| | Coefficients | Standard Error | tStat | P-value | Lower 95% |
| --- | --- | --- | --- | --- | --- |
| Intercept | −16.71847039 | 7.179010344 | −2.328798761 | 0.048252087 | −33.27330864 |
| Revenue | 1.452643014 | 0.216118353 | 6.721516208 | 0.00014936 | 0.954272876 |
| Assets | 1.034291795 | 0.769247372 | 1.344550312 | 0.215646057 | −0.739596974 |

From the output they see that the $F$ statistic is 23.09 and that the $p$ value is 0.0005. Since 0.0005 is less than their specified level of significance, 0.05, they reject $H_0$ and conclude that the model is significant. ∎

 **TRY IT NOW!**

### Order Filling

*Testing the Significance of the Model*

The mail-order company looking at factors related to the time to fill an order wants to know if its model is significant.

Write down the hypotheses that it needs to test.

The Excel output from the company's analysis is shown below:

SUMMARY OUTPUT

| Regression Statistics | |
| --- | --- |
| Multiple R | 0.915838993 |
| R Square | 0.838761061 |
| Adjusted R Square | 0.82696309 |
| Standard Error | 1.021396967 |
| Observations | 45 |

ANOVA

| | df | SS | MS | F | Significance F |
| --- | --- | --- | --- | --- | --- |
| Regression | 3 | 222.5057888 | 74.16859627 | 71.09366969 | 2.71039E-16 |
| Residual | 41 | 42.7733223 | 1.043251763 | | |
| Total | 44 | 265.2791111 | | | |

| | Coefficients | Standard Error | tStat | P-value | Lower 95% |
| --- | --- | --- | --- | --- | --- |
| Intercept | 0.215224599 | 0.55856837 | 0.385314691 | 0.701996141 | -0.912827729 |
| Number of Items | 0.281391801 | 0.029191616 | 9.639473395 | 4.26275E-12 | 0.222438105 |
| Locations | 1.259340956 | 0.103175681 | 12.20579253 | 3.10059E-15 | 1.050973327 |
| Assets | 0.019087782 | 0.047667374 | 0.400437032 | 0.690913642 | -0.077178487 |

Using the output, locate the values for MSR and MSE and their degrees of freedom.

Use the values of MSR and MSE to calculate the value of the $F$ statistic and compare it to the value in the table.

At the 0.05 level of significance, what is the critical value for the test?

What can the mail-order company conclude as a result of its test? Verify your answer by using the $p$ value from the printout.

At this point you should realize that just because the result of a statistical analysis is *statistically significant* does not mean that it will be useful or meaningful for purposes of decision making. If you have a lot of data, it might result in significant but essentially useless results.

The results of the significance test for a multiple regression model tell us that the *model* as a whole is significant, specifically that at least one of the coefficients is not zero. The *F* test tells us that the amount of variation in the dependent variable that is explained by the model is significantly larger than the amount left unexplained. These results lead to some questions; in particular, which coefficients are nonzero and how much of the variation in the dependent variable has been accounted for by the model? These questions can be answered by looking at some additional results of the analysis.

## 12.3.2 The Coefficient of Multiple Determination

In the simple linear regression model you learned that it is sometimes useful to look at the amount of the variation in $Y$ that can be explained by its association with $X$ (SSR) as a percentage of the total variation in $Y$ (SST). This measure is called the coefficient of determination, $R^2$, which can be interpreted as the amount by which the variation in $Y$ *could be reduced* if the value of $X$ is held constant. This type of interpretation is particularly useful in manufacturing control applications where the objective is to reduce the variation in the output (product) by controlling the variation in the input (raw materials, machine settings).

In multiple regression models there is an analogous measure, called the **coefficient of multiple determination, $R^2$**.

> The **coefficient of multiple determination, $R^2$**, is a measure of the percentage of the variation in the dependent variable, $Y$, that can be accounted for by the complete set of independent variables, $X_1$, $X_2$, . . . , $X_k$, in the model.

The coefficient of multiple determination is calculated in the same way as the coefficient of determination from the simple linear model:

*Formula for coefficient of multiple determination*

$$R^2 = \frac{\text{SSR}}{\text{SST}} \, 100\%$$

## EXAMPLE 12.9 Texas Real Estate

### *The Coefficient of Multiple Determination*

The group of planners looking at the selling prices of homes has found that their regression model with five independent variables is statistically significant. They would like a better idea of how much of the variation in selling price can be accounted for by the set of variables that they chose. They look at the output from the model for the value of the coefficient of multiple determination:

```
Regression Analysis

The regression equation is
1994 Avg. Value of Home Sold = - 140352 + 1.64 Avg. Household Income
     + 39.2 Total Instructional Expenditure + 90 Mean SAT Score
     - 47820 Population Diversity - 1132 1993/94 Avg. Violent Crime

Predictor      Coef       StDev          T        P
Constant     -140352     112958      -1.24    0.222
Avg. Hou      1.6377      0.3121      5.25    0.000
Total In       39.16      21.78       1.80    0.081
Mean SAT        89.6      113.9       0.79    0.437
Populati     -47820       62143      -0.77    0.447
1993/94       -1132        2147      -0.53    0.601

S = 37958      R-Sq = 75.1%      R-Sq(adj) = 71.6%
```

From the output they see that the value of $R^2$ is 75.1% which means that the set of five variables that they chose accounts for 75.1% in the variation of the average selling price of a house. The value makes them feel good about the set of variables that they chose, but at the same time they wonder if there is a better set of variables that they could use. ∎

At this point you may be thinking that "more is better" applies to the multiple regression model. Certainly, the more variables we add to the model the larger SSR will get and therefore $R^2$ will increase. This is true in an absolute sense—Numerically adding variables will never decrease SSE, so that as you add variables, $R^2$ will increase—but that does not mean that the model is better.

One of the objectives of modeling in general, and in the multiple regression model in particular, is to create a model that does a good job of predicting the value of the dependent variable, *with as small a set of input variables as possible.* There are several reasons for this. As the number of variables in a model increases, so does the complexity of the model. Complex models are harder to maintain and harder to explain to the people who want to use them. Also, as you increase the number of independent variables in a model, you increase the number of ways in which these variables may interact *with each other.* This results in unstable models—models that are very sensitive to small changes in the inputs. For these reasons, more is not necessarily better.

Another measure that is used in looking at multiple regression models is called the **adjusted $R^2$**.

> The **adjusted $R^2$** is the value of the coefficient of multiple determination adjusted to reflect the number of variables in the model.

Although it is true that as you increase the number of variables SSR will always increase, you also know that increasing the number of variables also *decreases* the number of degrees of freedom in the model. It is possible that adding a variable will increase SSR, and thereby decrease SSE, in such a way that the MSE actually increases! This happens because although the value you are dividing into (SSE) decreases, the value you are dividing by (degrees of freedom) decreases even more. The adjusted $R^2$ may actually decrease as variables are added to the model! We will talk about the adjusted $R^2$ further when we discuss model building.

## EXAMPLE 12.10   Phillips' Petroleum

### *The Coefficient of Multiple Determination*

The financial analysts who created a model for shareholders' equity want to see how the variables they selected account for the variation in shareholders' equity. They look at the output from their computer analysis and find

SUMMARY OUTPUT

| *Regression Statistics* | |
| --- | --- |
| Multiple R | 0.923228385 |
| R Square | 0.852350651 |
| Adjusted R Square | 0.815438314 |
| Standard Error | 1.067662859 |
| Observations | 11 |

From this output they see that the two variables in the model account for about 85% of the variation in shareholders' equity. They notice that the value of the adjusted $R^2$ is about 82%, but they are not sure how this number should be interpreted. ∎

*TRY IT NOW!*

### Order Filling
*The Coefficient of Multiple Determination*

The mail-order company looking at factors related to the time to fill an order wants to know if the set of independent variables that it selected does a good job of accounting for the variation in the time to fill an order. The relevant output from the model is shown below:

SUMMARY OUTPUT

| Regression Statistics | |
|---|---|
| Multiple R | 0.915838993 |
| R Square | 0.838761061 |
| Adjusted R Square | 0.82696309 |
| Standard Error | 1.021396967 |
| Observations | 45 |

Using the output, find the value of the coefficient of multiple determination.

If you were the manager of the company would you be satisfied with this model? Why or why not?

Look at the value of adjusted $R^2$. What do you think this value might be telling the company?

## 12.3.3 Testing the Individual Coefficients

In the simple linear regression model you learned that one way to test whether a model is significant is to perform a $t$ test on the slope coefficient. This tests the hypotheses

$$H_0: \quad \beta_1 = 0$$
$$H_A: \quad \beta_1 \neq 0$$

In the simple linear model, the $t$ test for $\beta_1$ and the ANOVA test are equivalent. In the multiple regression model the ANOVA tests all of the coefficients simultaneously. The result of rejecting the null hypothesis in this test of significance is that we know that at least one of the coefficients is nonzero. While this tells us, to some extent, whether we are on the right track, it does not tell us whether all of the variables in the model are useful or necessary in predicting the value of the dependent variable. It is certainly possible that in a model with three independent variables, only one of the coefficients is nonzero and that the other two variables are not adding any information to the model. If this were the case, it would be nice to be able to identify which variables contribute to the model and which variables do not.

This process is actually the beginning of a complex process called "model building," which we will discuss later in the chapter. At this point we will simply look at how to do a simple assessment of the value of each of the individual variables in the model.

We can test whether individual coefficients in a multiple regression model are equal to zero using a $t$ test like the one used in the simple linear model. The major difference in this test is that we are testing whether the coefficient of the particular variable is equal to zero *assuming that all other variables remain in the model*. This assumption is critical because it does not allow you to simply toss out all of the variables that do not prove to be significant. To test whether the coefficient of some variable $X_i$ in the model is different from zero we test the hypotheses

$$H_0: \quad \beta_i = 0$$
$$H_A: \quad \beta_i \neq 0$$

using a $t$ test. The test statistic is given by

$$t = \frac{b_i}{s_{b_i}}$$

*Formula for test statistic for individual coefficients*

Most computer packages give the information necessary to perform this test as part of their standard output. It is important that you understand exactly what it tells you so that you do not make errors in formulating your model.

## EXAMPLE 12.11 Texas Real Estate

**t *Tests for Individual Coefficients***

The planners who are looking at the average selling price of homes in the Dallas/Fort Worth area are pleased with the results of their model, but they wonder if all of the variables in the model are really contributing to the final result. They decide to look at the computer output again and see if they can obtain any additional information about the model. The relevant portion of the output is shown below:

```
Regression Analysis

The regression equation is
1994 Avg. Value of Home Sold = - 140352 + 1.64 Avg. Household Income
      + 39.2 Total Instructional Expenditure + 90 Mean SAT Score
      - 47820 Population Diversity - 1132 1993/94 Avg. Violent Crime

Predictor      Coef      StDev        T        P
Constant     -140352    112958     -1.24    0.222
Avg. Hou      1.6377     0.3121     5.25    0.000
Total In      39.16      21.78      1.80    0.081
Mean SAT      89.6       113.9      0.79    0.437
Populati     -47820      62143     -0.77    0.447
1993/94       -1132       2147     -0.53    0.601

S = 37958      R-Sq = 75.1%     R-Sq(adj) = 71.6%
```

The planners figure that what they want to do is perform five individual $t$ tests to see whether the individual coefficients are equal to zero. They decide to do this at the 0.05 level of significance and look at the $p$ values from the output. They see that only one variable, *average household income*, has a $p$ value less than 0.05. They reason that this means that the coefficient of $X_1$ is not zero. Using the same approach, they see that the result of the other four tests is that the coefficients of those variables, $\beta_2$ through $\beta_5$, are not nonzero.

What does this mean? Can the planners simply drop the other four variables from the model? What would happen if they did? ∎

## EXAMPLE 12.12   Phillips' Petroleum

### *Looking at Individual Coefficients*

Although the model for predicting shareholders' equity had only two variables in it, the analysts still wondered whether both variables were necessary. They decide to look at the relevant output again:

SUMMARY OUTPUT

| Regression Statistics | |
|---|---|
| Multiple R | 0.923228385 |
| R Square | 0.852350651 |
| Adjusted R Square | 0.815438314 |
| Standard Error | 1.067662859 |
| Observations | 11 |

ANOVA

| | df | SS | MS | F | Significance F |
|---|---|---|---|---|---|
| Regression | 2 | 52.64353179 | 26.32176589 | 23.09121324 | 0.000475254 |
| Residual | 8 | 9.119231848 | 1.139903981 | | |
| Total | 10 | 61.76276364 | | | |

| | Coefficients | Standard Error | tStat | P-value | Lower 95% |
|---|---|---|---|---|---|
| Intercept | −16.71847039 | 7.179010344 | −2.328798761 | 0.048252087 | −33.27330864 |
| Revenue | 1.452643014 | 0.216118353 | 6.721516208 | 0.00014936 | 0.954272876 |
| Assets | 1.034291795 | 0.769247372 | 1.344550312 | 0.215646057 | −0.739596974 |

They see that the $t$ statistic for *revenue* is 6.72 and $p$ value is essentially zero, indicating that the coefficient of that variable is nonzero. They also see that the $t$ statistic for *assets* is 1.34 and that the $p$ value for that test is 0.22. This would lead to the conclusion that the coefficient of the *assets* variable is zero.

Should the analysts drop the variable? At this point they are unsure about what to do. ∎

What do we learn from testing the individual coefficients in a model? We do learn something about the utility of the variables. Clearly, some of the variables are important and should stay in the model, but what about the others? It would appear that we cannot throw them all out because of the assumption, but what happens if we do? If we don't throw them all out, which ones should we discard?

## EXAMPLE 12.13   Texas Real Estate

### *Adjusting the Model*

The planners looking at the real estate model decide to drop all of the variables with low $t$ statistics out of the model. They are pretty sure this is not the right thing to do, but they want to see what happens if they take the most drastic approach.

```
Regression Analysis

The regression equation is
1994 Avg. Value of Home Sold = - 1167 + 2.09 Avg. Household Income

Predictor       Coef        StDev           T          P
Constant        -1167        14281        -0.08      0.935
Avg. Hou        2.0894      0.2084        10.03      0.000


S = 38118       R-Sq = 72.1%      R-Sq(adj) = 71.3%


Analysis of Variance

Source       DF           SS            MS          F          P
Regression    1  1.46103E+11  1.46103E+11    100.56      0.000
Error        39  56665134929   1452952178
```

The first thing they notice is that the model itself has changed. The intercept value has decreased from $-140{,}352$ dollars to $-1167$ dollars. This is quite a change. In addition, the coefficient of the variable *average household income* has increased from 1.64 to 2.09.

Interestingly, the value of $R^2$ has not changed much, from 75.1 to 72.1%. It certainly appears that the variables that were dropped were not of any use, but the planners are still not convinced they should drop them all. ∎

The answer to the question of dropping variables is not a simple yes or no. The $t$ statistics and/or $p$ values for the individual coefficients give a hierarchy for the variables in the model. With all variables in the model, we can consider the variable with the smallest $t$ value (or largest $p$ value) to be contributing the least to the model and we can eliminate that variable. But the simple approach stops there. If we drop a variable then the *entire model changes*. It is necessary to rerun the analysis and get a new model and look at those results in the same way we did the original model.

### EXAMPLE 12.14   Phillips' Petroleum

*Looking at Individual Coefficients*

The financial analysts from Phillips' Petroleum decide to see what the effect of dropping assets from the model is, so they rerun the model as a simple linear model. The results of that analysis are shown below:

SUMMARY OUTPUT

| Regression Statistics | |
|---|---|
| Multiple R | 0.904978067 |
| R Square | 0.818985302 |
| Adjusted R Square | 0.798872558 |
| Standard Error | 1.114548637 |
| Observations | 11 |

ANOVA

|  | df | SS | MS | F | Significance F |
|---|---|---|---|---|---|
| Regression | 1 | 50.58279565 | 50.58279565 | 40.71971956 | 0.000128074 |
| Residual | 9 | 11.17996798 | 1.242218665 | | |
| Total | 10 | 61.76276364 | | | |

|  | Coefficients | Standard Error | tStat | P-value | Lower 95% |
|---|---|---|---|---|---|
| Intercept | −7.795059065 | 2.857364461 | −2.728059081 | 0.023298434 | −14.25887147 |
| Revenue | 1.437769522 | 0.225313329 | 6.381200479 | 0.000128074 | 0.928074973 |

The first thing they look at is the model itself. Without including assets as a variable, the new model is

$$\hat{y} = -7.80 + 1.44x$$

The intercept term has changed considerably, but the coefficient for the revenue variable is essentially the same.

The analysts also notice that the $R^2$ value has changed. Without assets in the model, the model accounts for only 81.9% of the variation in shareholders' equity, while the two-variable model had an $R^2$ of 85.2%. The analysts also notice that the new $R^2$ is close to what the adjusted $R^2$ for the two-variable model was. They wonder how to interpret this.

At this point they cannot decide what to do. They like the model with the larger $R^2$ value, but it seems pointless to leave the variable in if it is not contributing significantly to the model. They decide to try to find a model that has a higher $R^2$ by looking for other independent variables that might prove to be significant. ∎

You can see that there are no easy answers here. In the two examples we looked at, dropping the variables with low $t$ values had different impacts. In one case the model changed but the value of $R^2$ did not, while in the other case the $R^2$ value dropped but the model coefficients remained essentially the same. In other cases both the model and the $R^2$ values can change.

 ***TRY IT NOW!***

## Order Filling

### *Testing Individual Regression Coefficients*

The mail-order company looking at factors related to the time to fill an order wants to know how the individual variables contribute to the model. It decides to look at the computer analysis again:

SUMMARY OUTPUT

| Regression Statistics | |
|---|---|
| Multiple R | 0.915838993 |
| R Square | 0.838761061 |
| Adjusted R Square | 0.82696309 |
| Standard Error | 1.021396967 |
| Observations | 45 |

ANOVA

| | df | SS | MS | F | Significance F |
|---|---|---|---|---|---|
| Regression | 3 | 222.5057888 | 74.16859627 | 71.09366969 | 2.71039E-16 |
| Residual | 41 | 42.7733223 | 1.043251763 | | |
| Total | 44 | 265.2791111 | | | |

| | Coefficients | Standard Error | tStat | P-value | Lower 95% |
|---|---|---|---|---|---|
| Intercept | 0.215224599 | 0.55856837 | 0.385314691 | 0.701996141 | −0.912827729 |
| Number of Items | 0.281391801 | 0.029191616 | 9.639473395 | 4.26275E-12 | 0.222438105 |
| Locations | 1.259340956 | 0.103175681 | 12.20579253 | 3.10059E-15 | 1.050973327 |
| Experience | 0.019087782 | 0.047667374 | 0.400437032 | 0.690913642 | −0.077178487 |

Look at the coefficients of each of the three variables in the model and perform the appropriate hypothesis tests.

Which variable(s) have nonzero coefficients? Which variable(s) have coefficients that are equal to zero?

As a result of these tests, what recommendation would you make?

The questions faced as a result of looking at the individual model coefficients are not ones that have simple answers. They give rise to a group of techniques called "model-building techniques," which try to find the "best" model from a set of input variables. By "best" we generally mean the model with the smallest number of variables that explains the largest portion of the variation in the dependent variable.

## 12.3.4 Exercises—Learning It!

**12.5** Consider the model on traffic fatalities for the different states. The variable $X_1$ represents the number of licensed drivers, $X_2$ is the number of registered vehicles, and $X_3$ is the number of vehicle miles driven annually. The output from the computer analysis is shown below:

*Requires Exercise 12.1*

SUMMARY OUTPUT

| Regression Statistics | |
|---|---|
| Multiple R | 0.982548538 |
| R Square | 0.96540163 |
| Adjusted R Square | 0.963193224 |
| Standard Error | 154.5407481 |
| Observations | 51 |

ANOVA

| | df | SS | MS | F | Significance F |
|---|---|---|---|---|---|
| Regression | 3 | 31321046.9 | 10440348.97 | 437.1485021 | 2.54274E-34 |
| Residual | 47 | 1122493.613 | 23882.84282 | | |
| Total | 50 | 32443540.51 | | | |

| | Coefficients | Standard Error | tStat | P-value | Lower 95% |
|---|---|---|---|---|---|
| Intercept | 51.7481659 | 30.43306219 | 1.700392999 | 0.095666076 | −9.475200509 |
| X Variable 1 | 0.06294764 | 0.048829545 | 1.289130172 | 0.20366155 | −0.035284642 |
| X Variable 2 | −0.211896991 | 0.055989427 | −3.784589385 | 0.000435566 | −0.324533083 |
| X Variable 3 | 0.029349954 | 0.003525079 | 8.326041167 | 8.34285E-11 | 0.022258416 |

(a) Set up the hypotheses to test whether the model as a whole is significant.
(b) Use the output to test the hypotheses at the 0.05 level of significance.
(c) What is the coefficient of multiple determination for this model?
(d) Do you think that the model does a good job of explaining the variation in the number of traffic fatalities in a state? Why or why not?
(e) Set up the hypotheses to test each regression coefficient individually.
(f) Perform the tests at the 0.05 level of significance.
(g) What can you conclude about the coefficient of variable $X_2$?

(h) What are your conclusions about $X_1$? About $X_3$?

(i) Would you recommend dropping any of these variables from the model? Why or why not?

*Requires Exercise 12.2*

**12.6** The analysts looking at the model on compliance with mandatory seat-belt laws is wondering whether their model is statistically significant. They look at the output from the model:

```
Regression Analysis

The regression equation
Percent Compliance = 0.416 + 0.00187 Fine + 0.0194 Years In Effect

47 cases used 4 cases contain missing values

Predictor        Coef       StDev          T          P
Constant       0.41558     0.05153       8.06      0.000
Fine           0.001874    0.001050      1.78      0.081
Years In       0.019397    0.004672      4.15      0.000


S = 0.09682     R-Sq = 30.8%     R-Sq(adj) = 27.7%


Analysis of Variance

Source        DF         SS          MS          F          P
Regression     2     0.183552    0.091776      9.79      0.000
Error         44     0.412431    0.009373
Total         46     0.595983


Source        DF      Seq SS
Fine           1     0.021952
Years In       1     0.161600
```

(a) Set up the hypotheses to test whether the model is significant.
(b) At the 0.01 level of significance, is the model they tested significant? What does this mean?
(c) What is the value of $R^2$ for this model? Do you think that the model does a good job of explaining the variation in compliance with the seat-belt laws? Why or why not?
(d) Set up the hypotheses to test for each of the regression coefficients individually and perform the test at the 0.05 level of significance.
(e) What are your conclusions from the tests on individual coefficients? What recommendations would you make about including the variables currently under consideration?

*Requires Exercise 12.3*

*CRIME.XXX*

**12.7** Consider the problem of the committee that is looking at the issue of serious crimes in some large cities.

(a) Set up the hypotheses to test whether the multiple regression model is significant.
(b) Use the output from your analysis to perform the test at the 0.05 level of significance. What is your conclusion?
(c) What percentage of the variation in the number of serious crimes in a city is accounted for by the set of variables in the model? Based on this number, do you think that the set of variables chosen by the committee does a good job of explaining the number of serious crimes in a large city?
(d) Set up the hypotheses to test the individual regression coefficients.
(e) At the 0.01 level of significance, what are your conclusions? What recommendations would you make about the set of variables currently in the model?

*Requires Exercise 12.4*

*INTERNET.XXX*

**12.8** Look at the data on Internet usage again and consider the two-variable model that you developed.

(a) At the 0.05 level of significance, what can you conclude about the significance of your model?
(b) What percentage of the variation in hours of Internet usage is explained by the set of variables in your model? Based on this percentage, do you think your model is a good one? Why or why not?

(c) Set up the hypotheses and test the individual regression coefficients in your model at the 0.05 level of significance. What are your conclusions?

(d) Based on the results of the tests on individual coefficients, what would you do next? Why?

**12.9** In the problem about number of traffic fatalities the model was rerun, dropping the data on number of licensed drivers that had the lowest $t$ statistic. The output is shown below:

SUMMARY OUTPUT

*Requires Exercises 12.1, 12.5*

| Regression Statistics | |
|---|---|
| Multiple R | 0.981925801 |
| R Square | 0.964178279 |
| Adjusted R Square | 0.962685707 |
| Standard Error | 155.6025568 |
| Observations | 51 |

ANOVA

| | df | SS | MS | F | Significance F |
|---|---|---|---|---|---|
| Regression | 2 | 31281357.04 | 15640678.52 | 645.9845512 | 1.99303E-35 |
| Residual | 48 | 1162183.473 | 24212.15568 | | |
| Total | 50 | 32443540.51 | | | |

| | Coefficients | Standard Error | tStat | P-value | Lower 95% |
|---|---|---|---|---|---|
| Intercept | 46.03605608 | 30.31563375 | 1.518558262 | 0.135433408 | −14.91757512 |
| X Variable 2 | −0.162799336 | 0.041321977 | −3.939776022 | 0.000263617 | −0.245882691 |
| X Variable 3 | 0.029996183 | 0.003513227 | 8.538070709 | 3.42795E-11 | 0.02293237 |

(a) Write down the equation of the new two-variable model.

(b) Compare the new model to the model with three variables. How much does the model change when number of licensed drivers is dropped?

(c) Compare the value of $R^2$ for both models. What does this make you think about the decision to drop number of licensed drivers from the model?

(d) Would you consider the two-variable model a good model? Why or why not?

(e) Based on the value of $R^2$ would you be satisfied with this model or would you want to consider other variables?

**12.10** The lobbyists looking at compliance with seat-belt laws decided to drop the variable associated with the amount of the fine for noncompliance from the model. The output from the modified model is shown below:

**Regression Analysis**

*Requires Exercises 12.2, 12.6*

```
The regression equation is
Percent Compliance = 0.462 + 0.0188 Years In

48 cases used 3 cases contain missing values

Predictor       Coef        StDev           T         P
Constant     0.46203      0.04443       10.40     0.000
Years In     0.018846     0.004673       4.03     0.000


S = 0.09806      R-Sq = 26.1%       R-Sq(adj) = 24.5%


Analysis of Variance

Source       DF          SS           MS         F       P
Regression    1      0.15642      0.15642     16.27   0.000
Error        46      0.44231      0.00962
Total        47      0.59873
```

(a) Write down the equation for the new model and compare it to the model with two independent variables. Do you think that the coefficients have changed significantly? Why or why not?

(b) Test the significance of the new model at the 0.05 level of significance.

(c) Compare the value of $R^2$ from the new model to $R^2$ from the old model. What does this make you think about the decision to drop the variable associated with the fine from the model?

(d) If you were among the group of lobbyists would you be satisfied with this model? Why or why not? Can you suggest any other variables that might be considered?

## Discovery Exercise 12.1
## FINDING THE BEST MODEL

The Office of Institutional Planning at a university in the Far West is interested in understanding what factors influence the graduation rate, that is, the percentage of entering freshmen who actually graduate from the university. The university planners have collected data on 46 universities with traits similar to theirs and have selected three variables that they think are related to the graduation rate: 25th percentile combined SATs of accepted students, the acceptance rate (percentage of students who apply that are accepted by the university), and educational expenditure per student ($) by the university. A sample of the data is shown below:

| Rank/School Name | 1995 Actual Graduation Rate (%) | SAT/ACT 25th | Acceptance Rate (%) | Education Expenditures per Student |
|---|---|---|---|---|
| Indiana University—Bloomington | 59 | NA | 80 | 9,713 |
| SUNY—Binghamton | 74 | NA | 40 | 9,080 |
| Allegheny University | 52 | 786 | 61 | 33,270 |
| Univ. of California—Riverside | 56 | 910 | 78 | 13,403 |
| Oregon State University | 53 | 949 | 89 | 10,182 |
| University of Hawaii—Manoa | 55 | 960 | 65 | 13,360 |
| New Jersey Inst. of Technology | 63 | 970 | 67 | 14,030 |

The entire data set is found in the file GRADRATE.XXX.

(a) Which variable do you think will have the greatest effect on graduation rate? Explain why you chose this variable.

(b) Find a simple linear model that predicts graduation rate using the variable that you think is most important. What is the value of $R^2$ for your model? Is the model significant?

(c) Do this again using the other two independent variables, and fill in the table below:

| Independent Variable | $R^2$ | $p$ Value |
|---|---|---|
| | | |
| | | |
| | | |

(d) Which one-variable model do you think is the best? Why?

(e) How many different models with two independent variables could you find? List them all.

(f) Find the multiple regression models for all of the two-variable models. Fill in the table below:

| Independent Variables | $R^2$ | $p$ Value |
|---|---|---|
|  |  |  |
|  |  |  |
|  |  |  |

(g) Which two-variable model do you think is best? Why? Does the two-variable model you think is best contain the variable from the best one-variable model? Would you expect it to?

(h) Find the multiple regression model that predicts graduation rate using all three independent variables. What is $R^2$ for this model? Is it significant?

(i) Now, fill in the table below for each of the "best" models that you have found.

| Number of Independent Variables | Independent Variables | $R^2$ | $p$ Value |
|---|---|---|---|
| 1 |  |  |  |
| 2 |  |  |  |
| 3 |  |  |  |

(j) If you were going to use a model to predict graduation rate, which of these three models would you choose? Why?

# 12.4 BUILDING A MULTIPLE REGRESSION MODEL

At the end of the last section we saw that the problem of deciding which variables to include in the multiple regression model is not an easy one. The objective is to find the "best" model, but that is not really well defined. Two somewhat conflicting criteria are involved in the definition: The model should do an acceptable job of predicting or explaining the dependent variable, and the model should use the smallest possible set of independent variables.

The task of deciding which variables should be included/excluded from the model must be done in a systematic way. If variables are dropped from a model, then the entire analysis must be redone. To keep the analysis from being chaotic and unproductive, several methods known as **"model building techniques"** have been developed.

> **Model building techniques** are methods used for identifying the best multiple regression model from a set of independent variables. These methods include Forward Selection, Backward Elimination, Stepwise Regression, and All Possible Regressions.

This section focuses on the issues of choosing the set of independent variables to be considered in a model as well as some of the model-building techniques available in most statistical software packages. We will not go into the technical details of most of the techniques since they can be quite complex, but rather we discuss the ideas behind the techniques and how to interpret the results.

## 12.4.1 Choosing a Set of Independent Variables

After reading the examples in this chapter and doing some of the exercises presented so far, you may be wondering how and why the people doing the analyses chose the variables they did. In many cases, you might have chosen an entirely different set of variables! The choice of the dependent variable is usually quite straightforward. There is a problem to be solved or a decision to be made that involves the variable in question. The choice for the set of independent variables is not quite as easy. Many factors are involved in the decision procedure.

One thing that can heavily impact the decision process is the experience and intuition of the "experts" involved. It is difficult, if not impossible, to solve a problem or make a decision if you know nothing about the subject. Often the group involved in the problem will brainstorm to find a list of variables that might be related to the dependent variable. Once this list is established and prioritized, other factors influence the final choice.

A factor that must be considered in the decision process is whether the data for the study will come from existing data or whether they will be collected especially for this analysis. When the data are to come from existing sources, availability and cost are two important considerations. Sometimes variables that the group considers important might be too expensive to obtain, or might not even be available. In this case, similar variables that are less desirable might have to be included. Occasionally, the set of available data is overwhelming and decisions have to be made about which variables to include.

### EXAMPLE 12.15   Texas Real Estate

*Choosing the Set of Input Variables*

The planners looking at the selling price of homes in different locations in the Dallas/Fort Worth area had 56 different variables on their list of possible variables. They were able to obtain data for all of the locations of interest for 20 of the 56 variables. They felt that 20 variables were too many to work with, so after some discussion they decided to try to find a set of 5 that they felt were most important. The question was how should they choose the 5 variables?

Of course, everyone in the group had opinions about which variables to include, and there was some overlap, but in the end the planners decided on a more objective approach. They knew that the correlation coefficient measured the strength of the relationship between two variables, and so they decided to calculate the correlation coefficient for each of the 20 variables with the average selling price of a home. They would try to pick the 5 variables with the highest correlation. The correlations are shown in the table below:

| | | | |
|---|---|---|---|
| Avg. household income | 0.84885 | Square miles | − 0.25142 |
| Percent passing AA | 0.63989 | Population density | 0.20480 |
| Percent taking SAT | 0.51699 | Property tax rate | − 0.20097 |
| Mean SAT | 0.50495 | 1994 pop | − 0.19425 |
| Total instructional exp. | 0.48006 | Distance to Dallas/Fort Worth | − 0.12979 |
| Population diversity | − 0.44283 | Annual insurance | − 0.10825 |
| 1993/94 violent crime | − 0.42840 | Percent owner occupied | 0.09989 |
| Hazardous waste sites | − 0.28200 | Median age | − 0.09370 |
| Appreciation of sale price | 0.27094 | Park land | − 0.05087 |
| 1993/94 nonviolent crime | − 0.26457 | Annual growth rate | 0.01887 |

They looked at the 5 variables that were most highly correlated with the average selling price of a home and decided not to use all 5 of them. The reason for this was that four out of the five variables were related to education, and they wanted some diversity in the variable set. They decided to take two of the education-related variables, and agreed on mean SAT scores and total instructional expenditures. ■

When data come from existing sources it is often tempting to include every bit of data that is available. This approach can be counterproductive and can lead to overly complex and unstable models. We will talk more about these and other problems with models in the next section of the chapter. If a model from a subset of available data is not acceptable, it is always possible to add more variables to the set under consideration.

When data are going to be collected as part of an analysis it is wise to include as many possible variables as you can afford. If you collect data by means of a survey or an experiment, then usually the only way to add variables to the study is to repeat the entire data collection procedure. This can be time-consuming and expensive, if not impossible.

 *TRY IT NOW!*

### GPA

*Choosing the Independent Variables*

Many studies have been done on what factors are related to a college student's grade point average (GPA). These studies often focus on precollege factors, such as high school performance, SAT or ACT scores, and general socioeconomic factors such as family income and race. Students know that once they are in college, many other factors influence their GPA. *(continued)*

You are going to try to find a model relating current factors to GPA. Make a list of all variables that you think are related to a college student's GPA.

From this list, identify what you think are the five most important variables.

How would you go about gathering the data you need to do your study?

Once the set of variables is chosen and the analysis begins, statistical software is used to actually find the "best" model from the set of variables under consideration.

### 12.4.2 Forward Selection and Backward Elimination

The simplest of the model-building techniques are Forward Selection and Backward Elimination. The techniques are very well described by their names. In forward selection, a model is built by adding variables one at a time, until adding variables does not cause the model to improve. In the backward elimination model, all variables are included at first and then dropped one by one until the model begins to degrade. There are several different ways to define improvement and degradation of a model. We will focus on the most commonly used methods.

Forward selection starts by running all possible one-variable models and selecting the "best" one. There are several equivalent measures of "best": largest $R^2$, smallest MSE, largest $F$ statistic, and largest $t$ statistic. After the one-variable model is found, each of the remaining unchosen variables is combined with the current model. Again, the "best" of these is chosen. Most computer packages define "best" in terms of something called the partial $F$ statistic, which is equivalent to the increase in the MSR, the decrease in MSE, or the highest $t$ statistic. The technique continues adding variables until the change in the model is not significant at some specified level.

Backward elimination works the same way as forward selection, except that it starts with the full model and drops the "worst" (using the same criteria) until dropping a variable causes the model to become significantly worse.

## EXAMPLE 12.16   Texas Real Estate

### Forward Selection

The planners looking at the real estate problem decided to use forward selection to try to find the "best" model for their set of five variables. The computer output from the procedure is shown and enlarged below:

```
* * * *  M U L T I P L E   R E G R E S S I O N  * * * *

Listwise Deletion of Missing Data

Equation Number 1    Dependent Variable. .   @994 AVG   1994 Avg. Value of Home

Block Number  1.  Method: Forward   Criterion   PIN  .0500    AVG. HOU TOTAL IN MEAN SAT @993 94   POPULATI

Variable(s) Entered on Step Number 1. .     AVG. HOU Avg. Household Income

Multiple R          .84885      Analysis of Variance
R Square            .72054                    DF     Sum of Squares       Mean Square
Adjusted R Square   .71338      Regression     1   146103347775.79370   146103347775.794
Standard Error 38117.60981      Residual      39    56665134929.23070   1452952177.67258

                    F =      100.55620      Signif F =  .0000
```

```
----------------- Variables in the Equation -----------------

Variable              B       SE B       Beta      T  Sig T

AVG. HOU        2.089442    .208365    .848848   10.028   .0000
(Constant) -1166.758282 14280.72621             -.082    .9353
```

```
------------- Variables not in the Equation -------------

Variable     Beta In  Partial  Min Toler      T   Sig T

TOTAL IN     .140535  .239924   .814503    1.523   .1359
MEAN SAT     .072320  .115670   .714894     .718   .4772
@993 94     -.066667 -.113157   .805119    -.702   .4869
POPULATI    -.049250 -.081890   .772608    -.507   .6154
```

```
----------------- Variables in the Equation -----------------

Variable              B       SE B       Beta      T  Sig T

AVG. HOU        2.089442    .208365    .848848   10.028   .0000
(Constant) -1166.758282 14280.72621              0.082   .9353
```

```
--------------- Variables not in the Equation -------------

Variable     Beta In  Partial  Min Toler      T   Sig T

TOTAL IN     .140535  .239924   .814503    1.523   .1359
MEAN_SAT     .072320  .115670   .714894     .718   .4772
@993_94_    -.066667 -.113157   .805119    -.702   .4869
POPULATI    -.049250 -.081890   .772608    -.507   .6154
```

From the output they saw that only one variable was included in the final model. They notice that the next variable for consideration is total instructional expenditure, which had the second highest $t$ statistic from the original model.

The model is $\hat{y} = -1167 + 2.09$ average household income. ■

*TRY IT NOW!*

## Order Filling
### *Forward Selection*

The mail-order company wants to use a standard method to find the best model from its set of three variables. It decides on forward selection because this method is the easiest to understand. The relevant portions of the computer output are shown below:

```
Method:  Forward        Criterion   PIN  .0500        NUMBER_O LOCATION EXPERIEN

Variable(s) Entered on Step Number  1..     LOCATION   Locations

Multiple R              .68766        Analysis of Variance
R Square               .47287                       DF   Sum of Squares   Mean Square
Adjusted R Square      .46061        Regression      1      125.44303     125.44303
Standard Error        1.80333        Residual       43      139.83608       3.25200

                                     F =      38.57409      Signif F =  .0000

------------------ Variables in the Equation ------------------

Variable              B         SE B       Beta        T   Sig T

LOCATION         1.116792    .179814    .687657      6.211  .0000
(Constant)       2.650287    .704624                 3.761  .0005

          Variable(s) Entered on Step Number  2..     NUMBER_O  Number of Items

Multiple R              .91549        Analysis of Variance
R Square               .83813                       DF   Sum of Squares   Mean Square
Adjusted R Square      .83042        Regression      2      222.33850     111.16925
Standard Error        1.01114        Residual       42       42.94061       1.02240

                                     F =     108.73411      Signif F =  .0000

------------------ Variables in the Equation ------------------

Variable              B         SE B       Beta        T   Sig T

NUMBER_O          .281325    .028898    .610427      9.735  .0000
LOCATION         1.256149    .101834    .773465     12.335  .0000
(Constant)        .338773    .460945                 .735  .4665

END PIN =     .050 Limits reached.
```

What variable is added on the first step of the procedure? What is its coefficient? What is the value of $R^2$ for the first model considered?

What variable is added on the second step? What is its coefficient? What does $R^2$ change to after this step?

Does the coefficient of the first variable change when the second variable is added? If so, what is the new value?

Write down the final model.

---

Forward selection and backward elimination are straightforward techniques, but there is no guarantee that they will find the "best" model. In fact, as you can see, many combinations of variables are not even considered as part of the process. For this reason, a technique called stepwise regression, which is a variation on these methods, is often used.

## 12.4.3 Stepwise Regression

Stepwise regression is actually a combination of the forward selection and backward elimination methods. It starts the same way as forward selection, by running all of the one-variable models and selecting the variable with the largest $F$ or $t$ statistic. It then adds variables one at a time to the current model using the same type of criteria as forward selection. The difference occurs when three variables have been included in the model. Stepwise regression includes a routine that considers dropping previously added variables from the model using criteria similar to the ones used by backward elimination. In this way the method considers combinations of variables that were never considered in either of the other methods.

For example, suppose that five variables are being considered for addition in the model and that the first variable chosen is $X_2$, the second is $X_1$, and the third is $X_5$. Up to this point the stepwise procedure is the same as forward selection. However, at this point the stepwise regression procedure considers dropping either $X_1$ or $X_2$ from the model, looking at the possible combinations of $X_1$ and $X_5$ or $X_2$ and $X_5$. While the second of these combinations, $X_2$ and $X_5$, was considered in the second iteration, the combination $X_1$ and $X_5$ was not. In this way, stepwise regression considers many more combinations of variables than the other two methods.

If it sounds complicated, it is. Many people make errors in using the stepwise regression technique because they do not understand it well enough. The method is very sensitive to the choice of significance levels chosen for adding and dropping variables. Small changes in these values can lead to completely different models being picked as the "best" model. If people are not familiar with the technique or with the software package they are using, they can get easily confused. Minitab requires that the user specify the criteria for adding and dropping variables from the model in terms of an $F$ statistic and defaults to the same value, 4.00 for each. SPSS allows the user to choose between significance levels (defaults to 0.05 and 0.10) or $F$ statistics (defaults to 3.84 and 2.71).

Clearly, the defaults and options are different among the packages and in many cases will produce different results. A powerful mainframe statistical package called SAS defaults to 0.15 for both significance levels. Without knowing what they are doing, users can get very different results!

### EXAMPLE 12.17   Texas Real Estate

***Stepwise Regression***

Knowing that forward selection does not consider many different combinations of variables, the group studying the real estate problem decided to use stepwise regression to see if it resulted in a different model. The output from the Minitab procedure is given below. The parts in ***bold italic*** indicate responses made by the user to questions from the procedure.

```
Stepwise Regression

  F-to-Enter:        4.00     F-to-Remove:        4.00

  Response is 1994 Avg on  5 predictors, with N =    41

      Step        1
Constant    -1167

Avg. Hou     2.09
T-Value     10.03

S           38118
R-Sq        72.05
  More? (Yes, No, Subcommand, or Help)
SUBC>y

  No variables entered or removed

  More? (Yes, No, Subcommand, or Help)
SUBC>no
```

The output is not very helpful, but after looking at it, the planners found that one variable, average household income, was added to the model and that its coefficient was 2.09. The $R^2$ from the model was 72.05.

In this case, the model from stepwise regression is the same as that from forward selection. ∎

The output from the stepwise regression procedure varies among software packages. The output from Minitab is interactive, requiring the user to make decisions at several points, and very lengthy, since it gives the results of each iteration. The output from SPSS is slightly less involved, with only the end results output. The major portion of the output contains information on the variables included in the model and those considered but not included. If you want the additional information available with standard regression output, such as predicted values and residuals, you need to rerun the model with the common regression routine.

 ***TRY IT NOW!***

### Order Filling

***Testing Individual Regression Coefficients***

The mail-order company looking at the model for order filling decides to use stepwise regression to see if it finds a model that is different from the one found using forward selection. The Minitab results (with the interactive parts edited out) are given here.

```
Stepwise Regression

F-to-Enter:        4.00     F-to-Remove:        4.00

Response is    Time   on   3 predictors, with N =    45

        Step        1
Constant    2.650

Location     1.12
T-Value      6.21

S            1.80
R-Sq        47.29

        Step        2
Constant    0.3388

Location     1.26
T-Value     12.34

Number o    0.281
T-Value      9.74

S            1.01
R-Sq        83.81

No variables entered or removed
```

From the output, how many iterations did the procedure take?

Which variable was added to the model first? What is its coefficient?

Which variable was added second? What is its coefficient?

Did the coefficient for the first variable change? If so, what is its coefficient after the second variable is added?

What is the final model from the stepwise procedure?

## 12.4.4 All Possible Regressions

One method of finding the "best" regression model that has been largely overlooked in the past is the method known as all possible regressions or all subsets. Practitioners have hesitated to use this technique because of the number of calculations that must be done. In the past, the technique has proved to be too computer-intensive for most situations. With the advent of powerful personal computers, this is no longer an issue and the technique is becoming widely used.

The method does exactly what its name states. Given a set of candidate variables, it runs regression models for all of the possible combinations of variables and then identifies the "best" model for each different size. For example, if there were five variables under consideration for a model, the all subsets method will run all models with one variable, all possible combinations of two variables, all three-variable combinations, etc. It then chooses, according to one or several criteria, the best one-variable model, the best two-variable model, and so on. The user is then free to choose among the models presented based on other decision criteria, such as cost, ease of access to the data, and personal preferences.

The reason that the technique is so computer-intensive is the way the number of possible combinations grows with respect to the number of variables. If there are $k$ variables under consideration, then there are $2^k - 1$ combinations to consider. Table 12.1 gives you an idea of how large this number gets.

The number of possible models for 5 variables is not unreasonable, even for 10 it is manageable, but doubling the number of variables from 10 to 20 multiplies the result by a factor of 1000!

All possible regression usually uses several criteria to define "best" for each level of model. Two of the criteria are ones you are already familiar with, $R^2$ and adjusted $R^2$. The third criterion is called the $C_p$ **statistic,** which is a measure of the total squared error. The quantity $p$ is the number of *terms* in the model being evaluated, including the intercept:

$$C_p = \frac{SSE_p}{MSE_{ALL}}(n - 2p)$$

The $C_p$ statistic uses the SSE from the model being evaluated and the MSE from the "full" model, the model that contains all of the potential variables. Using $C_p$, the "best model" with $p$ terms will result in $C_p = p$. For this model the estimates of the coefficients are unbiased.

**TABLE 12.1   Number of subsets**

| Number of Variables | Number of Subsets | Result |
|---|---|---|
| 1 | $2^1 - 1$ | 1 |
| 2 | $2^2 - 1$ | 3 |
| 3 | $2^3 - 1$ | 7 |
| 4 | $2^4 - 1$ | 15 |
| 5 | $2^5 - 1$ | 31 |
| ⋮ | ⋮ | ⋮ |
| 10 | $2^{10} - 1$ | 1,023 |
| ⋮ | ⋮ | ⋮ |
| 20 | $2^{20} - 1$ | 1,048,575 |

**EXAMPLE 12.18**   Texas Real Estate

*All Subsets*

Although the results of forward selection and stepwise regression procedures both came up with the same "best" model, the planners were not sure they wanted a one-variable model. They decided to use the all possible regression method. The Minitab output from the procedure is shown below. In this case, the output was the best two models for each size considered.

```
Best Subsets Regression

Response is 1994 Avg

                                    A P   M T
                                    v o l e o
                                    g p 9 a t
                                    . u 9 n a
                                      l 3   l
                                    H a / S
               R-Sq                 o t 9 A I
Vars  R-Sq    (adj)   C-p       S   u i 4 T n

   1  72.1    71.3    2.3     38118  X
   1  25.5    23.6   67.9     62238        X
   2  73.7    72.3    2.1     37488  X        X
   2  72.4    71.0    3.8     38357  X     X
   3  74.5    72.5    2.8     37351  X X      X
   3  74.3    72.2    3.2     37561  X      X X
   4  74.9    72.1    4.3     37575  X X    X X
   4  74.7    71.9    4.6     37742  X    X X X
   5  75.1    71.6    6.0     37958  X X X X X
```

The planners found the results of this model to be quite helpful. They saw that the best one-variable model was the same one found by the forward and stepwise procedures. By looking at the value of adjusted $R^2$, they saw that by going from a three-variable model to a four-variable model the value actually decreased from 72.5 to 72.1, indicating that adding the variable mean SAT scores provided no additional information. They also saw that there was a two-variable model that was the best on all criteria and that according to the $C_p$ index was better than the one-variable model. This model had both average household income and total instructional expenditure included. ■

One of the benefits of the all possible regressions procedure is that, for reasonably sized sets of input variables, it can be done with any software package that does regression analysis. The evaluation criteria $R^2$ and adjusted $R^2$ are included with normal regression output; the exception is $C_p$, which is often omitted but easy to calculate from the information available. The user simply needs to carefully list and run all of the combinations of independent variables.

 *TRY IT NOW!*

## Order Filling

### *All Possible Regressions*

The mail-order company is pretty sure that it has identified the best model, but it decides to use the all possible regressions method to make sure that it is not missing something. The output from the analysis is shown on page 632. *(continued)*

```
          Best Subsets Regression

          Response is Time
                                              N L E
                                              u o x
                                              m c p
                                              b a e
                                              e t r
                                              r i i
                          R-Sq                    o e
          Vars  R-Sq     (adj)    C-p      S   o n n

            1   47.3     46.1    93.0   1.8033   X
            1   25.2     23.4   149.3   2.1486 X
            2   83.8     83.0     2.2   1.0111 X X
            2   47.3     44.8    94.9   1.8239     X X
            3   83.9     82.7     4.0   1.0214 X X X
```

Is there a one-variable model that is best on all criteria? If so, what is it?

What is the best two-variable model?

Is there any benefit from moving to a three-variable model? Why or why not?

What model do you recommend the company use? Why?

---

The relative ease with which all possible regressions can be done has made the use of the forward, backward, and stepwise procedures less popular. These latter techniques are still well used to reduce an extremely large set of input variables down to a smaller set so that all subsets can be used.

## 12.5  CHECKING MODEL ADEQUACY

Just as in the simple linear model, some assumptions in the model must be checked to ensure that the model is appropriate. A model may appear to give good predictions and have a large value of $R^2$ and still not be a good multiple regression model. In this section, we discuss some of the problems that can occur in a multiple regression model.

---

ANS. YES, THE ONE WITH LOCATION, THE ONE WITH LOCATION AND NUMBER OF ITEMS; NO, ADJ. $R^2$ DECREASES; TWO VARIABLE, BEST USING ALL METHODS.

## 12.5.1 Residual Analysis

In the linear model some of the assumptions involved the error term in the model. Remember that the multiple regression model is given by

$$y = \beta_0 + \beta_1 x_1 + \beta_2 x_2 + \cdots + \beta_k x_k + \varepsilon$$

and that $\varepsilon$ is the error term. The basic assumptions about the error term $\varepsilon$ are

1. It has a mean value of zero ($\mu_\varepsilon = 0$).
2. For every value of $X$, the standard deviation, $\sigma$, of $\varepsilon$ is the same.
3. The distribution of $\varepsilon$ is normal.
4. The error terms for the different observations are not correlated with each other.

These assumptions also hold for the multiple regression model. In the simple linear model we used residual plots to tell us whether the assumptions were violated. These same plots help in the multiple regression model.

A plot of the residuals versus the predicted values gives information about whether the relationship is linear and whether the error terms have equal variances. If the plot of the residuals versus the $\hat{y}$ values shows a pattern, then the relationship between $Y$ and the set of predictor variables might be nonlinear. Similarly, if the amount of scatter in the residuals varies as the level of $Y$ varies, then the assumption of equal variances may be violated.

In addition to these plots, in multiple regression models it is also useful to plot the residuals against each of the independent variables to obtain additional information. If there are any problems indicated by the plot of the residuals versus the predicted values, you have information about a problem with the model as a whole. Plots against the individual variables can pinpoint whether the problem exists with all or just specific variables. If the problem exists with just one or two variables, the analyst can look at alternative models that do not contain these variables. Although the residual plots can be done in any package that creates scatter plots, Excel and Minitab include the plots as part of the output options in their regression routines.

To decide whether the assumption that the residuals are normally distributed is valid, you can use a normal probability plot or a histogram, just as you did in the simple linear model.

### EXAMPLE 12.19   Texas Real Estate

*Residual Plots*

The real estate planners have decided to use the two-variable model identified by all possible regressions. They decide to plot the residuals from the model against each of the independent variables. The plots are shown below:

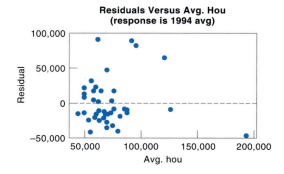

Both plots exhibit more variation on the high side than on the low side. In the plot against the total instructional expenditures it appears that the variance of the residuals increases as the variable increases. In the second plot, against average household income, the problem is not as severe, although it is more difficult to evaluate because there are some isolated values at the high end of the variable range.

A normal plot of the residuals was also done:

**Normal Probability Plot of the Residuals**
**(response is 1994 avg)**

From the plot it appears that the normality assumption is not valid because the points do not fall in a straight line.

From the diagnostic plots, it appears that the "best" model they have found violates some of the regression model assumptions. ■

At this point you might be thinking that you would be really upset if, after doing all that work and finding the "best" model, there is a problem with it. You might wonder why you didn't check these things before all that work. In fact, modeling is not a sequential procedure. It is an iterative procedure from which a solution gradually evolves. Checking earlier would involve a lot more checks than might be necessary with the final, candidate model.

 *TRY IT NOW!*

## Order Filling

### All Possible Regressions

Now that it has a model it likes, the mail-order company would like to make sure that the model does not violate any assumptions of the multiple regression model. The company creates a set of residual plots, which are shown below:

Look at the plot of residuals versus the locations. Does it appear that there is a problem with the assumption of equal variances?

Look at the plot of residuals versus the number of items. Does it appear that there is a problem with the assumption of equal variances?

The company also created a normal probability plot of the residuals:

**Normal Probability Plot**

From this plot, what can you say about the assumption of normality?

## 12.5.2 Values With Large Influence

Another problem that we encountered in the simple linear model was the problem of data points that exerted a large influence on the model. This could happen when the value of the dependent variable did not agree with the model or when an observation had an $X$ value that was distant from the other data points. Since there was only one independent variable, we could use plots such as scatter plots and boxplots to help identify these observations.

The problem of observations exerting a large influence can also happen in the multiple regression model, but it is more difficult to detect. This is because the influence might come from a combination of values of the independent variables and we lose the element of visualization that we had in the one-variable model. In multiple regression it is necessary to rely on other measures to identify observations with large influence. As we saw with the linear model, Minitab as part of the output generates a warning about observations that exert a large influence on the model either because of a large residual or because of $X$ values that are significantly larger.

One method that has emerged as a way to detect influential values uses a statistic called the **Cook's distance, D.**

> The **Cook's distance** method compares the values of the regression coefficients with all observations to the values when the $i$th observation is removed from the model.

If the $i$th observation is influencing the model then the difference in coefficients between the two models will be large and the corresponding value of the Cook's distance is also large. A definition of large that is suggested, is to compare $D$ to an $F$ statistic of a tail area of 0.50 and with $k + 1$ degrees of freedom in the numerator and $n - k - 1$ in the denominator, $F_{0.5,k+1,n-k-1}$. A rougher rule of thumb is to compare the Cook's distance to 1. If, in either case, $D$ exceeds the critical value, then the observation in question should be removed from the model. Outputting Cook's distance is an option of the regression routine in many statistical packages.

## EXAMPLE 12.20   Texas Real Estate

### Residual Plots

Looking at their data, the real estate planners wonder if one of the locations might be exerting a large influence on the model. In particular, they remember that when they made a boxplot of average home values, Highland Park was an outlier. They decide to look at the regression model again and to calculate Cook's distance for their model and see if they should eliminate those data. They specify for Minitab to calculate Cook's distances and also to output warnings about the data. The warning output is shown below:

```
Unusual Observations
Obs   Avg. Hou   1994 Avg      Fit   StDev Fit   Residual    St Resid
  1      49803     204800    108753      7458       96047        2.61R
 21     196697     361100    410877     28165      -49777       -2.01RX
 25      42483      71800    117684     20993      -45884       -1.48 X
 38      84007     269926    175521      7392       94405        2.57R
 39      87854     269926    181703      7868       88223        2.41R
 40     115368     324990    253062     15077       71928        2.10R
R denotes an observation with a large standardized residual
X denotes an observation whose X value gives it large influence
```

The planners notice that observation 21 has both a large residual and a large $X$ value influence. They check and find that it is indeed Highland Park. Minitab outputs Cook's distance for Highland Park to be 1.749 and compares it to $F_{0.5,3,38} = 0.803$. Since Cook's distance is so much larger than the $F$ value, the planners will have to think about redoing the analysis without that bit of data. None of the other Cook's distances exceed the critical value. ∎

## 12.5.3 Multicollinearity

A problem that occurs in the multiple regression model that is not a factor in the simple linear model is known as **multicollinearity.**

> A multiple regression model has **multicollinearity** when variables in the set of independent variables are correlated with each other.

It is not at all unusual to find that some of the variables in the set of independent variables are related to each other. This is especially likely to happen when the data used for a model come from existing sources. When data come from experimentation it is possible to control elements that might cause multicollinearity. When data come from existing sources, this is not easy.

Suppose you were trying to find a model to predict the amount of money that a family spends on food, and that in your set of independent variables you had family income, house value, number of children, family size, and level of education of the head of the household. It would seem almost certain that some of these variables would be related to each other. It is also possible that variables in the set are related to other variables that are not included in the model. When this happens, the results of a multiple regression analysis may be affected.

The main effect of multicollinearity on the model is instability of the regression coefficients. When two variables in the set of independent variables are highly correlated, then the coefficients of the model are not stable. This means that when you remove one of the variables and rerun the model the coefficients will change—sometimes even changing sign from negative to positive! If this is the case, then interpretation of the coefficients is not meaningful. We know that the coefficient in the model represents the change in the dependent variable when the variable of interest is changed *and all other variables are held constant.* If two variables are highly correlated then it may not be possible to vary one variable without also having the other variables change too.

Another problem that occurs when multicollinearity is present in a model is that the individual coefficients have large standard errors and therefore small $t$ statistics. It is possible that a model might test significant as a whole, but that no individual coefficient will be significantly different from zero. This is certainly a contradiction and an indication that something is wrong!

There are several complex statistical procedures to find out whether a model has multicollinearity, but some simple procedures work well too. One method for detecting multicollinearity is to calculate the correlations between every pair of independent variables and look for large values near 1 or $-1$. If it appears that some variables are highly correlated, then the analyst may wish to concentrate on models that do not include both variables. Some statistical packages, such as Minitab, will warn the user when variables are highly correlated and in cases where the problem is severe, will refuse to include a variable in the model.

### EXAMPLE 12.21  Texas Real Estate

#### Checking for Multicollinearity

The planners doing the real estate model know that their data came from existing sources and so might exhibit multicollinearity. They decide to run correlations for the set of five variables that they are considering.

```
Correlations (Pearson)

         Avg. Hou Populati  1993/94 Mean SAT
Populati -0.477
1993/94  -0.441    0.436
Mean SAT  0.534   -0.400   -0.208
Total In  0.431    0.073   -0.163    0.138
```

From the correlation matrix the planners see that the highest correlation is 0.534, which is not high at all. The correlation between the two variables used in the model is 0.434. Hence, they know that they do not have to worry about multicollinearity. ∎

A way to examine the effects of the multicollinearity in a model is to run the model with and without one of the variables. If the coefficients change to an unacceptable degree, for example, a complete change in sign from positive to negative, then another analysis method should be used. One method that produces coefficients that are not affected by multicollinearity is called ridge regression.

*TRY IT NOW!*

### Order Filling
*Checking for Multicollinearity*

The mail-order company wants to check to see if the model it is thinking about using has any problems with multicollinearity. It calculates the correlation between the two variables that are in the final model and finds that the correlation is $-0.141$.

Based on this information do you think that multicollinearity is a problem in their model? Why or why not?

Although multicollinearity causes big problems with the regression coefficients, the overall ability of the model to predict values of the dependent variable is not affected. If the users of the model are not looking at the actual equation but only at the predictors, the model may still be useful.

### 12.5.4 Using the Model

We have discussed some of the problems that can occur in a multiple regression model. Although none of the problems make the model unusable, we must understand the impact on the decision-making process. The biggest problem in terms of the usefulness of the model is instability of the regression coefficients. When the coefficients are unstable, then they really do not represent the change in the dependent variable that results from a corresponding change in one of the independent variables. If you try to use the model to bring about such a change you may find that the result does not even resemble what you expect.

Problems with coefficient instability or error result from multicollinearity, overspecified models, and influential observations. Unless the multicollinearity is severe (variables correlated with a correlation coefficient of 0.9999 and above), instability is not much of a problem.

Overspecification of a model occurs when the "more is better" approach is taken to modeling. If you add enough variables to a model you can get an almost perfect fit to the set of input data you are using. The problem is that the perfect fit is just for that particular set of data. If you vary the data set at all you will find that the model varies wildly, which makes it almost useless for "what if" analysis and scenario management. Overspecification can be avoided by using one of the techniques designed to find the "best" regression model.

Observations that exert a large influence on the model also cause the coefficients of the model to be inaccurate. Often dropping the observations that are erroneous will solve the problem.

It is useful to check the stability of the model you find. One simple way to do this is to run the regression with the selected independent variables on subsets of the original data. If you have a large enough data set then you could divide it into two randomly chosen subsets and run the model on both subsets. If the resulting models are not similar, then the model may have problems with stability. If you do not have a large enough data set to run two separate analyses then you can choose two or more random samples of a fixed size from the set of data and run the regression on each subset. Again, if the models do not agree, then there is a problem that you must be aware of before using the model to make decisions.

## 12.6 MULTIPLE REGRESSION MODELS IN EXCEL

The tools used in Excel for multiple regression models are the same ones that are used for the simple linear model. The only difference is that the data range for the $X$ variables will cover more than one column. It is *very* important, however, that the $X$ variable columns be adjacent to each other. If they are not, you will have to rearrange the worksheet before you do the analysis.

The output section includes information about each of the regression coefficients. If you choose the residual plots option, you will get a plot of residuals versus each independent variable. This is also true for the line fit plots. These plots allow you to look at the assumptions of the multiple regression model and decide whether it is appropriate.

Excel does not have any tools for finding the "best" regression model. These are sophisticated statistical tools, and Excel is not really a statistics software package. If you want to find the best model using Excel, you will have to use the All Possible Regressions approach.

# CHAPTER 12 SUMMARY

Finding a multiple regression model is easy. Finding a multiple regression model that is useful for the purposes of decision making is really an art. Modeling is an iterative process for which there is no single correct answer. The answer you choose depends on what you know and understand about the problem to be solved and how you intend to use the model.

There are several steps to the modeling process: identifying potential independent variables, collecting data, and finding a potential model are the beginning steps. Once a potential model is found, the process of *model building* takes place to find a "best" model. The objective of this process is to find a model that does an acceptable job of explaining or predicting the dependent variable with as few independent variables as possible. Some of the model-building techniques used are *Forward Selection, Backward Elimination, Stepwise Regression,* and *All Possible Regressions.* Finally, once the "best" model is identified, it must be checked for problems such as *violation of assumptions, influential observations,* and *multicollinearity* before it can be used for decision-making purposes.

## Key Terms

| Term | Definition | Page Reference |
|---|---|---|
| Adjusted $R^2$ | The **adjusted $R^2$** is the value of the coefficient of multiple determination adjusted to reflect the number of variables in the model. | 611 |
| Coefficient of multiple determination $R^2$ | The **coefficient of multiple determination, $R^2$,** is a measure of the percentage of the variation in the dependent variable, $Y$, that can be accounted for by the complete set of independent variables, $X_1, X_2, \ldots, X_k$, in the model. | 610 |
| Cook's distance | The **Cook's distance** method compares the values of the regression coefficients with all observations to the values when the $i$th observation is removed from the model. | 636 |
| Input variable | The set of independent variables, $X_1, X_2, \ldots, X_k$, are referred to as the **input variables.** | 594 |
| Model-building techniques | **Model-building techniques** are methods used for identifying the best multiple regression model from a set of independent variables. These methods include forward selection, backward elimination, stepwise regression, and all possible regressions. | 622 |
| Multicollinearity | A multiple regression model has **multicollinearity** when variables in the set of independent variables are correlated with each other. | 636 |

| Term | Definition | Page Reference |
|---|---|---|
| Multiple regression model | The true relationship between the independent variable $Y$ and the set of independent variables, $X_1, X_2, \ldots, X_k$, the **multiple regression model,** can be described by the equation: $$y = \beta_0 + \beta_1 x_1 + \beta_2 x_2 + \cdots + \beta_k x_k + \varepsilon$$ | 594 |
| Output variable | The dependent variable, $Y$, is often referred to as the **output variable.** | 594 |

## Key Formulas

| Term | Formula | Page Reference |
|---|---|---|
| F test statistic | $F = \dfrac{MSR}{MSE}$ | 606 |
| F critical value | $F_{\alpha,k,n-k-1}$ | 607 |
| $R^2$ | $R^2 = \dfrac{SSR}{SST} 100\%$ | 610 |
| t test statistic | $t = \dfrac{b_i}{s_{b_i}}$ | 613 |

# CHAPTER 12 EXERCISES

## Learning It

12.11 The analysts looking at the model for equity ($/share) at Phillips Petroleum decided that the model they had was not adequate for their needs. They decide to try the model with a different set of independent variables: debt ($ billions) and number of employees (thousands). The data and output of the regression analysis are given below:

| Year | Equity | Debt | Employees | Year | Equity | Debt | Employees |
|---|---|---|---|---|---|---|---|
| 1986 | 7.55 | 6.2 | 21.8 | 1992 | 10.37 | 3.8 | 21.4 |
| 1987 | 7.08 | 5.8 | 22.5 | 1993 | 10.28 | 3.2 | 19.4 |
| 1988 | 8.69 | 4.9 | 21.0 | 1994 | 11.29 | 3.1 | 18.4 |
| 1989 | 8.74 | 4.0 | 21.8 | 1995 | 12.16 | 3.1 | 17.4 |
| 1990 | 10.51 | 3.9 | 22.4 | 1996 | 16.15 | 3.0 | 17.2 |
| 1991 | 10.61 | 4.0 | 22.7 | | | | |

SUMMARY OUTPUT

| Regression Statistics | |
|---|---|
| Multiple R | 0.848198989 |
| R Square | 0.719441525 |
| Adjusted R Square | 0.649301907 |
| Standard Error | 1.471736507 |
| Observations | 11 |

ANOVA

| | df | SS | MS | F | Significance F |
|---|---|---|---|---|---|
| Regression | 2 | 44.43469688 | 22.21734844 | 10.2572774 | 0.006195745 |
| Residual | 8 | 17.32806676 | 2.166008345 | | |
| Total | 10 | 61.76276364 | | | |

| | Coefficients | Standard Error | tStat | P-value | Lower 95% |
|---|---|---|---|---|---|
| Intercept | 25.01669053 | 4.991469246 | 5.011889145 | 0.001037315 | 13.50633437 |
| Debt | −1.187745989 | 0.571806762 | −2.077180733 | 0.07142751 | −2.506335599 |
| Employees | −0.47922578 | 0.303483518 | −1.579083382 | 0.152970523 | −1.179060481 |

(a) Write down the equation of the regression model and interpret the coefficients.

(b) Use the regression equation to predict shareholders' equity for each year from 1986 through 1996.

(c) Calculate the residuals and compare the actual data to the predicted values. Which year has the largest residual? Which has the smallest?

(d) Do you think that the model does a good job of predicting shareholders' equity? Why or why not?

(e) Set up the hypotheses to test the significance of the overall model.

(f) At the 0.05 level of significance, what is your conclusion about the model?

(g) What is the value of $R^2$ for this model? What does it mean?

(h) Compare this model to the model using revenue and assets developed in the chapter. Which model do you think is better? Why?

*SALES.XXX*

**12.12** Analysts at a company that produces small appliances are looking at sales of 24 food preparation products in a medium-sized city in the Midwest. They have noticed that sales in this city have not been meeting forecast values for several months and want to look at the problem in more detail. They have collected data on monthly sales ($), advertising expenditure ($), number of competing products available, number of discount opportunities (sales, coupons, etc.) offered during the month, and the warranty period of the item. A portion of the data is shown below:

| Sales ($) | Advertising ($) | Number of Competitors | Discounts | Warranty (years) |
|---|---|---|---|---|
| 4565 | 459 | 1 | 1 | 2.00 |
| 4896 | 545 | 0 | 0 | 0.25 |
| 4480 | 472 | 2 | 2 | 1.00 |
| 4300 | 482 | 3 | 3 | 2.00 |
| 3502 | 435 | 3 | 3 | 0.25 |
| 4413 | 499 | 3 | 3 | 1.00 |
| 5868 | 604 | 0 | 0 | 1.00 |
| 4527 | 501 | 1 | 1 | 1.00 |
| 3849 | 370 | 3 | 3 | 1.00 |

(a) Consider the problem that needs to be solved. What do you think is the dependent variable in this problem? Which are the independent variables?

(b) As advertising expenditures increase, what do you think the effect on sales will be? Will they increase or decrease? How do you think that each of the other independent variables will affect monthly sales?

(c) Use a computer package to find the multiple regression model you identified in part (a).

(d) Write down the regression equation and interpret the coefficients in the model. How did the actual coefficients agree with your expectation of their effect?

(e) Use the software package to calculate the predicted sales for each of the products in the data set and compare them to the actual sales.

(f) Make a plot of predicted versus actual sales.

(g) How well do you think the model predicts monthly sales?

(h) Set up the hypotheses and test the significance of the model using $\alpha = 0.05$.

(i) What is the value of $R^2$ for this model and what does it mean?

*NATLPARK.XXX*

**12.13** The Department of the Interior is looking at state parks in the United States. It is wondering what variables might affect revenues and has assembled data for each state consisting of the number of acres, number of day visitors, number of overnight visitors, and operating expenses for 1992. A portion of the data is shown below:

| State | Acres (1000) | Day Visitors (1000) | Overnight Visitors (1000) | Revenues ($1000) | Operating Expenses ($1000) |
|---|---|---|---|---|---|
| Alabama | 50 | 5,174 | 1094 | 23,427 | 28,778 |
| Alaska | 3169 | 5,613 | 708 | 1,063 | 5,345 |
| Arizona | 33 | 1,250 | 500 | 3,734 | 10,400 |

(*continued*)

| State | Acres (1000) | Day Visitors (1000) | Overnight Visitors (1000) | Revenues ($1000) | Operating Expenses ($1000) |
|---|---|---|---|---|---|
| Arkansas | 47 | 6,063 | 662 | 12,046 | 21,503 |
| California | 1323 | 61,347 | 5954 | 73,563 | 149,223 |
| Colorado | 340 | 8,947 | 575 | 13,373 | 12,266 |
| Connecticut | 173 | 6,730 | 323 | 3,498 | 7,300 |
| Delaware | 13 | 3,336 | 283 | 4,176 | 9,712 |

SOURCE: *Statistical Abstract of the United States 1996.*

(a) Find a multiple regression model that predicts revenue as a function of the number of acres of park land, the number of day visitors, and operating expenses.

(b) Interpret the coefficients of the model.

(c) Use the model to predict the revenues for each state.

(d) Calculate the residuals and compare them to the actual data.

(e) Which state has the largest residual? Which state has the smallest?

(f) What is the value of $R^2$ for this model? Do you think that the model does a good job of predicting revenues for state parks? Why or why not?

12.14 A national chain of women's clothing stores with locations in large shopping malls   *MALL.XXX*
thinks that it can do a better job of planning store renovations and expansions if it understands what variables impact sales. It plans a small pilot study on stores in 24 different mall locations. The data it collects consist of monthly sales, store size (sq ft), number of linear feet of window display, number of competitors located in mall, size of the mall (sq ft), and distance to nearest competitor (ft). A sample of the data is shown below:

| Store | Sales | Size | Windows | Competitors | Mall Size | Nearest Competitor |
|---|---|---|---|---|---|---|
| 1 | 4453 | 3860 | 39 | 12 | 943,700 | 227 |
| 2 | 4770 | 4150 | 41 | 15 | 532,500 | 142 |
| 3 | 4821 | 3880 | 39 | 15 | 390,500 | 263 |
| 4 | 4912 | 4000 | 39 | 13 | 545,500 | 219 |
| 5 | 4774 | 4140 | 40 | 10 | 329,600 | 232 |
| 6 | 4638 | 4370 | 48 | 14 | 802,600 | 257 |

(a) What is the dependent variable in this model? Which variables are the independent variables?

(b) How do you think that the size of a store will impact sales? How do you think the other independent variables will impact sales?

(c) Use a computer software package to find a multiple regression model for the data.

(d) Interpret the values of the coefficients in the model. Do they agree with how you expected them to behave?

(e) Set up the hypotheses to test whether the model as a whole is significant. At the 0.05 level of significance, what is your conclusion?

(f) Use the model to predict monthly sales for each of the stores in the study. Calculate the residuals.

(g) Which store had the largest residual? Which had the smallest?

(h) Plot the residuals versus the actual values. Do you think that the model does a good job of predicting monthly sales? Why or why not?

## Thinking About It!

12.15 The financial analysts looking at a model for shareholders' equity at Phillips Petroleum   *Requires Example 12.2,*
have obtained similar data from Mobil Oil. They would like to construct models for Mobil   *Exercise 12.11*
similar to the two that they have already constructed for Phillips. The data are not avail-   *MOBIL.XXX*
able for all of the years of interest. The data for Mobil are shown at the top of page 644.

| Year | Assets ($ million) | Revenue ($ million) | Shareholder Equity ($/share) | Total Debt ($ million) | Employees (1000) |
|------|------|------|------|------|------|
| 1986 | NA | 44,936 | 32.86 | NA | 70.9 |
| 1987 | NA | 51,678 | 36.46 | NA | 68.1 |
| 1988 | NA | 54,740 | 38.19 | NA | 69.4 |
| 1989 | NA | 56,388 | 39.84 | NA | 67.9 |
| 1990 | NA | 64,774 | 42.44 | NA | 67.3 |
| 1991 | 25,464 | 63,311 | 43.74 | 8229 | 67.5 |
| 1992 | 25,075 | 64,456 | 41.06 | 8520 | 63.7 |
| 1993 | 25,037 | 63,975 | 42.74 | 8027 | 61.9 |
| 1994 | 25,503 | 67,383 | 42.61 | 7727 | 58.5 |
| 1995 | 24,850 | 75,370 | 44.71 | 6756 | 50.4 |
| 1996 | 27,479 | 81,503 | 47.62 | 6581 | 43.0 |

SOURCE: *Mobil Oil Annual Report,* World Wide Website: www.mobiloil.com.

(a) Using the data that are available, create a model for Mobil that relates shareholder equity to assets and revenue.

(b) In what ways might you expect the model for Mobil to be similar to the one for Phillips? In what ways might they be different?

(c) Compare the model for Mobil to the one found in Example 12.2 for Phillips Petroleum. Do the models agree with your answer to part (b)?

(d) Find a model for Mobil that relates shareholder equity to total debt and number of employees.

(e) Compare the model for Mobil to the one for Phillips from Exercise 12.11. How are they similar? How are they different?

(f) Which of the two models for Mobil do you think does a better job of predicting shareholder equity?

*MILES.XXX*

**12.16** Automobile salespeople want to understand the amount of driving done by a household so that they can plan advertising campaigns. They propose to use data from a survey sponsored by the Department of Transportation to help them understand what factors affect the annual number of miles driven by a household.

(a) What variables do you think might influence the number of miles driven annually by a household? Why did you choose these variables?

As part of the study, data were extracted from a data base containing the survey information. The data used consisted of 85 households with two or more people for a small city in the Northeast. The variables in the study were total miles driven in the previous 12 months, number of drivers in the household, number of vehicles in the household, number of children ages 5 to 19 in the household, total number of trips taken in the 24-hour period prior to the survey, and total number of miles driven in the same 24-hour period.

(b) Do you see any problems with these data? If so, what are they? Suggest any additional variables that you think would help to eliminate or clarify the problems.

(c) A portion of the data available from the Department of Transportation is shown below:

| HHVMILES | DRVRCNT | VEHCOUNT | NUM_KIDS | DTVCNT_H | DTVMILH |
|------|------|------|------|------|------|
| 0 | 2 | 0 | 2 | 3 | 3 |
| 20,975 | 2 | 2 | 1 | 6 | 35 |
| 23,000 | 3 | 2 | 0 | 7 | 99 |
| 36,000 | 1 | 1 | 0 | 2 | 70 |
| 6,327 | 2 | 2 | 2 | 2 | 10 |
| 27,423 | 2 | 2 | 0 | 5 | 58 |
| 7,058 | 1 | 2 | 0 | 0 | 0 |
| 24,365 | 1 | 3 | 1 | 4 | 10 |
| 101,313 | 1 | 2 | 1 | 4 | 13 |
| 2,519 | 2 | 2 | 0 | 2 | 6 |

The 6 variables are defined as follows:

| HHVMILES | *total annual miles for all vehicles in household* |
| DRVRCNT | *number of drivers in household* |
| VEHCOUNT | *number of vehicles in household* |
| NUM_KIDS | *number of children in household age 5 through 21* |
| DTVCNT_H | *number of travel day vehicle trips for the household* |
| DTVMILH | *travel day vehicle miles for household* |

The travel day is the 24-hour period prior to the survey.

(d) Use a computer software package to find the multiple regression model that predicts total annual vehicle miles using the other five variables as the set of independent variables.

(e) Write down the model and interpret the coefficients. Do they make sense?

(f) At the 0.05 level of significance what can you say about the model?

(g) What is the value of $R^2$ for this model? Do you think that the model will be useful for predicting total annual vehicle miles for a household? Why or why not?

**12.17** The planners looking at store characteristics and sales want to know more about the model that they have found.    *Requires Exercise 12.14*

(a) Find and interpret the value of $R^2$ for this model.

(b) Do you think that this model will be useful in helping the planners? Why or why not?

(c) Set up the hypotheses to test the individual regression coefficients. At the 0.05 level of significance, what are your conclusions?

(d) If you were going to drop just one variable from the model, which one would you choose? Why?

**12.18** The analysts looking at financial data from Phillips Petroleum are not satisfied with either of the two models that they have developed. They want to combine the two models and determine if there is a better model available.    *Requires Exercise 12.11 PHILLIPS.XXX*

(a) Based on the results of the previous models, which two variables do you think will be most helpful in predicting shareholder equity? Why?

(b) Use the all possible regressions method to find the best one-, two-, three-, and four-variable models.

(c) From the list of best models for each size, pick the model that you think should be used to predict shareholder equity.

(d) Write a memo to the Chief Financial Officer of the company explaining your model and justifying your choice.

**12.19** Look at the output from the analysis of the monthly sales data for the small appliances manufacturer. There were four variables in the original analysis, but the company wants to know if they were all important in the model.    *Requires Exercise 12.12*

(a) Set up the hypotheses to test whether the individual coefficients are different from zero and perform the tests at the 0.05 level of significance.

(b) What are your conclusions from this test?

(c) Consider the variables that have coefficients that are zero. Which of these would you recommend dropping from the model? Why?

(d) Rerun the analysis, dropping the variable you identified in part (c). How does this change the model as a whole?

(e) Does dropping this variable change the contribution of the other variables? If so, how?

(f) Would you consider dropping another variable? If so, which one?

(g) If you answered yes to part (f), then rerun the analysis dropping that variable.

(h) Based on all of your analyses, what model do you think the company should use?

(i) Does this approach to finding a model resemble any of the model building techniques you learned about? If so, which one?

**12.20** The store planners for the women's clothing chain want to find the best model that they can for understanding what store characteristics impact monthly sales.    *Requires Exercises 12.14, 12.17*

(a) Use stepwise regression or all possible regressions to find the best model for the data.

(b) Analyze the model you have identified to determine whether it has any problems.

(c) Write a memo reporting your findings to your boss. Identify the strengths and weaknesses of the model you have chosen.

*Requires Exercise 12.12, 12.19*

**12.21** Look again at the data on monthly sales of small appliances.

(a) Use stepwise regression or all possible regressions to find the best model from the set of four variables.

(b) Compare the results of the formal procedure to the model you chose in Exercise 12.19. Are the two models the same? Would you expect them to be?

(c) Consider the two "best" models you have. If they are different, which one would you choose as your model? Why?

*Requires Exercise 12.13*

**12.22** Look at the results of the analysis on the State Park data collected by the Department of the Interior.

(a) Set up the hypotheses for testing individual regression coefficients. At the 0.01 level of significance, what are your conclusions?

(b) Drop any of the variables whose coefficients tested to be zero from the model and rerun the analysis.

(c) How does the model change?

(d) Are you satisfied with this model or do you think that additional analysis needs to be done to find the best model?

(e) Perform any additional analyses you think are necessary to find the best model for predicting revenues.

*Requires Exercise 12.16*

**12.23** Look again at the model for predicting the total annual miles driven by a household.

(a) At the 0.05 level of significance, what can you say about the individual regression coefficients?

(b) Which variables would you consider dropping from the model? Why?

(c) Consider what the model is trying to predict. Do you think that households with no vehicles should be included in these data? Why or why not?

(d) Identify the observations that have no vehicles and drop them from the data. Rerun the analysis. Does this change anything?

(e) Again, think about what the model is trying to predict. What other data values do not represent the population of interest? Why?

(f) Drop any observations that correspond to your answer to part (e) and rerun the model. How do the results compare with the other two models? Do you feel more confident with this model? Why or why not?

*Requires Exercises 12.12, 12.19, 12.21*

**12.24** Look at the model that you decided on for the monthly small appliance sales.

(a) Prepare a set of residual plots for the model. Does the model appear to violate any of the assumptions of the multiple regression model? If so, what are the problems?

(b) Calculate the correlations between each pair of independent variables. Does there appear to be multicollinearity in the model? If so, what would you recommend doing?

(c) Divide the data set in half using some random selection process and run the model on each subset. Compare the models. Do the coefficients of the model appear stable? Why or why not?

(d) Use either Cook's distance or some other measure available in your software to determine whether any of the observations are exerting a strong influence on the model.

(e) Drop these observations from the data and rerun the analysis. How does this change the model?

(f) Prepare a memo describing your findings and suggesting a model to use for predicting revenue. Be sure to include any limitations of the model you suggest.

**12.25** Consider the model that you feel is currently best for predicting total annual miles for a household:

*Requires Exercises 12.16, 12.23*

(a) Prepare a set of residual plots for the model. Does the model appear to violate any of the assumptions of the multiple regression model? If so, what are the problems?

(b) Calculate the correlations between each pair of independent variables. Does there appear to be multicollinearity in the model? If so, what would you recommend doing?

(c) Use either Cook's distance or some other measure available in your software to determine whether any of the observations are exerting a strong influence on the model.

(d) Drop these observations from the data and rerun the analysis. How does this change the model?

(e) Prepare a memo describing your findings and suggesting a model to use for predicting revenue. Be sure to include any limitations of the model you suggest.

## Doing It!

**12.26** As we saw in the example about Texas Real Estate, there are many factors that might affect home prices. A group looking at the city of Boston collected data on a number of variables believed to be predictors of home prices. The variables are

| | |
|---|---|
| CRIM | per capita crime rate by town |
| ZN | proportion of residential land zoned for lots over 25,000 sq ft |
| INDUS | proportion of nonretail business acres per town |
| CHAS | Charles River dummy variable ($=1$ if tract bounds river, 0 otherwise) |
| NOX | nitric oxides concentrations (parts per 10 million) |
| RM | average number of rooms per dwelling |
| AGE | proportion of owner-occupied units built prior to 1940 |
| DIS | weighted distances to five Boston employment centers |
| RAD | index of accessibility to radial highways |
| TAX | full-value property-tax rate per $10,000 |
| PTRATIO | pupil–teacher ratio by town |
| LSTAT | % lower status of population |
| MEDV | median value of owner-occupied homes in $1000s |

| CRIM | ZN | INDUS | CHAS | NOX | RM | AGE | DIS | RAD | TAX | PTRATIO | LSTAT | MEDV |
|---|---|---|---|---|---|---|---|---|---|---|---|---|
| 0.00632 | 18.0 | 2.31 | 0 | 0.538 | 6.575 | 65.2 | 4.0900 | 1 | 296 | 15.3 | 4.98 | 24.0 |
| 0.02731 | 0.0 | 7.07 | 0 | 0.469 | 6.421 | 78.9 | 4.9671 | 2 | 242 | 17.8 | 9.14 | 21.6 |
| 0.02729 | 0.0 | 7.07 | 0 | 0.469 | 7.185 | 61.1 | 4.9671 | 2 | 242 | 17.8 | 4.03 | 34.7 |
| 0.03237 | 0.0 | 2.18 | 0 | 0.458 | 6.998 | 45.8 | 6.0622 | 3 | 222 | 18.7 | 2.94 | 33.4 |
| 0.06905 | 0.0 | 2.18 | 0 | 0.458 | 7.147 | 54.2 | 6.0622 | 3 | 222 | 18.7 | 5.33 | 36.2 |
| 0.02985 | 0.0 | 2.18 | 0 | 0.458 | 6.430 | 58.7 | 6.0622 | 3 | 222 | 18.7 | 5.21 | 28.7 |
| 0.08829 | 12.5 | 7.87 | 0 | 0.524 | 6.012 | 66.6 | 5.5605 | 5 | 311 | 15.2 | 12.43 | 22.9 |

(a) Calculate the correlation coefficients of each variable with the median value of owner-occupied homes (MEDV).

(b) Select a set of no more than 5 variables to use as the set of independent variables.

(c) Using the variables you selected, find the best model for predicting the median value of homes.

(d) Test to see whether your model is appropriate.

(e) Summarize your findings in a memo to the group.

# CHAPTER *13*

# TIME SERIES AND FORECASTING

## TAKE ME OUT TO
## THE BALL GAME

American Pastime, Inc. is a company that provides high-quality entertainment to communities. With a clear focus on families, American Pastime plans to offer people of all ages an affordable opportunity to view exciting Double A baseball in a clean and safe environment.

To plan for this facility, American Pastime must forecast ticket sales. This information will help the company determine the number of ticket windows and determine the ordering/inventory plan for the food and souvenir items to be sold in such a way that costs are minimized.

The facility will consist of a state-of-the-art 200,000-square-foot stadium. Seating for approximately 6000 fixed seats for baseball attendees will be available: 4200 chair reserved seats and 1800 bench with back general admission seats. There will be 61 home games.

The population figures for the past 10 years for the area are available to you. The data are divided into two groups: the number of people in families with children (FWC) and the number of people in the age group from 20 to 40 with no children (NC). These two groups are your target population. Once you have the forecast for the two population groups, you must predict the number of people who will attend each game. Past experience shows that the attendance rate is 1.91 per 1000 for the FWC group and 1.75 per 1000 for the NC group.

## 13.1 CHAPTER OBJECTIVES

The problem of forecasting ticket sales described above is similar to many business problems that require the prediction of some variable. For example, you may wish to predict sales of a new or existing product, the closing price of a stock, the yield of an agricultural crop, the attendance at a meeting, or the donations to a nonprofit organization.

Clearly, these variables depend on many other factors. Let's consider the daily sales of loaves of bread sold at your local convenience store. The manufacturer would like to know precisely how many loaves of bread will be sold each day. Then the manufacturer could supply exactly that number and know that it will not lose any customers due to a stock out and it will not be left with loaves of bread at the end of the day that cannot be sold. But the number of loaves of bread sold per day probably depends on a number of factors, such as the number of loaves sold the previous day, the weather, the traffic, and others things. We could try to build a regression model treating the number of loaves of bread sold as the dependent variable and the others as independent variables. However, in many instances such as this one there are few data available on the independent variables. Hence, it is often impossible to build a *causal* model.

In these cases, the only alternative is to use a time series forecasting model. The main idea behind any time series forecasting model is that we can predict future values by studying the behavior of the past values. There is not a sense of cause and effect in these models but simply an attempt to *see patterns, repetitive or systematic behavior, and/or trends in the data.* The following simple example may help you understand what this means.

Suppose you observe the time that Mr. Russell goes to work every morning for 10 mornings. The data are shown below:

| Monday | 8:02 AM | Monday | 7:59 AM |
|--------|---------|--------|---------|
| Tuesday | 9:15 AM | Tuesday | 9:20 AM |
| Wednesday | 7:55 AM | Wednesday | 8:10 AM |
| Thursday | 9:08 AM | Thursday | 9:09 AM |
| Friday | 8:06 AM | Friday | 8:05 AM |

A quick look at these times would tell you that on Mondays, Wednesdays, and Fridays, Mr. Russell leaves somewhere around 8 AM, whereas on Tuesdays and Thursdays he leaves shortly after 9 AM. We could use this pattern to predict his departure times for the next few days. We are banking on the fact that Mr. Russell, like most of us, is a creature of habit and therefore the future looks much like the past in terms of general patterns. We do not know what is causing this pattern and we cannot know the cause by simply examining these data. In a time series method, we do not need to know what is causing the difference in his departure times on MWF versus TuTh.

Typically, when you are looking at time series data, you have several observations of the variable of interest (the dependent variable) observed over time. So, in effect, time becomes the independent variable. The graph shown in Figure 13.1 displays the number of loaves of bread sold at a typical convenience store for the last 25 days.

*Remember: The independent variable is displayed on the x axis and the dependent variable on the y axis.*

This scatter plot depicts the number of loaves sold along the y axis and the number of the day along the x axis. It is the same scatter plot technique that you learned in Chapter 3 except now the data points are connected. The days are arbitrarily numbered beginning with day 1. To predict the number of loaves of bread to be sold on day 26 we should exploit the information in the scatter plot. This means we should try to identify any patterns that exist in the sales over time.

This small example illustrates the nature of **time series** data.

**FIGURE 13.1** Sample scatter plot of loaves sold by time

A *time series* is a set of observations of a variable at regular time intervals, such as yearly, monthly, weekly, daily, etc.

There are many different ways of extracting the information in a time series in order to predict the value of the variable in the next time period or beyond. In this chapter you will learn how to analyze time series data using four different methods: Simple Moving Average, Weighted Moving Average, Exponential Smoothing, and Regression. These are the most commonly used techniques. The moving average and exponential smoothing techniques are appropriate for stationary time series. This simply means there is no significant upward or downward long-term trend in the data. The technique of regression is appropriate for nonstationary time series, where there is some upward or downward trend over time. This is the same basic technique you learned in Chapter 11 only with the variable time as the independent variable now.

*Time series models are used to **extrapolate** past behavior in order to predict future behavior.*

This chapter covers the following material:

- Getting Started With Time Series Data
- Simple Moving Average Models
- Weighted Moving Average Models
- Exponential Smoothing Models
- Regression Models

## 13.2  GETTING STARTED WITH TIME SERIES DATA

### 13.2.1 Time Series Notation

To study time series data we must introduce some general notation. Consistent with the notation from regression, we will label the variable that we are trying to predict with the letter $Y$. Since each observation is taken at a particular *time,* we will subscript $Y$ with the letter $t$. Thus, the data in a time series are labeled

$$y_1, y_2, \ldots, y_t$$

where

$y_1$ is the observation of the variable at time period 1

$y_2$ is the observation of the variable at time period 2

$y_t$ is the observation of the variable at time period $t$

The observation that is the oldest in terms of the time that it was observed compared to the present is labeled $y_1$. For the bread example, the daily sales 25 days ago is the oldest observation and is therefore labeled $y_1$. The second oldest observation is labeled $y_2$ and so forth. With $t$ observations of the variable of interest, the current time is labeled time period $t$ and we are typically trying to predict the value for the next time period, $t + 1$. This prediction will be labeled $\hat{y}_{t+1}$. Remember that we used the notation $\hat{y}$ as the predicted value of $y$ from the regression model. In a similar manner, the caret in $\hat{y}_{t+1}$ tells you it is an estimate and the subscript $t + 1$ tells you that it is an estimate for the next time period, $t + 1$. Sometimes you may wish to predict beyond this point. In this case you may wish to predict the value of $y$ for time periods $t + 2$, $t + 3$, etc. These would be labeled $\hat{y}_{t+2}$, $\hat{y}_{t+3}$, etc.

Consider how the data would be labeled for the baseball problem.

### EXAMPLE 13.1  Baseball Attendance

*Population Data for the Past 10 Years*

The population data for the community surrounding the location of the new baseball stadium are shown below. Remember that FWC is the number of people in families with children and NC is the number of people in the age group from 20 to 40 with no children.

| Year | FWC | NC |
|------|-----|-----|
| 1989 | 1,098,909 | 564,790 |
| 1990 | 1,197,185 | 556,734 |
| 1991 | 1,169,860 | 653,134 |
| 1992 | 1,350,675 | 632,459 |
| 1993 | 1,335,213 | 642,387 |
| 1994 | 1,207,658 | 654,890 |
| 1995 | 1,379,723 | 657,238 |
| 1996 | 1,321,457 | 657,238 |
| 1997 | 1,387,692 | 717,903 |
| 1998 | 1,546,920 | 692,340 |

There are really two time series here: the time series consisting of observations of the number of people in families with children and those for the families with no children. For the moment, just consider the time series labeled FWC. The value from the year 1989 is the oldest so it is labeled $y_1$. The most recent observation is 1998 and so it is labeled $y_{10}$. Thus, the current time is $t = 10$ and we wish to predict the value for $t = 11$ or the year 1999. ∎

*The time period should be selected to fit the needs of the problem.*

Note that the time period could be every minute, every hour, every day, every week, every month, every year, or any other period that makes sense for the particular application you are working with. For the loaves of bread example in the previous section, the time period is every day since we observed daily sales. For the baseball ticket sales example, the population must be predicted for the next year based on the last 10 years of data. So the time period is every year. If you are predicting the closing price of a stock, the time period would be every day.

### 13.2.2 Displaying the Time Series

*Always display the data before deciding on a method of analysis.*

Once you have identified the data and labeled them properly, you should display them using a scatter plot. The $x$ axis should be time and the $y$ axis should be the variable of interest. Let's continue working with the FWC time series.

### EXAMPLE 13.2  Baseball Attendance

*Scatter Plot for FWC Time Series*

The scatter plot for the FWC time series is shown at the top of the next page.

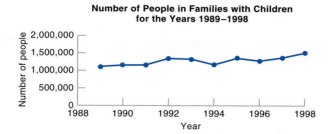

**Number of People in Families with Children for the Years 1989–1998**

Note that alternatively the $x$ axis could have been labeled year 1 to year 10. In this case, it is probably preferable to use the actual years for the $x$ axis, whereas for the bread data example, it was easiest to label the days from 1 to 25. ∎

After you plot the data, you should examine the plot to see if there are any obvious patterns or trends. In this case, the time series seems to have a slight upward trend but there are no regular patterns. Now you plot the scatter plot for the NC time series and examine it for patterns or trends.

 **TRY IT NOW!**

### Baseball Attendance
#### *Scatter Plot of NC Time Series*

Display the NC time series as a scatter plot.

Do you see any patterns or trends? If so, what are they?

The next two sections discuss simple and weighted moving average models of time series data.

## 13.2.3 Exercises—Learning It!

**13.1** Citing significant strides in math and science by women of all ethnic backgrounds, the College Board says that the male–female gap on Scholastic Assessment Test scores con-

ANS.

NO PATTERNS;
SLIGHT UPWARD TREND.

Number of People in Families with No Children

tinues to narrow among high school students. The average SAT scores for men and women from 1990 to 1994 are shown in the table below.

|      | Verbal | | Math | |
|------|--------|-------|------|-------|
|      | **Men** | **Women** | **Men** | **Women** |
| **1990** | 429 | 419 | 499 | 455 |
| **1991** | 426 | 418 | 497 | 453 |
| **1992** | 428 | 419 | 499 | 456 |
| **1993** | 428 | 420 | 502 | 457 |
| **1994** | 425 | 421 | 501 | 460 |

(a) Display each of the four time series on the same axis.

(b) Do you see any patterns?

**13.2** The graph below displays the high and low weekly closing price of John Wiley and Sons stock for the year 1994. Do you see any patterns in the time series?

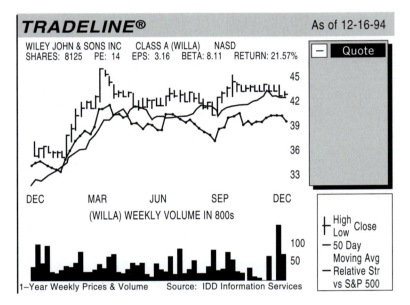

**13.3** The graph below displays the dropout rate for seniors in high school from 1971 to 1994. Describe the graph and note any patterns in the time series.

**13.4**    The table below shows the Indian sugar production in thousands of metric tons by year.

| Crop Year | Indian Sugar Production (Thousands of Metric Tons) | Crop Year | Indian Sugar Production (Thousands of Metric Tons) |
|---|---|---|---|
| 74–75 | 4,949 | 85–86 | 7,983 |
| 75–76 | 5,464 | 86–87 | 9,474 |
| 76–77 | 6,043 | 87–88 | 10,000 |
| 77–78 | 8,201 | 88–89 | 10,150 |
| 78–79 | 7,071 | 89–90 | 10,420 |
| 79–80 | 5,170 | 90–91 | 9,989 |
| 80–81 | 6,542 | 91–92 | 9,750 |
| 81–82 | 9,727 | 92–93 | 10,560 |
| 82–83 | 9,508 | 93–94 | 11,020 |
| 83–84 | 7,042 | 94–95 | 11,750 |
| 84–85 | 7,071 | | |

(a) Display this time series using a scatter plot.

(b) Do you notice any patterns?

**13.5**    The number of people living in the United States and the United Kingdom aged 15–64 is shown in the two time series below:

|  | 1990 | 1991 | 1992 | 1993 | 1994 | 1995 |
|---|---|---|---|---|---|---|
| USA | 164,654,840 | 166,150,720 | 167,660,448 | 169,184,152 | 170,721,960 | 172,274,000 |
| UK | 37,599,962 | 37,668,710 | 37,737,588 | 37,806,594 | 37,875,732 | 37,945,000 |

(a) Use a line graph to display these time series.

(b) Do you notice any patterns in the data? If so, is the pattern the same in both countries?

## *Discovery Exercise 13.1*
## LOOKING FOR PATTERNS AND TRENDS

Examine the time series displayed in the graphs below and identify any patterns and trends that you see. Use what you know about the particular data being collected to help you identify why the data look the way they do.

(a) This chart shows the average monthly price (cents per kilowatt·hour) of electricity across the United States from 1988 to 1996.

(b) The Bureau of Labor Statistics (BLS) reports data on commodity prices on a monthly basis. These data are reported on a regional and national basis. The graph at the top of page 656 displays the monthly average price (cents per pound) of white bread (pan) in the United States for the years 1980–1996.

*(continued)*

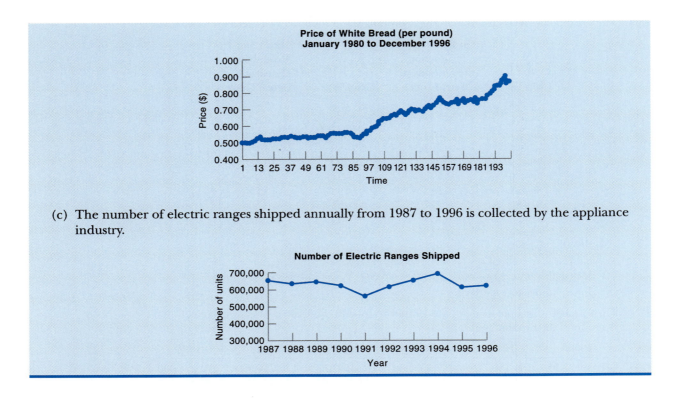

(c)  The number of electric ranges shipped annually from 1987 to 1996 is collected by the appliance industry.

## 13.3   SIMPLE MOVING AVERAGE MODELS

Let's look again at the problem on baseball attendance. If you had to forecast the number of people in families with children for 1999 right now, how would you use the data? If it helps, think about forecasting the number of loaves of bread to be sold on day 26. Remember you have data on only the past behavior of the variable of interest. Nothing else.

If you suggested finding the average of all of the data and using that as your forecast for time period $t + 1$, then you would be thinking along the right lines. Figure 13.2 redisplays the FWC time series with an additional line drawn at the average, $\bar{y}$.

The average FWC population over the past 10 years is 1,299,529. This, then, would become our forecast for 1999 if we decided to use the overall average as our forecast. By doing so we would be saying that we believe that the time series is station-

**FIGURE 13.2**   FWC time series with average displayed

ary and that all of the data, even the 10-year-old data, are equally valuable in constructing our forecast.

How good is this model? The graph of the straight line at the average value tells us that our forecast of the overall average is not terribly close to the data points that we have already observed. For instance, if we used the forecast of 1,299,529 to forecast for the year 1998 we would have been off by 1,546,920 − 1,299,529) = 247,391 people. Although there is not a strong trend in the data, it should be clear that the value from 1989 is not nearly as relevant in predicting the 1999 FWC population as the 1998 value. Perhaps we should not even use the 1989 value.

*We are trying to find a model that comes close to as many data points as possible.*

## 13.3.1 Calculating Simple Moving Averages

If you thought of this, then you have figured out what a moving average model is all about. Averaging the data was not a bad idea and, in fact, it is about the only thing you can do if you conclude that you have a stationary time series. But instead of averaging *all* of the data, we will average only the most recent observations. For example, we could average only the most recent 3 years as our forecast for the next year. In this case the predicted FWC population for 1999 would be calculated as follows:

$$\hat{y}_{11} = \frac{y_{10} + y_9 + y_8}{3}$$

$$= \frac{1,546,920 + 1,387,692 + 1,321,457}{3} = 1,418,690$$

*Remember that t, current time, is labeled t = 10 because you have 10 observations.*

Notice that this forecast is higher than the overall average forecast and ignores any data older than 3 years. Once the population for year 11 is observed then the forecast for year 12 would be calculated as follows:

$$\hat{y}_{12} = \frac{y_{11} + y_{10} + y_9}{3}$$

You can see that the observation from year 8 has been dropped and the newest observation, year 11, is included in the forecast calculation. Thus, the average "moves" along with time. This is why it is called a **moving average.**

*You can use any number period MA as long as the number of periods to be averaged doesn't exceed the number of periods for which you have data, namely t.*

> A *k*-period ***moving average*** is the average of the most recent *k* observations.

What we just calculated is called a 3-period moving average (MA), since we averaged the data from the most recent 3 time periods to get the forecast for the next period. You could instead use a 2-period moving average, a 4-period moving average or any number period moving average. In general, we will talk about a *k*-period moving average. The formula for a *k*-period moving average is given as

$$\hat{y}_{t+1} = \text{average of the most recent } k \text{ observations}$$

$$= \frac{y_t + y_{t-1} + \cdots + y_{t-k+1}}{k}$$

*Formula for a k-period moving average*

Don't let the notation confuse you; it is really just an average of *k* numbers.

## 13.3.2 Evaluating the Model

The next logical issue is to decide how to select the value of *k*. In other words, should we use a 2-period MA model, a 3-period MA model, or some other number period MA model? The right answer, of course, is that we should use the "best" model.

Ideally, we would like the forecasting model with zero error—that is, one that predicts perfectly. Recognizing that we will never find such a model, we look for a model with the smallest possible error. This is the same approach that we took in developing the "best" regression model. In regression, we used the model to predict the value of the dependent variable, $\hat{y}$, for each observation that we had. The predicted value, $\hat{y}$, was then compared to the actual value of $y$. This comparison took the form of a difference, $y - \hat{y}$. These differences were then squared and averaged to yield a

measure called the mean square error (MSE). The "best" regression model is the one with the smallest MSE. We will use precisely the same approach here.

Consider a 3-period moving average for the baseball attendance problem. We have 10 observations, so $t = 10$. To calculate the MSE for this 3-period MA model we must, in a sense, rewind the clock back to time $t = 4$ and use a 3-period MA to forecast for period 4, pretending for the moment that we don't know its value. The predicted value $\hat{y}_4$ can then be compared to the actual value, $y_4$, and the difference will become part of the mean square error calculation. Then we roll the clock forward to time $t = 5$, and use a 3-period MA to forecast for period 5 and so forth. It may seem odd to start the calculations of our predictions at time $t = 4$ when we have 3 data points prior to it. But you will quickly realize that to use a 3-period MA model, you must have 3 actual observations. So the earliest time period we can use to compare the model prediction to the actual value is for time period $t = 4$. In total we can compare the forecast to the actual values for time periods 4 through 10 or 7 time periods. We cannot use time period 11 in the calculation of the MSE because we don't yet know the actual value of the population for 1999 (time $t = 11$). The details are shown in the next example.

### EXAMPLE 13.3  Baseball Attendance

#### Calculating the MSE for a 3-Period MA Model for FWC

The steps for calculating the mean square error (MSE) for a 3-period moving average model are shown in the spreadsheet below. The column labeled FWC shows the time series data. The column labeled Prediction shows what the 3-period MA would have been for each of the time periods from 4 to 10. The column labeled Error is calculated by subtracting the predicted value from the actual value, $y_t - \hat{y}_t$. Finally, the column labeled Error Squared is calculated by squaring each of the errors. The MSE is found by averaging all of the values in the Error Squared column.

**3-Period Moving Average Model**

| Time | Year | FWC $y_t$ | Prediction $\hat{y}_t$ | Error $y_t - \hat{y}_t$ | Error Squared |
|------|------|-----------|------------------------|--------------------------|---------------|
| 1 | 1989 | 1,098,909 | | | |
| 2 | 1990 | 1,197,185 | | | |
| 3 | 1991 | 1,169,860 | | | |
| 4 | 1992 | 1,350,675 | 1,155,318 | 195,357 | 38,164,357,449 |
| 5 | 1993 | 1,335,213 | 1,239,240 | 95,973 | 9,210,816,729 |
| 6 | 1994 | 1,207,658 | 1,285,249 | − 77,591 | 6,020,363,281 |
| 7 | 1995 | 1,379,723 | 1,297,849 | 81,874 | 6,703,351,876 |
| 8 | 1996 | 1,321,457 | 1,307,531 | 13,926 | 193,933,476 |
| 9 | 1997 | 1,387,692 | 1,302,946 | 84,746 | 7,181,884,516 |
| 10 | 1998 | 1,546,920 | 1,362,957 | 183,963 | 33,842,385,369 |
| | | | | **MSE** | 14,473,870,385 |

The MSE is a rather large number in this case. This is nothing to be concerned about. It is due to the fact that the values in the time series are large, making the errors fairly large and making the errors squared even larger. Remember that like the variance calculation, the differences (errors) are squared before averaging them so that the positive and negative errors do not cancel each other out, leaving you with an inaccurate view of the usefulness of the model.

Notice that all but one of the errors are positive. Just like in regression, a pattern in the errors should make you suspicious. It usually means that there was a pattern in the original time series that you failed to model, in other words, a pattern you missed or you ignored. In this case, the positive errors tell you that your forecast from a 3-period MA model consistently underestimates the actual population. Because of this observation you consider using only 2 periods to forecast for the next period, a 2-period MA. You also begin to suspect that the assumption that this time series is stationary may be flawed. Since the forecast is almost always lagging behind (as evidenced by the positive errors), perhaps there is a long-term trend in the population for families with children. You saw a slight upward trend in Figure 13.2 but it wasn't clear at that point whether to consider the time series stationary or nonstationary.

*You should examine the errors for patterns.*

The formula for calculating the mean square error (MSE) for a *k*-period MA model is given below:

$$\text{MSE} = \frac{\sum_{i=k+1}^{t} (y_i - \hat{y}_i)^2}{t - k}$$

*Formula for MSE for k-period MA model*

Notice that the first difference, $y_{k+1} - \hat{y}_{k+1}$, in the numerator corresponds to the time period numbered $k + 1$. As we have already observed, this is the earliest time period for which you can calculate a *k*-period forecast. The last difference in the numerator corresponds to the current time period, labeled *t*. To get an average (or mean) square error, the numerator is divided by the number of squared errors being added together. For the baseball example we had 7 squared errors to be averaged in the column labeled Error Squared. In general, we will have $t - k$ values in this last column.

Now try the calculations for a 2-period MA model for the FWC time series.

 **TRY IT NOW!**

### Baseball Attendance
*A 2-Period MA Model*

Find the 2-period MA forecast for the number of people in families with children for 1999.

Complete the table below to find the MSE for the 2-period MA model for the FWC time series.

**2-Period Moving Average Model**

| Time | Year | FWC | Prediction | Error | Error Squared |
|------|------|-----------|------------|-------|---------------|
| 1 | 1989 | 1,098,909 | | | |
| 2 | 1990 | 1,197,185 | | | |
| 3 | 1991 | 1,169,860 | | | |
| 4 | 1992 | 1,350,675 | | | |
| 5 | 1993 | 1,335,213 | | | |
| 6 | 1994 | 1,207,658 | | | |
| 7 | 1995 | 1,379,723 | | | |
| 8 | 1996 | 1,321,457 | | | |
| 9 | 1997 | 1,387,692 | | | |
| 10 | 1998 | 1,546,920 | | | |

*(continued)*

How does the MSE for the 2-period MA model compare to the MSE for the 3-period MA?

---

Now we know that the 2-period MA has a smaller MSE than the 3-period MA. To see the difference in the performance of the 2-period MA model and the 3-period MA model, we can graph the original time series (FWC) and the 2 models on the same graph. This is shown in Figure 13.3.

*Note that a 10-period MA model would be an overall average. We would have no way to find an MSE for this model.*

But how do we know that we have the "best" model? Maybe a 4-period MA would be even better. The only way to be sure is to try all of the possible models. In this case that would mean models with $k = 2, 3, 4, 5, 6, 7, 8, 9, 10$. We have already considered the models for $k = 2, 3$, and 10. Clearly, if you have 30 observations in your time series you do not want to try all the possible values of $k$ from 2 to 30. Even with the "power of the spreadsheet," you would not want to do this. Worse yet, imagine if you had 100 observations in your time series! Usually you can try a few models and look for a pattern in the mean square error values. The MSEs for the 2-, 3-, and 4-period MA models are summarized in the table below.

| $k$ | MSE |
|---|---|
| 2 | 12,900,675,743 |
| 3 | 14,473,866,726 |
| 4 | 14,792,462,017 |

Thus it appears that the models that include more historical data (larger $k$ values) have higher MSEs. So, if we are going to use a moving average model, we should use a 2-period model and our estimate for the FWC population for the year 1999 is 1,467,306.

There is another measure that is sometimes used instead of the MSE to evaluate the goodness of a forecasting model. It is called the mean absolute deviation or MAD. You have seen this measure in Chapter 4 when we looked at ways to calculate a number that would estimate the spread or variability of a data set. In doing inferential statistics, the MSE or variance is preferred over the MAD because of its nice mathematical properties. In forecasting models, you are simply using the MSE to evaluate

**FIGURE 13.3**   FWC time series with a 2-period MA model and 3-period MA model

It is smaller.
MSE = 12,900,675,743
Ans. $\hat{y}_{11} = 1,467,306$

a model. Therefore, it really doesn't matter much if you use the MSE or MAD to evaluate models as long as you are consistent. That is, if you calculate the MSE for one model you must compare it to the MSE of a second model.

To calculate the MAD, you simply use the absolute value of the difference between the observed value and the predicted value instead of squaring the difference. The absolute value function eliminates the problem of the negative errors canceling out the positive errors. The formula is given below:

$$\text{MAD} = \frac{\sum_{i=k+1}^{t} |y_i - \hat{y}_i|}{t - k}$$

*Formula for mean absolute deviation (MAD) for k-period moving average*

The calculation of the MAD for the 3-period MA model of the FWC time series is shown in the next example. You will notice that all columns but the last column are the same as Example 13.3.

## EXAMPLE 13.4  Baseball Attendance

### Finding the MAD

The table below shows the details for calculating the MAD for the 3-period MA model for the FWC time series.

**3-Period Moving Average Model**

| Time | Year | FWC | Prediction | Error | Absolute Value of Error |
|---|---|---|---|---|---|
| 1 | 1989 | 1,098,909 | | | |
| 2 | 1990 | 1,197,185 | | | |
| 3 | 1991 | 1,169,860 | | | |
| 4 | 1992 | 1,350,675 | 1,155,318 | 195,357 | 195,357 |
| 5 | 1993 | 1,335,213 | 1,239,240 | 95,973 | 95,973 |
| 6 | 1994 | 1,207,658 | 1,285,249 | −77,591 | 77,591 |
| 7 | 1995 | 1,379,723 | 1,297,849 | 81,874 | 81,874 |
| 8 | 1996 | 1,321,457 | 1,307,531 | 13,926 | 13,926 |
| 9 | 1997 | 1,387,692 | 1,302,946 | 84,746 | 84,746 |
| 10 | 1998 | 1,546,920 | 1,362,957 | 183,963 | 183,963 |
| | | | | **MAD** | **104,776** |

The MAD is simply the average of the 7 values in the last column. ■

Like the MSE, the MAD itself is not a useful number. It is helpful as a way to help decide between different forecasting models. For the remainder of this chapter we will use the MSE to evaluate the models simply because it is more commonly used.

The models presented in this section are called **simple moving average models.** We have been referring to them as moving average models but should now note that they are *simple MA models* because the forecast is calculated as a simple average. This means that you simply add up all the values to be averaged and divide by *k*.

> A *simple moving average model* uses the simple average of the most recent *k* observations to predict for the next time period.

Each number in the average carries the same weight or importance in the averaging process. This is the type of average we have been working with throughout this book. However, it is possible to consider what is known as a weighted moving average. This is the subject of the next section.

## 13.3.3 Exercises—Learning It!

**13.6** Reconsider the baseball attendance shown in Example 13.1.

(a) Display the people with no children (NC) time series graphically using a line graph.

(b) Find the 2-period MA forecast for the population with no children (NC) for the year 1999.

(c) Find the MSE for the 2-period MA model for the NC time series.

(d) Find the 3-period MA forecast for the population with no children (NC) for the year 1999.

(e) Find the MSE for the 3-period MA model for the NC time series.

(f) Display the NC time series and the 2-period and 3-period MA models on the same graph.

*Requires Exercise 13.1*   **13.7** Analyze the math SAT scores.

(a) Find the 3-period MA forecast for 1995 math SAT scores for men.

(b) Find the MAD for this model.

(c) For the math SAT scores for women, find the 3-period MA forecast for 1995.

(d) Find the MAD for this model.

**13.8** The price for purchasing a half-gallon of ice cream in a typical U.S. city is one of many time series collected by the Bureau of Labor Statistics. The monthly prices for 1995 and 1996 are shown below:

|     | 1995  | 1996  |
| --- | ----- | ----- |
| Jan | 2.659 | 2.665 |
| Feb | 2.55  | 2.673 |
| Mar | 2.686 | 2.752 |
| Apr | 2.629 | 2.728 |
| May | 2.634 | 2.825 |
| Jun | 2.649 | 2.827 |
| Jul | 2.665 | 2.851 |
| Aug | 2.667 | 2.966 |
| Sep | 2.683 | 3.041 |
| Oct | 2.735 | 3.077 |
| Nov | 2.61  | 2.978 |
| Dec | 2.675 | 2.94  |

(a) Display this time series as a line graph.

(b) Find the 3-period, 4-period, and 5-period MA forecast for January 1997.

(c) Find the MSE for the 3-period, 4-period, and 5-period MA models.

(d) Which model yields the smallest MSE?

(e) Find the MAD for the 3-period, 4-period, and 5-period MA models.

(f) Which model yields the smallest MAD?

(g) Based on your analysis, which model would you recommend be used and why?

**13.9** The health of the construction industry is often an indication of the health of the economy. The data below show the number of bricks produced by quarter for 1995 and 1996.

| Year | Quarter | Brick Production |
| ---- | ------- | ---------------- |
| 1995 | I       | 1,800,252        |
| 1995 | II      | 1,849,733        |
| 1995 | III     | 1,770,466        |
| 1995 | IV      | 1,586,484        |
| 1996 | I       | 1,496,798        |
| 1996 | II      | 1,836,651        |
| 1996 | III     | 1,689,123        |
| 1996 | IV      | 1,654,901        |

(a) Display this time series as a line graph.

(b) Find the 4-period MA forecast for January 1997.

(c) Find the MSE for the 4-period MA model.

**13.10** Hospitals have been paying increased attention to customer service. To determine staffing needs in the emergency room, the number of patients seen daily by the ER of a city hospital is shown below. The city has about 150,000 residents. One month of data are shown below.

| Day | 1 | 2 | 3 | 4 | 5 | 6 | 7 | 8 | 9 | 10 | 11 | 12 | 13 | 14 | 15 | 16 | 17 | 18 | 19 | 20 | 21 | 22 | 23 | 24 | 25 | 26 | 27 | 28 | 29 | 30 |
|---|---|---|---|---|---|---|---|---|---|---|---|---|---|---|---|---|---|---|---|---|---|---|---|---|---|---|---|---|---|---|
| No. Patients | 43 | 47 | 41 | 43 | 45 | 44 | 34 | 33 | 35 | 37 | 38 | 43 | 41 | 42 | 36 | 30 | 35 | 38 | 45 | 42 | 43 | 34 | 39 | 30 | 42 | 45 | 38 | 33 | 38 | 39 |

(a) Display this time series as a line graph.
(b) Find the 7-period MA forecast for the next day.
(c) Find the MSE for the 7-period MA model.

# 13.4   WEIGHTED MOVING AVERAGES

In the previous section you saw that for one time series there are many possible forecasting models. In fact, all of the models considered in the last section were simple moving average models. In this section we will see that by using unequal weights in the calculation of the average, there will be an infinite number of models of this type. These are called **weighted moving average models.**

> A *weighted moving average model* is a moving average model with unequal weights.

You can quickly see that, as in multiple regression analyses, building a forecasting model is partly a science and partly an art. You must use your eye to discern patterns in the data and your knowledge of the application area to decide what type of models to consider. Fortunately, with the aid of the software, we can often consider many different possible models and pick the one with the smallest MSE as our model of choice.

## 13.4.1 Calculating Weighted Moving Averages

Reconsider the 3-period simple moving average model for the FWC time series. The forecast for 1999 (time period 11) is calculated as follows:

$$\hat{y}_{11} = \frac{y_{10} + y_9 + y_8}{3} = 1,418,690$$

To clearly see the weight that each value has in this calculation let's rewrite the equation as

$$\hat{y}_{11} = \tfrac{1}{3} y_{10} + \tfrac{1}{3} y_9 + \tfrac{1}{3} y_8$$

Now you can easily see that each value is weighted the same, $\frac{1}{3}$, in the formula for a 3-period *simple* moving average.

If we consider a 3-period *weighted* moving average, we will allow the weights to be different from each other. For example, we might decide to weight the most recent observation by $\frac{1}{2}$ (50%), the next most recent observation by $\frac{1}{4}$ (25%), and the third most recent observation by $\frac{1}{4}$ (25%). The forecast for 1999 would then be calculated as

$$\hat{y}_{11} = \tfrac{1}{2} y_{10} + \tfrac{1}{4} y_9 + \tfrac{1}{4} y_8$$
$$= (0.50)(1,546,920) + (0.25)(1,387,692) + (0.25)(1,321,457)$$
$$= 1,450,747$$

What have we accomplished by doing this? The forecast for 1999 calculated this way is larger than the forecast from the 3-period simple MA model. By weighting the most recent observation, $y_{10}$, more heavily than the older data we are saying that it is more important to our forecast. In this case, this argument makes sense, since we observed a slight upward tendency in the time series. Thus, the most recent observation should carry a greater weight in the forecast for the next period.

The only rule that needs to be observed as you pick the weights is that the sum of the weights must be 1 and each weight must be a positive number between 0 and 1. We will use the term $w_t$ to represent the weight to be used for the observation from time period $t$. The general formula for a 3-period weighted moving average is then

*Formula for 3-period weighted moving average*

$$\hat{y}_{t+1} = w_t \, y_t + w_{t-1} \, y_{t-1} + w_{t-2} \, y_{t-2}$$

where

$$0 \leq w_t, w_{t-1}, w_{t-2} \leq 1$$
$$w_t + w_{t-1} + w_{t-2} = 1$$

## 13.4.2 Finding the Best Model

Many other possible weights could be used for the 3-period weighted moving average. For instance, maybe our forecasting model would be improved if we used $w_t = 0.60$, $w_{t-1} = 0.30$, and $w_{t-2} = 0.10$. How can we tell? We can find the MSE for each model and see which one yields the lowest MSE. The MSE for two weighted MA models are shown in the next example.

### EXAMPLE 13.5  Baseball Attendance

***MSE Calculation for Two Different 3-Period Weighted MA Models***

The table below shows the calculation of the MSE for two different 3-period weighted MA models for the FWC time series.

**3-Period Weighted Moving Average Model**

| Time | Year | FWC | Prediction | Error | Error Squared |
|------|------|-----|------------|-------|---------------|
|      |      |     | Model 1: Weights: 0.50, 0.25, 0.25 | | |
| 1 | 1989 | 1,098,909 | | | |
| 2 | 1990 | 1,197,185 | | | |
| 3 | 1991 | 1,169,860 | | | |
| 4 | 1992 | 1,350,675 | 1,158,954 | 191,721 | 36,756,941,841 |
| 5 | 1993 | 1,335,213 | 1,267,099 | 68,114 | 4,639,516,996 |
| 6 | 1994 | 1,207,658 | 1,297,740 | −90,082 | 8,114,766,724 |
| 7 | 1995 | 1,379,723 | 1,275,301 | 104,422 | 10,903,954,084 |
| 8 | 1996 | 1,321,457 | 1,325,579 | −4,122 | 16,990,884 |
| 9 | 1997 | 1,387,692 | 1,307,574 | 80,118 | 6,418,893,924 |
| 10 | 1998 | 1,546,920 | 1,369,141 | 177,779 | 31,605,372,841 |
|    |      |           | **MSE** | | 14,065,205,328 |

| Time | Year | FWC | Prediction | Error | Error Squared |
|------|------|-----|------------|-------|---------------|
|      |      |     | Model 2: Weights: 0.60, 0.30, 0.10 | | |
| 1 | 1989 | 1,098,909 | | | |
| 2 | 1990 | 1,197,185 | | | |
| 3 | 1991 | 1,169,860 | | | |
| 4 | 1992 | 1,350,675 | 1,170,962 | 179,713 | 32,296,762,369 |
| 5 | 1993 | 1,335,213 | 1,281,082 | 54,131 | 2,930,165,161 |
| 6 | 1994 | 1,207,658 | 1,323,316 | −115,658 | 13,376,772,964 |

*(continued)*

| Time | Year | FWC | Prediction | Error | Error Squared |
|---|---|---|---|---|---|
| | | | Model 2: Weights: 0.60, 0.30, 0.10 | | |
| 7 | 1995 | 1,379,723 | 1,260,226 | 119,497 | 14,279,533,009 |
| 8 | 1996 | 1,321,457 | 1,323,653 | − 2,196 | 4,822,416 |
| 9 | 1997 | 1,387,692 | 1,327,557 | 60,135 | 3,616,218,225 |
| 10 | 1998 | 1,546,920 | 1,367,025 | 179,895 | 32,362,211,025 |
| | | | MSE | | 14,123,783,596 |

From Example 13.5 we learn that model 1 using weights of 0.50, 0.25, and 0.25, has a smaller MSE than model 2, but the 2-period simple MA has a lower MSE than both of these models. Does this mean that any 3-period weighted average will have a higher MSE than the 2-period simple MA model? We can't say this for sure but it seems that by including one more observation in the model we raise the MSE by quite a bit even with the weighted models. Perhaps we should consider a weighted 2-period model. Try one and see for yourself.

 *TRY IT NOW!*

## Baseball Attendance

### A 2-Period Weighted MA Model

Find the 2-period weighted MA forecast for the number of people in families with children for 1999. Weight the most recent observation by 0.75 and the second most recent observation by 0.25.

Complete the table below to find the MSE for the 2-period weighted MA model for the FWC time series.

**2-Period Weighted Moving Average Model**

| Time | Year | FWC | Prediction | Error | Error Squared |
|---|---|---|---|---|---|
| | | | Weights: 0.75, 0.25 | | |
| 1 | 1989 | 1,098,909 | | | |
| 2 | 1990 | 1,197,185 | | | |
| 3 | 1991 | 1,169,860 | | | |
| 4 | 1992 | 1,350,675 | | | |
| 5 | 1993 | 1,335,213 | | | |
| 6 | 1994 | 1,207,658 | | | |
| 7 | 1995 | 1,379,723 | | | |
| 8 | 1996 | 1,321,457 | | | |
| 9 | 1997 | 1,387,692 | | | |
| 10 | 1998 | 1,546,920 | | | |

Is this model better than the simple 2-period moving average?

*Some software packages allow you to optimize the weights for a MA model. For example, in Excel you can use the Solver feature of the Tools menu to find the set of weights that gives you the smallest MSE for a 3-period weighted MA model.*

***Remember*** *all the weights must be positive fractions and they must sum to one.*

The discussion in the previous paragraph is an illustration of the process by which you arrive at a "good" forecasting model. You try some models, look for patterns in the MSE, and arrive at what you feel is the "best" model. Most of the time you will not be guaranteed that there is no better model for the time series.

The formula for a *k-period weighted moving average* is shown below. It is simply a generalization of the 3-period weighted moving average formula that we have already seen.

$$\hat{y}_{t+1} = \text{weighted average of the most recent } k \text{ observations}$$
$$= w_t y_t + w_{t-1} y_{t-1} + \cdots + w_{t-k+1} y_{t-k+1}$$

where

$$0 \le w_t, w_{t-1}, \ldots, w_{t-k+1} \le 1$$
$$w_t + w_{t-1} + \cdots + w_{t-k+1} = 1$$

The formulas for calculating the MSE and the MAD are the same as for the simple moving average model.

## 13.4.3 Exercises—Learn It!

*Requires Exercise 13.6*

**13.11** For the baseball attendance problem:

(a) Find the 2-period weighted MA forecast for population with no children (NC) for the year 1999. Use $w_t = 0.80$ and $w_{t-1} = 0.20$.

(b) Find the MSE for the weighted 2-period MA model. How does it compare with the MSE for the simple 2-period MA model?

(c) Find the 3-period weighted MA forecast for population with no children (NC) for the year 1999. Use $w_t = 0.60$, $w_{t-1} = 0.20$, $w_{t-2} = 0.20$.

(d) Find the MSE for the 3-period weighted MA model. How does it compare with the MSE for the simple 2-period MA model?

(e) Display the NC time series and the 2-period and 3-period weighted MA models on the same graph.

*Requires Exercise 13.4*

**13.12** Use a 4-period weighted moving average to forecast for the 95–96 crop year of Indian sugar.

(a) Use weights of $w_t = 0.50$, $w_{t-1} = 0.20$, $w_{t-2} = 0.15$, $w_{t-3} = 0.15$.

(b) Find the MSE for this model.

(c) Try a different set of weights. Did you reduce the MSE?

**13.13** A small pizza restaurant is interested in using past data to estimate the sales for the next month. The management plans to use the sales forecast to determine the amount of cheese, tomato sauce, and other ingredients to order. The sales pattern for the past 2 years is shown below:

| Month | 1995 | 1996 |
|---|---|---|
| Jan | $5,838.90 | $12,958.07 |
| Feb | $5,226.89 | $12,731.63 |
| Mar | $6,326.66 | $13,901.36 |
| Apr | $6,815.22 | $13,232.64 |
| May | $6,665.17 | $14,223.89 |
| Jun | $10,801.43 | $13,440.71 |
| Jul | $10,171.44 | $13,751.10 |
| Aug | $6,047.26 | $15,324.82 |
| Sep | $4,840.70 | $11,738.72 |
| Oct | $5,431.87 | $14,185.08 |
| Nov | $5,536.20 | $14,607.72 |
| Dec | $5,944.26 | $15,167.76 |

(a) Display the time series as a line graph.

(b) Use a 3-period moving average with weights of $w_t = 0.75$, $w_{t-1} = 0.15$, and $w_{t-2} = 0.10$ to forecast for the next month.

(c) Find the MSE for this model.

(d) Use a 3-period weighted moving average with $w_t = 0.89$ and $w_{t-1} = 0.04$ and $w_{t-2} = 0.07$ to forecast for the next month.

(e) Find the MSE for this model.

(f) Compare the MSE for both models. Which one is smaller? What does this tell you?

(g) Examine the line graph of this time series and explain why you should have expected this.

**13.14** The cotton and wool outlook is published monthly by the Economic Research Service, U.S. Department of Agriculture, Washington, D.C. The number of 480-lb bales in the world is shown below for 1995 and 1996:

| | 1995 | 1996 |
|---|---|---|
| Jan | 29.94 | 28.89 |
| Feb | 30.14 | 28.92 |
| Mar | 30.72 | 28.75 |
| Apr | 29.62 | 28.87 |
| May | 27.54 | 28.91 |
| Jun | 27.42 | 33.34 |
| Jul | 29.45 | 33.78 |
| Aug | 29.84 | 34.33 |
| Sep | 29.68 | 34.95 |
| Oct | 29.50 | 35.59 |
| Nov | 29.37 | 35.63 |
| Dec | 29.18 | 36.16 |

(a) Display this time series as a line graph.

(b) Use a 6-period simple moving average to predict the number of bales for January 1997.

(c) Find the MSE for this model.

(d) Use a 6-period weighted moving average to predict the number of bales for January 1997. Use $w_t = 0.50$, $w_{t-1} = 0.30$, $w_{t-2} = 0.10$, $w_{t-3} = 0.05$, $w_{t-4} = 0.03$, and $w_{t-5} = 0.02$.

(e) Find the MSE for this model.

(f) Which model do you recommend and why?

**13.15** The amount of money spent on alcohol is collected by the Bureau of Labor Statistics (BLS) using a consumer expenditure survey. The amount spent on alcohol for two different age groups is shown below:

| | Under 25 Years of Age | Between 25 and 34 Years Old |
|---|---|---|
| 1990 | $318,000 | $365,000 |
| 1991 | $252,000 | $370,000 |
| 1992 | $356,000 | $365,000 |
| 1993 | $304,000 | $307,000 |
| 1994 | $247,000 | $347,000 |
| 1995 | $277,000 | $299,000 |

(a) Display each of the time series as a line graph.

(b) Describe any differences you see in the two time series.

(c) Use a 3-period weighted moving average to predict the alcohol expenditure for each age group for 1996. Use $w_t = 0.60$, $w_{t-1} = 0.30$, and $w_{t-2} = 0.10$.

## 13.5 EXPONENTIAL SMOOTHING MODELS

In the last section we saw that it is often difficult to determine the "best" values for the weights to be used in the weighted moving average model. The **exponential smoothing model** to be discussed in this section is another averaging technique that allows you to use unequal weights. This technique specifies a formula for the assignment of these weights.

> An *exponential smoothing model* is an averaging technique that uses unequal weights. The weights applied to past observations decline in an exponential manner.

### 13.5.1   FORECASTING USING AN EXPONENTIAL SMOOTHING MODEL

*Weighted MA models use only k periods of historical data in the calculation of the forecast for the next period.*

*Remember that the weights must be positive fractions and sum to one.*

The exponential smoothing model is different from the weighted moving average model because *all* of the historical data in the time series are used to generate the forecast for the next period. It is similar to a weighted MA model because the forecast is a weighted average. The weights are assigned in such a way that the most recent observation, $y_t$, carries the largest weight. The second most recent observation, $y_{t-1}$, carries the second largest weight and the weights assigned to the other data points decrease *systematically*.

What does the word "systematically" mean? Generally, it means that there is some order or pattern. Let's see how we could assign the weights systematically. Use the FWC time series for the baseball attendance problem again. Suppose we decided to assign a weight of 0.70 to the most recent observation in the time series. Call this weight $\alpha$. So $w_{10} = \alpha = 0.70$. We know from the above discussion that the weight for the observation at time period 9 should be less than 0.70. If we calculate the weight for this second most recent time period as $\alpha(1 - \alpha)$, then we know it would always be less than $\alpha$. In this case $w_9 = 0.70(1 - 0.70) = 0.21$. The weight for the third most recent observation should be less than 0.21. It is calculated as $\alpha(1 - \alpha)^2 = (0.70)(1 - 70)^2 = 0.063$. If we continue with this pattern the weights to be used in forecasting for period 11 are shown in the table and in the associated graph of Figure 13.4. The table and the graph show that the weights decrease rather quickly as the age of the data increases. In fact, the graph looks like the graph of an exponential curve. Hence, the name given to this technique is exponential smoothing.

The definition of $\alpha$ is not the same as that used in hypothesis testing. We could use a different Greek letter to avoid any confusion with the earlier definition of $\alpha$,

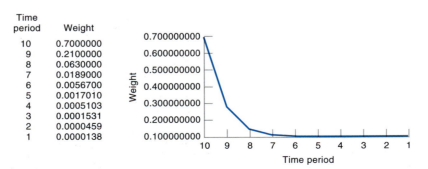

| Time period | Weight |
|---|---|
| 10 | 0.7000000 |
| 9 | 0.2100000 |
| 8 | 0.0630000 |
| 7 | 0.0189000 |
| 6 | 0.0056700 |
| 5 | 0.0017010 |
| 4 | 0.0005103 |
| 3 | 0.0001531 |
| 2 | 0.0000459 |
| 1 | 0.0000138 |

**FIGURE 13.4**   Weights for exponential smoothing model with $\alpha = 0.70$

but most forecasting books use $\alpha$ so we will do so as well. In this case, $\alpha$ is called the **smoothing constant.**

> The *smoothing constant, $\alpha$,* is the weight assigned to the most recent observation in an exponential smoothing model.

Let's see how to calculate the forecast for the FWC time series using a smoothing constant of $\alpha = 0.7$.

**EXAMPLE 13.6**   **Baseball Attendance**

*Exponential Smoothing Model*

The forecast for period 11 (FWC population for the year 1999) is calculated as a weighted average of all of the historical data using the weights shown in Figure 13.4:

$$\hat{y}_{11} = (0.7)(1{,}546{,}920) + (0.21)(1{,}387{,}692) + (0.063)(1{,}321{,}457)$$
$$+ \cdots + (0.0000138)(1{,}098{,}909)$$
$$= 1{,}493{,}651 \qquad \blacksquare$$

The general formula for the forecast for the next period, $t + 1$, is shown below.

$$\hat{y}_{t+1} = \sum_{i=1}^{t} w_i y_i$$
$$= \alpha y_t + \alpha(1 - \alpha)y_{t-1} + \alpha(1 - \alpha)^2 y_{t-2} + \cdots + \alpha(1 - \alpha)^n y_{t-n} + \cdots$$

*Formula for exponentially smoothed forecast for the next period*

Notice that the sum starts with the most recent observation weighted by $\alpha$ and includes data all the way back to the beginning of the time series. By examining the weights used when $\alpha = 0.70$, you can see that any data older than 3 time periods contributes virtually nothing. Why then should we bother to include these observations in the weighted average? The answer to that rather appropriate question has to do with a practical consideration. It can be shown that the exponentially smoothed forecast calculated using the formula above is algebraically equivalent to the following equation:

$$\hat{y}_{t+1} = \hat{y}_t + \alpha(y_t - \hat{y}_t)$$

*Alternative formula for computing the exponentially smoothed forecast for period $t + 1$*

It is not immediately obvious when you look at the formula that all of the data in the time series are actually being used. All of the data are rolled into the forecast made for time period $t$, $\hat{y}_t$. The new forecast $\hat{y}_{t+1}$ is equal to the previous forecast plus some adjustment. The adjustment is based on how much error there was in the previous forecast, that is, the difference between the forecast for period $t$ and the observation for period $t$, $(y_t - \hat{y}_t)$. If your previous forecast was "too high" then you need to adjust the forecast in the downward direction and the adjustment will be negative since $\hat{y}_t$ was larger than $y_t$. If, on the other hand, your previous forecast was "too low," then you need to make the forecast a bit larger and the adjustment will be positive since $y_t$ was bigger than $\hat{y}_t$.

## 13.5.2 Evaluating the Exponential Smoothing Model

The equation shown above is the best one to use to actually calculate the forecast using exponential smoothing. This is true because you need only the most recent forecast, $\hat{y}_t$, the most recent observation, $y_t$, and $\alpha$ to complete the computation. Let's see how to use this equation and find the MSE of the exponential smoothing model for the FWC time series.

**EXAMPLE 13.7** Baseball Attendance

*Forecast and MSE for Exponential Smoothing Model*

The first five columns of the table below display the year, the population for families with children, the forecast for each year, the error, and the adjustment to be made for the next forecast. The last column is used to find the mean square error.

| Year | FWC $y_t$ | Forecast $\hat{y}_t$ | Error, $y_t - \hat{y}_t$ | Alpha × Error $\alpha(y_t - \hat{y}_t)$ | Error Squared $(y_t - \hat{y}_t)^2$ |
|------|-----------|-----------------------|---------------------------|------------------------------------------|--------------------------------------|
| 1989 | 1,098,909 | 1,098,909 | 0 | 0 | 0 |
| 1990 | 1,197,185 | 1,098,909 | 98,276 | 68,793.20 | 9,658,172,176 |
| 1991 | 1,169,860 | 1,167,702 | 2,158 | 1,510.46 | 4,656,101 |
| 1992 | 1,350,675 | 1,169,213 | 181,462 | 127,023.64 | 32,928,580,838 |
| 1993 | 1,335,213 | 1,296,236 | 38,977 | 27,283.69 | 1,519,183,299 |
| 1994 | 1,207,658 | 1,323,520 | −115,862 | −81,103.39 | 13,424,000,588 |
| 1995 | 1,379,723 | 1,242,417 | 137,306 | 96,114.48 | 18,853,048,354 |
| 1996 | 1,321,457 | 1,338,531 | −17,074 | −11,951.86 | 291,524,175 |
| 1997 | 1,387,692 | 1,326,579 | 61,113 | 42,778.94 | 3,734,771,425 |
| 1998 | 1,546,920 | 1,369,358 | 177,562 | 124,293.28 | 31,528,204,498 |
| **Forecast** | | 1,493,651 | | **MSE** | 11,194,214,145 ■ |

*The observation for the first period becomes the forecast for the first period to get the technique started.*

Notice that to create the table shown in Example 13.7, we needed a forecast for period 1. Generally, you simply set the forecast for the first time period equal to the observation for the first time period. The effect of doing this is quickly washed out. As a result, the error for period 1 is always zero, making the second forecast precisely the same as the first forecast. Notice also that you cannot really forecast beyond period $t + 1$ because you do not know what the error will be for future forecasts. Since you have assumed a stationary time series (or you wouldn't be using simple exponential smoothing), your forecast for any period beyond $t + 1$ is the same as your forecast for period $t + 1$.

To compare the usefulness of the exponential smoothing model to the moving average and weighted moving average models that we have considered, we again use the mean square error (MSE). The MSE is computed in precisely the same manner as we have done before. It turns out that to calculate the forecast for $\hat{y}_{t+1}$ you need to calculate the error and so it is a simple matter of squaring these errors and averaging them to get the MSE.

**EXAMPLE 13.8** Baseball Attendance

*Evaluating the Exponential Smoothing Model*

In this case the MSE for the exponential smoothing model with a smoothing constant of 0.7 is smaller than for any other model we have tried for this time series. The graph shown below clearly depicts this as the time series and the forecasts are very close to each other for most periods.

Exponential Smoothing Model for FWC Time Series With $\alpha = 0.70$

Perhaps the MSE could be made even smaller with a different smoothing constant. Try it and see for yourself!

 ## *TRY IT NOW!*

### Baseball Attendance
#### *Exponential Smoothing Model*

Find the forecast for the number of people in families with children for 1999 using an exponential smoothing model with a smoothing constant of 0.6.

Complete the table below to find the MSE for the exponential smoothing model ($\alpha = 0.6$) for the FWC time series.

| Year | FWC | Forecast | Error | Alpha × Error | Error Squared |
|------|-----|----------|-------|---------------|---------------|
| 1989 | 1,098,909 | | | | |
| 1990 | 1,197,185 | | | | |
| 1991 | 1,169,860 | | | | |
| 1992 | 1,350,675 | | | | |
| 1993 | 1,335,213 | | | | |
| 1994 | 1,207,658 | | | | |
| 1995 | 1,379,723 | | | | |
| 1996 | 1,321,457 | | | | |
| 1997 | 1,387,692 | | | | |
| 1998 | 1,546,920 | | | | |
| | **Forecast** | | | **MSE** | |

Is this a better model than the model with a 0.7 smoothing constant?

As you can see, the selection of the smoothing constant affects both the forecast and the MSE for the model. Since $\alpha$ can vary anywhere from 0 to 1, it is impossible to try all values of $\alpha$ to find the model with the least MSE. Just as we did with the weighted moving average models, we can vary $\alpha$ and observe what happens to the MSE and attempt to zero in on a good value of $\alpha$.

*Some software packages allow you to optimize the value of $\alpha$. For example, in Excel you can use the Solver feature of the Tools menu to find the value of $\alpha$ that gives you the smallest MSE.*

### 13.5.3 Exercises—Learn It!

**13.16** Use the baseball attendance for the population with no children (NC) time series.
  (a) Forecast the attendance for 1999 using an exponential smoothing model with a value of $\alpha = 0.90$.
  (b) Find the MSE for this model.
  (c) How does this model compare with the weighted moving average models and the simple moving average models you tried in Exercises 13.6 and 13.11?
  (d) Which model do you recommend and why?

*Requires Exercises 13.6, 13.11*

**13.17** Use an exponential smoothing model with $\alpha = 0.65$ to forecast for the 95–96 crop year of Indian sugar.

*Requires Exercises 13.4, 13.12*

(a) Find the MSE for this model.

(b) How does this model compare with the weighted moving average models and the simple moving average models you tried in Exercises 13.4 and 13.12?

(c) Which model do you recommend and why?

*Requires Exercise 13.13*     **13.18** Use an exponential smoothing model with $\alpha = 0.55$ to forecast the sales for the next month for the pizza restaurant.

(a) Find the MSE for this model.

(b) How does this model compare with the weighted moving average model developed in Exercise 13.13?

(c) Which model do you recommend and why?

*Requires Exercise 13.10*     **13.19** Use an exponential smoothing model to forecast the number of patients seen in the emergency room for the next day. The data are shown in Exercise 13.10. Use $\alpha = 0.85$.

**13.20** The local liquor store has a bottle redemption center attached to it. The management is trying to determine staffing needs for the redemption center. The number of bottles brought into the redemption center varies from day to day. The data from the last 6 days are shown below:

| | |
|------|---------|
| Mon | 148,454 |
| Tues | 180,000 |
| Wed | 147,053 |
| Thurs | 152,197 |
| Fri | 150,583 |
| Sat | 122,366 |

(a) Use an exponential smoothing model with $\alpha = 0.75$ to predict the number of bottles to be returned on the following Monday; the store is closed on Sunday.

(b) Find the MSE for this model.

# 13.6   REGRESSION MODELS

*Remember: A stationary time series is one that exhibits no long-term trend.*

All of the models that have been considered thus far in the chapter are models that could be used if you have a time series that has no trends. But what if there is a long-term trend in the data, either upward or downward? In this case you should use a technique that explicitly models the trend. You are already familiar with one such technique, regression. There are other techniques that can be used to model trends in the data. Two of these are second-order moving averages and double exponential smoothing. These are extensions of the techniques covered in the previous sections. However, the most commonly used technique when a trend is present is the technique of regression. This is the only one that will be covered in this chapter.

## 13.6.1 Finding the Linear Regression Model

The first step to analyzing time series data is to display the data using a scatter plot. After you construct a scatter plot and visually examine the time series data, you may observe a clear upward or downward linear trend. In this case you should try a regression model. However, it is not always crystal clear that there is a trend in the data. This is the case with the FWC time series we have been working with. There appears to be a slight upward trend, but is that enough to warrant the use of regression? You may decide, as we have done, to use a stationary technique for such data. Upon examination of the errors (or residuals as they are called in regression) you may see a pattern. If there is a pattern, there probably is some more information in the time series data that you have not explicitly modeled. In this case you may also try a regression model.

In Chapter 11, you learned all about the technique of simple linear regression. The same technique is used to model time series data. Recall that the general model for a simple linear regression model is

$$y = \beta_0 + \beta_1 x + \varepsilon$$

where

$y$ is the dependent variable

$x$ is the independent variable

$\beta_0$ is the theoretical, unknown $y$ intercept of the line

$\beta_1$ is the theoretical, unknown slope of the line

$\varepsilon$ are independent with a normal distribution with a mean of zero and a constant standard deviation

Using observations of $x$ and $y$, estimates of $\beta_0$ and $\beta_1$ can be computed. The prediction equation is given as

$$\hat{y} = b_0 + b_1 x$$

where

$b_0$ is the estimate of the $y$ intercept

$b_1$ is the estimate of the slope

When using regression to model time series data, the independent variable is time and the dependent variable is the variable you are interested in forecasting. The prediction model thus becomes

$$\hat{y} = b_0 + b_1 t$$

*It makes more sense to label the independent variable as t when using regression to model a time series.*

Typically, you number the first time period as $t = 1$, the second time period as $t = 2$, and so forth. This is what we have done with the FWC time series. It would not be wrong to use the actual years for the independent variable in this case and you would get the same results for the slope ($b_1$) but different intercepts ($b_0$). You will get the same forecast as long as you use the correct number for time in the prediction equation.

Let's check to see that this is the case.

## EXAMPLE 13.9  Baseball Attendance

### Regression Model for FWC Time Series

The output from running regression in Excel is shown on page 674. The column of numbers containing the population for families with children (FWC) was identified as the dependent (or $y$) values and the column of numbers ranging from 1 to 10 was identified as the range for the independent variable (or $x$ values).

SUMMARY OUTPUT

| Regression Statistics | |
|---|---|
| Multiple R | 0.8485124 |
| R Square | 0.7199733 |
| Adjusted R Square | 0.68497 |
| Standard Error | 73836.86 |
| Observations | 10 |

ANOVA

| | df | SS | MS | F | Significance F |
|---|---|---|---|---|---|
| Regression | 1 | 1.12138E+11 | 1.12E+11 | 20.56871 | 0.001911048 |
| Residual | 8 | 43615055164 | 5.45E+09 | | |
| Total | 9 | 1.55753E+11 | | | |

| | Coefficients | Standard Error | t Stat | P-value |
|---|---|---|---|---|
| Intercept | 1096755.1 | 50440.17795 | 21.74368 | 2.11E−08 |
| X Variable 1 | 36868.012 | 8129.17074 | 4.535273 | 0.001911 |

Extracting the values for $b_0$ and $b_1$ from this output, we can write the regression model as

$$\hat{y}_t = b_0 + b_1 t$$
$$= 1{,}096{,}755.1 + 36{,}868.012t$$

The forecast for time period 11 (year 1999) is calculated by using the regression model and substituting the value of 11 for $t =$ time. This calculation is shown below:

$$\hat{y}_{11} = 1{,}096{,}755.1 + (36{,}868.012)(11)$$
$$= 1{,}502{,}303$$

If we used the actual years as the independent variable values instead, we would get the following results:

| | Coefficients |
|---|---|
| Intercept | −72196853 |
| X Variable 1 | 36868.012 |

The model is

$$\hat{y}_t = -72{,}196{,}853 + 36{,}868.012t$$

Notice that the estimate of the slope, $b_1$, is the same but the intercept term, $b_0$, is different. The same forecast is obtained as long as we use the year 1999 as the value for time to be predicted:

$$\hat{y}_{1999} = -72{,}196{,}853 + (36{,}868.012)(1999)$$
$$= 1{,}502{,}303$$                                           ■

It is clearly possible to forecast more than one period into the future. If we wished to estimate the population for the year 2000 we would simply use the year 2000 (or time $t = 12$) as the value for time in the regression model. Thus,

$$\hat{y}_{2000} = -72{,}196{,}853 + (36{,}868.012)(2000)$$
$$= 1{,}539{,}171.279$$

The farther out in time that you try to forecast, the less reliable your forecast will be since you have no additional values.

It should be noted that, by using a regression model to forecast for the next period or beyond, you are extrapolating. You were warned not to do this in the chapter on regression. However, in analyzing time series data the objective is to make a prediction for the next time period or beyond that will always be outside the range of the data. You are always extrapolating the past into the future when you work with time series data. The assumption for any time series model is that the future "looks" like the past. Without this assumption you cannot do anything with the historical data. So it is not that you were unduly cautioned in the chapter on regression but that the objective is now slightly different.

*A word on extrapolation*

Now that you have seen that there is no difference in the forecast resulting from different ways of coding the time values, we will always number the values of time starting with the value $t = 1$. This is often the easiest thing to do because sometimes your time variable is a calendar date, an hour of the day, or a day of the week.

*Remember to label time starting with $t = 1$.*

## Try It Now!

### Baseball Attendance
#### *Regression Model*

Find the regression model and forecast for the number of people with no children (NC) for 1999 (time $t = 11$).

The data are shown below:

| Time | Year | NC |
|------|------|---------|
| 1 | 1989 | 564,790 |
| 2 | 1990 | 556,734 |
| 3 | 1991 | 653,134 |
| 4 | 1992 | 632,459 |
| 5 | 1993 | 642,387 |
| 6 | 1994 | 654,890 |
| 7 | 1995 | 657,238 |
| 8 | 1996 | 657,238 |
| 9 | 1997 | 717,903 |
| 10 | 1998 | 692,340 |

## 13.6.2 Evaluating the Regression Model

We have been using the mean square error to evaluate the moving average and exponential smoothing models. This is easily done for the regression model because any software package that you use to run the regression model will calculate the pre-

ANS. $b_0 = 563,461.53$
$b_1 = 14,445.412$
$\hat{y}_{11} = 722,361$

dicted values and the residuals for each value of $y_t$. These residuals can then be squared and averaged to get the MSE. These values are shown in the next example.

### EXAMPLE 13.10 Baseball Attendance

***Finding the MSE for the Regression Model***

The first three columns of the table below were generated by Excel as part of the output of regression. The last column is simply the residuals (or errors) squared and then this column is averaged to get the MSE.

RESIDUAL OUTPUT

| Observation | Predicted Y | Residuals | Residuals Squared |
|---|---|---|---|
| 1 | 1133623.1 | −34714.14545 | 1205071895 |
| 2 | 1170491.2 | 26693.84242 | 712561223.4 |
| 3 | 1207359.2 | −37499.1697 | 1406187728 |
| 4 | 1244227.2 | 106447.8182 | 11331137996 |
| 5 | 1281095.2 | 54117.80606 | 2928736933 |
| 6 | 1317963.2 | −110305.2061 | 12167238484 |
| 7 | 1354831.2 | 24891.78182 | 619600802.1 |
| 8 | 1391699.2 | −70242.2303 | 4933970918 |
| 9 | 1428567.2 | −40875.24242 | 1670785443 |
| 10 | 1465435.3 | 81484.74545 | 6639763742 |
| **MSE** | | | **4,361,505,516** |

A comparison of the MSE for the regression model to the MSE values for all of the other models of this time series that we have tried shows that the regression model is the best. This makes sense since we did see a slight upward trend in the data when we examined the scatter plot and we saw a pattern in the errors when we used a moving average model (all but one were positive).

In addition to using the MSE as a way to evaluate the regression model, we can also use the methods learned in Chapter 11. Specifically the value of $R^2$ should be examined and a hypothesis test for the significance of the slope term should be done. The next example serves as a reminder of these methods using the FWC time series.

### EXAMPLE 13.11 Baseball Attendance

***$R^2$ and Test for Significance for the Regression Model***

The information needed to evaluate the regression model is shown in the output in Example 13.9. The coefficient of determination for the regression model is $R^2 = 72\%$. Thus, 72% of the variability in the $y$ values is explained by the independent variable of time.

The hypothesis test for the slope is

$$H_0: \quad \beta_1 = 0$$
$$H_A: \quad \beta_1 \neq 0$$

The $t$-test statistic is shown in the coefficient table as 4.535273 with a $p$ value of 0.0019. The low $p$ value (less than 0.01) tells us to reject the null hypothesis and conclude that the slope is nonzero. This tells us that the model is useful.

## *TRY IT NOW!*

### Baseball Attendance

#### *Evaluating the Regression Model for the NC Time Series*

Evaluate the regression model for the number of people in families with no children (NC). Find the MSE, the value of $R^2$ and test for the significance of the slope term.

## 13.6.3 Exercises—Learning It!

**13.21** The Bureau of Labor Statistics (BLS) reports data on commodity prices on a monthly basis. These data are reported on a regional and national basis. The following graph displays the monthly average price (cents per pound) of white bread in the United States for the years 1980–1996.

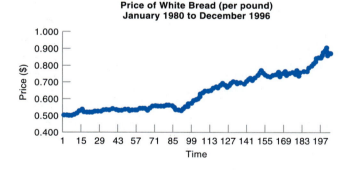

**Price of White Bread (per pound)
January 1980 to December 1996**

The output from running regression in Excel is shown below:

|  | Coefficients | Standard Error | t Stat | P-value |
|---|---|---|---|---|
| Intercept | 0.456250652 | 0.004516911 | 101.0094 | 7.1E−175 |
| Time | 0.001819793 | 3.82101E−05 | 47.62591 | 8.7E−112 |

(a) Predict the average price of bread for the next 3 months (time = 205, 206, 207).

(b) Test the significance of the slope for this model.

(c) Is the graph consistent with the results of your hypothesis test? Explain why or why not.

**13.22** The way companies do business is changing dramatically because of technology. The values shown below are the number of FAX machines per 1000 people for the years 1992–1995 for the United States (USA) and the United Kingdom (UK).

| | 1992 | 1993 | 1994 | 1995 |
|---|---|---|---|---|
| **USA** | 36 | 44 | 54 | 65 |
| **UK** | 17 | 22 | 26 | 31 |

(a) Find the slope and intercept for the regression model for the USA data.
(b) Use the slope and intercept to predict the number of FAX machines per 1000 people in the USA in 1996.
(c) Find the slope and intercept for the regression model for the UK data.
(d) Use the slope and intercept to predict the number of FAX machines per 1000 people in the UK in 1996.
(e) Comment on the rate of increase of FAX machines in the USA compared to the UK.

**13.23** The appliance industry continues to grow and change rapidly. The number of camcorders shipped for the years 1987 through 1996 are shown in the table below.

| Year | Number Shipped |
|---|---|
| 1987 | 1,604,153 |
| 1988 | 2,044,045 |
| 1989 | 2,286,326 |
| 1990 | 2,961,691 |
| 1991 | 2,864,395 |
| 1992 | 2,814,979 |
| 1993 | 3,088,427 |
| 1994 | 3,208,651 |
| 1995 | 3,560,497 |
| 1996 | 3,634,038 |

(a) Display these data as a line graph.
(b) Find the slope and intercept of the linear model that best fits these data.
(c) Use the slope and intercept to predict the number of camcorders shipped in 1997.
(d) Interpret the slope in terms of this application.
(e) Who might use this model and for what purpose?

*Requires Exercises 13.13, 13.18*

**13.24** Reconsider the sales of pizzas shown in Exercise 13.13.
(a) Find the slope and intercept of the regression model for these time series data.
(b) Compare the MSE for the regression model to the MSE for the weighted moving average model and the exponential smoothing model.

## 13.7 TIME SERIES ANALYSIS IN EXCEL

You learned how to do one of the forecasting methods described in this chapter when you learned about the linear regression model. Excel also has analysis tools for Moving Average and Exponential Smoothing models. In this section, we will look at two methods for the Moving Average model. For the Exponential Smoothing model, we will use the Excel data analysis tool.

Look at the problem of predicting the number of families with children that will attend Double A baseball games in a new stadium that is being built. The data collected are population data from the area from 1989 to 1998. The population figures are for two groups, families with children (FWC) and people 20–40 years old with no children (NC), both groups the stadium would like to attract. The data, in an Excel worksheet, are shown in Figure 13.5.

| | A | B | C |
|---|---|---|---|
| 1 | **Year** | **FWC** | **NC** |
| 2 | 1989 | 1,098,909 | 564,790 |
| 3 | 1990 | 1,197,185 | 556,734 |
| 4 | 1991 | 1,169,860 | 653,134 |
| 5 | 1992 | 1,350,675 | 632,459 |
| 6 | 1993 | 1,335,213 | 642,387 |
| 7 | 1994 | 1,207,658 | 654,890 |
| 8 | 1995 | 1,379,723 | 657,238 |
| 9 | 1996 | 1,321,457 | 657,238 |
| 10 | 1997 | 1,387,692 | 717,903 |
| 11 | 1998 | 1,546,920 | 692,340 |

**FIGURE 13.5**   Population data for baseball stadium

## 13.7.1 Moving Average Models in Excel

It is possible to analyze a moving average forecasting model in Excel using the **Add Trendline** option with a graph. To do this, you must first create a graph of the time series you want to analyze. The type of graph you want to create in Excel is a scatter plot, which you learned about in Chapter 5. Select the range that contains your data and make a scatter plot of the data. It should look similar to the one in Figure 13.6.

**FIGURE 13.6**   Scatter plot of FWC population data

Once the chart is created, follow these steps:

1. Click on the chart to select it, and click on any point on the line to select the data series. When you click on the chart to select it, a new option, **Chart,** is added to the menu bar.

2. From the **Chart** menu, select **Add T̲rendline.** The dialog box shown in Figure 13.7 opens.

**FIGURE 13.7**   The Add Trendline dialog box

**3.** Click on the Trend/Regression type box for **Moving average.** You can specify the number of periods you want to use in the model by entering the value in the textbox labeled **Period:** Type in 3, since we want a three-period moving average for this example. Click **OK;** a new line for the Moving Average model is inserted into the chart, as shown in Figure 13.8.

**FIGURE 13.8**    Chart with moving average trendline

Although this tool is simple to use, it is limited. The trendline appears on the chart, but you do not get the actual predicted values from the model anywhere in the worksheet; thus, you will be unable to calculate the error. Also, there is a trendline model only for the Moving Average model, and not for Exponential Smoothing. For these reasons, you might want to use the data analysis tools for time series. You can do this as follows:

**1.** From the list of Analysis Tools, select **Moving Average** and the dialog box shown in Figure 13.9 opens.

**FIGURE 13.9**    Moving Average dialog box

**2.** Position the cursor in the textbox labeled **Input Range:** and highlight the range that contains the data. Highlight only the actual data, not the column with the time periods. Put a check in the **Labels in First Row** box if the range contains a label.

3. Position the cursor in the textbox for **Interval:** and type in the number of periods you want to use for the moving average.

4. Now, position the cursor in the box labeled **Output Range:** and enter the location in the worksheet where you want the predicted values from the model to be located. Excel automatically skips the first value and enters NA for any values that cannot be predicted. For example, in a three-period moving average, the first value that can be predicted is $y_4$. To line up the predicted values with the actual data, position the cursor in the cell adjacent to the *second* data value. You cannot have the output from this tool placed in a new worksheet or new workbook.

*We do not use the standard errors for the predicted values, so there is no reason to include them.*

5. Finally put a check in the box for **Chart Output** and click **OK.** The worksheet will update to show the predicted values from the moving average model and a graph of the actual data and the predictions from the model, as shown in Figure 13.10.

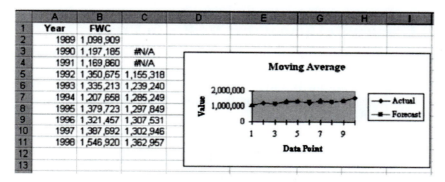

**FIGURE 13.10**  Output from moving average tool

If you are going to look at several models and try to choose the best model, then you will need to calculate the MSE for the model. To do this, we will use formulas in the worksheet:

1. First you must calculate the error for each prediction. Just as we saw in regression analysis, the errors provide valuable information about how well the predicted values match the actual data. Position the cursor in the cell adjacent to the first predicted value and type in a formula to find the difference between the data value and the predicted value.

*Be sure to remember the equal sign to tell Excel you are entering a formula.*

2. Next, you need to square the deviations. Position the cursor in the cell adjacent to the first error, and type in the formula "=E5^2".

3. Copy these formulas to the appropriate cells for the other data values and predicted values.

*Note: Your cell references should be the ones for your own data locations.*

4. In the cell directly below the squared errors, you will need to type in the rest of the formula for the MSE. That is, you must add up all of the squared deviations and divide by the number of values, $t - k$, in this case, 7. We can use the **SUM** and **COUNT** functions to do this, as shown in Figure 13.11.

=SUM(E5:E11)/COUNT(E5:E11)

**FIGURE 13.11**  Summing the squared deviations

The MSE will now appear in the cell just below the squared deviations. Figure 13.12 shows the worksheet with all of the calculated values.

| | A | B | C | D | E |
|---|---|---|---|---|---|
| 1 | Year | FWC | | Error | Squared Error |
| 2 | 1989 | 1,098,909 | | | |
| 3 | 1990 | 1,197,185 | #N/A | | |
| 4 | 1991 | 1,169,860 | #N/A | | |
| 5 | 1992 | 1,350,675 | 1,155,318 | 195,357 | 38,164,357,449 |
| 6 | 1993 | 1,335,213 | 1,239,240 | 95,973 | 9,210,816,729 |
| 7 | 1994 | 1,207,658 | 1,285,249 | -77,591 | 6,020,415,008 |
| 8 | 1995 | 1,379,723 | 1,297,849 | 81,874 | 6,703,406,459 |
| 9 | 1996 | 1,321,457 | 1,307,531 | 13,926 | 193,924,192 |
| 10 | 1997 | 1,387,692 | 1,302,946 | 84,746 | 7,181,884,516 |
| 11 | 1998 | 1,546,920 | 1,362,957 | 183,963 | 33,842,262,727 |
| 12 | | | | MSE | 14473866726 |
| 13 | | | | | |

**FIGURE 13.12**    Table with errors and MSE

## 13.7.2 Exponential Smoothing Models in Excel

The simplest way to analyze a time series using an Exponential Smoothing model in Excel is to use the data analysis tool. This tool works almost exactly like the one for Moving Average, except that you will need to input the value of $\alpha$ instead of the number of periods, $k$. The dialog box for the Exponential Smoothing analysis tool is shown in Figure 13.13.

**FIGURE 13.13**    The Exponential Smoothing dialog box

Once you have entered the data range and the damping factor, $\alpha$, and indicated what output you want and a location, the analysis is the same as the one for the Moving Average model.

# CHAPTER 13 SUMMARY

A lot of time series data are collected on virtually every aspect of life. The government, businesses, and even individuals keep track of many variables over periods of time. This type of data contains information on patterns and trends that can be used to forecast the behavior of the variable for the future. According to Peter Bernstein in his book entitled *Against the Gods: The Remarkable Story of Risk,* the idea that defines the boundary between modern times and the past is the mastery of risk. This means that the future is more than a whim of the gods and that men and women are not passive. Bernstein says that the ability to define what may happen in the future and to choose among alternatives lies at the heart of contemporary societies.

This chapter has provided you with a brief introduction to the topic of forecasting. You have learned some of the basic terminology and techniques in this interesting discipline. You learned how to use moving averages and weighted moving averages to predict future values of a numeric variable. You also saw an additional application of the technique of regression. Finally, in this chapter you learned how to evaluate the model in order to choose the "best" model.

## Key Terms

| Term | Definition | Page Reference |
|---|---|---|
| Time series | A **time series** is a set of observations of a variable at regular time intervals, such as yearly, monthly, weekly, or daily. | 651 |
| A *k*-period moving average | A *k*-period moving average is the average of the most recent *k* observations. | 657 |
| Simple moving average model | A **simple moving average model** uses the simple average of the most recent *k* observations to predict for the next time period. | 661 |
| Weighted moving average model | A **weighted moving average model** is a moving average model with unequal weights. | 663 |
| Exponential smoothing model | An **exponential smoothing model** is an averaging technique that uses unequal weights. The weights applied to past observations decline in an exponential manner. | 668 |
| Smoothing constant | The **smoothing constant, $\alpha$,** is the weight assigned to the most recent observation in an exponential smoothing model. | 669 |

## Key Formulas

| Term | Formula | Page Reference |
|---|---|---|
| **Simple moving average forecast** | $\hat{y}_{t+1}$ = average of the most recent $k$ observations $$= \frac{y_t + y_{t-1} + \cdots + y_{t-k+1}}{k}$$ | 657 |
| **Mean square error (MSE)** | $$MSE = \frac{\sum_{i=k+1}^{t} (y_i - \hat{y}_i)^2}{t - k}$$ | 659 |
| **Mean absolute deviation (MAD)** | $$MAD = \frac{\sum_{i=k+1}^{t} |y_i - \hat{y}_i|}{t - k}$$ | 661 |
| **Weighted moving average forecast** | $\hat{y}_{t+1}$ = weighted average of the most recent $k$ observations $$= w_t y_t + w_{t-1} y_{t-1} + \cdots + w_{t-k+1} y_{t-k+1}$$ | 666 |
| **Exponential smoothing forecast** | $$\hat{y}_{t+1} = \alpha y_t + \alpha(1 - \alpha)y_{t-1} + \alpha(1 - \alpha)^2 y_{t-2} \\ + \cdots + \alpha(1 - \alpha)^n y_{t-n} + \cdots$$ | 669 |

# CHAPTER 13 EXERCISES

## Learning It!

*Datafile: OZONE.XXX*

**13.25** A number of cities in the United States have had severe ozone problems caused by the burning of fossil fuels. The problem has been severe in cities such as Denver and Los Angeles (LA). The average monthly ozone rate in LA from 1979 to 1996 is given in the datafile. The first few lines of this datafile are shown below:

| | |
|---|---|
| Jan 1979 | 2.7 |
| Feb | 2 |
| Mar | 3.6 |
| Apr | 5 |
| May | 6.5 |
| Jun | 6.1 |

(a) Find the MSE for a 12-month simple moving average model of these data.

(b) Find the MSE for an exponential smoothing model of these data with $\alpha = 0.90$.

(c) Find the MSE for the regression model of these data.

(d) Select the model with the lowest MSE and forecast the average monthly ozone rate for January 1997.

**13.26** Most hospitals monitor length of stay of patients. These data help administrators to effectively use resources and staff. The average length of stay (in days) by month for a medium-sized New England city hospital are shown below:

| Month | Length of Stay |
|---|---|
| Jan 1996 | 7.15 |
| Feb | 4.97 |
| Mar | 5.3 |
| Apr | 5.9 |
| May | 5.62 |
| Jun | 5.55 |
| Jul | 4.69 |
| Aug | 5.16 |
| Sep | 4.85 |
| Oct | 5.4 |
| Nov | 8.12 |
| Dec | 5.7 |

(a) Display these data as a line graph.

(b) Find the MSE for a 3-month simple moving average model of these data.

(c) Find the MSE for a 3-month weighted moving average model using $w_t = 0.70$, $w_{t-1} = 0.20$, $w_{t-2} = 0.10$.

(d) Find the MSE for an exponential smoothing model of these data with $\alpha = 0.80$.

(e) Select the model with the lowest MSE and forecast the average length of stay for January 1997.

**13.27** Technology is affecting virtually every aspect of our lives. An increasing number of school districts across the United States are utilizing CD-ROM technology in educating children. The percentage of school districts using CD-ROM technology is shown below:

| Year | % Using CD-ROM |
|------|------|
| 1989 | 3.0 |
| 1990 | 4.5 |
| 1991 | 9.1 |
| 1992 | 11.4 |
| 1993 | 19.0 |
| 1994 | 42.6 |
| 1995 | 58.7 |

(a) Display this time series as a line graph.

(b) Which of the forecasting techniques would you recommend for these data and why?

(c) Develop a forecast for the percentage of school districts using CD-ROM technology in 1996 using the technique you identified in part (b).

**13.28** Virtually all medical research involves the collection and analysis of data. There are many job opportunities in the field of biostatistics. The American Statistical Association (ASA) did a salary survey of faculty teaching in Biostatistics Departments. The data shown below are the median starting salaries for assistant professors from 1990 to 1996:

| | |
|------|------|
| 1990 | $45,000 |
| 1991 | $45,000 |
| 1992 | $49,440 |
| 1993 | $48,000 |
| 1994 | $49,578 |
| 1995 | $50,250 |
| 1996 | $53,250 |

(a) Display these data as a line graph.

(b) Find the MSE for a 2-month moving average model.

(c) Find the MSE for an exponential smoothing model with $\alpha = 0.70$.

(d) Using the model with the smaller MSE, predict the starting salary for assistant professors of Biostatistics in 1997.

**13.29** The data presented in the file RIVER.XXX show the average monthly flow of the Elbe River in Canada from 1982 to 1996. The Department of Natural Resources monitors the water flow to detect any shifts in the ecology of the region. The first few lines of the data file are shown below:        *Datafile: RIVER.XXX*

| | |
|------|------|
| Jan 1982 | 170 |
| Feb | 376 |
| Mar | 658 |
| Apr | 311 |
| May | 205 |
| Jun | 385 |

(a) Find the MSE for a 12-month moving average model.

(b) Find the MSE for a 12-month weighted moving average model with the following weights: 0.34, 0.10, 0.11, 0, 0, 0, 0, 0, 0.02, 0.14, 0.25, 0.04. These weights are the optimal weights for a 12-month moving average model.

(c) Find the MSE for an exponential smoothing model with $\alpha = 0.55$. This is the optimal value of $\alpha$ for an exponential smoothing model.

(d) Find the MSE for a regression model.

(e) Using the model with the smallest MSE, predict the average monthly river flow for January 1997.

*Datafile: FIRES.XXX* **13.30** Each year during the "dry" season, a number of western states experience forest fires. Some of these fires are ignited by natural elements such as lightning, while others are started by arson. The FIRES.XXX file shows the number of acres lost in forest fires by year from 1941 to 1996. The first few lines of this datafile are shown below:

| | |
|---|---|
| 1941 | 1,009,000 |
| 1942 | 1,475,000 |
| 1943 | 856,000 |
| 1944 | 3,027,000 |
| 1945 | 4,271,000 |
| 1946 | 3,126,000 |
| 1947 | 1,115,000 |

(a) Find the MSE for a 3-year, 6-year, and 9-year moving average model.

(b) Using the model with the smallest MSE, forecast the number of acres to be lost in 1997.

## Thinking About It!

*Requires Exercise 13.20* **13.31** The forecast for the bottle redemption center done in Exercise 13.20 is based on one week of data. You now have two more weeks of bottle returns as shown below:

| | Week 1 | Week 2 | Week 3 |
|---|---|---|---|
| Mon | 148,454 | 154,737 | 133,633 |
| Tues | 180,000 | 183,968 | 180,202 |
| Wed | 147,053 | 137,972 | 151,757 |
| Thurs | 152,197 | 132,140 | 136,850 |
| Fri | 150,583 | 135,596 | 156,593 |
| Sat | 122,366 | 153,452 | 140,539 |

(a) What do you notice when you look at the graph for 3 weeks of data?

(b) What suggestions do you have for forecasting the number of returns for week 4?

*Requires Exercise 13.13* **13.32** When you talk to the owner of the pizza restaurant you discover that in January 1996 they began a delivery service to local businesses. What do you recommend be done with the data prior to January 1996 in building a forecasting model?

**13.33** Frequently when you are asked to analyze time series data, data are missing for some of the time periods. Think about how you would handle missing data. What suggestions do you have for dealing with missing data?

*Requires Exercise 13.26* **13.34** When you speak to the administrators of the hospital in Exercise 13.26 you discover that the length of stay data are collected by groups of physicians. Ten physician groups are tracked at this hospital. Suggest ways in which these data could be misused to the detriment of the patients.

**13.35** In most of the exercises in this chapter we have selected the model with the smallest MSE as the "best" model.

(a) If we used the MAD as our criterion instead of the MSE, would we always get the same "best" model?

(b) What characteristics in the dataset would lead you to the same "best" model regardless of whether you used the MSE or the MAD?

(c) What characteristics in the dataset would lead you to different "best" models depending on whether you used the minimum MSE or the minimum MAD as your criterion?

## Doing It!

**13.36** Due to population and economic development, the power consumption in Nigeria has steadily increased over the past several years. The data in the file POWER.XXX show the average monthly power consumption from January 1989 to December 1996 in kilowatt-hours. A portion of the datafile is shown below:

*Datafile: POWER.XXX*

| | |
|---|---|
| Jan 1989 | 56,634 |
| Feb | 56,427 |
| Mar | 65,170 |
| Apr | 63,847 |
| May | 56,864 |
| Jun | 67,288 |

(a) Display these data as a line graph.

(b) Find the MSE for a 12-month moving average model.

(c) Find the MSE for a 12-month optimized moving average model.

(d) Find the MSE for an exponential smoothing model with an optimized value of $\alpha$.

(e) Find the MSE for a regression model.

(f) Using the model with the smallest MSE, forecast the consumption for 1997.

**13.37** The daily opening, high, low, and closing prices of live hogs traded on the commodities market are shown in the datafile HOGS.XXX from 1 July 1996 to 31 December 1996.

*Datafile: HOGS.XXX*

| LH_M97 | Open | Hi | Lo | Close |
|---|---|---|---|---|
| 1-Jul-96 | 77.775 | 77.775 | 77.775 | 77.775 |
| 2-Jul-96 | 78.000 | 78.800 | 78.050 | 78.300 |
| 3-Jul-96 | 78.250 | 79.000 | 78.000 | 79.000 |
| 5-Jul-96 | 79.000 | 79.600 | 78.800 | 79.600 |
| 8-Jul-96 | 79.450 | 79.450 | 79.000 | 79.400 |
| 9-Jul-96 | 79.050 | 79.250 | 78.600 | 79.125 |
| 10-Jul-96 | 78.975 | 79.100 | 78.975 | 79.100 |

(a) Display each of the 4 time series on one graph.

(b) What patterns do you see in the data? Do all 4 series "move" in the same manner?

(c) Find the model that minimizes the MSE for each of the 4 series. Consider only a 6-day moving average, a 6-day weighted moving average with optimal weights, an exponential smoothing model with optimal $\alpha$, and a regression model.

(d) Find the model that minimizes the MAD for each of the 4 series. Consider only a 6-day moving average, a 6-day weighted moving average with optimal weights, an exponential smoothing model with optimal $\alpha$, and a regression model.

(e) Summarize your findings and recommend a model to forecast the opening, high, low, and closing prices for 2 January 1997. Why is no forecast needed for 1 January 1997?

# Part VI
# ADDITIONAL TOPICS

# CHAPTER 14

# EXPERIMENTAL DESIGN AND ANOVA

## THE AIRSPACE PROBLEM

Have you ever pulled the first tissue out of the box and had it tear? Have you ever opened a box of tissues and in trying to get one tissue out, ended up with several tissues? The problem causing both of these annoying things to happen is that there is not enough airspace in the box. This problem became an important issue for a large manufacturer of tissues when an unusually high number of complaints were registered.

Airspace is defined to be the amount of space between the top of the tissues and the top of the box. It is measured in millimeters and it should be at least 9 mm. Even if there is 9 mm of airspace when the box is manufactured, there may not be enough airspace by the time the customer opens the box. This is due to a phenomenon called "growback." As the box sits in the warehouse or on the supermarket shelf, the tissues, which were heavily compressed when they were put into the box, begin to expand or "grow back." Thus, the airspace is reduced.

The tissue company needs to estimate how much growback will occur in order to determine how much airspace should be left in the box at the time of manufacturing. A portion of the data is shown below:

| Position | Time | Airspace |
|----------|------|----------|
| 1 | 1 | 23 |
| 1 | 1 | 25 |
| 1 | 1 | 23 |
| 1 | 1 | 23 |
| 1 | 1 | 23 |

# 14.1 CHAPTER OBJECTIVES

*A simple random sample is a sample that has been selected in such a way that all members of the population have an equal chance of being selected.*

Throughout most of this book we have assumed that the samples have been selected at random from the population we are studying. These types of samples are called simple random samples. They were first defined in Chapter 2. The reasons we have studied primarily simple random samples are twofold. First, very often data are collected this way. Second, data analysis is simplest when the samples are simple random samples, so it makes sense to present these samples first when learning the data analysis techniques.

However, we are now at a point where it makes sense to think about *how* the sample is selected in order to get the most information for our money. Remembering that a statistical experiment is any action whose outcomes are recordable data, it is time to learn how to set up or design that statistical experiment. In doing so we will specify how the data are to be collected. You may recall that in Chapter 2 we recognized the fact that the amount of resources we have available does constrain the size of our sample. However, the formulas for determining the sample size that we learned about in Section 7.7 do not explicitly take this into account. So, it seems that we have, in effect, ignored the fact that we have limited resources. This chapter attempts to address that issue.

By now, we know that the techniques used to analyze data must take into account the type of data being analyzed and how the data were collected. Thus, this chapter also introduces a technique to analyze data that result from a designed experiment. This technique is called Analysis of Variance, which is abbreviated ANOVA.

In particular, this chapter covers the following topics:

- Motivation for Using a Designed Experiment
- Analysis of Data From One-Way Designs
- Assumptions of ANOVA
- Analysis of Data From Blocked Designs
- Analysis of Data From Two-Way Designs
- Other Types of Experimental Designs

# 14.2 MOTIVATION FOR USING A DESIGNED EXPERIMENT

To help the tissue company, you might suggest that the company take a random sample of tissue boxes and measure the amount of airspace in the box. Let's follow that idea for a minute. When should you measure the airspace? If you suggested measuring the airspace at the time that the tissue box is manufactured, then you are partway to the right answer. These are called in-process data, since they are taken at the time the box is manufactured.

Suppose there is an average of 12 mm of airspace in the tissue boxes right after they are manufactured. Is this enough to ensure that, even after growback, there is an average of 9 mm of airspace by the time the customer opens the tissue box? It looks like we need to check the airspace in the box at some future point in time, say 2 weeks after the box is made. Can we use the same boxes and measure the airspace 2 weeks later? Clearly, we cannot since the boxes need to be opened to measure the airspace right after they are manufactured. This is an example of destructive inspection. The product is, in a sense, destroyed once we open it and measure the airspace. We certainly cannot sell opened tissue boxes! We need to use another set of boxes that have been sitting on the shelf for 2 weeks.

Initially, it looks as if we could use the two population tests of Chapter 10 to compare the mean airspace of two populations: the population of boxes right after manufacturing is completed and the population of boxes that have been sitting on the shelf for 2 weeks. If these samples were selected randomly from the two populations then this would work. What might happen if we use this approach? Well, the boxes that we check 2 weeks after manufacturing might have been manufactured on a different day, under different conditions than those we check right after manufacturing. How can we tell if the differences in airspace are due to different manufacturing conditions or the "growback" during the 2-week time delay? If we decided that the boxes checked after 2 weeks should be selected at the same time as the ones checked immediately then we will have addressed this issue. We would have samples that are, in fact, comparable. They are matched in terms of the manufacturing conditions. This is an example of the simplest type of experimental design.

Actually, you have already been introduced to the idea of a designed experiment in Section 10.7. In that section, we recognized that in some situations, other characteristics will influence the variable being observed. One example we looked at was the effect on sales of a new product display. Clearly, the size of the store influences sales and to do a fair comparison we compared stores of similar size. The stores were matched on the size characteristic. These types of data were called paired data or dependent samples. The technique to analyze these types of data was a paired $t$ test. It was presented in Chapter 10 because in that chapter we were learning how to draw conclusions about two population parameters.

*Remember: In dependent samples the members of one sample are identical to or matched with members in the other sample.*

If we were to just compare the average airspace in process to the average airspace 2 weeks later then we could use a $t$ test. But how do we know that 2 weeks is the right time to check the airspace? Perhaps we should examine the airspace at several different times. This is precisely what was done at the tissue company. A total of 4 cases were sampled, one case every hour for 4 hours. One case was used to obtain in-process measurements. The other 3 cases were saved for observations taken after 24 hours, 2 weeks, and 4 weeks. Each case contained 24 cartons of tissue boxes or 120 tissue boxes.

A portion of the data is shown in Figure 14.1. The variable *Time* indicates when the measurement was taken and is coded as follows: 1 = in process, 2 = after 24 hours, 3 = after 2 weeks, and 4 = after 4 weeks. The variable *Airspace* is the measurement of the airspace measured in millimeters. Figure 14.2 shows the sample mean and sample standard deviation for each of the 4 time periods.

| Time | Airspace |
|------|----------|
| 1 | 23 |
| 1 | 25 |
| 1 | 23 |
| 1 | 23 |
| 1 | 23 |

**FIGURE 14.1**  A portion of the tissue company dataset

| Time | Sample Mean | Sample Standard Deviation |
|------|-------------|---------------------------|
| 1 | 21.01 | 1.53 |
| 2 | 17.46 | 1.48 |
| 3 | 15.78 | 2.08 |
| 4 | 15.02 | 2.32 |

**FIGURE 14.2**  Summary statistics for the tissue company

As you can see, we are now comparing the mean airspace of four different samples to draw a conclusion about four different populations: the population of boxes at the time of manufacturing (in process), the population of boxes that have been sitting on the shelf for 24 hours, the population of boxes that have been sitting on the shelf for 2 weeks, and the population of boxes that have been sitting on the shelf for 4 weeks. The average amount of airspace seems to be decreasing the longer the tissue box sits on the shelf. But we know from our work in earlier chapters that just because the sample means are different that does not automatically mean that the population means are different. We would like to be able to tell the tissue company if the mean airspace is the same for all four of these populations. Thus, our null and alternative hypotheses are as follows:

$$H_0: \quad \mu_1 = \mu_2 = \mu_3 = \mu_4$$

$H_A:$    At least one of the population means is different from the others.

We need a technique that extends the work we did in Chapter 10 beyond two populations. If the population means test different, that is, if we reject the null hypothesis, then we would like a model to explain how the airspace decreases as the box sits on the shelf for a longer period of time. Both of these issues are addressed in the next section.

# 14.3 ANALYSIS OF DATA FROM ONE-WAY DESIGNS

## 14.3.1 One-Way Designs: The Basics

The tissue company is interested in comparing the characteristics of four populations that differ on the basis of one **factor:** the amount of time that has elapsed since the box was manufactured. In this experiment the factor of interest has four different **levels** or possible values.

> A *factor* is a variable that can be used to differentiate one group or population from another. It is a variable that may be related to he variable of interest. A *level* is one of several possible values or settings that the factor can assume.

The variable that is being studied is referred to as the **response variable.** This is just a further description of the term "variable" which you have been working with throughout this text. The descriptor word "response" in front of the word variable indicates that what you are measuring or observing may respond to the experimental conditions, that is, the setting of the factor(s).

> The *response variable* is a quantitative variable that you are measuring or observing.

Many situations involve populations that differ on the basis of one factor. In fact, all of our work in Chapter 10 involved comparing populations that were different on the basis of one factor and that factor had only two levels. In Chapter 2 we said that a qualitative variable is often used to divide a large population into two smaller groups for comparisons. A commonly used qualitative variable is gender, which clearly has two levels: male and female. Thus, using the techniques of Chapter 10, we would compare the average salary of males and females. The factor would be gender and we would be comparing average male salaries to average female salaries.

The natural extension of this situation is to handle a factor that has more than two levels. For example, a manufacturer of diapers might wish to study the differences in the amount of fluid the diaper can absorb when different filler materials are used. Here, the factor is the filler material. If the diaper company is considering six different materials for the filler, then there are six levels of this factor. The response variable is the amount of fluid the diaper can absorb. A university career office might wish to study differences in starting salaries for different majors. Here, the factor is the academic major and the response variable is starting salary. If the career office is studying engineering, business, and humanities majors, then there are three levels of this factor. A soft drink manufacturer might be interested in studying the taste test of 4 different versions of a new drink. In this case, the factor is the version of the drink and there are 4 levels. The response variable is the taste test score. A medical researcher may wish to study the amount of time it takes for five different experimental drugs to work. In this case, the factor is the drug and there are five different levels. The response variable is the time it takes for the drug to work. A manufacturer may wish to study the life of products made at 3 different plants. In this case the plant location is the factor and there are three levels. The response variable is the life of the product.

Notice that, in each of these situations, the response variable is a quantitative variable (airspace, absorbency, salary, tastiness, time to work, product life) and the factor takes on one of several possible values. These are all examples of **one-way** or **completely randomized designs.**

> An experiment has a *one-way* or *completely randomized design* if there are several different levels of one factor being studied and the objects or people being observed/measured are randomly assigned to one of the levels of the factor.

The term *one-way* refers to the fact that the groups differ with regard to the one factor being studied. The term *completely randomized* refers to the fact that individual observations are assigned to the groups in a random manner. For example, in the case of the soft drink manufacturer, people would be assigned to sample one of the 4 versions of the drink in a random manner.

 **TRY IT NOW!**

## One-Way Designs
### *Designing a Simple Study*

Select a population that you might be interested in studying and identify a quantitative variable that you might wish to analyze.

Now, specify a factor that might be of interest. This should be some characteristic that you think might influence the variable you are analyzing.

*(continued)*

Indicate the various levels of the factor.

---

## 14.3.2 Understanding the Total Variation

When the data are the result of a one-way design and certain assumptions are met, then the proper tool to analyze the data is called **analysis of variance (ANOVA).**

> *Analysis of variance (ANOVA)* is the technique used to analyze the variation in the data to determine if more than two population means are equal.

R. A. Fisher originated the technique of ANOVA in England in connection with agricultural experiments. These experiments studied the yield of crops when the farmland was treated differently. In this case the factor was the treatment of the farmland and the response variable was the yield of the crop. The word **treatment** has remained a part of the language of designed experiments. It need not refer to treatment of the farmland, but more generally it refers to a particular setting or combination of settings of the factor(s) being studied.

> A *treatment* is a particular setting or combination of settings of the factor(s) being studied.

Your initial reaction to the name of this technique is most likely that it seems to be misnamed. After all, we are really interested in determining if the groups, which correspond to differing levels of the factor of interest, have different means. What does the variance have to do with this technique and why are we analyzing the variance and not examining the means? This is completely normal thinking at this point. Interestingly enough, it is through an analysis and breakdown of what is causing the variation that we see in the data that we reach conclusions about the group means! Let's see how this works.

*Remember that the units of measure on a variance calculation are the original units squared.*

Suppose we think about *all* of the data that were collected for the tissue company—that is, 120 observations per time period times the 4 time periods, or 480 observations of airspace. We are interested in deciding if the airspace changes, so we might ask the question, How different are these airspace measurements from each other? We know from our work in descriptive statistics that we measure how spread out the data values are by using a measure of spread or variation. Thus, we could calculate the variance of the entire dataset. The variance is calculated using the formula for the sample variance that we learned in Chapter 4:

*Sample variance formula*

$$s^2 = \frac{\sum_{i=1}^{n}(x_i - \bar{x})^2}{n - 1}$$

In this case $n = 480$ and the value of $\bar{x}$, the overall average for the dataset, is found to be 17.32 mm. Since the data were collected in 4 different groups corresponding to the 4 levels of the factor time, it is convenient to label each observation with a double subscript rather than just a single subscript ranging from 1 to 480. We use the letter $c$ for the number of different levels of the factor. In this case $c = 4$. Each observation can be written as $x_{ij}$, where the first subscript, $i$, tells you how the observation is numbered within the group and the second subscript, $j$, tells you what group the observation is in.

### EXAMPLE 14.1   Airspace Data

**Setting Up the Notation**

In the case of the tissue company, $x_{11}$ would refer to the first observation in the first group, which was the in-process data. Similarly, $x_{14}$ would refer to the first observation in the fourth group, the data collected 4 weeks after the time of manufacturing.

■

 **TRY IT NOW!**

### Airspace Data
**Getting Used to the ANOVA Notation**

Using the information in Figure 14.1, what is the value of $x_{31}$?

What notation would be used to refer to the 120th observation of the data taken 2 weeks after the time of manufacturing?

What is the range of values for the first subscript for group 1?

What is the range of values for the second subscript?

Carrying the benefits of this notation further, we notice that the number of observations in each group does not have to be the same. Thus, $n$, which is the total number of observations in the experiment, can be written as follows:

$$n = n_1 + n_2 + \cdots + n_j + \cdots + n_c$$

where each $n_j$ is the sample size of group $j$ and $c$ is the number of different groups or levels of the factor.

## EXAMPLE 14.2    Airspace Data

### Sample Size

For the tissue company, there are 4 time periods or 4 levels of the factor, so $c = 4$. There are 120 observations at each time period so $n_1 = n_2 = n_3 = n_4 = 120$ and

$$n = 120 + 120 + 120 + 120 = 480 \qquad \blacksquare$$

Finally the **overall mean** is relabeled $\bar{\bar{x}}$. This is read as "x double bar." For the tissue company $\bar{\bar{x}}$ is 17.32 mm. It is based on all 480 of the observations and therefore is sometimes called the **grand mean.**

> The **grand mean** or the **overall mean** is the sample average of all the observations in the experiment. It is labeled $\bar{\bar{x}}$.

We can now rewrite the variance calculation as follows:

*Formula for sample variance with double subscript notation*

$$s^2 = \frac{\sum\limits_{j=1}^{c} \sum\limits_{i=1}^{n_j} (x_{ij} - \bar{\bar{x}})^2}{n - 1}$$

This formula yields the same result as the formula for the variance we learned in Chapter 4. The numerator is simply rewritten to reflect the fact that each observation belongs to some level of the factor.

The numerator of this formula is called the **total variation** or **Sum of Squares Total (SST).**

> The **total variation** or **sum of squares total (SST)** is a measure of the variability in the entire data set considered as a whole.

SST is calculated as follows:

*Formula for sum of squares total (SST)*

$$\sum_{j=1}^{c} \sum_{i=1}^{n_j} (x_{ij} - \bar{\bar{x}})^2$$

The technique of ANOVA breaks down this total variation into several components and examines the contribution each component makes to the total variation. So instead of analyzing the sample variance, we will actually be analyzing the total variation, which is the numerator of the sample variance.

You could calculate SST by hand, but it can be a long and tedious calculation, even for small datasets. The next example shows you the first few terms of the calculation of SST for the tissue company.

## EXAMPLE 14.3    Airspace Data

### Calculation of SST for the Tissue Company

Using the information in Figure 14.1 and the fact that $\bar{\bar{x}} = 17.32$ mm, we can set up the first five terms of the calculation of SST for the tissue company:

| Time | Airspace, $x_{ij}$ | $(x_{ij} - \bar{\bar{x}})^2$ |
|------|------|------|
| 1 | 23 | $(23 - 17.32)^2 = 32.3$ |
| 1 | 25 | $(25 - 17.32)^2 = 59.0$ |
| 1 | 23 | $(23 - 17.32)^2 = 32.3$ |
| 1 | 23 | $(23 - 17.32)^2 = 32.3$ |
| 1 | 23 | $(23 - 17.32)^2 = 32.3$ |

The calculation shown in column 3 of the table would be done for all 480 observations and then totaled. This would give you a value of SST = 4249.87. ∎

Any statistical package that you use will calculate SST for you and produce what is known as an ANOVA table. You should recall that you have been introduced to SST and ANOVA tables in Chapter 11, Regression Analysis. The formula for SST given in Chapter 11 is written like the numerator of the variance formula presented above, before we introduced the double subscript notation. It measures the same thing as our formula for SST here. The ANOVA tables produced by Excel and MINITAB are shown in the next example.

## EXAMPLE 14.4 Airspace Data

***Output from Excel and Minitab***

ANOVA table from Excel and Minitab for the tissue company.

ANOVA

| Source of Variation | SS | df | MS | F | P-value | F crit |
|------|------|------|------|------|------|------|
| Between Groups | 2554.75 | 3 | 851.58333 | 239.13025 | 1.342E-94 | 2.6236364 |
| Within Groups | 1695.116667 | 476 | 3.5611695 | | | |
| Total | 4249.866667 | 479 | | | | |

**Excel output:** SST is shown in the line labeled Total under the column Source of Variation.

```
Analysis of Variance for Airspace

Source    DF      SS        MS       F       P
Time       3    2554.75    851.58   239.13   0.000
Error    476    1695.12      3.56
Total    479    4249.87
```

**Minitab output:** SST is shown in the line labeled Total under the column heading Source. ∎

Other values are printed in the ANOVA table besides SST. The next section explains what the other numbers mean and how to use them.

## 14.3.3 Components of Total Variation

If you look at the ANOVA tables shown in Example 14.4 you will see that there is a column labeled SS in both the Excel and the Minitab outputs. The SS label stands for sum of squares. So far we have seen what the SST or total sum of squares term means and how it is calculated. A small amount of detective work leads you to believe that the two numbers above SST add up to SST. This is exactly correct. The technique of ANOVA focuses on the relative sizes of these components of the total variation.

To start, we will use the Excel output terminology and then relate the discussion to the Minitab output. Looking at the words in the first column of the ANOVA table,

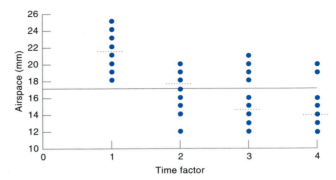

**FIGURE 14.3**  Dotplot of airspace by time for the tissue company

we can see that one of the components of SST is called the **Between Groups Variation** and the other component is called the **Within Groups Variation.** Let's try to understand what each of these components measures.

The between groups variation is also called the **Sum or Squares Between** or the **Sum of Squares Among** and it measures how much of the total variation comes from actual differences in the treatments. The dotplot shown in Figure 14.3 displays the sample average for each of the four time treatments. These are called **treatment means.**

> A *treatment mean* is the average of the response variable for a particular treatment.

*Note: The term "pretty different" will be quantified once we get a handle on the concept.*

By examining the graph we can see that the treatment means are "pretty different" from the grand mean of 17.32 mm. Thus, we might expect that a substantial portion of the total variation is due to the fact that the airspace measurements were taken at four different times. This is what the between groups variation component measures.

In computing this component, you look at differences between the treatment means ($\bar{x}_j$, shown as dotted lines) and the overall mean ($\bar{\bar{x}}$, shown as a solid line). Those differences are weighted by the sample size of each group ($n_j$).

> *Between Groups Variation* measures how different the individual treatment means are from the overall grand mean. It is often called the *sum of squares between* or the *sum of squares among (SSA).*

The formula for sum of squares among (SSA) is

*Formula for sum of squares among (SSA)*

$$SSA = \sum n_j(\bar{x}_j - \bar{\bar{x}})^2$$

Let's verify the calculation of SSA for the tissue company.

### EXAMPLE 14.5  Airspace Data

*Calculation of SSA*

The calculation of SSA for the tissue company is shown in the following table:

| Column A Time | Column B Sample Mean | Column C (Sample Mean − Grand Mean)$^2$ | Column D Column C × Group Sample Size |
|---|---|---|---|
| 1 | 21.01 | 13.63 | 1635.41 |
| 2 | 17.46 | 0.02 | 2.41 |
| 3 | 15.78 | 2.35 | 282.13 |
| 4 | 15.02 | 5.29 | 634.80 |
| | | SSA | 2554.75 |

The calculation of SSA is not as long and tedious as that of SST because there are only $c$ terms, but SSA is always part of an ANOVA table regardless of what statistical package you are using. There is really little value in calculating it by hand. Since you are learning the technique of ANOVA for the first time, it may help you to understand what the number SSA measures. For this reason it is time to try your hand at a small example.

 **TRY IT NOW!**

### Career Office

*Calculation of SSB*

The career office is interested in studying starting salaries for 3 different majors: Engineering, Business, and Humanities. The overall average starting salary was $\bar{\bar{x}} = \$28,200$. There were 30 students in each group and the averages are shown below. Find SSA.

| Column A Major | Column B Sample Mean | Column C (Sample Mean − Grand Mean)$^2$ | Column D Column C × Group Sample Size |
|---|---|---|---|
| Engineering | $38,100 | | |
| Business | $25,600 | | |
| Humanities | $20,900 | | |
| Total | | | |

Notice that the between group variation is labeled with the word "time" in the Minitab output under the column Source (of variation). This is because the label "time" was used when the data were entered into Minitab to indicate that time was the factor being studied.

The second component of the total variation is labeled "within groups variation" in the Excel output and it is labeled "error" in the Minitab output. These are the most common labels used to identify this component of variation. This component is also called experimental error, which explains why the word "error" is used in the Minitab output. Once again, this is not an error indicating that you have made a mistake but rather an indication of the fact that you are studying a sample.

If you take another look at the dotplot shown in Figure 14.3 you can see the within groups variation there as well. Concentrate for the moment on time period 1 or one group. Clearly, the depth of airspace in the boxes checked at the time of manufacturing differ from each other and from the treatment mean for that time period ($\bar{x}_1 = 21.01$ mm). It is this variation that is captured in the within groups component of the total variation. The difference between each observation and the mean of the group it is in is squared and these squared differences are accumulated into the **Sum of Squares Within,** which is also called the **Sum of Squares Error (SSE).**

> Within groups variation measures the variability in the measurements within the groups. It is often called sum of squares within or *sum of squares error (SSE).*

The sum of squares error is calculated as follows:

*Formula for sum of squares error (SSE)*

$$\textbf{SSE} = \sum_{j=1}^{c} \sum_{i=1}^{n_j} (x_{ij} - \bar{x}_j)^2$$

Let's set up the first few terms for the calculation of SSE for the tissue company. To show the detailed calculation of SSE for the tissue company we would have to show all 120 squared differences. It is not particularly helpful to do this.

### EXAMPLE 14.6  Airspace Data

***First 5 Terms of SSE***

Using the information in Figure 14.1 and the fact that $\bar{x}_1 = 21.01$ mm, we can set up the first five terms of the calculation of SSE for the tissue company:

SSE $= (23 - 21.01)^2 + (25 - 21.01)^2 + (23 - 21.01)^2 + (23 - 21.01)^2$
$\qquad + (23 - 21.01)^2 + \cdots$

We know from the computer output that SSE $= 1695.12$. ∎

Since SSA and SSE are the only two components of SST and you already know two of these numbers you can find SSE by simple subtraction. Thus,

$$SSE = SST - SSA$$

This formula shows us that SSE measures whatever variation is not a result of the fact that the data were collected in were several different groups.

### EXAMPLE 14.7  Airspace Data

***Finding SSE by Subtraction***

We can find SSE for the airspace data by subtraction:

$$SSE = SST - SSA = 4249.87 - 2554.75 = 1695.12$$ ∎

Let's summarize what we know so far about the airspace dataset. From the dot-plot and the relative size of SSA compared to SSE, it seems that much of the total variation in the data is due to the time factor being studied. To draw the inference that the means of the four populations are in fact different, we must perform a hypothesis test using the information in the ANOVA table. This is discussed in the next section.

Before moving on, one more example is presented. This example was first introduced in Chapter 9. The same scenario was reconsidered in Section 10.4 when you learned how to compare the mean of 2 populations. However, there are really 3 populations in this situation. Now you have the right tool to do the complete analysis.

## EXAMPLE 14.8 Tissue Strengths

### Finding SST, SSA, and SSE

You might remember from Chapter 9 that customer complaints about the tissues tearing led the company to look at the strength of the tissues. Two tissue strength measurements were studied: machine-directional strength *(MDStrength)* and the cross-directional strength *(CDStrength)*. They are both measured in lb/ream. We will look at only the *MDStrength* in this example. The data were collected over 3 days and the company would like to know if the average *MDStrength* is the same for all 3 days.

In the complete dataset there are 75 observations per day. In this example we will look at only five observations per day so you can see the computational details. At the end of the chapter you will have the opportunity to do a complete analysis of this dataset.

The five MD strengths for each of the 3 days are shown below.

|  | Day 1 | Day 2 | Day 3 |
|---|---|---|---|
|  | 1006 | 951 | 993 |
|  | 994 | 994 | 1093 |
|  | 1032 | 1017 | 939 |
|  | 875 | 965 | 966 |
|  | 1043 | 966 | 992 |
| **Averages** | 990 | 978.6 | 996.6 |
| **Grand mean** | 988.4 | | |

A scatter plot of the data was created in Excel and is shown below:

There are $c = 3$ levels of the factor "day" and $n_1 = n_2 = n_3 = 5$.

The treatment means have been calculated as well as the grand mean and are shown in the table: $\bar{x}_1 = 990$ lb/ream, $\bar{x}_2 = 978.6$ lb/ream, $\bar{x}_3 = 996.6$ lb/ream, and $\bar{\bar{x}} = 988.4$ lb/ream.

Recall that the formula for the sum of squares total is

$$SST = \sum_{j=1}^{c} \sum_{i=1}^{n_j} (x_{ij} - \bar{\bar{x}})^2$$

The Excel spreadsheet on page 704 displays the 15 squared differences, which added together give you SST.

| Day | MDStrength | (MDStrength − Grand Mean) | (MDStrength − Grand Mean)$^2$ |
|-----|-----------|--------------------------|-------------------------------|
| 1 | 1006 | 17.6 | 309.76 |
| 1 | 994 | 5.6 | 31.36 |
| 1 | 1032 | 43.6 | 1,900.96 |
| 1 | 875 | − 113.4 | 12,859.56 |
| 1 | 1043 | 54.6 | 2,981.16 |
| 2 | 951 | − 37.4 | 1,398.76 |
| 2 | 994 | 5.6 | 31.36 |
| 2 | 1017 | 28.6 | 817.96 |
| 2 | 965 | − 23.4 | 547.56 |
| 2 | 966 | − 22.4 | 501.76 |
| 3 | 993 | 4.6 | 21.16 |
| 3 | 1093 | 104.6 | 10,941.16 |
| 3 | 939 | − 49.4 | 2,440.36 |
| 3 | 966 | − 22.4 | 501.76 |
| 3 | 992 | 3.6 | 12.96 |
| | | **SST** | **35,297.6** |

Recalling that the formula for SSA or sum of squares among is

$$SSA = \Sigma n_j(\bar{x}_j - \bar{\bar{x}})^2$$

and  $\bar{x}_1 = 990$ lb/ream,  $\bar{x}_2 = 978.6$ lb/ream,  $\bar{x}_3 = 996.6$ lb/ream,  and  $\bar{\bar{x}} = 988.4$ lb/ream, we can easily compute this component as follows:

$$SSA = (5)(990 - 988.4)^2 + (5)(978.6 - 988.4)^2 + (5)(996.6 - 988.4)^2 = 829.2$$

Finally, the sum of squares error or SSE can be found by subtracting SSA from SST:

$$SSE = 35{,}297.6 - 829.2 = 34{,}468.4$$

Alternatively, SSE can be calculated using the formula

$$SSE = \sum_{j=1}^{c} \sum_{i=1}^{n_j} (x_{ij} - \bar{x}_j)^2$$

*If you are using Excel to generate the ANOVA table you must have the data organized in columns as displayed at the beginning of this example.*

The details of the calculation using the formula for SSE are shown below in an Excel spreadsheet.

| | Day 1 | SSE calculation | Day 2 | SSE calculation | Day 3 | SSE calculation | |
|---|-------|-----------------|-------|-----------------|-------|-----------------|---|
| | 1006 | 256 | 951 | 761.76 | 993 | 12.96 | |
| | 994 | 16 | 994 | 237.16 | 1093 | 9292.96 | |
| | 1032 | 1764 | 1017 | 1474.56 | 939 | 3317.76 | |
| | 875 | 13225 | 965 | 184.96 | 966 | 936.36 | |
| | 1043 | 2809 | 966 | 158.76 | 992 | 21.16 | |
| **Averages** | 990 | | 978.6 | | 996.6 | | |
| **Sum of SSE contribution** | | 18070 | | 2817.2 | | 13581.2 | 34468.4 |
| | | | | | | | **SSE** |

All of our computations are confirmed in the following ANOVA table.

**ANOVA**

| Source of Variation | SS | df | MS | F | P value | F crit |
|---|---|---|---|---|---|---|
| Between groups | 829.2 | 2 | 414.6 | 0.1443409 | 0.867073 | 3.8852903 |
| Within groups | 34,468.4 | 12 | 2872.3667 | | | |
| Total | 35,297.6 | 14 | | | | |

■

Clearly, you want to use a statistics package or a spreadsheet package that has some statistical tools in order to find SST, SSA, and SSE. Even this small example ($n = 15$) took two pages to perform by hand.

The next sections explain what the other numbers in the ANOVA table mean and how to do the hypothesis test to determine if any of the population means are different from the others.

## 14.3.4 The Mean Square Terms in the ANOVA Table

Remember that the technique of ANOVA analyzes the variances to make an inference about the equality of the population means. What we have found in the previous sections is the numerator of three variances. To turn these sums of squares into variances we must divide by a number that is typically one less than the number of observations in the sample. This number is called the degrees of freedom. It is typically labeled "df" or "DF." You have already encountered this term in conjunction with the $t$ distribution, the $F$ distribution, and the chi-square distribution.

The most obvious formula is the degrees of freedom associated with SST. Since this term measures the total variability in the dataset and uses all $n$ observations, it has $n - 1$ degrees of freedom associated with it. Following this line of thinking, we know that there are $c$ levels of the factor being compared and therefore $c$ terms are added together to find SSA. Thus, SSA has $c - 1$ degrees of freedom. This leaves us with $n - c$ degrees of freedom for SSE. There are two ways to think about why this is correct. First, just like the SS column, the degrees of freedom column must add up correctly. If the column labeled degrees of freedom must total $n - 1$ and we have used $c - 1$ degrees of freedom for SSA, then that leaves us with $(n - 1) - (c - 1)$ or $n - c$ degrees of freedom for SSE. Another way to think about this is to realize that each of the $c$ levels contributes $n_j - 1$ degrees of freedom to SSE. Summing these, we get the right number for the degrees of freedom for SSE:

$$\sum_{j=1}^{c} (n_j - 1) = n - c$$

If we divide each of the sum of square (SS) terms by the appropriate degrees of freedom we will have three variances or **mean square terms.**

The *mean square among* is labeled MSA. The *mean square error* is labeled MSE and the *mean square total* is labeled *MST*.

The formulas for the mean squares are

$$\text{MSA} = \frac{\text{SSA}}{c - 1} \qquad \text{MSE} = \frac{\text{SSE}}{n - c} \qquad \text{MST} = \frac{\text{SST}}{n - 1}$$

*Mean square formulas*

By looking at the output from Excel and Minitab you can see that these values are found in the column labeled MS. Notice that typically only MSA and MSE are printed.

**EXAMPLE 14.9** Airspace Data

*Calculating MSA and MSE*

For the airspace data the mean square calculations are

$$\text{MSA} = \frac{2554.75}{3} = 851.58$$

$$\text{MSE} = \frac{1695.12}{476} = 3.56$$

 TRY IT NOW!

**Tissue Strengths**

*Finding MSA and MSE*

Find MSA and MSE for the tissue strength data shown in Example 14.8. You should do the calculations using the formula and then find those values in the computer output.

## 14.3.5 Testing the Hypothesis of Equal Means

So far we have examined the first 4 columns of the ANOVA table. Typically, there are 2 more columns in the ANOVA table. These are labeled $F$ and $P$ or $p$ value. Excel provides one additional column labeled $F$ crit, which stands for the critical value of the $F$ distribution. These columns are based on the mean square column and allow us to do the hypothesis test that we set out to do, that is, test to see if at least one of the population means is different.

In general, the null and alternative hypotheses for a one-way designed experiment are shown below:

$H_0$:  $\mu_1 = \mu_2 = \mu_3 = \cdots = \mu_c$

$H_A$:  At least one of the population means is different from the others.

**EXAMPLE 14.10** Airspace Data

*Null and Alternative Hypotheses*

For the airspace data there are $c = 4$ levels of the factor, so the null and alternative hypotheses are as follows:

$H_0$:  $\mu_1 = \mu_2 = \mu_3 = \mu_4$

$H_A$:  At least one of the population means is different from the others.

Think for a moment about what it means if the null hypothesis is true. In the case of the airspace data, if the null hypothesis is true then the amount of time the box sits on the shelf does not affect the average airspace in the tissue box. In this case we say that there is no *treatment effect*.

If there is no treatment effect, then we conclude that the factor does not affect the variable being studied. Even in this case, we would not expect every observation in the dataset to be identical because there is some natural variation inherent in the process. Remember that SSE measures the variation within the groups and so MSE is certainly an estimate of this unknown population variability, $\sigma^2$.

If the null hypothesis is true and there is no treatment effect, then not only is MSE an estimate of $\sigma^2$, but all three of the mean square values, MSA, MSE, and MST, are estimates of the natural variability in the data. In this case we have 3 estimates of the same parameter, $\sigma^2$. We know from our work in Section 10.9 how to test for equality of variances of two normally distributed populations. To do this we used an $F$ test. We use this same $F$ test here to see if the MSA and MSE are actually two different estimates of the same parameter. If MSA and MSE test equal, then we will conclude that there is no treatment effect and the population means are equal. We will fail to reject the null hypothesis.

The formula for the $F$ test statistic from Section 10.9 is calculated by taking the ratio of the two sample variances: $F = s_1^2/s_2^2$. In ANOVA, MSA and MSE are our two sample variances. So the $F$ statistic is calculated as

$$F = \frac{\text{MSA}}{\text{MSE}}$$

*Formula for F test statistic for ANOVA*

The hypothesis test is easily done by determining if the $F$ value is in the rejection region. Excel provides the critical $F$ value as part of its ANOVA output. It is labeled $F$ crit. If the $F$ test statistic is "too large," that is, larger than the cutoff value shown in $F$ critical, then you conclude that MSA and MSE are not estimates of the same number and there is a treatment effect. If the software you are using does not provide you with the critical $F$ value, you can look it up in the $F$ table in the Appendix. The $F$ statistic has $c - 1$ degrees of freedom in the numerator (from the MSA term) and $n - c$ degrees of freedom in the denominator (from the MSE term).

*The value for F critical is obtained from the F table and the procedure for reading this table was explained in Section 10.9.*

An easier way to perform the hypothesis test is to use the $p$ value. Remember that we used $p$ values to do all of the hypotheses tests in this book. Both Minitab and Excel provide you with a $p$ value to use. Most statistical packages will typically output both the $F$ test statistic and the corresponding $p$ value as part of the ANOVA table.

Now we can complete the airspace data analysis.

## EXAMPLE 14.11 Airspace Data

### The Hypothesis Test

The $F$ test statistic for the airspace data is

$$F = \frac{\text{MSA}}{\text{MSE}} = \frac{851.58}{3.56} = 239.13$$

Looking at the ANOVA table from Excel or Minitab shown in Example 14.4, we find this $F$ value printed under the label F. This means that the ratio of the two mean squares is about 240:1, or MSA is 240 times larger than MSE. Clearly, the ratio is not even close to 1 and these two values are not likely to be estimates of the same number.

The $p$ value of 0.000 shown in the Minitab output and the $p$ value of $1.342 \times 10^{-94}$ (equivalent to a decimal point followed by 93 zeros and then 1342) confirms this line of thinking and tells us to reject the null hypothesis.

Thus, we conclude that at least one of the population means is different from the others and there is a treatment effect—the time the box sits on the shelf does affect the amount of airspace in the tissue box.

That's it! We have looked at all of the numbers in the ANOVA table and learned what they measure, how to calculate them, and how to use them to draw an inference about the population means. You should have noticed that the technique of ANOVA ties together a number of concepts and techniques that have been developed in earlier chapters. For this reason, the technique of ANOVA is in a sense a "capstone" technique for this book. It is a very commonly used tool and it is an interesting technique since it analyzes the variability in the data to see if there are any differences in the population means. There are some assumptions that we have glossed over up to this point and it is now time to take a look at these assumptions. People tend to use ANOVA without checking to see if it is actually appropriate, which could lead to erroneous conclusions. Section 14.4 discusses the assumptions necessary to use ANOVA.

Before moving on you should complete the analysis for the tissue strength data that was started in Example 14.8.

 **TRY IT NOW!**

### Tissue Strength Data
*Completing the Analysis*

Set up the null and the alternative hypotheses for the tissue strength example.

Using the ANOVA table shown in Example 14.8, find the *F* statistic and decide if you should reject the null hypothesis or fail to reject it.

What do you conclude about the variable *MDStrength* and what is your recommendation to the company?

### 14.3.6 A Summary of One-Way Designs

In this section you have learned how to examine a set of data that has resulted from a one-way designed experiment. There is a variable of interest and you wish to know if a particular factor affects the average level of this variable. So, you have one factor, which can be "set" to various values or levels. The variable of interest is repeatedly observed or measured with the factor set at each of the levels.

Once you have the data you quickly see that not all of the values of the variable are the same! No surprise here. The technique of ANOVA allows you to decide what is causing the variation that you see. Is the variation largely due to the fact that you have treated the observations differently by setting the factor at different levels? If so, you will conclude there is a treatment effect. Or is the variation largely due to natural or inherent variation in the variable being studied? If so, you will conclude that there is no treatment effect. You break the total variability in the data into two components and then examine the relative sizes of these two estimates of variability.

All of the computations necessary to carry out a one-way ANOVA are summarized in the ANOVA table shown below:

| Source of Variation | SS | df | MS | F | p Value |
|---|---|---|---|---|---|
| Between groups | SSA | $c - 1$ | $MSA = \dfrac{SSA}{c - 1}$ | $F = \dfrac{MSA}{MSE}$ | |
| Within groups | SSE | $n - c$ | $MSE = \dfrac{SSE}{n - c}$ | | |
| **Total** | **SST** | $n - 1$ | | | |

You can calculate each of the values in the ANOVA table by hand, but it is much more likely that you will be using a statistical software package or a spreadsheet package that has some data analysis tools built-in. The calculations are done from left to right across the columns of the table with the $p$ value being the final value to be calculated. Based on the $p$ value you make a decision to reject or fail to reject the null hypothesis. If you fail to reject the null hypothesis, then you conclude that the factor you studied does *not* affect the average value of the variable. If you reject the null hypothesis, then you conclude that there is a treatment effect and at least one of the population means is different from the others. In this case, although you cannot tell at this point which population mean(s) are different, you do know that the factor you have investigated is important to the response variable. This is one of the major focuses of the technique of ANOVA. It tells you whether anything interesting is going on with respect to this factor. It tells you whether you need to further investigate this factor or if it can simply be ignored because it does not affect the average level of the variable being studied.

## 14.3.7 Building the Model for the Response Variable

Remember when we started looking at the airspace data at the beginning of the chapter we wanted to be able to tell the tissue company if the mean airspace differed when the box was left on the shelf different amounts of time. We said that if the population means were different, then we wanted to be able to tell the tissue company how the airspace changes as the box sits on the shelf. Now that we know there is a treatment effect let's consider how to estimate the airspace.

Just as the overall variability in the data can be split into component parts so can a single observation be divided into its component parts. Each observation can be thought of as follows:

$$\text{Response} = \text{Grand mean} + \text{Treatment effect} + \text{Error}$$

Notice that if you concluded that there was no treatment effect (by failing to reject $H_0$), then each response would be equal to the overall mean plus some random variation called error.

We must use the data to estimate the grand mean and the treatment effect. As you will see in Section 14.4 on the assumptions of ANOVA, the error term is assumed to have an average of zero. The grand mean can be estimated by $\bar{\bar{x}}$ and the treatment effect is the adjustment that must be made to the grand mean to predict a response. The treatment effect can be estimated by taking the difference between the treatment mean and the overall mean, $\bar{x}_j - \bar{\bar{x}}$. Thus, the estimate of the response is

$$\text{Response} = \bar{\bar{x}} + (\bar{x}_j - \bar{\bar{x}}) + \text{Error} = \bar{x}_j + \text{Error}$$

or just the treatment mean + error.

### EXAMPLE 14.12   Airspace Data

*A Prediction Model*

For the airspace data consider $x_{11} = 23$ mm. Writing this in terms of its components gives us

$$\text{Response} = \text{Grand mean} + \text{Treatment effect} + \text{Error}$$
$$23 = 17.32 + (21.01 - 17.32) + \text{error}$$
$$23 = 17.32 + 3.69 + 1.99$$

Our model would predict an average of 21.01 mm of airspace in process. In this case we would be in error by 1.99 mm. ∎

 **TRY IT NOW!**

### Tissue Strength Data
*Prediction Model*

For the tissue strength data, write $x_{11}$ in terms of its components.

Using the first treatment mean as your prediction, what is your error for this particular observation?

## 14.3.8 The Next Step—Multiple Comparisons

*Remember that your goal is to understand how the airspace changes as the tissue box sits on the shelf.*

    If you have rejected the null hypothesis of equal population means, as we have in the case of the airspace data, you can say that there is sufficient evidence in the data to state that not all the population means are the same. Clearly, this is only the first step. To decide what type of experiment to run next to further investigate the effect this factor has on the variable of interest, we should try to learn a little bit more from this dataset.

    Let's see what tools we have already learned that might help us with our detective work. We started off by simply looking at the sample mean for each of the levels

ANS. $1006 = 988.4 + (990 - 988.4) + 16$

ERROR $= 16$ LB/REAM

```
Analysis of Variance

Source    DF        SS        MS       F        P
Time       3   2554.75    851.58   239.13    0.000
Error    476   1695.12      3.56
Total    479   4249.87

                                 Individual 95% CIs For Mean
                                 Based on Pooled StDev
Level      N      Mean     StDev   -------+---------+---------+---------
  1      120    21.008     1.526                                  (-*-)
  2      120    17.458     1.483                      (*-)
  3      120    15.783     2.083         (-*-)
  4      120    15.017     2.319   (-*-)
                                   -------+---------+---------+---------
Pooled StDev =    1.887              16.0      18.0      20.0
```

**FIGURE 14.4**  Confidence intervals for airspace data

of the factor. But we know from our work in Chapter 7 that the sample mean is a point estimate and is never likely to actually "hit" the population mean right on the money. To get an idea of how far off our point estimate was likely to be, we constructed confidence intervals for the population mean. It seems that this might be a useful tool at this point. In fact, most statistical software packages calculate the individual sample means and the corresponding confidence intervals when you run ANOVA. This portion of the Minitab output for the airspace data is shown in Figure 14.4.

By this point in the text, you have seen that you can learn a great deal by examining graphs and noticing patterns. As you examine these confidence intervals you can see that some of them overlap and some do not. For example, the confidence interval for the mean airspace in process (population or level 1) seems particularly different from the confidence interval for the mean airspace after 4 weeks (level 4). However, the confidence intervals for the mean airspace for levels 3 (2 weeks) and 4 (4 weeks) overlap a bit. Although this is not a formal statistical test, you can certainly intelligently speculate about whether there is one population mean that is causing the $F$ value in ANOVA to be large (leading to a rejection of the null hypothesis) or whether all of the means are different from each other. In the case of the airspace data it seems that there is a great difference between the in-process data and the 24-hour data. Perhaps it would be a good idea to look at the differences in the sample means. Table 14.1 shows the amount of growback (in mm) in the tissues from the time of manufacturing until 4 weeks after manufacturing. These values are easily found by subtracting pairs of sample means. For example, the difference between $\bar{x}_1$ and $\bar{x}_2$ is $(21.01 - 17.46 \text{ mm}) = 3.55 \text{ mm}$.

This table tells us that the airspace decreases 3.55 mm between the time of manufacturing (level 1) and 24 hours later (level 2), it decreases an additional 1.68 mm between the 24-hour mark and 2 weeks later, and it decreases only an additional 0.76 mm between 2 weeks and 4 weeks. In total, the tissues "grow back" about 6 mm from the time of manufacturing until the customer opens the box 4 weeks later. Clearly, our idea that the most of the growback occurs during the first 24 hours after manufacturing seems to be correct. In fact, the tissues are greatly compressed at the time

**TABLE 14.1  Airspace Data**

| Level | 1 | 2 | 3 |
|-------|------|------|------|
| 1 | | | |
| 2 | 3.55 | | |
| 3 | 5.23 | 1.68 | |
| 4 | 5.99 | 2.44 | 0.76 |

of manufacturing to get them into the box, so perhaps focusing on the first 24 hours is a good approach for your next experiment.

This analysis also points out how much easier it is to do data analysis when you are actually involved in the manufacturing process or the situation that generated the data. If you understand the situation you are much more likely to be able to make sense of the data and make relevant recommendations.

The intuitive, visual approach we have taken here works well at pointing you in the right direction and will not differ much from the conclusions you could draw from formal techniques if the sample sizes are equal, as they were in the airspace data. In fact, the most common formal statistical test done at this point uses the difference between two of the sample means as the center of a confidence interval, which estimates the difference in the corresponding population means. Several techniques could be used to do this. Alternatively, a hypothesis test of the difference in the two population means can be done. It might seem that a $t$ test or a paired $t$ test from Chapter 10 would work here. It is natural to think so since we now wish to compare two population means. Although the setup of the null and alternative hypotheses is the same here, unfortunately we cannot use either a $t$ test or a paired $t$ test to do the test. The reason for this has to do with the fact that you are making multiple pairwise comparisons from a single dataset and in addition to that you are doing the comparison after the data have been analyzed.

There are several techniques that could be used and most statistical software packages will have several options for doing the pairwise comparisons. For example, if you are using Minitab then you can use any one of four techniques to do the pairwise comparisons. Sometimes these techniques lead to conflicting conclusions. Many issues need to be addressed to properly explain these techniques and they will not be addressed here. However, by examining the confidence intervals for the mean of each level of the factor, looking at how they overlap and how much the sample means change as you change levels of the factor, you will detect a good deal about how the factor influences the variable you are studying. This will help guide the design of your next experiment.

### 14.3.9 Exercises—Learning It!

**14.1**   A diaper company is considering 3 different filler materials for their disposable diapers. Eight diapers were tested with each of the 3 filler materials and 24 toddlers were randomly given a diaper to wear. As the child played, fluid was injected into the diaper every 10 minutes until the product failed (leaked). The amount of fluid (in grams) at the time of failure was recorded for each diaper. The data are shown below:

| Material 1 | Material 2 | Material 3 |
|------------|------------|------------|
| 791 | 809 | 828 |
| 789 | 818 | 814 |
| 796 | 803 | 855 |
| 802 | 781 | 844 |
| 810 | 813 | 847 |
| 790 | 808 | 848 |
| 800 | 805 | 836 |
| 790 | 811 | 873 |

(a) What is the response variable and what is the factor?

(b) How many levels of the factor are being studied?

(c) Is there any difference in the average amount of fluid the diaper can hold using the 3 different filler materials? If so, which ones are different?

(d) What is your recommendation to the company and why?

*Datafile:*
*DISKFAIL.XXX*

**14.2**   In Chapter 4 you looked at a large company that sells software. The company had received complaints that the disks provided by the company failed to work properly after extended use. The company has decided to investigate the complaint. Data are collected on the time to failure for disks from their current supplier and two alternative suppliers. The data are shown below:

| Current Supplier | Alternative 1 Supplier | Alternative 2 Supplier |
|---|---|---|
| 486 | 489 | 508 |
| 490 | 489 | 510 |
| 491 | 491 | 517 |
| 491 | 492 | 506 |
| 494 | 492 | 515 |
| 494 | 492 | 520 |
| 496 | 492 | 503 |
| 498 | 493 | 524 |
| 498 | 493 | 515 |
| 502 | 494 | 503 |
| 504 | 495 | 507 |
| 505 | 496 | 515 |
| 506 | 497 | 509 |
| 507 | 497 | 518 |
| 508 | 497 | 495 |
| 510 | 498 | 499 |
| 514 | 499 | 510 |
| 515 | 502 | 503 |
| 527 | 503 | 533 |
| 498 | 505 | 517 |

(a) What is the response variable and what is the factor?

(b) How many levels of the factor are being studied?

(c) Is there any difference in the average time to failure of the disks from the 3 different suppliers? If so, which ones are different?

(d) What is your recommendation to the company and why?

14.3 Grading homework is a real problem. It takes an enormous amount of time and many students do not do a very good job or copy answers from other students or the back of the book. A teacher of elementary statistics decided to conduct a study to determine what effect grading homework had on her students' exam scores. She taught 3 sections of Elementary Statistics and randomly assigned each class one of three conditions: (1) no homework given, (2) homework given but not collected, and (3) homework given, collected, and graded. After the first exam, she collected the data (exam scores). They are shown below:

*Datafile:*
*HOMEWORK.XXX*

| No Homework | Homework, Not Collected | Homework, Collected |
|---|---|---|
| 69 | 73 | 83 |
| 69 | 63 | 97 |
| 92 | 68 | 72 |
| 84 | 79 | 79 |
| 79 | 57 | 84 |
| 84 | 68 | 76 |
| 76 | 72 | 91 |
| 63 | 74 | 76 |
| 76 | 49 | 83 |
| 82 | 84 | 88 |
| 89 | 79 | 91 |
| 72 | 71 | 96 |
| 72 | 80 | 68 |
| 65 | 74 | 99 |
| 73 | 71 | 89 |
| 47 | 63 | 80 |

(*continued*)

| No Homework | Homework, Not Collected | Homework, Collected |
|---|---|---|
| 92 | 88 | 79 |
| 71 | 83 | 91 |
| 83 | 89 | 83 |
| 81 | 82 | 83 |
| 92 | 69 | 76 |
| 80 | 92 | 90 |
| 64 | 79 | 79 |
| 72 | 81 | 67 |
| 84 | 76 | 86 |
| 79 | 81 | 86 |
| 74 | 81 | 82 |
| 81 | 75 | 84 |

(a) What is the response variable and what is the factor?

(b) How many levels of the factor are being studied?

(c) Is there any difference in the average exam scores using the 3 different approaches to homework? If so, which ones are different?

(d) What is your recommendation to the teacher and why?

**14.4** Two tissue strength measurements were studied: machine-directional strength (MD-Strength) and the cross-directional strength (CDStrength). They are both measured in lb/ream. We will look at only the CDStrength in this exercise. The data were collected over 3 days and the company would like to know if the average CDStrength is the same for all 3 days.

In the complete dataset there are 75 observations per day. In this example we will look at only five observations per day. At the end of the chapter you will have the opportunity to do a complete analysis of this dataset.

The five CD strengths for each of the 3 days are shown below:

| Day | CDStrength | Day | CDStrength | Day | CDStrength |
|---|---|---|---|---|---|
| 1 | 422 | 2 | 436 | 3 | 473 |
| 1 | 448 | 2 | 468 | 3 | 440 |
| 1 | 423 | 2 | 419 | 3 | 441 |
| 1 | 435 | 2 | 458 | 3 | 443 |
| 1 | 445 | 2 | 459 | 3 | 400 |

(a) Display the data as a dotplot.

(b) Find the grand mean and the treatment means.

(c) Mark the grand mean and the treatment means on the dotplot.

(d) Examine the first observation for day 1: 422 lb/ream. Find the difference between the first treatment mean and the grand mean. This is the treatment effect for the first treatment. On the dotplot mark off this difference.

(e) Continue to examine the first observation for day 1: 422 lb/ream. Find the difference between this observation and the first treatment mean and the grand mean. This is the error for this observation. On the dotplot mark off this difference.

(f) Analyze these data using one-way ANOVA. Is there a significant treatment effect?

*Datafile: BALLDESN.XXX*

**14.5** The sports industry is a large and competitive market. A manufacturer of golf balls is considering 4 new ball designs. A sample of 36 balls from each of the 4 models is tested and the distance the balls carry (in yards) is recorded. The balls are hit by a machine. The data are shown below.

| M1 Model | M2 Model | M3 Model | M4 Model |
|---|---|---|---|
| 257 | 256 | 244 | 250 |
| 255 | 255 | 243 | 255 |
| 256 | 258 | 241 | 251 |
| 255 | 257 | 243 | 249 |

*(continued)*

| M1 Model | M2 Model | M3 Model | M4 Model |
|---|---|---|---|
| 255 | 257 | 240 | 250 |
| 256 | 257 | 249 | 251 |
| 255 | 258 | 244 | 251 |
| 258 | 258 | 249 | 248 |
| 252 | 256 | 248 | 247 |
| 256 | 258 | 243 | 253 |
| 253 | 258 | 242 | 250 |
| 254 | 256 | 245 | 255 |
| 260 | 259 | 253 | 250 |
| 258 | 255 | 250 | 256 |
| 258 | 259 | 251 | 260 |
| 257 | 257 | 249 | 259 |
| 259 | 256 | 249 | 252 |
| 257 | 255 | 249 | 253 |
| 255 | 255 | 255 | 257 |
| 256 | 257 | 249 | 255 |
| 259 | 260 | 251 | 251 |
| 261 | 255 | 250 | 255 |
| 260 | 261 | 248 | 253 |
| 257 | 262 | 250 | 256 |
| 260 | 258 | 258 | 262 |
| 262 | 260 | 258 | 258 |
| 258 | 260 | 257 | 263 |
| 260 | 256 | 258 | 259 |
| 258 | 255 | 254 | 256 |
| 258 | 256 | 256 | 261 |
| 258 | 257 | 255 | 258 |
| 256 | 251 | 254 | 258 |
| 260 | 253 | 255 | 256 |
| 260 | 254 | 260 | 263 |
| 261 | 254 | 254 | 264 |
| 257 | 255 | 253 | 257 |

(a) What is the response variable and what is the factor?

(b) How many levels of the factor are being studied?

(c) Is there any difference in the average distance the ball carries?

(d) What is your recommendation to the golf manufacturer and why?

# 14.4 ASSUMPTIONS OF ANOVA

We have already mentioned that ANOVA is a very commonly used technique. Unfortunately, many times ANOVA is used when it is not the appropriate technique. This may be partly due to the fact that it is relatively easy to run ANOVA using a statistical package and most software tools simply do the calculations but do not check that it is the appropriate technique. This is your job as the data analyst. As with many of the tools presented in this book, the tool of ANOVA gives you valid conclusions if your data meet certain assumptions. If your data do not meet these assumptions and you use ANOVA, you could easily draw the wrong conclusions. Thus, you should always check that the three major assumptions of ANOVA are met *before* you use this technique.

The three major assumptions of ANOVA are as follows:

**1.** The errors are random and independent of each other.

**2.** Each population has a normal distribution.

**3.** All of the populations have the same variance.

### 14.4.1 Assumption About the Errors

The first assumption talks about the behavior of the errors. What are these errors? The term error refers to the difference between any observed value and the sample mean of its group. We looked at these differences when we developed the formula for SSE in Section 14.3.3. Remember that SSE is found by adding up the squared differences of the form $x_{ij} - \bar{x}_j$. Each term in the sum for SSE is considered an "error." Thus, the first assumption says that these differences should be random and the error for one observation should not influence the error for another observation. If you have randomly selected a sample from each of the populations then typically this assumption is met. The most common situation in which this assumption is violated is when you have time series data. In this case, the observation at time period $t$ may influence the observation at time period $t + 1$ (in fact, it is precisely this dependency that you exploit when you build a time series model). If the observations are dependent, then the errors will be dependent. A violation of this assumption could seriously affect the conclusions reached using ANOVA. If you are suspicious of a bias in the sampling process or if you have time-dependent data, then you should consult a more advanced textbook.

### 14.4.2 Assumptions About the Underlying Populations

The second and third assumptions of ANOVA refer to the behavior of the variable being collected. The second assumption says that the variable should have a normal distribution in each of the populations (each level of the factor). You can check this assumption formally using a chi-square goodness of fit test, which is the subject of the next chapter. At this point you should visually check the shape of the data in each sample by displaying a histogram of the data.

### EXAMPLE 14.13    Airspace Data

*Looking at Histograms for Normality*

Histograms for all 4 levels of the factor "time" for the airspace data are shown below:

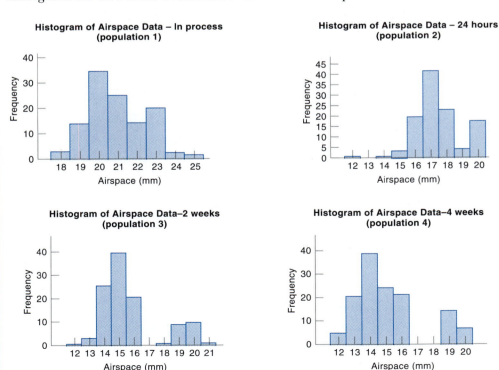

A look at these graphs tells us that we may be in violation of the assumption of normality for at least one of the populations. ■

Another way to test for the assumption of normality is to use a normal probability plot. You were introduced to this concept in the discussion of the assumptions of simple linear regression in Section 11.6.1. A normal probability plot for the in-process data is shown in Figure 14.5. It was generated by Minitab.

*Remember: When the data are normally distributed, the plot will be a straight line.*

Average: 17.4583
StDev: 1.48322
N: 120

Anderson-Darling Normality Test
A. Squared: 4.758
P. Value: 0.000

**FIGURE 14.5**  Normal probability plot of 24-hour airspace data

Fortunately, the technique of ANOVA is not particularly sensitive to violations of the normality assumption. This means that if your data follow a distribution that is not extremely different from a normal distribution, then your conclusions from ANOVA are probably fine, especially for a large sample size.

The third assumption of ANOVA also has to do with the behavior of the distribution of the variable in each of the populations. Once you have checked to see that the second assumption has been met, or at least is not severely violated, then you must check to see that the amount of variability in each of the populations is the same. That is, the variance within each population should be equal. This is necessary in order to combine the data from the various samples to get an estimate of the inherent variability measured by SSE.

*Remember: You are using ANOVA to test equality of means.*

At a minimum, you should compare the size of the sample variances and do a visual check of the data by looking at a boxplot of the data to see if the spread in each sample looks about the same. Neither of these approaches constitutes a test for equal variances and, of course, there are such tests, but they will certainly identify glaring violations of this assumption.

## EXAMPLE 14.14   Airspace Data

*Checking for Equal Variances Using a Boxplot*

The sample variances for the airspace data are shown in the table below:

| Time | Sample Standard Deviation | Sample Variance |
|------|---------------------------|-----------------|
| 1 | 1.53 | 2.34 |
| 2 | 1.48 | 2.19 |
| 3 | 2.08 | 4.33 |
| 4 | 2.32 | 5.38 |

The boxplots of airspace by time are shown on page 718.

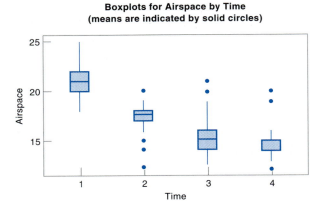

The largest sample variance (5.38) is almost 2.5 times larger than the smallest sample variance (2.19). This represents a large difference in the variances given the size of this sample. These data may be in violation of the third assumption of ANOVA. ■

There are formal statistical procedures to test the assumption of equal variances. The statistical software you are using will most likely have more than one way to test this assumption. If the data are in violation of the third assumption, you will need to use an appropriate data transformation to normalize the data and stabilize the variance. Alternatively, you could use a nonparametric technique.

## 14.4.3 Exercises—Learning It!

*Requires Exercise 14.1*      **14.6**   Check the ANOVA assumptions for the disposable diaper company interested in testing different materials.

(a) Is there any reason to believe that the errors are not independent?

(b) Construct histograms of the amount of fluid in the diapers at the time of failure for each of the 3 different materials. Do they look normally distributed?

(c) Calculate and compare the size of the sample variances and do a visual check of the data by looking at boxplots of the data to see if the spread in each sample looks about the same.

(d) Comment on the validity of the assumptions of ANOVA for these data.

*Requires Exercise 14.2*      **14.7**   Check the ANOVA assumptions for the software company.

(a) Is there any reason to believe that the errors are not independent?

(b) Construct histograms of the time to failure for each of the 3 different suppliers. Do they look normally distributed?

(c) Calculate and compare the size of the sample variances and do a visual check of the data by looking at boxplots of the data to see if the spread in each sample looks about the same.

(d) Comment on the validity of the assumptions of ANOVA for these data.

*Requires Exercise 14.3*      **14.8**   Check the ANOVA assumptions for the statistics teacher investigating the effect of homework on exam scores.

(a) Is there any reason to believe that the errors are not independent?

(b) Construct histograms of the exam scores for each of the 3 different methods. Do they look normally distributed?

(c) Calculate and compare the size of the sample variances and do a visual check of the data by looking at boxplots of the data to see if the spread in each sample looks about the same.

(d) Comment on the validity of the assumptions of ANOVA for these data.

*Requires Exercise 14.5*      **14.9**   Check the ANOVA assumptions for the golf ball manufacturer.

(a) Is there any reason to believe that the errors are not independent?

(b) Construct histograms of the distance the balls carried for the 4 different ball designs. Do they look normally distributed?

(c) Calculate and compare the size of the sample variances and do a visual check of the data by looking at boxplots of the data to see if the spread in each sample looks about the same.

(d) Comment on the validity of the assumptions of ANOVA for these data.

# 14.5 ANALYSIS OF DATA FROM BLOCKED DESIGNS

## 14.5.1 Motivation for Block Designs

In this chapter we have looked at two issues for the tissue company: the effect of time on the airspace in the box and the strength of the tissues being manufactured as measured by the variable *MDStrength*. In both of these situations there was one factor of interest and the company was able to collect a sample of *comparable* observations from each population.

Often there is one factor of interest but the observations in the sample may not be comparable due to some other factor. Consider the effect on sales of product display. One factor is being studied here: the product display. Suppose we are considering 3 different product displays; then there are 3 levels for this factor. To use a one-way ANOVA we would need to collect sales data from a sample of stores using display method 1, a sample of stores using display method 2, and a sample of stores using display method 3. This is not a problem, but the question that arises is, Are the sales from these stores different due to some reason other than the display method being used? It is likely that the answer to this question is yes. The stores are probably different in size and in different geographic locations. These factors of size and location would not matter if they did not affect the variable we were studying: sales. If they did not affect sales, then we could happily use a one-way ANOVA, randomly assigning stores to use one of the 3 product display methods. But they most likely do affect sales. We are not going to find a sample of like size stores in the same location! So we cannot get a sample of *comparable* stores to sample.

In this example, it seems as though we actually have 3 factors: product display method, size, and location of the store. But really we are interested only in the effect of the product display method on sales. The other 2 factors are in a sense "nuisance factors." They need to be taken into account, but we are not directly interested in their effect on sales at this point. The factors size and location need to be blocked out so that we may see the effect of the product display on sales. Otherwise, there is no way to tell if the differences we see in sales are a result of the differing product displays or the differing size and/or location of the store.

There are two possible ways to handle the data collection for this example. We could find 3 stores that are similar in size and location (large city, small city, rural, etc.) and randomly assign a product display method to each store. Then we would need to find another 3 stores that are similar in size and location and do the same thing. Each set of 3 stores becomes what is called a **block.** We would use as many blocks as feasible given the amount of time, the amount of money, and the availability of stores.

Alternatively, we could observe a set of 20 stores for 3 weeks. During the first week, each store would randomly be assigned to use one of the 3 product display methods, during the second week, the store would use a second product display method, and during the third week, the store would use the remaining product display method. In this case each store is a **block.**

> A *block* is a group of objects or people that have been matched. An object or person can be matched with itself, meaning that repeated observations are taken on that object or person and these observations form a block.

If the realities of data collection lead you to use blocks, then you must take this into account in your analysis. Your experimental design is called a **randomized block design.** Instead of using a one-way ANOVA you must use a block ANOVA.

> An experiment has a *randomized block design* if several different levels of one factor are being studied and the objects or people being observed/ measured have been matched. Each object or person is randomly assigned to one of the $c$ levels of the factor.

## 14.5.2 Partitioning the Total Variation

The analysis of data from a block design is conceptually the same as that of a one-way design. You essentially still have just one factor of interest. You are still interested in deciding if there is a treatment effect. That is, you are interested in seeing if the population means are all equal or if at least one is different. The null and alternative hypotheses remain as follows:

$$H_0: \quad \mu_1 = \mu_2 = \mu_3 = \cdots = \mu_c$$
$$H_A: \quad \text{At least one of the means is different.}$$

Like the approach we took with data from a one-way design, the idea is to take the total variability as measured by SST and break it down into its components. With a block design there is one additional component: the variability between the blocks. It is called the **sum of squares blocks** and is labeled **SSBL.**

> The *sum of squares blocks* measures the variability between the blocks. It is labeled *SSBL.*

The difference between each observation and the mean of its block is squared and these squared differences are accumulated into the sum of squares blocks. This component measures the portion of the variation that is attributable to the fact that the data were collected in blocks. Since we used blocks because we thought that there were some "interfering factors," we would expect this component to be of significant size. If it is not, then we probably worried about factors that do not, in fact, affect the variable of interest.

For a block design, the variation we see in the data is due to one of three things: the level of the factor, the block, or the error. Thus, the total variation is divided into three components:

*Formula for SST in terms of other components in a block design*

$$SST = SSA + SSBL + SSE$$

The terms SSA and SSE have the same formulas and definition as for a one-way ANOVA. The effect of partitioning the total variation into another component is to reduce the error term, SSE. Recall that SSE is used to find MSE, which becomes the denominator of the $F$ test statistic. If you can reduce SSE by the use of blocks, then the MSE term will be smaller, making the $F$ test statistic larger. Since "big" $F$ values lead you to reject the null hypothesis of equal means, the use of blocks allows you to see differences in the treatment means that you might not have seen otherwise.

### 14.5.3 Using the ANOVA Table in a Block Design

Consider a randomized block design with $r$ blocks and $c$ groups. There are a total of $n = rc$ observations. The ANOVA table for such a block design looks just like the ANOVA table for a one-way design with an additional row. It is shown below:

| Source of Variation | SS | df | MS | F | p Value |
|---|---|---|---|---|---|
| Between groups | SSA | $c - 1$ | $\text{MSA} = \dfrac{\text{SSA}}{c - 1}$ | $F = \dfrac{\text{MSA}}{\text{MSE}}$ | |
| Between blocks | SSBL | $r - 1$ | $\text{MSBL} = \dfrac{\text{SSBL}}{r - 1}$ | | |
| Within groups | SSE | $(r - 1)(c - 1)$ | $\text{MSE} = \dfrac{\text{SSE}}{(r - 1)(c - 1)}$ | | |
| Total | SST | $rc - 1$ | | | |

The last 2 columns of the ANOVA table are used to do the hypothesis test of the equality of the population means. The hypothesis test is easily done by determining if the $F$ value is in the rejection region. If the $F$ test statistic is "too large," that is, larger than the critical $F$ value from the table, then you conclude that MSA and MSE are not, in fact, estimates of the same number and there is a treatment effect. If the software you are using does not provide you with the critical $F$ value, you can look it up in the $F$ table in Appendix A. The $F$ statistic has $(c - 1)$ degrees of freedom in the numerator (from the MSA term) and $(r - 1)(c - 1)$ degrees of freedom in the denominator (from the MSE term).

Let's see how this works by following the sales and product display example.

### EXAMPLE 14.15    Product Display Location

*A Block Design*

A national chain of stores is investigating the effect of the product display on sales of the product. The analysts are interested in 3 product display locations: on the shelf (group 1), at the end of the aisle (group 2), and at the entrance to the store (group 3). They have matched stores with respect to size and location and have created 10 blocks of stores. The stores in the blocks were randomly assigned a product display location and the monthly sales were recorded. The data are shown below:

| Block | Shelf | End of Aisle | Front of Store |
|---|---|---|---|
| 1 | 4457 | 4500 | 5800 |
| 2 | 4400 | 4370 | 5290 |
| 3 | 4310 | 4300 | 5000 |
| 4 | 4600 | 4400 | 5600 |
| 5 | 5000 | 5000 | 6000 |
| 6 | 4500 | 4500 | 5300 |
| 7 | 4700 | 5100 | 5900 |
| 8 | 4590 | 4280 | 5460 |
| 9 | 8510 | 8670 | 10600 |
| 10 | 6470 | 6500 | 8410 |
| **Average** | **5153.7** | **5162** | **6336** |

The ANOVA table and the group sample means and confidence intervals generated by Minitab are shown below:

```
Analysis of Variance for Sales

Source      DF        SS          MS
Display      2     9253927     4626964
Block        9    60738440     6748716
Error       18     1371972       76221
Total       29    71364339
```

```
                          Individual 95% CI
Display                       Mean    ------+---------+---------+---------+-----
Shelf                         5154        (----*---)
End of Aisle      5162     (----*----)
Front of Store                6336                                   (---*----)
                                       ------+---------+---------+---------+-----
                                        5200      5600      6000      6400
```

Looking at the sample means and the confidence intervals for the individual population means, we can see that locating the product at the front of the store generates considerably more sales.

To complete the hypothesis test we need to find the $F$ value and then use the $F$ table to find the rejection region. Minitab does not calculate the $F$ or $p$ values for a block design.

The null and alternative hypotheses are shown below:

$$H_0: \quad \mu_1 = \mu_2 = \mu_3$$
$$H_A: \quad \text{At least one population mean is different.}$$

The $F$ statistic is MSA/MSE = 4,626,964/76,221 = 60.7. The critical value for the rejection region is found from the $F$ table with 2 $(c - 1)$ and 18 $(r - 1)(c - 1)$ degrees of freedom. If $\alpha = 0.05$, then the critical $F$ value is 3.55. Clearly, $F = 60.7$ is larger than the critical $F$ value and we reject $H_0$.

The null hypothesis is rejected and we conclude that the location of the product display affects the average sales. Our scatter plot tells us that it is probably group 3 (front of store) that has the different mean, since the other two sample means are quite similar and the confidence intervals overlap substantially. ■

All of the assumptions of one-way ANOVA pertain to the block design. Thus, you should be sure to check these assumptions for your dataset.

## *TRY IT NOW!*

### Participative Management
#### *Block Design*

In Chapter 10 you looked at the number of sick days used by employees in the past 12 months by employees who were traditionally managed and those who were involved in a participative management style. Management is now considering a third approach which is a mixture of the old structure with the team-based approach. The study involves 75 employees matched by age, type of work, and gender.

Use the Minitab output shown below to determine if there are any differences in the mean number of sick days used by the different populations.

```
Analysis of Variance for Sick Day

Source    DF        SS         MS
Style      2     57.680    28.840
Block     24    336.213    14.009
Error     48     24.987     0.521
Total     74    418.880


                    Individual 95% CI
Style    Mean    -------+---------+---------+---------+----
Part.    4.72      (----*----)
Hybrid   6.56                                (----*----)
Trad     6.60                                (----*----)
                  -------+---------+---------+---------+----
                      4.80      5.40      6.00      6.60
```

## 14.5.4 Benefits of Blocking

The purpose of blocking is to allow you to see a treatment effect with a smaller sample size. However, it is often difficult to set up the blocks and so it is interesting to examine whether the effort to set up the blocks is worth it. The following formula gives you a measure of the relative efficiency of the block design compared to a one-way design:

$$\text{Relative efficiency} = \frac{(r-1)\text{MSBL} + r(c-1)\text{MSA}}{(n-1)\text{MSE}}$$

*Formula for relative efficiency of block design*

The terms in this formula are all found in the ANOVA table from a block design. The value of $n$ is the total sample size. The relative efficiency value tells you by what factor the sample size would have to be increased to see the treatment effect without using a block design. Clearly, if the relative efficiency is a low number such as 1.5, then perhaps it is not worth the effort to set up the blocks. Let's see how this formula works for the product display location.

---

ANS. $F = 55.35$; $F_{\text{crit}} = 3.23$ USING 2 AND 40 DEGREES OF FREEDOM AND $\alpha = 0.05$; REJECT $H_0$. AT LEAST ONE GROUP IS DIFFERENT. FROM THE CONFIDENCE INTERVALS, IT APPEARS THAT PARTICIPATIVE GROUP IS DIFFERENT FROM THE OTHER TWO.

**EXAMPLE 14.16**   **Product Display Location**

*Efficiency of Block Design*

The ANOVA table for the product display location exercise is shown below:

| Source | DF | SS | MS |
|--------|----|----|----|
| Display | 2 | 9,253,927 | 4,626,964 |
| Block | 9 | 60,738,440 | 6,748,716 |
| Error | 18 | 1,371,972 | 76,221 |
| **Total** | **29** | **71,364,339** | |

The relative efficiency of this design is calculated as follows:

$$\text{Relative efficiency} = \frac{(10 - 1)6,748,716 + 10(3 - 1)4,626,964}{(30 - 1)76,221} = 69.3$$

This number tells us that we would have needed 69 times as many observations to see the treatment effect if we did not use a block design. Clearly, the blocking was worth it in this case. ∎

## TRY IT NOW!

**Participative Management**

*Efficiency of Block Design*

Calculate the relative efficiency of the block design for the sick days data. Comment on the usefulness of the blocking.

## 14.5.5 Exercises—Learning It!

*Datafile: AMPS1.XXX*

**14.10** A manufacturer of electronic components was concerned about the reliability of one of its products. The company wished to see if the input current that the component would draw from a power source such as a battery changed over time. Twenty components were pretested at the time of manufacture, 9 days later, and 58 days later. The data (in milliamps) are shown below:

| Number | Pretest | 9 Days | 58 Days |
|--------|---------|--------|---------|
| 1 | 26.1 | 27.0 | 27.5 |
| 2 | 26.2 | 27.4 | 27.8 |
| 3 | 26.3 | 27.8 | 28.5 |
| 4 | 27.4 | 28.1 | 28.3 |
| 5 | 27.8 | 29.5 | 29.8 |
| 6 | 26.8 | 27.9 | 27.8 |
| 7 | 28.2 | 28.9 | 29.3 |
| 8 | 28.2 | 29.3 | 29.3 |

*(continued)*

| Number | Pretest | 9 Days | 58 Days |
|---|---|---|---|
| 9 | 27.1 | 28.1 | 28.9 |
| 10 | 26.9 | 27.9 | 28.0 |
| 11 | 27.8 | 28.9 | 29.0 |
| 12 | 27.4 | 28.8 | 28.8 |
| 13 | 27.1 | 28.0 | 27.7 |
| 14 | 27.4 | 29.0 | 29.5 |
| 15 | 26.5 | 27.7 | 27.8 |
| 16 | 27.0 | 27.6 | 27.7 |
| 17 | 28.3 | 29.1 | 29.6 |
| 18 | 28.2 | 29.1 | 29.1 |
| 19 | 27.0 | 27.5 | 28.0 |
| 20 | 28.1 | 29.7 | 30.4 |

(a) What is the factor and how many levels are there in this design?
(b) What is the blocking factor?
(c) Is there a significant difference in the power drawn over time?
(d) Calculate the relative efficiency of the blocking. Is the blocking effective?

**14.11** Many techniques for finding the "best" or optimal solution to a problem require that you provide the computer software with a starting solution. There are often different ways to get this starting solution. Four algorithms for generating starting solutions are run on 30 different problems. The number of iterations required to move from the starting solution to the optimal solution was recorded. A portion of the data is shown below: *Datafile: ALGORTH.XXX*

| Problem | Method | Iteration |
|---|---|---|
| 1 | 1 | 5 |
| 2 | 1 | 4 |
| 3 | 1 | 4 |
| 4 | 1 | 4 |
| 5 | 1 | 2 |
| 6 | 1 | 2 |

(a) Is there any difference in the number of iterations required to find the optimal solution that is due to the difference in the method used to generate the starting solution?
(b) Which method would you recommend and why?

**14.12** No. 1 Foods contracts with colleges and universities to provide the food services for the school. It has 3 different cafeteria layouts, which it is evaluating in terms of the time (in minutes) it takes for a "typical" student to get his/her food. To do a fair comparison No. 1 Foods has blocked on the size of the school. Is there a difference in the average time spent getting food for the 3 different layouts? If so, which layout is best? The data are shown below:

| | Layout 1 | Layout 2 | Layout 3 |
|---|---|---|---|
| 1 | 6.5 | 2.9 | 6.9 |
| 2 | 5.8 | 4.5 | 6.1 |
| 3 | 5.8 | 4.1 | 7.0 |
| 4 | 6.7 | 4.2 | 7.4 |
| 5 | 6.4 | 5.1 | 6.9 |
| 6 | 5.8 | 4.7 | 7.7 |
| 7 | 6.2 | 5.2 | 7.8 |
| 8 | 5.7 | 4.9 | 8.1 |
| 9 | 6.9 | 4.8 | 7.8 |
| 10 | 5.0 | 4.3 | 8.2 |
| 11 | 5.8 | 4.2 | 7.1 |
| 12 | 4.8 | 4.1 | 7.4 |
| 13 | 6.7 | 4.7 | 7.2 |
| 14 | 6.6 | 4.3 | 7.6 |
| 15 | 4.7 | 4.4 | 7.5 |

**14.13** Should you listen to classical music, watch TV, or just have silence when you study? To answer this question a group of 10 undergraduate students were randomly selected from a large university. For one semester they listened to classical music while they studied, for the next semester they watched TV while they studied, and for the third semester they studied in the library. Their GPA for each semester was recorded. The data are shown below:

| Student | Music | TV | Library |
|---|---|---|---|
| 1 | 3.62 | 3.35 | 3.23 |
| 2 | 3.38 | 3.23 | 3.07 |
| 3 | 3.09 | 2.24 | 2.70 |
| 4 | 2.69 | 1.88 | 2.41 |
| 5 | 3.29 | 2.31 | 3.01 |
| 6 | 2.97 | 2.75 | 2.82 |
| 7 | 2.60 | 1.98 | 2.14 |
| 8 | 3.56 | 3.26 | 3.07 |
| 9 | 3.50 | 2.74 | 3.23 |
| 10 | 3.74 | 3.19 | 3.55 |

(a) What is the response variable?
(b) What is the factor? How many levels are there?
(c) What is the blocking factor?
(d) Use block ANOVA to determine if there are any differences in the average GPA scores for the three treatment groups.
(e) What other factors would you suggest be included in a subsequent study?

*Datafile:*
*GRADSAL.XXX*

**14.14** What can I expect as a starting salary if I major in Business? Engineering? Education? This is a commonly asked question. The career office at a university decided to see if there were any differences in the starting salaries of these 3 majors. Analysts selected 50 students from last year's graduating class and recorded the starting salary for each student. To do a fair comparison, the students were matched on the basis of the overall grade point average. The data are shown below:

| Business | Engineering | Education |
|---|---|---|
| 22521 | 25698 | 20577 |
| 22482 | 25423 | 20609 |
| 22822 | 25662 | 20483 |
| 22503 | 25512 | 20531 |
| 22300 | 25453 | 20530 |
| 22418 | 25322 | 20710 |
| 22574 | 25715 | 20290 |
| 22545 | 25792 | 20730 |
| 22406 | 25499 | 20215 |
| 22521 | 25532 | 20491 |
| 22249 | 25543 | 20512 |
| 22398 | 25509 | 20540 |
| 22144 | 25451 | 20570 |
| 22615 | 25504 | 20662 |
| 22399 | 25367 | 20564 |
| 22565 | 25335 | 20500 |
| 22512 | 25461 | 20385 |
| 22469 | 25650 | 20672 |
| 22532 | 25399 | 20510 |
| 22516 | 25618 | 20412 |
| 22558 | 25481 | 20680 |
| 22414 | 25592 | 20587 |
| 22608 | 25479 | 20719 |
| 22327 | 25442 | 20390 |
| 22699 | 25192 | 20324 |
| 22545 | 25501 | 20418 |

(*continued*)

| Business | Engineering | Education |
|----------|-------------|-----------|
| 22431 | 25761 | 20593 |
| 22622 | 25514 | 20482 |
| 22467 | 25208 | 20430 |
| 22064 | 25740 | 20309 |
| 22388 | 25743 | 20452 |
| 22805 | 25867 | 20689 |
| 22600 | 25622 | 20600 |
| 22446 | 25386 | 20523 |
| 22667 | 25568 | 20647 |
| 22655 | 25604 | 20298 |
| 22500 | 25340 | 20441 |
| 22285 | 25601 | 20374 |
| 22218 | 25466 | 20601 |
| 22471 | 25394 | 20328 |
| 22176 | 25479 | 20713 |
| 22403 | 25489 | 20004 |
| 22293 | 25072 | 20548 |
| 22670 | 25627 | 20610 |
| 22531 | 25250 | 20577 |
| 22208 | 25412 | 20228 |
| 22550 | 25434 | 20527 |
| 22483 | 25204 | 20413 |
| 22501 | 25409 | 20754 |
| 22810 | 25396 | 20439 |

(a) What is the response variable?

(b) What is the factor? How many levels are there?

(c) What is the blocking factor?

(d) Use block ANOVA to determine if there are any differences in the average starting salary for the three groups.

(e) What other factors would you suggest be included in a subsequent study?

**14.15** Surely you have heard of the "freshman ten," referring to the average weight gain of 10 lb by students during the freshman year. Does the same phenomenon occur to graduate students? A sample of 15 students was taken and their weight (in lb) was recorded in September, February of the following year, July of the following year, and then July two years later. The data are shown below:

*Datafile: WEIGHT.XXX*

| Person | Sept | Feb | July | July, 1 Year Later |
|--------|------|-----|------|--------------------|
| 1 | 110 | 115 | 115 | 111 |
| 2 | 140 | 144 | 146 | 141 |
| 3 | 165 | 169 | 171 | 165 |
| 4 | 123 | 128 | 131 | 124 |
| 5 | 158 | 167 | 169 | 159 |
| 6 | 180 | 187 | 187 | 181 |
| 7 | 173 | 179 | 179 | 173 |
| 8 | 178 | 185 | 190 | 180 |
| 9 | 195 | 203 | 207 | 196 |
| 10 | 145 | 154 | 158 | 145 |
| 11 | 136 | 137 | 137 | 137 |
| 12 | 170 | 179 | 180 | 170 |
| 13 | 166 | 168 | 171 | 168 |
| 14 | 180 | 184 | 185 | 181 |
| 15 | 130 | 136 | 138 | 130 |

A new student of statistics decided to use a one-way ANOVA for these data and the results from Minitab are shown on page 728.

Analysis of Variance for Weight

| Source | DF | SS | MS | F | P |
|--------|-----|-------|-----|------|-------|
| Month  | 3  | 634   | 211 | 0.36 | 0.784 |
| Error  | 56 | 33108 | 591 |      |       |
| Total  | 59 | 33742 |     |      |       |

(a) Using this ANOVA table, what would you conclude about the average weights at the four times the data were collected?

(b) Explain why this analysis is incorrect and why it must be analyzed as a block design.

(c) What are the blocks?

(d) The Minitab output for the block ANOVA is shown below. What can you conclude about the average weights for the 4 time periods?

Analysis of Variance for Weight

| Source | DF | SS | MS |
|--------|-----|----------|---------|
| Person | 14 | 32990.73 | 2356.48 |
| Month  | 3  | 633.87   | 211.29  |
| Error  | 42 | 117.13   | 2.79    |
| Total  | 59 | 33741.73 |         |

# Discovery Exercise 14.1
# THE BENEFITS OF BLOCKING

In manufacturing electronic products such as loudspeakers, it is important that connections are strong and hold so that they do not disconnect. The same is true for commercial heaters. One of the customers of a manufacturer of heaters complained that the pull poundage was too low. Thus, the customer did not have confidence in the terminal connection. New wires with the terminals attached were made and a pull test was done on a sample of 25. The terminals were connected and pulled apart. The pull poundage at which the terminals would disconnect from each other was recorded. Then the same terminals were reconnected and the test was repeated a total of 6 times. The object was to determine if the pull poundage changed over time. The minimum pull poundage required by the customer was 5 lb. The data (datafile HEATERS.XXX) are shown below:

| Trial 1 | Trial 2 | Trial 3 | Trial 4 | Trial 5 | Trial 6 |
|---------|---------|---------|---------|---------|---------|
| 11.0 | 12.3 | 9.8  | 12.3 | 10.8 | 8.5  |
| 8.5  | 10.8 | 7.5  | 11.0 | 8.0  | 8.0  |
| 9.5  | 10.0 | 9.0  | 9.0  | 9.0  | 7.8  |
| 10.0 | 7.5  | 8.0  | 9.8  | 10.5 | 8.5  |
| 10.5 | 7.5  | 10.3 | 9.0  | 9.0  | 9.8  |
| 11.8 | 11.8 | 10.0 | 9.0  | 9.5  | 10.3 |
| 11.5 | 10.0 | 14.0 | 11.3 | 14.0 | 13.0 |
| 10.0 | 9.5  | 10.0 | 8.0  | 8.3  | 8.0  |
| 11.3 | 12.0 | 11.0 | 11.3 | 10.5 | 10.3 |
| 10.3 | 10.5 | 9.8  | 9.8  | 11.8 | 10.5 |
| 11.0 | 15.0 | 13.5 | 11.5 | 9.8  | 10.0 |
| 11.3 | 10.8 | 9.8  | 10.3 | 10.0 | 9.5  |
| 10.5 | 10.0 | 10.5 | 10.5 | 10.0 | 9.5  |
| 11.8 | 10.8 | 10.0 | 11.8 | 9.8  | 9.5  |
| 10.0 | 10.0 | 9.5  | 11.8 | 9.0  | 9.3  |
| 11.5 | 11.0 | 11.0 | 9.8  | 9.0  | 9.8  |
| 10.5 | 9.0  | 12.5 | 11.3 | 11.0 | 11.8 |

*(continued)*

| Trial 1 | Trial 2 | Trial 3 | Trial 4 | Trial 5 | Trial 6 |
|---------|---------|---------|---------|---------|---------|
| 11.3 | 11.0 | 9.5 | 9.0 | 10.0 | 8.0 |
| 11.0 | 11.3 | 11.3 | 10.0 | 11.3 | 9.5 |
| 11.0 | 8.0 | 10.5 | 7.0 | 7.0 | 12.0 |
| 11.8 | 10.3 | 12.3 | 11.8 | 12.5 | 11.0 |
| 12.0 | 12.3 | 11.5 | 12.0 | 12.3 | 11.0 |
| 11.5 | 12.0 | 11.0 | 10.0 | 11.5 | 11.5 |
| 12.0 | 9.5 | 12.5 | 9.5 | 9.0 | 9.3 |
| 10.0 | 7.5 | 10.5 | 10.3 | 9.5 | 9.0 |

### Part I: One-Way ANOVA

Use a one-way ANOVA to analyze these data.

(a)  What is the response variable and what is the factor?

(b)  How many levels of the factor are being studied?

(c)  Is there any difference in the average pull poundage among the trials? If so, which ones are different?

(d)  What is your recommendation to the company and why?

### Part II: Block Design

Since the same terminals were connected and pulled apart 6 times, each row is actually a block. Reanalyze these data using a block design.

(a)  Is there any difference in the average pull poundage among the trials? If so, which ones are different?

(b)  Explain why you got a different answer when you analyzed the data as a one-way design.

(c)  Calculate the sample size needed to see the treatment effect using a randomized one-way design.

(d)  Now, what is your recommendation to the company?

# 14.6  ANALYSIS OF DATA FROM TWO-WAY DESIGNS

After learning about block designs and thinking a bit more about the results of the airspace analysis, you wonder if there might be some other factor that influences the airspace in the box (something other than the length of time that the box sits on the shelf). You return to talk to the people who work in manufacturing to ask them about how the tissue boxes are made.

## 14.6.1 Motivation for a Factorial Design Model

In Figure 14.6 (page 730) you see a picture of a hardroll, which is a large roll of raw paper stock. Each box of 250 tissues is made from 25 hardrolls, each of which has 10 different slit positions. Four such slit positions are shown in the diagram. One tissue is taken from each slit and they are pressed into a tissue box and sealed. A hardroll lasts for 4 hours in production.

The manufacturing personnel feel that the mean airspace might differ by position in the hardroll from which the tissues were made. This hypothesis was proposed because of two facts:

**FIGURE 14.6**   Schematic of a hardroll

1. As the hardroll sits in the warehouse, the outside collects moisture. Thus boxes made from tissues taken from the outside of the roll might have less growback.

2. The core (very center) of the roll is very compressed (a hardroll weighs 1000 pounds). Thus, boxes of tissues made from the core have the "life stretched out of them."

As a result of this discussion you have identified another potential factor, the position in the hardroll from which the box was produced. This is not a blocking factor because you are interested in knowing whether this factor affects the airspace in the box. Hence, it is not a nuisance factor. You return to look at your data and realize that a total of 4 cases of tissue boxes were sampled, one case every hour for four hours. One case was used to obtain in-process measurements. The other 3 cases were saved for observations taken after 24 hours, 2 weeks, and 4 weeks. Each case contained 24 cartons of tissue boxes or 120 tissue boxes. This means that the tissues that were allowed to sit on the shelf for 24 hours were made from a different part of the hardroll than the ones that sat on the shelf for 2 weeks or 4 weeks. You cannot tell from this dataset if the position in the hardroll is indeed a significant factor because you did not account for this factor in your experimental design.

Based on this new information you suggest the following sampling plan:

1. Sample 5 times through the hardroll. Since a hardroll lasts for 4 hours the 5 observations were taken at the beginning and once an hour thereafter. This results in observations at the top, the core, and 3 in the middle of the hardroll.

2. Each sample consists of a case of tissues (24 cartons or 120 tissue boxes).

3. A total of 4 cases is sampled every hour. One was used to obtain in-process measurements. The other 3 cases are saved for observations taken after 24 hours, 2 weeks, and 4 weeks.

Each airspace measurement is taken on a tissue box made from a certain position in the hardroll and allowed to sit on the shelf a certain amount of time before measurement. Thus, we have two factors: position and time. The position indicates from what part of the hardroll the sample was taken. The values for position are 1, 2, 3, 4, and 5. Position 1 indicates that it was taken from the outside of the roll and position 5 indicates it was taken from the core. The factor time indicates when the measurement was taken:  1 = in process,  2 = after 24 hours,  3 = after 2 weeks,  and 4 = after 4 weeks.

This experimental design is an example of a **factorial design with two factors.** The two factors do not need to have the same number of levels but we will assume that the sample size is the same for each combination of the levels. For the airspace data this means that the number of tissue boxes sampled from position 1 and tested in process is the same as the number of tissue boxes sampled from position 3 and allowed to sit 2 weeks on the shelf and this is the same as the number of tissue boxes sampled from each of the 5 positions and allowed to sit for each of 4 different time possibilities. In this design, a sample of 1 case (24 cartons or 120 boxes) is selected from each of the $5 \times 4 = 20$ populations. This is referred to as **equal replication.**

Each observation within a population is referred to as a **replicate.** If you have unequal sample sizes, you must refer to a more advanced text.

> An experimental design is called a *factorial design with two factors* if there are several different levels of two factors being studied. The first factor is called *factor A* and there are $r$ levels of factor A. The second factor is called *factor B* and there are $c$ levels of factor B.

## EXAMPLE 14.17   Airspace Data

*Two-Factor Design*

For the airspace data, we can label the position factor as factor A and the time factor as factor B. There are $r = 5$ levels of factor A: positions 1, 2, 3, 4, and 5. Position 1 refers to tissues made from the very outside of the hardroll and position 5 refers to tissues made from the core of the hardroll. There are $c = 4$ levels of the factor B, as we had before. ∎

> The design is said to have *equal replication* if the same number of objects or people being observed/measured are randomly selected from each population. The population is described by a specific level for each of the two factors. Each observation is called a *replicate.* There are $n'$ observations or replicates observed from each population. There are $n = n'rc$ observations in total.

## EXAMPLE 14.18   Airspace Data

*Number of Replicates*

For the airspace data, one case or 120 tissue boxes is selected for each combination of position and time. There are 20 combinations of position and time ($rc = (5)(4) = 20$). Thus, $n'$ is 120 and there are $(20)(120) = 2400$ observations in total. ∎

The tissue company wishes to know if there a difference in airspace due to position in the hardroll and if there a difference in airspace by time tested. To answer these questions we must extend our analysis technique to handle this second factor.

## 14.6.2 Partitioning the Variation

The underlying concept in analyzing data from a two-factor design is the same as the approach we took in analyzing data from a one-way design. Remember that we looked at the relative sizes of the components of variation. For a two-factor design the variation we see in the data is due to one of four things: factor A, factor B, the interaction between factor A and B, and error. The idea is to take the total variability as measured by SST and break it down into 4 component sum of squares.

The names and labels for these are **sum of squares due to factor A (SSA), sum of squares due to factor B (SSB), sum of squares due to the interacting effect of A and B (SSAB),** and **sum of squares error (SSE).** Although the names are a bit different from those we used in a one-way ANOVA, they are an extension of the same concept.

The **sum of squares due to factor A** is labeled SSA. It measures the squared differences between the mean of each level of factor A and the grand mean.

The **sum of squares due to factor B** is labeled **SSB**. It measures the squared differences between the mean of each level of factor B and the grand mean.

The **sum of squares due to the interacting effect of A and B** is labeled **SSAB**. It measures the effect of combining factor A and factor B.

The **sum of squares error** is labeled **SSE**. It measures the variability in the measurements within the groups.

Thus, the total variation is divided into four components:

*Formula for SST in terms of other components in a two-way design*

$$SST = SSA + SSB + SSAB + SSE$$

The terms SSA and SSE have the same formulas and definition as for a one-way ANOVA. The SSB and SSAB terms are calculated in a similar fashion. If you are using a two-way experimental design, you will need to use software to do the calculations. The formulas for the sum of squares terms are long and tedious and any statistical software will generate the ANOVA table for a two-way design.

## 14.6.3 Using the ANOVA Table in a Two-Way Design

Consider a factorial design with $r$ levels of factor A and $c$ levels of factor B with $n'$ replicates in each combination of factor A and B. There are a total of $n = n'rc$ observations. The ANOVA table for such a design looks just like the ANOVA table for a one-way design with two additional rows. It is shown below:

| Source of Variation | SS | df | MS | F | p Value |
|---|---|---|---|---|---|
| Factor A | SSA | $r - 1$ | $MSA = \dfrac{SSA}{r - 1}$ | $F = \dfrac{MSA}{MSE}$ | |
| Factor B | SSB | $c - 1$ | $MSBL = \dfrac{SSB}{c - 1}$ | $F = \dfrac{MSB}{MSE}$ | |
| Interaction of A and B | SSAB | $(r - 1)(c - 1)$ | $MSAB = \dfrac{SSAB}{(r - 1)(c - 1)}$ | $F = \dfrac{MSAB}{MSE}$ | |
| Error | SSE | $rc(n' - 1)$ | $MSE = \dfrac{SSE}{rc(n' - 1)}$ | | |
| **Total** | **SST** | $rcn' - 1$ | | | |

The last 2 columns of the ANOVA table are used to do the hypothesis tests. In a two-way ANOVA, three hypothesis tests should be done. (1) To test the hypothesis of no difference due to factor A we would have the following null and alternative hypotheses:

$H_0$:    There is no difference in the population means due to factor A.

$H_A$:    There is a difference in the population means due to factor A.

This hypothesis test is easily done by determining if the $F$ value in the row of the ANOVA table corresponding to factor A is in the rejection region. If the $F$ test statistic is "too large," that is, larger than the critical $F$ value from the table, then you conclude that factor A is significant. If the software you are using does not provide you with the critical $F$ value, you can look it up in the table of $F$ distribution values. The $F$ statistic has $(r - 1)$ degrees of freedom in the numerator (from the MSA term) and $rc(n' - 1)$ degrees of freedom in the denominator (from the MSE term).

(2) To test the hypothesis of no difference due to factor B we would have the following null and alternative hypotheses:

$H_0$: There is no difference in the population means due to factor B.

$H_A$: There is a difference in the population means due to factor B.

This hypothesis test is easily done by determining if the $F$ value in the row of the ANOVA table corresponding to factor B is in the rejection region. If the $F$ test statistic is "too large," that is, larger than the critical $F$ value from the table, then you conclude that factor B is significant. If the software you are using does not provide you with the critical $F$ value, you can look it up in the $F$ distribution table. The $F$ statistic has $c - 1$ degrees of freedom in the numerator (from the MSB term) and $rc(n' - 1)$ degrees of freedom in the denominator (from the MSE term).

(3) To test the hypothesis of no difference due to the interaction of factors A and B we would have the following null and alternative hypotheses:

$H_0$: There is no difference in the population means due to the interaction of factors A and B.

$H_A$: There is a difference in the population means due to the interaction of factors A and B.

This hypothesis test is easily done by determining if the $F$ value in the row of the ANOVA table corresponding to the interaction of factors A and B is in the rejection region. If the $F$ test statistic is "too large," that is, larger than the critical $F$ value from the table, then you conclude that the interaction is significant. If the software you are using does not provide you with the critical $F$ value, you can look it up in the $F$ distribution table. The $F$ statistic has $(r - 1)(c - 1)$ degrees of freedom in the numerator (from the MSAB term) and $rc(n' - 1)$ degrees of freedom in the denominator (from the MSE term). The interaction effect is discussed further in the next section. But first let's look at the airspace data.

## EXAMPLE 14.19   Airspace Data

### Two-Way ANOVA

The output from Minitab for the airspace data is shown below. Factor A is position and factor B is time.

```
Two-Way Analysis of Variance

Analysis of Variance for Airspace

Source          DF          SS          MS
Position         4      5942.98     1485.74
Time             3     12773.75     4257.92
Interaction     12       865.52       72.13
Error         2380      1667.08        0.70
Total         2399     21249.33
```

To test the hypothesis of no difference due to factor A we would have the following null and alternative hypotheses:

$H_0$:   There is no difference in the population means due to position.

$H_A$:   There is a difference in the population means due to position.

The $F$ value is MSA/MSE = 1485.74/0.70 = 2122.5. With such a large $F$ value we really don't need an $F$ critical to tell us that our $F$ value is bigger than $F$ critical. However, for completeness the table $F$ value is found for $\alpha = 0.05$ with 4 degrees of freedom in the numerator and 2380 degrees of freedom in the denominator. $F$ critical = 2.37, so, clearly, position is a significant factor.

To test the hypothesis of no difference due to factor B we would have the following null and alternative hypotheses:

$H_0$:   There is no difference in the population means due to time.

$H_A$:   There is a difference in the population means due to time.

The $F$ value is MSB/MSE = 4257.92/0.70 = 6082.7. With such a large $F$ value we really don't need an $F$ critical to tell us that our $F$ value is bigger than $F$ critical. However, for completeness the table $F$ value is found for $\alpha = 0.05$ with 3 degrees of freedom in the numerator and 2380 degrees of freedom in the denominator. $F$ critical = 2.60 so, clearly, time is a significant factor.

To test the hypothesis of no difference due to the interaction of factors A and B we would have the following null and alternative hypotheses:

$H_0$:   There is no difference in the population means due to the interaction of position and time.

$H_A$:   There is a difference in the population means due to the interaction of position and time.

The $F$ value is MSAB/MSE = 72.13/0.70 = 103.0. With such a large $F$ value we really don't need an $F$ critical to tell us that our $F$ value is bigger than $F$ critical. However, for completeness the table $F$ value is found for $\alpha = 0.05$ with 12 degrees of freedom in the numerator and 2380 degrees of freedom in the denominator. $F$ critical = 1.75, so, clearly, the interaction is significant.

At this point we know from the analysis that both position in the hardroll from which the tissue box is made and the length of time the box sits on the shelf affect the average airspace in the box. We also know that these two factors interact. ■

 **TRY IT NOW!**

## Participative Management
### *Two-Way Design*

In a previous "Try It Now" you looked at the number of sick days used in the past 12 months by employees who were traditionally managed, employees who were involved in a participative management style, and employees who were involved in a management style that was a mixture of the old structure with the team-based approach. Management is wondering whether the employee's department is a factor as well.

Use the accompanying Minitab output to determine if there are any differences in the mean number of sick days by management style and by department and if there is any interaction effect.

**Analysis of Variance for Sick Day**

| Source | DF | SS | MS |
|---|---|---|---|
| Department | 4 | 298.187 | 74.547 |
| Mgt Style | 2 | 51.440 | 25.720 |
| Interaction | 8 | 4.293 | 0.537 |
| Error | 60 | 42.000 | 0.700 |
| Total | 74 | 395.920 | |

## 14.6.4 Understanding the Interaction Effect

In the previous section we performed a hypothesis test to determine if the two factors had a significant interaction effect. The easiest way to understand this effect is to look at a graph of the sample averages for each of the possible combinations of the two factors. The line graph shown in Figure 14.7 displays the 20 sample means for airspace.

From this graph you can see that the mean airspace decreases the longer the box sits on the shelf, regardless of from what position in the hardroll the box was made. However, the line connecting the sample means for those made from position 2 crosses the line connecting the sample means for those made from position 5. This indicates that the amount of growback in the tissue box between the 24-hour mark and the 2-week mark depends on where in the hardroll the tissue came from. Thus, the airspace behavior is affected by the interaction of the time on the shelf and the position in the hardroll from which it was made. This is what we expected based on our hypothesis test.

The information needed to look at the interaction effect graphically is the average of the observed variable for each group. Such information can be generated by using a statistical software package. Sometimes it is part of the output of ANOVA and sometimes it is a separate command.

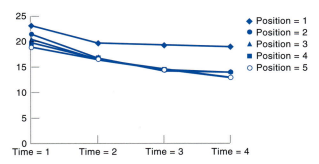

**FIGURE 14.7**  Average airspace based on time on shelf for different positions

**EXAMPLE 14.20**   Airspace Data

*Sample Means for All Combinations for Position and Time Tested*

In Minitab the analysis of means command generates output as shown below for the airspace data.

```
Rows: Position  Columns: Time

            1          2          3          4          All
1         120        120        120        120         480
        23.00      19.75      19.58      19.33       20.42
        0.648      0.523      0.705      0.473       1.613

2         120        120        120        120         480
        20.83      16.88      15.08      14.13       16.73
        0.803      1.171      0.762      0.668       2.713

3         120        120        120        120         480
        20.25      17.08      14.54      14.08       16.49
        0.664      1.120      0.916      0.705       2.605

4         120        120        120        120         480
        19.29      16.67      15.04      13.29       16.07
        0.679      0.853      0.938      0.938       2.371

5         120        120        120        120         480
        21.67      16.92      14.67      14.25       16.88
        1.286      0.816      0.690      0.882       3.096

All       600        600        600        600        2400
        21.01      17.46      15.78      15.02       17.32
        1.521      1.478      2.076      2.311       2.976

Cell Contents—

    Airspace:N

        Mean

        StDev
```

If there were no interaction effect, the lines connecting the sample means would be parallel. This would tell us that the growback behavior of the tissues in the box was the same no matter what position in the hardroll the tissues were made from. The graph might look something like the one shown in Figure 14.8.

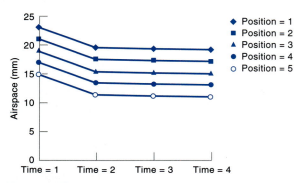

**FIGURE 14.8**   Hypothetical average airspace if there was no interaction effect

## 14.6.5 Exercises—Learning It!

14.16 The manufacturer of batteries is designing a battery to be used in a device that will be subjected to extremes in temperature. The company has a choice of 3 materials to use in the manufacturing process. An experiment is designed to study the life of the battery when it is made from materials A, B, and C and is exposed to temperatures of 15, 70, and 125°F. For each combination of material and temperature, 4 batteries are tested. The lifetimes in hours of the batteries are shown below:

| | Temperature | | |
|---|---|---|---|
| | **15°F** | **70°F** | **125°F** |
| Material A | 130,155,74,180 | 34,40,80,75 | 20,70,82,58 |
| Material B | 150,188,159,126 | 126,122,106,115 | 25,70,58,45 |
| Material C | 138,110,168,160 | 174,120,150,139 | 96,104,82,60 |

(a) Calculate the average life for each of the 3 materials.

(b) Calculate the average life for each of the 3 temperatures.

(c) Calculate the average life for each of the 9 treatment groups.

(d) Plot the 9 treatment means on a graph with temperature factor on the *x* axis, and the life of the battery in hours on the *y* axis. Use a different color for each of the 3 materials and connect the averages for those of the same material. What do you speculate about the interaction effect based on the graph?

(e) Confirm your suspicions by doing a two-way ANOVA and testing to see if there is a significant interaction effect.

(f) What material do you recommend to this manufacturer and why?

14.17 The golf ball data that you analyzed as a one-way ANOVA in Exercise 14.5 were actually collected at three different times of day. To see if the time of day is a factor, reanalyze the data using a two-way ANOVA. The first 12 observations of each model were taken early in the morning, the second 12 observations were taken mid-day, and the last 12 observations were taken late in the afternoon.    *Requires Exercise 14.5*

(a) Is there any difference in the average distance the ball carries by design?

(b) Is there any difference in the average distance the ball carries by time of day?

(c) Is there any interaction effect? If so, interpret this in terms of the application.

(d) Check the assumptions of ANOVA for these data.

14.18 The company that tested the power supplies in Exercise 14.10 wished to see if the input current that the component would draw from a power source, such as a battery, changed over time. Twenty components were pretested at the time of manufacture, 9 days later, and 58 days later. The company was also concerned about the possible differences in the product performance from lot to lot. The entire experiment was repeated for a sample taken from another lot. The data (in milliamps) are shown below:    *Requires Exercise 14.10* *Datafile: AMPS2.XXX*

| Number | Pretest | 9 Days | 58 Days |
|---|---|---|---|
| 1 | 25.6 | 27.3 | 28.5 |
| 2 | 27.9 | 30.0 | 30.5 |
| 3 | 25.7 | 27.1 | 27.8 |
| 4 | 25.9 | 27.6 | 29.1 |
| 5 | 25.2 | 26.6 | 27.5 |
| 6 | 26.1 | 27.9 | 29.4 |
| 7 | 26.5 | 28.8 | 29.4 |
| 8 | 29.9 | 32.1 | 33.2 |
| 9 | 27.9 | 29.1 | 30.3 |
| 10 | 27.3 | 29.8 | 32.0 |
| 11 | 25.1 | 27.2 | 28.5 |
| 12 | 25.7 | 27.2 | 27.9 |
| 13 | 27.0 | 28.3 | 29.2 |
| 14 | 25.6 | 26.2 | 26.8 |
| 15 | 26.2 | 27.1 | 27.6 |

*(continued)*

| Number | Pretest | 9 Days | 58 Days |
|--------|---------|--------|---------|
| 16 | 26.5 | 28.3 | 28.9 |
| 17 | 24.7 | 25.8 | 26.8 |
| 18 | 29.5 | 33.1 | 34.7 |
| 19 | 25.3 | 27.2 | 28.2 |
| 20 | 24.9 | 26.2 | 27.4 |

(a) With the additional lot this can now be considered a two-factor design. What is factor A? Factor B? How many replicates are there?

(b) Analyze both lots as a two-factor ANOVA.

(c) Is there any difference in the amount of power drawn from the battery over time?

(d) Is there any difference in the amount of power drawn from the battery by lot?

(e) Is there any interaction effect? If so, interpret this in terms of the application.

*Datafile:*
*MOUSE1.XXX*

14.19 With the increased use of computers at all levels of the organization, many managers are finding themselves faced with new and different problems. One manager wished to study the effect of the employees' efficiency measured by how long it took to complete a particular task (in seconds). Twenty employees were timed at the beginning of the day and at 2, 4, and 6 hours after the employee had been doing intensive work. In each case the employee used his/her right hand to complete the task 20 times and the left hand 20 times. The data are shown below.

| Beginning of Day | | After 2 h Intensive Work | | After 4 h Intensive Work | | After 6 h Intensive Work | |
|------|------|------|------|------|------|------|------|
| Right | Left | Right | Left | Right | Left | Right | Left |
| 67 | 162 | 84 | 63 | 52 | 89 | 57 | 114 |
| 64 | 86 | 78 | 71 | 53 | 53 | 53 | 61 |
| 69 | 88 | 74 | 86 | 56 | 122 | 71 | 61 |
| 88 | 99 | 91 | 111 | 66 | 88 | 61 | 61 |
| 72 | 83 | 70 | 99 | 59 | 93 | 73 | 53 |
| 80 | 88 | 73 | 88 | 77 | 116 | 50 | 43 |
| 85 | 113 | 86 | 121 | 64 | 104 | 53 | 56 |
| 116 | 69 | 71 | 66 | 62 | 53 | 80 | 111 |
| 77 | 88 | 76 | 56 | 54 | 58 | 63 | 38 |
| 78 | 101 | 76 | 56 | 65 | 119 | 41 | 40 |
| 68 | 73 | 61 | 48 | 71 | 96 | 63 | 58 |
| 51 | 61 | 62 | 51 | 92 | 104 | 41 | 41 |
| 54 | 76 | 94 | 123 | 71 | 146 | 53 | 84 |
| 75 | 53 | 63 | 56 | 50 | 81 | 63 | 58 |
| 71 | 71 | 70 | 83 | 71 | 53 | 61 | 136 |
| 64 | 61 | 63 | 63 | 58 | 58 | 46 | 93 |
| 86 | 63 | 66 | 86 | 77 | 73 | 68 | 71 |
| 98 | 101 | 71 | 63 | 53 | 56 | 64 | 68 |
| 103 | 86 | 53 | 61 | 81 | 136 | 49 | 73 |
| 91 | 71 | 81 | 171 | 70 | 83 | 70 | 74 |

(a) Is there any difference in the average time to complete the task after differing amounts of computer intensive work?

(b) Is there any difference in the average amount of time to complete the task for the right hand versus the left hand?

(c) Is there a significant interaction effect?

(d) Display the treatment means to get some additional insight into the interaction effect. What does the graph tell you that the hypothesis test does not tell you?

(e) What is your recommendation to this manager?

14.20 A manufacturer of adhesive products designed an experiment to compare a new adhesive product to a competitor's product. The adhesive product, or glue, is used by automobile manufacturers. The response variable was the strength of the glue measured by tensile strength in pounds per square inch (psi). The ability to adhere to oil-contaminated surfaces under different humidity conditions was studied. There were 2 levels for factor A: no oil or oil. Oil contamination was applied by hand dipping the samples in an oil solution and allowing them to air dry at room temperature for 2 hours. There were 2 levels for factor B: 50% humidity and 90% humidity. Three samples were tested for each of the combinations of factor A and factor B. The tensile values (psi) for the new product are shown below:

| Humidity | No Oil | Oil |
|----------|--------|-----|
| 50% | 175,100,175 | 43,42,44 |
| 90% | 95,115,85 | 95,105,116 |

(a)  Does the product behave significantly differently if the surface is oil contaminated?

(b)  Does the product behave significantly differently at different humidity levels?

(c)  Is there any significant interaction effect present?

# 14.7  OTHER TYPES OF EXPERIMENTAL DESIGNS

Often you wish to investigate more than 2 factors. The analysis technique of ANOVA extends directly to what is known as a factorial design with more than 2 factors. The interaction terms become a bit more complicated and the required sample size increases as you consider additional factors. However, the basic approach that we have taken works and the concepts you have learned extend directly to handle more than 2 factors.

We can see from our work in the previous section that the number of observations that you need increases very quickly as you increase the number of factors. For the airspace data we ended up with 20 groups from 2 factors. This means your total sample size increases. Suppose, for example, that the tissue company wished to consider a third factor which had 3 levels. We would have (20)(3) or 60 groups to study. If we selected a sample of 5 from each of these populations we would need a total sample of size of 300. You often do not have the resources to collect this many data.

With the addition of a third factor you could perform more hypothesis tests. You would have 3 hypothesis tests for the effect of the individual factors; 3 hypothesis tests to test for interactions between 2 of the variables (factor A and factor B; factor B and factor C; factor A and factor C), and 1 hypothesis test to check for interaction between all 3 variables. Although there are more tests to be performed, each one is done using an $F$ test just as we have seen in this chapter.

The two-way interaction terms are often relatively easy to interpret, as we have seen in the previous section. However, the higher order interactions are typically difficult to interpret and often do not tell you much. Thus, instead of using a full factorial design as we have done with two factors, we use a technique that helps keep the sample size down while still giving information about the main effect of the factors on the variable we are studying. These are called fractional factorial designs. In the case of the tissue company faced with 60 groups to study, the fractional design may call for a sample to be selected from only 30 groups. Thus, half the number of groups is studied, hence the term fractional design. You want to select the 30 groups to be studied judiciously so that you get information on the 3 factors and some of the important interaction effects. You lose information on some of the higher order interactions, but at least you have information on the major factors individually and some of the two-way interactions. This is better than not doing the experiment at all because of cost reasons. If you are in a situation where you wish to study more than 2 factors you should consult an advanced text on experimental design.

## 14.8    ANOVA IN EXCEL

Although Excel is not really a statistical software package, it does have some built-in tools for performing the basics of Analysis of Variance. We will look at how to use these tools for one- and two-way designs.

### 14.8.1 One-Way Analysis of Variance in Excel

To perform a one-factor ANOVA using Excel, the data must be in a worksheet with the data for each factor level in separate, adjacent columns. A portion of the airspace data is shown in Figure 14.9.

|     | A      | B      | C      | D      |
|-----|--------|--------|--------|--------|
| 1   | Time 1 | Time 2 | Time 3 | Time 4 |
| 2   | 23     | 20     | 19     | 19     |
| 3   | 25     | 19     | 19     | 19     |
| 4   | 23     | 20     | 20     | 19     |
| 5   | 23     | 20     | 20     | 19     |
| 6   | 23     | 20     | 19     | 19     |
| 7   | 23     | 19     | 20     | 19     |
| 8   | 23     | 20     | 20     | 19     |
| 9   | 23     | 20     | 19     | 19     |
| 10  | 22     | 20     | 19     | 19     |
| 11  | 22     | 20     | 20     | 20     |
| 12  | 23     | 20     | 19     | 20     |
| 13  | 23     | 19     | 19     | 19     |
| 14  | 22     | 20     | 19     | 19     |
| 15  | 22     | 20     | 21     | 19     |
| 16  | 24     | 20     | 20     | 19     |
| 17  | 23     | 20     | 18     | 20     |
| 18  | 23     | 20     | 19     | 20     |

**FIGURE 14.9**    Airspace data in Excel

To perform the analysis to see whether there is a difference in mean airspace for the different times, we will use the **ANOVA: Single Factor** tool from the **Tools > Data Analysis** menu. The dialog box shown in Figure 14.10 opens.

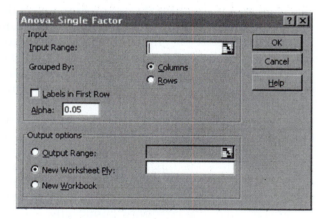

**FIGURE 14.10**    ANOVA: Single Factor dialog box

The input is similar to most Excel data analysis tools.

1.  Enter the range that contains the data in the **Input Range:** textbox. If your data range contains labels, click on the textbox for **Labels in First Row.**

2.  Since each factor is in a different column, click the radio button for **Grouped By: Columns.** This is the default value.

3. In the textbox for **Alpha,** enter the level of significance that you want to use for the test.

4. Finally, indicate where you want Excel to put the output from the analysis.

5. Click **OK;** the output will appear in the location specified. The output for the airspace data is shown in Figure 14.11. As with the output from the regression analysis, you might want to select **Format > Column > AutoFit Selection** while the output is still highlighted.

| F | G | H | I | J | K | L |
|---|---|---|---|---|---|---|
| Anova: Single Factor | | | | | | |
| | | | | | | |
| SUMMARY | | | | | | |
| Groups | Count | Sum | Average | Variance | | |
| Time 1 | 120 | 2521 | 21.00833333 | 2.327661064 | | |
| Time 2 | 120 | 2095 | 17.45833333 | 2.199929972 | | |
| Time 3 | 120 | 1894 | 15.78333333 | 4.339215686 | | |
| Time 4 | 120 | 1802 | 15.01666667 | 5.377871148 | | |
| | | | | | | |
| | | | | | | |
| ANOVA | | | | | | |
| Source of Variation | SS | df | MS | F | P-value | F crit |
| Between Groups | 2554.75 | 3 | 851.5833333 | 239.1302467 | 1.34182E-94 | 2.623636419 |
| Within Groups | 1695.116667 | 476 | 3.561169468 | | | |
| | | | | | | |
| Total | 4249.866667 | 479 | | | | |

**FIGURE 14.11**   Output from ANOVA on airspace data

Excel provides some summary statistics for each factor and the ANOVA table as output. From the ANOVA table, you see that the $F$ value for **Between Groups** is 239.13, compared to the critical value of 2.6236. The $p$ value of the test is zero for all practical purposes. Thus, the conclusion is that there is a difference in the mean airspace values for the different time periods.

Excel does not provide any tools for doing multiple comparisons to determine which levels are different, nor does it provide any diagnostic plots for checking assumptions. Since the fitted values are not provided, there is no way to find the residuals and perform the analysis yourself.

## 14.8.2 Two-Way ANOVA Designs in Excel

The data analysis tools in Excel allow you to analyze two-way designs for situations where there are no replications, and for situations where there are replications. The

| | A | B | C | D |
|---|---|---|---|---|
| 1 | Store | Shelf | End of Aisle | Front of Store |
| 2 | 1 | 4457 | 4500 | 5800 |
| 3 | 2 | 4400 | 4370 | 5290 |
| 4 | 3 | 4310 | 4300 | 5000 |
| 5 | 4 | 4600 | 4400 | 5600 |
| 6 | 5 | 5000 | 5000 | 6000 |
| 7 | 6 | 4500 | 4500 | 5300 |
| 8 | 7 | 4700 | 5100 | 5900 |
| 9 | 8 | 4590 | 4280 | 5460 |
| 10 | 9 | 8510 | 8670 | 10600 |
| 11 | 10 | 6470 | 6500 | 8410 |

**FIGURE 14.12**   Display/Sales data

no-replication tool is similar to the method you would use to analyze a blocked design, whereas the replications tool allows you to analyze a full two-way design.

To analyze a blocked design, we will look at the data on the effects of display type on sales. The data should be in a spreadsheet with the factors (display type) in columns and the blocks (stores) in rows, as shown in Figure 14.12 (see page 741).

1. From the analysis tools list, select **ANOVA: Two Factor Without Replication** and the dialog box shown in Figure 14.13 opens.

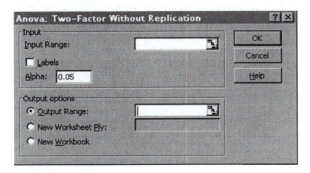

**FIGURE 14.13**   Two-way without replication dialog box

2. Enter the worksheet range for the data and fill in the other textboxes with the appropriate information.

3. Click **OK;** the output will be placed in the location specified. The output for the store display data is shown in Figure 14.14.

| Anova: Two-Factor Without Replication | | | | | | |
|---|---|---|---|---|---|---|
| **SUMMARY** | Count | Sum | Average | Variance | | |
| 1 | 3 | 14757 | 4919 | 582583 | | |
| 2 | 3 | 14060 | 4686.667 | 273233.3 | | |
| 3 | 3 | 13610 | 4536.667 | 161033.3 | | |
| 4 | 3 | 14600 | 4866.667 | 413333.3 | | |
| 5 | 3 | 16000 | 5333.333 | 333333.3 | | |
| 6 | 3 | 14300 | 4766.667 | 213333.3 | | |
| 7 | 3 | 15700 | 5233.333 | 373333.3 | | |
| 8 | 3 | 14330 | 4776.667 | 374233.3 | | |
| 9 | 3 | 27780 | 9260 | 1353100 | | |
| 10 | 3 | 21380 | 7126.667 | 1235433 | | |
| | | | | | | |
| Shelf | 10 | 51537 | 5153.7 | 1782646 | | |
| End of Aisle | 10 | 51620 | 5162 | 1970196 | | |
| Front of Store | 10 | 63360 | 6336 | 3148316 | | |
| | | | | | | |
| ANOVA | | | | | | |
| Source of Variation | SS | df | MS | F | P-value | F crit |
| Rows | 60738440.03 | 9 | 6748716 | 88.5418 | 4.73E-13 | 2.456282 |
| Columns | 9253927.267 | 2 | 4626964 | 60.70484 | 9.97E-09 | 3.554561 |
| Error | 1371972.067 | 18 | 76220.67 | | | |
| | | | | | | |
| Total | 71364339.37 | 29 | | | | |

**FIGURE 14.14**   Output for blocked design

The output includes summary information for each factor level and block in addition to the ANOVA table. We see from the $p$ values that both the blocks (located in the Rows) and the display types (located in the Columns) were significant.

The method for analyzing a full two-way design with replications is very similar to the method for the one-way and two-way without replications. The only difference is in the way Excel expects the data to be stored in the worksheet.

Excel expects to find the data in table format, with the levels of one factor in columns and the levels of the other factors in rows. Each replicate (observation) for a pair of levels is placed in a cell under the previous one. An explanation of this data setup from the Excel documentation is shown in Figure 14.15.

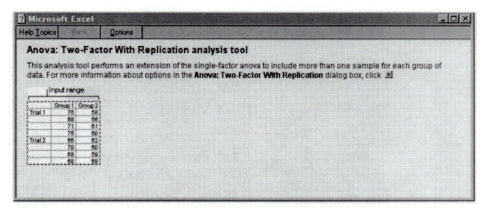

**FIGURE 14.15**   Data setup for two-way with replications

A subset of the airspace data is shown in Figure 14.16. In this example there are 24 boxes taken from each of five positions. The airspace is measured at four different times.

|  | A | B | C | D | E | F |
|---|---|---|---|---|---|---|
| 1 |  |  | Time1 | Time2 | Time 3 | Time 4 |
| 2 |  | Position 1 | 23 | 20 | 19 | 19 |
| 3 |  |  | 25 | 19 | 19 | 19 |
| 4 |  |  | 23 | 20 | 20 | 19 |
| 5 |  |  | 23 | 20 | 20 | 19 |
| 6 |  |  | 23 | 20 | 19 | 19 |
| 7 |  |  | 23 | 19 | 20 | 19 |
| 8 |  |  | 23 | 20 | 20 | 19 |
| 9 |  |  | 23 | 20 | 19 | 19 |
| 10 |  |  | 22 | 20 | 19 | 19 |
| 11 |  |  | 22 | 20 | 20 | 20 |
| 12 |  |  | 23 | 20 | 19 | 20 |
| 13 |  |  | 23 | 19 | 19 | 19 |
| 14 |  |  | 22 | 20 | 19 | 19 |
| 15 |  |  | 22 | 20 | 21 | 19 |
| 16 |  |  | 24 | 20 | 20 | 19 |
| 17 |  |  | 23 | 20 | 18 | 20 |
| 18 |  |  | 23 | 20 | 19 | 20 |
| 19 |  |  | 23 | 20 | 20 | 19 |
| 20 |  |  | 23 | 18 | 20 | 20 |
| 21 |  |  | 23 | 20 | 20 | 19 |
| 22 |  |  | 23 | 20 | 20 | 19 |
| 23 |  |  | 23 | 20 | 21 | 20 |
| 24 |  |  | 24 | 20 | 20 | 20 |
| 25 |  |  | 23 | 19 | 19 | 20 |
| 26 |  | Position 2 | 20 | 18 | 14 | 14 |
| 27 |  |  | 20 | 17 | 15 | 14 |
| 28 |  |  | 21 | 17 | 15 | 14 |
| 29 |  |  | 20 | 16 | 16 | 14 |
| 30 |  |  | 21 | 17 | 14 | 15 |
| 31 |  |  | 21 | 18 | 15 | 14 |
| 32 |  |  | 20 | 17 | 15 | 14 |

**FIGURE 14.16**   Airspace data for two-way analysis

The completed dialog box for the two-way with replications analysis tool is shown in Figure 14.17.

**FIGURE 14.17**   Two-way with replications dialog box

The only difference in the input is that you need to tell Excel how many replications there are by filling in the textbox labeled **Rows per sample:**. In this case there were 24. You *must* highlight the row with the first factor labels and the column with the second factor labels. Excel does not give you a choice here. The output is shown in Figure 14.18. It is very similar to the output from the other ANOVA tools.

**Anova: Two-Factor With Replication**

| SUMMARY | Time1 | Time2 | Time 3 | Time 4 | Total |
|---|---|---|---|---|---|
| **Position 1** | | | | | |
| Count | 24 | 24 | 24 | 24 | 96 |
| Sum | 552 | 474 | 470 | 464 | 1960 |
| Average | 23 | 19.75 | 19.5833333 | 19.3333333 | 20.4166667 |
| Variance | 0.43478261 | 0.2826087 | 0.51449275 | 0.23188406 | 2.6245614 |
| **Position 2** | | | | | |
| Count | 24 | 24 | 24 | 24 | 96 |
| Sum | 500 | 405 | 362 | 339 | 1606 |
| Average | 20.8333333 | 16.875 | 15.0833333 | 14.125 | 16.7291667 |
| Variance | 0.66666667 | 1.41847826 | 0.60144928 | 0.46195652 | 7.42061404 |
| **Position 3** | | | | | |
| Count | 24 | 24 | 24 | 24 | 96 |
| Sum | 486 | 410 | 349 | 338 | 1583 |
| Average | 20.25 | 17.0833333 | 14.5416667 | 14.0833333 | 16.4895833 |
| Variance | 0.45652174 | 1.29710145 | 0.86775362 | 0.51449275 | 6.84199561 |
| **Position 4** | | | | | |
| Count | 24 | 24 | 24 | 24 | 96 |
| Sum | 463 | 400 | 361 | 319 | 1543 |
| Average | 19.2916667 | 16.6666667 | 15.0416667 | 13.2916667 | 16.0729167 |
| Variance | 0.47644928 | 0.75362319 | 0.91123188 | 0.91123188 | 5.6683114 |
| **Position 5** | | | | | |
| Count | 24 | 24 | 24 | 24 | 96 |
| Sum | 520 | 406 | 352 | 342 | 1620 |
| Average | 21.6666667 | 16.9166667 | 14.6666667 | 14.25 | 16.875 |
| Variance | 1.71014493 | 0.6884058 | 0.49275362 | 0.80434783 | 9.66842105 |
| **Total** | | | | | |
| Count | 120 | 120 | 120 | 120 | |
| Sum | 2521 | 2095 | 1894 | 1802 | |
| Average | 21.0083333 | 17.4583333 | 15.7833333 | 15.0166667 | |
| Variance | 2.32766106 | 2.19992997 | 4.33921569 | 5.37787115 | |

**ANOVA**

| Source of Variation | SS | df | MS | F | P-value | F crit |
|---|---|---|---|---|---|---|
| Sample | 1188.59583 | 4 | 297.148958 | 409.963072 | 3.869E-150 | 2.39132447 |
| Columns | 2554.75 | 3 | 851.583333 | 1174.89128 | 3.56E-215 | 2.62429012 |
| Interaction | 173.104167 | 12 | 14.4253472 | 19.9020037 | 4.7221E-35 | 1.77324111 |
| Within | 333.416667 | 460 | 0.72481884 | | | |
| Total | 4249.86667 | 479 | | | | |

**FIGURE 14.18**   Output from airspace two-way design

# *CHAPTER 14 SUMMARY*

In this chapter you have learned the tool of analysis of variance (ANOVA), which is used to draw inferences about more than 2 population means. It directly extends the two-population tests to compare means which you learned in Chapter 10. You typically find yourself comparing more than 2 populations when you are trying to determine what factors influence the variable you are studying.

Regardless of the specific number of factors you are investigating, the technique of ANOVA partitions the variation in the data into components and compares the relative sizes of these components. So it is through an analysis of the variation in the data that we learn something about the means. It is a clever approach and one that draws on many of the basic concepts and tools that you have learned in earlier chapters.

The results of the hypothesis tests that you can perform will guide you in your design of subsequent experiments as you try to understand the factors that cause your variable of interest to vary. Ultimately you are interested in building a model to predict or control this variation. Remember that unexpected variation is the enemy according to Deming, and customers are demanding higher quality in products and services. It is only by understanding variation that we can produce quality products.

## **Key Terms**

| Term | Definition | Page Reference |
|---|---|---|
| **Response variable** | The **response variable** is a quantitative variable that you are measuring or observing. | 694 |
| **Factor** | A **factor** is a variable that can be used to differentiate one group or population from another. | 694 |
| **Level** | A **level** is one of several possible values or settings that the factor can assume. | 694 |
| **One-way or completely randomized design** | An experiment has a **one-way** or **completely randomized design** if several different levels of one factor are being studied and the objects or people being observed/measured are randomly assigned to one of the $c$ levels of the factor. | 695 |
| **ANOVA** | **Analysis of variance (ANOVA)** is the technique used to analyze the variation in the data to determine if more than two population means are equal. | 696 |
| **Treatment** | A **treatment** is a particular setting or combination of settings of the factor(s) being studied. | 696 |

*(continued)*

| Term | Definition | Page Reference |
|---|---|---|
| **Grand mean** | The **grand mean** or the **overall mean** is the sample average of all the observations in the experiment. It is labeled $\bar{\bar{x}}$. | 698 |
| **Sum of squares total (SST)** | The **total variation** or **sum of squares total (SST)** is a measure of the variability in the entire data set considered as a whole. | 698 |
| **Treatment mean** | A **treatment mean** is the average of the response variable for a particular treatment. | 700 |
| **Sum of squares among (SSA)** | **Between groups variation** measures how different the individual treatment means are from the overall grand mean. It is often called the **sum of squares between** or the **sum of squares among (SSA).** | 700 |
| **Sum of squares error (SSE)** | Within groups variation measures the variability in the measurements within the groups. It is often called **sum of squares within** or **sum of squares error (SSE).** | 702 |
| **Mean squares** | The **mean square among** is labeled **MSA**. The **mean square error** is labeled **MSE** and the **mean square total** is labeled **MST**. | 705 |
| **Block** | A **block** is a group of objects or people that have been matched. An object or person can be matched with itself, meaning that repeated observations are taken on that object or person and these observations form a block. | 720 |
| **Randomized block design** | An experiment has a **randomized block design** if there are several different levels of one factor being studied and the objects or people being observed/measured have been matched. Each object or person is randomly assigned to one of the $c$ levels of the factor. | 720 |
| **Sum of squares blocks (SSBL)** | The **sum of squares blocks** measures the variability between the blocks. It is labeled **SSBL.** | 720 |
| **Factorial design with two factors** | An experimental design is called a **factorial design with two factors** if there are several different levels of two factors being studied. The first factor is called **factor A** and there are $r$ levels of factor A. The second factor is called **factor B** and there are $c$ levels of factor B. | 731 |
| **Replicate** | The design is said to have **equal replication** if the same number of objects or people being observed/measured are randomly selected from each population. The population is described by a specific level for each of the two factors. Each observation is called a **replicate.** There are $n'$ observations or replicates observed from each population. There are $n = n'rc$ observations in total. | 731 |
| **Sum of squares due to factor A (SSA)** | The **sum of squares due to factor A** is labeled **SSA.** It measures the squared differences between the mean of each level of factor A and the grand mean. | 732 |
| **Sum of squares due to factor B (SSB)** | The **sum of squares due to factor B** is labeled **SSB.** It measures the squared differences between the mean of each level of factor B and the grand mean. | 732 |
| **Sum of squares due to the interacting effect of A and B (SSAB)** | The **sum of squares due to the interacting effect of A and B** is labeled **SSAB.** It measures the effect of combining factor A and factor B. | 732 |

## Key Formulas

| Term | Formula | Page Reference |
|------|---------|----------------|
| SST, one-way design | $\sum_{j=1}^{c} \sum_{i=1}^{n_j} (x_{ij} - \bar{\bar{x}})^2$ | 698 |
| SSA, one-way design | $SSA = \sum n_j(\bar{x}_j - \bar{\bar{x}})^2$ | 700 |
| SSE, one-way design | $\sum_{j=1}^{c} \sum_{i=1}^{n_j} (x_{ij} - \bar{x}_j)^2$ | 702 |
| MSA, one-way design | $\dfrac{SSA}{c-1}$ | 705 |
| MSE, one-way design | $\dfrac{SSE}{n-c}$ | 705 |
| MST, one-way design | $\dfrac{SST}{n-1}$ | 705 |
| F statistic, one-way design | $\dfrac{MSA}{MSE}$ | 707 |
| SST, block design | $SST = SSA + SSBL + SSE$ | 720 |
| Relative efficiency of block design | $\dfrac{(r-1)MSBL + r(c-1)MSA}{(n-1)MSE}$ | 723 |
| SST, two-way design | $SST = SSA + SSB + SSAB + SSE$ | 732 |

# CHAPTER 14 EXERCISES

## Learning It!

**14.21** A manufacturer of watches is concerned about the defective rate. There have been numerous customer complaints. The watches are made on 3 shifts and the company wishes to see if there is any difference in the average percent defective on the 3 shifts. The defective rate for each shift for 10 days is recorded and shown below:

|        | Shift 1 | Shift 2 | Shift 3 |
|--------|---------|---------|---------|
| Day 1  | 0.020   | 0.037   | 0.036   |
| Day 2  | 0.016   | 0.032   | 0.032   |
| Day 3  | 0.018   | 0.028   | 0.037   |
| Day 4  | 0.021   | 0.029   | 0.033   |
| Day 5  | 0.020   | 0.031   | 0.038   |
| Day 6  | 0.022   | 0.034   | 0.035   |
| Day 7  | 0.019   | 0.028   | 0.034   |
| Day 8  | 0.019   | 0.032   | 0.027   |
| Day 9  | 0.019   | 0.031   | 0.036   |
| Day 10 | 0.021   | 0.030   | 0.034   |

(a) What type of design is this?

(b) Is there any difference in the percent defective by shift? If so, which one(s)?

(c) Check the assumptions of ANOVA for these data.

**14.22** A manufacturer of dishwashers was concerned about the consistency of measuring the resistance of the heater in the dishwasher product from one meter to another. A sample of 30 heaters was selected. The resistance of each heater was measured by 4 meters. The data (in ohms) are shown on page 748.

*Datafile: DISHWASH.XXX*

| Sample | Meter 1 | Meter 2 | Meter 3 | Meter 4 |
|--------|---------|---------|---------|---------|
| 1 | 14.5 | 14.4 | 14.3 | 14.63 |
| 2 | 14.8 | 14.8 | 14.7 | 14.98 |
| 3 | 14.8 | 14.8 | 14.7 | 14.93 |
| 4 | 14.6 | 14.6 | 14.5 | 14.80 |
| 5 | 15.8 | 15.7 | 15.6 | 15.95 |
| 6 | 14.6 | 14.6 | 14.5 | 14.80 |
| 7 | 14.8 | 14.8 | 14.7 | 14.99 |
| 8 | 14.4 | 14.4 | 14.3 | 14.62 |
| 9 | 14.6 | 14.5 | 14.4 | 14.71 |
| 10 | 14.3 | 14.2 | 14.2 | 14.43 |
| 11 | 14.3 | 14.3 | 14.2 | 14.49 |
| 12 | 14.6 | 14.6 | 14.5 | 14.75 |
| 13 | 14.3 | 14.2 | 14.2 | 14.44 |
| 14 | 15.0 | 14.9 | 14.8 | 15.11 |
| 15 | 14.5 | 14.5 | 14.4 | 14.68 |
| 16 | 14.7 | 14.7 | 14.5 | 14.86 |
| 17 | 14.6 | 14.6 | 14.5 | 14.76 |
| 18 | 14.7 | 14.6 | 14.5 | 14.81 |
| 19 | 14.7 | 14.7 | 14.6 | 14.85 |
| 20 | 14.6 | 14.5 | 14.4 | 14.71 |
| 21 | 14.5 | 14.4 | 14.3 | 14.58 |
| 22 | 14.4 | 14.3 | 14.3 | 14.50 |
| 23 | 15.3 | 15.3 | 15.2 | 15.44 |
| 24 | 14.4 | 14.4 | 14.3 | 14.54 |
| 25 | 14.5 | 14.4 | 14.3 | 14.58 |
| 26 | 14.8 | 14.7 | 14.6 | 14.86 |
| 27 | 14.6 | 14.5 | 14.4 | 14.72 |
| 28 | 14.5 | 14.4 | 14.3 | 14.57 |
| 29 | 14.4 | 14.3 | 14.2 | 14.50 |
| 30 | 14.6 | 14.6 | 14.5 | 14.75 |

(a) What type of design is this?

(b) Is there any difference in the measurements taken by the different operators? If so, which one(s)?

(c) Check the assumptions of ANOVA for these data.

*Requires Exercise 14.22*   **14.23** The dishwasher company wondered whether it mattered if the meters had time to warm up before being used. In addition to the 4 meters used in Exercise 14.22, another meter of the same type as meter 4 was allowed a 2-hour warm-up period before being used. The readings on this meter for the same 30 sample products are shown below:

| | | | | | |
|---|---|---|---|---|---|
| Samples 1–5 | 14.62 | 15.00 | 14.95 | 14.78 | 15.93 |
| Samples 6–10 | 14.79 | 15.01 | 14.61 | 14.70 | 14.45 |
| Samples 11–15 | 14.48 | 14.76 | 14.45 | 15.09 | 14.67 |
| Samples 16–20 | 14.86 | 14.77 | 14.79 | 14.84 | 14.71 |
| Samples 21–25 | 14.58 | 14.50 | 15.45 | 14.55 | 14.59 |
| Samples 26–30 | 14.85 | 14.72 | 14.58 | 14.51 | 14.76 |

(a) Rerun the ANOVA including these data as a 5th treatment.

(b) Do any of your conclusions change?

*Datafile:*
*ONHOLD.XXX*   **14.24** Many on-line service companies have a technical support telephone service to assist users who are having difficulty getting connected or using the service. Recently there has been a concern about the amount of time that a customer has to wait "on hold" before hearing a human voice. In an effort to decide which on-line service to purchase, a study was conducted. Thirty calls were made to 4 on-line services during 3 times of day: early morning (6–8 AM), midday (noon to 2 PM), and late afternoon (4–6 PM). The amount of time spent on hold in minutes was recorded. A portion of the data is shown in the next table.

| | Company 1 | Company 2 | Company 3 | Company 4 |
|---|---|---|---|---|
| **Early AM** | 11 | 11 | 7 | 3 |
| | 8 | 11 | 9 | 8 |
| | 7 | 11 | 7 | 6 |
| | 8 | 9 | 9 | 9 |
| | 8 | 9 | 8 | 11 |
| | 7 | 6 | 8 | 9 |
| | 8 | 9 | 5 | 9 |

(a) Is there any difference in the average time on hold among the four companies?

(b) Is there any difference in the average amount of time on hold by time of day?

(c) Is there any interaction between the time of day and the four companies?

(d) If you could use more than one of the companies, which company would you use during each of the three time periods? Explain your reasoning.

(e) If you cannot use more than one company, which company would you select? Explain your reasoning.

(f) What other factors might influence the time on hold?

**14.25** One of the benefits of implementing a Total Quality Management program is the decreased time it takes to go from product design to market sales. This is often referred to as the cycle time. An automobile company recently implemented a TQM program companywide. The cycle times (in months) for 3 new cars before TQM and one year after TQM are shown below:

| Before TQM | After TQM |
|---|---|
| 26, 29, 22 | 17, 9, 13 |

(a) Based on these data, what can you conclude about the average cycle time before and after the implementation of TQM?

(b) What else would you like to know about these products to be certain that the reduction in cycle time is a result of the TQM program and not something else?

**14.26** There has been increasing concern about the declining reading and math levels of American students. A study was conducted to see the effect of income on four response variables: behavior problems index, reading score, math score, and vocabulary score. All three of the scores were from 0 to 100. The behavior index ranges from 0 to 100 with low scores indicating fewer behavior problems. One hundred children aged 5 to 7 were tested from families with varying income levels. Three income levels were used: an annual income of $15,000 or less, an annual income of $25,000, and an annual income of $40,000. The results are found in the datafile named SCORES.xxx. *Datafile: SCORES.XXX*

(a) For each of the response variables, test to see if there is a difference in the average score for the 3 levels of the income factor.

(b) Have any of the assumptions of ANOVA been violated?

(c) What recommendations can you make based on your analysis?

**14.27** The manager from Exercise 14.19 is also concerned about the accuracy of her employees after many hours of intensive computer work. At the same time as the data on time to complete the task was taken, a measure of accuracy (% correct) was also recorded. These data are shown below: *Requires Exercise 14.19* *Datafile: MOUSE2.XXX*

| Beginning of Day | | After 2 h Intensive Work | | After 4 h Intensive Work | | After 6 h Intensive Work | |
|---|---|---|---|---|---|---|---|
| **Right** | **Left** | **Right** | **Left** | **Right** | **Left** | **Right** | **Left** |
| 95.88 | 89.70 | 93.60 | 90.78 | 87.96 | 82.54 | 79.88 | 92.38 |
| 91.94 | 90.00 | 98.00 | 94.17 | 96.39 | 87.96 | 84.70 | 86.58 |
| 92.79 | 85.68 | 90.10 | 94.17 | 98.59 | 95.76 | 93.68 | 94.17 |
| 95.76 | 95.53 | 97.00 | 89.70 | 96.00 | 91.00 | 93.92 | 95.00 |
| 91.46 | 81.97 | 87.96 | 96.84 | 84.35 | 83.24 | 97.76 | 96.84 |
| 92.72 | 90.15 | 95.53 | 90.78 | 96.39 | 95.00 | 74.68 | 84.48 |
| 93.60 | 92.72 | 96.84 | 92.72 | 94.90 | 94.61 | 86.40 | 89.23 |

(*continued*)

| Beginning of Day | | After 2 h Intensive Work | | After 4 h Intensive Work | | After 6 h Intensive Work | |
|---|---|---|---|---|---|---|---|
| **Right** | **Left** | **Right** | **Left** | **Right** | **Left** | **Right** | **Left** |
| 94.61 | 83.45 | 80.76 | 91.00 | 94.61 | 87.96 | 99.00 | 90.00 |
| 99.00 | 98.00 | 95.53 | 92.19 | 79.38 | 79.60 | 94.61 | 86.11 |
| 97.76 | 88.30 | 92.00 | 88.82 | 89.70 | 85.00 | 76.65 | 82.91 |
| 100.00 | 93.60 | 96.39 | 97.17 | 93.68 | 96.84 | 98.59 | 94.17 |
| 89.18 | 89.23 | 80.90 | 85.58 | 98.59 | 90.00 | 81.13 | 90.00 |
| 87.47 | 97.00 | 85.68 | 94.61 | 92.93 | 91.75 | 86.96 | 95.88 |
| 94.17 | 96.39 | 93.29 | 91.94 | 85.68 | 94.90 | 92.72 | 83.00 |
| 91.40 | 91.94 | 96.00 | 84.97 | 93.60 | 87.27 | 91.94 | 94.90 |
| 95.53 | 84.19 | 94.90 | 90.15 | 90.57 | 89.70 | 80.69 | 94.61 |
| 93.68 | 96.39 | 94.34 | 89.00 | 96.39 | 84.70 | 91.75 | 87.79 |
| 94.90 | 84.44 | 92.19 | 87.79 | 87.96 | 95.53 | 95.00 | 95.00 |
| 97.76 | 94.17 | 95.00 | 91.40 | 94.90 | 93.92 | 85.58 | 82.97 |
| 97.00 | 93.29 | 98.59 | 91.40 | 95.53 | 90.57 | 96.84 | 88.69 |

(a) Is there any difference in the average accuracy level after differing amounts of computer intensive work?

(b) Is there any difference in the average accuracy level to complete the task for the right hand versus the left hand?

(c) Is there a significant interaction effect?

(d) What is your recommendation to this manager?

**14.28** In an effort to reduce crime, several cities have implemented a plan that encourages members of the police force to live in the city they patrol. These police officers have been given the opportunity to purchase homes at half-price, the other half of the cost being funded by the city budget. To see if this program has had the desired effect, the yearly number of murders in 5 cities has been recorded the year before the program was initiated, one year after the program started, and two years after the program was initiated. The data are shown below:

| City | Year Before Program Started | 1 Year After Program Started | 2 Years After Program Started |
|---|---|---|---|
| Los Angeles | 44 | 43 | 43 |
| Chicago | 35 | 34 | 33 |
| Detroit | 41 | 39 | 38 |
| Springfield | 21 | 20 | 19 |
| Philadelphia | 29 | 28 | 27 |

(a) What type of experimental design was used in this situation?

(b) Has the program been effective? Why or why not?

(c) Have any of the assumptions of ANOVA been violated?

(d) A government publication states that the average number of murders has dropped from 34 to 31.8 for the cities that have adopted this program. The publication uses this statement to propose expanding this program to other cities. Do you agree with this analysis? Why or why not? Should this program be adopted by other cities or should it be scrapped?

*Requires Exercise 14.20*   **14.29** The adhesive company in Exercise 14.20 was interested in comparing its new product with that of its competitor. The same experimental conditions were used to test the competitor's product. The tensile values (psi) for the competitor's product are shown below:

| Humidity | No Oil | Oil |
|---|---|---|
| 50% | 437,437,450 | 9,10,6 |
| 90% | 115,115,50 | 105,87,105 |

(a) Does the competitor's product behave significantly differently if the surface is oil contaminated?

(b) Does the competitor's product behave significantly differently at different humidity levels?

(c) Is there a significant interaction effect for the competitor's product?

(d) Summarize your findings about the new product and the competitor's product in terms of the response to oil-contaminated surfaces and varying levels of humidity. Should this company continue to develop the new product?

## Thinking About It!

**14.30** Upon further thought about the dishwasher company, you have decided to compare just 2 of the columns of data: meter 4 and meter 4 after a warm-up period.

*Requires Exercises 14.22, 14.23*

  (a) Use a paired $t$ test to do this comparison. Is there any difference in the readings if you let the meter warm up?

  (b) Run the appropriate ANOVA using just meter 4 and meter 4 after warm-up as the dataset. Does the ANOVA indicate any difference in the treatment means?

  (c) Should you have expected the same answer using the paired $t$ test and the ANOVA? Explain why or why not.

  (d) Take the value of the $t$ statistic from the paired test and square it. Compare it to the $F$ statistic from the ANOVA. What do you notice? Do you think this will always happen or is this just a coincidence for this dataset?

**14.31** Reconsider the teacher of statistics who was wondering whether to assign and grade homework. The teacher used 3 different classes to investigate the 3 different possible ways of dealing with homework. The teacher used a one-way design thinking that students randomly select sections of a course to take and so they should be comparable. Is there any reason to think that a block design should have been used in this situation instead of the one-way design that was used? Explain why or why not.

*Requires Exercise 14.3*

**14.32** When the diaper company tested absorbency the product was actually placed on a child and the child was allowed to play during the testing. As the child played, fluid was injected into the diaper every 10 minutes until the product failed (leaked). The amount of fluid (in grams) at the time of failure was recorded for each diaper. This experiment was set up as a one-way design. Can you think of any reason why a block design might be better?

*Requires Exercise 14.1*

**14.33** In any ANOVA you test the null hypothesis of equal means using an $F$ statistic. If you fail to reject the null hypothesis and you construct a confidence interval for the difference in two of the treatment means, what number should you expect to find in the confidence interval? Explain why.

**14.34** In Exercise 14.18 you analyzed a dataset with two factors: time and lot. Since then, two more months have gone by and data have been collected on two more lots. Unfortunately, for the third lot the experiment was run only for the pretest and 9 days later. For the fourth lot, data were collected during the pretest and 9 days and 16 days later (instead of 58 days). The data are shown below:

*Requires Exercise 14.18*

| Number | Pretest | 9 Days | Number | Pretest | 9 Days | 16 Days |
|--------|---------|--------|--------|---------|--------|---------|
| 1  | 26.1 | 31.5 | 1  | 30.8 | 34.6 | 35.0 |
| 2  | 26.8 | 29.8 | 2  | 27.5 | 30.9 | 31.4 |
| 3  | 25.3 | 28.7 | 3  | 26.8 | 31.3 | 31.8 |
| 4  | 28.1 | 32.2 | 4  | 29.4 | 32.8 | 33.5 |
| 5  | 25.3 | 30.3 | 5  | 28.7 | 32.6 | 32.0 |
| 6  | 27.3 | 31.7 | 6  | 24.0 | 26.5 | 26.9 |
| 7  | 26.5 | 30.4 | 7  | 26.4 | 30.9 | 31.2 |
| 8  | 24.8 | 28.0 | 8  | 28.3 | 31.0 | 31.2 |
| 9  | 26.1 | 28.6 | 9  | 30.0 | 32.4 | 32.8 |
| 10 | 26.2 | 29.1 | 10 | 25.0 | 27.3 | 27.4 |
| 11 | 25.3 | 27.4 | 11 | 26.8 | 30.3 | 30.8 |
| 12 | 28.0 | 31.6 | 12 | 26.7 | 30.9 | 31.2 |
| 13 | 25.5 | 29.1 | 13 | 26.1 | 27.9 | 28.2 |
| 14 | 24.8 | 28.2 | 14 | 27.6 | 31.5 | 31.7 |
| 15 | 27.2 | 31.0 | 15 | 27.4 | 29.7 | 30.0 |
| 16 | 25.1 | 27.1 | 16 | 26.3 | 29.0 | 29.5 |
| 17 | 26.2 | 28.2 | 17 | 28.0 | 31.5 | 32.0 |
| 18 | 26.8 | 29.5 | 18 | 25.3 | 27.1 | 27.5 |
| 19 | 27.3 | 29.4 | 19 | 26.4 | 28.3 | 28.4 |
| 20 | 27.1 | 28.8 | 20 | 27.2 | 30.0 | 30.7 |

(a) Explain why you cannot simply add two more levels to the lot factor and reanalyze the entire dataset using ANOVA.

(b) What analysis do you suggest be done to use these additional data?

**14.35** You have decided to use your newly acquired data analysis skills in your coaching of a 6th-grade girls' basketball team. You want to see if there is any difference in the average number of points your team scores when you use different defense strategies. You suggest the following design: use "man-to-man" defense strategy for the first third of the season and record the number of points scored for each game. Switch to full-court press strategy for the second third of the season and record the number of points scored for each game. During the last third of the season use a mixture of defense strategies.

(a) Will you be able to tell which strategy is best from the data collected from this experiment?

(b) Suggest an alternative experimental design.

*Requires Exercise 14.15*   **14.36** In Exercise 14.15 you concluded that the average weights for the students were different at the four different times the data were collected. Suppose you now learned that these students were enrolled in a one-year weekend MBA program starting in September of one year and ending a year later and that they were all working at full-time jobs during this time.

(a) What might be causing the weight differences that you observed?

(b) Support your theory by taking another look at this dataset and analyzing it differently.

*Requires Exercises 14.19,*   **14.37** In Exercises 14.19 and 14.27 you analyzed the amount of time to complete a task and
*14.27*   the accuracy level after varying amounts of time of intensive computer work. What other factors would you suggest be studied in a subsequent experiment on this subject?

## Doing It!

**14.38** A manufacturing company that makes paper products is having trouble with its paper towel line. When a roll of paper towels is manufactured, the last step before the roll is put in a case is to wrap it with the poly wrapper in which the consumer sees it on the store shelf. A large amount of product is scrapped because it is not being wrapped properly.

The problem appears to be that the diameters of the towel rolls are too large and the wrapper does not go all the way around and seal properly. There are many factors in the manufacturing process that would appear to affect roll diameter, but the engineers involved were not sure exactly how the different machine settings really impacted roll diameter. In fact, they were not convinced that all of the factors made a difference! The towel machine team designed a study to determine the effect that different machine settings actually had on roll diameter. The specifications for the diameter of the roll of towels is 5.35 inches. Several different people on the team expressed a concern that some settings that would result in a good roll diameter might have an adverse impact on another towel roll characteristic, roll firmness.

If a towel is not firm its wrapping will also be affected, and if it is too firm consumers will still react negatively. Fixing the roll diameter problem at the expense of firmness was not an option.

The team decided to look at two machine factors:

**1.** Embosser roll gap: the mechanism that puts the pattern in the towel

**2.** Speed: the speed at which the machine winds the rolls of towels

The machine was run at two different settings for speed and two different sizes of the embosser roll gap. The response variables were diameter of the roll of paper towels in inches and firmness measured on a specialized scale. Lower numbers indicate softer rolls, while larger numbers indicate firmer rolls. A portion of the dataset is shown below:

| Speed | Embosser | Diameter | Firmness |
|-------|----------|----------|----------|
| 0 | 0 | 5.43307 | 0.271667 |
| 0 | 0 | 5.39370 | 0.336667 |
| 0 | 0 | 5.47244 | 0.260333 |
| 0 | 0 | 5.39370 | 0.261333 |
| 0 | 0 | 5.43307 | 0.297000 |
| 1 | 0 | 5.39370 | 0.312000 |
| 1 | 0 | 5.47244 | 0.294667 |

*(continued)*

| Speed | Embosser | Diameter | Firmness |
|-------|----------|----------|----------|
| 1 | 0 | 5.43307 | 0.342333 |
| 1 | 0 | 5.47244 | 0.306000 |
| 1 | 0 | 5.43307 | 0.365667 |

The variable *Speed* is a 0–1 variable that indicates the machine speed. A 0 indicates that the slower machine speed was used and a 1 indicates that the faster machine speed was used. The variable *Embosser* is a 0–1 variable that indicates the size of the embosser roll gap. A 0 indicates that the smaller gap measurement was used, while a 1 indicates that the larger gap measurement was used.

*Datafile: TOWEL.XXX*

(a) Is there a difference in the diameter of the paper towel roll due to the speed at which the machine is run?

(b) Is there a difference in the diameter of the paper towel roll due to the size of the embosser roll gap?

(c) Is there any interaction of the two factors with respect to the diameter of the paper towel roll?

(d) Is there a difference in the firmness of the paper towel roll due to the speed at which the machine is run?

(e) Is there a difference in the firmness of the paper towel roll due to the size of the embosser roll gap?

(f) Is there any interaction of the two factors with respect to the firmness of the paper towel roll?

(g) Check the ANOVA assumptions for both response variables.

(h) Prepare a report for management explaining the effects that each of the machine settings have on *Diameter* and *Firmness*. Indicate which settings result in acceptable values for roll diameter. Make a recommendation on machine settings, if you can.

**14.39** In several of the worked examples in this chapter we looked at a portion of a dataset measuring tissue strength. The management of the tissue company was concerned about complaints involving sheets tearing on removal. One possibility was the lack of airspace in the box. We have studied that dataset extensively in this chapter. Another possible explanation is that the tissues are not strong enough. Two tissue strength measurements were studied: machine-directional strength *(MDStrength)* and the cross-directional strength *(CDStrength)*. They are both measured in lb/ream. The data were collected over 3 days and the company would like to know if the average *MDStrength* and the average *CDStrength* are the same for all 3 days. There are 75 observations per day. A portion of the dataset is shown below:

| Day | MDStrength | CDStrength | Day | MDStrength | CDStrength |
|-----|-----------|-----------|-----|-----------|-----------|
| 1 | 1006 | 422 | 1 | 962 | 464 |
| 1 | 994 | 448 | 1 | 973 | 472 |
| 1 | 1032 | 423 | 1 | 1036 | 489 |
| 1 | 875 | 435 | 1 | 1084 | 440 |
| 1 | 1043 | 445 | | | |

*Datafile: TISSUE.XXX*

The specifications indicate that *MDStrength* should have a mean of 1000 lb/ream and a standard deviation of 50 lb/ream. The specifications indicate that *CDStrength* should be normally distributed with a mean of 450 and a standard deviation of 25 lb/ream.

(a) Is there any difference in the average *MDStrength* for the 3 days? If so, which day(s) are different?

(b) Is the *MDStrength* running to specification?

(c) Is there any difference in the average *CDStrength* for the 3 days? If so, which day(s) are different?

(d) Is the *CDStrength* running to specification?

(e) Check the assumptions of ANOVA for both response variables.

(f) Using the results of your analysis, prepare a report to management telling them about the current tissue manufacturing process. Make recommendations to them on whether the process needs to be adjusted or whether it is running to target. Remember, if it is running to target they will be considering making changes to the specifications to reduce customer complaints about dispensing.

# CHAPTER *15*

# *THE ANALYSIS OF QUALITATIVE DATA*

## COLLEGE DRINKING

For many students, consumption of alcohol is part of the "college experience." Some colleges and universities attract students based on their reputation as "party schools." Students do not often think about the ramifications of their drinking until they experience problems related to it, such as failing classes or flunking out entirely, unwanted pregnancies, or encounters with law enforcement. University administrations are often unaware of or in denial of the problems and do not have programs in place to help students address the problems of drinking and alcohol-related health behaviors.

The problem of excessive alcohol consumption in college is one that has been documented many times, in many different ways. A random national survey involving 140 colleges and universities, conducted by Harvard University, looked at the problems connected with *binge drinking,* which is defined as the consumption of five or more drinks during one episode of drinking for men and consumption of four or more drinks during one episode of drinking for women. The study found that as the number of binge drinking episodes for an individual increased, the number of alcohol-related problems experienced increased as well. The types of problems that the two studies examined included having a hangover, missing class, having unplanned sex, and driving while intoxicated.

A senior Public Health major at a university in the Northeast conducted a survey similar to the Harvard survey to examine the problem of binge drinking on the officially "alcohol-free" campus. The survey was administered to students registered in a Public Health course entitled "Wellness." Although the sample is not really a random sample of the student population, the course was one that is required of all students at the University. The study was limited to students 25 years of age or under and participation was voluntary.

The study resulted in 221 usable responses. Data were collected in four different areas: information about drinking habits (4 questions), problems experienced related to the students' drinking (15 questions), problems experienced because of other students' drinking (8 questions), and some demographic information (9 questions).

A sample of some of the data is shown below:

| Five | Four | Three | Last | Binger | Hangover | Miss | Behind |
|------|------|-------|------|--------|----------|------|--------|
| 0 | 0 | 0 | 2 | 1 | 1 | 1 | 1 |
| 0 | 0 | 0 | 4 | 1 | 1 | 1 | 1 |
| 0 | 1 | 1 | 3 | 2 | 1 | 1 | 1 |
| 0 | 0 | 0 | 3 | 1 | 1 | 1 | 1 |
| 0 | 0 | 0 | 3 | 1 | 1 | 1 | 1 |
| 0 | 0 | 0 | 4 | 1 | 1 | 1 | 1 |

(The variable names are abbreviated and are explained in the text as needed.)

## 15.1   CHAPTER OBJECTIVES

You have learned that while *qualitative data* are important in understanding the results of statistical analyses, there are not many statistical tools that deal specifically with these types of data. You learned some graphical techniques for displaying qualitative data in Chapters 3 and 5 and you learned how to calculate probabilities for these variables in Chapter 6. In Chapters 9 and 10 you learned some techniques for testing hypotheses about population proportions, which arise from *qualitative data.*

This chapter deals primarily with the analysis of qualitative data. Almost all of the techniques that are used to analyze qualitative data are known as *chi-square tests.* In this chapter you will learn to use chi-square tests for

- testing whether a particular probability model fits a set of data (goodness of fit test)
- testing equality of proportions for more than two populations
- testing whether two qualitative variables are dependent or independent

## 15.2   TEST FOR GOODNESS OF FIT

In Chapters 9 and 10 you learned several different hypothesis tests that *assumed* that the data came from a population that was normally distributed. At the time you may have wondered how you could know if the assumption was valid for a particular set of data. In Chapter 11 you saw a normal probability plot, which you used informally to determine whether the residuals were normally distributed.

There are some other, *informal,* ways to check to see if the assumption of normality is valid, such as creating a histogram, dotplot, or boxplot of the data and seeing if the shape resembles that of the normal distribution. This is not a bad idea, but whenever we can substantiate an *eyeball* test with a formal statistical technique, we are more secure in the decisions we make.

One method of testing to see if data come from a population with a certain distribution is to perform a test called a **chi-square goodness of fit test.**

> The *chi-square goodness of fit test* checks to see how well a set of data fit the model for a particular probability distribution.

## 15.2.1 The Chi-Square Test

You know by now that every hypothesis test has

- a set of hypotheses, $H_0$ and $H_A$
- a test procedure that defines how the data are collected and processed
- a test statistic that has a certain sampling distribution
- one or more critical values
- a decision or conclusion that can be made as a result of the test

For a chi-square goodness of fit test, the hypotheses are

$H_0$:  The data come from a population with a specific probability distribution (normal, binomial, etc.).

$H_A$:  The data do not come from a population with the specified distribution.

*The chi-square goodness of fit test is always a one-tail or one-sided test*

    The test procedure is to collect a set of sample data and to create a frequency histogram for the data. The test then compares the **observed** frequency distribution of the data to the frequency distribution that would be **expected** if the null hypothesis were true.

> The *observed frequencies* are the actual number of observations that fall into each class in a frequency distribution or histogram.

> The *expected frequencies* are the number of observations that should fall into each class in a frequency distribution under the hypothesized probability distribution.

### EXAMPLE 15.1  College Drinking

#### *Choosing the Hypothesized Distribution*

Because the data collected for the survey on binge drinking were not a random sample, the researcher was interested in knowing how well the sample represented the actual student population. The researcher thought that the distribution of students among the four classes—freshman, sophomore, junior, and senior—was approximately uniform. That is, the probability distribution for the students he expected was

| Class | Freshman | Sophomore | Junior | Senior |
|---|---|---|---|---|
| **Probability** | 25% | 25% | 25% | 25% |

This type of distribution is known as a **uniform distribution.** ∎

> A *uniform distribution* is one in which each outcome or class of outcomes is equally likely to occur.

### EXAMPLE 15.2 College Drinking

***Setting Up the Goodness of Fit Test***

In this situation, the hypotheses we wish to test are

$H_0$: The distribution of students is uniform over the classes.

$H_A$: The distribution of students is not uniform.

Because of a mixup in the early administration of the survey, the question about class was left off some of the surveys. As a result there were only 171 responses to this question. The data are shown below:

| Class | Observed Frequency |
|-------|--------------------|
| Freshman | 86 |
| Sophomore | 36 |
| Junior | 30 |
| Senior | 19 |
| **Total ($n$)** | **171** |

From simply looking at the data it might be reasonable to wonder whether the distribution is uniform, since the number of freshmen is larger than that of the other classes combined. ■

Remember that we often test hypotheses when observed data seem unusual or inconsistent with what we expect.

### EXAMPLE 15.3 Quality Problems

***The Chi-Square Goodness of Fit Test***

A company that manufactures CD jewel cases contracts with various computer software manufacturers to provide a product that is 10% defective. One of the software manufacturers is wondering about the quality of the product it receives and decides to check on the quality of the cases.

Since the company has just received a shipment of product it decides to sample 1000 boxes of the jewel cases and take 5 jewel cases from each. The reason for checking 5 jewel cases is that this is consistent with the company's current acceptance sampling procedures. The company will inspect and count the number of defective cases. If the vendor is conforming to the contract, then the data collected on the number of defective cases should have a binomial distribution with $\pi = 0.10$ (or 10%).

The hypotheses the company wishes to test are

$H_0$: The distribution of defective cases is binomial with $n = 5$ and $\pi = 0.10$.

$H_A$: The distribution of defective cases is NOT binomial with $n = 5$ and $\pi = 0.10$.

It decides to perform the test at the 0.05 level of significance. ■

You might be wondering why we would want to do this, since you already know how to do hypothesis tests about proportions. This is a good question. The reason is tied to the *information* we are trying to obtain. If the software manufacturer tests the hypothesis $\pi = 0.10$ (against $\pi \neq 0.10$ or $\pi > 0.10$), then the conclusion will only be about the percent defective over the 5000 jewel cases in the sample. Depending on the length of time over which the product was manufactured, an average percent defective may not be a good indication of what is happening.

The chi-square goodness of fit test will give information about the distribution of the number of defective cases in a sample of size 5, which may enable the software manufacturer to obtain additional information about when or how the population changed. Remember that when you learned about the binomial probability distribution, one of the assumptions we made was that $\pi$, the proportion of successes in the population, remains constant over time. In a manufacturing process this is not always a safe assumption, since the process may shift over time. If the process changes but *averages* 0.10 over time, the hypothesis test about proportions might not be able to identify the problem. If we test the distribution as well as the value of $\pi$, we may be able to identify a problem. You will see what we mean as the example progresses.

 **TRY IT NOW!**

### Seat-Belt Usage

*Setting Up the Hypotheses for the Goodness of Fit Test*

Analysts for insurance companies assume that the number of drivers who wear seat belts is a binomial random variable with $\pi = 0.70$. To test this assumption they decide to set up checkpoints and sample 10 drivers every 2 hours. Set up the hypotheses to perform an appropriate chi-square goodness of fit test.

In a chi-square goodness of fit test, we are always trying to decide whether the data we observed fit a particular probability distribution model. We can use this model to determine how many data points we would *expect* to fall in each class of the frequency distribution since we know the theoretical probabilities of observing any of the possible values.

### EXAMPLE 15.4   College Drinking

*Finding the Expected Frequencies*

If the distribution of students over the classes is really uniform, then we would expect there to be the same number of students in each class; that is,

$$(25\%)(171) = 42.75 \text{ students in each year}$$

If we add a column to the frequency table we see that indeed the observed frequency does differ from what we expected to happen.

| Class | Observed Frequency | Expected Frequency |
|---|---|---|
| Freshman | 86 | 42.75 |
| Sophomore | 36 | 42.75 |
| Junior | 30 | 42.75 |
| Senior | 19 | 42.75 |
| **Total (n)** | **171** | **171** |

We can see the deviation from what is expected visually by looking at the frequency histograms for both the observed and expected distributions.

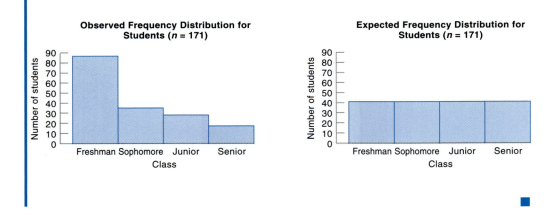

In the case of the uniform distribution, there is not much calculation involved in finding the expected frequencies of the probability distribution. For other probability distributions, you will need to do some additional calculations to find the expected frequencies.

## EXAMPLE 15.5   Quality Problems

### *Calculating the Expected Frequencies*

The software manufacturer who is wondering about the quality of the jewel cases it purchases from an outside vendor has collected the data and obtained a frequency distribution and histogram:

| Number of Defective Cases | Frequency |
|:---:|:---:|
| 0 | 485 |
| 1 | 340 |
| 2 | 127 |
| 3 | 48 |
| 4 | 0 |
| 5 | 0 |
| **Total** | **1000** |

**Defective CD Jewel Cases**

The data have the right shape for a binomial distribution with a small value of $\pi$ but there seem to be too many crates that have a high number of defective cases.

The next step is to determine the expected frequencies for the binomial distribution with $n = 5$ and $\pi = 0.10$. Looking in the binomial tables, the manufacturer finds that the probability distribution is

| $x$ | 0 | 1 | 2 | 3 | 4 | 5 |
|:---:|:---:|:---:|:---:|:---:|:---:|:---:|
| $p(x)$ | 0.590 | 0.328 | 0.073 | 0.008 | 0.000 | 0.000 |

The *expected* frequency for each class of the frequency distribution, $e_i$, is calculated by multiplying the probability for that class by the sample size:

$$e_i = np_i$$

where $n$ is the sample size of the data, and $p_i$ is the probability that the value in the $i$th class will occur. For this problem, $n = 1000$, so we have

| $x$ | $np_i$ |
|---|---|
| 0 | $(1000)(0.590) = 590$ |
| 1 | $(1000)(0.328) = 328$ |
| 2 | $(1000)(0.073) = 73$ |
| 3 | $(1000)(0.008) = 8$ |
| 4 | $(1000)(0.000) = 0$ |
| 5 | $(1000)(0.000) = 0$ |

Putting it all together in one table, we see that there is definitely a difference between the observed and expected frequencies.

| Number of Defective Cases | Observed Frequency | Expected Frequency |
|---|---|---|
| 0 | 485 | 590 |
| 1 | 340 | 328 |
| 2 | 127 | 73 |
| 3 | 48 | 8 |
| 4 | 0 | 0 |
| 5 | 0 | 0 |
| **Total** | **1000** | **999** |

*The reason for the discrepancy in the totals is rounding. If the binomial tables had 4 decimal places you could find the expected frequencies to 1 decimal place and avoid the problem.*

## TRY IT NOW!

### Seat-Belt Usage

*Calculating the Expected Frequencies*

The insurance analysts collect data for 1000 samples of 10 drivers and obtain the frequency distribution shown in the table. Find the expected frequency distribution for the data if the distribution is really binomial with $n = 10$ and $\pi = 0.70$.

| Number Wearing Seat Belts ($x$) | Observed Frequency | $p(x)$ | Expected Frequency |
|---|---|---|---|
| 0 | 0 | | |
| 1 | 0 | | |
| 2 | 1 | | |
| 3 | 6 | | |
| 4 | 33 | | |
| 5 | 116 | | |
| 6 | 213 | | |
| 7 | 275 | | |
| 8 | 216 | | |
| 9 | 119 | | |
| 10 | 21 | | |
| **Total** | **1000** | | |

*For the moment, don't worry about rounding if the expected frequency column does not add to 1.*

(*continued*)

Create frequency histograms for both the observed and the expected frequency distributions. At this point, does it appear that the observed data conform to the binomial distribution with $n = 10$ and $\pi = 0.70$? Why or why not?

Once we have collected the data, formed a hypothesis, and found the expected frequencies for the hypothesized distribution, we need to find a way to quantify the results of the test. The next section describes the way that this is done.

### 15.2.2 The Chi-Square Statistic

You know from previous work that often it is departure from what we expect to happen that leads us to perform a hypothesis test. The question is, how can we quantify the "deviation" from what is expected, and how will we know when what we observe deviates by more than it should or more than is likely by pure chance? To answer these questions we need to define a test statistic that measures the deviation of interest, and has a sampling distribution whose behavior we understand. This test statistic is known as the **chi-square statistic** and is calculated as

*Chi-square statistic*

$$\chi^2 = \frac{\sum_{i=1}^{k}(o_i - e_i)^2}{e_i}$$

where

$o_i$ = observed frequency in the $i$th class of the frequency distribution

$e_i$ = expected frequency in the $i$th class of the frequency distribution

$k$ = the number of classes in the frequency distribution

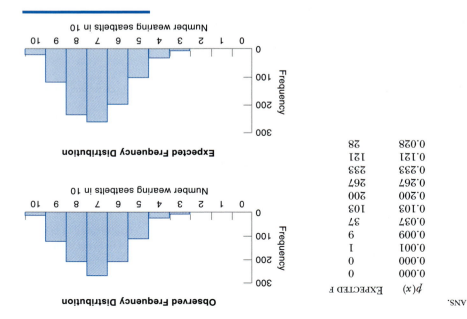

| $p(x)$ | EXPECTED F |
| --- | --- |
| 0.000 | 0 |
| 0.000 | 0 |
| 0.001 | 1 |
| 0.009 | 9 |
| 0.037 | 37 |
| 0.103 | 103 |
| 0.200 | 200 |
| 0.267 | 267 |
| 0.233 | 233 |
| 0.121 | 121 |
| 0.028 | 28 |

ANS.

This test statistic has a chi-square distribution with $k - p - 1$ degrees of freedom, where $p$ = the number of parameters of the theoretical distribution that were estimated from the data.

*Don't worry too much about the value of p right now. It will explained in detail later on when we need it.*

This statistic is a little different from any that you have encountered before. We are trying to measure the amount by which the observed frequency distribution deviates from the expected frequency distribution. Clearly, looking at the difference between the two quantities, $o_i - e_i$, is a good start. Just as clearly, though, if you simply add the deviations they will cancel each other out, since some will be positive and some will be negative. So, just as we did when we learned about the variance and the standard deviation, we square the quantities so that they are all positive. Adding the squared deviations together is a measure of *total* deviation. Dividing by $e_i$ is similar to calculating the *percent error* instead of the *absolute error.*

## EXAMPLE 15.6    College Drinking

### Calculating the Chi-Square Statistic

For the example about the distribution of students, we can calculate the chi-square statistic by extending the frequency table that we have developed:

| Class | Observed Frequency ($o_i$) | Expected Frequency ($e_i$) | $o_i - e_i$ | $\dfrac{(o_i - e_i)^2}{e_i}$ |
|---|---|---|---|---|
| Freshman | 86 | 42.75 | 43.25 | 43.76 |
| Sophomore | 36 | 42.75 | −6.75 | 1.07 |
| Junior | 30 | 42.75 | −12.75 | 3.80 |
| Senior | 19 | 42.75 | −23.75 | 13.19 |
| Total ($n$) | 171 | 171 | | 61.82 |

You see that the value of the test statistic is 61.82. This is a measure of the *total* deviation of the observed frequencies from the expected frequencies. The question now is, what does this tell us about the distribution of students? ■

## EXAMPLE 15.7    Quality Problems

### Calculating the Chi-Square Statistic

From the data that were collected it appears that there are definitely differences between the observed data and what would be expected if indeed the population is binomial with $\pi = 0.10$ and $n = 5$. To quantify the total difference we need to calculate the chi-square test statistic:

| Number of Defective Cases | Observed Frequency | Expected Frequency | $o_i - e_i$ | $\dfrac{(o_i - e_i)^2}{e_i}$ |
|---|---|---|---|---|
| 0 | 485 | 590.5 | −105.5 | 18.85 |
| 1 | 340 | 328.1 | 11.9 | 0.43 |
| 2 | 127 | 73.0 | 54 | 39.95 |
| 3 | 48 | 8.1 | 39.9 | 196.54 |
| 4 | 0 | 0.5 | −0.5 | 0.5 |
| 5 | 0 | 0.0 | 0.0 | 0.0 |
| Total | 1000 | 1000.2 | −0.2 | 256.27 |

*To avoid large errors from rounding, the probabilities were calculated with a computer package and the expected frequencies were calculated to 1 decimal place.*

From the calculation of the chi-square statistic we see that the measure of total deviation between the expected and observed frequencies is 256.77. We also see that the classes in the frequency distribution for values of 2 and 3 contribute the most to the total.

*TRY IT NOW!*

### Seat-Belt Usage
*Calculating the Chi-Square Statistic*

Fill in the table below to calculate the value of the chi-square statistic for the data obtained by the insurance analysts.

| Number Wearing Seat Belts | Observed Frequency | $p(x)$ | Expected Frequency | $o - e$ | $\dfrac{(o-e)^2}{e}$ |
|:---:|:---:|:---:|:---:|:---:|:---:|
| 0 | 0 | 0.000 | 0 | | |
| 1 | 0 | 0.000 | 0 | | |
| 2 | 1 | 0.001 | 1 | | |
| 3 | 6 | 0.009 | 9 | | |
| 4 | 33 | 0.037 | 37 | | |
| 5 | 116 | 0.103 | 103 | | |
| 6 | 213 | 0.200 | 200 | | |
| 7 | 275 | 0.267 | 267 | | |
| 8 | 216 | 0.233 | 233 | | |
| 9 | 119 | 0.121 | 121 | | |
| 10 | 21 | 0.028 | 28 | | |
| **Total** | **1000** | **0.999** | **999** | | |

Once we have quantified the difference between the observed and expected frequency distributions, we are ready to determine whether or not the difference is significant.

### 15.2.3 The Critical Value and the Decision Rule

We now have a way to measure the deviation of the observed data from what is expected, the chi-square statistic, and we know that this test statistic has a $\chi^2$ distribution with $k - 1 - p$ degrees of freedom. All that remains of the test is finding the critical value and making a decision about our hypothesis.

To find the appropriate critical value we will need a level of significance for the test, $\alpha$. Chi-square goodness of fit tests are always one-sided and upper-tailed tests. This is because we are looking at the *total deviation* and trying to see if that *exceeds*

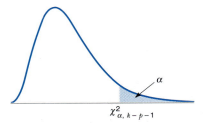

**FIGURE 15.1** The upper tail of a chi-square distribution

some reasonable level. To find the critical value we look up $\chi^2_{\alpha,k-p-1}$ for the appropriate values of $\alpha$ and $k - p - 1$ as shown in Figure 15.1.

To make a decision about our hypothesis we compare the value of the test statistic to the critical value. Since it is an upper-tail test, if the chi-square test statistic is greater than the $\chi^2$ value from the table we will reject $H_0$ and conclude that the data did not come from the hypothesized distribution. If the chi-square statistic is less than the critical value, we fail to reject $H_0$ and conclude that there is no evidence that the hypothesized distribution is not appropriate.

There are two things we must know to find the critical value for the test: the level of significance of the test, $\alpha$, and the degrees of freedom, $k - p - 1$.

### EXAMPLE 15.8   College Drinking

***Performing the Chi-Square Test***

For the problem with the student distribution, suppose we would like to perform the test at the 0.05 level of significance. This would mean that there is a 5% chance that the test will conclude that the data collected did not come from the uniform distribution when in fact it did.

To determine the number of degrees of freedom, we first look at the number of classes in the frequency distribution, $k$. In this case, $k = 4$. The value $p$ represents the number of population parameters we estimated from the data. In this case we did not use the data to estimate any parameters. Thus, the number of degrees of freedom for the test is $4 - 0 - 1 = 3$. From the chi-square table, we look up $\chi^2_{0.05,3}$ and find that the critical value is 7.82.

To make a decision, we compare the chi-square test statistic, 61.82, to the critical value of 7.82. Since 61.82 is larger than 7.82 (the test statistic falls beyond or outside the critical value) we reject $H_0$. That is, the conclusion of the test is that the data do not come from a population that has a uniform distribution. This indicates that the survey was not evenly spread out over the four years. ■

### EXAMPLE 15.9   Quality Problems

***Performing the Chi-Square Test***

The software company found the value of the chi-square statistic for the data on the defective jewel cases to be 256.27. To decide what this means, it needs to find the critical value for the test, compare the test statistic to the critical value, and make a decision.

To find the critical value the company analysts need to know $\alpha$ and the degrees of freedom for the test. They have already decided to test at the 0.05 level of significance. The number of degrees of freedom for the test is calculated to be

$$\text{Degrees of freedom} = k - p - 1 = 6 - 0 - 1 = 5$$

Remember that $p$ is the number of parameters of the hypothesized distribution that were estimated from the data. The binomial distribution has 2 parameters, $n$ and $\pi$, both of which were given or assumed, not estimated.

> This critical value, $\chi^2_{0.05,5}$, is 11.07. Since 256.27 is clearly beyond the critical value, the software company can reject $H_0$ and conclude that the number of defective jewel cases in a sample of 5 does not have a binomial distribution with $\pi = 0.10$. That is, the data do not fit the assumed probability model. ■

 ## TRY IT NOW!

### Seat-Belt Usage
*Finding the Critical Value and Performing the Test*

The insurance analysts decide that they want to test the goodness of fit hypotheses at the 0.01 level of significance.

How many degrees of freedom will the critical value for the test have?

Find the critical value for the test.

Based on the chi-square test statistic and the critical value, what can you conclude about the distribution of the number of people in a sample of size 10 that wear seat belts?

One of the things that happens as we learn more and more statistical techniques is that we lose sight of the fact that we are doing the analyses to help us identify problems, that is, to understand why things happen and to make informed decisions. After the chi-square goodness of fit test, what do we know and how can we use the information?

### EXAMPLE 15.10   College Drinking

*After the Test is Over. . .*

As a result of the chi-square test, the researcher who administered the survey on binge drinking knows that the distribution of students in his sample is not uniformly distributed over the four classes. This *means* that his sample *may not be representative of the population he is trying to study* and that he will have to be careful about any conclusions or generalizations he makes.

In this case, the researcher began to think a little more about his null hypothesis. Does it really make sense that students are uniformly distributed over the four classes? After some thought he realized that if you consider the problems of student retention (loss of students from transfer or dropout) and the transfer in of students from other universities and community colleges, it is more likely that the true percentage of students in each class decreases as the class level increases. He contacted the university administrator who keeps track of such things and found out that the historical distribution of students is the one in the table below:

| Class | Expected Distribution (%) | Expected Frequency |
|---|---|---|
| Freshman | 45.2 | 77.3 |
| Sophomore | 18.6 | 31.8 |
| Junior | 18.0 | 30.8 |
| Senior | 18.2 | 31.1 |
| **Total ($n$)** | **100** | **171** |

The researcher then decided to compare the actual frequency distribution to the new distribution obtained from the university and redo the chi-square test. The comparison is shown in the table below:

| Class | Observed Frequency ($o_i$) | Expected Frequency ($e_i$) | $o - e$ | $\dfrac{(o - e)^2}{e}$ |
|---|---|---|---|---|
| Freshman | 86 | 77.3 | 8.7 | 0.98 |
| Sophomore | 36 | 31.8 | 4.2 | 0.55 |
| Junior | 30 | 30.8 | − 0.8 | 0.02 |
| Senior | 19 | 31.1 | − 12.1 | 4.71 |
| **Total ($n$)** | **171** | **171.0** | | **6.26** |

Since nothing else changed, the critical value of the test is still $\chi^2_{0.05,3} = 7.82$. The value of the test statistic, 6.26, is not outside the critical value so there is not sufficient evidence to reject $H_0$. This means that there is no reason to say that the sample did not come from a population with the hypothesized distribution and so it is reasonable to assume that the sample is representative of the student population of the university *as defined by the university administration.* ∎

One of the benefits of the chi-square test is that it lets us proceed with confidence in our analysis. Knowing that the sample does indeed represent the population of interest gives much more credibility to any conclusions drawn from the analysis.

## EXAMPLE 15.11   Quality Problems

### After the Test Is Over. . .

The company that used the outside vendor as a source of CD jewel cases knows as a result of the test that the number of defective cases in a sample of size 5 does not have a binomial distribution with $\pi = 0.10$. The company does *not* know if the distribution is binomial with a different value of $\pi$, or if some other factor makes the binomial distribution inappropriate. This is the time to use some common sense, knowledge of how the data were collected, and descriptive and graphical statistics to see if the company can identify the problem.

There are many things that the company analysts could do, depending on how they collected the data and what variables they recorded, but in this case they decided to plot the data in the order in which the cases were produced. Each sample that they took came from a different box and they recorded the lot number and production data along with the data on the number of defective cases. A time plot of the data indicated that there was an increase in the number of defectives in the sample starting at about the 750th sample. Since the first plot was very dense the analysts decided to also plot every fifth sample so that they could see the shift more clearly.

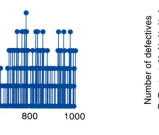

As a result of the chi-square test and the plot of the data they learned that the percent defective had apparently changed during production. This is why the binomial distribution was not correct.   ■

## 15.2.4 Testing for Normality and Other Considerations

The results of the chi-square goodness of fit test are similar to the test you learned for testing whether the variances of two populations are equal. The test does not result in a definitive answer that the hypothesized distribution is correct, just as the hypothesis test to see if two variances are equal does not tell you that the variances are equal. Rather, if you reject $H_0$, it will tell you when the hypothesized distribution is *not* correct. When you are using a chi-square test to test the assumption of normality, you will either decide that the assumption of normality is not appropriate, or that since there is no evidence to the contrary, the assumption is a reasonable one.

### EXAMPLE 15.12   Jar Weights

*Testing for Normality*

The quality specifications for a 15.0-ounce jar of peanut butter state that the weights of the jars should be normally distributed with a mean of 15.1 oz. The Quality Focus Group for the product line selects a random sample of 500 jars to determine whether the specifications are being met. It decides to test at the 0.05 level of significance.

The hypotheses for the test are

$H_0$:   The jar weights are normally distributed with $\mu = 15.1$ oz.

$H_A$:   The jar weights are not normally distributed with $\mu = 15.1$ oz.

The group collects the data and uses a computer package to create a frequency distribution and histogram.

| Weights (<) | Observed Frequency |
|---|---|
| 14.75 | 3 |
| 14.80 | 5 |
| 14.85 | 34 |
| 14.90 | 43 |
| 14.95 | 79 |
| 15.00 | 83 |
| 15.05 | 92 |
| 15.10 | 77 |
| 15.15 | 56 |
| 15.20 | 20 |
| 15.25 | 5 |
| 15.30 | 1 |
| 15.35 | 1 |
| 15.40 | 1 |
| **Total** | **500** |

**Observed Frequency for 500 Peanut Butter Jar Weights**

The data are symmetric, although this does not mean that they are definitely normally distributed.

    To calculate the expected frequency for each class the analysts need to calculate the probability that a randomly selected jar will fall into the weight range defined by each class. For the first class they need to find $P(14.70 < X \le 14.75)$.

    Remember from Chapter 6 that you need to turn the two values into standard normals ($Z$'s) and look them up. But to do that you need $\mu$ and $\sigma$! The specifications for the jars give only a value for $\mu$, so the analysts will have to use the data to estimate $\sigma$.

    Using the same software package, they find that the sample standard deviation, $s$, is 0.102 oz. So, to find the probability of interest they use

$$\text{For } 14.70: \quad Z = \frac{14.70 - 15.10}{0.102} = -3.92$$

$$\text{For } 14.75: \quad Z = \frac{14.75 - 15.1}{0.102} = -3.43$$

Looking up the $Z$ values in the table and subtracting the results they find that $P(14.70 < X \le 14.75) = 0.0003 - 0.0000 = 0.0003$. When they were doing the calculations, they noticed that there was a discrepancy between the total for the expected frequency and the total of 500 observations. Looking at the table, they realized that this was probably because there were a few additional classes on the high side that had nonzero probabilities. They expanded the table to include these classes.

    To find the expected number of jars out of the 500 that would weigh between 14.70 and 14.75 oz the analysts multiplied the probability by 500 and obtained $(0.0003)(500) = 0.15$. They found the probabilities and expected frequencies for all of the classes:

*The software package creates classes the same way you have learned. For example, 14.80 is the same as $14.75 < x \le 14.80$.*

| Weight (<) | Frequency | Observed $P(a < X < b)$ | Expected Frequency | $\dfrac{(o_i - e_i)^2}{e_i}$ |
|---|---|---|---|---|
| 14.75 | 3 | 0.0003 | 0.1 | 84.10 |
| 14.80 | 5 | 0.0013 | 0.7 | 28.13 |
| 14.85 | 34 | 0.0055 | 2.7 | 355.99 |
| 14.90 | 43 | 0.0179 | 8.9 | 130.33 |
| 14.95 | 79 | 0.0458 | 22.9 | 137.71 |
| 15.00 | 83 | 0.0927 | 46.4 | 28.93 |
| 15.05 | 92 | 0.1486 | 74.3 | 4.23 |

(*continued*)

| Weight (<) | Observed Frequency | Expected P(a < X < b) | Frequency | $\dfrac{(o_i - e_i)^2}{e_i}$ |
|---|---|---|---|---|
| 15.10 | 77 | 0.1879 | 94.0 | 3.07 |
| 15.15 | 56 | 0.1879 | 94.0 | 15.36 |
| 15.20 | 20 | 0.1486 | 74.3 | 39.66 |
| 15.25 | 5 | 0.0927 | 46.4 | 36.91 |
| 15.30 | 1 | 0.0458 | 22.9 | 20.92 |
| 15.35 | 1 | 0.0179 | 8.9 | 7.03 |
| 15.40 | 1 | 0.0055 | 2.7 | 1.11 |
| 15.45 | 0 | 0.0013 | 0.7 | 0.67 |
| 15.50 | 0 | 0.0003 | 0.1 | 0.13 |
| **Total** | **500** | **1.0000** | **500** | **894.28** |

Looking at the chi-square statistic, there would appear to be some kind of problem. The data may not be *exactly* normally distributed, but they are not that far off. Why is the value of the test statistic so large?  ■

Before we proceed any further with the example we need to discuss some of the problems with the chi-square goodness of fit test. The problems are related to the number of observations that are used to create the frequency distribution.

You may remember from Chapter 3 that a fairly large sample size is required to obtain a good frequency distribution. If your frequency histogram is highly variable due to lack of data, the frequencies in some classes will have greater differences from the expected frequencies.

*Remember, dividing by numbers close to 0 causes the result to be large.*

Another related problem is that when the total sample size is small, the expected frequencies for each class will also be small. Since the chi-square statistic divides by the expected frequency, this can inflate the value of the statistic artificially and lead to rejection of $H_0$ when, in fact, it is true. For this reason, it is recommended that the chi-square test not be used if the expected frequency in any cell is less than 5. If this is not the case, it is possible to combine adjacent cells to get the expected frequencies above 5, but this results in a loss of degrees of freedom.

### EXAMPLE 15.13   Jar Weights

***Collapsing Classes when the Expected Frequencies Are <5***

Looking at the table of the observed and expected frequencies, the Quality Focus Group for the product called in a statistician to help them. The statistician explained that there are too many classes with very small expected frequencies that are causing the chi-square statistic to be inflated. The group decides to collapse cells on both ends of the distribution to get the expected frequencies above 5. The table was adjusted and the chi-square statistic recalculated:

| Weight (<) | Observed Frequency | P(a < X < b) | Expected Frequency | $\dfrac{(o_i - e_i)^2}{e_i}$ |
|---|---|---|---|---|
| 14.75 | 3 ⎫ | 0.0003 | 0.1 ⎫ | |
| 14.80 | 5 ⎬ 85 | 0.0013 | 0.7 ⎬ 12.4  423.89 | |
| 14.85 | 34 ⎪ | 0.0055 | 2.7 ⎪ | |
| 14.90 | 43 ⎭ | 0.0179 | 8.9 ⎭ | |
| 14.95 | 79 | 0.0458 | 22.9 | 137.71 |
| 15.00 | 83 | 0.0927 | 46.4 | 28.93 |
| 15.05 | 92 | 0.1486 | 74.3 | 4.23 |
| 15.10 | 77 | 0.1880 | 94.0 | 3.07 |
| 15.15 | 56 | 0.1880 | 94.0 | 15.36 |

| Weight (<) | Observed Frequency | $P(a < X < b)$ | Expected Frequency | $\dfrac{(o_i - e_i)^2}{e_i}$ |
|---|---|---|---|---|
| 15.20 | 20 | 0.1486 | 74.3 | 39.66 |
| 15.25 | 5 | 0.0927 | 46.4 | 36.91 |
| 15.30 | 1 | 0.0458 | 22.9 | 20.92 |
| 15.35 | 1 | 0.0179 | 8.9 | |
| 15.40 | 1 | 0.0055 | 2.7 | |
| 15.45 | 0 | 0.0013 | 0.7 | |
| 15.50 | 0 | 0.0003 | 0.1 | |
| | 2 | | 12.4 | 8.78 |
| **Total** | **500** | **1.0000** | **500** | **719.46** |

Clearly the chi-square statistic is still large, but it has been considerably reduced.

To complete the test, the Quality Focus Group found the critical value for the test. It calculated the degrees of freedom to be

$$\text{Degrees of freedom} = k - p - 1 = 10 - 1 - 1 = 8$$

Since it wanted to test at $\alpha = 0.05$, it used $\chi^2_{0.05,8} = 15.51$.

Since the value of the test statistic, 719.46, is beyond the critical value of 15.51, the group rejects $H_0$ and concludes that the jar weights are not normally distributed with a mean of 15.1 oz and a standard deviation of 0.102 oz. ∎

*After collapsing classes there were only 10 classes left. Also, the normal distribution has two parameters, $\mu$ and $\sigma$, and the data were used to estimate one of them, $\sigma$.*

Again, we need to look at the conclusion of the chi-square goodness of fit test and realize that, in fact, the conclusion comes as a package deal. In the previous example the conclusion is that the data do not come from a normal distribution with a mean of 15.1 oz. It is not necessarily true that the data are not normally distributed—they might be normally distributed with a *different mean* than the one hypothesized. It is also not necessarily true that the mean is not 15.1 oz; the mean could be 15.1 oz, but the distribution *might not be normal*. To really know what is going on here, the Quality Focus Group might want to perform a hypothesis test about the value of the mean to see if that is the problem, or rerun the chi-square test using the data to estimate $\mu$.

## 15.2.5 Exercises—Learning It!

15.1   The administration of a university has been using the following distribution to classify the ages of their students:

| Age Group | Estimated % of Student Population |
|---|---|
| Less than 18 | 2.7 |
| 18–19 | 29.9 |
| 20–24 | 53.4 |
| Older than 24 | 14 |

A recent student survey provided the following data on age of students:

| Age Group | Frequency |
|---|---|
| Less than 18 | 6 |
| 18–19 | 118 |
| 21–24 | 102 |
| Older than 24 | 26 |

(a) Set up a table that compares the expected and observed frequencies for each group.

(b) Based on the table, do you think that the data represent the estimated distribution?

(c) Set up the hypotheses for the chi-square goodness of fit test.

(d) Perform the goodness of fit test at the 0.05 level of significance.

(e) Based on the chi-square test, is the estimated age distribution that the university is using correct?

**15.2**   As part of a survey on the use of Office Suites Software, the company doing the polling wanted to know whether its population was uniformly distributed over the following age distribution: under 25, 25 to 44, 45 and up. The company looked at the data it had collected so far and found the following distribution:

| Age Group | Number of Respondents |
|---|---|
| Under 25 | 73 |
| 25 to 44 | 61 |
| 45 and up | 66 |
| **Total** | **200** |

(a) Based on the data, do you think that the respondents are uniformly distributed over the age categories?

(b) Set up the hypotheses to test whether the data are uniformly distributed over the age categories.

(c) Find the expected frequency distribution and perform the chi-square goodness of fit test.

(d) At the 0.05 level of significance, would you say that the respondents were uniformly distributed over the age groups?

**15.3**   The Transit Authority in a large city estimates that 80% of business commuters get a seat for their entire commute. It decided to take a random sample of its subway system over the course of a 12-week period to see if its estimate is correct. It set up an exit poll at the most common destination station and asked 15 randomly selected commuters whether or not they got a seat for the entire commute. The data the Transit Authority obtained are as follows:

| Number of Commuters in 15 Who Got Seats | Count |
|---|---|
| 0 | 0 |
| 1 | 0 |
| 2 | 0 |
| 3 | 0 |
| 4 | 0 |
| 5 | 0 |
| 6 | 0 |
| 7 | 0 |
| 8 | 1 |
| 9 | 3 |
| 10 | 10 |
| 11 | 21 |
| 12 | 31 |
| 13 | 20 |
| 14 | 9 |
| 15 | 5 |
| **Total** | **100** |

(a) Why is the number of commuters in 15 who got seats a binomial random variable?

(b) Set up the hypotheses to test whether the sample data come from a binomial distribution with $n = 15$ and $\pi = 0.80$.

(c) Find the expected frequencies for the hypothesized distribution.

(d) Perform the chi-square goodness of fit test at the 0.01 level of significance.

(e) Is it reasonable to assume that the data come from a binomial distribution with $n = 15$ and $\pi = 0.80$?

**15.4**   One of the assumptions of the small sample ($t$) hypothesis test about a mean is that the underlying population is normally distributed. The accompanying data represent the diameter of the holes in washers that were purchased by a company. The specification on the washers is that the mean diameter of the hole be 0.5000 in.

| | | | |
|---|---|---|---|
| 0.5053 | 0.5098 | 0.4606 | 0.4606 |
| 0.4711 | 0.4627 | 0.4800 | 0.4800 |
| 0.4672 | 0.5642 | 0.5495 | 0.5495 |
| 0.4672 | 0.5346 | 0.5745 | 0.5745 |
| 0.5340 | 0.3767 | 0.3933 | 0.3933 |

(a) Make a frequency distribution and a histogram for the data. Do not use more than 5 classes.

(b) From the histogram, would you be willing to believe that the data are normally distributed?

(c) Set up the hypotheses to test that the data come from a normal distribution with a mean of 0.5000 and a standard deviation of 0.0600 in.

(d) Find the expected frequency distribution of the data.

(e) At the 0.05 level of significance, does it appear reasonable to assume that the data are normally distributed?

15.5   A large banking corporation believes that 80% of the loan applications it receives are approved within 24 hours. It decides to take a random sample of 10 loan applications every day for 3 months and record the number of the applications that are approved within 24 hours. The following data are obtained:

| Number of Loan Applications in 10 Approved in 24 Hours | Frequency |
|:---:|:---:|
| 4 | 1 |
| 5 | 5 |
| 6 | 11 |
| 7 | 19 |
| 8 | 27 |
| 9 | 18 |
| 10 | 7 |
| **Total** | **88** |

(a) Set up the necessary hypotheses to test whether the data come from a binomial distribution with $n = 10$ and $\pi = 0.80$.

(b) Find the expected frequency distribution for the data.

(c) At the 0.05 level of significance, is it reasonable to assume that the number of loan applications that are approved in 24 hours has a binomial distribution with $\pi = 0.80$?

# 15.3   TESTING PROPORTIONS FROM MORE THAN TWO POPULATIONS

In Chapter 10 you learned how to compare parameters from two different populations. In particular, you learned that you can use a $Z$ test to determine whether the population proportions for two populations are equal. What if you are interested in comparing the population proportions for *more than two populations?*

Using a chi-square test, it is possible to compare proportions for more than two populations. Although it would be possible to compare a set of $c$ populations by using the $Z$ test you learned in Chapter 10 to test all of the possible pairs of populations, this is not a good thing to do. The reason for doing the single test is that each hypothesis test you do has a type I error probability associated with it. When you do multiple tests there is an increased chance that you will decide that two population proportions are different when in fact they are not. The single test controls the probability of making this mistake.

## 15.3.1 Testing Proportions for More Than Two Populations

In general, some characteristic of importance (a success) is defined for several different populations. We want to determine whether the proportion of successes in each population is the same. That is, the general set of hypotheses we wish to test is

$$H_0: \quad \pi_1 = \pi_2 = \cdots = \pi_c$$
$$H_A: \quad \text{At least one } \pi_i \text{ is different.}$$

To test these hypotheses for two populations you take a sample from each population and count the number of times that the characteristic of interest occurs. The data collection for more than two populations is the same.

In Chapter 5 you learned how to summarize data that involved two qualitative variables—the contingency table. For this test the two variables are the population that the sample item comes from ($i = 1, 2, \ldots, c$) and whether the item is a success (the characteristic is present) or a failure (the characteristic is not present). An example of the contingency table is shown in Table 15.1.

**TABLE 15.1   Data Table for Testing Equality of Proportions**

| Number of | Population 1 | 2 | ... | c | Totals |
|---|---|---|---|---|---|
| Successes | $s_1$ | $s_2$ | ... | $s_c$ | $s$ |
| Failures | $f_1$ | $f_2$ | ... | $f_c$ | $f$ |
| Totals | $n_1$ | $n_2$ | ... | $n_c$ | $n$ |

The proportion of successes for each population can be calculated as

$$p_i = \frac{s_i}{n_i}$$

### EXAMPLE 15.14   College Drinking

***Setting Up a Chi-Square Test of Proportions***

For the purposes of the study on binge drinking, students were classified as non-drinkers, non-binge drinkers, infrequent binge drinkers, and frequent binge drinkers. One question of interest to the researcher was whether there was a difference in academic responsibility among the three different categories of drinkers. He decided to look at the question of whether the proportion of people who missed class because of drinking is the same for each of the three groups. He created the contingency table shown below:

| Miss Class | Type of Drinker Non-binge | Infrequent | Frequent | Total |
|---|---|---|---|---|
| Never | 77 | 39 | 31 | 147 |
| Once or More | 7 | 19 | 39 | 65 |
| Total | 84 | 58 | 70 | 212 |

The populations in this case are the different types of drinkers and the characteristic of interest is whether the student missed class. The hypotheses are

$$H_0: \quad \pi_N = \pi_I = \pi_F$$
$$H_A: \quad \text{At least one of the } \pi_i \text{ is different.}$$

From the data you can see that $p_N = 0.083$, $p_I = 0.328$, and $p_F = 0.557$. The proportions certainly appear to be different but because of sampling error, a statistical test is necessary. ■

The hypotheses are equivalent to asking the question, Is missing class related to the amount of drinking by a student? This is similar to the question that we asked when looking at regression models. The difference is that the variables are *qualitative*.

### EXAMPLE 15.15   Comparing Machine Quality

*Setting up a Chi-Square Test of Proportions*

Suppose that a manufacturing facility had three different machines, all of which produced the same product. In an attempt to better understand their product, the management of the facility wants to know if the percent defective product is the same for all machines. That is, they wish to test the following hypotheses:

$$H_0: \quad \pi_1 = \pi_2 = \pi_3$$
$$H_A: \quad \text{At least one of the } \pi_i \text{ is different.}$$

The statistical tool used to test these types of hypotheses is a *chi-square test* similar to the one used to test goodness of fit.

The manufacturing facility collected finished product from the end of the production line. For each product analysts collected data on the machine and whether the product was defective. They assembled the data into a contingency table, which is shown below:

| Number of | Machine 1 | 2 | 3 | Totals |
|---|---|---|---|---|
| Defective | 12 | 22 | 14 | 48 |
| Good | 188 | 148 | 196 | 532 |
| Totals | 200 | 170 | 210 | 580 |

From the data the analysts can estimate the proportion defective for each machine, $p_1 = 0.06$, $p_2 = 0.13$, and $p_3 = 0.07$. Clearly, the numbers are different, but the analysts know enough about sampling error to know that they cannot tell for sure without a statistical test. ■

Contingency tables like the one in the example above are tedious to do by hand for large data sets, but are easily done using computer software packages like Minitab or Excel.

 **TRY IT NOW!**

### Technical Support

*Setting Up the Chi-Square Test for Proportions*

A company that sells computer software has three different locations set up to provide customers with technical support for their products. The support representatives keep a log for each call to technical support, and as part of that log, they record whether the problem was resolved successfully.

The company analysts are interested in knowing whether the percentage of calls that are successfully resolved is the same for each location. They randomly select logs from each location and collect data on

(*continued*)

the number of calls that result in a successful resolution of the problem. The data are summarized in the table below:

| Number of | Location 1 | 2 | 3 | Totals |
|---|---|---|---|---|
| Successful calls | 257 | 264 | 283 | 804 |
| Unsuccessful calls | 43 | 86 | 97 | 226 |
| Totals | 300 | 350 | 380 | 1030 |

Set up the hypotheses for the software company.

Calculate the proportion of successfully resolved calls for each location. Based *solely* on these numbers, do you think that the proportion of successfully resolved calls for all three locations is the same?

## 15.3.2 Description of the Test

In Section 15.1 you learned that a chi-square test compares *expected and observed frequencies*. The observed frequencies are the raw data that are collected and the expected frequencies are the frequencies that would be predicted if the null hypothesis were true. The data in Table 15.1 can be analyzed using exactly this method.

If the null hypothesis, $H_0$, is true, then the data for all of the populations being sampled can be combined to give one overall estimate of the true proportion of successes, $\pi$. This overall estimate, $p$, can then be used to predict, for example, the number of defectives that the manufacturing facility should *expect* to see in the sample from each machine. In general, the estimate for $\pi$ is calculated by counting all of the successes observed and dividing by the total number of objects sampled. That is, the **overall proportion of successes** is given by

*Overall proportion of successes*

$$p = \frac{s_1 + s_2 + \cdots + s_c}{n_1 + n_2 + \cdots + n_c} = \frac{s}{n}$$

and the **expected number of successes** in the sample from population $i$ is

*Expected number of successes*

$$e_i = \frac{s}{n}\, n_i = \frac{s n_i}{n} = p n_i$$

These are all equivalent ways of calculating the expected frequency. For the chi-square test on proportions the last expression is probably easiest to use. For other chi-square tests, people find the second expression simplest.

Since we know the sample size for each population and the expected number of successes in the sample, the easiest way to find the expected number of failures is by subtraction.

## EXAMPLE 15.16   College Drinking

*Calculating the Expected Frequencies*

For the college drinking survey the researcher needed to estimate $\pi$, the overall proportion of students who miss class because of drinking. The value of $\pi$ is estimated assuming that the null hypothesis is true. The overall proportion of students who miss class because of drinking is

$$p = \frac{7 + 19 + 39}{84 + 58 + 70} = \frac{65}{212} = 0.307$$

The expected number for each group is calculated as

$$e_N = (0.307)(84) = 25.8$$
$$e_I = (0.307)(58) = 17.8$$
$$e_F = (0.307)(70) = 21.5$$

The next step was to calculate the number of students who did not miss class for each group:

| | |
|---|---|
| Non-binge drinkers: | $84 - 25.8 = 58.2$ |
| Infrequent binge drinkers: | $58 - 17.8 = 40.2$ |
| Frequent binge drinkers: | $70 - 21.5 = 48.5$ |

To get a better picture, the observed and expected frequencies were put into a single table as shown below. The expected frequencies are below the observed frequencies and are in parentheses.

| Miss Class | Type of Drinker | | | Total |
|---|---|---|---|---|
| | Non-binge | Infrequent | Frequent | |
| Never | 77 | 39 | 31 | 147 |
| | (58.2) | (40.2) | (48.5) | |
| Once or More | 7 | 19 | 39 | 65 |
| | (25.8) | (17.8) | (21.5) | |
| Total | 84 | 58 | 70 | 212 |

## EXAMPLE 15.17   Comparing Machine Quality

*Calculating the Expected Frequencies*

For the manufacturing facility that wanted to compare machine quality, the estimate for $\pi$, the overall proportion defective, is found to be

$$p = \frac{12 + 22 + 14}{200 + 170 + 210} = \frac{48}{580} = 0.083$$

The expected number of defectives for each machine are

$$e_1 = (0.083)(200) = 16.6$$
$$e_2 = (0.083)(170) = 14.1$$
$$e_3 = (0.083)(210) = 17.4$$

The table at the top of page 778 shows the observed and expected frequency of defectives for each machine. The expected frequencies are below the observed frequencies and enclosed in parentheses.

| Number of | Machine 1 | Machine 2 | Machine 3 | Totals |
|---|---|---|---|---|
| Defective | 12 (16.6) | 22 (14.1) | 14 (17.4) | 48 |
| Good | 188 (183.4) | 148 (155.9) | 196 (192.6) | 532 |
| Totals | 200 | 170 | 210 | 580 |

## TRY IT NOW!

### Technical Support
***Calculating the Expected Frequencies***

The data for the computer software company interested in its technical support locations are

| Number of | Location 1 | Location 2 | Location 3 | Totals |
|---|---|---|---|---|
| Successful calls | 257 | 264 | 283 | 804 |
| Unsuccessful calls | 43 | 86 | 97 | 226 |
| Totals | 300 | 350 | 380 | 1030 |

*As with the goodness of fit test, the expected frequencies might not add up to the total because of rounding.*

Estimate $\pi$, the percentage of calls that are resolved successfully, assuming that the three locations are the same.

Use the overall proportion of successful calls to find the expected frequency of successful calls for each location.

### 15.3.3 Performing the Test

Once we have the expected frequency for each population, all that remains to be done is to calculate the chi-square test statistic, find the critical value for the test, and make a decision.

We have already defined the chi-square test statistic in general and have shown how it is used for the goodness of fit test. The application of the chi-square statistic to testing proportions is exactly the same:

$$\chi^2 = \sum_{\text{all cells}} \frac{(o_i - e_i)^2}{e_i}$$

*Numbering of the cells is not critical, although generally we number cells in a table row by row.*

where

$o_i$ is the observed frequency in the $i$th cell of the table

$e_i$ is the expected frequency in the $i$th cell of the table

The degrees of freedom for a chi-square test involving a table are given by

$$\text{Degrees of freedom} = (r - 1)(c - 1)$$

where $r$ is the number of rows in the table and $c$ is the number of columns.

### EXAMPLE 15.18    College Drinking

*Performing a Chi-Square Test for Equality of Proportions*

The data on missing class and drinking that the researcher has summarized so far are shown again below:

| Miss Class | Type of Drinker Non-binge | Infrequent | Frequent | Total |
|---|---|---|---|---|
| Never | 77 | 39 | 31 | 147 |
| | (58.2) | (40.2) | (48.5) | |
| Once or more | 7 | 19 | 39 | 65 |
| | (25.8) | (17.8) | (21.5) | |
| Total | 84 | 58 | 70 | 212 |

To perform the chi-square test at a 0.05 level of significance, the researcher needs to calculate the test statistic and find the critical value for comparison. The test statistic is calculated as shown:

$$\frac{(77 - 58.2)^2}{58.2} + \frac{(39 - 40.2)^2}{40.2} + \frac{(31 - 48.5)^2}{48.5} + \frac{(7 - 25.8)^2}{25.8} + \frac{(19 - 17.8)^2}{17.8}$$
$$+ \frac{(39 - 21.5)^2}{21.5} = 40.45$$

To find the critical value he needs to first find the degrees of freedom of the test, $(r - 1)(c - 1)$:

$$(2 - 1)(3 - 1) = 2$$

The critical value is then $\chi^2_{0.05,2} = 5.99$.

To finish the test he compares the test statistic of 40.45 to the critical value and finds that since 40.45 is outside the critical value the decision is to reject $H_0$ and conclude that the proportion of students who miss class is not the same for each population. That is, missing class *is related* to the amount that a student drinks. ∎

Software packages make it easy to perform the calculations for the expected frequencies, and some software will display these values as part of their output, as shown in Figure 15.2 on page 780.

```
               Non-Binge     Infrequent     Frequent      All

Never             77             39             31          147
                58.25          40.22          48.54        147.0

Once or            7             19             39           65
more             25.75          17.78          21.46         65.0

All               84             58             70          212
                84.00          58.00          70.00        212.0


Chi-Square = 40.484, DF = 2, P-Value = 0.00


Cell Contents—
                      Count
                      Exp Freq
```

**FIGURE 15.2**   Minitab output from chi-square analysis

**EXAMPLE 15.19**   **Comparing Machine Quality**

***Performing a Chi-Square Test for Equality of Proportions***

The manufacturing facility has collected and summarized its data and calculated the expected number of defectives for each machine:

| Number of | **Machine** | | | |
|---|---|---|---|---|
| | **1** | **2** | **3** | **Totals** |
| Defective | 12 | 22 | 14 | **48** |
| | (16.6) | (14.1) | (17.4) | |
| Good | 188 | 148 | 196 | **532** |
| | (183.4) | (155.9) | (192.6) | |
| Totals | **200** | **170** | **210** | **580** |

It wants to perform the test at the 0.05 level of significance. It calculates the chi-square statistic as

$$\frac{(12-16.6)^2}{16.6} + \frac{(22-14.1)^2}{14.1} + \frac{(14-17.4)^2}{17.4} + \frac{(188-183.4)^2}{183.4} + \frac{(148-155.9)^2}{155.9}$$
$$+ \frac{(196-192.6)^2}{192.6} = 6.94$$

Since the level of significance is 0.05 and the degrees of freedom is $(2-1)(3-1) = 2$, the critical value for the test is $\chi^2_{0.05,2} = 5.99$.

Comparing the test statistic of 6.94 to the critical value of 5.99, the manufacturer rejects $H_0$ and concludes that the proportion defective for the three machines is not the same. ∎

# TRY IT NOW!

## Technical Support

***Performing the Chi-Square Test for Proportions***

The computer software company with the different technical support locations wants to complete the test to determine whether the percentage of successfully resolved calls is the same at all three locations. It wants to test at the 0.01 level of significance.

Calculate the value of the chi-square test statistic and complete the test.

Is the proportion of successfully resolved calls the same at each location?

## 15.3.4 Using the Results of the Chi-Square Test for Proportions

You might be wondering (and with good reason) just how you can *use* the chi-square test for equality of proportions to gain useful information or to make decisions. The results of the chi-square test are similar to the results of analysis of variance that you learned in Chapter 14. When the test does not lead to rejection of the null hypothesis, there is not really anything else to do. However, when the test leads to rejection of the null hypothesis, you conclude that at least one of the populations is different from the others. Does it mean that one of the populations is different from all of the others? Does it mean that each population is different from every other population? How useful is this kind of conclusion?

Remember that statistical analyses are not answers in themselves, but rather tools that can be used to identify problems or aid in decision making. The key words here are *identify* and *aid*. If you have done a chi-square test for proportions, then it is likely that you suspect that something is going on that is causing the populations to be different. The results of the test *verify* that your suspicions are justified. The test result tells you that further investigation is warranted.

How might you go about determining exactly which of the populations are the same and which are different? Certainly, you will know that the two samples that are farthest apart are probably different, but how about the others in between? How do they compare? For ANOVA there are formal techniques for testing which means are different. For chi-square tests, the techniques are much less formal.

Often, descriptive and graphical techniques can help. You might simply plot the sample proportions for each population on the same graph and see how they compare.

## EXAMPLE 15.20   College Drinking

### After the Chi-Square Test

As a result of the chi-square test, the researcher knows that at least one of the populations is different from the others with respect to missing classes. The problem is that that is *all* he knows. The chi-square test does not tell him whether all three populations are different from each other or if the two binge drinking categories are similar but different from the non-bingers.

He decides to plot the proportions for each sample on a graph. From the graph it appears that all three groups are different from each other. The chi-square test verifies his original ideas, but if he wants more definitive conclusions he might need to use a designed experiment.

**Proportion of Students Who Miss Class**

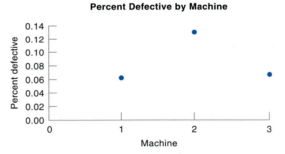

As you can see from the previous example, the answer is not often obvious from a plot. If the sample sizes are relatively the same you could find confidence intervals for each of the samples and compare them.

### EXAMPLE 15.21   Comparing Machine Quality

*After the Chi-Square Test*

Now that the manufacturing facility knows that at least one of its machines is different from the others, it needs to try to identify which one(s) are different so that it can determine where the problem lies. The company decides to plot the proportion defective for each machine:

**Percent Defective by Machine**

From the plot it certainly appears that machines 1 and 3 are comparable, while machine 2 has a much higher proportion defective. Just to be sure the company decides to calculate and plot 95% confidence intervals for the proportion defectives:

$$\text{Machine 1:} \quad 0.06 \pm 1.96 \sqrt{\frac{(0.06)(0.94)}{200}} = 0.06 \pm 0.0329 = (0.0271, 0.0929)$$

*Note: There has been some rounding in these calculations.*

$$\text{Machine 2:} \quad 0.13 \pm 1.96 \sqrt{\frac{(0.13)(0.87)}{170}} = 0.13 \pm 0.0504 = (0.0796, 0.1804)$$

$$\text{Machine 3:} \quad 0.07 \pm 1.96 \sqrt{\frac{(0.07)(0.93)}{210}} = 0.07 \pm 0.0337 = (0.0363, 0.1037)$$

**Confidence Intervals for Percent Defective**

From the plot they see that the confidence intervals for machines 1 and 3 almost completely overlap, while the confidence interval for machine 2 barely overlaps the other two intervals. Thus, the company is pretty sure that machine 2 is different from the other two and that it produces a higher proportion of defective items. ■

An important thing to remember when doing the chi-square test for proportions, or any statistical test, is that the test is not the *answer* or *solution* to the problem. It is simply verification that something is *causing* the differences that you observed. Once the statistical analysis is done, it always requires knowledge of the situation to identify the causes and correct the problem if that is needed.

## 15.3.5 Exercises—Learning It!

**15.6** In an experiment to study the attitude of voters concerning term limitations in Congress, voters in Indiana, Ohio, and Kentucky were polled with the following results:

| Opinion | Indiana | Kentucky | Ohio |
|---|---|---|---|
| Support | 82 | 107 | 93 |
| Do not support | 97 | 66 | 74 |

(a) Set up the hypotheses to test whether the proportion of voters who support congressional term limits is the same for all three states.

(b) Calculate the proportion of voters that support congressional term limits for each state individually. Based on these values, do you think there is a difference in the proportions?

(c) Calculate the overall proportion of voters who support term limits for Congress.

(d) Calculate the expected frequencies for each cell and find the value of the chi-square test statistic.

(e) At the 0.05 level of significance, is there a difference in the proportion of voters who support congressional term limits among the three states?

**15.7** In a survey about satisfaction with local phone service, those respondents who rated their current service as excellent and those who rated it Poor–Very Poor were asked to classify their current local service provider. The results are given in the table below:

| | Type of Company | | | | |
|---|---|---|---|---|---|
| Current Service Source | Long Distance | Local Phone | Power | Cable TV | Cellular Phone |
| Excellent | 264 | 444 | 131 | 215 | 198 |
| Poor–Very Poor | 1394 | 1318 | 485 | 431 | 572 |

(a) Set up the hypotheses to test whether the proportion of people who rated their company as excellent is the same for each type of company.

(b) Calculate the overall proportion of people who rate their current phone service as excellent.

(c) Find the expected frequency for each cell and calculate the chi-square test statistic.

(d) If you wanted to perform the test at the 0.05 level of significance, what would the critical value of the test be?

(e) At the 0.05 level of significance, is there a difference in the proportion of people who rate their local phone service as excellent among the different types of companies?

**15.8** A computer magazine surveyed its readers to determine how likely it was that people who planned to purchase new computers in the near future would buy a portable/notebook or desktop model. The results are tabulated below:

| | When Purchase Will Be Made | | |
|---|---|---|---|
| Type of Computer | 0–3 months | 3–6 Months | 6–12 Months |
| Notebook/Portable | 34 | 156 | 258 |
| Desktop | 56 | 346 | 128 |

(a) Suppose that you were interested in determining whether there is a relationship between the type of computer that a person is planning to buy and when the person plans to make the purchase. Set up the hypotheses to test this. Which variable represents the population?

(b) Calculate the overall proportion of people who plan to buy a notebook/portable computer.

(c) Use this value to find the expected frequency for each cell.

(d) Calculate the value of the chi-square test statistic.

(e) At the 0.01 level of significance, is there a relationship between the type of computer that will be purchased and the time when the purchase will be made?

15.9   In a survey on the use of Office Suite software, 201 people who responded to the survey were asked about their experience with such software. The people who were administering the survey were interested in determining whether females had different experience with such software than males. The data they compiled are shown in the table:

| | | *Experience With Office Suite Software* | | |
|---|---|---|---|---|
| **Gender** | **Use It Now** | **Don't Use It Now, But Have in Past** | **Know Someone Who Uses It** | **No Experience With It** |
| Female | 46 | 3 | 3 | 4 |
| Male | 118 | 12 | 5 | 10 |

(a) Set up the hypotheses to determine whether experience with Office Suite software is related to gender.

(b) Calculate the overall proportion of female respondents in the survey.

(c) Use the overall proportion to calculate the expected frequencies for each cell.

(d) If you want to do this test at the 0.10 level of significance, what is the critical value?

(e) At the 0.10 level of significance, are gender and experience with Office Suite software related?

# 15.4   THE CHI-SQUARE TEST FOR INDEPENDENCE

In Chapter 5 you looked at bivariate data. For qualitative data you learned how to summarize the data using a contingency table and for quantitative data you learned how to look at relationships between two variables. You learned that two qualitative variables can be dependent, meaning there is a relationship between the variables, or independent, meaning that there is no relationship between the variables. You also learned how to identify the type of relationship that exists (linear, nonlinear) and to quantify a linear relationship in the form of an equation.

In the previous section you learned that it is also possible for there to be a relationship between two qualitative variables. For example, suppose the Human Resources department of a large company collected data on employees' job levels and the type of medical coverage that the employees chose. Both of these variables, job level and type of medical coverage, are qualitative, but it makes sense that there might be a relationship between the two. We might expect that people with higher level, higher paying jobs would have better health plan coverage. How can we determine whether such a relationship exists? That is, how can we determine whether two qualitative variables are dependent or independent?

The chi-square test for proportions involved a contingency table that had 2 rows, successes and failures, and a column for each population. This test can be extended to look at two qualitative variables where each row corresponds to a different value of the first variable and each column corresponds to a different value of the second variable.

## 15.4.1 Probability and Independence

By now you are quite familiar with the idea that we rarely know the populations that we are studying, and instead we use data to *estimate* or *approximate* different characteristics of these populations. You learned in Chapter 6 that data collected over time are often used as *estimates* of the probabilities that different events will occur. These probabilities are known as *empirical* probabilities.

When we collect data on two qualitative variables we can display the data using a contingency table. The rows of the table represent the possible categories for one of the variables and the columns of the table represent the categories of the other variable. We can use the frequencies in the table to estimate different probabilities about the elements of the population. We can estimate whether an element will have a certain characteristic (event A), a particular pair of characteristics (event A AND B), or one or more of the characteristics (event A OR B).

### EXAMPLE 15.22   Health-Care Plans

*Classifying Events*

Suppose that the Human Resources department of the company mentioned previously looked at the records of 300 employees and collected data on two characteristics: Job Classification and Health Plan. The data are summarized below:

| Job Classification | Physician Network | HMO | No Coverage | Total |
|---|---|---|---|---|
| Salaried professional | 35 | 12 | 3 | 50 |
| Salaried clerical | 21 | 67 | 12 | 100 |
| Hourly | 6 | 112 | 32 | 150 |
| Totals | 62 | 191 | 47 | 300 |

*(Type of Coverage spans Physician Network, HMO, No Coverage)*

*We are assuming here that the sample is representative of the employees of the company and we are using the data to estimate the probabilities for the company as a whole.*

If we define A as the event that a person in the company selects the HMO for medical coverage, and B as the event that a person in the company is an hourly worker, then

$$P(A) = \frac{191}{300} \qquad P(B) = \frac{150}{300} \qquad P(A \text{ AND } B) = \frac{112}{300}$$

## Definition of Independence in Probability

When you looked at bivariate relationships in Chapters 5 and 11, you were trying to decide whether the value of one quantitative variable depended on another or if they were independent. We can also use the word **independent** in terms of events and probability.

> Two events are *independent* if the probability that one event occurs in any given trial of an experiment is not affected or changed by the occurrence of the other event.

In probability, two events, A and B, are independent *exactly when*

$$P(A \text{ AND } B) = P(A) \times P(B)$$

*Formula for independent events*

The phrase *exactly when* means that the statement can be used in both directions. That is, if we know two events are independent, then to find the probability that both will occur we can multiply the individual probabilities together. It also means that if we know that the probability that both occur is equal to the product of the individual probabilities, we can conclude that the events are independent. We will use this fact in the chi-square test for independence.

### EXAMPLE 15.23   Health-Care Plans

*Independence and Probability*

In the example about the Human Resources department we can check to see if events A and B are independent by comparing the quantity P(A AND B) to the quantity P(A) × P(B):

$$P(A) \times P(B) = \left(\frac{191}{300}\right)\left(\frac{150}{300}\right) = 0.318 \quad \text{and} \quad P(A \text{ AND } B) = \frac{112}{300} = 0.373$$

Thus, events A and B are not independent *in this sample*. We know that when we use sample data the probabilities are *exactly correct* only when applied to the sample. Does it mean that the two events A and B are not independent in the population? Not really.     ■

You know from your study of sampling error and sampling distributions that if the company were to take another sample of 300 employees the results would not be identical. If both samples are representative of the population of interest, then the results should be similar, but there will be some sampling error. We need a test for independence that takes into account the fact that we are dealing with *sample data* and not population values. In the next section we see how the chi-square test that we have been studying can be used to accomplish this task.

## 15.4.2 Description of the Test

We would like to decide whether or not two qualitative variables are related. The hypotheses that we will test are

$H_0$:   The two variables are independent of each other.

$H_A$:   The two variables are not independent.

When we collect data on two qualitative variables and summarize them using a contingency table, the number in each cell of the table is the *observed* frequency associated with one particular category of the first variable *and* one particular category of the second variable. A general example of the contingency table for two variables is shown in Table 15.2.

Variable 1 has $r$ possible categories and variable 2 has $c$ possible categories. The value $o_{12}$ is the number of sample elements that were in category 1 of variable 1 and category 2 of variable 2.

**TABLE 15.2   Contingency Table for Test for Independence**

| | | Variable 2 | | | |
|---|---|---|---|---|---|
| **Variable 1** | **1** | **2** | **. . .** | **c** | **Totals** |
| 1 | $o_{11}$ | $o_{12}$ | . . . | $o_{1c}$ | $r_1$ |
| 2 | $o_{21}$ | $o_{22}$ | . . . | $o_{2c}$ | $r_2$ |
| ⋮ | ⋮ | ⋮ | ⋮ | ⋮ | |
| $r$ | $o_{r1}$ | $o_{r2}$ | . | $o_{rc}$ | $r_r$ |
| **Totals** | $c_1$ | $c_2$ | . . . | $c_c$ | $n$ |

## EXAMPLE 15.24   College Drinking

***Setting Up the Contingency Table***

In addition to questions used to classify drinking habits, the survey on drinking asked students about where they lived: residence halls or dormitories, fraternity or sorority houses, other university housing, or off campus house or apartment. In analyzing the results of the survey, the researcher asked whether the amount of drinking by students depends on where they live. His hypotheses were

$H_0$:   The amount that a student drinks is independent of where he or she lives.

$H_A$:   The amount that a student drinks depends on where he or she lives.

He compiled a contingency table of the data:

| Where They Live | Type of Drinker Non-binge | Infrequent | Frequent | Total |
|---|---|---|---|---|
| Residence hall or dormitory | 35 | 25 | 46 | **106** |
| Fraternity or sorority | 0 | 1 | 0 | **1** |
| Other university housing | 0 | 2 | 1 | **3** |
| Off-campus house or apartment | 49 | 30 | 24 | **103** |
| **Total** | **84** | **58** | **71** | **213** ∎ |

To do a chi-square test, we must find the expected frequency for each cell in the table and compare it to the observed frequency. You remember from the other chi-square tests that we have done that we find the expected frequencies by *assuming* that $H_0$ is true and calculating the probability for each cell under that assumption. For the test of independence, the null hypothesis is that the two variables *are independent*. To find the probabilities for each value of a variable, you use the row or column totals, the same way you did in Chapter 6 when you learned about probability. That is, for row $i$,

$$P(i) = \frac{r_i}{n}$$

and for column $j$,

$$P(j) = \frac{c_j}{n}$$

Remember that the definition of independence works two ways. If we know that two events are independent, then we can multiply their probabilities together to find the probability that they will both occur. That is, when we have independence for each cell in the contingency table, the probability that variable 1 will have category $i$ and variable 2 will have category $j$ is

$$P(i \text{ AND } j) = \frac{r_i}{n} \frac{c_j}{n}$$

To find the **expected frequency** for any cell we multiply the probability by the sample size, so

$$e_{ij} = \left( \frac{r_i}{n} \frac{c_j}{n} \right) n = \frac{r_i c_j}{n}$$

*Formula for expected frequency in a cell*

**EXAMPLE 15.25**   College Drinking

*Finding the Expected Frequencies*

The data on where a student lives and the type of drinker that he or she is are shown again below:

|  | Type of Drinker | | | |
| Where They Live | Non-binge | Infrequent | Frequent | Total |
|---|---|---|---|---|
| Residence hall or dormitory | 35 | 25 | 46 | 106 |
| Fraternity or sorority | 0 | 1 | 0 | 1 |
| Other university housing | 0 | 2 | 1 | 3 |
| Off-campus house or apartment | 49 | 30 | 24 | 103 |
| **Total** | **84** | **58** | **71** | **213** |

The calculations for the expected frequencies for each cell are

$$e_{11} = \frac{(106)(84)}{213} = 41.80 \quad e_{12} = \frac{(106)(58)}{213} = 28.86 \quad e_{13} = \frac{(106)(71)}{213} = 35.33$$

$$e_{21} = \frac{(1)(84)}{213} = 0.39 \quad e_{22} = \frac{(1)(58)}{213} = 0.27 \quad e_{23} = \frac{(1)(71)}{213} = 0.33$$

$$e_{31} = \frac{(3)(84)}{213} = 1.18 \quad e_{32} = \frac{(3)(58)}{213} = 0.82 \quad e_{33} = \frac{(3)(71)}{213} = 1.00$$

$$e_{41} = \frac{(103)(84)}{213} = 40.62 \quad e_{42} = \frac{(103)(58)}{213} = 28.05 \quad e_{43} = \frac{(103)(71)}{213} = 34.33$$

The expected frequencies are shown with the observed frequencies in the table below:

|  | Type of Drinker | | | |
| Where They Live | Non-binge | Infrequent | Frequent | Total |
|---|---|---|---|---|
| Residence hall or dormitory | 35 | 25 | 46 | 106 |
|  | (41.80) | (28.86) | (35.33) |  |
| Fraternity or sorority | 0 | 1 | 0 | 1 |
|  | (0.39) | (0.27) | (0.33) |  |
| Other university housing | 0 | 2 | 1 | 3 |
|  | (1.18) | (0.82) | (1.00) |  |
| Off-campus house or apartment | 49 | 30 | 24 | 103 |
|  | (40.62) | (28.05) | (34.33) |  |
| **Total** | **84** | **58** | **71** | **213** ∎ |

Remember that in any chi-square test the expected frequency for each cell or class should be at least five. When this is not the case, the categories must be collapsed or joined. For the goodness of fit test, the data are quantitative, so you always join classes that are adjacent. In the test for proportions or the test for independence this is not the case. The categories should be collapsed on a *logical* basis, that is, one that preserves the intent of the question.

**EXAMPLE 15.26**   College Drinking

*Adjusting the Contingency Table*

When he looked at the contingency table on type of drinker and residence, the researcher saw that two of the categories for residence did not have expected frequencies of at least 5. He knew that this meant he should collapse some of the categories.

His first thought was to combine the two low-frequency categories (Fraternity or sorority and Other university housing) into a single category designated Other university housing, but he realized that this would still not raise the expected values above 5.

His final decision was to collapse all of the on-campus housing into a single category so that he met the requirements of the chi-square test. The modified contingency table is shown below:

|  | Type of Drinker | | | |
| Where They Live | Non-binge | Infrequent | Frequent | Total |
|---|---|---|---|---|
| On campus | 35 (43.38) | 28 (29.95) | 47 (36.67) | 110 |
| Off-campus house or apartment | 49 (40.62) | 30 (28.05) | 24 (34.33) | 103 |
| **Total** | **84** | **58** | **71** | **213** ■ |

## EXAMPLE 15.27  Health Care Plans

### Setting up the Contingency Table

For the Human Resources department, the hypotheses to be tested are

$H_0$:  Choice of health-care plan and job classification are independent.

$H_A$:  Choice of health-care plan and job classification are not independent.

For the employee/health plan data the expected frequencies are calculated as follows:

$$e_{11} = \frac{(50)(62)}{300} = 10.3 \quad e_{12} = \frac{(50)(191)}{300} = 31.8 \quad e_{13} = \frac{(50)(47)}{300} = 7.8$$

$$e_{21} = \frac{(100)(62)}{300} = 20.7 \quad e_{22} = \frac{(100)(191)}{300} = 63.7 \quad e_{23} = \frac{(100)(47)}{300} = 15.7$$

$$e_{31} = \frac{(150)(62)}{300} = 31.0 \quad e_{32} = \frac{(150)(191)}{300} = 95.5 \quad e_{33} = \frac{(150)(47)}{300} = 23.5$$

The expected frequencies together with the original data are shown below. There are clearly discrepancies between the observed and expected frequencies, but it is not clear if they are simply the result of random variation or if they are indicative that $H_0$ is false.

|  | Type of Coverage | | | |
| Job Classification | Physician Network | HMO | No Coverage | Total |
|---|---|---|---|---|
| Salaried professional | 35 (10.3) | 12 (31.8) | 3 (7.8) | 50 |
| Salaried clerical | 21 (20.7) | 67 (63.7) | 12 (15.7) | 100 |
| Hourly | 6 (31.0) | 112 (95.5) | 32 (23.5) | 150 |
| **Totals** | **62** | **191** | **47** | **300** ■ |

As in the case with the goodness of fit test and the test for proportions, it is important to remember that the expected frequency in each cell should be at least 5. If this is not the case, then categories for one or the other or both of the variables need to be combined. Since the categories are not numerical, the categories should be grouped on similarities.

*TRY IT NOW!*

### Drinking Survey
*Setting Up the Contingency Table for a Test for Independence*

The Public Health student who did the study on drinking also collected data on the number of times that the student drove while intoxicated in the last two weeks (coded). The contingency table for the usable responses is given below:

| Class | Number of Times Drive While Intoxicated | | | |
| --- | --- | --- | --- | --- |
| | Not at All | Once | Twice or More | Total |
| Freshman | 72 | 5 | 9 | 86 |
| Sophomore | 19 | 8 | 9 | 36 |
| Junior | 16 | 8 | 6 | 30 |
| Senior | 8 | 4 | 7 | 19 |
| **Total** | **115** | **25** | **31** | **171** |

The university is interested in knowing if the number of times a student drove while intoxicated is related to his or her class in school. It feels that this information will help target student audiences for programs on drinking and driving.

   Set up the hypotheses that the university should test.

Calculate the expected frequencies for each cell and put them in the appropriate location in the table above.

Are any of the expected frequencies less than 5? If so, can you suggest a logical way to combine categories to avoid this problem?

## 15.4.3 Performing the Test for Independence

Once we have the expected and observed frequencies for each cell in the table, nothing new needs to be learned to perform the test. The chi-square test statistic is calculated in the same way as for the test of proportions:

$$\chi^2 = \sum_{\text{all cells}} \frac{(o_{ij} - e_{ij})^2}{e_{ij}}$$

and the degrees of freedom are calculated as

$$(r - 1)(c - 1)$$

### EXAMPLE 15.28   College Drinking

*Performing the Chi-Square Test for Independence*

The researcher looking at whether where students live is related to the amount that they drink calculated the chi-square statistic for the revised contingency table:

|  | *Type of Drinker* | | | |
| Where They Live | Non-binge | Infrequent | Frequent | Total |
|---|---|---|---|---|
| On campus | 35 | 28 | 47 | **110** |
|  | (43.38) | (29.95) | (36.67) |  |
| Off-campus house or apartment | 49 | 30 | 24 | **103** |
|  | (40.62) | (28.05) | (34.33) |  |
| **Total** | **84** | **58** | **71** | **213** |

$$\frac{(35 - 43.38)^2}{43.38} + \frac{(28 - 29.95)^2}{29.95} + \cdots + \frac{(30 - 28.05)^2}{28.05} + \frac{(24 - 34.33)^2}{34.33}$$

$$= 1.62 + 0.13 + \cdots + 0.14 + 3.11 = 9.63$$

He decided to do the test at the 0.05 level of significance and calculated the degrees of freedom of the test to be $(2 - 1)(3 - 1) = 2$. The critical value of the test is $\chi^2_{0.05,2} = 5.99$. Since the test statistic of 9.63 is outside the critical value, he rejects $H_0$ and concludes that the variables residence and type of drinker are not independent. That is, the amount of drinking that students do depends on where they live. ∎

### EXAMPLE 15.29   Health Care Plans

*Performing the Chi-Square Test for Independence*

The Human Resources department that was looking at the relationship between employee job classification and type of health plan chosen calculated the chi-square statistic for its data as

$$\frac{(35 - 10.3)^2}{10.3} + \frac{(12 - 31.8)^2}{31.8} + \cdots + \frac{(112 - 95.5)^2}{95.5} + \frac{(32 - 23.5)^2}{23.5}$$

$$= 58.88 + 12.36 + \cdots + 2.85 + 3.07 = 101.35$$

The analysts decide to perform the test with $\alpha = 0.05$. The degrees of freedom for the test are $(3 - 1)(3 - 1) = 4$, and they find the critical value to be 9.49. Since the chi-square test statistic is definitely outside the critical value, they reject $H_0$ and conclude that the type of job and health plan chosen are not independent; that is, they are related. ∎

 **TRY IT NOW!**

### Drinking Survey

*Performing the Chi-Square Test for Independence*

The university that is looking at the relationship between class year and drinking and driving wants to perform the test at the 0.05 level of significance.

*(continued)*

Calculate the value of the chi-square statistic for the data.

*Don't forget to collapse the categories so that there are no cells with expected values below 5.*

Find the critical value and perform the test.

Are class and drinking and driving independent?

## 15.4.4 Exercises—Learning It!

**15.10** A report by the Department of Justice on rape victims reports on interviews with 3721 victims. The attacks were classified by age of the victim and the relationship of the victim to the rapist. The results of the study are given below:

| | | Relationship of Rapist | |
|---|---|---|---|
| Age of Victim | Family | Acquaintance or Friend | Stranger |
| Under 12 | 153 | 167 | 13 |
| 12 to 17 | 230 | 746 | 172 |
| Over 17 | 269 | 1232 | 739 |

(a) Set up the hypotheses to test whether age of victim and relationship of rapist are independent.

(b) Calculate the expected frequencies for each cell.

(c) How many degrees of freedom will the chi-square test for indpendence have? Using a level of significance of 0.01, what is the critical value for the test?

(d) Calculate the value of the chi-square test statistic.

(e) Is the age of the victim independent of the relationship of the rapist?

**15.11** A company that manufactures cardboard boxes for packaging cereals wants to determine whether the type of defect that a particular box has is related to the shift on which it was produced. It compiles the following data. In each case, if a box had multiple defects the most serious defect was recorded.

| | | Type of Defect | |
|---|---|---|---|
| Shift | Printing | Rips/Tears | Size |
| 1 | 55 | 60 | 85 |
| 2 | 58 | 63 | 79 |
| 3 | 89 | 63 | 48 |

(a) Set up the appropriate hypotheses for the test.

(b) Calculate the expected frequencies for each cell and calculate the value of the chi-square test statistic.

(c) How many degrees of freedom does the test have?

(d) At the 0.05 level of significance are defect type and shift related?

15.12 A company that depends heavily on advertising for selling its products wants to know if its various advertising media have different effectiveness relative to the age of the customer. The company uses its warranty return cards to collect the following information:

| Type of Advertisement | Age of Customer 21–30 | 31–40 | 41–50 | Over 50 | Totals |
|---|---|---|---|---|---|
| Store display | 21 | 28 | 8 | 6 | 63 |
| Catalog | 8 | 5 | 1 | 1 | 15 |
| Magazine | 1 | 23 | 8 | 1 | 33 |
| Newspaper | 18 | 14 | 2 | 4 | 38 |
| Totals | 48 | 70 | 19 | 12 | 149 |

(a) Set up the hypotheses to test whether type of advertisement and age of customer are independent.

(b) What population will the test apply to?

(c) Calculate the expected frequencies for each cell.

(d) How many degrees of freedom will the test have?

(e) Without collapsing any cells, calculate the value of the chi-square test statistic.

(f) At the 0.05 level of significance, are type of advertisement and age of customer independent?

(g) Do you think that categories should have been collapsed? Why or why not?

(h) What categories would you recommend collapsing to fix the problem?

15.13 The people who did the survey on Office Suite software were also interested in whether experience with such software was related to age. They created a contingency table to look at this question:

| Experience | Age Group Under 25 | 25 to 44 | 45 and Up |
|---|---|---|---|
| Use it now | 11 | 51 | 101 |
| Don't use it now, but have in past | 2 | 2 | 11 |
| Know someone who uses it | 2 | 2 | 4 |
| No experience | 2 | 3 | 10 |

(a) Set up the hypotheses to test whether experience with Office Suite software is independent of age group.

(b) How many degrees of freedom will the chi-square test have?

(c) Calculate the expected frequencies for each cell.

(d) Without collapsing any categories, calculate the value of the chi-square test statistic.

(e) At the 0.05 level of significance, is experience with Office Suite software independent of age?

(f) Are the cells with frequencies less than 5 a factor in this case? Why or why not? Do you think that you should redo the test after collapsing categories?

# 15.5   THE CHI-SQUARE TEST FOR INDEPENDENCE IN EXCEL

Excel provides the tools needed to do a chi-square test for independence, although they are not found as a single analysis tool. You are already familiar with one of them, the pivot table. In this section, we will explain how to use Excel to perform a chi-square test for independence.

## 15.5.1 Creating the Contingency Table in Excel

In Chapter 5, you learned how to create a contingency table for two qualitative variables in Excel. This contingency table provides the backbone for the chi-square test for independence in Excel. We will look at the data from the binge drinking survey, and test whether where a student lives is independent of the type of binge drinker. A small portion of the data in an Excel worksheet is shown in Figure 15.3.

| | A | B | C |
|---|---|---|---|
| 1 | BINGER | MISS | LIVE |
| 2 | 1 | 1 | 4 |
| 3 | 1 | 1 | 4 |
| 4 | 2 | 1 | 1 |
| 5 | 1 | 1 | 4 |
| 6 | 1 | 1 | 4 |
| 7 | 1 | 1 | 1 |
| 8 | 1 | 1 | 4 |
| 9 | 3 | 1 | 4 |
| 10 | 1 | 1 | 4 |
| 11 | 1 | 1 | 4 |
| 12 | 2 | 2 | 1 |
| 13 | 3 | 2 | 4 |
| 14 | 0 | 1 | 4 |
| 15 | 3 | 1 | 1 |
| 16 | 3 | 3 | 1 |

**FIGURE 15.3**   Binge drinking data

The first step is to create a contingency table in Excel using the **Pivot table** tool. From the **Data** menu, select **Pivot Table Report** and follow the steps of the pivot table wizard to create the table. In this example, we will put the variable for type of binge drinker (BINGER) in the columns and the variable for where a student lives (LIVE) in the rows. At step 4 of 4, after you indicate where you want the table to be placed, do not select **Finish.** Instead, click **Options...** and the dialog box shown in Figure 15.4 opens.

**FIGURE 15.4**   Pivot table options dialog box

Make sure that the checkbox next to **For empty cells, show** is checked, and in the textbox next to it, type "0". To perform a chi-square test, Excel will not accept empty cells in the contingency table. Click **OK** and then **Finish.** The contingency table for the data is shown in Figure 15.5.

Since we are interested only in students who are drinkers, we do not want the Binger column for 0, which indicates that the student does not drink. To eliminate this column, double click on the field button BINGER and the pivot table field dialog box shown in Figure 15.6 opens.

| O | P | Q | R | S | T |
|---|---|---|---|---|---|
| Count of BINGER | BINGER | | | | |
| LIVE | 0 | 1 | 2 | 3 | Grand Total |
| 1 | 11 | 35 | 25 | 46 | 117 |
| 2 | 0 | 0 | 1 | 0 | 1 |
| 3 | 0 | 0 | 2 | 1 | 3 |
| 4 | 28 | 49 | 30 | 24 | 131 |
| Grand Total | 39 | 84 | 58 | 71 | 252 |

**FIGURE 15.5**   Contingency table for binge drinking data

**FIGURE 15.6**   Pivot table field dialog box

At the bottom of the box under **Hide Items:,** highlight "0" and click **OK.** The pivot table will adjust to hide this column. To work with this table, use **Paste Special > Values** to make a copy in a different location in the worksheet. You can then enter titles that are more descriptive and change format features.

## 15.5.2 Creating the Table of Expected Frequencies in Excel

The contingency table provides the actual frequencies for each cell. To perform the chi-square test for independence, we also need the expected frequencies for each cell. You can use formulas in Excel to calculate these. Excel expects to find these frequencies in a separate table, directly below the table of actual frequencies.

Remember that the expected frequency for any cell is calculated by

$$e_{ij} = \left(\frac{r_i}{n} \times \frac{c_j}{n}\right) n = \frac{r_i c_j}{n}$$

where $r_i$ is the row total, $c_j$ is the column total, and $n$ is the total number of observations.

1. Copy the row and column headings of the pivot table to a location just below it. This will define the table for the expected frequencies.
2. Position the cursor in the first cell of the table, for non-binge drinkers who live in dorms or residence halls, and type "=$S20*$P$24/$S$24". In this formula, S20 is the location of the row total for dorms and residence halls, P24 is the location of the column total for non-binge drinkers, and S24 is the location of the grand total. You must use relative references if you are going to copy the for-

mula to the rest of the cells and not retype them. The relative references in this formula will allow you to copy the formula to the rest of the cells in this column. To copy to the rest of the cells in a row, make the first reference $S$20 and the second P$24. The table of expected frequencies showing the formulas in each cell is found in Figure 15.7.

| 26 | | Type of Drinker | | |
|---|---|---|---|---|
| 27 | LIVE | Non-Binge | Infrequent | Frequent |
| 28 | Residence Hall or Dormitory | =$S$20*P$24/$S$24 | =$S$20*Q$24/$S$24 | =$S$20*R$24/$S$24 |
| 29 | Fraternity or Sorority | =$S$21*P$24/$S$24 | =$S$21*Q$24/$S$24 | =$S$21*R$24/$S$24 |
| 30 | Other University Housing | =$S$22*P$24/$S$24 | =$S$22*Q$24/$S$24 | =$S$22*R$24/$S$24 |
| 31 | Off Campus House or Apartment | =$S$23*P$24/$S$24 | =$S$23*Q$24/$S$24 | =$S$23*R$24/$S$24 |
| 32 | | | | |

**FIGURE 15.7**   Table of expected frequency formulas

## 15.5.3 Performing the Chi-Square Test

Excel has a function, **CHITEST,** that performs the chi-square test for independence. The function accepts the data ranges for the observed and expected values as input and returns the $p$ value of the test as output.

To perform the test for the binge drinking data:

1. Position the cursor in the cell where you want the $p$ value of the test to be located, and start the function wizard. Choose **Statistical** as the function category and **CHITEST** as the function name. The **CHITEST** dialog box opens as shown in Figure 15.8.

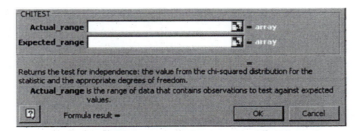

**FIGURE 15.8**   The CHITEST dialog box

2. Position the cursor in the textbox for **Actual_range** and highlight the range of cells that contain the observed frequencies. Be careful not to highlight the row and column totals.

3. Do the same thing for the **Expected_range,** highlighting the location of the expected frequencies.

4. Click **OK;** the result of the test will appear in the designated cell. In this case, the value returned is 0.01744.

If the test is performed at the 0.05 level of significance, we conclude that type of drinker and where a student lives are not independent.

From the examples in this chapter, you know that there was a problem with this test because of cells with expected frequencies less than 5. You can use the pivot table to combine rows and columns and then perform the test again. To combine rows or columns in a pivot table, use the **Group** command found on the pivot table toolbar. You can also access this command by highlighting the rows or columns you want to combine, right clicking the mouse, and selecting **Group and Outline > Group** from the menu.

# CHAPTER 15 SUMMARY

The chi-square test involves comparing observed and expected frequencies for different classes of data. The test is quite versatile and can be used to test *goodness of fit, equality of proportions for more than two populations,* and *independence of qualitative variables.*

As is true with *any* statistical tool, the results of a chi-square test do not *solve* a problem or make a decision for the user. They simply point out when further action is indicated and when it is not. Solving the real problem requires knowledge of the situation and sometimes further data collection and analysis.

## Key Terms

| Term | Definition | Page Reference |
|---|---|---|
| Chi-square goodness of fit test | The **chi-square goodness of fit test** checks to see how well a set of data fits the model for a particular probability distribution. | 756 |
| Expected frequencies, $e_i$ | The **expected frequencies** are the number of observations that should fall into each class in a frequency distribution under the hypothesized probability distribution. | 757 |
| Independent events | Two events are independent if the probability that one event occurs is not affected or changed by the occurrence of the other event. | 785 |
| Observed frequencies, $o_i$ | The **observed frequencies** are the actual number of observations that fall into each class in a frequency distribution or histogram. | 757 |
| Uniform distribution | A **uniform distribution** is one in which each outcome or class of outcomes is equally likely to occur. | 757 |

## Key Formulas

| Term | Formula | Page Reference |
|---|---|---|
| Chi-square critical value | $\chi^2_{\alpha,\, k-p-1}$ | 765 |
| Chi-square statistic | $\chi^2 = \dfrac{\sum\limits_{i=1}^{k}(o_i - e_i)^2}{e_i}$ | 762 |
| Degrees of freedom for goodness of fit test | $k - p - 1$ | 763 |

*(continued)*

| Term | Formula | Page Reference |
|------|---------|----------------|
| **Degrees of freedom for test for independence** | $(r-1)(c-1)$ | 791 |
| **Expected frequency for test for independence** | $e_{ij} = \left(\dfrac{r_i}{n}\dfrac{c_j}{n}\right)n = \dfrac{r_i c_j}{n}$ | 787 |
| **Expected frequency for test for proportions** | $e_i = \dfrac{s}{n}n_i = \dfrac{sn_i}{n} = pn_i$ | 776 |
| **Probability for independent events** | $P(A \text{ AND } B) = P(A) \times P(B)$ | 785 |
| **Overall proportion of success** | $p = \dfrac{s_1 + s_2 + \cdots + s_c}{n_1 + n_2 + \cdots + n_c} = \dfrac{s}{n}$ | 776 |

# CHAPTER 15 EXERCISES

## Learning It!

**15.14** A random sample of 69 Porsche drivers were asked how many miles they had driven their vehicle in the past calendar year. A frequency table for the data is given below:

| Number of Miles | Observed Frequency |
|-----------------|--------------------|
| $0 < x \le 4000$ | 20 |
| $4000 < x \le 8000$ | 22 |
| $8000 < x \le 12{,}000$ | 14 |
| $12{,}000 < x \le 16{,}000$ | 6 |
| $16{,}000 < x \le 20{,}000$ | 1 |
| $20{,}000 < x \le 24{,}000$ | 4 |
| $24{,}000 < x \le 28{,}000$ | 0 |
| $28{,}000 < x \le 32{,}000$ | 2 |

(a) Set up the hypotheses to test whether the number of miles driven by Porsche drivers is normally distributed with a mean of 7500 miles with a standard deviation of 6500 miles.

(b) Find the expected frequencies for each cell.

(c) Calculate the chi-square test statistic. If necessary, collapse cells so that the expected frequency for each category is at least 5.

(d) At the 0.05 level of significance, what can you conclude about the number of miles driven annually by Porsche drivers?

**15.15** To begin a study comparing the monthly salaries of the auditors for a particular company to accountants' salaries in the industry in general, a random sample of 30 auditors was taken and their monthly salaries recorded. The data are shown here:

| | | | | |
|---|---|---|---|---|
| 2832 | 3032 | 3122 | 3208 | 3325 |
| 2843 | 3050 | 3123 | 3224 | 3328 |
| 2875 | 3050 | 3128 | 3231 | 3375 |
| 2878 | 3087 | 3130 | 3233 | 3396 |
| 2995 | 3093 | 3133 | 3237 | 3429 |
| 3010 | 3096 | 3175 | 3239 | 3450 |

(a) Create a frequency distribution for the data using six classes and display the data graphically.

(b) Do you think that the assumption that the data are normally distributed is at least reasonable? Why or why not?

(c) Set up the hypotheses to test whether the data come from a normal population that has a mean of $3150 and a standard deviation of $165.

(d) Perform the chi-square goodness of fit test with a level of significance of 0.05.

(e) What does the test tell you about the data?

**15.16** The problem of workplace violence is growing. A survey of 600 full-time American workers on workplace violence concentrated on those respondents who were victims of harassment, threat of physical violence, and actual physical violence. In follow-up interviews the victims were asked to identify the major effect that the violence had on them. The data are shown below:

| Major Effect on Worker | Type of Violence | | |
|---|---|---|---|
| | Harassment | Threat | Physical |
| Psychological | 56 | 28 | 15 |
| Disrupted work life | 39 | 13 | 12 |
| Physical injury or sickness | 15 | 5 | 10 |
| No negative effect | 5 | 7 | 6 |

(a) Suppose that the group that conducted the study were interested in whether the type of violence that a person experienced is related to the effect of the violence on their life. Set up the hypotheses for this test.

(b) Find the expected frequencies for each cell.

(c) Without collapsing any categories, how many degrees of freedom will the test have?

(d) At the 0.05 level of significance, are type of violence and effect on a worker independent?

**15.17** In a survey sponsored by the National Cancer Institute (NCI) and the American Cancer Society in 1993, adults (people 18 years of age and older) in eight states were surveyed on their attitudes toward smoking in public places. One question of interest was whether the respondent favored banning smoking in fast-food restaurants. The data collected are shown in the table below:

| Ban Smoking in Fast-Food Restaurants | Louisiana | Missouri | New Jersey | Ohio | Oklahoma | South Carolina | Texas | Washington |
|---|---|---|---|---|---|---|---|---|
| Yes | 129 | 124 | 133 | 130 | 107 | 210 | 204 | 272 |
| No | 146 | 124 | 128 | 128 | 145 | 161 | 201 | 159 |

(a) Set up the hypotheses to test whether the proportion of adults who favor banning smoking in fast-food restaurants is the same for all eight states.

(b) Calculate the overall proportion of adults who favor a ban on smoking in fast-food restaurants.

(c) Use the overall proportion to find the expected frequencies for each cell.

(d) Calculate the chi-square statistic.

(e) At the 0.05 level of significance, do the data indicate that the proportion of adults who favor banning smoking in fast-food restaurants is the same for all eight states?

**15.18** In the survey about using Office Suite software the researchers were interested in finding whether the proportion of people who installed the software themselves was different for different age groups. They created the contingency table shown below:

| Install Software | Age Group | | |
|---|---|---|---|
| | Under 25 | 25 to 44 | 45 and Up |
| Yes | 8 | 45 | 94 |
| No | 9 | 14 | 31 |

(a) Set up the hypotheses to test whether the proportion of people who install their own Office Suite software is the same for all three age groups.

(b) Use the data to calculate the overall proportion of people who install their own Office Suite software.

(c) Find the expected frequencies for each cell and calculate the chi-square statistic.

(d) At the 0.05 level of significance, what can you conclude about the proportion of people who install their own Office Suite software?

**15.19** A software company that was looking at the time to failure of the diskettes it uses decides to look at its two suppliers of the product. The specifications for the product state that the life of the diskette should be normally distributed with a mean of 500 hours and a standard deviation of 5 hours. The data for each of the suppliers are shown below:

| Supplier A | | | | | Supplier B | | | | |
|---|---|---|---|---|---|---|---|---|---|
| 474 | 492 | 498 | 504 | 511 | 487 | 492 | 495 | 497 | 499 |
| 486 | 492 | 500 | 505 | 511 | 488 | 492 | 495 | 497 | 499 |
| 489 | 494 | 501 | 506 | 512 | 489 | 492 | 495 | 497 | 499 |
| 490 | 494 | 501 | 507 | 513 | 489 | 492 | 495 | 497 | 501 |
| 490 | 494 | 501 | 508 | 514 | 489 | 493 | 495 | 498 | 502 |
| 490 | 495 | 502 | 508 | 515 | 491 | 493 | 496 | 498 | 503 |
| 491 | 496 | 502 | 509 | 517 | 491 | 494 | 496 | 498 | 503 |
| 491 | 498 | 504 | 509 | 519 | 491 | 494 | 496 | 498 | 505 |
| 491 | 498 | 504 | 510 | 528 | 492 | 494 | 496 | 499 | 506 |

(a) Create a frequency distribution and a histogram for each supplier.

(b) From the histogram, does it appear that the data for each supplier are normally distributed?

(c) Do a chi-square goodness of fit test for each supplier to see if it is reasonable to say that the diskettes meet the specifications. Use a level of significance of 0.05.

**15.20** As part of an annual survey by the Department of Transportation, randomly selected households are asked, among other questions, about the number of vehicles in the household and the availability of public transportation. The results for people who lived in urban areas are tabulated below:

| Public Transportation? | *Number of Vehicles* | | | | | |
|---|---|---|---|---|---|---|
| | **0** | **1** | **2** | **3** | **4** | **5+** |
| Yes | 974 | 2592 | 2427 | 687 | 194 | 59 |
| No | 98 | 378 | 540 | 199 | 58 | 23 |

(a) Set up the hypotheses to determine whether the number of vehicles in a household and the availability of public transportation are independent.

(b) How many degrees of freedom will the critical value for the test have?

(c) Calculate the expected frequencies and compute the test statistic.

(d) At the 0.05 level of significance, what can you conclude about the number of vehicles in a household and the availability of public transportation for households in urban areas?

**15.21** In a survey of adults who were currently smokers done in 1992 the following data were collected on gender and education level:

| | *Education Level* | | | |
|---|---|---|---|---|
| Gender | Less Than 12 Years | 12 Years | 13 to 15 Years | 16 Years or More |
| Male | 359 | 546 | 293 | 173 |
| Female | 364 | 662 | 392 | 175 |

(a) Set up the hypotheses to test whether the proportion of female smokers is the same for each level of education.

(b) Calculate the overall percentage of women who smoke.

(c) Calculate the expected frequency for each cell.

(d) At the 0.05 level of significance, what can you say about the proportion of women who smoke for the different levels of education?

## Thinking About It!

**15.22** One of the big controversies during the Vietnam War era was whether the draft lottery was truly random. Many people thought that the lottery was biased against people who were born in certain months of the year. The accompanying data show the number of birth dates in each month that were chosen in the first half of the draft lottery (there are 366 possible birth dates, so these are the first 183 birth dates that were chosen).

| Month | Number of Birth Dates in the First Half of the Draft | Month | Number of Birth Dates in the First Half of the Draft |
|---|---|---|---|
| January | 13 | July | 14 |
| February | 12 | August | 18 |
| March | 9 | September | 19 |
| April | 11 | October | 13 |
| May | 14 | November | 21 |
| June | 14 | December | 25 |

(a) Just by looking at the data, do you think that the claim was worth investigating? Why or why not?

(b) At the 0.05 level of significance do the data indicate that the distribution of the first 183 draft dates was something other than random? (*Hint:* Think about what random would mean in this situation.)

**15.23** Consider the company that is looking at the relationship between type of advertising media and age of customer. The data are shown again below:   *Requires Exercise 15.12*

| Type of Advertisement | Age of Customer 21–30 | 31–40 | 41–50 | Over 50 | Totals |
|---|---|---|---|---|---|
| Store display | 21 | 28 | 8 | 6 | 63 |
| Catalog | 8 | 5 | 1 | 1 | 15 |
| Magazine | 1 | 23 | 8 | 1 | 33 |
| Newspaper | 18 | 14 | 2 | 4 | 38 |
| Totals | 48 | 70 | 19 | 12 | 149 |

(a) Create a revised contingency table that collapses categories and fixes the problem of cells with expected frequencies less than 5.

(b) Perform the chi-square test for independence again.

(c) Compare the results of this test with the results of the test where categories were not collapsed.

**15.24** Consider the data from the problem about the diskette suppliers.   *Requires Exercise 15.19*

(a) Why is doing a chi-square goodness of fit test for each supplier to see if they meet specifications different than doing a test to see whether the mean for each supplier is 500 hours and a test to see if the variance for each supplier is 25 hours$^2$?

(b) Just because you reject the hypothesis that the data are normally distributed with a mean of 500 hours and a standard deviation of 5 hours, does that mean that the data are *not* normally distributed?

(c) Calculate the sample mean and standard deviation for each supplier from the data provided.

(d) Do a chi-square test for each supplier to see if the data are normally distributed using the sample means and standard deviations.

(e) How does this change the test procedure itself?

(f) How do the results of these tests compare to the results of the previous tests?

**15.25** Look at the results of the chi-square test on proportions for the data on smoking in fast-food restaurants.   *Requires Exercise 15.17*

(a) Prepare a plot of the proportion of adults who favor banning smoking in fast-food restaurants by state.

(b) When you reject $H_0$ you decide that at least one population (in this case state) is different. Use your plot to come to a more informative conclusion.

**15.26** Look at the data on the number of miles driven by the sample of Porsche drivers.   *Datafile: PORSCHE.XXX*

(a) Make a boxplot of the data.

(b) Do the data look normally distributed?

(c) Delete the outliers and replot the data. Did this change your opinion about the normality of the data? Why or why not?

(d) Calculate the mean and standard deviation of the data without the outliers.

(e) Perform a chi-square goodness of fit test to determine whether the data come from a normally distributed population with the calculated mean and standard deviation.

(f) What are your conclusions?

## Doing It!

**15.27** The Public Health student who was looking at the issue of college drinking administered a survey that asked questions on four different areas: drinking habits, problems related to drinking, problems related to other people's drinking, and living habits. A portion of the data from the survey and an explanation of the variables are given below:

| Five | Four | Three | Last | Binger | Hangover | Miss | Behind | Regret | Forget |
|---|---|---|---|---|---|---|---|---|---|
| 0 | 0 | 0 | 2 | 1 | 1 | 1 | 1 | 1 | 1 |
| 0 | 0 | 0 | 4 | 1 | 1 | 1 | 1 | 1 | 1 |
| 0 | 1 | 1 | 3 | 2 | 1 | 1 | 1 | 1 | 1 |
| 0 | 0 | 0 | 3 | 1 | 1 | 1 | 1 | 1 | 1 |
| 0 | 0 | 0 | 3 | 1 | 1 | 1 | 1 | 1 | 1 |
| 0 | 0 | 0 | 4 | 1 | 1 | 1 | 1 | 1 | 1 |

| Argue | Engage | Protect | Damage | Police | Injured | Medical | Drive | DWI | Ride |
|---|---|---|---|---|---|---|---|---|---|
| 1 | 1 | 1 | 1 | 1 | 1 | 1 | 1 | 1 | 1 |
| 1 | 1 | 1 | 1 | 1 | 1 | 1 | 1 | 1 | 1 |
| 1 | 1 | 1 | 1 | 1 | 1 | 1 | 1 | 1 | 1 |
| 1 | 1 | 1 | 1 | 1 | 1 | 1 | 1 | 1 | 1 |
| 1 | 1 | 1 | 1 | 1 | 1 | 1 | 3 | 1 | 3 |
| 1 | 1 | 1 | 1 | 1 | 1 | 1 | 1 | 1 | 1 |

| Age | Gender | Fulltime | Greek | Live | Roommate | Race | Class |
|---|---|---|---|---|---|---|---|
| 18 | 1 | Y | N | 4 | P | 2 | 1 |
| 62 | 2 | Y | N | 4 | S | 1 | |
| 18 | 2 | Y | N | 1 | R | 1 | 1 |
| 31 | 1 | Y | N | 4 | R | 1 | |
| 19 | 2 | Y | N | 4 | P | 1 | 2 |
| 20 | 2 | Y | N | 1 | R | 1 | 2 |

For the purposes of this survey a "drink" means any of the following:

12-ounce can or bottle of beer
4-ounce glass of wine
12-ounce bottle or can of wine cooler
1 oz (shot) of liquor straight or in a mixed drink

Binge drinking is defined as

The consumption of five or more drinks in one episode of drinking for males.
The consumption of four or more drinks in one episode of drinking for females.

The variables *Five, Four,* and *Three* refer to the two-week period just before the survey was administered and answer the questions indicated. They are coded as follows:

0   None
1   Once
2   Twice
3   Three to five times
4   Six to nine times
5   Ten or more times

- *Five*   "How many times have you had five or more drinks in a row?"
  *Four*   "How many times have you had four drinks in a row (but no more than that)?"
  *Three*  "How many times have you had three drinks in a row (but no more than that)?"

- The variable *Last* answers the question When did you have your last drink? The answers are coded as follows:

0   I never had a drink
1   Not in the past year
2   More than 30 days ago, but less than a year ago
3   More than one week ago, but less than 30 days ago
4   Within the last week

- The variable *Binger* is defined as the type of drinker that a student is based on his or her answer to the first four questions.

0   Nondrinker
1   Non-binge drinker
2   Infrequent binge drinker
3   Frequent binge drinker

The next 15 variables answer questions that start with the statement "Since the beginning of the school year, how often has your drinking caused you to. . . " The answers are coded as follows:

0   Not at all
1   Once
2   Twice or more

- *Hangover*   ". . . have a hangover?"
- *Miss*   ". . . miss a class?"
- *Behind*   ". . . get behind in school work?"
- *Regret*   ". . . do something you later regret?"
- *Forget*   ". . . forget where you were or what you did?"
- *Argue*   ". . . argue with friends?"
- *Engage*   ". . . engage in unplanned sexual activity?"
- *Protect*   ". . . not use protection when you had sex?"
- *Damage*   ". . . damage property?"
- *Police*   ". . . get into trouble with campus or local police?"
- *Injured*   ". . . get hurt or injured?"
- *Medical*   ". . . require medical treatment for an alcohol overdose?"
- *Drive*   ". . . drive after drinking alcohol?"
- *DWI*   ". . . drive after having five or more drinks?"
- *Ride*   ". . . ride with a driver who was high or drunk?"

- The variable *Age* is the age of the student.

- The variable *Gender* is the gender of the student and is coded as

1   Male
2   Female

- The variable *Fulltime* is Y if the student is full-time and N if the student is part-time.

- The variable *Greek* is Y if the student is a member of a fraternity or sorority and N if not.

- The variable *Live* records where the student lives and is coded as follows:

1   Residence hall or dormitory
2   Fraternity or sorority house
3   Other university housing
4   Off-campus house or apartment

- The variable *Roommate* describes with whom the student lives and is coded as follows:

A   Alone
R   Roommate(s) or housemate(s)
S   Spouse
P   Parent(s) or other relative(s)
O   Significant other
C   Children

- The variable *Race* describes the race or ethnic category of the student and is coded as

  | Blank | Hispanic (separate question) |
  |---|---|
  | 1 | White |
  | 2 | Black/African American |
  | 3 | Asian/Pacific Islander |
  | 4 | Native American/Native Alaskan |
  | 5 | Other |

- The variable *Class* describes the student's year at the university and is coded as

  | 1 | Freshman |
  |---|---|
  | 2 | Sophomore |
  | 3 | Junior |
  | 4 | Senior |

(a) Look at the variable *Drive,* which refers to whether a student drove after drinking. Create a contingency table for this variable and the variable *Binger.*

(b) At the 0.05 level of significance, is whether a student drove after drinking independent of the type of drinker he or she is?

(c) If you exclude nondrinkers and look at only those students who drink, does the answer change?

(d) From the variables *Live* and *Roommate* derive a way to classify students as residents/commuters and to separate commuters living at home from commuters living with nonrelatives.

(e) Create contingency tables that will allow you to look at the relationship between how often a student missed class *(Miss)* and resident/commuter students. Would you expect these two variables to be related? Why or why not?

(f) Do a chi-square test to determine whether these two variables are related.

(g) Look at the variables *Miss* and *Gender* and do the same analysis.

(h) Look at the variables *Binger* and *Age.* Is there a relationship between them?

(i) From the variable *Age* create a new variable that classifies students as above and below the legal drinking age (assume that it is age 21 in the state of interest). How is this new variable related to the type of drinker that a student is? to *Drive?* to *Miss?*

(j) Is there a relationship between *Race* and *Binger?*

(k) Consider all of the other variables relating to behavior after drinking. Determine whether these variables are related to the type of drinker that a student is.

(l) Consider any other variables and relationships that you think are important and investigate them. Prepare a report for the university administration that describes the drinking behaviors of the students at the university.

**15.28** The Chamber of Commerce that is studying the credit problems of small businesses asked the businesses 3 questions to classify their business and 7 questions related to the issue of credit problems. A portion of the datafile and an explanation of each variable are given below:

| Size | Employees | Nature | Problem | Understd | Concern | Call | Loan | Collateral | Access |
|---|---|---|---|---|---|---|---|---|---|
| 2 | 2 | 1 | 1 | 2 | 1 | 2 | 2 | 0 | 2 |
| 1 | 2 | 3 | 1 | 2 | 2 | 0 | 2 | 2 | 1 |
| 4 | 3 | 1 | 2 | 1 | 2 | 2 | 2 | 2 | 0 |
| 1 | 1 | 1 | 2 | 1 | 2 | 2 | 2 | 2 | 2 |
| 1 | 2 | 1 | 2 | 0 | 0 | 0 | 0 | 0 | 0 |
| 3 | 1 | 5 | 2 | 0 | 2 | 2 | 2 | 2 | 1 |
| 3 | 3 | 2 | 2 | 1 | 1 | 2 | 2 | 2 | 1 |
| 2 | 2 | 3 | 2 | 1 | 2 | 2 | 2 | 2 | 2 |
| 3 | 5 | 2 | 2 | 1 | 2 | 2 | 2 | 1 | 2 |

- The variable *Size* refers to the annual sales of the company and is coded as follows:

  | 1 | Under $1 million | 4 | $11–20 million |
  |---|---|---|---|
  | 2 | $1–5 million | 5 | Over $20 million |
  | 3 | $6–10 million | | |

- The variable *Employees* refers to the number of employees that the company currently employs. This variable was coded as

  | | |
  |---|---|
  | 1 | 0–5 employees |
  | 2 | 6–10 employees |
  | 3 | 11–50 employees |
  | 4 | 51–150 employees |
  | 5 | 151–250 employees |
  | 6 | Over 250 employees |

- The variable *Nature* refers to the type of business and is coded as

  | | |
  |---|---|
  | 1 | Manufacturing |
  | 2 | Retail |
  | 3 | Service |
  | 4 | Real Estate |
  | 5 | Other |

The next seven variables contain the response to the questions or statements indicated and are coded as follows:

| | |
|---|---|
| 1 | Yes |
| 2 | No |

- *Problem*     "Are you experiencing credit-related problems?"

- *Understd*     "The bank understands my problems."

- *Concern*     "I am concerned that my note might be recalled."

- *Call*     "The bank is planning to recall my loan."

- *Loan*     "The bank has called my loan."

- *Collateral*     "The bank has demanded more collateral."

- *Access*     "Access to credit is affecting my business."

(a) Look at the variable *Nature,* which refers to the type of business. Is the sample of businesses that were surveyed by the Chamber of Commerce uniformly distributed over the different types of companies?

(b) Answer the question posed in part (a) for *Size* and *Employees.*

(c) Create a contingency table for the variables *Problem* and *Nature.* Is the proportion of companies that are having credit-related problems different for the different types of companies?

(d) Answer the same question for the variable *Nature* and each of the variables *Understd* and *Access.*

(e) Investigate the relationship between the variables *Size* and *Access* and *Size* and *Problem.*

(f) Investigate the relationship between the variables *Employees* and *Access* and *Employees* and *Problem.*

(g) Is there a relationship between the size of the company and the type of business that the company does?

(h) Is there a relationship between the size of the company and the number of employees in the company?

(i) Look at the variables related to loan recall, *Concern, Call,* and *Loan.* Investigate whether these variables are related to *Size, Employees,* and *Nature.*

(j) Look at any other relationships that you think might be useful to the Chamber of Commerce.

(k) Prepare a report for the Chamber of Commerce about your findings.

# CHAPTER 16

# IMPROVING AND MANAGING QUALITY

## THE PLASTICS COMPANY

One of the quality characteristics of plastic is the melt flow (min) at 200°F. This measures the viscosity of the liquid. The plastic is poured into molds to make razors, pens, disposable lighters, and many other familiar products. The melt flow indicates how the plastic will fill the molds. The plastics company monitors the process that manufactures the plastic by regularly measuring the melt flow and plotting the data on a control chart. By analyzing the data, the company determines if the plastic is ready for sale. If the plastic does not have the correct melt flow, then air bubbles may form when the plastic is poured into a mold, causing underfilling or overfilling of the mold.

The plastics company sells its product to other manufacturers to be used in making such consumer products as automobile bumpers and pens. The company that purchases the plastic is also concerned about the melt flow of the material it is purchasing and so it checks a sample of the raw material. The melt flow results for five tests done by the manufacturer and the customer are shown in the portion of the dataset displayed below:

| Manufacturer | Customer |
|---|---|
| 10.2 | 8.7 |
| 9.4 | 9.2 |
| 9.5 | 9.2 |
| 9.4 | 8.8 |
| 9.9 | 9.1 |

Clearly, the manufacturer and the customer are not getting the same results and something must be done to resolve these differences.

## 16.1   CHAPTER OBJECTIVES

The story of the plastics company detailed above is an example of the increased attention being paid to quality by manufacturers and customers. This book has repeatedly emphasized the importance of data analysis as a tool for improving quality both in the manufacturing environment and more recently in service industries, such as health care, and the entertainment industry. In the face of increased competition, companies must continuously monitor and improve the products and services they are marketing. To do this data must be collected, analyzed, and used to improve the products or services being sold. This idea is known as continuous improvement.

There are many definitions of the word quality, but in this chapter we will focus on the idea that a quality product or service should have the same characteristics regardless of when it was made, what shift it was made on, who was working, or which day it is. That is, the product or service should have predictable characteristics. Control charts are one of a group of tools known as the "7 Tools of Quality," which are used to deliver products or services that are consistent.

In particular, this chapter will cover:

- Supplier/Customer Relations—A New Paradigm
- The Language of Control Charts
- Types of Control Charts
- Individual or *x*-charts
- Xbar- and *R*-charts
- The *p*-chart

## 16.2   SUPPLIER/CUSTOMER RELATIONS—A NEW PARADIGM

As a result of the quality movement, the relationship between supplier and customer has changed dramatically. Formerly, the job of the purchasing agent was to purchase raw materials as inexpensively as possible without being much concerned with quality. The performance of the purchasing agent was judged on how inexpensively he/she was able to purchase raw material. But we all know the old adage "garbage in, garbage out." The quality movement has helped companies see the importance of changing this paradigm and treating the steps involved in bringing the product to market as a **process.**

> A *process* is a series of actions that changes inputs to outputs.

You are familiar with many processes. The process of getting ready for class involves all the steps you take to change yourself into a student prepared for a day of school. This might involve showering, packing the needed books and notebooks into your backpack, completing the homework due that day, and grabbing your keys and ID so you can eat and get back into your room!

Manufacturers use processes to turn raw materials into products. Sometimes those products are products that we buy directly, such as a car or a computer. Sometimes they are products that are sold to another company to be used to manufacture another item that we may buy directly or might in turn be sold to another manufacturer. The plastic made by the company described in the opening scenario, for example, is manufactured and then sold to another company to be made into car bumpers that are then sold to car manufacturers.

You may be familiar with more processes than you think. Services such as getting admitted to a hospital are processes, too; the way you sign up for courses for the upcoming semester is a process; and the way your paycheck is created is a process.

Certainly, if the inputs to the process are not of high quality then the outputs will not be either. This realization has changed the way companies work with suppliers. In particular, the job of purchasing is now to build partnerships with suppliers. This idea of a partnership means that the company will show some loyalty to the supplier and not change suppliers month to month based solely on price and the supplier will in turn work with the company to improve the quality of the inputs. In many cases, purchasing managers visit companies to see the quality initiatives that the supplier has in place. Both sides can benefit from the team approach. The supplier has a more committed customer and the customer can trust the quality of the inputs.

# 16.3  THE LANGUAGE OF CONTROL CHARTS

Two important points need to be made about processes. First, no two items produced by a process are the same. And second, variability is to be expected in the output of all processes. What does this mean? It means that no matter how hard we try, we cannot completely eliminate variation. So, no two sheets of plastic will be exactly the same, no two golf balls will be exactly the same, no two tissue boxes will be exactly the same, and no two experiences at a hospital check-in will be exactly the same. However, we can identify the sources of variation.

If you think about where the variation in a product or service comes from you might identify the following sources: people, machines, materials, methods, measurement, and the environment. These are considered to be the 6 major sources of variation. A **control chart** is a graphical tool used to monitor process variation. Walter Shewhart of Bell Laboratories originally developed control charts during the 1920s.

> A *control chart* is a graph that allows us to study variation in a product or service.

## EXAMPLE 16.1  The Soup Company

*Sources of Variation*

Suppose you are filling cans of soup and there should be 10.5 oz of soup in each can. Consider each of the 6 sources of variation for the process of filling the cans with soup.

*People:* Different operators of the machines might do the filling.

*Machines:* Different machines might be used for the process.

*Materials:* Different ingredients might be used to make the soup.

*Methods:* The way in which the cans are filled might differ.

*Measurement:* More than one scale might be used to check the ounces.

*Environment:* The temperature of the plant might vary from day to day.  ■

## TRY IT NOW!

### Coffee and a Bagel

*Identifying Possible Sources of Variation*

Suppose you go to a donut shop for coffee and a bagel. Consider the service and quality of your food. Using the 6 categories of sources of variation, identify where variation might come from.

### 16.3.1 Changes in the Distribution of a Variable

To get a better handle on describing variation and determining the causes of the variation we need to rely on some of the concepts developed in Chapters 6 and 7. Consider the process of filling the cans of soup again. The variable of interest is the number of ounces in the can. The values that this variable can assume and their corresponding likelihood can be described by a probability distribution. We studied the normal distribution in Chapter 6 and learned that many quantitative variables can be described by this bell shaped distribution. Any single observation of the variable, such as the ounces of soup in one can, is governed by the underlying distribution. The distribution can change.

Figure 16.1 shows the number of ounces in cans of soup. The measurements are taken every hour. There are four different scenarios depicted in this figure: (a) the underlying distribution is not changing over time; (b) the mean of the distribution is increasing over time; (c) the variance of the distribution is increasing; and (d) there was a shift in the mean of the distribution at one point in time.

(a) Distribution not changing

(b) Mean of the distribution is increasing

(c) Variance of the distribution is increasing

(d) Shift in the mean at time 7

**FIGURE 16.1**   Different possibilities for changes in the distribution of the variable

Control charts help us detect situations (b) – (d). You can see by looking at the graphs that it is not easy to do. If the normal distributions were not displayed it would be difficult to see the changes that have occurred. This is because you cannot eliminate variation. Therefore, it is difficult to tell if the variation you are seeing in the variable is what is reasonable to expect or if something has happened to the process that needs attention.

> A process whose distribution does not change over time is said to be in a state of *statistical control* or *in control.* If it does change over time, it is said to be *out of statistical control* or *out of control.*

Figure 16.1b shows what the observations could look like if the mean of the process, in this case the average amount of soup going into the cans, was gradually increasing. This could happen as a machine part wears out. Figure 16.1c shows what the observations could look like if the variance in the process increases, causing the soup amount dispensed into the can to vary more. This could happen if the person running the machine or providing the service gets tired or fatigued. Figure 16.1d shows what the observations might look like if there was a change in the raw materials at some point in time. In this case there is not a gradual shifting of the mean but a jump in the mean, perhaps due to a different batch of raw materials or a new operator.

## 16.3.2 Types of Variation

As you can see from Figure 16.1a, the output of a process that is in control still exhibits variation but there is no pattern to the data. The data exhibit random behavior. Such variation is said to be due to **common causes of variation.**

> *Common causes of variation* are the methods, machines, people, and environment that are an inherent part of a process. They are part of the design of the process.

To reduce variability in a stable process, you must eliminate some common cause of variation by changing the process in some way. This is the job of management.

Out of control processes exhibit variation that is the result of both common causes and **special causes of variation.**

> *Special causes of variation* are events or actions that are not part of the process design.

Control charts are one of the tools used to detect the presence of special causes of variation.

### EXAMPLE 16.2    The Plastics Company

*Common and Special Causes of Variation*

Consider the plastics company. Possible common causes of variation in the melt flow of the plastic might be poor lighting in the plant or inconsistent raw materials. Possible special causes of variation might be the accidental setting of the controls of a machine incorrectly or a negligent supplier.

### 16.3.3 Exercises—Learn It!

**16.1**   A fast-food chain monitors the amount of time it takes between the customer placing the order and the customer receiving the order. Identify some potential sources of variation in this time variable.

**16.2**   The percentage of free throw shots that a player makes varies from game to game. Identify some causes of variation.

**16.3**   The diameter of a golf ball varies from ball to ball. Identify some causes of variation.

**16.4**   The strength of a tissue varies from tissue to tissue. Identify some causes of variation.

**16.5**   The average grade in an introductory statistics course varies from semester to semester. Identify some causes of variation.

## 16.4   TYPES OF CONTROL CHARTS

In the opening problem statement at the beginning of the chapter, you read that the machine operators plot data on control charts. There are several different types of control charts. As we have emphasized in this book on many occasions, the type of data you have dictates the tools that you can use. The same is true for control charts. The type of control chart you use depends on whether you have quantitative or qualitative data and also on the sample size you use each time you collect data.

The following decision tree appeared in the February 1998 issue of *Quality Digest*. The article was entitled "When in Doubt, Get the *x*-chart Out" by Thomas Pyzdek. The decision tree gives us a nice way to look at the types of control charts and how to decide which type to use.

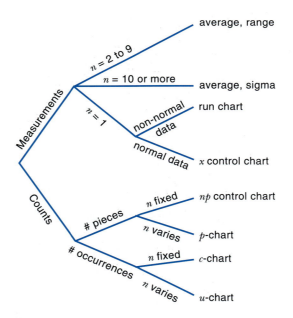

The first decision you need to make is whether you have quantitative (measurement) data or qualitative (counts) data. Following the quantitative data branch at the top of the tree, your next decision is with regard to the sample size that is used each time data are collected. If you are observing single observations each time you collect data, then you must decide if you have normal or non-normal data. The next two sections cover *x*-charts, Xbar-charts, and *R*-charts (referred to as "average, range" in the tree diagram).

Following the bottom branch of the decision tree when you have qualitative data, the next decision point involves deciding if you are counting the number of units that have a defect or the number of defects per unit. In the first case a unit is either defective or not. For instance, there is either an error in the bill or not. The latter situation occurs when you are counting, for instance, the number of errors in each bill. Once you have established which situation you have, then you must decide if your sample size is fixed or if it varies from time to time. The last section of this chapter covers *np* and *p* control charts. This book does not cover the other types of control charts, but they are based on the same concepts.

# 16.5  INDIVIDUAL OR *x*-CHARTS

In this section we learn about individual or *x*-charts. They are called individual charts because individual values ($n = 1$) of the variable are plotted. The theory of the individual control chart is based on the normal distribution.

## 16.5.1 Constructing an *x*-chart

An **x-chart** is a graph that plots the sample values in chronological order along the *x* axis. There are three lines on a control chart that tell the operator whether the process is in control.

> An **x-chart** is a control chart used to monitor individual observations of a variable over time. Time is plotted on the *x* axis and the quality characteristic being monitored is plotted on the *y* axis. The graph also has 3 lines drawn parallel to the *x* axis. These are called the center line, the upper control limit, and the lower control limit.

These lines are constructed using the following steps:

**Step 1:** The center line (CL) of the chart is drawn at the theoretical or desired value for the mean of the distribution, $\mu$.

*Steps for constructing an x-chart*

**Step 2:** The upper control limit (UCL) of the chart is a horizontal line drawn at a value that is three standard deviations above the center line (called the 3 sigma limit), $\mu + 3\sigma$.

**Step 3:** The lower control limit (LCL) of the chart is a horizontal line drawn at a value that is three standard deviations below the center line (called the 3 sigma limit), $\mu - 3\sigma$.

## EXAMPLE 16.3  The Plastics Company

*Finding CL, UCL, and LCL*

Suppose the desired value for the mean melt flow is $\mu = 9.5$ and the standard deviation is 0.60. The center line (CL) is drawn parallel to the *x* axis at the value $y = 9.5$. The upper control limit (UCL) of the chart is drawn parallel to the *x* axis at the value $y = 9.5 + 3(0.6) = 11.3$. The lower control limit (LCL) of the chart is drawn parallel to the *x* axis at the value of $y = 9.5 - 3(0.6) = 7.7$. These lines are shown in the graph at the top of page 814.

Once the upper and lower control limits have been established, the observations of the quality characteristic, such as the melt flow, are plotted on the graph. If the process is working properly, that is, if the only variation is due to natural variation, then the observations should fall between the upper and lower control limits with no discernable pattern.

The completed $x$-chart for the melt flow is shown in the next example.

### EXAMPLE 16.4   The Plastics Company

#### $x$-Chart for Melt Flow

## TRY IT NOW!

### The Plastics Company
#### $x$-Chart for the Customer Data

Use the same upper and lower control chart limits but plot the first 10 melt flows that the customer observed. They are as follows:

|       |     |     |     |     |     |     |     |     |     |
|-------|-----|-----|-----|-----|-----|-----|-----|-----|-----|
| 8.7   | 9.2 | 9.2 | 8.8 | 9.1 | 8.8 | 9.2 | 9.4 | 8.6 | 8.3 |

ANS.

In the *Try It Now* that you just completed, you should see that all the melt flows fall between the upper and lower limits, and that they all fall below the center line. This is a pattern that cannot be ignored. To resolve the differing results the customer and the manufacturer will have to engage in conversation and utilize another of the "7 Tools of Quality" known as a fishbone diagram. The fishbone diagram is a tool to examine the major and secondary causes of a problem. It gets its name from the fact that it looks like the skeleton of a fish. Since the intent of this chapter is to introduce the basic idea of control charts, we will leave fishbone diagrams to be covered in a course on quality management.

## 16.5.2 False Alarms

You might wonder why the control limits are set 3 standard deviations above and below the mean. Would it be better to set the limits at 2 standard deviations? The answer to this question lies in our understanding of the normal distribution.

We know from our study of the normal distribution (see Figure 7.6) that 99.73% of all values should fall within 3 standard deviations of the target or population mean. Thus, the probability that a value will fall outside the control limits is 0.0027 or 27 in 10,000 if the process is working properly. If we observe a value outside the limits, then the process should be examined, because it is highly unlikely that we would see values outside these limits if the process is working properly. It may seem to you that we would be better off setting the limits at only 2 standard deviations above and below the target mean. Then we would detect problems sooner. That part of your thinking is correct but there is a trade-off.

The flip side of this issue is that no matter how small, there is some nonzero chance (in this case 0.0027) of observing values outside the limits when the process really is working properly. Using our control chart, we would flag these observations as unusual and conclude that there was some special cause of the variation. It is possible, however, that there is nothing unusual happening and that it was just chance (albeit a small chance) that the values would be outside the control chart limits. This is known as a false alarm. If we were to drop the limits to 2 standard deviations above and below the mean, we would increase the chance of a false alarm to 0.0456 or 4.56%. This is the probability of observing values above or below two standard deviation limits if the mean has not shifted. It is generally felt that 4.56% is too great a chance of a false alarm, although clearly the cost of making a false alarm versus the cost of missing a problem should be factored into the decision.

## 16.5.3 Exercises—Learn It!

**16.6**   A manufacturer of golf balls monitors the weight of its golf balls. Suppose the target weight is 45 g and the target standard deviation is 0.15 g.

(a) Construct the *x*-chart and plot the following observations on the chart:

45.3    45.2    45.2    45.3    45.5    45.3    45.2    45.1    45.3    45.2    45.3

(b) Are any of the points out of control?

**16.7**   A soda machine is supposed to dispense an average of 8 oz of soda with a standard deviation of 0.03 ounce into cups. A representative from the company checks the machines in use every 2 weeks. The following data show the actual soda dispensed over the last 25 weeks.

| | | | | |
|---|---|---|---|---|
| 8.09 | 8.01 | 7.98 | 7.97 | 7.94 |
| 8.07 | 8.01 | 7.98 | 7.97 | 7.92 |
| 8.06 | 8.00 | 7.98 | 7.96 | 7.91 |
| 8.06 | 7.99 | 7.98 | 7.95 | 7.89 |
| 8.05 | 7.99 | 7.98 | 7.95 | 7.89 |

(a) Construct the *x*-chart for the soda machine company.

(b) Plot the observations on the chart.

(c) Are any points out of control?

16.8 You are interested in monitoring the number of minutes you can work out before you get fatigued. You think you are increasing your stamina but are not sure if what you see in the data is just natural variation. You have recorded the following number of minutes for your workout over the past 30 days. The first column of data shows the first 10 days, the second column the next 10 days, and the third column the last 10 days.

| | | |
|---|---|---|
| 61.54 | 51.13 | 66.37 |
| 66.02 | 65.29 | 78.42 |
| 62.08 | 58.77 | 73.83 |
| 55.20 | 59.46 | 66.32 |
| 49.93 | 66.15 | 78.80 |
| 55.76 | 64.89 | 68.86 |
| 58.32 | 69.12 | 77.45 |
| 60.55 | 72.04 | 76.73 |
| 53.83 | 69.11 | 68.48 |
| 55.53 | 69.95 | 64.13 |

(a) If your average has been 60 minutes with a standard deviation of 5 minutes, construct the *x*-chart for these data.

(b) Does it appear that your process is in control?

(c) If not, does that mean you are increasing your stamina?

16.9 A company that manufactures tissues is concerned about the cross-directional strength of its tissues. The target mean is 450 lb/ream with a standard deviation of 25 lb/ream. The following data were taken over 15 hours:

| | | |
|---|---|---|
| 422 | 464 | 437 |
| 448 | 472 | 396 |
| 423 | 489 | 432 |
| 435 | 440 | 475 |
| 445 | 441 | 439 |

(a) Construct the *x*-chart for the tissue company.

(b) Plot the observations on the chart.

(c) Are any points out of control?

16.10 A mail-order company monitors the amount of time that customers are on hold before being serviced. The company randomly monitors one call every hour. The target on-hold time is 1 minute with a standard deviation of 0.5 minute. The following on-hold times (in minutes) were recorded for one day:

| | |
|---|---|
| 0.82 | 0.57 |
| 0.60 | 0.78 |
| 1.22 | 0.03 |
| 1.31 | 0.81 |
| 0.84 | 0.99 |
| 0.77 | 0.45 |

(a) Construct the *x*-chart for the mail-order company.

(b) Plot the observations on the chart.

(c) Are any points out of control?

# 16.6 XBAR- AND *R*-CHARTS

Although the individual *x*-charts developed in the previous section are useful, it is more common to monitor the average of a sample of observations. These are called

Xbar-charts and they are used to detect the presence of special causes of variation. The Xbar-chart is almost always used with the *R*-chart and so they are presented together in this section.

## 16.6.1 Xbar-charts

The **Xbar-chart** is constructed the same way that an *x*-chart is constructed: There is a center line and an upper and lower control limit. Instead of plotting individual observations ($n = 1$) on the *y* axis, the average of *n* observations is plotted over time. So the variable being monitored is $\overline{X}$.

> The *Xbar-chart* is a control chart used to monitor the mean of a variable over time. Time is plotted on the *x* axis and the sample means are plotted on the *y* axis. The graph also has 3 lines drawn parallel to the *x* axis. These are called the center line, the upper control limit, and the lower control limit.

From our study of the central limit theorem (Chapter 7) we know that the standard deviation of $\overline{X}$ is called the standard error and is labeled $\sigma_{\overline{X}}$. The standard error is $\sigma/\sqrt{n}$. We also know that the mean of $\overline{X}$ is labeled $\mu_{\overline{X}}$ and is the same as the population or target mean, $\mu$. Using this information, we arrive at the following steps for constructing an Xbar-chart:

**Step 1:**  The center line (CL) of the chart is drawn at the theoretical or desired value for the mean of the distribution, $\mu$.

*Steps for constructing an Xbar chart*

**Step 2:**  The upper control limit (UCL) of the chart is a horizontal line drawn at a value that is three standard deviations above the center line (called the 3 sigma limit), $\mu + 3\,\sigma/\sqrt{n}$.

**Step 3:**  The lower control limit (LCL) of the chart is a horizontal line drawn at a value that is three standard deviations below the center line (called the 3 sigma limit), $\mu - 3\,\sigma/\sqrt{n}$.

The values for $\mu$ and $\sigma$ come from the specifications of the manufacturing process or they may be estimated by $\overline{X}$ and *s* from sample data. You might remember the diaper company that we looked at in Chapter 7. The company sampled 5 diapers every hour and measured the diaper weight in grams. Let's see how to set up the Xbar-chart for the diaper company.

## EXAMPLE 16.5  The Diaper Company

*Calculation of Upper and Lower Control Limits for Xbar-Chart*

For the diaper weight the target mean is 55 g with a target standard deviation of 0.55 g. Using these values in the formulas for UCL and LCL, you get

$$\text{UCL} = 55 + (3)(0.55)/\sqrt{5}$$
$$= 55 + 0.738 = 55.738 \text{ g}$$

and

$$\text{LCL} = 55 - (3)(0.55)/\sqrt{5}$$
$$= 55 - 0.738 = 54.262 \text{ g}$$

The chart setup looks like this:

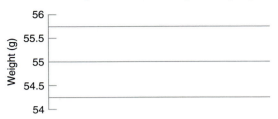

**Setup for Xbar–The Diaper Company**

## EXAMPLE 16.6   The Diaper Company

***Control Chart for Average Diaper Weight***

Plotting the sample averages for the weight for 52 hours gives you the following chart:

**Xbar-chart for the Diaper Company**

Clearly none of the sample means are outside the control limits in this case.     ■

Sometimes we do not have the values of $\mu$ and $\sigma$ from the specifications or we wish to develop the control chart independently of these values. In this case we use the overall mean, $\overline{\overline{X}}$, for the center line. The upper and lower control limits are calculated as follows:

*Steps for constructing Xbar chart when $\mu$ and $\sigma$ are unavailable*

$$\text{UCL} = \overline{\overline{X}} + A_2\overline{R}$$
$$\text{Center line} = \overline{\overline{X}}$$
$$\text{LCL} = \overline{\overline{X}} - A_2\overline{R}$$

*Remember that the sample range is the Max − Min.*

where $\overline{R}$ is the average of the sample ranges, and $A_2$ is a constant that depends on the sample size, $n$.

The value of $A_2\overline{R}$ is an estimate of $(3\sigma/\sqrt{n})$. You can see this when you know that

$$\hat{\sigma}_{\overline{X}} = \frac{\overline{R}}{d_2\sqrt{n}} \quad \text{and} \quad A_2 = \frac{3}{d_2\sqrt{n}}$$

so

$$A_2\overline{R} = \frac{3}{d_2\sqrt{n}}\,\overline{R} = 3\left(\frac{\overline{R}}{d_2\sqrt{n}}\right) = 3\hat{\sigma}_{\overline{X}} = 3\text{ sigma}$$

Table 16.1 shows the value of $A_2$ for a variety of different sample sizes, $n$. A complete table of $A_2$ values is shown in Table 7 of the Appendix.

**TABLE 16.1  Values of A₂**

| n | $A_2$ |
|---|-------|
| 2 | 1.88 |
| 3 | 1.023 |
| 4 | 0.729 |
| 5 | 0.577 |
| 6 | 0.483 |
| 7 | 0.419 |
| 8 | 0.373 |
| 9 | 0.337 |

The use of $A_2\overline{R}$ as an estimate of $(3\sigma/\sqrt{n})$ was developed to simplify the calculations. It is easier to find the range of a sample than the standard deviation. Despite the fact that the need for ease of calculation is considerably less important with the technology that we have today, these formulas continue to be favored by industry. The following example gives the observations for the first 3 samples of diaper weights. Remember that each sample is of $n = 5$ observations. The sample average and range are shown for each of these samples and the values of $\overline{\overline{X}}$ and $\overline{R}$ are shown if we were to construct the control chart based on only these 3 samples.

## EXAMPLE 16.7  The Diaper Company

### Limits for Xbar-chart from the Data

In this example the upper and lower control chart limits will be constructed just from these 3 samples so you can see where the values come from. The value of $A_2$ is 0.577 since $n = 5$.

| Sample | Diaper | Weight | Sample | Diaper | Weight | Sample | Diaper | Weight |
|--------|--------|--------|--------|--------|--------|--------|--------|--------|
| 1 | 1 | 55.87 | 2 | 1 | 54.85 | 3 | 1 | 54.40 |
| 1 | 2 | 55.35 | 2 | 2 | 54.79 | 3 | 2 | 56.44 |
| 1 | 3 | 54.50 | 2 | 3 | 54.65 | 3 | 3 | 54.11 |
| 1 | 4 | 53.97 | 2 | 4 | 55.56 | 3 | 4 | 54.67 |
| 1 | 5 | 54.29 | 2 | 5 | 55.82 | 3 | 5 | 54.56 |

| | | | | | | | | |
|---|---|---|---|---|---|---|---|---|
| Average | | 54.80 | | | 55.13 | | | 54.84 |
| Range | | 1.90 | | | 1.17 | | | 2.33 |
| Grand average | | 54.92 | | | | | | |
| Average range | | 1.80 | | | | | | |

$$\text{UCL} = \overline{\overline{X}} + A_2\overline{R} = 54.92 + (0.577)1.80 = 55.96$$
$$\text{Center line} = \overline{\overline{X}} = 54.92$$
$$\text{LCL} = \overline{\overline{X}} - A_2\overline{R} = 54.92 - (0.577)1.80 = 53.88$$ ∎

Consider a different example. The sleeve shown in Figure 16.2 on page 820 contains three inside diameter key characteristics. They are all machined on the same lathe but with different tools. Each inside diameter is a different size. The customer requires stability of the lathe process.

The time required to manufacture a sleeve is 3 minutes. The measurements are taken every hour from three consecutive sleeves. The following table shows the data collected for 9 samples.

| Sample | Sleeve | Characteristic | | |
|--------|--------|--------|--------|--------|
| | | **a** | **b** | **c** |
| 1 | 1 | 0.8500 | 1.1494 | 1.2494 |
| 1 | 2 | 0.8512 | 1.1494 | 1.2489 |
| 1 | 3 | 0.8501 | 1.1491 | 1.2494 |
| 2 | 1 | 0.8507 | 1.1498 | 1.2494 |
| 2 | 2 | 0.8498 | 1.1491 | 1.2488 |
| 2 | 3 | 0.8510 | 1.1499 | 1.2497 |
| 3 | 1 | 0.8509 | 1.1496 | 1.2494 |
| 3 | 2 | 0.8507 | 1.1498 | 1.2490 |
| 3 | 3 | 0.8506 | 1.1497 | 1.2496 |
| 4 | 1 | 0.8504 | 1.1499 | 1.2849 |
| 4 | 2 | 0.8502 | 1.1496 | 1.2943 |
| 4 | 3 | 0.8512 | 1.1497 | 1.2500 |
| 5 | 1 | 0.8512 | 1.1500 | 1.2490 |
| 5 | 2 | 0.8499 | 1.1498 | 1.2496 |
| 5 | 3 | 0.8508 | 1.1497 | 1.2497 |
| 6 | 1 | 0.8511 | 1.1498 | 1.2493 |
| 6 | 2 | 0.8507 | 1.1501 | 1.2497 |
| 6 | 3 | 0.8505 | 1.1498 | 1.2493 |
| 7 | 1 | 0.8512 | 1.1498 | 1.2490 |
| 7 | 2 | 0.8506 | 1.1494 | 1.2498 |
| 7 | 3 | 0.8500 | 1.1501 | 1.2496 |
| 8 | 1 | 0.8506 | 1.1498 | 1.2495 |
| 8 | 2 | 0.8506 | 1.1501 | 1.2497 |
| 8 | 3 | 0.8512 | 1.1500 | 1.2492 |
| 9 | 1 | 0.8510 | 1.1501 | 1.2495 |
| 9 | 2 | 0.8518 | 1.1498 | 1.2498 |
| 9 | 3 | 0.8499 | 1.1503 | 1.2492 |

**FIGURE 16.2**  Three key characteristics of sleeve inside diameter

---

## EXAMPLE 16.8  The Sleeve Company

### Xbar-chart

Let's construct the Xbar-chart for sleeve characteristic a using the data shown in the table above. First we must calculate the average and range for each of the 9 samples. Then we can calculate $\overline{\overline{X}}$ and $\overline{R}$ using the sample averages and sample ranges:

| | Sample Averages | | | Sample Ranges | | |
|--------|--------|--------|--------|--------|--------|--------|
| **Sample** | **a** | **b** | **c** | **a** | **b** | **c** |
| 1 | 0.8504 | 1.1493 | 1.2492 | 0.0012 | 0.0003 | 0.0005 |
| 2 | 0.8505 | 1.1496 | 1.2493 | 0.0012 | 0.0008 | 0.0009 |

| Sample | Sample Averages | | | Sample Ranges | | |
|---|---|---|---|---|---|---|
| | a | b | c | a | b | c |
| 3 | 0.8507 | 1.1497 | 1.2493 | 0.0003 | 0.0002 | 0.0006 |
| 4 | 0.8506 | 1.1497 | 1.2764 | 0.0010 | 0.0003 | 0.0443 |
| 5 | 0.8506 | 1.1498 | 1.2494 | 0.0013 | 0.0003 | 0.0007 |
| 6 | 0.8508 | 1.1499 | 1.2494 | 0.0006 | 0.0003 | 0.0004 |
| 7 | 0.8506 | 1.1498 | 1.2495 | 0.0012 | 0.0007 | 0.0008 |
| 8 | 0.8508 | 1.1500 | 1.2495 | 0.0006 | 0.0003 | 0.0005 |
| 9 | 0.8509 | 1.1501 | 1.2495 | 0.0019 | 0.0005 | 0.0006 |
| Overall average | 0.8507 | 1.1498 | 1.2524 | 0.0010 | 0.0004 | 0.0055 |

$$\overline{\overline{X}} = \text{average of the 9 sample means} = 0.850663$$
$$\overline{R} = \text{average of the 9 sample ranges} = 0.001033$$
$$\text{UCL} = \overline{\overline{X}} + A_2\overline{R} = 0.850663 + 1.023(0.001033) = 0.8517$$
$$\text{Center line} = \overline{\overline{X}} = 0.850663$$
$$\text{LCL} = \overline{\overline{X}} - A_2\overline{R} = 0.850663 - 1.023(0.001033) = 0.8496$$

Finally, the graph is created by plotting the control limits and the sample averages for each of the 9 samples:

**Xbar-chart for Characteristic a**

# TRY IT NOW!

## The Sleeve Company

### *Xbar-chart for Characteristic b*

Using the sample averages shown in Example 16.8, create the Xbar-chart for characteristic b for the manufacturer of the sleeves.

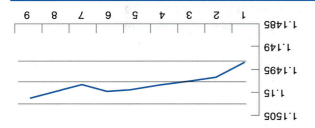

ANS.

## 16.6.2 R-charts

At the beginning of the chapter we looked at 4 different scenarios for a soup can filling process. The 4 scenarios showed a stable process, two ways that the mean of the process could change, and one illustration of a change in the process variability. We learned that control charts are used to help detect changes in the mean of the distribution and changes in the variance of the distribution. The Xbar-chart developed in the previous section is used to detect changes in the mean of the distribution. In this section we look at a control chart to detect changes in the variability of the distribution. It is called the **R-chart.**

> The **R-chart** is a control chart used to monitor the variation of a process over time. Time is plotted on the $x$ axis and the sample range, $R = $ Max $-$ Min, of the quality characteristic being monitored is plotted on the $y$ axis. The graph also has 3 lines drawn parallel to the $x$ axis. These are called the center line, the upper control limit, and the lower control limit.

You might wonder why we don't plot the sample standard deviations in an effort to detect changes to the process variability. This is a good question since we have repeatedly said that the standard deviation is the most commonly used measure of variability. There is, in fact, the $s$-chart, which is a plot of the sample standard deviations over time. However, for small sample sizes ($n = 9$ or less), the $s$-chart and the $R$-chart provide about the same information and $R$-charts are widely used by industry. For this reason, we will look only at $R$-charts.

The logic used to construct the $R$-chart is the same as that used to construct the $x$-chart and the Xbar-chart. There is a center line and upper and lower control limits that are set 3 standard deviations above and below the center line. Instead of plotting sample averages on the chart, sample ranges are plotted. To construct the upper and lower control limits we must estimate the standard deviation of $R$. The most commonly used estimate is $d_3(\overline{R}/d_2)$, where $d_3$ and $d_2$ are constants that depend on the size of the sample. They are found in Table 7 of the Appendix. Using this estimate for the standard deviation of $R$, the procedure for setting up the $R$-chart is as follows:

*Steps for constructing an R-chart*

**Step 1:**   The center line (CL) of the chart is drawn at the average of all the sample ranges, $\overline{R}$.

**Step 2:**   The upper control limit (UCL) of the chart is a horizontal line drawn at a value that is three standard deviations above the center line (called the 3 sigma limit), $\overline{R} + 3d_3(\overline{R}/d_2)$.

**Step 3:**   The lower control limit (LCL) of the chart is a horizontal line drawn at a value that is three standard deviations below the center line *(called the 3 sigma limit)*, $\overline{R} - d_3(\overline{R}/d_2)$.

By doing some algebra on the formulas for the upper and lower control limits, you can rewrite them as

*Simpler steps for constructing an R-chart*

$$\text{UCL} = \overline{R}\left(1 + \frac{3d_3}{d_2}\right) = \overline{R}\,D_4$$

$$\text{LCL} = \overline{R}\left(1 - \frac{3d_3}{d_2}\right) = \overline{R}\,D_3$$

In these formulas $D_4 = 1 + 3d_3/d_2$ and $D_3 = 1 - 3d_3/d_2$. They have been calculated from $d_3$ and $d_2$ and can be found in the same table in the Appendix, Table 7. Using these formulas for the upper and lower control limits makes the calculation much easier.

### EXAMPLE 16.9   The Diaper Company

*R-chart*

The sample ranges for 20 hours of data from the diaper company are shown below. Recall that the diaper company samples 5 diapers every hour and checks the weight in grams.

| Sample | Range | Sample | Range |
|--------|-------|--------|-------|
| 1 | 1.9 | 11 | 0.81 |
| 2 | 1.17 | 12 | 1.36 |
| 3 | 2.33 | 13 | 1.44 |
| 4 | 0.65 | 14 | 2.11 |
| 5 | 1.86 | 15 | 1.62 |
| 6 | 0.86 | 16 | 1.92 |
| 7 | 1.46 | 17 | 1.2 |
| 8 | 1.64 | 18 | 1.2 |
| 9 | 2.37 | 19 | 1.55 |
| 10 | 1.83 | 20 | 0.93 |

The centerline is found by averaging the 20 sample ranges to get $\overline{R} = 1.511$ g. The upper control limit is $\overline{R}\,D_4 = (1.511)(2.114) = 3.193$. The lower control limit is $\overline{R}\,D_3 = (1.511)(0) = 0$.

   Plotting these lines and the sample ranges gives you the *R*-chart shown below:

**R-chart for the Diaper Company**

■

## TRY IT NOW!

### The Sleeve Company

*R-chart for Characteristic b*

Using the sample ranges shown in Example 16.8, create the *R*-chart for characteristic b for the manufacturer of the sleeves.

ANS.

### 16.6.3 Guidelines for Sampling

The data required for both the Xbar-chart and the R-chart come from sampling $n$ observations periodically over time. To collect such data you must decide the sample size, $n$, and the frequency with which samples are to be taken (once an hour, every 10 minutes, etc.). These samples are often referred to in the quality literature as **rational subgroups.**

There are some guidelines for selecting rational subgroups. Basically you want the sample observations in a subgroup to be similar to each other since you are finding the sample average and the sample range. Thus, you usually select consecutively produced items or consecutively served customers. Typically, sample sizes of 4 to 10 are used.

To determine the frequency of sampling you need to think about what the potential sources of variation might be. Then you need to sample with a frequency that will allow the samples to differ from each other. For example, suppose you were manufacturing tubes of glue and the tubes were filled 8 at a time. If the timing of your sampling always got you a sample of the 8th tube, then you could be missing some variation simply because you have not looked at the other tube positions in the filling process.

### 16.6.4 Exercises—Learn It!

**16.11** Set up the Xbar- and R-charts for characteristic c of the sleeve company.

**16.12** An instructor wonders if the average grade in the introductory statistics course is a stable process. The average course grade for the past 10 semesters is shown below as well as the range for each semester. The average and ranges are based on classes of size 15.

| Semester | Average grade | Range |
|---|---|---|
| 1 | 2.92 | 3.2 |
| 2 | 2.68 | 2.5 |
| 3 | 3.06 | 3.1 |
| 4 | 3.32 | 3 |
| 5 | 3.30 | 2 |
| 6 | 3.43 | 4 |
| 7 | 2.45 | 3.5 |
| 8 | 2.94 | 3.1 |
| 9 | 3.27 | 2.9 |
| 10 | 2.73 | 2.9 |

(a) Construct the Xbar-chart.

(b) Construct the R-chart.

(c) Has there been any change to the mean or variability of the grading process over the past 10 semesters? Explain why or why not.

*Datafile:*
*FASTFOOD.XXX*

**16.13** A fast-food chain is concerned about putting the "fast" back into fast food. The variable to be monitored is the time between when the customer finishes placing his order and when he leaves the register after receiving his food. Five customers are sampled every hour during the peak hours and the following times in minutes were recorded.

| Hour 1 | Hour 2 | Hour 3 | Hour 4 | Hour 5 | Hour 6 | Hour 7 | Hour 8 | Hour 9 | Hour 10 |
|---|---|---|---|---|---|---|---|---|---|
| 4.81 | 3.74 | 4.88 | 4.91 | 5.06 | 5.40 | 3.94 | 4.64 | 4.85 | 4.44 |
| 4.77 | 4.03 | 5.27 | 3.93 | 4.64 | 3.23 | 4.92 | 4.46 | 5.33 | 4.71 |
| 4.62 | 4.39 | 4.11 | 3.97 | 4.30 | 4.28 | 4.64 | 4.68 | 4.86 | 4.95 |
| 4.61 | 4.29 | 5.07 | 4.50 | 4.82 | 4.71 | 4.44 | 4.19 | 3.93 | 4.74 |
| 3.59 | 3.21 | 4.18 | 5.17 | 4.48 | 4.31 | 5.24 | 4.33 | 5.57 | 4.50 |

(a) Construct the Xbar- and R-charts for this process.

(b) Has there been any change to the process mean over time?

(c) Has there been any change to the process variability over time?

(d) If the process is in control, why might customers still be unhappy with service times?

(e) What recommendations would you make to the fast-food chain?

**16.14** The golf ball company has decided to sample 4 golf balls every 2 hours to check the weight of the balls. The sample averages and sample ranges for 24 samples are shown below:

| Sample | Xbar | Range |
|--------|-------|-------|
| 1 | 45.02 | 0.09 |
| 2 | 45.09 | 0.01 |
| 3 | 45.03 | 0.00 |
| 4 | 44.98 | 0.16 |
| 5 | 44.93 | 0.12 |
| 6 | 44.99 | 0.08 |
| 7 | 45.04 | 0.08 |
| 8 | 45.08 | 0.10 |
| 9 | 44.95 | 0.12 |
| 10 | 44.98 | 0.04 |
| 11 | 45.03 | 0.00 |
| 12 | 44.98 | 0.04 |
| 13 | 45.04 | 0.11 |
| 14 | 44.98 | 0.04 |
| 15 | 45.02 | 0.18 |
| 16 | 45.01 | 0.04 |
| 17 | 44.97 | 0.20 |
| 18 | 44.98 | 0.06 |
| 19 | 44.97 | 0.15 |
| 20 | 44.94 | 0.22 |
| 21 | 44.95 | 0.32 |
| 22 | 44.99 | 0.16 |
| 23 | 44.96 | 0.03 |
| 24 | 45.03 | 0.07 |

(a) Construct the Xbar- and _R_-charts for the golf ball weights. The company has a target mean of 45 g and a target standard deviation of 0.1 g.

(b) What recommendations would you make to the company?

**16.15** A company that manufactures tissues has discovered that the amount of airspace left in the box at the time of production impacts the customer complaints about the first tissue tearing. The target airspace at the time of production is 20 mm with a standard deviation of 1 mm. A sample of 4 tissue boxes is checked every hour. The sample means and ranges for 20 hours are shown below:

| Sample | Xbar | Range |
|--------|-------|-------|
| 1 | 21.96 | 2.10 |
| 2 | 18.63 | 1.92 |
| 3 | 21.63 | 2.21 |
| 4 | 20.34 | 2.67 |
| 5 | 20.46 | 3.20 |
| 6 | 21.85 | 2.51 |
| 7 | 19.18 | 2.46 |
| 8 | 20.17 | 1.81 |
| 9 | 20.07 | 2.27 |
| 10 | 19.23 | 1.78 |
| 11 | 17.94 | 0.90 |
| 12 | 17.24 | 2.67 |
| 13 | 16.85 | 1.15 |
| 14 | 16.96 | 2.78 |
| 15 | 14.45 | 2.06 |
| 16 | 17.91 | 2.39 |
| 17 | 17.66 | 2.23 |
| 18 | 17.93 | 1.97 |
| 19 | 17.29 | 1.61 |
| 20 | 18.20 | 1.53 |

(a) Construct the Xbar- and _R_- charts for the airspace data.

(b) What can you conclude about the packaging process from these data?

## 16.7  THE *p*-CHART

The tree diagram shown in Section 16.4 indicates that the first question you must ask to decide which control chart to use has to do with whether you have quantitative or qualitative data. The control charts we have presented so far in this chapter have dealt with quantitative data such as melt flow, weight, and diameter. However, very often we wish to monitor the number of defects, the number of errors, etc. In this case a product or service either has the attribute of interest (e.g., a defect or an error) or it does not. For such qualitative or attribute data, the most commonly used control chart is called **the *p*-chart.**

> The ***p*-chart** is a control chart used to monitor the proportion, *p*, of items that have a particular attribute or characteristic over time. Time is plotted on the *x* axis and the sample proportion, *p* = number in the sample with the attribute/*n*, is plotted on the *y* axis. The graph also has 3 lines drawn parallel to the *x* axis. These are called the center line, the upper control limit, and the lower control limit.

There are many situations in which you might use the *p*-chart, for example, the proportion of invoices processed with at least one error, the proportion of restaurant customers that have to wait more than 10 minutes for a table, the proportion of free throw shots missed, the proportion of bytes transmitted incorrectly. In each of these situations you are interested in monitoring the proportion of items or people that have these attributes.

The underlying assumption in situations like these is that the proportion of the population that has the attribute, $\pi$, remains constant over time. Thus, the population distribution is a binomial distribution, which you learned about in Chapter 6. To monitor $\pi$, samples of size *n* are selected over time. The number of items in the sample that have the attribute you are monitoring is observed and labeled *x*. This number is divided by the sample size, *n*, to get a sample proportion, *p*:

$$p = \frac{x}{n}$$

If the process is a stable process then there will, of course, be common variation in the sample proportion, *p*. If the process is unstable, then the sample proportion will have both common and special variation. The *p*-chart is used to detect this latter situation. Variation in *p* is displayed in the *p*-chart and when the sample proportion goes outside the control chart limits or exhibits a pattern, then the process needs to be examined.

### 16.7.1 Constructing the *p*-chart

To determine the control chart limits we need to know the mean and standard deviation of the sample proportion, *p*. You might recall from Chapter 7 that if the sample size is large enough the estimate *p* has a normal distribution with

- a mean equal to population proportion, $\pi$, and
- a standard deviation of

$$\sqrt{\frac{\pi(1 - \pi)}{n}}$$

Using this information, we can construct the upper and lower control chart limits in the same basic manner as we have done for the other charts. The population proportion, $\pi$, could be known if there is a target proportion, but most of the time $\pi$ must be estimated from the data. This is analogous to the situation we found ourselves in for the Xbar-chart when the population mean was unknown. In that case we used $\overline{\overline{X}}$ to estimate $\mu$. In this case we will use $\overline{p}$ to estimate $\pi$. The estimate $\overline{p}$ is found by averaging all the sample proportions. This is equivalent to totaling the number of items with the attribute and dividing by the total number of items sampled:

$$\overline{p} = \frac{\Sigma x}{\Sigma n}$$

*Formula for $\overline{p}$*

The center line and the upper and lower control limits are found using the following steps:

**Step 1:** The center line (CL) of the chart is drawn at the average of all the sample proportions, $\overline{p}$.

*Steps for constructing a p-chart*

**Step 2:** The upper control limit (UCL) of the chart is a horizontal line drawn at a value that is three standard deviations above the center line (called the 3 sigma limit),

$$\overline{p} + 3\sqrt{\frac{\overline{p}(1 - \overline{p})}{n}}$$

**Step 3:** The lower control limit (LCL) of the chart is horizontal line drawn at a value that is three standard deviations below the center line (called the 3 sigma limit),

$$\overline{p} - 3\sqrt{\frac{\overline{p}(1 - \overline{p})}{n}}$$

Once the upper and lower control chart limits have been graphed, the sample proportions are graphed over time. An example will help illustrate the construction of the *p*-chart.

## EXAMPLE 16.10   Software Company

### *p*-chart for Proportion of Defective Disks

A manufacturer of computer disks is concerned about the proportion of defective disks. Each day a sample of 50 disks are tested and the number of defectives is reported. The proportion of defects found for the past 20 days is shown below:

| **Days 1–10** | 0.00 | 0.02 | 0.04 | 0.02 | 0.02 | 0.00 | 0.04 | 0.02 | 0.06 | 0.02 |
| **Days 11–20** | 0.02 | 0.02 | 0.00 | 0.00 | 0.04 | 0.02 | 0.00 | 0.02 | 0.02 | 0.04 |

The center line is at $\overline{p}$ = average of all the sample proportions = 0.021. The upper control limit is at

$$\overline{p} + 3\sqrt{\frac{\overline{p}(1 - \overline{p})}{n}} = 0.021 + 3\sqrt{\frac{0.021(1 - 0.021)}{50}} = 0.021 + 3(0.0203) = 0.082$$

The lower control limit is at

$$0.021 - 3\sqrt{\frac{0.021(1 - 0.021)}{50}} = 0.021 - 3(0.0203) = 0$$

(since the proportion cannot be negative). The center line, the control limits, and the sample proportions are shown in the *p*-chart at the top of page 828.

*p*-chart for Software Company

All of the sample proportions are in between the control limits, so we conclude that the process is stable. ■

Notice that the sample size remains constant from sample to sample in a *p*-chart. If you have a varying sample size, then you must use what is called the *np*-chart. This is very similar to the *p*-chart but accommodates the changing sample size.

## TRY IT NOW!

### The Basketball Player

***p*-chart for Proportion of Successful Free Throws**

A basketball player is monitoring his free throw percentage. He keeps track of the number of baskets made in 50 practice shots on each day. His data are shown below for the past 15 days of practice:

| Practice Session | Shots Made | Practice Session | Shots Made |
|:---:|:---:|:---:|:---:|
| 1 | 37 | 9 | 38 |
| 2 | 34 | 10 | 33 |
| 3 | 35 | 11 | 30 |
| 4 | 35 | 12 | 41 |
| 5 | 29 | 13 | 37 |
| 6 | 32 | 14 | 34 |
| 7 | 39 | 15 | 37 |
| 8 | 36 | | |

Create the *p*-chart to monitor the percentage of free throw shots that he makes.

Is this player consistent?

## 16.7.2 Exercises—Learn It!

16.16 The emergency room at a major hospital is concerned about the number of patients who must wait more than 1 hour to be seen by a doctor. For the past 35 days the hospital has sampled 30 patients a day and recorded the number of patients who had to wait more than one hour to be seen. The data are shown below:

*Datafile:*
*PATIENT.XXX*

| Day | Day of the week | Number of patients who waited more than 1 hour |
|-----|-----------------|-----------------------------------------------|
| 1 | Monday | 19 |
| 2 | Tuesday | 17 |
| 3 | Wednesday | 19 |
| 4 | Thursday | 22 |
| 5 | Friday | 19 |
| 6 | Saturday | 25 |
| 7 | Sunday | 15 |
| 8 | Monday | 14 |
| 9 | Tuesday | 12 |
| 10 | Wednesday | 19 |
| 11 | Thursday | 15 |
| 12 | Friday | 15 |
| 13 | Saturday | 28 |
| 14 | Sunday | 17 |
| 15 | Monday | 17 |
| 16 | Tuesday | 18 |
| 17 | Wednesday | 12 |
| 18 | Thursday | 15 |
| 19 | Friday | 17 |
| 20 | Saturday | 22 |
| 21 | Sunday | 17 |
| 22 | Monday | 17 |
| 23 | Tuesday | 16 |
| 24 | Wednesday | 12 |
| 25 | Thursday | 15 |
| 26 | Friday | 11 |
| 27 | Saturday | 27 |
| 28 | Sunday | 20 |
| 29 | Monday | 20 |
| 30 | Tuesday | 15 |
| 31 | Wednesday | 20 |
| 32 | Thursday | 18 |
| 33 | Friday | 20 |
| 34 | Saturday | 21 |
| 35 | Sunday | 15 |

(a) Find the sample proportion who waited longer than 1 hour each day.

(b) Find the values for the center line and the upper and lower control limits.

(c) Construct the *p*-chart for these data.

(d) Is the process in control? If not, which points are out of control? Can you offer any explanation to the hospital?

16.17 A college registrar's office is auditing the process by which transfer credits are awarded. A sample of 50 student records is checked for each of the past 8 semesters. The number of records with errors is recorded. The data are shown below:

| Semester | Number with errors | Semester | Number with errors |
|----------|--------------------|---------|--------------------|
| 1 | 7 | 5 | 9 |
| 2 | 7 | 6 | 11 |
| 3 | 6 | 7 | 10 |
| 4 | 6 | 8 | 13 |

(a) Find the sample proportion of records with errors for each semester.

(b) Construct a *p*-chart for the registrar.

(c) Is this process under control? Why or why not?

**16.18** A manufacturer of T-shirts checks to be sure that the design printed on the shirt is straight. The company checks a sample of 100 shirts a day. The number of defects found for each of the past 20 days is shown below:

| Day | Number defective | Day | Number defective |
|-----|------------------|-----|------------------|
| 1 | 2 | 11 | 2 |
| 2 | 1 | 12 | 2 |
| 3 | 2 | 13 | 2 |
| 4 | 2 | 14 | 1 |
| 5 | 2 | 15 | 2 |
| 6 | 4 | 16 | 2 |
| 7 | 2 | 17 | 3 |
| 8 | 2 | 18 | 2 |
| 9 | 2 | 19 | 0 |
| 10 | 4 | 20 | 2 |

(a) Find the sample proportion defective for each day.

(b) Construct the *p*-chart for these data.

(c) Is the process in control? Explain why or why not.

**16.19** A mail-order company monitors a sample of 25 calls a day to see if a sales representative is following the company guidelines. The number of calls that have quality problems (e.g., the sales representative did not provide accurate information or did not treat the customer with respect) is recorded. The data are shown below:

| Day | Number of calls with quality problems | Day | Number of calls with quality problems |
|-----|---------------------------------------|-----|---------------------------------------|
| 1 | 4 | 11 | 3 |
| 2 | 3 | 12 | 2 |
| 3 | 3 | 13 | 2 |
| 4 | 3 | 14 | 4 |
| 5 | 4 | 15 | 3 |
| 6 | 4 | 16 | 2 |
| 7 | 1 | 17 | 4 |
| 8 | 3 | 18 | 2 |
| 9 | 3 | 19 | 3 |
| 10 | 2 | 20 | 2 |

(a) Find the sample proportion of calls with quality problems for each day.

(b) Construct a *p*-chart for these data.

(c) Is this process in control? Explain why or why not.

**16.20** A manufacturer of heaters for dishwashers performs a number of different quality checks on the heaters before they are shipped. The number of heaters with defects in samples of size 25 are recorded by shift. The data are shown below:

| Day | Shift 1 | Shift 2 | Shift 3 | Day | Shift 1 | Shift 2 | Shift 3 |
|-----|---------|---------|---------|-----|---------|---------|---------|
| 1 | 0 | 2 | 5 | 9 | 0 | 3 | 5 |
| 2 | 0 | 3 | 5 | 10 | 2 | 3 | 6 |
| 3 | 1 | 3 | 4 | 11 | 1 | 3 | 5 |
| 4 | 0 | 3 | 6 | 12 | 2 | 3 | 5 |
| 5 | 1 | 3 | 5 | 13 | 1 | 3 | 5 |
| 6 | 1 | 4 | 5 | 14 | 1 | 2 | 5 |
| 7 | 1 | 3 | 5 | 15 | 2 | 4 | 5 |
| 8 | 1 | 3 | 5 | | | | |

(a) Construct a *p*-chart for the proportion of defects produced during the day shift, shift 1.

(b) Construct a *p*-chart for the proportion of defects produced during the evening shift, shift 2.

(c) Construct a *p*-chart for the proportion of defects produced during the graveyard shift, shift 3.

(d) Which of the processes are under control?

# 16.8 CREATING CONTROL CHARTS IN EXCEL

Although Excel has no built-in tools for creating control charts, you can use several of the tools you already know about to create them. In this section we will detail the procedure for creating an Xbar-chart. The procedures for the other control charts are very similar.

## 16.8.1 Creating Subgroup Statistics in Excel

Suppose that we want to create an Xbar-chart for the diaper data. The variable measured was diaper weight for samples of size 5. The data are shown in Figure 16.3.

| | A | B | C | D |
|---|---|---|---|---|
| **1** | **Sample** | **Diaper** | **Weight** | **Bulk** |
| **2** | 1 | 1 | 55.87 | 0.419 |
| **3** | 1 | 2 | 55.35 | 0.380 |
| **4** | 1 | 3 | 54.50 | 0.365 |
| **5** | 1 | 4 | 53.97 | 0.406 |
| **6** | 1 | 5 | 54.29 | 0.360 |
| **7** | 2 | 1 | 54.85 | 0.397 |
| **8** | 2 | 2 | 54.79 | 0.405 |
| **9** | 2 | 3 | 54.65 | 0.393 |
| **10** | 2 | 4 | 55.56 | 0.413 |
| **11** | 2 | 5 | 55.82 | 0.367 |
| **12** | 3 | 1 | 54.40 | 0.397 |

**FIGURE 16.3**   Diaper weight data

To create the control chart, we need to calculate the mean and range for each sample. This is easily done using pivot tables.

1. Create a pivot table for the data by dragging the Sample field button to the Row area and the Weight field button to the Data area. Double click on Weight to open the Pivot Table Field box and change **Summarize by:** to Average. Finish making your selections and click **Finish.**

2. You cannot summarize the data in a pivot table using the range, but you can do it for the maximum and minimum values. Create pivot tables for the **Max** and **Min** of each sample and locate them adjacent to the one for average. After you have done this, the worksheet will look like the one in Figure 16.4 (page 832).

| | Sample | Diaper | Weight | Bulk | Average of Weight | | Max of Weight | | Min of Weight | |
|---|---|---|---|---|---|---|---|---|---|---|
| 1 | | | | | Sample | Total | Sample | Total | Sample | Total |
| 2 | 1 | 1 | 55.87 | 0.419 | | | | | | |
| 3 | 1 | 2 | 55.35 | 0.380 | 1 | 54.796 | 1 | 55.87 | 1 | 53.97 |
| 4 | 1 | 3 | 54.50 | 0.365 | 2 | 55.134 | 2 | 55.82 | 2 | 54.65 |
| 5 | 1 | 4 | 53.97 | 0.406 | 3 | 54.836 | 3 | 56.44 | 3 | 54.11 |
| 6 | 1 | 5 | 54.29 | 0.360 | 4 | 54.59 | 4 | 54.87 | 4 | 54.22 |
| 7 | 2 | 1 | 54.85 | 0.397 | 5 | 55.268 | 5 | 56.3 | 5 | 54.44 |
| 8 | 2 | 2 | 54.79 | 0.405 | 6 | 55.276 | 6 | 55.75 | 6 | 54.89 |
| 9 | 2 | 3 | 54.65 | 0.393 | 7 | 55.026 | 7 | 55.78 | 7 | 54.32 |
| 10 | 2 | 4 | 55.56 | 0.413 | 8 | 55.232 | 8 | 55.78 | 8 | 54.14 |
| 11 | 2 | 5 | 55.82 | 0.367 | 9 | 54.718 | 9 | 55.97 | 9 | 53.6 |
| 12 | 3 | 1 | 54.40 | 0.397 | 10 | 55.054 | 10 | 55.7 | 10 | 53.87 |
| 13 | 3 | 2 | 56.44 | 0.345 | 11 | 55.116 | 11 | 55.55 | 11 | 54.74 |
| 14 | 3 | 3 | 54.11 | 0.421 | 12 | 54.542 | 12 | 55.22 | 12 | 53.86 |
| 15 | 3 | 4 | 54.67 | 0.419 | 13 | 55.06 | 13 | 55.87 | 13 | 54.43 |

**FIGURE 16.4**   Data and summary pivot tables

Since we really don't need all of those columns labeled Sample, you can hide those columns to make the worksheet less crowded. To work easily with the pivot table columns, you might want to use **Copy > Paste Special > Value.**

3. Use a formula in the cell adjacent to the minimum value to calculate the range for each sample by subtracting the minimum weight value from the maximum.

## 16.8.2 Calculating the Control Limits in Excel

Now that you have the summary information for each sample, you need to calculate the values for the center line and upper and lower control limits. In this example, we will use the target specification of 55.00 g for the center line, but we will use the sample range data to find the upper and lower control limits. That is, we want to create a chart with a known population mean, but unknown standard deviation. Remember that the formulas for the control limits are:

$$LCL = \mu - A_2\overline{R}$$
$$UCL = \mu + A_2\overline{R}$$

The factor $A_2$ depends on the sample size. In this case, from the table in the Appendix for $n = 5$, the value of $A_2$ is 0.577.

1. Use a formula to find the average of the sample ranges, $\overline{R}$. The formula should be = AVERAGE(data range), where data range is the cell reference that contains the sample ranges. For this example, the value of $\overline{R}$ is 1.44.

2. Insert three new columns next to the column that contains the sample means and label them CL, LCL, and UCL.

*Note: You could use cell references instead of the actual values, but be sure you use absolute cell referencing.*

3. In the first cell under the label CL, type "55.00" and copy it to the cells next to each of the sample means.

4. In the first cell under the label LCL, type in the formula "= 55 − 0.577*1.44." Copy this formula down the column for each sample.

5. In the first cell under the label UCL, type in the formula "= 55 + 0.577*1.44", and copy this formula to the cells below.

Your worksheet should contain columns that look like the ones shown in Figure 16.5.

| Average of Weight | | | | | |
|---|---|---|---|---|---|
| Sample | Average | CL | LCL | UCL | |
| 1 | 54.796 | 55 | 54.16912 | 55.83088 | |
| 2 | 55.134 | 55 | 54.16912 | 55.83088 | |
| 3 | 54.836 | 55 | 54.16912 | 55.83088 | |
| 4 | 54.59 | 55 | 54.16912 | 55.83088 | |
| 5 | 55.268 | 55 | 54.16912 | 55.83088 | |
| 6 | 55.276 | 55 | 54.16912 | 55.83088 | |
| 7 | 55.026 | 55 | 54.16912 | 55.83088 | |
| 8 | 55.232 | 55 | 54.16912 | 55.83088 | |
| 9 | 54.718 | 55 | 54.16912 | 55.83088 | |
| 10 | 55.054 | 55 | 54.16912 | 55.83088 | |
| 11 | 55.116 | 55 | 54.16912 | 55.83088 | |
| 12 | 54.542 | 55 | 54.16912 | 55.83088 | |
| 13 | 55.06 | 55 | 54.16912 | 55.83088 | |
| 14 | 55.136 | 55 | 54.16912 | 55.83088 | |
| 15 | 54.73 | 55 | 54.16912 | 55.83088 | |
| 16 | 55.45 | 55 | 54.16912 | 55.83088 | |

**FIGURE 16.5**   Worksheet with CL, LCL, and UCL

## 16.8.3 Creating the Control Chart

In Excel you will use the chart wizard to create the control chart from the worksheet that you have prepared.

1. Highlight the columns in the worksheet that contain the sample means, CL, LCL, and UCL, and start the chart wizard.

2. From the list of **Chart types,** choose **Line** and **Lines with Markers** as the **Chart sub-type** as shown in Figure 16.6. Click **Next >** twice to continue.

**FIGURE 16.6**   Choosing a line chart

3. Add titles and other formatting changes, and click **Finish.** The chart will be placed on the worksheet.

If you want the control limits to be plotted without markers, you can accomplish this quite easily.

**4.** Click on the chart to select it, and then click on one of the three limits. The **Format Data Series** dialog box will open. Select the **Patterns** tab, and under **Marker,** click the radio button for **None.** The dialog box is shown in Figure 16.7.

**FIGURE 16.7**   Removing markers from lines

**5.** Repeat this for the other two control lines. The finished graph is shown in Figure 16.8.

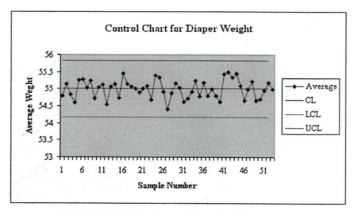

**FIGURE 16.8**   Control chart for diaper weight

The steps for creating other types of control charts such as $R$- and $p$-charts are very similar to the ones for the Xbar-chart. The only changes are the calculation of the summary values for each sample and the formulas for the control limits.

# *CHAPTER 16 SUMMARY*

Control charts are one of many quality tools being used with increasing frequency by both manufacturers and service companies. Customers are also beginning to partner with suppliers and in doing so are also monitoring the quality of their suppliers. This is a growing and dynamic area and we have just introduced the basics in this chapter.

In this chapter you learned the basics about constructing and using control charts. You have seen that there are many different kinds of charts but they all are constructed using the same logic. In all cases a sample of data is collected periodically over time to monitor some characteristic of a process. Depending on the process you are monitoring, you might observe a quantitative variable or you might observe the presence or lack of presence of some characteristic. Regardless of the situation, a control chart is used to detect the presence of special cause variation. If a process has some special cause variation, then one or more of the plotted points will go outside the upper and lower control limits.

The center line is located at the average value of the variable being plotted. The upper and lower control limits are typically set three standard deviations above and below the center line. This allows for detection of the presence of special variation without setting off false alarms very often. A false alarm occurs when you conclude that there is some special variation in the process when there is not any.

## Key Terms

| Term | Definition | Page Reference |
|---|---|---|
| **Common causes of variation** | **Common causes of variation** are the methods, machines, people, and environment that are an inherent part of a process. They are part of the design of the process. | 811 |
| **Control chart** | A **control chart** is a graph that allows us to study variation in a product or service. | 809 |
| **Out of statistical control** | If the distribution of a process changes over time, it is said to be **out of statistical control** or **out of control**. | 811 |
| **p-chart** | The **p-chart** is a control chart used to monitor the proportion, $p$, of items that have a particular attribute or characteristic over time. | 826 |
| **Process** | A **process** is a series of actions that changes inputs to outputs. | 808 |

(*continued*)

| Term | Definition | Page reference |
|---|---|---|
| **R-chart** | The **R-chart** is a control chart used to monitor the variation of a process over time. | 822 |
| **Special causes of variation** | **Special causes of variation** are events or actions that are not part of the process design. | 811 |
| **Statistical control** | A process whose distribution does not change over time is said to be in a state of **statistical control** or **in control.** | 811 |
| **Xbar-chart** | The **Xbar-chart** is a control chart used to monitor the mean of a variable over time. | 817 |
| **x-chart** | An **x-chart** is a graph that plots the sample values in chronological order along the $x$ axis. | 813 |

# CHAPTER 16   EXERCISES

## Learning It!

*Datafile:*
*MELTFLOW.XXX*

**16.21** A customer of the plastics company has collected its own data on the melt flow of the plastic it is purchasing. The customer data are shown below:

| | | | | |
|---|---|---|---|---|
| 8.7 | 9.3 | 9.8 | 10.0 | 8.4 |
| 9.2 | 8.8 | 7.8 | 9.4 | 8.6 |
| 9.2 | 8.9 | 9.2 | 9.9 | 9.1 |
| 8.8 | 9.0 | 9.5 | 9.5 | 9.5 |
| 9.1 | 8.3 | 9.1 | 9.0 | 9.5 |
| 8.8 | 8.9 | 9.5 | 9.1 | 6.5 |
| 9.2 | 7.6 | 9.0 | 9.6 | |
| 9.4 | 8.4 | 8.9 | 9.6 | |
| 8.6 | 8.7 | 9.9 | 8.3 | |
| 8.3 | 9.2 | 9.8 | 8.1 | |

(a) Create the *x*-chart for these data.

(b) Is the process in control? Explain why or why not.

**16.22** An amusement park is concerned about the average amount of time that patrons wait in line for a ride. The park manager has collected waiting times for samples of 5 consecutive customers every hour for 12 hours. The data are shown below:

| Hour | Customer | Waiting time | Hour | Customer | Waiting time | Hour | Customer | Waiting time |
|---|---|---|---|---|---|---|---|---|
| 1 | 1 | 4.70 | 4 | 1 | 10.72 | 7 | 1 | 7.04 |
| | 2 | 5.06 | | 2 | 10.19 | | 2 | 8.31 |
| | 3 | 4.85 | | 3 | 10.38 | | 3 | 7.34 |
| | 4 | 4.62 | | 4 | 9.77 | | 4 | 7.31 |
| | 5 | 4.58 | | 5 | 10.79 | | 5 | 7.10 |
| 2 | 1 | 5.31 | 5 | 1 | 10.19 | 8 | 1 | 7.04 |
| | 2 | 4.78 | | 2 | 9.80 | | 2 | 6.21 |
| | 3 | 5.15 | | 3 | 10.35 | | 3 | 6.72 |
| | 4 | 5.73 | | 4 | 10.33 | | 4 | 7.47 |
| | 5 | 5.37 | | 5 | 9.96 | | 5 | 6.46 |
| 3 | 1 | 5.14 | 6 | 1 | 10.11 | 9 | 1 | 5.36 |
| | 2 | 4.90 | | 2 | 10.35 | | 2 | 7.19 |
| | 3 | 4.55 | | 3 | 10.61 | | 3 | 7.14 |
| | 4 | 4.67 | | 4 | 9.87 | | 4 | 7.35 |
| | 5 | 5.20 | | 5 | 9.68 | | 5 | 7.00 |

(a) Create an appropriate control chart to monitor the average waiting time.

(b) Is this process in control? Explain why or why not.

16.23 A company that buys blank VHS tapes for video recording is concerned about the actual amount of time they are able to record. The tapes are rated at 120 minutes. The company tests 10 tapes each time it makes a purchase of tapes. The sample averages and sample ranges (in minutes) for the last 12 purchases are shown below:

| Sample | Xbar | Range |
|---|---|---|
| 1 | 122.39 | 2.71 |
| 2 | 116.96 | 5.05 |
| 3 | 122.49 | 4.16 |
| 4 | 118.11 | 1.26 |
| 5 | 119.23 | 5.76 |
| 6 | 120.47 | 4.91 |
| 7 | 117.66 | 5.31 |
| 8 | 117.50 | 4.25 |
| 9 | 119.14 | 3.64 |
| 10 | 124.00 | 5.91 |
| 11 | 116.54 | 4.49 |
| 12 | 114.30 | 4.31 |

(a) Using the manufacturer's target mean of 120 minutes, construct an Xbar-chart for these data.

(b) Is the process mean in control? Explain why or why not.

(c) Redo the control chart using $\overline{\overline{X}}$ instead of the target mean of 120 minutes. Is your conclusion different based on this chart?

(d) Construct an R-chart for these data.

(e) Is the process variability in control? Explain why or why not.

(f) Should this company look for a new supplier? Support your answer.

16.24 A donut shop is concerned about the percentage of the last batch of donuts that get thrown away at closing time each day. The percentages for the past 3 weeks are shown below:

| Day | Proportion of Donuts Tossed at Closing (%) | Day | Proportion of Donuts Tossed at Closing (%) | Day | Proportion of Donuts Tossed at Closing (%) |
|---|---|---|---|---|---|
| 1 | 15 | 8 | 15 | 15 | 16 |
| 2 | 14 | 9 | 15 | 16 | 15 |
| 3 | 18 | 10 | 13 | 17 | 8 |
| 4 | 13 | 11 | 18 | 18 | 13 |
| 5 | 14 | 12 | 12 | 19 | 12 |
| 6 | 13 | 13 | 18 | 20 | 13 |
| 7 | 14 | 14 | 12 | 21 | 12 |

(a) Construct a p-chart for these data. Assume that the last batch of donuts made for the day has 6 dozen donuts.

(b) Is this process under control? Explain why or why not.

16.25 A company that manufactures pens is concerned about the large number of customers who are complaining that the pens do not write and return them for a free pen. The company starts a systematic sampling process and tests 30 pens from each batch of pens produced. The number of pens that do not write is recorded and is shown below for the last 20 batches:

| Batch | Number Defective | Batch | Number Defective |
|---|---|---|---|
| 1 | 6 | 4 | 7 |
| 2 | 5 | 5 | 6 |
| 3 | 7 | 6 | 6 |

(continued)

| Batch | Number Defective | Batch | Number Defective |
|-------|------------------|-------|------------------|
| 7 | 6 | 14 | 6 |
| 8 | 6 | 15 | 6 |
| 9 | 5 | 16 | 5 |
| 10 | 7 | 17 | 6 |
| 11 | 7 | 18 | 6 |
| 12 | 7 | 19 | 6 |
| 13 | 6 | 20 | 5 |

(a) Construct a *p*-chart for the proportion of defective pens.

(b) Is the process under control? Explain why or why not.

## Thinking About It!

*Requires Exercise 16.20*

**16.26** For the manufacturer of dishwasher heaters that you considered in Exercise 16.20, you constructed 3 different *p*-charts. Combine the data into one data set respecting the time order. The number defective on day 1, shift 1, shift 2, shift 3, is followed by the number defective on day 2, shift 1, shift 2, shift 3, and so on.

(a) Construct a single *p*-chart for these data.

(b) Do you reach any different conclusions than you did when you did them as separate charts by shift?

(c) Which approach do you think makes more sense and why?

*Requires Exercise 16.25*

**16.27** The process that produces the pens in Exercise 16.25 is in control but customer complaints continue to pour in to the 800 number. What kind of variation exists in this process and how can it be reduced?

**16.28** Upon further investigation of the melt flow problem you discover that the plastic tested was produced on two different lines. How would you incorporate this new information into your analysis?

**16.29** A fast-food burger place discovers that the average customer waiting time is not a stable characteristic. Suggest some possible sources of special variation.

*Requires Exercise 16.17*

**16.30** Upon further investigation of the registrar's office in Exercise 16.17 you discover that there was a change in personnel at the beginning of semester 5. How does this new information help you?

**16.31** You have been keeping track of how long it takes you to drive to work every day. What type of control chart would you suggest for these data?

## Doing It!

*Datafile: DIAPER.XXX*

**16.32** The diaper company (from Chapter 7) has collected data on 5 diapers per hour for a total of 52 hours. Thus it has 260 rows of data. The company has hired you to help analyze these data. The data are located on the data disk in a file named DIAPER.XXX. A portion of the data is shown below:

| Sample | Diaper | Weight | Bulk |
|--------|--------|--------|------|
| 1 | 1 | 55.87 | 0.419 |
| 1 | 2 | 55.35 | 0.380 |
| 1 | 3 | 54.50 | 0.365 |
| 1 | 4 | 53.97 | 0.406 |

| Variable | Description |
|----------|-------------|
| Sample | Keeps track of the sequence in which the samples were taken and goes from 1 to 52. |
| Diaper | Keeps track of the individual diapers within a sample and goes from 1 to 5. |
| Weight | The diaper weight in grams. |
| Bulk | The diaper bulk in millimeters. |

(a) Construct the Xbar- and $R$-charts for the average weight and average bulk.

(b) What recommendations can you make to the diaper company?

# CHAPTER 17

# *PULLING IT ALL TOGETHER*

## THE GREAT DIVIDE

> *"The earnings gap between executives at the very top of corporate America and middle managers and workers has stretched into a vast chasm."*
> (Wall Street Journal, *11 April 1996*)

A *Wall Street Journal* article reports that in 1995 the heads of about 30 major companies received compensation that was 212 times higher than the pay of the average American employee. The widening gulf has ignited a political firestorm and CEOs are scurrying to explain their pay increases. Compensation of chief executives "has reached Marie Antoinette proportions," says Nell Minow, a principal at Lens Inc., an activist investment fund in Washington, D.C. "People are getting disgusted with it."

William M. Mercer Inc., New York compensation consultants, analyzed 350 proxy statements of the biggest U.S. businesses. The Securities and Exchange Commission (SEC) rules for proxy statements require disclosure of the compensation of the chief executive officer plus four or more other highly compensated executives. For each individual the following data were collected:

**Company name**

**Type of business**

**Standard Industrial Classification Code (SIC)**

**Number of employees—1995**

**Company revenue—1995**

**CEO's name**

**1995 Salary—base salary earned in 1995**

**1995 Bonus—annual bonus earned in 1995**

**Percent change from 1994**—change in salary/bonus from 1994 to 1995. Percent change data have been excluded for new CEOs and in cases where salary/bonus data for 1995 or 1994 are not valid.

**Long-term compensation: options gain**— gains from the exercise of stock options and/or stock appreciation rights during 1995

**Long-term compensation: other**—includes 1995 (1) value of shares of restricted stock or restricted stock units at grant, and (2) value of payouts under any other long-term incentive compensation plan.

The furor over the great pay divide is creating pressure for higher tax rates on the rich and new federal limits on executive compensation. One idea is to give tax breaks for businesses that treat their workers well—for example, by capping the highest-paid executive's pay at perhaps 50 times the lowest-paid employee's.

## 17.1   CHAPTER OBJECTIVES

This is the final chapter of the book. It is a good time to look back on the techniques that have been covered and "pull them all together." We have been building up to this point throughout the entire book.

Each of the chapters in this book has been developed around a business problem and each problem was chosen to introduce the specific techniques explained in the chapter. Typically, the *Doing It!* exercises at the end of each chapter presented you with the data set from the chapter opening business problem. In these exercises, you received some guidance in the analysis and knew to apply the techniques that had been presented in the chapter. At the end of Chapters 5 and 10 you have worked on data sets with less guidance. In these summary exercises you used techniques from more than one chapter of the book.

Now it is time to complete the course and to solve some problems without any guidance. This is, after all, what you will need to be able to do as you enter the workforce. You must use all that you have learned about data analysis to make some sense out of these data sets, see the information in the data, draw some conclusions, and make some recommendations. Hopefully you have completed the paradigm shift that we started talking about in Chapter 1 and you now see the world through a "statistical thinking" lens. It is this vision which will make you an important member of any team in the workplace.

Specifically, this chapter will include:

- Summary of the Techniques Covered in This Book
- How to Analyze the "Great Divide" Data Set
- Ethical Issues in Data Analysis

## 17.2   SUMMARY OF THE TECHNIQUES COVERED IN THIS BOOK

As you look back over the material covered in this book, try not to see the small numerical details of each of the techniques; instead, look for the unifying themes that underlie what you have learned. Take a global look at the book and use this summary to identify the major threads that tie the book together.

This book has been divided into 17 chapters. Now, let's take the chapters and group them into six major parts.

| Part | Chapters | Topic |
|---|---|---|
| Part I | Chapters 1 and 2 | *Getting Started* |
| | Main idea:   Sample as a snapshot of the population | |
| Part II | Chapters 3, 4, and 5 | *Exploratory Data Analysis* |
| | Main ideas:   Summarize the data in picture and numerical form | |
| | Look for the center, the spread, and trends | |
| Part III | Chapters 6 and 7 | *The Keys That Unlock the Door to Inferential Statistics* |
| | Main ideas:   Distributions and parameters | |
| | Link between probability and inferential statistics | |
| Part IV | Chapters 8, 9, and 10 | *Inferential Statistics* |
| | Main idea:   Use the sample to draw conclusions about the population | |
| Part V | Chapters 11, 12, and 13 | *Model Building* |
| | Main idea:   Use sample data to build a model to explain the relationship between two or more variables | |
| Part VI | Chapters 14, 15, 16, and 17 | *Additional Topics* |

We have managed to reduce the book to five main ideas! Now let's see if there are some major themes that tie these ideas together.

1. Early in the text we learned that even if we did our sampling perfectly, we would still undoubtedly have incomplete information. This is not because we made any mistakes but because of what is called *sampling error*. Sampling error is the difference between what the sample predicts and what is true for the population. It is due to the fact that the sample is only a piece of the population. No matter what you do in statistics, unless you take a census (examine the entire population), you will have some sampling error. We have learned that even if the sample data are "a bit" inconsistent with the hypothesized theory, this doesn't automatically mean the theory is wrong. It is possible and maybe even likely to find that 52% of a sample of employees feel the company has empowered them when in fact less than half of all of the employees feel this way.

2. Without a doubt, the concept of *variability* has been a common thread from the beginning of this book. One of the great quality gurus, W. Edwards Deming, said that variation is our greatest enemy. We have indeed seen this ourselves. In fact, we observed that if there were no variability in the population we could take a sample of size $n = 1$ and have perfect information. The *standard deviation*, which is the square root of the variance, is the yardstick by which we evaluate all differences when analyzing data.

3. Another theme that you may have noticed throughout the text is the focus on *relationships* between variables. Right from the beginning of descriptive statistics you compared graphs of variables and used scatter plots to look for relationships between variables. This theme was extended when you learned to do hypothesis tests to compare two or more population means or proportions. In one of the later chapters you learned how to build models that would allow you to exploit the relationship between two or more variables for prediction purposes.

4. *Decision making in any form should be based on a proper analysis of data.* To remain competitive in today's fast-changing business environment, more data than ever before are being collected. However, the data alone are useless. They must be properly analyzed to reveal the information they contain. The entire quality movement depends on the proper use of data to make informed business decisions.

So now we have reduced the entire book to four fundamental themes. Despite the many different techniques that have been covered and even some that have not been covered, there are really only a few fundamental concepts that you need to learn in statistics.

## 17.3 HOW TO ANALYZE THE "GREAT DIVIDE" DATA SET

The "story" presented at the beginning of this chapter is a real-life story. Your job is to analyze this data set to reveal the information they contain. The data are found in the file named DIVIDE.xxx on the data disk. This section will provide you with some ideas for your analysis. Before we look at the particulars of this data set, let's see if we can identify some basic steps that should be followed anytime we begin to analyze data to solve a problem.

The material in the text has been presented in the order that you would use the techniques.

*Datafile: DIVIDE.XXX*

**Step 1:** Explore the data visually using the graphical techniques of descriptive statistics. These include bar charts, pie charts, and histograms.

**Step 2:** Summarize the data numerically. This includes finding the mean, median, quartiles, range, and standard deviation (at a minimum).

**Step 3:** Look for relationships in the data visually and numerically. Visually you would examine scatter plots. Numerically, you should use cross-tabulation tables to examine averages and standard deviations for subsets of the data.

**Step 4:** On the basis of the results of steps 1–3, formulate some hypotheses that might help you answer your questions and solve the problem.

**Step 5:** Test the hypotheses, being sure to check that the assumptions of the test are not violated. For example, if the variables must be normally distributed for the test to be valid, then first run a chi-square goodness-of-fit test. If the technique requires you to assume a common variance, use an $F$ test to be sure this is a reasonable assumption.

**Step 6:** If there appears to be a pattern in the data over time, use a time series technique.

**Step 7:** If appropriate, estimate the parameters of a multiple regression model for the variable you are trying to predict. Be sure that you check the assumptions of the model. Determine if your model is useful by testing for the significance of each of the variables and the significance of the regression model.

**Step 8:** Draw some conclusions from your analysis and make some recommendations.

*First, examine the data visually using the graphical techniques of descriptive statistics.*

Let's look at the details of implementing these eight steps for the "great divide" data set. At step 1, you should visually explore the data. You should consider using pie charts and bar charts for the qualitative variables such as type of business and SIC code. You should consider using histograms for the quantitative variables such as 1995 salary, 1995 bonus, number of employees, and annual revenue.

*Next, calculate numerical descriptors for the quantitative variables.*

At step 2, for each of the quantitative variables you should calculate summary statistics including the mean, median, quartiles, range, and standard deviation.

At step 3 you should start to look for patterns, trends, and relationships among the variables. You should also calculate some summary statistics for subsets of the data. For example, you might wish to calculate the average CEO salary for all "Energy" companies. For these data you might construct scatter plots to see if

*Look for relationships in the data.*

- salary and/or bonus is related to type of business
- salary and/or bonus is related to annual revenue of the company
- salary and/or bonus is related to size of the company as measured by the number of employees.

*Formulate hypotheses based on your analysis of the sample.*

Step 4 is the point at which you summarize what you have learned from your exploratory data analysis. Remember that the tools you have used thus far in the procedure simply describe the sample. At this step you formulate some hypotheses in an attempt to use the sample to answer questions about the population related to the problem you are trying to solve.

*Use hypothesis testing and ANOVA to determine if the differences in the sample are significant.*

At step 5, you should consider using $t$ tests to test for differences in the average salary. For example, you could divide the companies into two groups: companies with less than 500 employees and those with more than 500 employees. Then you could test to see if the average salary is different for the two groups. To see if salary is related to the type of business, you would use ANOVA to see if the average CEO salaries are equal across the various types of businesses. If they are not, determine which ones are different.

Step 6 proposes that you examine the data over time when appropriate. For this data set, you essentially have 2 years worth of data because you have the percent change in salary and bonus from 1994 to 1995. This is not really enough data to do any time series analysis but you could look for patterns in the percent changes by type of business, SIC code, or size of the company.

*Use time series analysis if there appears to be a pattern in a variable over time.*

At step 7 you should try to make some predictions. For this data set, you should consider trying to predict the salary of the CEO based on the type of business, size of the company, SIC code, and annual revenue of the company. To do this you would use a multiple regression model and determine which factors are significant.

*Use regression to develop a prediction model.*

At step 8, you should reach some conclusions and make some recommendations on the basis of your analysis. Perhaps the "great divide" is most pronounced in large companies or in a particular type of business. This type of conclusion might influence the type of legislation that would help solve the problem.

*Draw some conclusions.*

# 17.4 ETHICAL ISSUES IN DATA ANALYSIS

In this text we have cautioned you numerous times to present the data fairly and accurately. You have seen that you can distort the message of the data by altering the scale of the graphs, by changing the level of $\alpha$ at which you perform the hypothesis test, by withholding data, and other ways.

The American Statistical Association (ASA) is one of the professional associations that statisticians belong to. The ASA is concerned about the misuse of data and has published the following "Ethical Guidelines for Statistical Practice":

1. Statisticians have a public duty to maintain integrity in their professional work, particularly in the application of statistical skills to problems where private interests may inappropriately affect the development or application of statistical knowledge. For these reasons, statisticians should:
   - present their findings and interpretations honestly and objectively
   - avoid untrue, deceptive, or undocumented statements
   - disclose any financial or other interests that may affect, or appear to affect, their professional statements

2. Recognizing that collecting data for a statistical inquiry may impose a burden on respondents, that it may be viewed by some as an invasion of privacy, and that it often involves legitimate confidentiality considerations, statisticians should:
   - collect only the data needed for the purpose of their inquiry
   - inform each potential respondent about the general nature and sponsorship of the inquiry and the intended uses of the data
   - establish their intentions, where pertinent, to protect the confidentiality of information collected from respondents, strive to ensure that these intentions realistically reflect their pledges of confidentiality and their limitations to the respondents
   - ensure that the means are adequate to protect confidentiality to the extent pledged or intended, that processing and use of data conform with the pledges made, that appropriate care is taken with directly identifying information, that appropriate techniques are applied to control statistical disclosure

- ensure that, whenever data are transferred to other persons or organizations, this transfer conforms with the established confidentiality pledges, and require written assurance from the recipients of the data that the measures employed to protect confidentiality will be at least equal to those originally pledged

3. Recognizing that statistical work must be visible and open to assessment with respect to quality and appropriateness in order to advance knowledge, and that such assessment may involve an explanation of the assumptions, methodology, and data processing used, statisticians should:
   - delineate the boundaries of the inquiry as well as the boundaries of the statistical inferences which can be derived from it
   - emphasize that statistical analysis may be an essential component of an inquiry and should be acknowledged in the same manner as other essential components
   - be prepared to document data sources used in an inquiry, known inaccuracies in the data, and steps taken to correct or refine the data, statistical procedures applied to the data, and the assumptions required for their application
   - make the data available for analysis by other responsible parties with appropriate safeguards for privacy concerns
   - recognize that the selection of a statistical procedure may to some extent be a matter of judgment and that other statisticians may select alternative procedures
   - direct any criticism of a statistical inquiry to the inquiry itself and not to the individuals conducting it

4. Recognizing that a client or employer may be unfamiliar with statistical practice and be dependent upon the statistician for expert advice, statisticians should:
   - make clear their qualifications to undertake the statistical inquiry at hand
   - inform a client or employer of all factors that may affect or conflict with their impartiality
   - accept no contingency fee arrangements
   - fulfill all commitments in any inquiry undertaken
   - apply statistical procedures without concern for a favorable outcome
   - state clearly, accurately, and completely to a client the characteristics of alternate statistical procedures along with the recommended methodology and the usefulness and implications of all possible approaches
   - disclose no private information about or belonging to any present or former client without the client's approval

# CHAPTER 17 SUMMARY

In this chapter we have taken a global look at the techniques you learned in this book. Despite the large number of different techniques and many formulas presented in this book, we looked at the "bigger picture." We have grouped the chapters into five main parts and found four themes that tie all of our work in this book together.

With the increasing emphasis on quality and increasing international competition, there is clearly a need for business people to use data to make informed business decisions. You have learned the basic tools to allow you to see the world through a statistical paradigm. It is this lens that will help you advance in your career.

At whatever level you find yourself in an organization, you may be called upon to analyze data. You have learned how to do this and the final section of this book reminds you that this analysis must always be done ethically and fairly. This has been a constant point throughout the discussions in this book and it is appropriate to end the book in this way. Lives have been lost and saved by fair data analysis. You must use your statistical paradigm responsibly.

## CHAPTER 17 EXERCISES

### Exercises—Thinking About It!

**17.1** After you received your master's degree in statistics last year you took the only statistical job you could find near your fiancée. It is with a large pharmaceutical corporation, where you are now the only statistician. You spend most of your time supporting some randomized clinical trials initiated during the tenure of your predecessor, who left on very short notice and for reasons referred to rather vaguely in terms of "mutual agreement." The company has been losing ground to the competition, and management is now counting on new product development to "save the company."

The protocol for the largest of the ongoing trials, now coming to a close, clearly states the null hypothesis to be that cancer patients treated by a new procedure will have the same 2-year survival as those treated by the standard Regimen X, but no statistical method for the comparison is specified. Your analysis produces a two-tail $p$-value of 0.08. Your boss says that switching the hypothesis to a one-tail test yields a $p$-value of 0.05 and the data clearly support the alternative hypothesis. He is prepared to recommend that the CEO approve a large budget (including your salary) to further develop this new procedure.

What do you do now? OK, what do you *really* do now?

**17.2**   You have been hired by an attorney to "prove" that Company ZZZ does not practice age discrimination. The specifics of the case are that Mr. Oldstone, who had worked for the company for over 20 years, lost his job a year ago. Mr. Oldstone is suing Company ZZZ for 1 million dollars claiming age discrimination. The company claims that it reorganized and that Mr. Oldstone's position was eliminated in the reorganization. Since that time Ms. Youngsville has been hired to perform tasks similar in nature to those formerly performed by Mr. Oldstone. Ms. Youngsville is a recent college graduate and is paid $30,000 less than Mr. Oldstone's salary. You have access to the company's human resource records. An initial review of the data seems to support Mr. Oldstone's claim. However, you are being paid to disprove this claim. You are a poor, starving graduate student and you and your spouse are expecting your first child in 3 months. You also know that there are several other companies with similar suits against them and they are watching the results of this trial. You know that you can probably manipulate the highly disorganized human resource data to support the company's claim that it does not practice age discrimination.

What do you do now? OK, what do you *really* do now?

### Exercise—Doing It!

**17.3**   Complete the analysis of the "great divide" data set by following the suggestions that were outlined for each of the eight steps in Section 17.3.

## *REALLY DOING IT*
## *Responding to the Customer:*
## *Pulling It All Together*

Congratulations! You have completed your introductory statistics course and been hired by a company that is interest in applying for ISO (International Standards Organization) certification. You know that this certification requires the company to document all its processes, including the process by which it obtains and uses customer feedback to continuously improve.

The company has provided the customers with a fax number and an e-mail address to report complaints or problems. When a complaint comes in, the company must respond to the customer. Your job is to analyze the complaint data collected when customers call to complain about the product.

Your boss is most concerned about the time it takes to respond to a customer complaint. The response time is the number of minutes from receipt of the fax or e-mail until first contact with the customer. Clearly, lower response times are desirable, with all response times targeted at less than one-half day (240 minutes). The data include the number of staff personnel scheduled for that day, whether that type of problem had been encountered previously (1 = yes; 0 = no), how the issue was received (via fax or e-mail), whether the issue statement was transferred into a written resolution log, and, finally, the volume (or number) of issues electronically recorded for that day. Information is also provided on the coded volume level (Cvol = 1 if Volume ≤ 25; Cvol = 2 if 26 < Volume ≤ 50; Cvol = 3 for [51, 75]; Cvol = 4 for [76, 100]; Cvol = 5 for >100). A small portion of the data set is shown here.

| Run | Staff | Exper | Type | Log | Volume | Cvol | RespTime |
|-----|-------|-------|------|-----|--------|------|----------|
| 1 | 5 | 1 | E | Y | 114 | 5 | 97 |
| 2 | 8 | 1 | E | Y | 54 | 3 | 39 |
| 3 | 6 | 1 | F | Y | 36 | 2 | 54 |
| 4 | 9 | 1 | F | Y | 66 | 3 | 55 |
| 5 | 10 | 0 | E | Y | 125 | 5 | 125 |
| 6 | 5 | 1 | E | Y | 121 | 5 | 108 |
| 7 | 8 | 0 | F | Y | 69 | 3 | 115 |
| 8 | 6 | 1 | F | Y | 40 | 2 | 60 |

The data are found in a file named CUSTOMER.xxx. Looking at the data, you notice that some of the variables are qualitative variables, some are quantitative variables, and some of the variables are coded responses. You know that the type of data will certainly influence what techniques you use to analyze the data.

Your boss would like you to

1. Describe the data to him. This would include graphs and numerical summaries when appropriate.
2. Look for relationships in the data. Check to see if the response time to the customer is related to any of the other variables and, if so, how are they related.
3. Test to see what factors impact the response time.
4. Look for patterns or trends in the complaints over time.

Prepare a report for your boss addressing these issues and make recommendations based on your findings.

## *REALLY DOING IT*
## *Too Much Scrap: Pulling It All Together*

A manufacturing company that makes paper products has been having trouble with its paper towel line. When a roll of paper towels is manufactured, the last step before the roll is put in a case is to wrap it with plastic wrapper that the customer sees on the store shelf. A large amount of the product is being scrapped because it is not being wrapped properly.

Because of your data analysis skills, you have been assigned to be a member of the team formed to solve this problem. Clearly the first thing to do is to decide why the towels are not being wrapped properly. To help you understand the problem, you decide to collect data on the towels that have been manufactured over a 3-day period. This includes information regarding the date the roll was manufactured, the shift during which the roll was manufactured, the diameter of the roll, and whether or not the roll was scrapped. A small portion of the data set is shown below:

| Day | Shift | Diameter | Scrapped |
|-----|-------|----------|----------|
| 1 | A | 5.35433 | N |
| 1 | A | 5.51181 | Y |
| 1 | A | 5.31496 | N |
| 1 | A | 5.35433 | N |
| 1 | A | 5.35433 | N |
| 1 | A | 5.27559 | N |

The data are found in a data file named ROLLS.xxx. It seems to you that the roll cannot be properly wrapped if the diameter of the roll is too large. Use graphs, numerical summaries, and hypothesis testing to determine if there is any difference in the average diameter of the rolls that are scrapped and those that are not scrapped. Also investigate whether or not there is any difference in the average roll diameter over time or by shift.

After doing this analysis, you ask the engineers on the team what factors affect the roll diameter. They were not sure but they helped you understand the manufacturing process. You learn that there are several machine settings that might affect the roll diameter. They are:

1. Embosser roll gap: the mechanism that puts the pattern in the towel
2. Draw roll gap: the opening through which the towel material is pulled into the winders
3. Speed: the speed at which the machine winds the rolls of towels

The engineers also told you to be sure to pay attention to the firmness of the towel roll. They warned you that in fixing the diameter problem, you might cause a different problem. If the towel roll is not firm that will also affect wrapping, and if it is too firm consumers will perceive it as stiff and complain.

The machine was run at different settings for each factor and rolls of towels were taken from the end of the production line. The towel diameter and roll firmness were measured. A small portion of the data set is shown below:

| Drawroll | Speed | Embosser | Diameter | Firmness |
|----------|-------|----------|----------|----------|
| 0 | 0 | 0 | 5.43307 | 0.271667 |
| 0 | 0 | 0 | 5.39370 | 0.336667 |
| 0 | 0 | 0 | 5.47244 | 0.260333 |
| 0 | 0 | 0 | 5.39370 | 0.261333 |
| 0 | 0 | 0 | 5.43307 | 0.297000 |
| 0 | 1 | 0 | 5.39370 | 0.312000 |
| 0 | 1 | 0 | 5.47244 | 0.294667 |

The data are found in a file named FACTORS.xxx.

Prepare a report for management explaining the effects that each of the machine settings has on roll diameter and firmness. Management also wishes to know whether the rolls meet the specifications for the mean roll diameter set at 5.35 inches and the maximum roll diameter of 5.50 inches. Indicate which settings result in the best values for roll diameter. Make a recommendation on machine settings if you can.

Before your report is completed, someone on the team reminds you that the roll diameter was measured to the nearest millimeter and converted to inches because the product specifications for the paper towel product are given in English measures, not metric. You remember your statistics teacher pointing out the fact that statistically different and practically different are *not* necessarily the same. Perhaps you should reconsider your conclusions.

# APPENDIX *A*

# *STATISTICAL TABLES*

### TABLE I  Random Number Table

| Row # | 1 | 2 | 3 | 4 | 5 | 6 | 7 | 8 | 9 |
|---|---|---|---|---|---|---|---|---|---|
| 1 | 094632795 | 711501513 | 537971597 | 562758635 | 410398128 | 182794408 | 773761503 | 455139927 | 132682754 |
| 2 | 033413186 | 653475420 | 289063704 | 485441982 | 460744361 | 328703833 | 289612212 | 569540556 | 620271271 |
| 3 | 297556368 | 658953044 | 738968017 | 414437050 | 296126017 | 075254187 | 702140315 | 467039889 | 762226273 |
| 4 | 472960570 | 785645638 | 574817322 | 817883255 | 976076280 | 843373358 | 118284363 | 445336907 | 327380271 |
| 5 | 256883707 | 716249997 | 378236162 | 467694224 | 193707682 | 380141891 | 605807481 | 180164558 | 473854769 |
| 6 | 179451522 | 878902420 | 602450872 | 987686989 | 686677180 | 242196303 | 517640224 | 691116863 | 275385608 |
| 7 | 894964682 | 704841116 | 241902107 | 750429362 | 794778197 | 693242123 | 316755091 | 193593484 | 913974355 |
| 8 | 738120861 | 744470405 | 873393138 | 758824215 | 394004646 | 496696605 | 006936567 | 163371217 | 727267920 |
| 9 | 803156944 | 653387115 | 716335974 | 835667154 | 066959782 | 908783760 | 165946696 | 735683921 | 894672507 |
| 10 | 187636922 | 953598780 | 481536873 | 055734541 | 493193305 | 566923120 | 435549770 | 007706188 | 839596393 |
| 11 | 102021077 | 286953643 | 851411058 | 132935798 | 745770831 | 187026467 | 363837178 | 791264282 | 107184709 |
| 12 | 734254842 | 133959443 | 113708008 | 443989454 | 786207141 | 772432741 | 682053431 | 048076059 | 617648837 |
| 13 | 757417865 | 524596578 | 240504889 | 544970942 | 340054233 | 417544234 | 302126745 | 003333205 | 250247568 |
| 14 | 227658086 | 233543943 | 487116060 | 577966118 | 524453480 | 934483237 | 367425608 | 431112250 | 536516890 |
| 15 | 746794759 | 557361146 | 826373105 | 870360802 | 412399571 | 804914923 | 128067420 | 659566961 | 452000520 |
| 16 | 235158119 | 336776002 | 728424416 | 086967212 | 040966064 | 335090111 | 985461873 | 832921870 | 461741235 |
| 17 | 944558037 | 787700710 | 060386364 | 635482046 | 558143223 | 600009181 | 448499754 | 064172342 | 713707601 |
| 18 | 658344036 | 271853277 | 275251035 | 744269244 | 877186509 | 130398637 | 367142231 | 846275675 | 485443650 |
| 19 | 179411521 | 104680475 | 020354893 | 576185422 | 778014690 | 380931445 | 886031872 | 320231466 | 062555147 |
| 20 | 865570814 | 699503925 | 628956988 | 683503622 | 276170341 | 744494133 | 081246804 | 523527226 | 198219562 |
| 21 | 984902072 | 022065717 | 504274676 | 136174524 | 195356906 | 027900159 | 809382340 | 381669407 | 544140648 |
| 22 | 139846732 | 496390379 | 582502144 | 768571665 | 177715615 | 830320391 | 105937107 | 329901920 | 618226629 |
| 23 | 492146502 | 493503513 | 138813631 | 479880385 | 684082619 | 010963692 | 268892703 | 552334849 | 488002392 |
| 24 | 739727377 | 916314641 | 263944162 | 861966588 | 286459084 | 491798049 | 760316559 | 837966446 | 371951811 |
| 25 | 075823838 | 115491339 | 547215506 | 007869049 | 323138362 | 193432798 | 361574944 | 787418390 | 016648846 |
| 26 | 915745558 | 104176259 | 349828840 | 546922404 | 266406684 | 490531595 | 336155799 | 242136076 | 641061181 |
| 27 | 136001350 | 309685268 | 986533618 | 587428568 | 052231831 | 422269870 | 302793461 | 564542482 | 915031158 |
| 28 | 691703257 | 926306032 | 988266746 | 716231516 | 519662016 | 986665536 | 993015206 | 066999095 | 731533696 |
| 29 | 154222042 | 316334873 | 963715901 | 044966315 | 846937935 | 104409586 | 768790545 | 113341348 | 108261519 |
| 30 | 983823702 | 641345385 | 203912928 | 219869690 | 208288383 | 861497163 | 149918954 | 160034550 | 759951927 |
| 31 | 277515731 | 805241329 | 549047139 | 285206828 | 534033122 | 130094940 | 970748730 | 798208639 | 485614463 |
| 32 | 566308464 | 543263173 | 711363354 | 339738940 | 051286779 | 714375200 | 698531722 | 072971345 | 762369710 |
| 33 | 533605194 | 994619567 | 798813607 | 079914804 | 405016946 | 275797011 | 801942743 | 814124918 | 033457635 |
| 34 | 374266385 | 237398626 | 014653680 | 107885763 | 848594153 | 093210516 | 751461171 | 121583622 | 388598493 |
| 35 | 739563239 | 736604709 | 251789737 | 480977217 | 432264262 | 975146204 | 639768152 | 455086460 | 573742841 |
| 36 | 539277994 | 536590953 | 293592699 | 279474008 | 525803109 | 281596944 | 199856046 | 646139218 | 124051051 |
| 37 | 609535944 | 455877404 | 251255783 | 334162937 | 523110770 | 461537085 | 359043224 | 423641232 | 223047648 |
| 38 | 254030985 | 962503045 | 747584829 | 988588554 | 631976076 | 901087130 | 961891746 | 149014870 | 557453130 |
| 39 | 869881662 | 489992047 | 240739861 | 875737562 | 237409010 | 135678267 | 196964820 | 343397286 | 364892121 |
| 40 | 935509137 | 382168564 | 659392779 | 628617853 | 533473897 | 569590209 | 333032014 | 937163707 | 460780426 |
| 41 | 727184789 | 651476235 | 562537081 | 259345598 | 118701307 | 970343244 | 678458590 | 189103174 | 840164700 |
| 42 | 076139028 | 276588587 | 329963439 | 184510242 | 904140612 | 761154973 | 175127287 | 520167477 | 242700207 |
| 43 | 181087637 | 087629199 | 028292894 | 460181436 | 623140518 | 207937371 | 238398056 | 136009368 | 581975565 |
| 44 | 514095654 | 401875869 | 095986936 | 976620836 | 391483115 | 713574093 | 679457157 | 184527765 | 553593737 |
| 45 | 804743163 | 129181317 | 005547120 | 712455031 | 948648814 | 909230622 | 839276576 | 726704039 | 427115217 |
| 46 | 492640056 | 038991626 | 749280637 | 430162677 | 414656226 | 603291802 | 983746203 | 756647413 | 333907575 |
| 47 | 105925664 | 168586200 | 348119547 | 829517480 | 244539107 | 448715108 | 154400559 | 634475954 | 701530403 |
| 48 | 688908833 | 510541646 | 706386776 | 935447659 | 798110618 | 127019341 | 889979390 | 455625169 | 283128630 |
| 49 | 806445894 | 067701203 | 566577304 | 808746117 | 093933115 | 198530698 | 142531634 | 042491555 | 776838859 |
| 50 | 249997369 | 047395403 | 102944245 | 987149692 | 239682871 | 971259345 | 193515078 | 797533485 | 459099813 |

## TABLE 2   Binomial Probability Tables

$P(X = x)$

**n = 5**                                                                π

| x | 0.05 | 0.10 | 0.20 | 0.25 | 0.30 | 0.40 | 0.50 | 0.60 | 0.70 | 0.75 | 0.80 | 0.90 | 0.95 |
|---|------|------|------|------|------|------|------|------|------|------|------|------|------|
| 0 | 0.774 | 0.590 | 0.328 | 0.237 | 0.168 | 0.078 | 0.031 | 0.010 | 0.002 | 0.001 | 0.000 | 0.000 | 0.000 |
| 1 | 0.204 | 0.328 | 0.410 | 0.396 | 0.360 | 0.259 | 0.156 | 0.077 | 0.028 | 0.015 | 0.006 | 0.000 | 0.000 |
| 2 | 0.021 | 0.073 | 0.205 | 0.264 | 0.309 | 0.346 | 0.313 | 0.230 | 0.132 | 0.088 | 0.051 | 0.008 | 0.001 |
| 3 | 0.001 | 0.008 | 0.051 | 0.088 | 0.132 | 0.230 | 0.313 | 0.346 | 0.309 | 0.264 | 0.205 | 0.073 | 0.021 |
| 4 | 0.000 | 0.000 | 0.006 | 0.015 | 0.028 | 0.077 | 0.156 | 0.259 | 0.360 | 0.396 | 0.410 | 0.328 | 0.204 |
| 5 | 0.000 | 0.000 | 0.000 | 0.001 | 0.002 | 0.010 | 0.031 | 0.078 | 0.168 | 0.237 | 0.328 | 0.590 | 0.774 |

**n = 10**                                                                π

| x | 0.050 | 0.100 | 0.200 | 0.250 | 0.300 | 0.400 | 0.500 | 0.600 | 0.700 | 0.750 | 0.800 | 0.900 | 0.950 |
|---|-------|-------|-------|-------|-------|-------|-------|-------|-------|-------|-------|-------|-------|
| 0 | 0.599 | 0.349 | 0.107 | 0.056 | 0.028 | 0.006 | 0.001 | 0.000 | 0.000 | 0.000 | 0.000 | 0.000 | 0.000 |
| 1 | 0.315 | 0.387 | 0.268 | 0.188 | 0.121 | 0.040 | 0.010 | 0.002 | 0.000 | 0.000 | 0.000 | 0.000 | 0.000 |
| 2 | 0.075 | 0.194 | 0.302 | 0.282 | 0.233 | 0.121 | 0.044 | 0.011 | 0.001 | 0.000 | 0.000 | 0.000 | 0.000 |
| 3 | 0.010 | 0.057 | 0.201 | 0.250 | 0.267 | 0.215 | 0.117 | 0.042 | 0.009 | 0.003 | 0.001 | 0.000 | 0.000 |
| 4 | 0.001 | 0.011 | 0.088 | 0.146 | 0.200 | 0.251 | 0.205 | 0.111 | 0.037 | 0.016 | 0.006 | 0.000 | 0.000 |
| 5 | 0.000 | 0.001 | 0.026 | 0.058 | 0.103 | 0.201 | 0.246 | 0.201 | 0.103 | 0.058 | 0.026 | 0.001 | 0.000 |
| 6 | 0.000 | 0.000 | 0.006 | 0.016 | 0.037 | 0.111 | 0.205 | 0.251 | 0.200 | 0.146 | 0.088 | 0.011 | 0.001 |
| 7 | 0.000 | 0.000 | 0.001 | 0.003 | 0.009 | 0.042 | 0.117 | 0.215 | 0.267 | 0.250 | 0.201 | 0.057 | 0.010 |
| 8 | 0.000 | 0.000 | 0.000 | 0.000 | 0.001 | 0.011 | 0.044 | 0.121 | 0.233 | 0.282 | 0.302 | 0.194 | 0.075 |
| 9 | 0.000 | 0.000 | 0.000 | 0.000 | 0.000 | 0.002 | 0.010 | 0.040 | 0.121 | 0.188 | 0.268 | 0.387 | 0.315 |
| 10 | 0.000 | 0.000 | 0.000 | 0.000 | 0.000 | 0.000 | 0.001 | 0.006 | 0.028 | 0.056 | 0.107 | 0.349 | 0.599 |

**n = 15**                                                                π

| x | 0.050 | 0.100 | 0.200 | 0.250 | 0.300 | 0.400 | 0.500 | 0.600 | 0.700 | 0.750 | 0.800 | 0.900 | 0.950 |
|---|-------|-------|-------|-------|-------|-------|-------|-------|-------|-------|-------|-------|-------|
| 0 | 0.463 | 0.206 | 0.035 | 0.013 | 0.005 | 0.000 | 0.000 | 0.000 | 0.000 | 0.000 | 0.000 | 0.000 | 0.000 |
| 1 | 0.366 | 0.343 | 0.132 | 0.067 | 0.031 | 0.005 | 0.000 | 0.000 | 0.000 | 0.000 | 0.000 | 0.000 | 0.000 |
| 2 | 0.135 | 0.267 | 0.231 | 0.156 | 0.092 | 0.022 | 0.003 | 0.000 | 0.000 | 0.000 | 0.000 | 0.000 | 0.000 |
| 3 | 0.031 | 0.129 | 0.250 | 0.225 | 0.170 | 0.063 | 0.014 | 0.002 | 0.000 | 0.000 | 0.000 | 0.000 | 0.000 |
| 4 | 0.005 | 0.043 | 0.188 | 0.225 | 0.219 | 0.127 | 0.042 | 0.007 | 0.001 | 0.000 | 0.000 | 0.000 | 0.000 |
| 5 | 0.001 | 0.010 | 0.103 | 0.165 | 0.206 | 0.186 | 0.092 | 0.024 | 0.003 | 0.001 | 0.000 | 0.000 | 0.000 |
| 6 | 0.000 | 0.002 | 0.043 | 0.092 | 0.147 | 0.207 | 0.153 | 0.061 | 0.012 | 0.003 | 0.001 | 0.000 | 0.000 |
| 7 | 0.000 | 0.000 | 0.014 | 0.039 | 0.081 | 0.177 | 0.196 | 0.118 | 0.035 | 0.013 | 0.003 | 0.000 | 0.000 |
| 8 | 0.000 | 0.000 | 0.003 | 0.013 | 0.035 | 0.118 | 0.196 | 0.177 | 0.081 | 0.039 | 0.014 | 0.000 | 0.000 |
| 9 | 0.000 | 0.000 | 0.001 | 0.003 | 0.012 | 0.061 | 0.153 | 0.207 | 0.147 | 0.092 | 0.043 | 0.002 | 0.000 |
| 10 | 0.000 | 0.000 | 0.000 | 0.001 | 0.003 | 0.024 | 0.092 | 0.186 | 0.206 | 0.165 | 0.103 | 0.010 | 0.001 |
| 11 | 0.000 | 0.000 | 0.000 | 0.000 | 0.001 | 0.007 | 0.042 | 0.127 | 0.219 | 0.225 | 0.188 | 0.043 | 0.005 |
| 12 | 0.000 | 0.000 | 0.000 | 0.000 | 0.000 | 0.002 | 0.014 | 0.063 | 0.170 | 0.225 | 0.250 | 0.129 | 0.031 |
| 13 | 0.000 | 0.000 | 0.000 | 0.000 | 0.000 | 0.000 | 0.003 | 0.022 | 0.092 | 0.156 | 0.231 | 0.267 | 0.135 |
| 14 | 0.000 | 0.000 | 0.000 | 0.000 | 0.000 | 0.000 | 0.000 | 0.005 | 0.031 | 0.067 | 0.132 | 0.343 | 0.366 |
| 15 | 0.000 | 0.000 | 0.000 | 0.000 | 0.000 | 0.000 | 0.000 | 0.000 | 0.005 | 0.013 | 0.035 | 0.206 | 0.463 |

TABLE 2  BINOMIAL PROBABILITY TABLES          **A3**

**n = 20**                                                                         π

| x | 0.050 | 0.100 | 0.200 | 0.250 | 0.300 | 0.400 | 0.500 | 0.600 | 0.700 | 0.750 | 0.800 | 0.900 | 0.950 |
|---|-------|-------|-------|-------|-------|-------|-------|-------|-------|-------|-------|-------|-------|
| 0 | 0.358 | 0.122 | 0.012 | 0.003 | 0.001 | 0.000 | 0.000 | 0.000 | 0.000 | 0.000 | 0.000 | 0.000 | 0.000 |
| 1 | 0.377 | 0.270 | 0.058 | 0.021 | 0.007 | 0.000 | 0.000 | 0.000 | 0.000 | 0.000 | 0.000 | 0.000 | 0.000 |
| 2 | 0.189 | 0.285 | 0.137 | 0.067 | 0.028 | 0.003 | 0.000 | 0.000 | 0.000 | 0.000 | 0.000 | 0.000 | 0.000 |
| 3 | 0.060 | 0.190 | 0.205 | 0.134 | 0.072 | 0.012 | 0.001 | 0.000 | 0.000 | 0.000 | 0.000 | 0.000 | 0.000 |
| 4 | 0.013 | 0.090 | 0.218 | 0.190 | 0.130 | 0.035 | 0.005 | 0.000 | 0.000 | 0.000 | 0.000 | 0.000 | 0.000 |
| 5 | 0.002 | 0.032 | 0.175 | 0.202 | 0.179 | 0.075 | 0.015 | 0.001 | 0.000 | 0.000 | 0.000 | 0.000 | 0.000 |
| 6 | 0.000 | 0.009 | 0.109 | 0.169 | 0.192 | 0.124 | 0.037 | 0.005 | 0.000 | 0.000 | 0.000 | 0.000 | 0.000 |
| 7 | 0.000 | 0.002 | 0.055 | 0.112 | 0.164 | 0.166 | 0.074 | 0.015 | 0.001 | 0.000 | 0.000 | 0.000 | 0.000 |
| 8 | 0.000 | 0.000 | 0.022 | 0.061 | 0.114 | 0.180 | 0.120 | 0.035 | 0.004 | 0.001 | 0.000 | 0.000 | 0.000 |
| 9 | 0.000 | 0.000 | 0.007 | 0.027 | 0.065 | 0.160 | 0.160 | 0.071 | 0.012 | 0.003 | 0.000 | 0.000 | 0.000 |
| 10 | 0.000 | 0.000 | 0.002 | 0.010 | 0.031 | 0.117 | 0.176 | 0.117 | 0.031 | 0.010 | 0.002 | 0.000 | 0.000 |
| 11 | 0.000 | 0.000 | 0.000 | 0.003 | 0.012 | 0.071 | 0.160 | 0.160 | 0.065 | 0.027 | 0.007 | 0.000 | 0.000 |
| 12 | 0.000 | 0.000 | 0.000 | 0.001 | 0.004 | 0.035 | 0.120 | 0.180 | 0.114 | 0.061 | 0.022 | 0.000 | 0.000 |
| 13 | 0.000 | 0.000 | 0.000 | 0.000 | 0.001 | 0.015 | 0.074 | 0.166 | 0.164 | 0.112 | 0.055 | 0.002 | 0.000 |
| 14 | 0.000 | 0.000 | 0.000 | 0.000 | 0.000 | 0.005 | 0.037 | 0.124 | 0.192 | 0.169 | 0.109 | 0.009 | 0.000 |
| 15 | 0.000 | 0.000 | 0.000 | 0.000 | 0.000 | 0.001 | 0.015 | 0.075 | 0.179 | 0.202 | 0.175 | 0.032 | 0.002 |
| 16 | 0.000 | 0.000 | 0.000 | 0.000 | 0.000 | 0.000 | 0.005 | 0.035 | 0.130 | 0.190 | 0.218 | 0.090 | 0.013 |
| 17 | 0.000 | 0.000 | 0.000 | 0.000 | 0.000 | 0.000 | 0.001 | 0.012 | 0.072 | 0.134 | 0.205 | 0.190 | 0.060 |
| 18 | 0.000 | 0.000 | 0.000 | 0.000 | 0.000 | 0.000 | 0.000 | 0.003 | 0.028 | 0.067 | 0.137 | 0.285 | 0.189 |
| 19 | 0.000 | 0.000 | 0.000 | 0.000 | 0.000 | 0.000 | 0.000 | 0.000 | 0.007 | 0.021 | 0.058 | 0.270 | 0.377 |
| 20 | 0.000 | 0.000 | 0.000 | 0.000 | 0.000 | 0.000 | 0.000 | 0.000 | 0.001 | 0.003 | 0.012 | 0.122 | 0.358 |

**n = 25**                                                                         π

| x | 0.050 | 0.100 | 0.200 | 0.250 | 0.300 | 0.400 | 0.500 | 0.600 | 0.700 | 0.750 | 0.800 | 0.900 | 0.950 |
|---|-------|-------|-------|-------|-------|-------|-------|-------|-------|-------|-------|-------|-------|
| 0 | 0.277 | 0.072 | 0.004 | 0.001 | 0.000 | 0.000 | 0.000 | 0.000 | 0.000 | 0.000 | 0.000 | 0.000 | 0.000 |
| 1 | 0.365 | 0.199 | 0.024 | 0.006 | 0.001 | 0.000 | 0.000 | 0.000 | 0.000 | 0.000 | 0.000 | 0.000 | 0.000 |
| 2 | 0.231 | 0.266 | 0.071 | 0.025 | 0.007 | 0.000 | 0.000 | 0.000 | 0.000 | 0.000 | 0.000 | 0.000 | 0.000 |
| 3 | 0.093 | 0.226 | 0.136 | 0.064 | 0.024 | 0.002 | 0.000 | 0.000 | 0.000 | 0.000 | 0.000 | 0.000 | 0.000 |
| 4 | 0.027 | 0.138 | 0.187 | 0.118 | 0.057 | 0.007 | 0.000 | 0.000 | 0.000 | 0.000 | 0.000 | 0.000 | 0.000 |
| 5 | 0.006 | 0.065 | 0.196 | 0.165 | 0.103 | 0.020 | 0.002 | 0.000 | 0.000 | 0.000 | 0.000 | 0.000 | 0.000 |
| 6 | 0.001 | 0.024 | 0.163 | 0.183 | 0.147 | 0.044 | 0.005 | 0.000 | 0.000 | 0.000 | 0.000 | 0.000 | 0.000 |
| 7 | 0.000 | 0.007 | 0.111 | 0.165 | 0.171 | 0.080 | 0.014 | 0.001 | 0.000 | 0.000 | 0.000 | 0.000 | 0.000 |
| 8 | 0.000 | 0.002 | 0.062 | 0.124 | 0.165 | 0.120 | 0.032 | 0.003 | 0.000 | 0.000 | 0.000 | 0.000 | 0.000 |
| 9 | 0.000 | 0.000 | 0.029 | 0.078 | 0.134 | 0.151 | 0.061 | 0.009 | 0.000 | 0.000 | 0.000 | 0.000 | 0.000 |
| 10 | 0.000 | 0.000 | 0.012 | 0.042 | 0.092 | 0.161 | 0.097 | 0.021 | 0.001 | 0.000 | 0.000 | 0.000 | 0.000 |
| 11 | 0.000 | 0.000 | 0.004 | 0.019 | 0.054 | 0.147 | 0.133 | 0.043 | 0.004 | 0.001 | 0.000 | 0.000 | 0.000 |
| 12 | 0.000 | 0.000 | 0.001 | 0.007 | 0.027 | 0.114 | 0.155 | 0.076 | 0.011 | 0.002 | 0.000 | 0.000 | 0.000 |
| 13 | 0.000 | 0.000 | 0.000 | 0.002 | 0.011 | 0.076 | 0.155 | 0.114 | 0.027 | 0.007 | 0.001 | 0.000 | 0.000 |
| 14 | 0.000 | 0.000 | 0.000 | 0.001 | 0.004 | 0.043 | 0.133 | 0.147 | 0.054 | 0.019 | 0.004 | 0.000 | 0.000 |
| 15 | 0.000 | 0.000 | 0.000 | 0.000 | 0.001 | 0.021 | 0.097 | 0.161 | 0.092 | 0.042 | 0.012 | 0.000 | 0.000 |
| 16 | 0.000 | 0.000 | 0.000 | 0.000 | 0.000 | 0.009 | 0.061 | 0.151 | 0.134 | 0.078 | 0.029 | 0.000 | 0.000 |
| 17 | 0.000 | 0.000 | 0.000 | 0.000 | 0.000 | 0.003 | 0.032 | 0.120 | 0.165 | 0.124 | 0.062 | 0.002 | 0.000 |
| 18 | 0.000 | 0.000 | 0.000 | 0.000 | 0.000 | 0.001 | 0.014 | 0.080 | 0.171 | 0.165 | 0.111 | 0.007 | 0.000 |
| 19 | 0.000 | 0.000 | 0.000 | 0.000 | 0.000 | 0.000 | 0.005 | 0.044 | 0.147 | 0.183 | 0.163 | 0.024 | 0.001 |
| 20 | 0.000 | 0.000 | 0.000 | 0.000 | 0.000 | 0.000 | 0.002 | 0.020 | 0.103 | 0.165 | 0.196 | 0.065 | 0.006 |
| 21 | 0.000 | 0.000 | 0.000 | 0.000 | 0.000 | 0.000 | 0.000 | 0.007 | 0.057 | 0.118 | 0.187 | 0.138 | 0.027 |
| 22 | 0.000 | 0.000 | 0.000 | 0.000 | 0.000 | 0.000 | 0.000 | 0.002 | 0.024 | 0.064 | 0.136 | 0.226 | 0.093 |
| 23 | 0.000 | 0.000 | 0.000 | 0.000 | 0.000 | 0.000 | 0.000 | 0.000 | 0.007 | 0.025 | 0.071 | 0.266 | 0.231 |
| 24 | 0.000 | 0.000 | 0.000 | 0.000 | 0.000 | 0.000 | 0.000 | 0.000 | 0.001 | 0.006 | 0.024 | 0.199 | 0.365 |
| 25 | 0.000 | 0.000 | 0.000 | 0.000 | 0.000 | 0.000 | 0.000 | 0.000 | 0.000 | 0.001 | 0.004 | 0.072 | 0.277 |

## TABLE 3 Standard Normal Table

$P(Z < z)$

### Second Decimal Place

| z | 0.00 | 0.01 | 0.02 | 0.03 | 0.04 | 0.05 | 0.06 | 0.07 | 0.08 | 0.09 |
|---|------|------|------|------|------|------|------|------|------|------|
| -3.9 | 0.0000 | 0.0000 | 0.0000 | 0.0000 | 0.0000 | 0.0000 | 0.0000 | 0.0000 | 0.0000 | 0.0000 |
| -3.8 | 0.0001 | 0.0001 | 0.0001 | 0.0001 | 0.0001 | 0.0001 | 0.0001 | 0.0001 | 0.0001 | 0.0001 |
| -3.7 | 0.0001 | 0.0001 | 0.0001 | 0.0001 | 0.0001 | 0.0001 | 0.0001 | 0.0001 | 0.0001 | 0.0001 |
| -3.6 | 0.0002 | 0.0002 | 0.0001 | 0.0001 | 0.0001 | 0.0001 | 0.0001 | 0.0001 | 0.0001 | 0.0001 |
| -3.5 | 0.0002 | 0.0002 | 0.0002 | 0.0002 | 0.0002 | 0.0002 | 0.0002 | 0.0002 | 0.0002 | 0.0002 |
| -3.4 | 0.0003 | 0.0003 | 0.0003 | 0.0003 | 0.0003 | 0.0003 | 0.0003 | 0.0003 | 0.0003 | 0.0002 |
| -3.3 | 0.0005 | 0.0005 | 0.0005 | 0.0004 | 0.0004 | 0.0004 | 0.0004 | 0.0004 | 0.0004 | 0.0003 |
| -3.2 | 0.0007 | 0.0007 | 0.0006 | 0.0006 | 0.0006 | 0.0006 | 0.0006 | 0.0005 | 0.0005 | 0.0005 |
| -3.1 | 0.0010 | 0.0009 | 0.0009 | 0.0009 | 0.0008 | 0.0008 | 0.0008 | 0.0008 | 0.0007 | 0.0007 |
| -3.0 | 0.0013 | 0.0013 | 0.0013 | 0.0012 | 0.0012 | 0.0011 | 0.0011 | 0.0011 | 0.0010 | 0.0010 |
| -2.9 | 0.0019 | 0.0018 | 0.0018 | 0.0017 | 0.0016 | 0.0016 | 0.0015 | 0.0015 | 0.0014 | 0.0014 |
| -2.8 | 0.0026 | 0.0025 | 0.0024 | 0.0023 | 0.0023 | 0.0022 | 0.0021 | 0.0021 | 0.0020 | 0.0019 |
| -2.7 | 0.0035 | 0.0034 | 0.0033 | 0.0032 | 0.0031 | 0.0030 | 0.0029 | 0.0028 | 0.0027 | 0.0026 |
| -2.6 | 0.0047 | 0.0045 | 0.0044 | 0.0043 | 0.0041 | 0.0040 | 0.0039 | 0.0038 | 0.0037 | 0.0036 |
| -2.5 | 0.0062 | 0.0060 | 0.0059 | 0.0057 | 0.0055 | 0.0054 | 0.0052 | 0.0051 | 0.0049 | 0.0048 |
| -2.4 | 0.0082 | 0.0080 | 0.0078 | 0.0075 | 0.0073 | 0.0071 | 0.0069 | 0.0068 | 0.0066 | 0.0064 |
| -2.3 | 0.0107 | 0.0104 | 0.0102 | 0.0099 | 0.0096 | 0.0094 | 0.0091 | 0.0089 | 0.0087 | 0.0084 |
| -2.2 | 0.0139 | 0.0136 | 0.0132 | 0.0129 | 0.0125 | 0.0122 | 0.0119 | 0.0116 | 0.0113 | 0.0110 |
| -2.1 | 0.0179 | 0.0174 | 0.0170 | 0.0166 | 0.0162 | 0.0158 | 0.0154 | 0.0150 | 0.0146 | 0.0143 |
| -2.0 | 0.0228 | 0.0222 | 0.0217 | 0.0212 | 0.0207 | 0.0202 | 0.0197 | 0.0192 | 0.0188 | 0.0183 |
| -1.9 | 0.0287 | 0.0281 | 0.0274 | 0.0268 | 0.0262 | 0.0256 | 0.0250 | 0.0244 | 0.0239 | 0.0233 |
| -1.8 | 0.0359 | 0.0351 | 0.0344 | 0.0336 | 0.0329 | 0.0322 | 0.0314 | 0.0307 | 0.0301 | 0.0294 |
| -1.7 | 0.0446 | 0.0436 | 0.0427 | 0.0418 | 0.0409 | 0.0401 | 0.0392 | 0.0384 | 0.0375 | 0.0367 |
| -1.6 | 0.0548 | 0.0537 | 0.0526 | 0.0516 | 0.0505 | 0.0495 | 0.0485 | 0.0475 | 0.0465 | 0.0455 |
| -1.5 | 0.0668 | 0.0655 | 0.0643 | 0.0630 | 0.0618 | 0.0606 | 0.0594 | 0.0582 | 0.0571 | 0.0559 |
| -1.4 | 0.0808 | 0.0793 | 0.0778 | 0.0764 | 0.0749 | 0.0735 | 0.0721 | 0.0708 | 0.0694 | 0.0681 |
| -1.3 | 0.0968 | 0.0951 | 0.0934 | 0.0918 | 0.0901 | 0.0885 | 0.0869 | 0.0853 | 0.0838 | 0.0823 |
| -1.2 | 0.1151 | 0.1131 | 0.1112 | 0.1093 | 0.1075 | 0.1056 | 0.1038 | 0.1020 | 0.1003 | 0.0985 |
| -1.1 | 0.1357 | 0.1335 | 0.1314 | 0.1292 | 0.1271 | 0.1251 | 0.1230 | 0.1210 | 0.1190 | 0.1170 |
| -1.0 | 0.1587 | 0.1562 | 0.1539 | 0.1515 | 0.1492 | 0.1469 | 0.1446 | 0.1423 | 0.1401 | 0.1379 |
| -0.9 | 0.1841 | 0.1814 | 0.1788 | 0.1762 | 0.1736 | 0.1711 | 0.1685 | 0.1660 | 0.1635 | 0.1611 |
| -0.8 | 0.2119 | 0.2090 | 0.2061 | 0.2033 | 0.2005 | 0.1977 | 0.1949 | 0.1922 | 0.1894 | 0.1867 |
| -0.7 | 0.2420 | 0.2389 | 0.2358 | 0.2327 | 0.2296 | 0.2266 | 0.2236 | 0.2206 | 0.2177 | 0.2148 |
| -0.6 | 0.2743 | 0.2709 | 0.2676 | 0.2643 | 0.2611 | 0.2578 | 0.2546 | 0.2514 | 0.2483 | 0.2451 |
| -0.5 | 0.3085 | 0.3050 | 0.3015 | 0.2981 | 0.2946 | 0.2912 | 0.2877 | 0.2843 | 0.2810 | 0.2776 |
| -0.4 | 0.3446 | 0.3409 | 0.3372 | 0.3336 | 0.3300 | 0.3264 | 0.3228 | 0.3192 | 0.3156 | 0.3121 |
| -0.3 | 0.3821 | 0.3783 | 0.3745 | 0.3707 | 0.3669 | 0.3632 | 0.3594 | 0.3557 | 0.3520 | 0.3483 |
| -0.2 | 0.4207 | 0.4168 | 0.4129 | 0.4090 | 0.4052 | 0.4013 | 0.3974 | 0.3936 | 0.3897 | 0.3859 |
| -0.1 | 0.4602 | 0.4562 | 0.4522 | 0.4483 | 0.4443 | 0.4404 | 0.4364 | 0.4325 | 0.4286 | 0.4247 |
| 0.0 | 0.5000 | 0.4960 | 0.4920 | 0.4880 | 0.4840 | 0.4801 | 0.4761 | 0.4721 | 0.4681 | 0.4641 |

TABLE 3   STANDARD  NORMAL  TABLE          **A5**

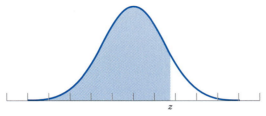

$P(Z < z)$

## Second Decimal Place

| z | 0.00 | 0.01 | 0.02 | 0.03 | 0.04 | 0.05 | 0.06 | 0.07 | 0.08 | 0.09 |
|---|---|---|---|---|---|---|---|---|---|---|
| 0.0 | 0.5000 | 0.5040 | 0.5080 | 0.5120 | 0.5160 | 0.5199 | 0.5239 | 0.5279 | 0.5319 | 0.5359 |
| 0.1 | 0.5398 | 0.5438 | 0.5478 | 0.5517 | 0.5557 | 0.5596 | 0.5636 | 0.5675 | 0.5714 | 0.5753 |
| 0.2 | 0.5793 | 0.5832 | 0.5871 | 0.5910 | 0.5948 | 0.5987 | 0.6026 | 0.6064 | 0.6103 | 0.6141 |
| 0.3 | 0.6179 | 0.6217 | 0.6255 | 0.6293 | 0.6331 | 0.6368 | 0.6406 | 0.6443 | 0.6480 | 0.6517 |
| 0.4 | 0.6554 | 0.6591 | 0.6628 | 0.6664 | 0.6700 | 0.6736 | 0.6772 | 0.6808 | 0.6844 | 0.6879 |
| 0.5 | 0.6915 | 0.6950 | 0.6985 | 0.7019 | 0.7054 | 0.7088 | 0.7123 | 0.7157 | 0.7190 | 0.7224 |
| 0.6 | 0.7257 | 0.7291 | 0.7324 | 0.7357 | 0.7389 | 0.7422 | 0.7454 | 0.7486 | 0.7517 | 0.7549 |
| 0.7 | 0.7580 | 0.7611 | 0.7642 | 0.7673 | 0.7704 | 0.7734 | 0.7764 | 0.7794 | 0.7823 | 0.7852 |
| 0.8 | 0.7881 | 0.7910 | 0.7939 | 0.7967 | 0.7995 | 0.8023 | 0.8051 | 0.8078 | 0.8106 | 0.8133 |
| 0.9 | 0.8159 | 0.8186 | 0.8212 | 0.8238 | 0.8264 | 0.8289 | 0.8315 | 0.8340 | 0.8365 | 0.8389 |
| 1.0 | 0.8413 | 0.8438 | 0.8461 | 0.8485 | 0.8508 | 0.8531 | 0.8554 | 0.8577 | 0.8599 | 0.8621 |
| 1.1 | 0.8643 | 0.8665 | 0.8686 | 0.8708 | 0.8729 | 0.8749 | 0.8770 | 0.8790 | 0.8810 | 0.8830 |
| 1.2 | 0.8849 | 0.8869 | 0.8888 | 0.8907 | 0.8925 | 0.8944 | 0.8962 | 0.8980 | 0.8997 | 0.9015 |
| 1.3 | 0.9032 | 0.9049 | 0.9066 | 0.9082 | 0.9099 | 0.9115 | 0.9131 | 0.9147 | 0.9162 | 0.9177 |
| 1.4 | 0.9192 | 0.9207 | 0.9222 | 0.9236 | 0.9251 | 0.9265 | 0.9279 | 0.9292 | 0.9306 | 0.9319 |
| 1.5 | 0.9332 | 0.9345 | 0.9357 | 0.9370 | 0.9382 | 0.9394 | 0.9406 | 0.9418 | 0.9429 | 0.9441 |
| 1.6 | 0.9452 | 0.9463 | 0.9474 | 0.9484 | 0.9495 | 0.9505 | 0.9515 | 0.9525 | 0.9535 | 0.9545 |
| 1.7 | 0.9554 | 0.9564 | 0.9573 | 0.9582 | 0.9591 | 0.9599 | 0.9608 | 0.9616 | 0.9625 | 0.9633 |
| 1.8 | 0.9641 | 0.9649 | 0.9656 | 0.9664 | 0.9671 | 0.9678 | 0.9686 | 0.9693 | 0.9699 | 0.9706 |
| 1.9 | 0.9713 | 0.9719 | 0.9726 | 0.9732 | 0.9738 | 0.9744 | 0.9750 | 0.9756 | 0.9761 | 0.9767 |
| 2.0 | 0.9772 | 0.9778 | 0.9783 | 0.9788 | 0.9793 | 0.9798 | 0.9803 | 0.9808 | 0.9812 | 0.9817 |
| 2.1 | 0.9821 | 0.9826 | 0.9830 | 0.9834 | 0.9838 | 0.9842 | 0.9846 | 0.9850 | 0.9854 | 0.9857 |
| 2.2 | 0.9861 | 0.9864 | 0.9868 | 0.9871 | 0.9875 | 0.9878 | 0.9881 | 0.9884 | 0.9887 | 0.9890 |
| 2.3 | 0.9893 | 0.9896 | 0.9898 | 0.9901 | 0.9904 | 0.9906 | 0.9909 | 0.9911 | 0.9913 | 0.9916 |
| 2.4 | 0.9918 | 0.9920 | 0.9922 | 0.9925 | 0.9927 | 0.9929 | 0.9931 | 0.9932 | 0.9934 | 0.9936 |
| 2.5 | 0.9938 | 0.9940 | 0.9941 | 0.9943 | 0.9945 | 0.9946 | 0.9948 | 0.9949 | 0.9951 | 0.9952 |
| 2.6 | 0.9953 | 0.9955 | 0.9956 | 0.9957 | 0.9959 | 0.9960 | 0.9961 | 0.9962 | 0.9963 | 0.9964 |
| 2.7 | 0.9965 | 0.9966 | 0.9967 | 0.9968 | 0.9969 | 0.9970 | 0.9971 | 0.9972 | 0.9973 | 0.9974 |
| 2.8 | 0.9974 | 0.9975 | 0.9976 | 0.9977 | 0.9977 | 0.9978 | 0.9979 | 0.9979 | 0.9980 | 0.9981 |
| 2.9 | 0.9981 | 0.9982 | 0.9982 | 0.9983 | 0.9984 | 0.9984 | 0.9985 | 0.9985 | 0.9986 | 0.9986 |
| 3.0 | 0.9987 | 0.9987 | 0.9987 | 0.9988 | 0.9988 | 0.9989 | 0.9989 | 0.9989 | 0.9990 | 0.9990 |
| 3.1 | 0.9990 | 0.9991 | 0.9991 | 0.9991 | 0.9992 | 0.9992 | 0.9992 | 0.9992 | 0.9993 | 0.9993 |
| 3.2 | 0.9993 | 0.9993 | 0.9994 | 0.9994 | 0.9994 | 0.9994 | 0.9994 | 0.9995 | 0.9995 | 0.9995 |
| 3.3 | 0.9995 | 0.9995 | 0.9995 | 0.9996 | 0.9996 | 0.9996 | 0.9996 | 0.9996 | 0.9996 | 0.9997 |
| 3.4 | 0.9997 | 0.9997 | 0.9997 | 0.9997 | 0.9997 | 0.9997 | 0.9997 | 0.9997 | 0.9997 | 0.9998 |
| 3.5 | 0.9998 | 0.9998 | 0.9998 | 0.9998 | 0.9998 | 0.9998 | 0.9998 | 0.9998 | 0.9998 | 0.9998 |
| 3.6 | 0.9998 | 0.9998 | 0.9999 | 0.9999 | 0.9999 | 0.9999 | 0.9999 | 0.9999 | 0.9999 | 0.9999 |
| 3.7 | 0.9999 | 0.9999 | 0.9999 | 0.9999 | 0.9999 | 0.9999 | 0.9999 | 0.9999 | 0.9999 | 0.9999 |
| 3.8 | 0.9999 | 0.9999 | 0.9999 | 0.9999 | 0.9999 | 0.9999 | 0.9999 | 0.9999 | 0.9999 | 0.9999 |
| 3.9 | 1.0000 | 1.0000 | 1.0000 | 1.0000 | 1.0000 | 1.0000 | 1.0000 | 1.0000 | 1.0000 | 1.0000 |

## TABLE 4   *t* Critical Values

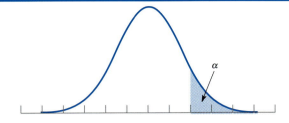

| Degrees of Freedom | Upper Tail Probability (α) | | | | | | | | |
|---|---|---|---|---|---|---|---|---|---|
| | 0.15 | 0.10 | 0.05 | 0.025 | 0.015 | 0.01 | 0.005 | 0.001 | 0.0005 |
| 1 | 1.963 | 3.078 | 6.314 | 12.706 | 21.205 | 31.821 | 63.657 | 318.309 | 1273.155 |
| 2 | 1.386 | 1.886 | 2.920 | 4.303 | 5.643 | 6.965 | 9.925 | 22.327 | 44.703 |
| 3 | 1.250 | 1.638 | 2.353 | 3.182 | 3.896 | 4.541 | 5.841 | 10.215 | 16.326 |
| 4 | 1.190 | 1.533 | 2.132 | 2.776 | 3.298 | 3.747 | 4.604 | 7.173 | 10.305 |
| 5 | 1.156 | 1.476 | 2.015 | 2.571 | 3.003 | 3.365 | 4.032 | 5.893 | 7.976 |
| 6 | 1.134 | 1.440 | 1.943 | 2.447 | 2.829 | 3.143 | 3.707 | 5.208 | 6.788 |
| 7 | 1.119 | 1.415 | 1.895 | 2.365 | 2.715 | 2.998 | 3.499 | 4.785 | 6.082 |
| 8 | 1.108 | 1.397 | 1.860 | 2.306 | 2.634 | 2.896 | 3.355 | 4.501 | 5.617 |
| 9 | 1.100 | 1.383 | 1.833 | 2.262 | 2.574 | 2.821 | 3.250 | 4.297 | 5.291 |
| 10 | 1.093 | 1.372 | 1.812 | 2.228 | 2.527 | 2.764 | 3.169 | 4.144 | 5.049 |
| 11 | 1.088 | 1.363 | 1.796 | 2.201 | 2.491 | 2.718 | 3.106 | 4.025 | 4.863 |
| 12 | 1.083 | 1.356 | 1.782 | 2.179 | 2.461 | 2.681 | 3.055 | 3.930 | 4.717 |
| 13 | 1.079 | 1.350 | 1.771 | 2.160 | 2.436 | 2.650 | 3.012 | 3.852 | 4.597 |
| 14 | 1.076 | 1.345 | 1.761 | 2.145 | 2.415 | 2.625 | 2.977 | 3.787 | 4.499 |
| 15 | 1.074 | 1.341 | 1.753 | 2.131 | 2.397 | 2.602 | 2.947 | 3.733 | 4.417 |
| 16 | 1.071 | 1.337 | 1.746 | 2.120 | 2.382 | 2.583 | 2.921 | 3.686 | 4.346 |
| 17 | 1.069 | 1.333 | 1.740 | 2.110 | 2.368 | 2.567 | 2.898 | 3.646 | 4.286 |
| 18 | 1.067 | 1.330 | 1.734 | 2.101 | 2.356 | 2.552 | 2.878 | 3.611 | 4.233 |
| 19 | 1.066 | 1.328 | 1.729 | 2.093 | 2.346 | 2.539 | 2.861 | 3.579 | 4.187 |
| 20 | 1.064 | 1.325 | 1.725 | 2.086 | 2.336 | 2.528 | 2.845 | 3.552 | 4.146 |
| 21 | 1.063 | 1.323 | 1.721 | 2.080 | 2.328 | 2.518 | 2.831 | 3.527 | 4.109 |
| 22 | 1.061 | 1.321 | 1.717 | 2.074 | 2.320 | 2.508 | 2.819 | 3.505 | 4.077 |
| 23 | 1.060 | 1.319 | 1.714 | 2.069 | 2.313 | 2.500 | 2.807 | 3.485 | 4.047 |
| 24 | 1.059 | 1.318 | 1.711 | 2.064 | 2.307 | 2.492 | 2.797 | 3.467 | 4.021 |
| 25 | 1.058 | 1.316 | 1.708 | 2.060 | 2.301 | 2.485 | 2.787 | 3.450 | 3.997 |
| 26 | 1.058 | 1.315 | 1.706 | 2.056 | 2.296 | 2.479 | 2.779 | 3.435 | 3.974 |
| 27 | 1.057 | 1.314 | 1.703 | 2.052 | 2.291 | 2.473 | 2.771 | 3.421 | 3.954 |
| 28 | 1.056 | 1.313 | 1.701 | 2.048 | 2.286 | 2.467 | 2.763 | 3.408 | 3.935 |
| 29 | 1.055 | 1.311 | 1.699 | 2.045 | 2.282 | 2.462 | 2.756 | 3.396 | 3.918 |
| 30 | 1.055 | 1.310 | 1.697 | 2.042 | 2.278 | 2.457 | 2.750 | 3.385 | 3.902 |
| 40 | 1.050 | 1.303 | 1.684 | 2.021 | 2.250 | 2.423 | 2.704 | 3.307 | 3.788 |
| 50 | 1.047 | 1.299 | 1.676 | 2.009 | 2.234 | 2.403 | 2.678 | 3.261 | 3.723 |
| 60 | 1.045 | 1.296 | 1.671 | 2.000 | 2.223 | 2.390 | 2.660 | 3.232 | 3.681 |
| 120 | 1.041 | 1.289 | 1.658 | 1.980 | 2.196 | 2.358 | 2.617 | 3.160 | 3.578 |
| **Z critical value** | | | | | | | | | |
| | 1.036 | 1.282 | 1.645 | 1.960 | 2.170 | 2.326 | 2.576 | 3.090 | 3.290 |
| **Level of Significance for a one-tailed test** | 0.15 | 0.10 | 0.05 | 0.025 | 0.015 | 0.01 | 0.005 | 0.001 | 0.0005 |
| **Level of Significance for a two-tailed test** | 0.30 | 0.20 | 0.10 | 0.05 | 0.03 | 0.02 | 0.01 | 0.002 | 0.001 |

TABLE 5  CHI-SQUARE  DISTRIBUTION  TABLE     **A7**

## TABLE 5  Chi-Square Distribution Table

The entries in this table give the critical values of $\chi^2$ for the specified number of degrees of freedom and areas in the right tail.

| Degrees of Freedom | Upper Tail Areas | | | | | | | | | | | |
|---|---|---|---|---|---|---|---|---|---|---|---|---|
| | 0.995 | 0.99 | 0.975 | 0.95 | 0.9 | 0.75 | 0.25 | 0.1 | 0.05 | 0.025 | 0.01 | 0.005 |
| 1 | 0.000 | 0.000 | 0.001 | 0.004 | 0.016 | 0.102 | 1.323 | 2.706 | 3.841 | 5.024 | 6.635 | 7.879 |
| 2 | 0.010 | 0.020 | 0.051 | 0.103 | 0.211 | 0.575 | 2.773 | 4.605 | 5.991 | 7.378 | 9.210 | 10.597 |
| 3 | 0.072 | 0.115 | 0.216 | 0.352 | 0.584 | 1.213 | 4.108 | 6.251 | 7.815 | 9.348 | 11.345 | 12.838 |
| 4 | 0.207 | 0.297 | 0.484 | 0.711 | 1.064 | 1.923 | 5.385 | 7.779 | 9.488 | 11.143 | 13.277 | 14.860 |
| 5 | 0.412 | 0.554 | 0.831 | 1.145 | 1.610 | 2.675 | 6.626 | 9.236 | 11.070 | 12.832 | 15.086 | 16.750 |
| 6 | 0.676 | 0.872 | 1.237 | 1.635 | 2.204 | 3.455 | 7.841 | 10.645 | 12.592 | 14.449 | 16.812 | 18.548 |
| 7 | 0.989 | 1.239 | 1.690 | 2.167 | 2.833 | 4.255 | 9.037 | 12.017 | 14.067 | 16.013 | 18.475 | 20.278 |
| 8 | 1.344 | 1.647 | 2.180 | 2.733 | 3.490 | 5.071 | 10.219 | 13.362 | 15.507 | 17.535 | 20.090 | 21.955 |
| 9 | 1.735 | 2.088 | 2.700 | 3.325 | 4.168 | 5.899 | 11.389 | 14.684 | 16.919 | 19.023 | 21.666 | 23.589 |
| 10 | 2.156 | 2.558 | 3.247 | 3.940 | 4.865 | 6.737 | 12.549 | 15.987 | 18.307 | 20.483 | 23.209 | 25.188 |
| 11 | 2.603 | 3.053 | 3.816 | 4.575 | 5.578 | 7.584 | 13.701 | 17.275 | 19.675 | 21.920 | 24.725 | 26.757 |
| 12 | 3.074 | 3.571 | 4.404 | 5.226 | 6.304 | 8.438 | 14.845 | 18.549 | 21.026 | 23.337 | 26.217 | 28.300 |
| 13 | 3.565 | 4.107 | 5.009 | 5.892 | 7.041 | 9.299 | 15.984 | 19.812 | 22.362 | 24.736 | 27.688 | 29.819 |
| 14 | 4.075 | 4.660 | 5.629 | 6.571 | 7.790 | 10.165 | 17.117 | 21.064 | 23.685 | 26.119 | 29.141 | 31.319 |
| 15 | 4.601 | 5.229 | 6.262 | 7.261 | 8.547 | 11.037 | 18.245 | 22.307 | 24.996 | 27.488 | 30.578 | 32.801 |
| 16 | 5.142 | 5.812 | 6.908 | 7.962 | 9.312 | 11.912 | 19.369 | 23.542 | 26.296 | 28.845 | 32.000 | 34.267 |
| 17 | 5.697 | 6.408 | 7.564 | 8.672 | 10.085 | 12.792 | 20.489 | 24.769 | 27.587 | 30.191 | 33.409 | 35.718 |
| 18 | 6.265 | 7.015 | 8.231 | 9.390 | 10.865 | 13.675 | 21.605 | 25.989 | 28.869 | 31.526 | 34.805 | 37.156 |
| 19 | 6.844 | 7.633 | 8.907 | 10.117 | 11.651 | 14.562 | 22.718 | 27.204 | 30.144 | 32.852 | 36.191 | 38.582 |
| 20 | 7.434 | 8.260 | 9.591 | 10.851 | 12.443 | 15.452 | 23.828 | 28.412 | 31.410 | 34.170 | 37.566 | 39.997 |
| 21 | 8.034 | 8.897 | 10.283 | 11.591 | 13.240 | 16.344 | 24.935 | 29.615 | 32.671 | 35.479 | 38.932 | 41.401 |
| 22 | 8.643 | 9.542 | 10.982 | 12.338 | 14.041 | 17.240 | 26.039 | 30.813 | 33.924 | 36.781 | 40.289 | 42.796 |
| 23 | 9.260 | 10.196 | 11.689 | 13.091 | 14.848 | 18.137 | 27.141 | 32.007 | 35.172 | 38.076 | 41.638 | 44.181 |
| 24 | 9.886 | 10.856 | 12.401 | 13.848 | 15.659 | 19.037 | 28.241 | 33.196 | 36.415 | 39.364 | 42.980 | 45.558 |
| 25 | 10.520 | 11.524 | 13.120 | 14.611 | 16.473 | 19.939 | 29.339 | 34.382 | 37.652 | 40.646 | 44.314 | 46.928 |
| 26 | 11.160 | 12.198 | 13.844 | 15.379 | 17.292 | 20.843 | 30.435 | 35.563 | 38.885 | 41.923 | 45.642 | 48.290 |
| 27 | 11.808 | 12.878 | 14.573 | 16.151 | 18.114 | 21.749 | 31.528 | 36.741 | 40.113 | 43.195 | 46.963 | 49.645 |
| 28 | 12.461 | 13.565 | 15.308 | 16.928 | 18.939 | 22.657 | 32.620 | 37.916 | 41.337 | 44.461 | 48.278 | 50.994 |
| 29 | 13.121 | 14.256 | 16.047 | 17.708 | 19.768 | 23.567 | 33.711 | 39.087 | 42.557 | 45.722 | 49.588 | 52.335 |
| 30 | 13.787 | 14.953 | 16.791 | 18.493 | 20.599 | 24.478 | 34.800 | 40.256 | 43.773 | 46.979 | 50.892 | 53.672 |
| 31 | 14.458 | 15.655 | 17.539 | 19.281 | 21.434 | 25.390 | 35.887 | 41.422 | 44.985 | 48.232 | 52.191 | 55.002 |
| 32 | 15.134 | 16.362 | 18.291 | 20.072 | 22.271 | 26.304 | 36.973 | 42.585 | 46.194 | 49.480 | 53.486 | 56.328 |
| 33 | 15.815 | 17.073 | 19.047 | 20.867 | 23.110 | 27.219 | 38.058 | 43.745 | 47.400 | 50.725 | 54.775 | 57.648 |
| 34 | 16.501 | 17.789 | 19.806 | 21.664 | 23.952 | 28.136 | 39.141 | 44.903 | 48.602 | 51.966 | 56.061 | 58.964 |
| 35 | 17.192 | 18.509 | 20.569 | 22.465 | 24.797 | 29.054 | 40.223 | 46.059 | 49.802 | 53.203 | 57.342 | 60.275 |
| 40 | 20.707 | 22.164 | 24.433 | 26.509 | 29.051 | 33.660 | 45.616 | 51.805 | 55.758 | 59.342 | 63.691 | 66.766 |
| 60 | 35.534 | 37.485 | 40.482 | 43.188 | 46.459 | 52.294 | 66.981 | 74.397 | 79.082 | 83.298 | 88.379 | 91.952 |
| 120 | 83.852 | 86.923 | 91.573 | 95.705 | 100.624 | 109.220 | 130.055 | 140.233 | 146.567 | 152.211 | 158.950 | 163.648 |

## TABLE 6 The $F$ Distribution Table

a. Area in the Right Tail under the $F$ Distribution Curve = .01

| Degrees of Freedom Denominator | \multicolumn{17}{c}{Degrees of Freedom - Numerator} |
|---|---|---|---|---|---|---|---|---|---|---|---|---|---|---|---|---|---|
| | 1 | 2 | 3 | 4 | 5 | 6 | 7 | 8 | 9 | 10 | 11 | 12 | 13 | 14 | 15 | 16 | 17 |
| 1 | 4052.185 | 4999.340 | 5403.534 | 5624.257 | 5763.955 | 5858.950 | 5928.334 | 5980.954 | 6022.397 | 6055.925 | 6083.399 | 6106.682 | 6125.774 | 6143.004 | 6156.974 | 6170.012 | 6181.188 |
| 2 | 98.502 | 99.000 | 99.164 | 99.251 | 99.302 | 99.331 | 99.357 | 99.375 | 99.390 | 99.397 | 99.408 | 99.419 | 99.422 | 99.426 | 99.433 | 99.437 | 99.441 |
| 3 | 34.116 | 30.816 | 29.457 | 28.710 | 28.237 | 27.911 | 27.671 | 27.489 | 27.345 | 27.228 | 27.132 | 27.052 | 26.983 | 26.924 | 26.872 | 26.826 | 26.786 |
| 4 | 21.198 | 18.000 | 16.694 | 15.977 | 15.522 | 15.207 | 14.976 | 14.799 | 14.659 | 14.546 | 14.452 | 14.374 | 14.306 | 14.249 | 14.198 | 14.154 | 14.114 |
| 5 | 16.258 | 13.274 | 12.060 | 11.392 | 10.967 | 10.672 | 10.456 | 10.289 | 10.158 | 10.051 | 9.963 | 9.888 | 9.825 | 9.770 | 9.722 | 9.680 | 9.643 |
| 6 | 13.745 | 10.925 | 9.780 | 9.148 | 8.746 | 8.466 | 8.260 | 8.102 | 7.976 | 7.874 | 7.790 | 7.718 | 7.657 | 7.605 | 7.559 | 7.519 | 7.483 |
| 7 | 12.246 | 9.547 | 8.451 | 7.847 | 7.460 | 7.191 | 6.993 | 6.840 | 6.719 | 6.620 | 6.538 | 6.469 | 6.410 | 6.359 | 6.314 | 6.275 | 6.240 |
| 8 | 11.259 | 8.649 | 7.591 | 7.006 | 6.632 | 6.371 | 6.178 | 6.029 | 5.911 | 5.814 | 5.734 | 5.667 | 5.609 | 5.559 | 5.515 | 5.477 | 5.442 |
| 9 | 10.562 | 8.022 | 6.992 | 6.422 | 6.057 | 5.802 | 5.613 | 5.467 | 5.351 | 5.257 | 5.178 | 5.111 | 5.055 | 5.005 | 4.962 | 4.924 | 4.890 |
| 10 | 10.044 | 7.559 | 6.552 | 5.994 | 5.636 | 5.386 | 5.200 | 5.057 | 4.942 | 4.849 | 4.772 | 4.706 | 4.650 | 4.601 | 4.558 | 4.520 | 4.487 |
| 11 | 9.646 | 7.206 | 6.217 | 5.668 | 5.316 | 5.069 | 4.886 | 4.744 | 4.632 | 4.539 | 4.462 | 4.397 | 4.342 | 4.293 | 4.251 | 4.213 | 4.180 |
| 12 | 9.330 | 6.927 | 5.953 | 5.412 | 5.064 | 4.821 | 4.640 | 4.499 | 4.388 | 4.296 | 4.220 | 4.155 | 4.100 | 4.052 | 4.010 | 3.972 | 3.939 |
| 13 | 9.074 | 6.701 | 5.739 | 5.205 | 4.862 | 4.620 | 4.441 | 4.302 | 4.191 | 4.100 | 4.025 | 3.960 | 3.905 | 3.857 | 3.815 | 3.778 | 3.745 |
| 14 | 8.862 | 6.515 | 5.564 | 5.035 | 4.695 | 4.456 | 4.278 | 4.140 | 4.030 | 3.939 | 3.864 | 3.800 | 3.745 | 3.698 | 3.656 | 3.619 | 3.586 |
| 15 | 8.683 | 6.359 | 5.417 | 4.893 | 4.556 | 4.318 | 4.142 | 4.004 | 3.895 | 3.805 | 3.730 | 3.666 | 3.612 | 3.564 | 3.522 | 3.485 | 3.452 |
| 16 | 8.531 | 6.226 | 5.292 | 4.773 | 4.437 | 4.202 | 4.026 | 3.890 | 3.780 | 3.691 | 3.616 | 3.553 | 3.498 | 3.451 | 3.409 | 3.372 | 3.339 |
| 17 | 8.400 | 6.112 | 5.185 | 4.669 | 4.336 | 4.101 | 3.927 | 3.791 | 3.682 | 3.593 | 3.518 | 3.455 | 3.401 | 3.353 | 3.312 | 3.275 | 3.242 |
| 18 | 8.285 | 6.013 | 5.092 | 4.579 | 4.248 | 4.015 | 3.841 | 3.705 | 3.597 | 3.508 | 3.434 | 3.371 | 3.316 | 3.269 | 3.227 | 3.190 | 3.158 |
| 19 | 8.185 | 5.926 | 5.010 | 4.500 | 4.171 | 3.939 | 3.765 | 3.631 | 3.523 | 3.434 | 3.360 | 3.297 | 3.242 | 3.195 | 3.153 | 3.116 | 3.084 |
| 20 | 8.096 | 5.849 | 4.938 | 4.431 | 4.103 | 3.871 | 3.699 | 3.564 | 3.457 | 3.368 | 3.294 | 3.231 | 3.177 | 3.130 | 3.088 | 3.051 | 3.018 |
| 21 | 8.017 | 5.780 | 4.874 | 4.369 | 4.042 | 3.812 | 3.640 | 3.506 | 3.398 | 3.310 | 3.236 | 3.173 | 3.119 | 3.072 | 3.030 | 2.993 | 2.960 |
| 22 | 7.945 | 5.719 | 4.817 | 4.313 | 3.988 | 3.758 | 3.587 | 3.453 | 3.346 | 3.258 | 3.184 | 3.121 | 3.067 | 3.019 | 2.978 | 2.941 | 2.908 |
| 23 | 7.881 | 5.664 | 4.765 | 4.264 | 3.939 | 3.710 | 3.539 | 3.406 | 3.299 | 3.211 | 3.137 | 3.074 | 3.020 | 2.973 | 2.931 | 2.894 | 2.861 |
| 24 | 7.823 | 5.614 | 4.718 | 4.218 | 3.895 | 3.667 | 3.496 | 3.363 | 3.256 | 3.168 | 3.094 | 3.032 | 2.977 | 2.930 | 2.889 | 2.852 | 2.819 |
| 25 | 7.770 | 5.568 | 4.675 | 4.177 | 3.855 | 3.627 | 3.457 | 3.324 | 3.217 | 3.129 | 3.056 | 2.993 | 2.939 | 2.892 | 2.850 | 2.813 | 2.780 |
| 26 | 7.721 | 5.526 | 4.637 | 4.140 | 3.818 | 3.591 | 3.421 | 3.288 | 3.182 | 3.094 | 3.021 | 2.958 | 2.904 | 2.857 | 2.815 | 2.778 | 2.745 |
| 27 | 7.677 | 5.488 | 4.601 | 4.106 | 3.785 | 3.558 | 3.388 | 3.256 | 3.149 | 3.062 | 2.988 | 2.926 | 2.872 | 2.824 | 2.783 | 2.746 | 2.713 |
| 28 | 7.636 | 5.453 | 4.568 | 4.074 | 3.754 | 3.528 | 3.358 | 3.226 | 3.120 | 3.032 | 2.959 | 2.896 | 2.842 | 2.795 | 2.753 | 2.716 | 2.683 |
| 29 | 7.598 | 5.420 | 4.538 | 4.045 | 3.725 | 3.499 | 3.330 | 3.198 | 3.092 | 3.005 | 2.931 | 2.868 | 2.814 | 2.767 | 2.726 | 2.689 | 2.656 |
| 30 | 7.562 | 5.390 | 4.510 | 4.018 | 3.699 | 3.473 | 3.305 | 3.173 | 3.067 | 2.979 | 2.906 | 2.843 | 2.789 | 2.742 | 2.700 | 2.663 | 2.630 |
| 40 | 7.314 | 5.178 | 4.313 | 3.828 | 3.514 | 3.291 | 3.124 | 2.993 | 2.888 | 2.801 | 2.727 | 2.665 | 2.611 | 2.563 | 2.522 | 2.484 | 2.451 |
| 60 | 7.077 | 4.977 | 4.126 | 3.649 | 3.339 | 3.119 | 2.953 | 2.823 | 2.718 | 2.632 | 2.559 | 2.496 | 2.442 | 2.394 | 2.352 | 2.315 | 2.281 |
| 120 | 6.851 | 4.787 | 3.949 | 3.480 | 3.174 | 2.956 | 2.792 | 2.663 | 2.559 | 2.472 | 2.399 | 2.336 | 2.282 | 2.234 | 2.191 | 2.154 | 2.119 |

TABLE 6   THE *F* DISTRIBUTION TABLE                    **A9**

## TABLE 6   *(Continued)*

| Degrees of Freedom Denominator | Degrees of Freedom - Numerator | | | | | | | | | | | | | | | |
|---|---|---|---|---|---|---|---|---|---|---|---|---|---|---|---|---|
| | 18 | 19 | 20 | 21 | 22 | 23 | 24 | 25 | 26 | 27 | 28 | 29 | 30 | 40 | 60 | 120 |
| 1 | 6191.432 | 6200.746 | 6208.662 | 6216.113 | 6223.097 | 6228.685 | 6234.273 | 6239.861 | 6244.518 | 6249.174 | 6252.900 | 6257.091 | 6260.350 | 6286.427 | 6312.970 | 6339.513 |
| 2 | 99.444 | 99.448 | 99.448 | 99.451 | 99.455 | 99.455 | 99.455 | 99.459 | 99.462 | 99.462 | 99.462 | 99.462 | 99.466 | 99.477 | 99.484 | 99.491 |
| 3 | 26.751 | 26.719 | 26.690 | 26.664 | 26.639 | 26.617 | 26.597 | 26.579 | 26.562 | 26.546 | 26.531 | 26.517 | 26.504 | 26.411 | 26.316 | 26.221 |
| 4 | 14.079 | 14.048 | 14.019 | 13.994 | 13.970 | 13.949 | 13.929 | 13.911 | 13.894 | 13.878 | 13.864 | 13.850 | 13.838 | 13.745 | 13.652 | 13.558 |
| 5 | 9.609 | 9.580 | 9.553 | 9.528 | 9.506 | 9.485 | 9.466 | 9.449 | 9.433 | 9.418 | 9.404 | 9.391 | 9.379 | 9.291 | 9.202 | 9.112 |
| 6 | 7.451 | 7.422 | 7.396 | 7.372 | 7.351 | 7.331 | 7.313 | 7.296 | 7.281 | 7.266 | 7.253 | 7.240 | 7.229 | 7.143 | 7.057 | 6.969 |
| 7 | 6.209 | 6.181 | 6.155 | 6.132 | 6.111 | 6.092 | 6.074 | 6.058 | 6.043 | 6.029 | 6.016 | 6.003 | 5.992 | 5.908 | 5.824 | 5.737 |
| 8 | 5.412 | 5.384 | 5.359 | 5.336 | 5.316 | 5.297 | 5.279 | 5.263 | 5.248 | 5.234 | 5.221 | 5.209 | 5.198 | 5.116 | 5.032 | 4.946 |
| 9 | 4.860 | 4.833 | 4.808 | 4.786 | 4.765 | 4.746 | 4.729 | 4.713 | 4.698 | 4.684 | 4.672 | 4.660 | 4.649 | 4.567 | 4.483 | 4.398 |
| 10 | 4.457 | 4.430 | 4.405 | 4.383 | 4.363 | 4.344 | 4.327 | 4.311 | 4.296 | 4.283 | 4.270 | 4.258 | 4.247 | 4.165 | 4.082 | 3.996 |
| 11 | 4.150 | 4.123 | 4.099 | 4.077 | 4.057 | 4.038 | 4.021 | 4.005 | 3.990 | 3.977 | 3.964 | 3.952 | 3.941 | 3.860 | 3.776 | 3.690 |
| 12 | 3.910 | 3.883 | 3.858 | 3.836 | 3.816 | 3.798 | 3.780 | 3.765 | 3.750 | 3.736 | 3.724 | 3.712 | 3.701 | 3.619 | 3.535 | 3.449 |
| 13 | 3.716 | 3.689 | 3.665 | 3.643 | 3.622 | 3.604 | 3.587 | 3.571 | 3.556 | 3.543 | 3.530 | 3.518 | 3.507 | 3.425 | 3.341 | 3.255 |
| 14 | 3.556 | 3.529 | 3.505 | 3.483 | 3.463 | 3.444 | 3.427 | 3.412 | 3.397 | 3.383 | 3.371 | 3.359 | 3.348 | 3.266 | 3.181 | 3.094 |
| 15 | 3.423 | 3.396 | 3.372 | 3.350 | 3.330 | 3.311 | 3.294 | 3.278 | 3.264 | 3.250 | 3.237 | 3.225 | 3.214 | 3.132 | 3.047 | 2.959 |
| 16 | 3.310 | 3.283 | 3.259 | 3.237 | 3.216 | 3.198 | 3.181 | 3.165 | 3.150 | 3.137 | 3.124 | 3.112 | 3.101 | 3.018 | 2.933 | 2.845 |
| 17 | 3.212 | 3.186 | 3.162 | 3.139 | 3.119 | 3.101 | 3.083 | 3.068 | 3.053 | 3.039 | 3.026 | 3.014 | 3.003 | 2.920 | 2.835 | 2.746 |
| 18 | 3.128 | 3.101 | 3.077 | 3.055 | 3.035 | 3.016 | 2.999 | 2.983 | 2.968 | 2.955 | 2.942 | 2.930 | 2.919 | 2.835 | 2.749 | 2.660 |
| 19 | 3.054 | 3.027 | 3.003 | 2.981 | 2.961 | 2.942 | 2.925 | 2.909 | 2.894 | 2.880 | 2.868 | 2.855 | 2.844 | 2.761 | 2.674 | 2.584 |
| 20 | 2.989 | 2.962 | 2.938 | 2.916 | 2.895 | 2.877 | 2.859 | 2.843 | 2.829 | 2.815 | 2.802 | 2.790 | 2.778 | 2.695 | 2.608 | 2.517 |
| 21 | 2.931 | 2.904 | 2.880 | 2.857 | 2.837 | 2.818 | 2.801 | 2.785 | 2.770 | 2.756 | 2.743 | 2.731 | 2.720 | 2.636 | 2.548 | 2.457 |
| 22 | 2.879 | 2.852 | 2.827 | 2.805 | 2.785 | 2.766 | 2.749 | 2.733 | 2.718 | 2.704 | 2.691 | 2.679 | 2.667 | 2.583 | 2.495 | 2.403 |
| 23 | 2.832 | 2.805 | 2.780 | 2.758 | 2.738 | 2.719 | 2.702 | 2.686 | 2.671 | 2.657 | 2.644 | 2.632 | 2.620 | 2.536 | 2.447 | 2.354 |
| 24 | 2.789 | 2.762 | 2.738 | 2.716 | 2.695 | 2.676 | 2.659 | 2.643 | 2.628 | 2.614 | 2.601 | 2.589 | 2.577 | 2.492 | 2.403 | 2.310 |
| 25 | 2.751 | 2.724 | 2.699 | 2.677 | 2.657 | 2.638 | 2.620 | 2.604 | 2.589 | 2.575 | 2.562 | 2.550 | 2.538 | 2.453 | 2.364 | 2.270 |
| 26 | 2.715 | 2.688 | 2.664 | 2.642 | 2.621 | 2.602 | 2.585 | 2.569 | 2.554 | 2.540 | 2.526 | 2.514 | 2.503 | 2.417 | 2.327 | 2.233 |
| 27 | 2.683 | 2.656 | 2.632 | 2.609 | 2.589 | 2.570 | 2.552 | 2.536 | 2.521 | 2.507 | 2.494 | 2.481 | 2.470 | 2.384 | 2.294 | 2.198 |
| 28 | 2.653 | 2.626 | 2.602 | 2.579 | 2.559 | 2.540 | 2.522 | 2.506 | 2.491 | 2.477 | 2.464 | 2.451 | 2.440 | 2.354 | 2.263 | 2.167 |
| 29 | 2.626 | 2.599 | 2.574 | 2.552 | 2.531 | 2.512 | 2.495 | 2.478 | 2.463 | 2.449 | 2.436 | 2.423 | 2.412 | 2.325 | 2.234 | 2.138 |
| 30 | 2.600 | 2.573 | 2.549 | 2.526 | 2.506 | 2.487 | 2.469 | 2.453 | 2.437 | 2.423 | 2.410 | 2.398 | 2.386 | 2.299 | 2.208 | 2.111 |
| 40 | 2.421 | 2.394 | 2.369 | 2.346 | 2.325 | 2.306 | 2.288 | 2.271 | 2.256 | 2.241 | 2.228 | 2.215 | 2.203 | 2.114 | 2.019 | 1.917 |
| 60 | 2.251 | 2.223 | 2.198 | 2.175 | 2.153 | 2.134 | 2.115 | 2.098 | 2.083 | 2.068 | 2.054 | 2.041 | 2.028 | 1.936 | 1.836 | 1.726 |
| 120 | 2.089 | 2.060 | 2.035 | 2.011 | 1.989 | 1.969 | 1.950 | 1.932 | 1.916 | 1.901 | 1.886 | 1.873 | 1.860 | 1.763 | 1.656 | 1.533 |

## TABLE 6   (Continued)

b. Area in the Right Tail under the $F$ Distribution Curve = .05

| Degrees of Freedom Denominator | 1 | 2 | 3 | 4 | 5 | 6 | 7 | 8 | 9 | 10 | 11 | 12 | 13 | 14 | 15 | 16 | 17 |
|---|---|---|---|---|---|---|---|---|---|---|---|---|---|---|---|---|---|
| 1 | 161.446 | 199.499 | 215.707 | 224.583 | 230.160 | 233.988 | 236.767 | 238.884 | 240.543 | 241.882 | 242.981 | 243.905 | 244.690 | 245.363 | 245.949 | 246.466 | 246.917 |
| 2 | 18.513 | 19.000 | 19.164 | 19.247 | 19.296 | 19.329 | 19.353 | 19.371 | 19.385 | 19.396 | 19.405 | 19.412 | 19.419 | 19.424 | 19.429 | 19.433 | 19.437 |
| 3 | 10.128 | 9.552 | 9.277 | 9.117 | 9.013 | 8.941 | 8.887 | 8.845 | 8.812 | 8.785 | 8.763 | 8.745 | 8.729 | 8.715 | 8.703 | 8.692 | 8.683 |
| 4 | 7.709 | 6.944 | 6.591 | 6.388 | 6.256 | 6.163 | 6.094 | 6.041 | 5.999 | 5.964 | 5.936 | 5.912 | 5.891 | 5.873 | 5.858 | 5.844 | 5.832 |
| 5 | 6.608 | 5.786 | 5.409 | 5.192 | 5.050 | 4.950 | 4.876 | 4.818 | 4.772 | 4.735 | 4.704 | 4.678 | 4.655 | 4.636 | 4.619 | 4.604 | 4.590 |
| 6 | 5.987 | 5.143 | 4.757 | 4.534 | 4.387 | 4.284 | 4.207 | 4.147 | 4.099 | 4.060 | 4.027 | 4.000 | 3.976 | 3.956 | 3.938 | 3.922 | 3.908 |
| 7 | 5.591 | 4.737 | 4.347 | 4.120 | 3.972 | 3.866 | 3.787 | 3.726 | 3.677 | 3.637 | 3.603 | 3.575 | 3.550 | 3.529 | 3.511 | 3.494 | 3.480 |
| 8 | 5.318 | 4.459 | 4.066 | 3.838 | 3.688 | 3.581 | 3.500 | 3.438 | 3.388 | 3.347 | 3.313 | 3.284 | 3.259 | 3.237 | 3.218 | 3.202 | 3.187 |
| 9 | 5.117 | 4.256 | 3.863 | 3.633 | 3.482 | 3.374 | 3.293 | 3.230 | 3.179 | 3.137 | 3.102 | 3.073 | 3.048 | 3.025 | 3.006 | 2.989 | 2.974 |
| 10 | 4.965 | 4.103 | 3.708 | 3.478 | 3.326 | 3.217 | 3.135 | 3.072 | 3.020 | 2.978 | 2.943 | 2.913 | 2.887 | 2.865 | 2.845 | 2.828 | 2.812 |
| 11 | 4.844 | 3.982 | 3.587 | 3.357 | 3.204 | 3.095 | 3.012 | 2.948 | 2.896 | 2.854 | 2.818 | 2.788 | 2.761 | 2.739 | 2.719 | 2.701 | 2.685 |
| 12 | 4.747 | 3.885 | 3.490 | 3.259 | 3.106 | 2.996 | 2.913 | 2.849 | 2.796 | 2.753 | 2.717 | 2.687 | 2.660 | 2.637 | 2.617 | 2.599 | 2.583 |
| 13 | 4.667 | 3.806 | 3.411 | 3.179 | 3.025 | 2.915 | 2.832 | 2.767 | 2.714 | 2.671 | 2.635 | 2.604 | 2.577 | 2.554 | 2.533 | 2.515 | 2.499 |
| 14 | 4.600 | 3.739 | 3.344 | 3.112 | 2.958 | 2.848 | 2.764 | 2.699 | 2.646 | 2.602 | 2.565 | 2.534 | 2.507 | 2.484 | 2.463 | 2.445 | 2.428 |
| 15 | 4.543 | 3.682 | 3.287 | 3.056 | 2.901 | 2.790 | 2.707 | 2.641 | 2.588 | 2.544 | 2.507 | 2.475 | 2.448 | 2.424 | 2.403 | 2.385 | 2.368 |
| 16 | 4.494 | 3.634 | 3.239 | 3.007 | 2.852 | 2.741 | 2.657 | 2.591 | 2.538 | 2.494 | 2.456 | 2.425 | 2.397 | 2.373 | 2.352 | 2.333 | 2.317 |
| 17 | 4.451 | 3.592 | 3.197 | 2.965 | 2.810 | 2.699 | 2.614 | 2.548 | 2.494 | 2.450 | 2.413 | 2.381 | 2.353 | 2.329 | 2.308 | 2.289 | 2.272 |
| 18 | 4.414 | 3.555 | 3.160 | 2.928 | 2.773 | 2.661 | 2.577 | 2.510 | 2.456 | 2.412 | 2.374 | 2.342 | 2.314 | 2.290 | 2.269 | 2.250 | 2.233 |
| 19 | 4.381 | 3.522 | 3.127 | 2.895 | 2.740 | 2.628 | 2.544 | 2.477 | 2.423 | 2.378 | 2.340 | 2.308 | 2.280 | 2.256 | 2.234 | 2.215 | 2.198 |
| 20 | 4.351 | 3.493 | 3.098 | 2.866 | 2.711 | 2.599 | 2.514 | 2.447 | 2.393 | 2.348 | 2.310 | 2.278 | 2.250 | 2.225 | 2.203 | 2.184 | 2.167 |
| 21 | 4.325 | 3.467 | 3.072 | 2.840 | 2.685 | 2.573 | 2.488 | 2.420 | 2.366 | 2.321 | 2.283 | 2.250 | 2.222 | 2.197 | 2.176 | 2.156 | 2.139 |
| 22 | 4.301 | 3.443 | 3.049 | 2.817 | 2.661 | 2.549 | 2.464 | 2.397 | 2.342 | 2.297 | 2.259 | 2.226 | 2.198 | 2.173 | 2.151 | 2.131 | 2.114 |
| 23 | 4.279 | 3.422 | 3.028 | 2.796 | 2.640 | 2.528 | 2.442 | 2.375 | 2.320 | 2.275 | 2.236 | 2.204 | 2.175 | 2.150 | 2.128 | 2.109 | 2.091 |
| 24 | 4.260 | 3.403 | 3.009 | 2.776 | 2.621 | 2.508 | 2.423 | 2.355 | 2.300 | 2.255 | 2.216 | 2.183 | 2.155 | 2.130 | 2.108 | 2.088 | 2.070 |
| 25 | 4.242 | 3.385 | 2.991 | 2.759 | 2.603 | 2.490 | 2.405 | 2.337 | 2.282 | 2.236 | 2.198 | 2.165 | 2.136 | 2.111 | 2.089 | 2.069 | 2.051 |
| 26 | 4.225 | 3.369 | 2.975 | 2.743 | 2.587 | 2.474 | 2.388 | 2.321 | 2.265 | 2.220 | 2.181 | 2.148 | 2.119 | 2.094 | 2.072 | 2.052 | 2.034 |
| 27 | 4.210 | 3.354 | 2.960 | 2.728 | 2.572 | 2.459 | 2.373 | 2.305 | 2.250 | 2.204 | 2.166 | 2.132 | 2.103 | 2.078 | 2.056 | 2.036 | 2.018 |
| 28 | 4.196 | 3.340 | 2.947 | 2.714 | 2.558 | 2.445 | 2.359 | 2.291 | 2.236 | 2.190 | 2.151 | 2.118 | 2.089 | 2.064 | 2.041 | 2.021 | 2.003 |
| 29 | 4.183 | 3.328 | 2.934 | 2.701 | 2.545 | 2.432 | 2.346 | 2.278 | 2.223 | 2.177 | 2.138 | 2.104 | 2.075 | 2.050 | 2.027 | 2.007 | 1.989 |
| 30 | 4.171 | 3.316 | 2.922 | 2.690 | 2.534 | 2.421 | 2.334 | 2.266 | 2.211 | 2.165 | 2.126 | 2.092 | 2.063 | 2.037 | 2.015 | 1.995 | 1.976 |
| 40 | 4.085 | 3.232 | 2.839 | 2.606 | 2.449 | 2.336 | 2.249 | 2.180 | 2.124 | 2.077 | 2.038 | 2.003 | 1.974 | 1.948 | 1.924 | 1.904 | 1.885 |
| 60 | 4.001 | 3.150 | 2.758 | 2.525 | 2.368 | 2.254 | 2.167 | 2.097 | 2.040 | 1.993 | 1.952 | 1.917 | 1.887 | 1.860 | 1.836 | 1.815 | 1.796 |
| 120 | 3.920 | 3.072 | 2.680 | 2.447 | 2.290 | 2.175 | 2.087 | 2.016 | 1.959 | 1.910 | 1.869 | 1.834 | 1.803 | 1.775 | 1.750 | 1.728 | 1.709 |

Degrees of Freedom - Numerator

.05

TABLE 6   THE *F* DISTRIBUTION TABLE    A11

## TABLE 6   (Continued)

Degrees of Freedom – Numerator

| Degrees of Freedom Denominator | 18 | 19 | 20 | 21 | 22 | 23 | 24 | 25 | 26 | 27 | 28 | 29 | 30 | 40 | 60 | 120 |
|---|---|---|---|---|---|---|---|---|---|---|---|---|---|---|---|---|
| 1 | 247.324 | 247.688 | 248.016 | 248.307 | 248.579 | 248.823 | 249.052 | 249.260 | 249.453 | 249.631 | 249.798 | 249.951 | 250.096 | 251.144 | 252.196 | 253.254 |
| 2 | 19.440 | 19.443 | 19.446 | 19.448 | 19.450 | 19.452 | 19.454 | 19.456 | 19.457 | 19.459 | 19.460 | 19.461 | 19.463 | 19.471 | 19.479 | 19.487 |
| 3 | 8.675 | 8.667 | 8.660 | 8.654 | 8.648 | 8.643 | 8.638 | 8.634 | 8.630 | 8.626 | 8.623 | 8.620 | 8.617 | 8.594 | 8.572 | 8.549 |
| 4 | 5.821 | 5.811 | 5.803 | 5.795 | 5.787 | 5.781 | 5.774 | 5.769 | 5.763 | 5.759 | 5.754 | 5.750 | 5.746 | 5.717 | 5.688 | 5.658 |
| 5 | 4.579 | 4.568 | 4.558 | 4.549 | 4.541 | 4.534 | 4.527 | 4.521 | 4.515 | 4.510 | 4.505 | 4.500 | 4.496 | 4.464 | 4.431 | 4.398 |
| 6 | 3.896 | 3.884 | 3.874 | 3.865 | 3.856 | 3.849 | 3.841 | 3.835 | 3.829 | 3.823 | 3.818 | 3.813 | 3.808 | 3.774 | 3.740 | 3.705 |
| 7 | 3.467 | 3.455 | 3.445 | 3.435 | 3.426 | 3.418 | 3.410 | 3.404 | 3.397 | 3.391 | 3.386 | 3.381 | 3.376 | 3.340 | 3.304 | 3.267 |
| 8 | 3.173 | 3.161 | 3.150 | 3.140 | 3.131 | 3.123 | 3.115 | 3.108 | 3.102 | 3.095 | 3.090 | 3.084 | 3.079 | 3.043 | 3.005 | 2.967 |
| 9 | 2.960 | 2.948 | 2.936 | 2.926 | 2.917 | 2.908 | 2.900 | 2.893 | 2.886 | 2.880 | 2.874 | 2.869 | 2.864 | 2.826 | 2.787 | 2.748 |
| 10 | 2.798 | 2.785 | 2.774 | 2.764 | 2.754 | 2.745 | 2.737 | 2.730 | 2.723 | 2.716 | 2.710 | 2.705 | 2.700 | 2.661 | 2.621 | 2.580 |
| 11 | 2.671 | 2.658 | 2.646 | 2.636 | 2.626 | 2.617 | 2.609 | 2.601 | 2.594 | 2.588 | 2.582 | 2.576 | 2.570 | 2.531 | 2.490 | 2.448 |
| 12 | 2.568 | 2.555 | 2.544 | 2.533 | 2.523 | 2.514 | 2.505 | 2.498 | 2.491 | 2.484 | 2.478 | 2.472 | 2.466 | 2.426 | 2.384 | 2.341 |
| 13 | 2.484 | 2.471 | 2.459 | 2.448 | 2.438 | 2.429 | 2.420 | 2.412 | 2.405 | 2.398 | 2.392 | 2.386 | 2.380 | 2.339 | 2.297 | 2.252 |
| 14 | 2.413 | 2.400 | 2.388 | 2.377 | 2.367 | 2.357 | 2.349 | 2.341 | 2.333 | 2.326 | 2.320 | 2.314 | 2.308 | 2.266 | 2.223 | 2.178 |
| 15 | 2.353 | 2.340 | 2.328 | 2.316 | 2.306 | 2.297 | 2.288 | 2.280 | 2.272 | 2.265 | 2.259 | 2.253 | 2.247 | 2.204 | 2.160 | 2.114 |
| 16 | 2.302 | 2.288 | 2.276 | 2.264 | 2.254 | 2.244 | 2.235 | 2.227 | 2.220 | 2.212 | 2.206 | 2.200 | 2.194 | 2.151 | 2.106 | 2.059 |
| 17 | 2.257 | 2.243 | 2.230 | 2.219 | 2.208 | 2.199 | 2.190 | 2.181 | 2.174 | 2.167 | 2.160 | 2.154 | 2.148 | 2.104 | 2.058 | 2.011 |
| 18 | 2.217 | 2.203 | 2.191 | 2.179 | 2.168 | 2.159 | 2.150 | 2.141 | 2.134 | 2.126 | 2.119 | 2.113 | 2.107 | 2.063 | 2.017 | 1.968 |
| 19 | 2.182 | 2.168 | 2.155 | 2.144 | 2.133 | 2.123 | 2.114 | 2.106 | 2.098 | 2.090 | 2.084 | 2.077 | 2.071 | 2.026 | 1.980 | 1.930 |
| 20 | 2.151 | 2.137 | 2.124 | 2.112 | 2.102 | 2.092 | 2.082 | 2.074 | 2.066 | 2.059 | 2.052 | 2.045 | 2.039 | 1.994 | 1.946 | 1.896 |
| 21 | 2.123 | 2.109 | 2.096 | 2.084 | 2.073 | 2.063 | 2.054 | 2.045 | 2.037 | 2.030 | 2.023 | 2.016 | 2.010 | 1.965 | 1.916 | 1.866 |
| 22 | 2.098 | 2.084 | 2.071 | 2.059 | 2.048 | 2.038 | 2.028 | 2.020 | 2.012 | 2.004 | 1.997 | 1.990 | 1.984 | 1.938 | 1.889 | 1.838 |
| 23 | 2.075 | 2.061 | 2.048 | 2.036 | 2.025 | 2.014 | 2.005 | 1.996 | 1.988 | 1.981 | 1.973 | 1.967 | 1.961 | 1.914 | 1.865 | 1.813 |
| 24 | 2.054 | 2.040 | 2.027 | 2.015 | 2.003 | 1.993 | 1.984 | 1.975 | 1.967 | 1.959 | 1.952 | 1.945 | 1.939 | 1.892 | 1.842 | 1.790 |
| 25 | 2.035 | 2.021 | 2.007 | 1.995 | 1.984 | 1.974 | 1.964 | 1.955 | 1.947 | 1.939 | 1.932 | 1.926 | 1.919 | 1.872 | 1.822 | 1.768 |
| 26 | 2.018 | 2.003 | 1.990 | 1.978 | 1.966 | 1.956 | 1.946 | 1.938 | 1.929 | 1.921 | 1.914 | 1.907 | 1.901 | 1.853 | 1.803 | 1.749 |
| 27 | 2.002 | 1.987 | 1.974 | 1.961 | 1.950 | 1.940 | 1.930 | 1.921 | 1.913 | 1.905 | 1.898 | 1.891 | 1.884 | 1.836 | 1.785 | 1.731 |
| 28 | 1.987 | 1.972 | 1.959 | 1.946 | 1.935 | 1.924 | 1.915 | 1.906 | 1.897 | 1.889 | 1.882 | 1.875 | 1.869 | 1.820 | 1.769 | 1.714 |
| 29 | 1.973 | 1.958 | 1.945 | 1.932 | 1.921 | 1.910 | 1.901 | 1.891 | 1.883 | 1.875 | 1.868 | 1.861 | 1.854 | 1.806 | 1.754 | 1.698 |
| 30 | 1.960 | 1.945 | 1.932 | 1.919 | 1.908 | 1.897 | 1.887 | 1.878 | 1.870 | 1.862 | 1.854 | 1.847 | 1.841 | 1.792 | 1.740 | 1.683 |
| 40 | 1.868 | 1.853 | 1.839 | 1.826 | 1.814 | 1.803 | 1.793 | 1.783 | 1.775 | 1.766 | 1.759 | 1.751 | 1.744 | 1.693 | 1.637 | 1.577 |
| 60 | 1.778 | 1.763 | 1.748 | 1.735 | 1.722 | 1.711 | 1.700 | 1.690 | 1.681 | 1.672 | 1.664 | 1.656 | 1.649 | 1.594 | 1.534 | 1.467 |
| 120 | 1.690 | 1.674 | 1.659 | 1.645 | 1.632 | 1.620 | 1.608 | 1.598 | 1.588 | 1.579 | 1.570 | 1.562 | 1.554 | 1.495 | 1.429 | 1.352 |

**TABLE 7**   **Quality Control Chart Constants**

| Number of Observations in Subgroup, $n$ | $A_2$ | $d_2$ | $d_3$ | $D_3$ | $D_4$ |
|---|---|---|---|---|---|
| 2 | 1.880 | 1.128 | 0.853 | 0.000 | 3.267 |
| 3 | 1.023 | 1.693 | 0.888 | 0.000 | 2.574 |
| 4 | 0.729 | 2.059 | 0.880 | 0.000 | 2.282 |
| 5 | 0.577 | 2.326 | 0.864 | 0.000 | 2.114 |
| 6 | 0.483 | 2.534 | 0.848 | 0.000 | 2.004 |
| 7 | 0.419 | 2.704 | 0.833 | 0.076 | 1.924 |
| 8 | 0.373 | 2.847 | 0.820 | 0.136 | 1.864 |
| .9 | 0.337 | 2.970 | 0.808 | 0.184 | 1.816 |
| 10 | 0.308 | 3.078 | 0.797 | 0.223 | 1.777 |
| 11 | 0.285 | 3.173 | 0.787 | 0.256 | 1.744 |
| 12 | 0.266 | 3.258 | 0.778 | 0.283 | 1.717 |
| 13 | 0.249 | 3.336 | 0.770 | 0.307 | 1.693 |
| 14 | 0.235 | 3.407 | 0.762 | 0.328 | 1.672 |
| 15 | 0.223 | 3.472 | 0.755 | 0.347 | 1.653 |
| 16 | 0.212 | 3.532 | 0.749 | 0.363 | 1.637 |
| 17 | 0.203 | 3.588 | 0.743 | 0.378 | 1.622 |
| 18 | 0.194 | 3.640 | 0.738 | 0.391 | 1.608 |
| 19 | 0.187 | 3.689 | 0.733 | 0.403 | 1.597 |
| 20 | 0.180 | 3.735 | 0.729 | 0.415 | 1.585 |
| 21 | 0.173 | 3.778 | 0.724 | 0.425 | 1.575 |
| 22 | 0.167 | 3.819 | 0.720 | 0.434 | 1.566 |
| 23 | 0.162 | 3.858 | 0.716 | 0.443 | 1.557 |
| 24 | 0.157 | 3.895 | 0.712 | 0.451 | 1.548 |
| 25 | 0.153 | 3.931 | 0.709 | 0.459 | 1.541 |

<br>

## APPENDIX *B*

# *EXPLANATION OF LARGE DATA SETS*

## THE INFORMATION THAT FOLLOWS WILL HELP YOU UNDERSTAND THE DATA FILE CARS.XXX.

### Explanation of the Variable Names in CARS

| | |
|---|---|
| CMSA | Household location—CMSA |
| HHFAMINC | Household family income category |
| HHLOC | MSA status |
| HHMSA | Household location—MSA |
| HHSIZE | Total number of persons in household |
| HH_RACE | Race of household reference person |
| HOUSEID | Household identification number |
| LIF_CYC | Family life cycle |
| MAKECODE | NASS Code for vehicle make |
| MSASIZE | Size of MSA/CMSA of household |
| MSTR_MON | Date of household master interview—Month |
| POPDNSTY | Population density of household ZIP code area |
| POVERTY | Household below, near, or above poverty level |
| REF_AGE | Age of household reference person |
| REF_EDUC | Education of household reference person |
| REF_SEX | Sex of household reference person |
| URBAN | Urbanized area indicator |
| URBNAREA | Urbanized area status |
| URBNSIZE | Size of urbanized area |
| VEH12MNT | Vehicle received in last 12 months |
| VEHHHOWN | Vehicle owned by household member |
| VEHMILES | Reported vehicle mileage previous 12 months |
| VEHNEW | Vehicle new or used when received |
| VEHTYPE | Vehicle type |
| VEHYEAR | Model year of vehicle |

# Explanation of Coded Values for Variables in CARS

### CMSA Household location — CMSA

| Value | Label |
|---|---|
| 1122 | Boston-Lawrence-Salem, MA-NH |
| 1282 | Buffalo-Niagara Falls, NY |
| 1602 | Chicago-Gary-Lake County, IL-IN-WI |
| 1642 | Cincinnati-Hamilton, OH-KY-IN |
| 1692 | Cleveland-Akron-Lorain, OH |
| 1922 | Dallas-Fort Worth, TX |
| 2082 | Denver-Boulder, CO |
| 2162 | Detroit-Ann Arbor, MI |
| 3282 | Hartford-New Britain-Middletown, CT |
| 3362 | Houston-Galveston-Brazoria, TX |
| 4472 | Los Angeles-Anaheim-Riverside, CA |
| 4992 | Miami-Fort Lauderdale, FL |
| 5082 | Milwaukee-Racine, WI |
| 5602 | New York-Northern New Jersey-Long Island, NY-NJ-CT |
| 6162 | Philadelphia-Wilmington-Trenton, PA-NJ-DE-MD |
| 6282 | Pittsburgh-Beaver Valley, PA |
| 6442 | Portland-Vancouver, OR-WA |
| 6482 | Providence-Pawtucket-Fall River, RI-MA |
| 7362 | San Francisco-Oakland-San Jose, CA |
| 7602 | Seattle-Tacoma, WA |

### HHFAMINC Household family income category

| Value | Label |
|---|---|
| 01 | Less than $5,000 |
| 02 | $5,000 to $9,999 |
| 03 | $10,000 to $14,999 |
| 04 | $15,000 to $19,999 |
| 05 | $20,000 to $24,999 |
| 06 | $25,000 to $29,999 |
| 07 | $30,000 to $34,999 |
| 08 | $35,000 to $39,999 |
| 09 | $40,000 to $44,999 |
| 10 | $45,000 to $49,999 |
| 11 | $50,000 to $54,999 |
| 12 | $55,000 to $59,999 |
| 13 | $60,000 to $64,999 |
| 14 | $65,000 to $69,999 |
| 15 | $70,000 to $74,999 |
| 16 | $75,000 to $79,999 |
| 17 | $80,000 or over |
| 98 | Not Ascertained |
| 99 | Refused |

### HHLOC MSA status

| Value | Label |
|---|---|
| 1 | In MSA central city |
| 2 | In MSA, not in central city |
| 3 | Not in MSA |

### HHSIZE Total number of persons in household

| Value | Label |
|---|---|
| 01 | One person in household |
| 02 | Two people in household |
| 03 | Three people in household |
| 04 | Four people in household |
| 05 | Five people in household |
| 06 | Six people in household |
| 07 | Seven people in household |
| 08 | Eight people in household |
| 09 | Nine people in household |
| 10 | Ten people in household |

### HH_RACE Race of household reference person

| Value | Label |
|---|---|
| 01 | White |
| 02 | Black |
| 03 | Other |
| 98 | Not Ascertained |
| 99 | Refused |

### HOUSEID Household identification number

(This is just an index number that is used to keep track of each interview.)

## LIF_CYC Family life cycle

| Value | Label |
|---|---|
| 01 | Single adult, no children |
| 02 | Two or more adults, no children |
| 03 | Single adult, youngest child age 0−5 |
| 04 | Two or more adults, youngest child age 0−5 |
| 05 | Single adult, youngest child age 6−15 |
| 06 | Two or more adults, youngest child age 6−15 |
| 07 | Single adult, youngest child age 16−21 |
| 08 | Two or more adults, youngest child age 16−21 |
| 09 | Single adult, retired, no children |
| 10 | Two or more adults, retired, no children |
| 98 | Not Ascertained |

## MAKECODE NASS code for vehicle make

| Value | Label | Value | Label | Value | Label |
|---|---|---|---|---|---|
| 001 | American Motors | 030 | Volkswagen | 048 | Subaru |
| 002 | Jeep (includes Kaiser-Jeep) | 031 | Alfa Romeo | 049 | Toyota |
| 003 | AM General | 032 | Audi | 050 | Triumph |
| 006 | Chrysler | 033 | Austin/Austin Healey | 051 | Volvo |
| 007 | Dodge | 034 | BMW | 052 | Mitsubishi |
| 008 | Imperial | 035 | Nissan/Datsun | 053 | Suzuki |
| 009 | Plymouth | 036 | Fiat | 054 | Acura |
| 010 | Eagle | 037 | Honda | 055 | Hyundai |
| 012 | Ford | 038 | Isuzu | 056 | Merkur |
| 013 | Lincoln | 039 | Jaguar | 057 | Yugo |
| 014 | Mercury | 040 | Lancia | 058 | Infiniti |
| 018 | Buick | 041 | Mazda | 059 | Lexus |
| 019 | Cadillac | 042 | Mercedes-Benz | 060 | Daihatsu |
| 020 | Chevrolet | 043 | MG | 069 | Other foreign |
| 021 | Oldsmobile | 044 | Peugeot | 084 | International Harvester/Navistar |
| 022 | Pontiac | 045 | Porsche | 089 | Other medium/heavy trucks and buses |
| 023 | GMC | 046 | Renault | 099 | Unknown |
| 024 | Saturn | 047 | Saab | 994 | Legitimate skip |
| 029 | Other domestic | | | | |

## MSASIZE Size of MSA/CMSA of household

| Value | Label |
|---|---|
| 01 | Less than 250,000 |
| 02 | 250,000 to 499,999 |
| 03 | 500,000 to 999,999 |
| 04 | 1,000,000 to 2,999,999 |
| 05 | 3,000,000 or more |
| 94 | Not in MSA |

### MSTR_MON Date of household master interview — Month

| Value | Label |
|---|---|
| 01 | January |
| 02 | February |
| 03 | March |
| 04 | April |
| 05 | May |
| 06 | June |
| 07 | July |
| 08 | August |
| 09 | September |
| 10 | October |
| 11 | November |
| 12 | December |
| 98 | Not ascertained |

### POPDNSTY Population density of household ZIP code area

| Value | Label |
|---|---|
| 01 | 0 to 99 |
| 02 | 100 to 249 |
| 03 | 250 to 499 |
| 04 | 500 to 749 |
| 05 | 750 to 999 |
| 06 | 1,000 to 1,999 and in MSA |
| 07 | 2,000 to 2,999 and in MSA |
| 08 | 3,000 to 3,999 and in MSA |
| 09 | 4,000 to 4,999 and in MSA |
| 10 | 5,000 to 7,499 and in MSA |
| 11 | 7,500 to 9,999 and in MSA |
| 12 | 10,000 to 49,999 and in MSA |
| 13 | 50,000 or more and in MSA |
| 14 | 1,000 or more and not in MSA |

### POVERTY Household below, near, or above poverty level

| Value | Label |
|---|---|
| 01 | Below poverty level |
| 02 | Near poverty level |
| 03 | Above poverty level |
| 98 | Not Ascertained |
| 99 | Refused |

### REF_AGE Age of household reference person

**Number except the following:**

| Value | Label |
|---|---|
| 077 | Reference person age 76 through 79 |
| 082 | Reference person age 80 through 84 |
| 088 | Reference person age 85 and over |
| 998 | Not Ascertained |
| 999 | Refused |

### REF_SEX Sex of household reference person

| Value | Label |
|---|---|
| 01 | Male |
| 02 | Female |
| 98 | Not Ascertained |
| 99 | Refused |

### REF_EDUC Education of household reference person

| Value | Label |
|---|---|
| 01 | First grade |
| 02 | Second grade |
| 03 | Third grade |
| 04 | Fourth grade |
| 05 | Fifth grade |
| 06 | Sixth grade |
| 07 | Seventh grade |
| 08 | Eighth grade |
| 09 | Ninth grade |
| 10 | Tenth grade |
| 11 | Eleventh grade |
| 12 | Twelfth grade |
| 13 | Technical school after high school |
| 21 | 1st (Freshman) year of college or equivalent |
| 22 | 2nd (Sophomore) year of college or equivalent |
| 23 | 3rd (Junior) year of college or equivalent |
| 24 | 4th (Senior) year of college or equivalent |
| 31 | One year of graduate school |
| 32 | Two or more years of graduate school |
| 98 | Not Ascertained |
| 99 | Refused |

# THE FOLLOWING INFORMATION WILL HELP YOU UNDERSTAND THE DATA FILE CONSPEXP.XXX.

## Explanation of Variable Names Used in the Consumer Expenditures Survey

| | |
|---|---|
| AGE2 | Age of Spouse |
| AGE_REF | Age of Ref. Person |
| ALCBEVCQ | Alcoholic Beverages This Quarter |
| ALCBEVPQ | Alcoholic Beverages Last Quarter |
| APPARCQ | Apparel and Services This Quarter |
| APPARPQ | Apparel and Services Last Quarter |
| AS_COMP1 | No. of Males Age 16 and Over in Cu |
| AS_COMP2 | No. of Females Age 16 and Over |
| AS_COMP3 | No. of Males Age 2 Through 15 |
| AS_COMP4 | No. of Females Age 2 Through 15 |
| AS_COMP5 | No. of Members Under Age 2 in Cu |
| BLS_URBN | Urban/rural |
| CUTENURE | Housing Tenure |
| EARNINCX | Cu Earned Inc. Before Taxes |
| EDUCA2 | Education of Spouse |
| EDUC_REF | Education of Ref. Person |
| ENTERTCQ | Entertainment This Quarter |
| FAMTFEDX | Fed Inc. Tax Deducted From Last Pay |
| FAM_SIZE | No. of Members in Cu |
| FDAWAYCQ | Food Away From Home This Quarter |
| FDAWAYPQ | Food Away From Home Last Quarter |
| FDHOMECQ | Food At Home This Quarter |
| FDHOMEPQ | Food At Home Last Quarter |
| FINCATAX | Cu Inc. After Taxes |
| FINDRETX | Money Placed in Individual Retirement Account (IRA) |
| FOODCQ | Total Food This Quarter |
| FOODPQ | Total Food Last Quarter |
| FRRETIRX | Soc. Sec. and Railroad Income |
| FSALARYX | Cu Wage and Salary Inc. Before Taxes |
| GASMOCQ | Gasoline and Motor Oil This Quarter |
| GASMOPQ | Gasoline and Motor Oil Last Quarter |
| HEALTHCQ | Health Care This Quarter |
| HEALTHPQ | Health Care Last Quarter |
| HOUSEQCQ | House furnishings, Equip. This Quarter |
| HOUSEQPQ | House furnishings, Equip. Last Quarter |
| INC_HRS1 | Wkly. Hrs. Reference Person Usually Worked |
| INC_HRS2 | Weekly Hrs. Spouse Usually Worked |
| NEWID | Cu Identification Number |
| NUM_AUTO | No. of Automobiles |
| PUBTRACQ | Pub. Transportation This Quarter |
| PUBTRAPQ | Pub. Transportation Last Quarter |
| RENTEQVX | Approx. Mthly. Rental Value of Home |
| SAVACCTX | Savings Accts. Balances At Banks |
| SECESTX | Est. Market Value of Securities |
| TRANSCQ | Transportation This Quarter |
| TRANSPQ | Transportation Last Quarter |
| UTILCQ | Utilities, Fuels, Public Services This Quarter |
| UTILPQ | Utilities, Fuels, Public Services Last Quarter |
| VEHQ | Number of Vehicles |

## Explanation of Code Values for Variables in Consumers Expenditure Survey

**BLS_URBN—Whether the location of the household is urban or rural**

| Value | Label |
|---|---|
| 1 | Urban |
| 2 | Rural |

**CUTENURE Type of ownership of house**

| Value | Label |
|---|---|
| 1 | Owned With Mortgage |
| 2 | Owned Without Mortgage |
| 3 | Owned Mort. Not Reported |
| 4 | Rented |
| 5 | Occ. Without Payment of Cash Rent |
| 6 | Stud. Housing |

## EDUCA2 and EDUC_REF Education level of head of household

| Value | Label |
|---|---|
| 1 | Elementary 1–8 Yrs. |
| 2 | High School, Less Than H.S. Grad. |
| 3 | High School Graduate |
| 4 | Coll., Less Than Coll. Grad. |
| 5 | Coll. Graduate |
| 6 | Graduate School |
| 7 | Never Att. School |

## URBAN Urbanized area indicator

| Value | Label |
|---|---|
| 1 | Household in urbanized area |
| 2 | Household not in urbanized area |

## URBNAREA Urbanized area status

| Value | Label |
|---|---|
| 1 | Urbanized, in MSA central city |
| 2 | Urbanized, not in MSA central city |
| 3 | Not in urbanized area |

## URBNSIZE Size of urbanized area

| Value | Label |
|---|---|
| 01 | 50,000 to 199,999 |
| 02 | 200,000 to 499,999 |
| 03 | 500,000 to 999,999 |
| 04 | 1,000,000 or more without subway/rail |
| 05 | 1,000,000 or more with subway/rail |
| 94 | Not in urbanized area |

## VEHYEAR Model year of vehicle

| Value | Label |
|---|---|
| 055 | 1919–1959 |
| 063 | 1960–1964 |
| 065 | 1965 |
| 066 | 1966 |
| 067 | 1967 |
| 068 | 1968 |
| 069 | 1969 |
| 070 | 1970 |
| 071 | 1971 |
| 072 | 1972 |
| 073 | 1973 |
| 074 | 1974 |
| 075 | 1975 |
| 076 | 1976 |
| 077 | 1977 |
| 078 | 1978 |
| 079 | 1979 |
| 080 | 1980 |
| 081 | 1981 |
| 082 | 1982 |
| 083 | 1983 |
| 084 | 1984 |
| 085 | 1985 |
| 086 | 1986 |
| 087 | 1987 |
| 088 | 1988 |
| 089 | 1989 |
| 090 | 1990 |
| 091 | 1991 |
| 994 | Legitimate skip |
| 998 | Not Ascertained |
| 999 | Refused |

## VEH12MNT Vehicle received in last 12 months

| Value | Label |
|---|---|
| 01 | Yes |
| 02 | No |
| 98 | Not ascertained |
| 99 | Refused |

## VEHHHOWN Vehicle owned by household member

| Value | Label |
|---|---|
| 01 | Yes |
| 02 | No |
| 98 | Not ascertained |
| 99 | Refused |

## VEHMILES Reported vehicle mileage previous 12 months

| Value | Label |
|---|---|
| 999994 | Legitimate skip |
| 999998 | Not ascertained |

## VEHNEW Vehicle new or used when received

| Value | Label |
|---|---|
| 01 | New |
| 02 | Used |
| 98 | Not ascertained |
| 99 | Refused |

## VEHTYPE Vehicle type

| Value | Label |
|---|---|
| 01 | Automobile (includes station wagon) |
| 02 | Passenger van |
| 03 | Cargo van |
| 04 | Pickup truck (including pickup with camper) |
| 05 | Other truck |
| 06 | RV or motor home |
| 07 | Motorcycle |
| 08 | Moped |
| 09 | Other |
| 98 | Not Ascertained |
| 99 | Refused |

# AN INTRODUCTION TO MICROSOFT EXCEL

This appendix will introduce you to the basics of Microsoft Excel 97, if you are not familiar with the program. If you are already familiar with Excel, you can simply refer to the sections on using Excel for statistics at the end of each chapter.

## C.1   STARTING MICROSOFT EXCEL

You can start Microsoft Excel 97 in several different ways:

- from the **Start > Programs** menu
- from the Microsoft Office shortcut bar

To open Excel from the Program menu, click on **Start > Programs** and then on the icon for Excel. Excel is usually located in a folder for MS Office 97 as shown in Figure C.1. If there is not a folder for Office 97, then Excel might be a separate icon on the program menu.

**FIGURE C.1**    Starting Excel from the Start button

If your computer has the Microsoft Office shortcut bar running, you will see an icon for Excel on the toolbar, as shown in Figure C.2. Clicking on the icon will launch Excel.

**FIGURE C.2**    The Microsoft Office toolbar

## C.2    THE EXCEL WORKBOOK

Excel starts by opening a new workbook. A workbook in Excel is a collection of work-sheets. Each worksheet is a grid with 256 columns and 65,536 rows. An Excel work-book, the menus, and toolbars are shown in Figure C.3. We will now look at the dif-ferent parts of the workbook and talk about their functions.

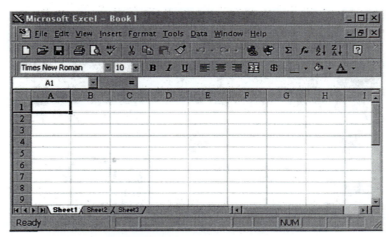

**FIGURE C.3**   The Excel workbook

### The Worksheet

The worksheet portion of the Excel workbook is where you will store your data and do all of your calculations. A worksheet consists of 256 columns labeled with letters and 65,536 rows, labeled with numbers. An individual location in a worksheet is called a **cell.** Each cell in a worksheet is identified by its **location.** Locations are de-scribed in terms of the column letter followed by the row number. In Figure C.4, you see a typical worksheet and its formula bar. The important parts of the worksheet are labeled.

**FIGURE C.4**   The Excel worksheet

In this worksheet, cell C4 is the **active cell.** A cell is made active by either clicking on it or by moving to the cell using the directional keys. The active cell will have a darker border around it. The location of the active cell is shown in the Cell/Range Reference and the contents of the active cell appear in the Cell Edit box.

A workbook consists of a collection of worksheets. You can move from sheet to sheet, by clicking on the **sheet tabs** at the bottom of the workbook. It is a good idea to give different names to the worksheets to keep your worksheet organized and so that you don't waste time looking for something important. To change the name of a worksheet, double-click on the sheet tab and type in the new name.

## The Excel Menu and Toolbars

The other portion of the workbook that you see when you use Excel consists of the Excel menu bar and some toolbars as shown in Figure C.5. You can specify exactly which toolbars Excel shows, but the ones shown by default are the standard and formatting toolbars.

**FIGURE C.5**   Excel menus and toolbars

The **Standard Toolbar** contains icons for often-used tasks from the Excel menu bar. Some of these tasks are **New, Open, Print, Print Preview,** and **Save** from the File menu, **Copy, Cut,** and **Paste** from the Edit menu, and **Sort** from the Data menu. The **Formatting Toolbar** includes icons for changing the appearance of entries in the worksheet.

## C.3   FILE OPERATIONS IN EXCEL

When you start Excel a new workbook automatically opens. After you have been working in Excel awhile you might want to save your work and to open your own workbooks the next time you start Excel.

## Opening an Existing Workbook in Excel

There are two basic ways to open workbook files in Excel. One way is simply to double-click on the name of the file in Windows Explorer or My Computer. Excel files have the file extension "**.xls**" after the name of the file. When you open the file this way, Excel will start automatically with the file as the open workbook.

If you have already started Excel, then you can open a file using the **File** menu. Click on the **File** in the menu bar and the menu shown in Figure C.6 (page C4) will appear.

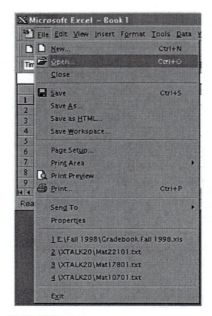

**FIGURE C.6**    The File menu in Excel

Choose **Open** and the dialog box shown in Figure C.7 opens.

**FIGURE C.7**    The Open File dialog box

Clicking on the arrow next to the **Look in:** box will bring up a list of possible locations for your file. Most of the time your files will be on a floppy disk or on one of the hard drives of the computer. If the file is located on a floppy disk, choose the $3\frac{1}{2}$ Floppy option. If it is located on some other drive such as your hard drive or a network drive, you should choose the appropriate disk letter.

  After you choose the drive that contains your file, a list of all Excel files on that drive will appear in the dialog box, as shown in Figure C.8. To open the file, click on the name of the file you want and choose **Open** or simply double-click on the file name.

**FIGURE C.8**   List of Excel files on the A: drive

## Saving a File in Excel

After you have done some work in Excel you will want to save your file, perhaps to work on at a later time or to hand in to your instructor. The process of saving a file in Excel is very similar to that of opening a file. Choose **File > Save** from the menu bar. If the workbook is a new one *that has not been saved before,* the **Save As** dialog box will open as shown in Figure C.9.

**FIGURE C.9**   The Save As dialog box

Use the **Save in:** list to choose the location where you want to save the file. In the **File name:** box, type in a name for the file that will help you remember what the file contains. Cute names, however funny, do not help you when you have ten files saved on the same disk! Click **Save** to save the file to the location you selected.

*This is only one way to save a file. You can also click on the floppy disk icon in the Standard Toolbar, or hit Ctrl+S.*

No computer or computer network is foolproof. Even if you are not finished with your work, you will want to save your file regularly. If you do not, and the system crashes, you will lose any work you have done. Once the file has been saved for the first time, selecting **File > Save** will automatically save the file to its previous location.

## C.4   PRINTING IN EXCEL

You will most likely want to print out some of your work in Excel. This can be done in several different ways.

To start printing, you can click on the printer icon in the **Standard Toolbar** or select **File > Print.** This will print your current worksheet using the current printing setup. You might not like the way your output appears if you do this. You can customize the way your printout appears by using the **Print Preview** option.

To use **Print Preview** click on the icon in the **Standard Toolbar** or choose **File >
Print Preview.** When you select Print Preview your current worksheet will be shown
exactly as it will appear on the printed page as shown in Figure C.10.

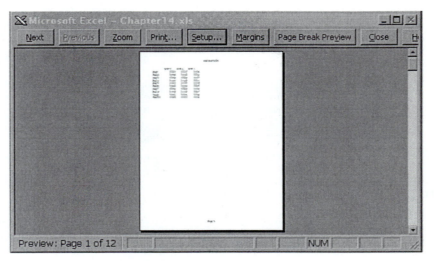

**FIGURE C.10**     The Print Preview dialog box

If you like the way it looks you can simply click **Print . . .** and it will be printed.
If you do not like the way it looks you can change several aspects of the printout by
selecting **Setup . . . .** The **Page Setup** dialog box opens. The dialog box has four
tabs: **Page, Margins, Header/Footer,** and **Sheet.** Each of these lets you control a different aspect of the printout.

The **Page** tab shown in Figure C.11 allows you to choose the page orientation,
portrait or landscape. Your choice will depend on how the data are located in your
spreadsheet. If you use a lot of columns and not many rows, you will want Landscape.
If you have many rows and not so many columns you will want Portrait.

**FIGURE C.11**     The Page tab

The other tabs in the Page Setup dialog box let you adjust other aspects about
how your selection will print. You can adjust the margins, add headers and footers
(or delete the default headers and footers that Excel uses), and choose whether or
not you want the gridlines on the sheet to print.

# C.5   OTHER USEFUL TIPS FOR EXCEL

## Selecting a Range

Many activities in Excel require you to select a range of cells in the worksheet. A range of cells might be in a single column or in several columns. There are several ways to select a range:

1. Make the first cell in the range active by moving the cursor to that cell. Press the left mouse button and drag the cursor to highlight the remaining cells in the range.
2. Click the first cell in the range, and then hold down SHIFT and click the last cell in the range. You can scroll to make the last cell visible.
3. To select non-adjacent ranges, select the first range and then hold down the CTRL button while you select the other range(s).

## Copying, Cutting, and Pasting

Some of the actions that you will perform most often in Excel are copying, cutting, and pasting. These commands are found in the **Edit** menu and allow you to manage the data in your worksheets. To copy a range of cells from one location in a workbook to another, you must first select the range of cells as just described. Then, from the **Edit** menu, select **Copy**, or click on the **Copy** icon on the toolbar. Move the cursor to the first cell where you want to place the copy and then select **Paste** from the edit menu or click on the **Paste** icon on the toolbar.

If you want to move a range of cells from one location to another, after selecting the range of cells, choose **Cut** from the **Edit** menu or click on the **Cut** icon on the toolbar. Then move the cursor to the destination cells and select **Paste**.

## Right Clicking

Clicking the right mouse button in Excel is a useful action. When you click the right mouse button, a pop-up context-sensitive menu appears. The commands on this menu are the ones that most likely apply to the action you are performing.

## Switching Among Workbooks

In Excel it is possible to keep more than one workbook open at a time. This is useful if you need to copy information or sheets from a workbook to another workbook. To switch among open workbooks in Excel, select **Window** from the menu bar. At the bottom of the menu is a list of all open workbooks, as shown in Figure C.12.

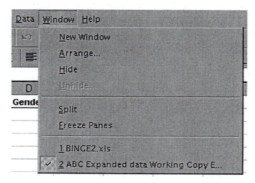

**FIGURE C.12**   The Window menu

To select a different workbook, click on the name of the workbook you want to use.

# INDEX